SELECTED TOPICS IN THE CHEMISTRY OF NATURAL PRODUCTS

SELECTED TOPICS IN THE CHEMISTRY OF NATURAL PRODUCTS

EDITED BY RAPHAEL IKAN
Emeritus Professor,
Hebrew University of Jerusalem, Israel

World Scientific
NEW JERSEY • LONDON • SINGAPORE • BEIJING • SHANGHAI • HONG KONG • TAIPEI • CHENNAI

Published by

World Scientific Publishing Co. Pte. Ltd.
5 Toh Tuck Link, Singapore 596224
USA office: 27 Warren Street, Suite 401-402, Hackensack, NJ 07601
UK office: 57 Shelton Street, Covent Garden, London WC2H 9HE

British Library Cataloguing-in-Publication Data
A catalogue record for this book is available from the British Library.

SELECTED TOPICS IN THE CHEMISTRY OF NATURAL PRODUCTS

ISBN-13 978-981-270-569-3
ISBN-10 981-270-569-4

Typeset by Stallion Press
Email: enquiries@stallionpress.com

Printed in Singapore by World Scientific Printers (S) Pte Ltd

To my dear wife Yael, our children, Amiram, Ariel, Arnon and Eliana, their spouses and our eleven lovely grandchildren — all natural products.

Contents

Preface

Natural products can be thought of as originating from mankind's curiosity about odor, taste and cures for diseases, both in terrestrial and marine environments.

The study of natural products has always been the starting point in the discipline of chemistry worldwide. This observation is due to the importance of organic compounds in agriculture, medicine and industry.

Nowadays, every student and scientist feel the need to acquire further knowledge in this fascinating field which reveals natural "secrets."

The book contains 15 chapters which cover the following topics: The origin and nature of natural products; plant–derived natural products in drug discovery and development: an overview; plant and brain cannabinoids: the chemistry of major new players in physiology; natural products as biomarker tracers for environmental and geological processes; toxins of marine invertebrates and microorganisms; enantiometric composition of odorous oxygenated monoterpenes in aromatic plants; recent trends of natural sweet substances from plants; natural products for pest management; natural products in mycelial microorganisms: impact of morphology; insects pheromones; natural products and related compounds of realized and potential use in treating neurodegenerative disease; phytotoxic compounds with calmodulin inhibitor properties from selected Mexican fungi and plants, potential anticancer natural products from plant-anticancer fungi; plant fungal endophytes: interactions, metabolites and biosyntheses; antibiotics.

The reviews were written by accomplished scientists, with extensive records in their fields. This book is devoted to multidisciplinary studies of various topics of natural products, which have always been fascinating and challenging for organic chemists.

Our wish is that this book will help in consolidating our understanding of various aspects of natural products.

The contributing authors are from Israel, Mexico, the United Kingdom and the United States of America.

Since the book covers many disciplines of natural products, I wish to complete this short Preface with a quote by Albert Szent-Györgi in his "Personal Reminiscences": "*In my hunt for the secret of life, I started research in histology. Unsatisfied by the information that cellular morphology could give me clues about life, I turned to physiology. Finding physiology too complex, I took up pharmacology. Still finding the situation too complicated, I turned to bacteriology. But bacteria were even more complex, so I descended to the molecular level. Studying chemistry and physical chemistry. After twenty years of work, I was led to conclude that to understand life we have to descend to the electronic level, and to the world of wave mechanics. But electrons are just electrons, and have no life at all. Evidently on the way I lost life; it had run out between my fingers . . .*"

This statement approves once again the importance of natural products in our daily activities and well being.

Raphael Ikan, CChem FRSC
Department of Organic Chemistry
Faculty of Sciences
Hebrew University of Jerusalem, Israel

Acknowledgements

The major credit for this scientific merit goes to the contributing authors and co-authors of the reviews and also the publisher, especially to Ms. Ang Ching Ting who was efficient and helpful.

Sincere thanks are due to Professor A Douglas Kinghorn, Chair of the Division of Medicinal Chemistry and Pharmacognosy, College of Pharmacy, the Ohio State University, Columbus, Ohio, and Editor-in-Chief of the *Journal of Natural Products*, for recommending some of the senior scientists (from various disciplines of natural products) as candidates for the preparation of the appropriate reviews for this book. Finally, I wish to express my deep appreciation and love to my wife, Yael, for her sincere devotion in corresponding with the authors on matters regarding their reviews.

Contributors

MARK BAHAR
Division of Medicinal Chemistry and Pharmacognosy
College of Pharmacy
The Ohio State University
Columbus, OH 43210
USA

YEHUDA BENAYAHU
School of Chemistry and Department of Zoology
Tel-Aviv University
Ramat Aviv 69978
Israel

SERGEI BRAUN
Institute of Life Sciences
The Hebrew University of Jerusalem
Safra Campus
Jerusalem, 91904
Israel

CHARLES CANTRELL
Natural Product Utilization Research
Agricultural Research Service
United States Department of Agriculture
University, MS 38677
USA

BERNARD CRAMMER
Department of Organic Chemistry
Laboratory of Natural Products
The Hebrew University of Jerusalem
Safra Campus
Jerusalem, 91904
Israel

YE DANG
Division of Medicinal Chemistry and Pharmacognosy
College of Pharmacy
The Ohio State University
Columbus, OH 43210
USA

FRANCK E. DAYAN
Natural Product Utilization Research
Agricultural Research Service
United States Department of Agriculture
University, MS 38677
USA

STEPHEN O. DUKE
Natural Product Utilization Research
Agricultural Research Service
United States Department of Agriculture
University, MS 38677
USA

JOSHUA N. FLETCHER
Division of Medicinal Chemistry and Pharmacognosy
College of Pharmacy
The Ohio State University
Columbus, OH 43210
USA

MAGESH J. GOUNDALKAR
Department of Chemistry
State University of New York
College of Environmental Science and Forestry
Syracuse, New York 13210
USA

A.A. LESLIE GUNATILAKA
Southwest Center for Natural Products Research and Commercialization
Office of Arid Lands Studies
College of Agriculture and Life Sciences
The University of Arizona
250 East Valencia Road
Tucson, Arizona 85706-6800
USA

LUMIR HANUŠ
Hebrew University Medical Faculty
School of Pharmacy
Department of Medicinal Chemistry and Natural Products
Jerusalem
Israel

PETER J. HOUGHTON
Pharmacognosy Research Laboratories
Pharmaceutical Sciences Research Division
Kings College London
150 Stamford St. London SE1 9NH
United Kingdom

MELANIE JAYNE HOWES
Jodrell Laboratory Royal
Botanic Gardens Kew
Richmond, Surrey TW9 3DS
United Kingdom

RAPHAEL IKAN
Department of Organic Chemistry
Laboratory of Natural Products
The Hebrew University of Jerusalem
Safra Campus
Jerusalem, 91904
Israel

YOEL KASHMAN
School of Chemistry
and Department of Zoology
Tel-Aviv University
Ramat Aviv, 69978
Israel

A. DOUGLAS KINGHORN
Division of Medicinal Chemistry and Pharmacognosy
College of Pharmacy
The Ohio State University
Columbus, OH 43210
USA

SRINIVAS KODALI
Merck & Co. Inc
P.O.Box 2000
R80Y-210; Rahway, NJ 07065
USA

SERGIO-MARTINEZ LUIS
Departamento de Farmacia, Facultad de
Quimica Universidad Nacional Autonoma de
Mexico, DF 04510, Mexico City
Mexico

RACHEL MATA
Departamento de Farmacia, Facultad de
Quimica Universidad Nacional Autonoma de
Mexico, DF 04510, Mexico City
Mexico

MARILYN T. MARRON
Southwest Center for Natural Products Research and Commercialization
Office of Arid Lands Studies
College of Agriculture and Life Sciences
The University of Arizona
250 East Valencia Road
Tucson, Arizona 85706-6800
USA

RAPHAEL MECHOULAM
Hebrew University Medical Faculty
School of Pharmacy
Department of Medicinal Chemistry and Natural Products
Jerusalem
Israel

KUMUDINI M. MEEPAGALA
Natural Product Utilization Research
Agricultural Research Service
United States Department of Agriculture
University, MS 38677
USA

ARACELI PEREZ-VASQUES
Departamento de Farmacia, Facultad de
Quimica Universidad Nacional Autonoma de
Mexico, DF 04510, Mexico City
Mexico

JOHN R. PORTER
Cell Biology and Biotechnology Program
Department of Biological Sciences
University of Sciences in Philadelphia
600 S. 43rd Street
Philadelphia, PA. 19104
USA

UZI RAVID
Department of Aromatic Plants
Agricultural Research Organization
P.O.Box 1021
Ramat Yishay, 30095
Israel

AGNES M. RIMANDO
Natural Product Utilization Research
Agricultural Research Service
United States Department of Agriculture
University, MS 38677
USA

KEVIN K. SCHRADER
Natural Product Utilization Research
Agricultural Research Service
United States Department of Agriculture
University, MS 38677
USA

BERND R.T. SIMONEIT
Department of Chemistry and
College of Oceanic and Atmospheric Sciences
Oregon State University
Corvallis, Oregon 97331
USA

NURHAYAT TABANCA
Natural Product Utilization Research
Agricultural Research Service
United States Department of Agriculture
University, MS 38677
USA

FRANCIS X. WEBSTER
Department of Chemistry
State University of New York
College of Environmental Science and Forestry
Syracuse, New York 13210
USA

JUN WANG
Merck & Co. Inc.
P.O.Box 2000
R80W-250; Rahway
NJ 07065
USA

DAVID E. WEDGE
Natural Product Utilization Research
Agricultural Research Service
United States Department of Agriculture
University, MS 38677
USA

Chapter 1

THE ORIGIN AND THE NATURE OF NATURAL PRODUCTS

Raphael Ikan

1.1 INTRODUCTION

Throughout history, mankind has always been interested in naturally occurring compounds from prebiotic, microbial, plants and animals sources. Various extracts of flowers, plants and insects have been used for isolating compounds whose task, color and odor could be used for various purposes. Many natural products, such as plant hormones, have a regulatory role, while others function as chemical defense against pests. The role of certain compounds is to act as chemical messengers, such as sex-attractants (pheromones) in insects, terrestrial and marine animals and humans. What is the origin of natural products?

It has been suggested that earth planet was created about 5000 million years ago and human beings were created about 100 000 years ago. In the beginning, God created the heaven and the earth. And God said: Let the earth put forth grass, herb yielding seed, and fruit tree bearing fruit after its kind, and God saw that it was good. And there was evening and there was morning on the third day. On the fifth day, he created the fish in the sea and the fowl in the air. In a sense, we are concerned with the third and the fifth days in this review. There are other ways in which the idea of creation has been represented. Recently, there has been a literary expression of Thornton Wilder in his book, *The Eighth Day*, "Nature never sleeps. The process of life never stands still. The Bible says that God created man in the sixth day and rested, but each one of these days was many millions of years long (Fig. 1)."[1]

1.2 HUMAN MEDICINAL AGENTS FROM PLANTS

The ancient Egyptians have described several useful preparations such as opium and castor oil. They also used "rotten bread" for treating infections which "resembles" our use of antibiotics produced by moulds and fungi. The Roman physician, Dioscorides, studied the medical uses of hundreds of plants and wrote the first systematic *materia medica* during the first century. He also described the medicinal properties of wines.

The Chinese are considered as leaders in using natural products for healing. The oldest compilation of Chinese herbs is Shen Nung Pen Ts'ao, which lists 385 materials. Pen Ts'ao Ma catalogue, written by Li-Chen during the Ming Dynasty, (1573–1620) mentions 1898 herbal drugs and 8160 prescriptions.

5267 medicinal herbs were used in China in 1979. One of the most famous herbs among them is the ginseng root, *Panax ginseng* used for health maintenance and the treatment of various diseases. Another popular folk drug is the extract of the Ginkgo tree, *Ginkgo biloba*, which can improve memory and sharpen mental alertness.

During the 17th century, the Jesuits brought with them from South America the bark of the China tree for the treatment of malaria. In 1820, Pelletia and Caventou isolated from the China tree the active compound, quinine. American Indians used the powerful hallucinogen, mescaline, for a long period. The Indian hemp plant, *Cannabis sativa*, has been used since 3000 BC, and it is also used as marijuana or hashish. Its constituent, Δ^1-THC (tetrahydrocannabinol) is responsible for its mind-altering effect. It was synthesized by Mechoulam and Gaoni.[2]

The rapid and impressive development of organic chemistry in the 19th century had a tremendous effect on the discoveries of natural products. Towards the end of the 19th century, microbiology has developed enormously. The antibiotic penicillin was first isolated by Alexander Fleming

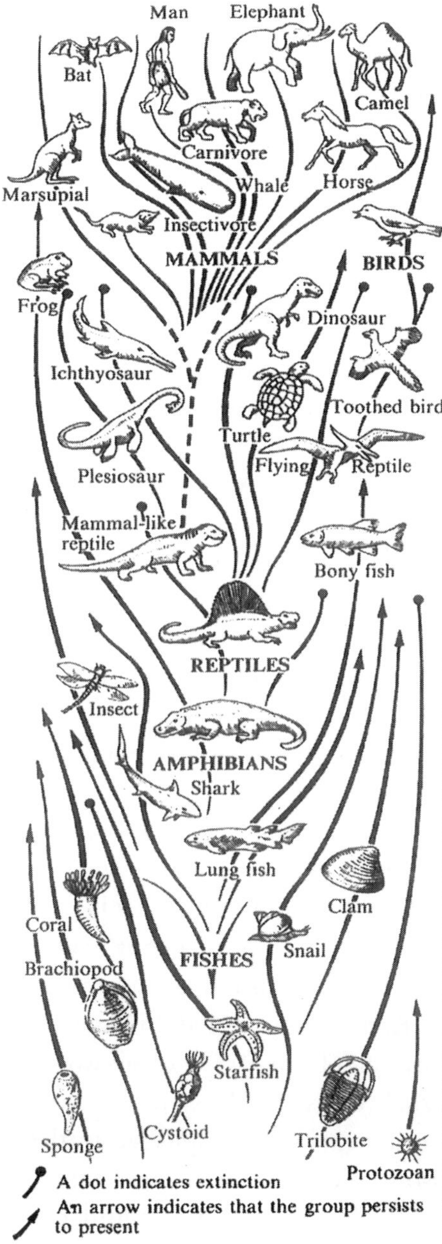

Years ago (Millons)	Durations	PERIODS	Eras
2	2	QUARTERNARY Recent Pleistocene	CAINOZOIC
65	63	TERTIARY Pliocene Miocene Oligocene Eocene Palaeocene	CAINOZOIC
135	70	CRETACEOUS	MESOZOIC
190	55	JURASSIC	MESOZOIC
225	35	TRIASSIC	MESOZOIC
280	55	PERMIAN	PALAEOZOIC
345	65	CARBONIFEROUS	PALAEOZOIC
395	50	DEVONIAN	PALAEOZOIC
430	35	SILURIAN	PALAEOZOIC
500	70	ORDOVICIAN	PALAEOZOIC
570	70	CAMBRIAN	PALAEOZOIC

Fig. 1. The palaeontological record of the Palaeozoic.[1]

in 1929. In 1944, Waxman isolated streptomycin which is used for the treatment of tuberculosis.

1.3 RECENT PROGRESS IN THE CHEMISTRY OF NATURAL PRODUCTS[3,4,5]

Among the recent outstanding contributions to the chemistry of natural products is the conformational analysis designed by Derek Barton. He used it for the structural determinations of many complex molecules such as β-amyrin and cycloartenol. Robert B. Woodward was involved in the structural determinations of penicillin, strychnine, patalin, terramycin, aureomycin and the synthesis of Vitamin B12.

The other important scientists include V. Prelog, L. Ruzicka, P. Plattner, O. Jeger, A. Eschenmoser, D. Arigoni, A. Dreiding, J. Dunitz and many more. Other famous investigators regarding the biosynthetic studies were A. Birch and R. Robinson who studied the biosynthesis of polyketides having C_6-C_3-C_6 backbones such as plant phenolics, polyene macrolides, terpenoids and alkaloids, sterols, fatty acids and prostaglandins (discovered in seminal fluids).

Otto Wallach (1847–1931) proposed the "isoprene rule" and many scientists were engaged in isoprenoid studies. Among them was Wieland, Windaus, Karrer, Kuhn, Butenandt and Ruzicka. Paul Karrer (1889–1971) established the foundation of carotenoid chemistry which was proceeded by Otto Isler and Hans Eugster. Adolf Butenandt (1903–1998) dealt with human sex hormones and obtained estrone from pregnant women, progesterone from sow ovaries and 50 mg of androsterone from 400 gallons of male urine. Together with Peter Karlson, he isolated 25 mg of ecdysone (insect and crustacean molting hormone) from 500 kg of silkworm larvae.

Studies on isolation from adrenal cortex and the synthesis of cortisone (in 28 steps), an anti-arthritic hormone, was accomplished in the 1940s by Woodward and others. Cortisone was used as an important military medicine during World War II. Carl Djerassi from Stanford University directed the research at the Syntex Laboratories, which led to the synthesis of the first oral contraceptive "pill" for women. Koji Mori is very active in the field of the synthesis of pheromones.

Extracts of toxic plants has been used for hunting and murder throughout the world for thousands of years. Thus, *Strychnos* and *Chondrodendion* (both containing strychnine) were used in arrow poisons. The Colombian arrow poison consists of toxins from the legs of frogs. When rye is infected by the fungus, *Claviceps porpurea*, the toxin ergotamine and a number of ergot alkaloids are produced. These compounds cause serious illnesses.

In 1952, Bloch and Woodward suggested a mechanism for the cyclization of squalene to cholesterol. In 1962, Francis Crick and James Watson described the double helix structure of proteins. Hodgkin determined the structure of vitamin B12 and of penicillin through collaboration between Woodward and Eschenmoser, involving postdoctoral fellows. In 1877, Alexander Fleming discovered penicillin which was active against tuberculosis.

The study of marine natural products has great possibilities for new products. Thus, Paul Scheuer from Hawaii studied bioactive compounds from mollusks and other marine sources. Luigi Minale, Raffaele Riccio and Maria Iorizzi from Italy, conducted a comprehensive research on marine steroidal glycosides. Joel Kashman from Tel-Aviv University investigated on the biologically active natural products from marine organisms.

Palytoxin, a most poisonous substance from the Hawaiian soft coral, *Palythoa toxica*, was isolated by Woodward. It contains about 64 chiral centers and seven double bonds, capable of E/Z isomerism, giving rise to the possibility of 2^{71} isomers (close to Avogadro's number). It was synthesized by Armstrong, *et al.*, in 1989. In 1996, a most toxic toxin, maitotoxin was isolated and its structures was determined by two Japanese groups. Its empirical formula is $C_{164}H_{256}O_{68}S_2NO_2$ and its molecular weight is 3422. Human medicinal agents from plants were described by Kinghorn and Balandrin.[6]

1.4 THE CLASSIFICATION OF NATURAL PRODUCTS

The classification of natural products may follow the four schemes below:

(1) Classification based on the molecular skeletal structure: Open-chain aliphatic, alicyclic and cycloaparaffinic, aromatic, benzenoid and heterocyclic.

(2) Classification based on physiological activity: The interest in natural products is frequently initiated by attempts to isolate and clarify a physiologically active principle of plant or animal origin. Actually, many medicines currently in use are natural products, e.g. alkaloids, such as morphine and penicillin G.
(3) Classification based on chemotaxonomy: The field of chemotaxonomy attempts to review plant constituents according to plant taxa. Namely, constituents are regarded as markers for evolution as well as the classification of plants.
(4) Classification based on biogenesis: It has been established that the primary synthetic process in nature is photosynthesis by which green plants utilize the energy of the sun for the production of organic compounds from carbon dioxide. The initial products of photosynthesis are carbohydrates. Further metabolic alterations lead to the formation of a pool of organic compounds of low molecular weight and simple structures such as carboxylic- and amino acids, which are vital for the living organisms. They form the synthetic starting materials for specific, genetically controlled, enzymatically catalyzed reactions that lead to the complex compounds that characterize the secondary metabolism of plants and mammals. The reaction pathway leading to a particular natural product is called the biosynthetic pathway and the corresponding event is known as biogenesis. Different plant and animal species can employ different biosynthetic pathways to produce the same metabolite. This feature can be employed in the classification of plants in terms of their chemotaxonomy.

Of the four major classes of biochemicals (carbohydrates, proteins, nucleic acids and lipids), experiments have shown that the first three classes could have arisen through prebiotic chemistry.[7] Although the biosynthesis of many natural products can be traced back to acetate (e.g. fatty acids, terpenes and polyketide biosynthesis) or amino acids (e.g. alkaloid biosynthesis), there are many whose biosynthetic origins are either obscure or result from a complex combination of pathways (Fig. 2).

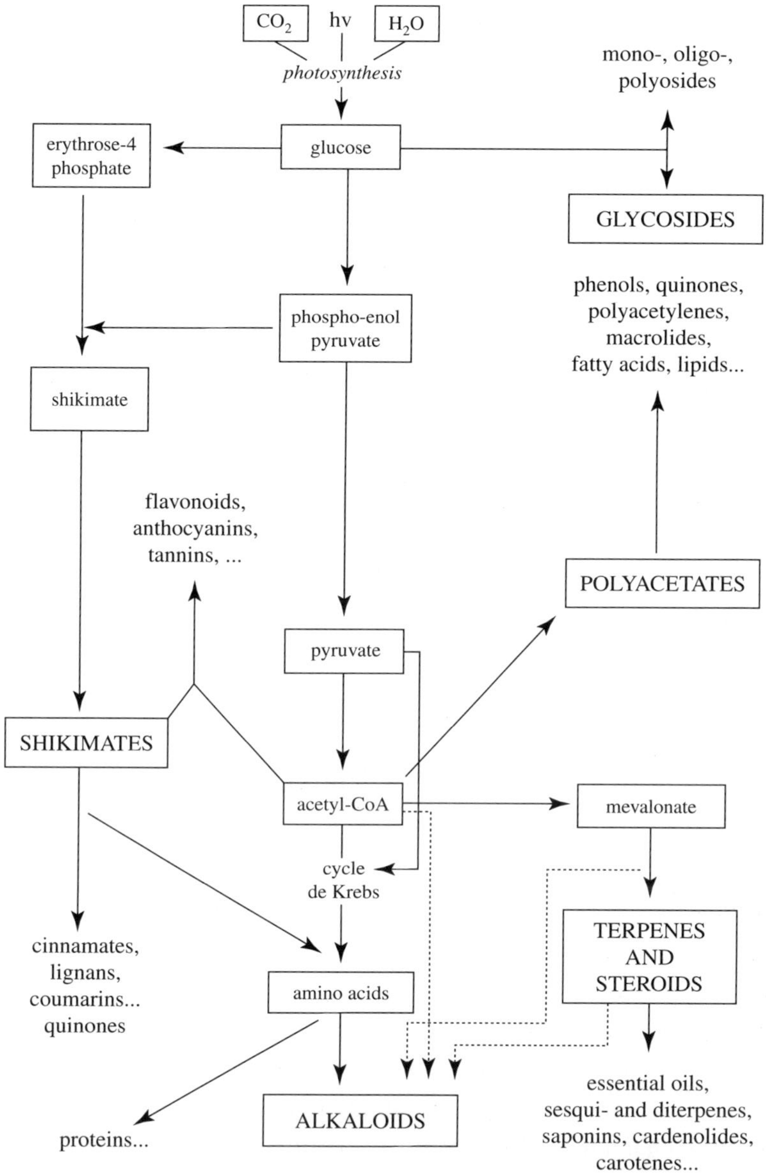

Fig. 2. Schematic presentation of natural products formation.[8]

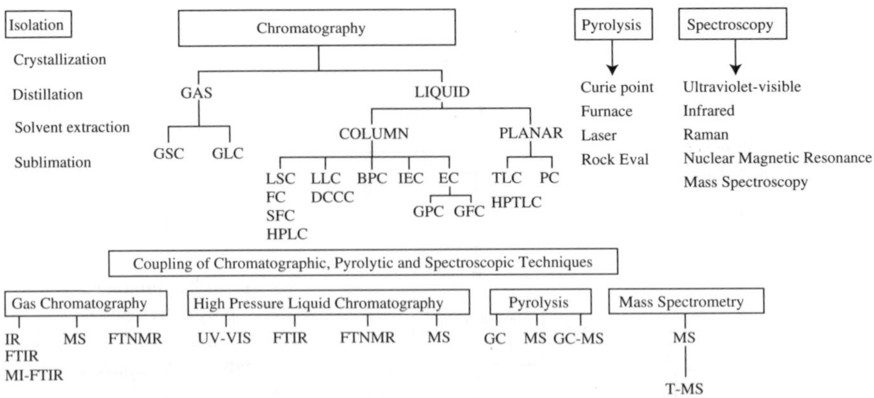

Chromatographic and spectroscopic techniques for detection and identification of organic compounds.

GC, gas chromatography; GLC, gas–liquid chromatography; GSC, gas–solid chromatography; TLC, thin layer chromatography; HPTLC, high performance thin layer chromatography; PC, paper chromatography; LSC, liquid–solid chromatography; FC, flash chromatography; SFC, supercritical fluid chromatography; LLC, liquid–liquid chromatography; DCCC, droplet counter current chromatography; PBC, bonded phase chromatography; HPLC, high pressure liquid chromatography; IEC, ion exchange chromatography; EC, exclusion chromatography; GPC, gel permeation chromatography; GFC, gel filtration chromatography; IR, infrared; UV, ultraviolet; NMR, nuclear magnetic resonance; MS, mass spectroscopy; FT, fourier transform; T-MS, Tandem mass spectroscopy; MI-FTIR, matrix isolation fourier transform infrared.

Fig. 3. Chromatographic and spectroscopic techniques.[9]

1.5 IDENTIFICATION OF NATURAL PRODUCTS BY CHROMATOGRAPHIC AND SPECTROSCOPIC METHODS

Recently, natural products chemistry has undergone explosive growth due to advances in isolation techniques, synthetic and biosynthetic approaches as well as spectroscopic and chromatographic methods.

The advent of computers and Fourier transform completely revolutionized the detection and identification of organic compounds. Modern automated instruments allow very small samples in the nanogram (10^{-9} g) range to be characterized in a very short time. The application of Fourier transform nuclear magnetic resonance (FTNMR) and Fourier transform infrared (FTIR) allows recovery of the sample in contrast to mass spectrometric (MS) determination which is a destructive but quite often a necessary technique.

Modern methods used to separate complex organic mixtures utilizing gas-liquid chromatography (GLC), high-pressure liquid chromatography (HPLC), and droplet counter-current (DCC) chromatography can separate samples rapidly and efficiently in the picogram range. This has been impossible until recently. Coupling the chromatographic instruments to spectrometers enables a partially automated analysis in an even shorter period of time. The following coupling of chromatographic instruments has been performed: GC-MS, GC-FTIR, GC-MI-FTIR, GC-UV-VIS, HPLC-MS, HPLC-FTIR, HPLC-FTNMR and MS-MS (Fig. 3).

These semiautomated systems of analyzing and characterizing small samples are vital to the natural product organic chemist and biochemist for the detection of highly active substances in extremely low concentrations in living organisms. A typical example is in the field of pheromones which includes insect sex attractants that differ quite markedly in many insects. Their concentration has often been found in the 10^{-9}–10^{-12} g range.

REFERENCES

1. Calvin M, *Chemical Evolution*, Clarendon Press, 1969.
2. Mechoulam R, Gaoni Y, A stereospecific synthesis of (-) delta-1-and (-) delta-6-tetrahydrocannabinol, *J Amer Chem Soc* **89**: 4532–4552, (1967).
3. Nakanishi K, A historical perspective of natural products chemistry in Barton D and Nakanishi K (eds.), *Comprehensive Natural Products Chemistry*, Vol. 2, XXI–XXXVIII, Elsevier Publishers, 1995.
4. Koskinen A, *Asymmetric Synthesis of Natural Products*, John Wiley and Sons, 1993.
5. Geissman TA, Croat DHG, *Organic Chemistry of Secondary Plant Metabolism*, Freeman, Cooper and Company, 1969.
6. Kinghorn AD, Balandrin MF (eds.), *Human medicinal agents*, Plenum Press, New York, 1993.
7. Jarvis BB, Miller JB, Natural products, complexity and evolution, in Romeo *et al.* (eds.), *Biochemical Diversity and Redundancy in Biological Interactions*, Plenum Press, New York, 1996.
8. Bruneton J, *Pharmocognosy, Phytochemistry, Medicinal Plants, Intersept* 2nd edn., pp. 4, 1999.
9. Ikan R, Crammer B, Organic Chemistry: Compound detection, in *Encyclopedia of Physical Sciences and Technology*, 3rd edn **11**: 453–496, 2002.

Chapter 2

PLANT-DERIVED NATURAL PRODUCTS IN DRUG DISCOVERY AND DEVELOPMENT: AN OVERVIEW

Mark Bahar, Ye Deng, Joshua N. Fletcher and A. Douglas Kinghorn

2.1 INTRODUCTION

For a large proportion of recorded human history, medicinal plants have been a major resource for treating and preventing disease. (For general references in this field, see[1–6]). Natural products, whether from terrestrial microbes and fungi, vertebrate animals, plants, or from marine fauna and flora, continue to be very valuable for this purpose.[7–17] In addition, small organic molecules from organisms are invaluable sources of prototype drug molecules ("lead compounds") and biochemical tools for probing fundamental cellular processes. The last fifty years have heralded great advances in the field of drug discovery in terms of understanding the molecular mechanisms underlying human disease, and natural products have played a part in elucidating these processes.[18–24]

When plant-derived natural products are examined specifically, it may be seen that they remain an invaluable but still incompletely exhausted resource for drug discovery purposes. Largely as a result of advances in natural product isolation techniques and spectroscopic methods for structural determination as well as the utilization of high-throughput bioassay systems, new drugs from higher plants are still being developed even now. In addition, biologically active compounds afforded from plants have unique structural types and possess novel mechanisms of action against many biological targets, as will be exemplified herein. This chapter serves as an update to a review on the same topic co-published by one of the present authors in 1993 in the edited volume *Human Medicinal Agents from Plants*.[25]

2.2 PLANT NATURAL COMPOUNDS AND DRUG DISCOVERY

A "paradigm shift" in drug discovery occurred in the early 19th century, marked by the isolation of pure bioactive entities from medicinal plants, beginning with the isoquinoline alkaloid, morphine.[3–5] The purification of plant drug molecules such as atropine, cocaine, codeine, digitoxin, and quinine, later in the same century, proved to be significant not only for the extensive medicinal uses of these isolates, but also for the crucial roles these molecules played in better understanding human disease and in the development of organic and medicinal chemistry.[26]

In the 20th century, additional important drugs were isolated from plants, including artemisinin, digoxin, paclitaxel, vinblastine, and vincristine. Also in this same century, as a result of advances made in pharmacology and a better understanding of human diseases at the molecular level, physicians and pharmacists gradually shifted from the use of plant extractives in prescriptions to pure naturally occurring and synthetically modified natural products, or totally synthetic compounds. Today, the effort to find new bioactive principles from medicinal plants may bring together scientists working in a diverse range of disciplines including biochemistry, botany, ethnobotany, medicinal chemistry, microbiology, molecular biology, organic chemistry, pharmaceutics, pharmacognosy, pharmacology, plant ecology, and taxonomy.[9,12–16] The advances in molecular biology are duly reflected in the complexity of bioassays employed in the medicinal plant drug discovery field, and also provide the "mode of action" information at the molecular level in a rapid and accurate fashion.[27] Preliminary *in vitro* experiments may then be followed up with a variety of *in vivo* bioassays.[28]

Terrestrial plants and other organisms are known to be the sources of a plethora of small organic molecules, representing considerable structural diversity, which may not be matched by the creativity of synthetic chemists.[15,29] When all currently known (both synthetic and natural) chemical entities are taken into consideration, the area of chemical space occupied by bioactive molecules is a relatively limited one.[15,29,30] When considered statistically, natural products have been found to account for much greater chemical diversity than compounds generated by both synthetic and combinatorial chemistry methods together. They also possess some unique structural differences, which afford them with greater drug-like qualities.[31–35] Accordingly, libraries of pure natural product, when combined with the advantages of combinatorial chemistry, offer an even more effective and reliable avenue for exploring the "more bioactive part" of chemical space when compared with a purely synthetic approach.[29,33,35]

From a chemical informatics perspective, natural products are significantly different from synthetic- and combinatorial chemistry-produced compounds in that they have more single bonds, and protonated amino and free hydroxyl groups, while having fewer aromatic rings.[36] Also, natural compounds possess a greater diversity of ring systems and tend to be more rigid than their synthetic counterparts.[32,34] Natural products also exhibit more chiral centers and fewer rotatable bonds than substances produced by synthesis.[32] The "rule of five," developed originally to guide the efforts of combinatorial chemistry, has spearheaded a new approach to drug discovery by defining the physicochemical characteristics a "drug-like" molecule should have.[37] In this respect, comparison studies have shown that natural products resemble proprietary drugs and show a high degree of "drug-likeness," compared to their synthetic counterparts.[34] The intricate biosynthetic processes that lead to bioactive secondary metabolites from organisms with bioactivity continue to be a major research interest and there has been much thought as to *how* and *why* these compounds are biosynthesized. The most widely accepted and satisfactory explanations refer to an evolutionary perspective, as will be described in the next two paragraphs.[30,38–40]

For survival purposes, all living organisms rely on an ability to transform and interconvert a diverse set of organic and inorganic compounds in order to utilize them as a source of energy and as their structural building blocks.[3] The presence of these crucial building molecules, the "primary metabolites" (amino acids, fatty acids, nucleosides, and sugars) can be considered synonymous with "life" since they are ubiquitous among all organisms. While the role of primary metabolism for the survival of species can be appreciated readily, the rationale for the presence of some other compounds with no apparent role in the internal economy of the producing organism is not so clearly evident. These small-molecule organic compounds of considerable structural diversity and typically a limited taxonomical distribution, the "secondary metabolites," are believed to provide the producing organisms with a survival advantage. Support for this may be inferred from the observation that organisms lacking immune systems (e.g. higher and lower plants, algae, fungi and microorganisms) generally show a high abundance of these compounds.[38] The term "natural product" is usually used interchangeably with "secondary metabolite" for the purposes of drug discovery from organisms.[3,38] It is found commonly that plants, like other organisms, tend to produce a series of analogues of a given structural type, rather than only one main secondary metabolite of a given class. It has been argued that over millions of years, the compounds providing a survival advantage are preserved (and even fine-tuned structurally through biosynthetic modification), while the not-so-active analogues are eliminated through evolutionary pressure.[30]

Plants interact chemically with other organisms such as insects, microorganisms, other plants, and even mammals as a result of their secondary metabolites, leading to a multitude of biological responses.[41,42] Clearly, since secondary metabolites are produced at the expense of the producing orgańism, it would be expected that this "chemical artillery" must offer the plant some advantage against its potential adversaries.[43] Secondary metabolites mostly exert their effects by acting through enzyme or protein interactions.[15,24,44] While some of these compounds act as substrates at the receptor level and mimic the endogenous substances in the target organism, others simply disrupt protein-protein interactions necessary for normal cell function (for examples, see[23,45–50]). Therefore, it might be concluded that it is this ability of secondary metabolites to interact with the physiology of other species that renders them as an impressive source of "evolutionarily fine-tuned" drug-like molecules.[30]

Drug discovery efforts from plants has been evolving continuously in response to a number of recent technological advances, such as the development of chromatographic methods that allow reproducible and fast purification steps for diverse compound classes; the availability of sensitive spectroscopic methods permitting the structural characterization of samples in microgram quantities; efficient chemical methods that permit the synthesis, derivatization, and optimization of bioactive lead compounds; and the wide accessibility of diverse sets of bioassays that provide the natural products chemist with critical information to target bioactive compounds from the early stages of separation studies.[11,17,28,33,35,51–54]

The standard practice in plant natural products isolation chemistry for drug discovery can be broken down into five steps, namely, organism collection, extraction, compound isolation, structure determination, and bioassay.[9,17,27,55–57] Plant samples are authenticated taxonomically and extracted with a solvent of choice and the resultant extracts are screened against pharmacologically relevant targets.[27,28,57] Once a positive result is achieved, isolation studies follow, guided by the relevant bioassay studies using the so-called "activity-guided fractionation" technique. When an active principle is isolated, thorough spectroscopic and spectrometric or X-ray crystallographic methods are employed as needed for the unambiguous assignment of its structure and configuration.[9,16,57] In the pharmaceutical industry and in other laboratories with considerable resources, many of the above-mentioned techniques are employed along with combinatorial and synthetic chemistry efforts, as well as computational modeling and chemical informatics studies coupled with specific high-throughput screening methods.[18,29,35,42,44,51–54,58–62]

2.3 EXAMPLES OF NATURALLY OCCURRING DRUGS OBTAINED FROM PLANTS

Despite the "synthetic revolution" in the pharmaceutical industry, it is readily apparent that even today, a large portion of the world's population, especially in developing countries, depends on medicinal plants as their primary source of medicine.[16,63] Fractionation studies on these traditionally used plant drugs very often reveal principles having biological activities consistent with their implicated usage.[63] In this section of the chapter, specific attention will be given to examples of plant-derived compounds that have played a role in the treatment of various diseases in their natural forms. A summary of some important plant-derived drugs used in their unmodified forms is shown in Table 1 and their structures are provided in Fig. 1.

Artemisinin ("qinghaosu") (**18**), a sesquiterpene lactone antimalarial compound with an endoperoxide group, discovered in the People's Republic of China as a constituent of *Artemisia annua* L., has created great interest in the biomedical community, owing to its unique mechanism of action on the heme complex. Artemisinin serves as an option for the treatment of chloroquine (**41**)-resistant malaria and is used in some Asian countries as an antimalarial.[64,65] However, the use of artemisinin as a single agent antimalarial is a potential risk since the malaria parasite may become resistant to this compound class.[66]

There is much interest in the medical applications of *Cannabis sativa* L. (marijuana). An oral spray consisting of the marijuana constituents, cannabidiol (CBD,**19**) and Δ^9-*trans*-tetrahydrocannabinol (THC, **20**), has been approved recently in Canada for the treatment of neuropathic pain associated with multiple sclerosis (MS), and it is possible that this drug will be approved elsewhere in the near future.[67]

In England, in the late 18th century, the purple foxglove (*Digitalis purpurea* L.) was shown to be efficacious in the treatment of dropsy, caused by congestive heart failure.[68] By the late 19th century, the cardiac glycoside digitoxin (**16**) was isolated from *D. purpurea*, while in the early 20th century, a similar cardiotonic glycoside, digoxin (**17**) was purified from the related species, *Digitalis lanata* Ehrh. These glycosides act by competing with K^+ ions for the enzyme ATPase. Both compounds (**16**) and (**17**) are still in use for their positive inotropic effects in cardiac patients.[69]

Galanthamine (**10**), an Amaryllidaceae-type alkaloid from *Galanthus woronowii* Losinsk and other species of this genus, has been approved in the last few years for the treatment of early-onset Alzheimer's disease.

Table 1. Some Examples of Major Plant-derived Natural Product Drugs

Compound or Class	Botanical Source	Therapeutic Category/Use
A. Alkaloids		
1. Belladonna-type solanaceous tropane alkaloids [Atropine (**1**), (−)-Hyoscyamine (**2**), Scopolamine ((−)-Hysoscine)(**3**)]	*Atropa belladonna* L. (belladonna), *Datura metel* L., *D. stramonium* L. (jimson weed), *Hyoscyamus niger* L. (henbane), *Mandragora officinarum* L. (European mandrake), and other solanaceous species	Anticholinergics (parasympatholytics)
2. *Catharanthus* (*Vinca*) alkaloids [Vinblastine (**4**), Vincristine (**5**)]	*Catharanthus roseus* (L.) G. Don (Madagascar periwinkle)	Anticancer
3. *Cinchona* alkaloids [Quinidine (**6**), Quinine (**7**)]	*Cinchona* spp. (*Cinchona* bark)	Cardiac antiarrhythmic Antimalarial
4. Cocaine (**8**)	*Erythroxylum coca* Lam. (coca leaves)	Local anesthetic
5. Colchicine (**9**)	*Colchicum autumnale* L. (autumn crocus), *Gloriosa superba* L. (glory lily)	Anti-gout
6. Galanthamine (**10**)	*Galanthus woronowii* Losinsk (snowdrop)	Anti-Alzheimer's disease
7. Nicotine (**11**)	*Nicotiana tabacum* L.	Smoking cessation (as transdermal patches)
8. Opium alkaloids [Codeine (**12**), Morphine (**13**)]	*Papaver somniferum* L. (opium poppy)	Antitussive Narcotic analgesic
9. Physostigmine (**14**)	*Physostigma venenosum* Balfour (Calabar bean)	Cholinergic (parasympathomimetic)
10. Pilocarpine (**15**)	*Pilocarpus jaborandi* Holmes (jaborandi), and related species	Cholinergic (parasympathomimetic)

(*Continued*)

Table 1. (Continued)

Compound or Class	Botanical Source	Therapeutic Category/Use
B. Steroidal glycosides		
1. *Digitalis* glycosides [Digitoxin (**16**), Digoxin (**17**)]	*Digitalis purpurea* L., *D. lanata* Ehrhart (foxgloves)	Cardiotonic glycosides
C. Terpenoids		
1. Artemisinin (**18**)	*Artemisia annua* L.	Antimalarial
2. Marijuana principles [Cannabidiol (**19**), Δ^9-*trans*-Tetrahydrocannabinol (**20**)]	*Cannabis sativa* L. (marijuana)	Analgesic (indicated for neuropathic pain in multiple sclerosis)
3. Paclitaxel (**21**)	*Taxus brevifolia* Nutt. (western or pacific yew)	Anticancer

Galanthamine exerts its activity through a competitive and reversible cholinesterase inhibitor action.[20,70–72]

Certain plants of the family Solanaceae, such as *Atropa belladonna* L., *Hyoscyamus niger* L., and *Datura stramonium* L., have been used medicinally for centuries in Europe because they contain tropane-type alkaloids.[9,26] For example, atropine (**1**) [a racemic mixture of (+)- and (−)-hyoscyamine (**2**)] and (−)-hyoscyamine are competitive antagonists at the muscarinic acetylcholine receptor site, leading to antispasmodic and antiallergic effects.[26] Scopolamine [(−)-hyoscine)] (**3**) is used in a transdermal patch for the prevention of motion sickness. Since these tropane alkaloids penetrate the blood-brain barrier, they also have psychoactive effects.[26]

Codeine (**12**) and morphine (**13**), both major active constituents of opium resin, the dried latex from the immature fruits of *Papaver somniferum* L., are pharmacologically important drugs as analgesic agents with considerable use today.[5,26,73,74] Like other opiates, morphine and codeine exert their actions on the opioid receptors.[73]

Nicotine (*Nicotiana tabacum* L.) (**11**), an agonist of the nicotinic-type acetylcholine receptor (nAChR), has long been known for its mechanism of action. Nicotine-like compounds offer much potential in the treatment

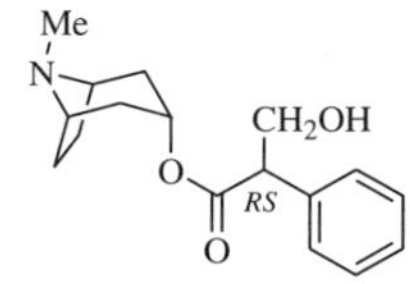

(1) Atropine

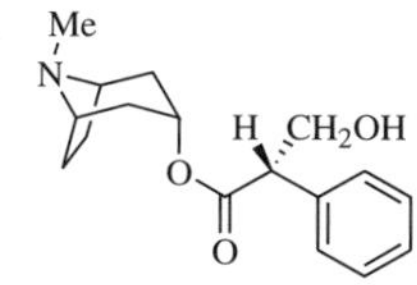

(2) (-)-Hyoscyamine

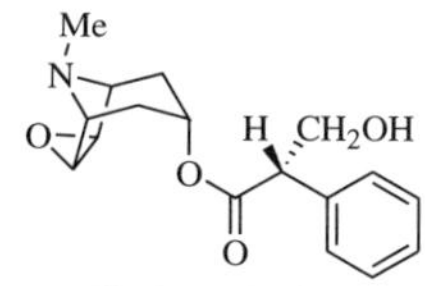

(3) Scopolamine [(-)-hyoscine]

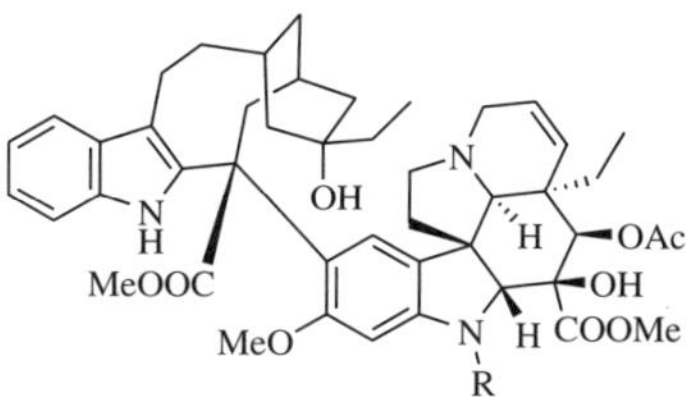

(4) R = CH_3 Vinblastine
(5) R = CHO Vincristine

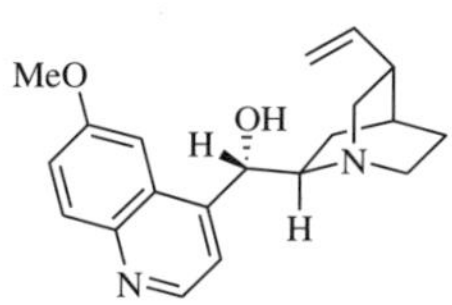

(6) Quinidine

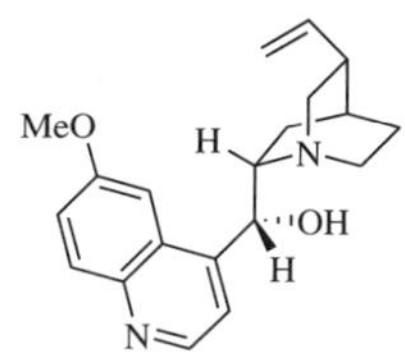

(7) Quinine

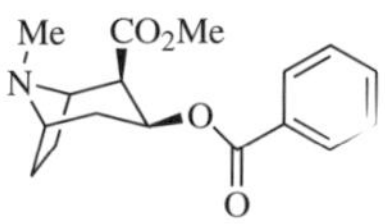

(8) Cocaine

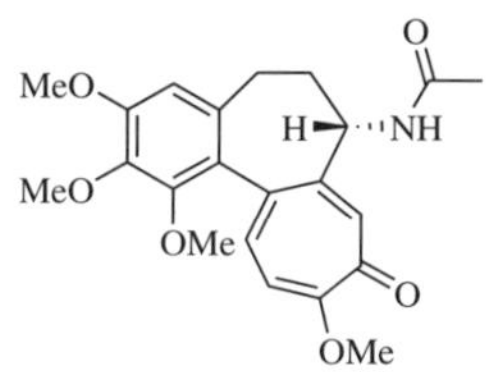

(9) Colchicine

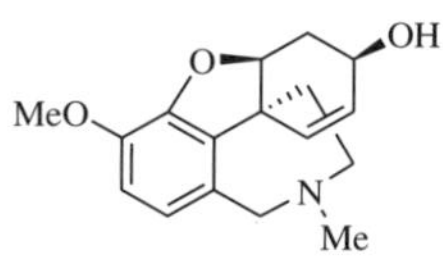

(10) Galanthamine

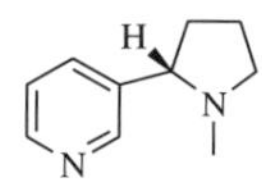

(11) Nicotine

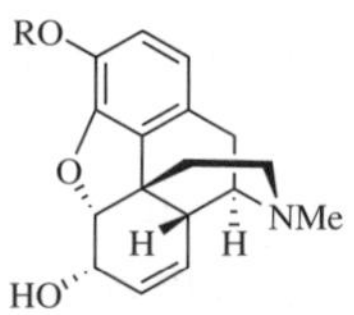

(12) R = Me Codeine
(13) R = H Morphine

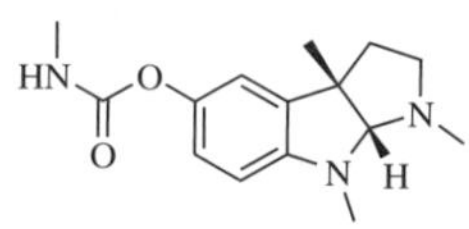

(14) Physostigmine

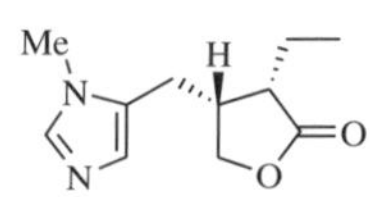

(15) Pilocarpine

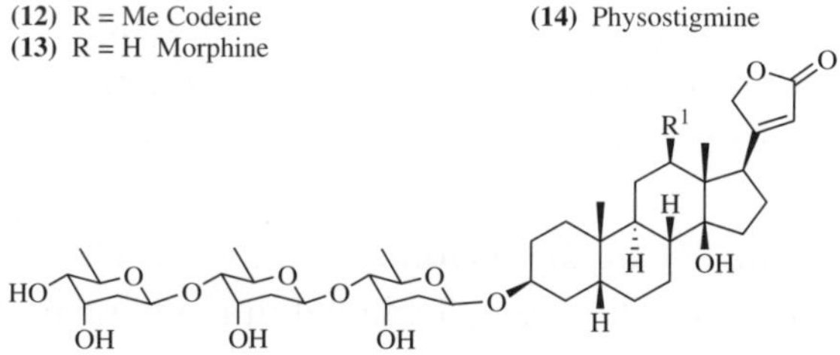

(16) R^1=H, Digitoxin
(17) R^1=OH Digoxin

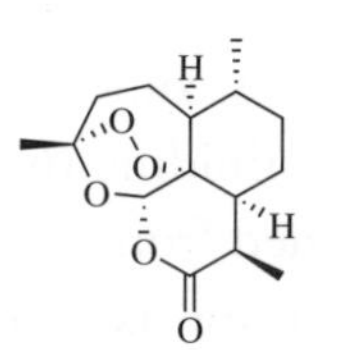

(18) Artemisinin

Fig. 1. Structures of major plant-derived natural product drugs.

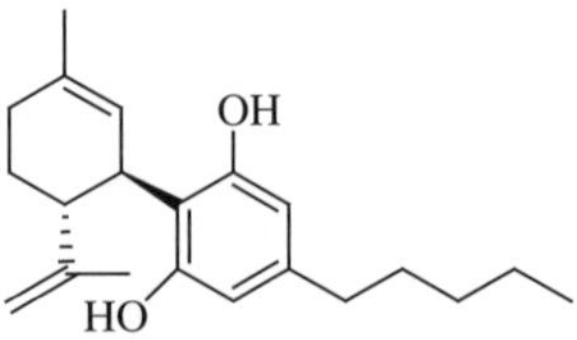

(**19**) Cannabidiol (CBD)

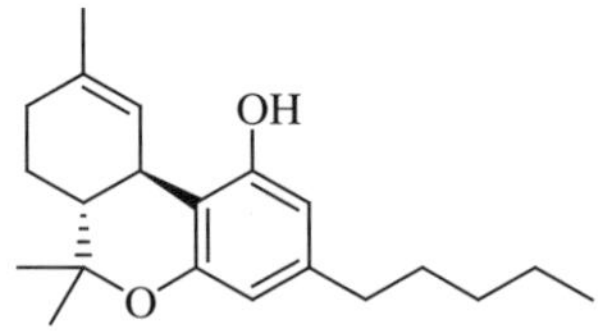

(**20**) Δ^9-*trans*-Tetrahydrocannabinol (THC)

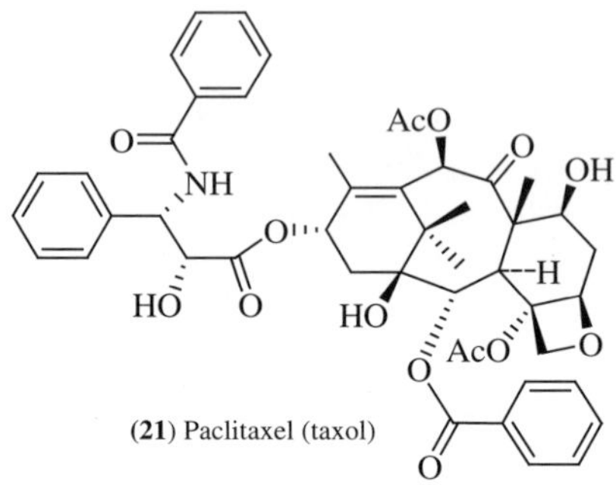

(**21**) Paclitaxel (taxol)

Fig. 1. (*Continued*)

of neurodegenerative diseases.[20,75] Nicotine is incorporated in transdermal patches employed for smoking cessation.[76]

Paclitaxel (**21**), formerly known as "taxol", is a nitrogen-containing diterpenoid compound isolated from the bark of *Taxus brevifolia* Nutt. (Pacific yew).[77] As an anticancer agent, paclitaxel acts as a tubulin stabilizer and leads to cell cycle arrest.[78] Since paclitaxel was originally isolated from the bark of the slow-growing species, *T. brevifolia*, sourcing was a major obstacle in the development of this drug and its introduction into the market.[78] However, as described later in this chapter, this has now been overcome.

Quinidine (**6**) and quinine (**7**) are diastereomeric quinoline alkaloids obtained from *Cinchona* spp. Quinidine (**6**) is included in many pharmacopeias for its antiarrhythmic effects.[2,3] Quinine was the first antimalarial drug and served as an effective remedy for this deadly infectious disease in colonial times, making European settlement in many tropical and subtropical parts of the world possible.[79] Owing to the development of resistance to synthetic antimalarials, quinine is still reverted to some extent for this purpose.[1–3]

Vinblastine (**4**) and vincristine (**5**) are closely related indole-dihydroindole dimers (bisindole alkaloids), isolated from *Catharanthus roseus* (L.) G. Don (formerly known as *Vinca rosea* L.), the Madagascar periwinkle. Both of these anticancer agents, known as vinca alkaloids in the medical literature, are specific binders of tubulin, leading to tubulin depolymerization and cell cycle arrest in the metaphase stage.[80–82]

2.4 EXAMPLES OF SEMISYNTHETIC DRUGS BASED ON LEAD COMPOUNDS FROM PLANTS

By the end of 19th century, the first semisynthetic and synthetic derivatives of plant secondary metabolites had begun to find use as drugs.[26] The fact that one of the earliest examples of these compounds utilized as a drug, acetylsalicylic acid (**44**), more commonly known as aspirin, still finds worldwide use, is testimony to the long-established potential of natural products as leads in drug discovery and design. From 1981 to 2002, semisynthetic compounds based on natural product leads represented 23% of the new drugs placed on the market, with an additional 24% of the new drugs from this same period being "natural product mimics" or utilizing a natural product "pharmacophore", the collective term for the groups on a molecule that are necessary for biological assay.[10] A new study providing updated figures on this topic has been published recently.[83] Synthetic optimization of natural product structures is used, for example, to enhance potency; increase bioavailability; remove unwanted side effects; solve sourcing problems; or to display biological properties that are opposite to that of the natural product through agonist/antagonist interactions. Examples of semisynthetic drugs of plant origin are included in Table 2, and the structures of some clinically significant drugs mentioned, along with their natural product lead compounds, are shown in Figs. 1 and 2.

Artemisinin (**18**) is a natural product for which many semisynthetic derivatives have been generated. The major rationale to produce these derivatives was to deal with the low aqueous solubility of artemisinin and its short half-life in plasma.[84] The lipid-soluble arteether (**22**) and artemether (**23**), and the water-soluble sodium artesunate (**24**), were designed for

Table 2. Examples of Semisynthetic Agents from Plant Natural Product Lead Compounds

Lead Compound	Compound	Class of Compound	Therapeutic Category / Use
Artemisinin (18)	Arteether (22)	Sesquiterpenoid	Antimalarial
Artemisinin (18)	Artemether (23)	Sesquiterpenoid	Antimalarial
Artemisinin (18)	Sodium artesunate (24)	Sesquiterpenoid	Antimalarial
Atropine (1)	Ipratropium bromide (25)	Tropane alkaloid	Asthma
Atropine (1)	Tiotropium bromide (26)	Tropane alkaloid	Chronic obstructive pulmonary disease
Camptothecin (27)	Irinotecan (28)	Quinoline alkaloid	Anticancer
Camptothecin (27)	Topotecan (29)	Quinoline alkaloid	Anticancer
(+)-Cytisine (30)	Varenicline (31)	Quinolizidine alkaloid	Smoking cessation
Khellin (32)	Cromolyn sodium (33)	Chromone derivative	Vernal keratoconjunctivitis, vernal conjunctivitis, and vernal keratitis
Leptospermone (34)	Nitisinone (35)	β-Triketone	Hereditary tyrosinemia type 1
Morphine (13)	Apomorphine (36)	Benzyltetrahydro-isoquinoline alkaloid	Parkinson's disease
Paclitaxel (21)	Docetaxel (37)	Diterpenoid	Anticancer
Podophyllotoxin (38)	Etoposide (39)	Lignan	Anticancer
Podophyllotoxin (38)	Teniposide (40)	Lignan	Anticancer
Quinine (7)	Chloroquine (41)	Quinoline alkaloid	Antimalarial
Salicin (42)	*p*-Aminosalicylic acid (43)	Salicylic acid derivative	Antituberculosis
Salicin (42)	Aspirin (44)	Salicylic acid derivative	Analgesic
Salicin (42)	Mesalamine (45)	Salicylic acid derivative	Ulcerative colitis
Thebaine (46)	Hydrocodone (47)	Benzyltetrahydro-isoquinoline alkaloid	Analgesic
Thebaine (46)	Naloxone (48)	Benzyltetrahydro-isoquinoline alkaloid	Antidote for opiate overdose
Thebaine (46)	Naltrexone (49)	Benzyltetrahydro-isoquinoline alkaloid	Alcohol and opiate dependence
Theophylline (50)	Aminophylline (51)	Purine alkaloid	Antiasthma
Vinblastine (4)	Vinorelbine (52)	Indole alkaloid	Anticancer

use in severe malaria cases.[84] Dihydroartemisinin is a metabolic product of the artemisinin derivatives, and shows greater potency than artemisinin (**18**).[84] Currently, the World Health Organization (WHO) recommends the following artemisinin-based oral combination therapies for the treatment of malaria: artemether (**23**) and lumefantrine; sodium

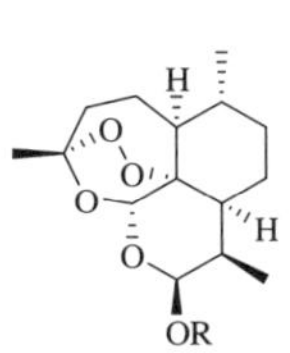

(**22**) R=Et Arteether
(**23**) R=Me Artemether

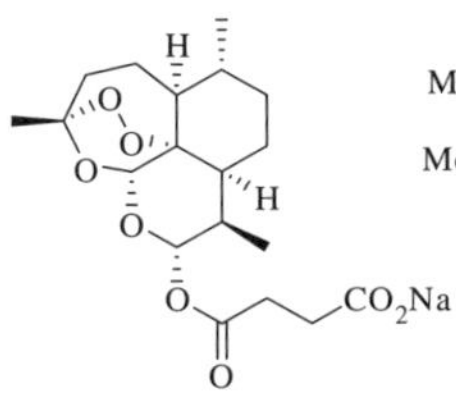

(**24**) Sodium artesunate

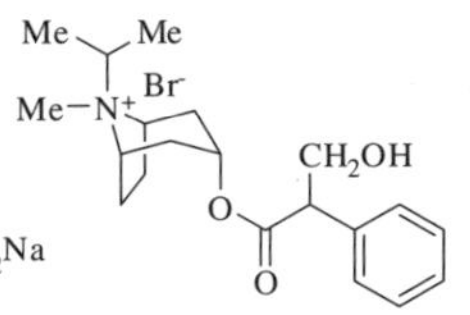

(**25**) Ipratropium bromide

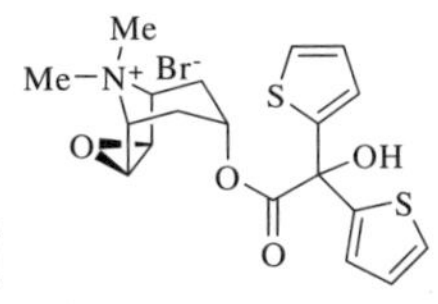

(**26**) Tiotropium bromide

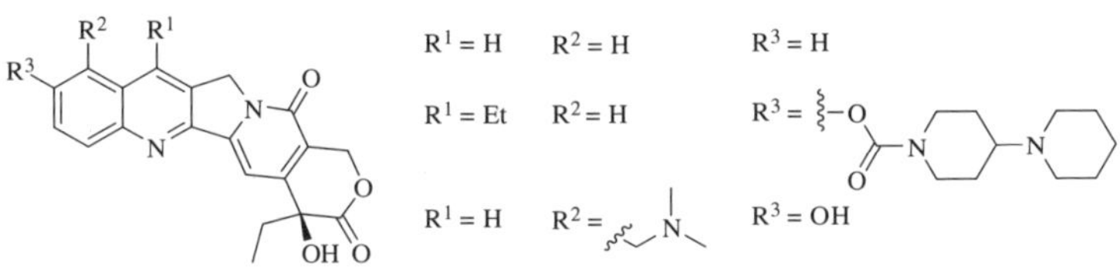

(**27**) Camptothecin
(**28**) Irinotecan
(**29**) Topotecan

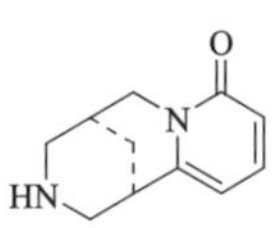

(**30**) (-)-cytisine

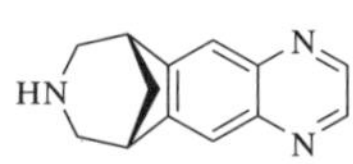

(**31**) Varenicline

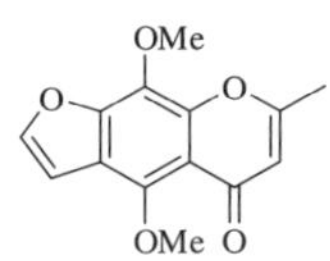

(**32**) Khellin

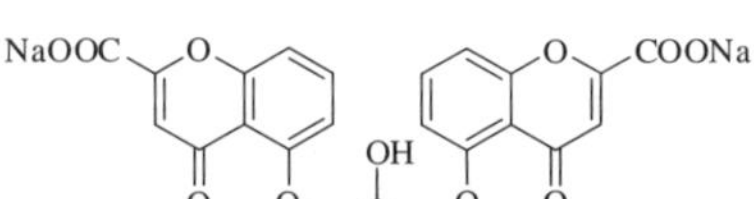

(**33**) Cromolyn sodium

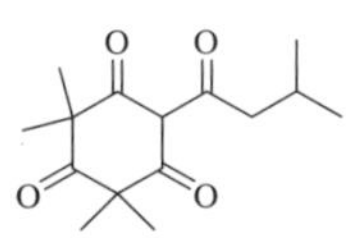

(**34**) Leptospermone

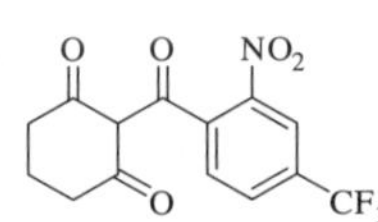

(**35**) Nitisinone

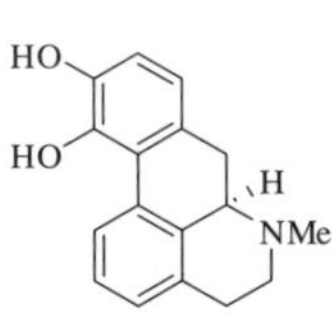

(**36**) Apomorphine

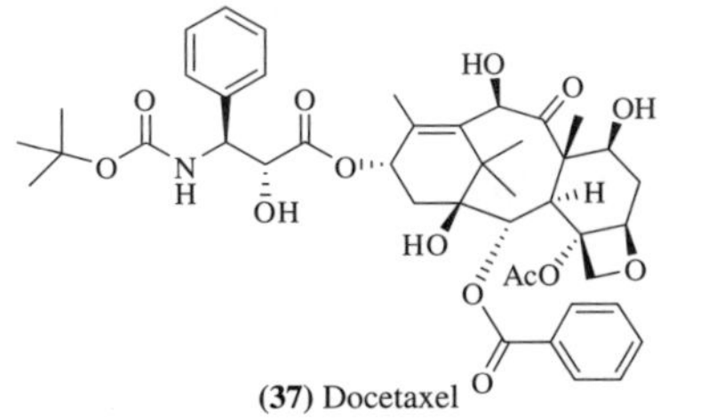

(**37**) Docetaxel

Fig. 2. Structures of plant natural product lead compounds and semisynthetic agents.

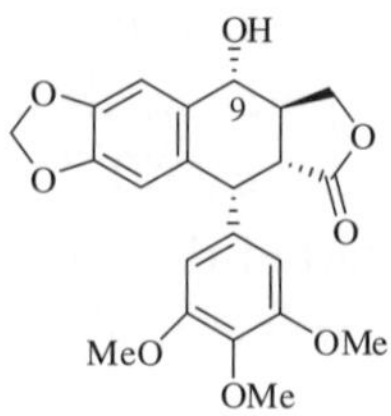

(38) Podophyllotoxin

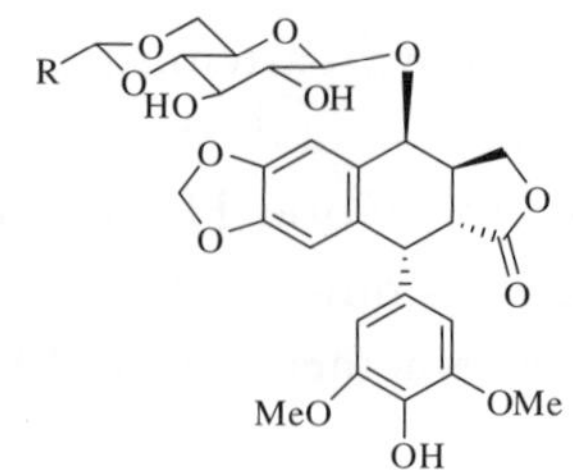

(39) R=Me Etoposide

(40) R= 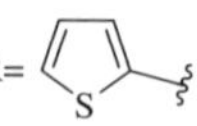 Teniposide

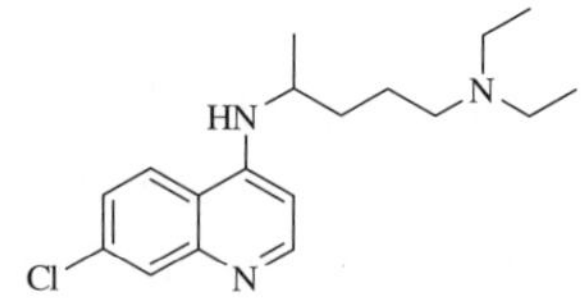

(41) Chloroquine

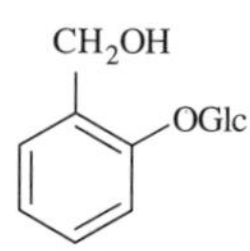

(42) Salicin

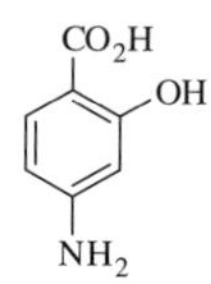

(43) *p*-Aminosalicylic acid

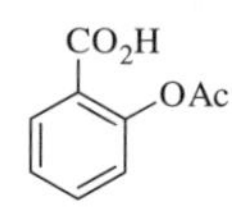

(44) Aspirin (acetylsalicylic acid)

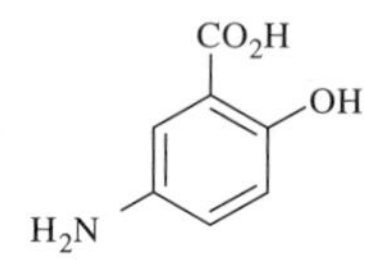

(45) Mesalamine

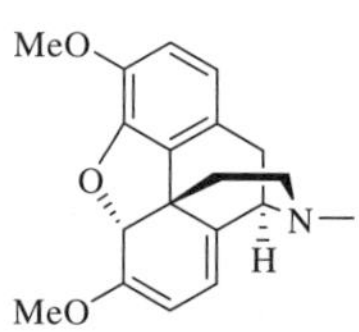

(46) Thebaine

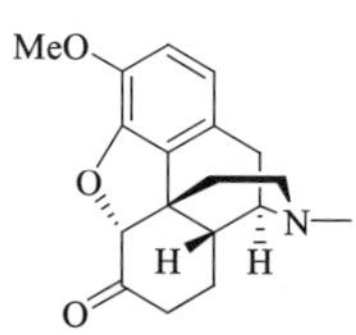

(47) Hydrocodone

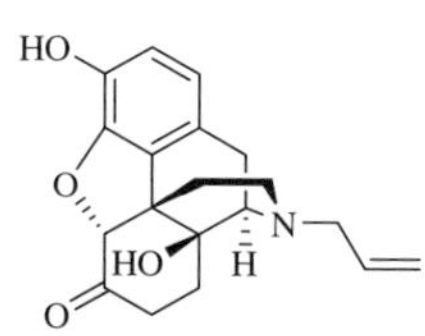

(48) Naloxone

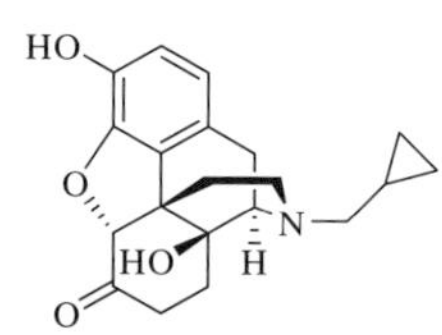

(49) Naltrexone

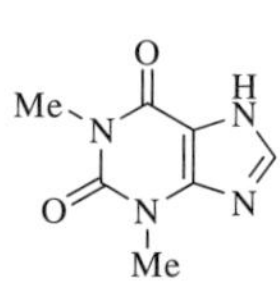

(50) Theophylline

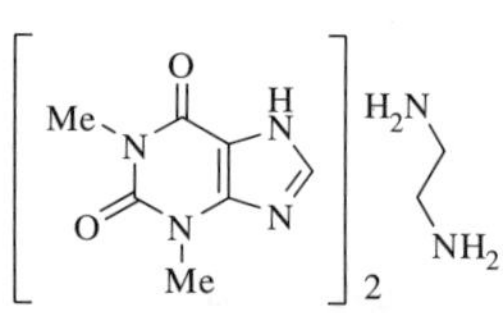

(51) Aminophylline [mixture of theophylline and ethylenediamine (2:1)]

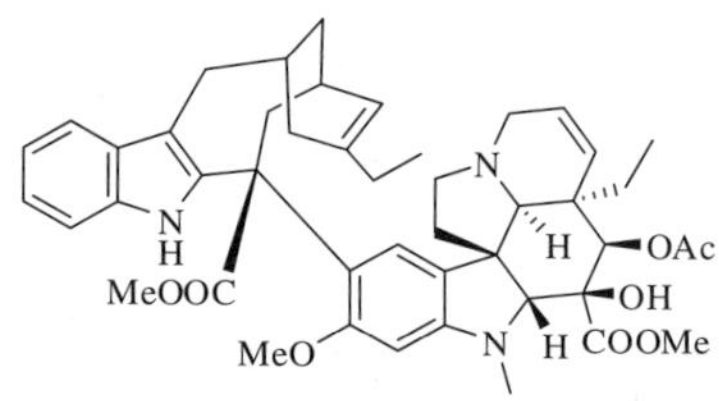

(52) Vinorelbine

Fig. 2. (*Continued*)

artesunate (24) and amodiaquine; sodium artesunate (24) and mefloquine; and sodium artesunate (24) and sulfadoxine–pyrimethamine.[85] Dihydroartemisinin-piperaquine has been found to be safe and effective in large trials in Asia, but is not yet recommended by WHO because of a lack of trials in Africa and South America, and also due to sourcing problems.[85] OZ277/RBx-11160 is a trioxolane with antimalarial activity inspired by the structure of artemisinin (**18**).[86] This drug candidate is currently in phase II clinical trials.[87]

Aspirin (**44**) has its roots in the use of salicylate-containing plants with analgesic, anti-inflammatory, and antipyretic properties.[26,88] In the 1800s, salicylaldehyde and salicin (**42**) were isolated from meadowsweet, *Filipendula ulmara* L. (Rosaceae) (originally named *Spirea ulmaria*), and methyl salicylate from the oil of wintergreen *Gaultheria procumbens* L. from each of these compounds, salicylic acid was produced.[26,88] Pure salicin (from *Salix* spp.) was used to treat rheumatism in Western medicine throughout the mid 1800s. However, salicylic acid, a salicin-like compounds, became widely available after its facile synthesis was devised.[26,88] Phenolic groups were found to irritate the stomach and therefore the phenolic hydroxyl group of salicylic acid was substituted with an acetyl group, creating acetylsalicylic acid or aspirin (**44**).[26,88] Two other semisynthetic derivatives of salicylic acid that are clinically significant today are *p*-aminosalicylic acid (**43**) and mesalamine (**45**), indicated in the treatment of tuberculosis and ulcerative colitis, respectively.[89,90]

Derivatives of atropine (**1**) represent a major set of drugs based on a natural product lead molecule. Indeed, synthetic analogues of atropine account for more approved drugs than any other plant-derived secondary metabolite. Chemical modification of atropine to find the effect of translocating the ester group about the tropane ring led to the discovery that the tropane ring was unnecessary and that all anticholinergic activity remained with the tropate ester linked to a tertiary amine by two or three carbons, allowing for many tropane derivatives to be made.[26] To provide an early example, homatropine was developed in 1884, and was found to dilate the pupils at a faster rate than atropine while paralyzing the muscles of the eye for a shorter duration, compared to atropine.[26] Ipratropium bromide (**25**), which is used to treat asthma, is a contemporary example of a clinically useful atropine derivative.[91] Further illustrating the potential of this class of compounds, atropine (**1**) has served as a lead molecule for tiotropium bromide (**26**), which was introduced in 2002 as a treatment for chronic obstructive pulmonary disorder (COPD).[92]

The *Catharanthus* (*Vinca*) alkaloids, vinblastine (**4**) and vincristine (**5**), are not only very useful anticancer agents, but are also *bona fide* lead compounds. A semisynthetic vinblastine (**4**) analogue, vinorelbine (**52**) is indicated for lung and mammary carcinomas.[81] Vinflunine was further developed from vinorelbine, and has shown promise in phase II clinical trials.[81,93]

Representatives of another important class of plant-derived semisynthetic compounds are the camptothecin (**27**) derivatives, irinotecan (**28**) and topotecan (**29**). Camptothecin (**27**) was originally discovered as an antileukemic agent in a mouse model when isolated from *Camptotheca acuminata* Decne.[78] Compounds (**28**) and (**29**) are both employed in cancer chemotherapy. These substances are important mechanistically because of their activity against the enzyme, topoisomerase I. These compounds were designed to address efficacy and toxicity concerns with the parent compound, camptothecin, and its sodium salt.[78]

(+)-Cytisine (**30**), from *Cytisus scoparius* L., has been found to be a selective and specific probe for the nicotinic-type acetylcholine receptor (nAChR) as a partial agonist and has served as the lead compound for varenicline (**31**), a drug recently approved by the FDA for smoking cessation.[94]

Initial observations on khellin (**32**), a furochromone-type compound from *Ammi visnaga* L., showed a mast-cell stabilizing effect. This led to the synthesis of many derivatives of this lead for the potential treatment of allergy and asthma, most notably the approved preventive antiasthma drug, cromolyn sodium (**33**). The precise molecular mechanism of this class of drugs is not yet clearly understood.[95]

Leptospermone (**34**), a representative of an important new class of herbicides from the bottlebrush plant, *Callistemon citrinus* (Curtis) Skeels, has been found to have an inhibitory effect on the enzyme, *p*-hydroxyphenylpyruvate dioxygenase (HPPD), involved in the synthesis of plastoquinone in plants. Nitisinone (**35**), a synthetic derivative of (**34**), has recently been introduced to the market for the treatment of hereditary tyrosinemia type 1 (HT-1), a severe genetic disease caused by a deficiency of fumaryl acetoacetate hydrolase (FAH).[96]

Morphine (**13**), from *Papaver somniferum* L., is an excellent example of a lead compound from which numerous synthetic derivatives have been produced.[26,97] Opiates, narcotic compounds isolated or chemically derived from the constituents of *P. somniferum*, have been used for various therapeutic effects including use as analgesics, antidiarrheals, antitussives, and sedatives.[97] One of the early morphine derivatives to be produced was apomorphine (**36**), which has found use recently as a therapy for Parkinson's disease.[26] The most widely prescribed drug in 2004 in the USA was a mixture containing acetaminophen and the semisynthetic opiate hydrocodone (**47**), derived from the alkaloid lead compound, thebaine (**46**).[74] Naloxone (**48**), and naltrexone (**49**) are opiates that are opioid receptor pure antagonists, with the former being used to prevent fatal overdoses of opiate narcotics, while the latter is indicated in overcoming alcohol dependence as well as opiate addiction.[26,98,99] Dextromethorphan is a nonnarcotic opiate derivative that is widely used as an over-the-counter antitussive.[100]

Two derivatives of podophyllotoxin (**38**), etoposide (**39**) and teniposide (**40**), are further examples of the use of a natural product as a drug lead. In 1890, podophyllotoxin was isolated from *Podophyllum peltatum* L. (American May Apple), and in 1948, it was shown to have antineoplastic activity in mice bearing tumors.[101,102] Then, hundreds of podophyllotoxin analogues were synthesized in the hope of finding a drug with a better therapeutic index than the lead compound, (**38**), which showed considerable toxicity.[102] The epipodophyllotoxin derivatives (differing from (**38**) by the inverted stereochemistry at C-9), etoposide (**39**) and teniposide (**40**), were products of this search. Interestingly, they have a different mechanism of action compared with podophyllotoxin, in that while the lead compound is a microtubule inhibitor, etoposide and teniposide inhibit the enzyme topoisomerase II.[102] Etopophos (etoposide phosphate) is a clinically used prodrug of etoposide (**39**), and has an increased water solubility compared to etoposide.[102]

As will be mentioned further in this chapter, the discovery of the anticancer drug, paclitaxel (**21**), was soon followed by sourcing issues due to the low yield of this compound in the source plant, *Taxus brevifolia* Nutt.[77] This led to the semisynthesis of paclitaxel (**21**), from a precursor molecule, 10-deacetylbaccatin III, readily available from the leaves of *Taxus baccata* L., a renewable source with high yields of this compound.[26,77] Docetaxel (**37**) is a second taxane class anticancer drug and is a semisynthetic derivative of paclitaxel (**21**).[103]

A plant-derived compound used in the treatment of asthma and COPD is the methylxanthine-type alkaloid, theophylline (**50**), found naturally in tea (*Camellia sinensis* Kuntze). This compound demonstrated higher activity when complexed with bases, as in its semisynthetic analogue aminophylline (**51**).[95]

2.5 PLANT-DERIVED COMPOUNDS USED AS "BIOCHEMICAL TOOLS"/"PHARMACOLOGICAL PROBES"

Until the mid-20th century, the discovery of bioactive entities from plants was largely the result of a trial and error process, without much understanding of their mechanism of action in vertebrate physiological systems. Opium, the dried latex of the capsules of *Papaver somniferum* L., for instance, has been used for its analgesic properties for several millennia, with the major active narcotic analgesic alkaloid morphine (**13**) being isolated at the beginning of 19th century.[104] Nevertheless, it was only in 1973 that the opioid receptors were characterized, thus helping to explain the mode of action of opiates.[73] In a similar fashion, plant secondary metabolites have helped define many receptor types and have served as invaluable tools to help elucidate biological processes. These types of compounds are known as "biochemical tools" or "pharmacological probes." The structures of the compounds mentioned in this section of the chapter are in Figs. 1–3. Compounds are described in turn, in terms of their activities on the CNS, DNA, proteins, enzymes, and miscellaneous molecular targets.

The isolation of pure plant natural products, such as, pilocarpine (**15**), physostigmine (**14**), (−)-hyoscyamine (**2**), and nicotine (**11**), and the fungal isolate, muscarine (**53**), have played a major role in the development of modern pharmacology through enabling the discovery of intricate receptor-ligand mechanisms, which could not be probed earlier. Natural products, including some traditionally known to have psychoactive properties, have provided many potent CNS-active agents.[20] Based on the contribution of plant psychoactive compounds in the understanding of neurochemical processes, *in silico* methods for screening the

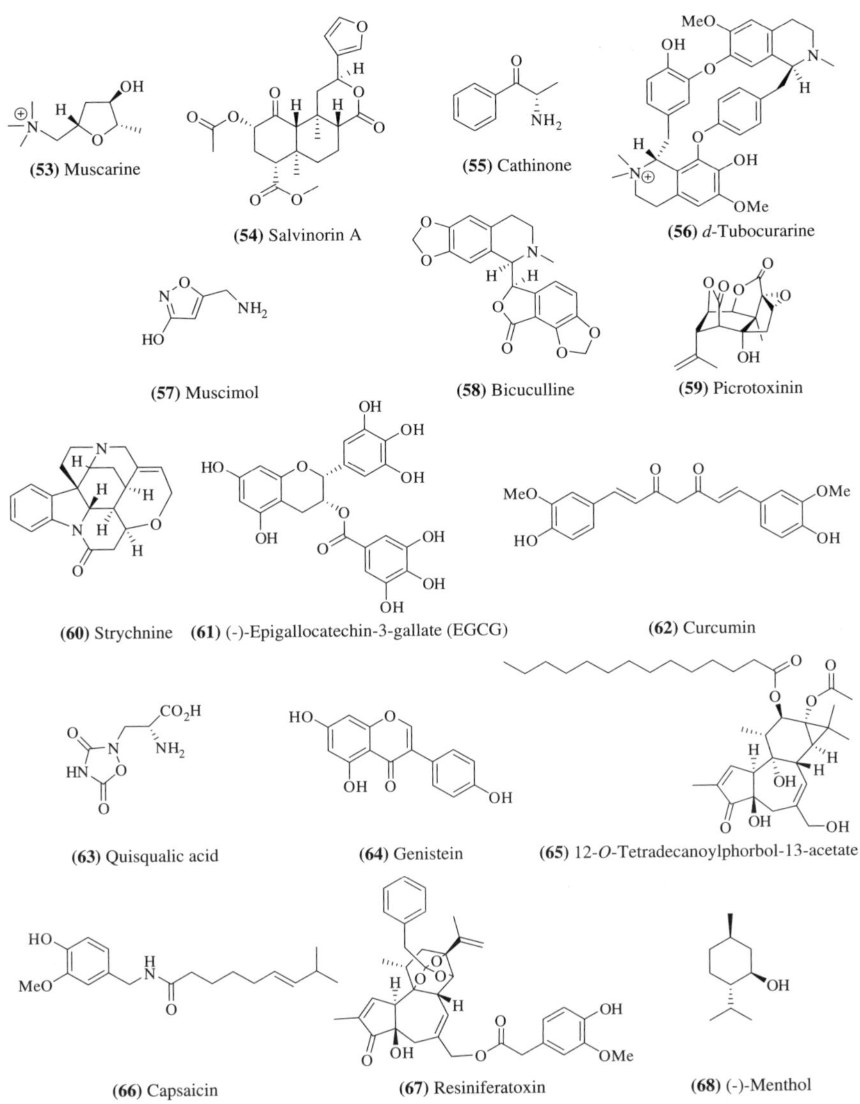

Fig. 3. Structures of "biological tools" ("pharmacological probes") from plants.

receptorome for plant-derived compounds have been suggested to further elucidate and define molecular mechanisms, along with methods to make them more compatible with modern assay systems, such as fluorescent modifications.[60,105]

The cannabinoid-type psychoactive principles of marijuana (*Cannabis sativa* L.), have been of interest to pharmacologists for many years.[106] The biological characterization of the major euphoriant principle, (−)-Δ^9-*trans*-tetrahydrocannabinol (THC) (**20**), has led to an understanding of the molecular mechanisms of these compounds, and ultimately enabled the characterization of the cannabinoid receptor (CB_1).[107] This finding was followed by the discovery of the endogenous ligands, and also another cannabinoid receptor subtype, CB_2. The first selective antagonist-inverse agonist at the CB_1 receptor, rimonabant, was recently approved in the European Union for the treatment of obesity and metabolic disorder.[108]

Cocaine (**8**), from *Erythroxylum coca* Lam., besides causing euphoria by inhibiting the dopamine transport protein (DAT) responsible for its recreational and illegal use, exerts a local anesthetic activity through blocking sodium channels and is still used as a probe for this target.[109]

The discovery of a potent and selective κ-opioid receptor agonist compound, salvinorin A (**54**), a hallucinogenic neoclerodane diterpenoid from *Salvia divinorum* Epling and Jativa, has created particular interest in recent years, since it is the first nonnitrogenous compound found to demonstrate this type of activity.[110,111]

Chiral protoalkaloids, such as cathinone (**55**), a psychotropic constituent of *Catha edulis* Forssk. (khat), have provided probes for studying the mechanistic properties of biogenic amine transporters, and afforded information regarding the effect of stereochemistry at the transportation level of these nitrogenous compounds.[112]

d-Tubocurarine (**56**), isolated from a South American plant used as an arrow poison, *Chondodendron tomentosum* Ruiz and Pav. (curare), while no longer employed as a skeletal muscle relaxant drug, is still used in pharmacological studies as a competitive agonist of nicotinic acetylcholine receptors (nAChR) and also as an agonist of the 5-hydroxytryptamine (5-HT_3) receptor.[48] Muscimol (**57**), a centrally acting principle of the mushroom *Amanita muscaria* (fly agaric), is known to be a potent $GABA_A$ receptor agonist and led to the development of a bicyclic analogue, gaboxadol, a drug currently in phase III clinical trials for the treatment of sleep disorders. Bicuculline (**58**) (*Dicentra cucullaria* Bernh.) and picrotoxinin (**59**) (*Cocculus indicus* Royle), on the other hand, serve as potent antagonists at $GABA_A$ receptors.[48] Strychnine (**60**), the toxic principle of *Strychnos nux-vomica* L., has proven to be useful in the studies on glycine (Gly) receptors as a competitive antagonist of this receptor type at nM levels.[48] Members of the chalcone subtype of flavonoids have provided useful tools for probing the p53/MDM2 system through the inhibition of ubiquitin ligase.[50] Quisqualic acid (**63**) (*Quisqualis indica* L.) is a potent agonist for the ionotopic glutamate (iGlu) receptors.[48]

Epigenetic modulation is important for the treatment of cancer and the understanding of cancer cell dynamics. DNA-methylation is regulated by the DNA-methyl transferase (DNMT) family of enzymes which constitute potential targets with antiproliferative outcome in cancer chemotherapy. (−)-Epigallocatechin-3-gallate (EGCG) (**61**) (*Camellia sinensis* Kuntze), and other flavan derivatives have been shown to be DNMT1 inhibitors at μM levels.[49] Curcumin (**62**) (*Curcuma longa* L.) has been found to act as a histone acetyl transferase (HAT) inhibitor, leading to transcriptional silencing.[50]

Podophyllotoxin (**38**) (*Podophyllum peltatum* L.), colchicine (**9**) (*Colchicum autumnale* L.), vinblastine (**4**), and vincristine (**5**) [*Catharanthus roseus* (L.) G. Don] are standard microtubule-destabilizing agents used in cancer research.[80,113] Paclitaxel (**21**), from *Taxus brevifolia* Nutt., acts as a promoter of stabilization of microtubules and causes mitotic arrest in an unusual fashion.[77,114]

Genistein (**64**), an isoflavone found in plants of the family Leguminosae, is an inhibitor of several protein tyrosine kinases (PTK) and is currently in phase II clinical trials for its potential as an angiogenesis inhibitor.[23,115] Genistein also has been utilized as a probe to identify binding sites for PTKs by observing the effect it demonstrates on cyclic-nucleotide-gated channels.[116]

Long-chain ester derivatives of phorbol, a tetracyclic diterpene from the seed oil of *Croton tiglium* L., including its most abundant representative, 12-*O*-tetradecanoylphorbol-13-acetate (**65**), are potent activators of protein kinase C (PKC) and are used as standard tumor promoters for the study of experimental carcinogenesis in animal models.[117]

Investigations on capsaicin (**66**), the vanillyl-group containing "hot" principle in chili peppers (*Capsicum* spp.), as well as the irritant compound resiniferatoxin (**67**) (*Euphorbia resinifera* Berg.), have led to the characterization of transient receptor potential (TRP) channels and the vanilloid receptors (TRPV1).[118] These receptors show the possibility of serving as new targets to assist with diseases that involve chronic pain.[119] While ligands acting on TRPV1 cause a burning sensation, compounds with a cooling effect, such as (−)-menthol (**68**) (*Mentha spicata* L.), have led to the characterization of cold sensors, most importantly the TRPM8 receptor, which also plays a role in the nociceptive process.[22]

2.6 STRATEGIES FOR MEDICINAL PLANT CONSERVATION

In spite of the advances in chemical synthesis, combinatorial chemistry, and biological engineering and genetic manipulation of microbes in the course of finding new small-molecule therapeutic leads, drug discovery from plants still plays an important role.[10] However, to date, only a small portion of all plant species have been phytochemically and/or pharmacologically investigated. Even from previously well studied medicinal plants, there is still a good chance of finding new bioactive constituents for potential drug development.[56] With the continual removal of forests for timber, farmland, and industrial zones, as well as the constantly increasing demand for medicinal plants worldwide, the survival of natural stands of many presently used and potential medicinal plants is threatened. This situation can be highlighted by several examples. The plant silphium (most likely a *Ferula* species), also known as silphion, which was widely used in ancient Libya as a perfume and a spice, as well as a drug for the treatment of gynecological and other ailments, disappeared from the market around 100 CE and became extinct, because of its unsustainable exploitation.[120] The Himalayan yew tree (*Taxus wallichiana* Zucc.) has been threatened to close to extinction in the Himalayan region of India due to deforestation and uncontrolled harvesting for the extraction of paclitaxel (**21**).[121,122] Recent research has also shown that with the increasing demand for huperzine A, used to improve mental performance, supplies of the source plants from the genus *Huperzia* have become endangered due to uncontrolled collection from the wild and concomitant difficulties in cultivation of these species.[123] The herb Goldenseal (*Hydrastis canadensis* L.) was formerly distributed widely throughout the northeastern USA and eastern Canada. After early settlers adopted the plant for its antibiotic, anti-inflammatory, and tonic effects, the populations of this plant have decreased dramatically. This trend has continued to the present day, since Goldenseal has become a popular botanical dietary supplement in the United States. In 1997, *H. canadensis* was listed in the Convention on International Trade in Endangered Species of Wild Fauna and Flora (CITES) Appendix II for protection purposes.[124]

In 1988, the International Consultation on the Conservation of Medicinal Plants was held in Chiang Mai, Thailand. It was convened by the World Conservation Union (IUCN), the World Health Organization (WHO), and the World Wildlife Fund (WWF). This conference led to the "Chiang Mai Declaration" that called for global efforts for the conservation and sustainable utilization of medicinal plants.[125] In 1993, WHO published Guidelines on the Conservation of Medicinal Plants, which provided a general framework for implementation of the Chiang Mai Declaration.[126] It is estimated by IUCN that more than 138 organizations are now devoted to the conservation of medicinal plants.[127] For example, the Medicinal Plant Specialist Group (MPSG) is a member of the IUCN Species Survival Commission. The organization was formed in 1994 with the purpose of increasing "awareness of conservation threats to medicinal plants and to promote conservation action" (MPSG home page, http://www.iucn.org/themes/ssc/sgs/mpsg/index.html). The World Bank has financed several programs in developing countries for the conservation and sustainable development of medicinal plants. These programs include the Kerala Forest Project in India, the Sri Lanka Medicinal Plants Project in Sri Lanka, and TRAMIL in Central America and the Caribbean.[125,128] Different conservation methods have been studied, with some of these applied in practice. Such methods include *in situ*/on-farm conservation like agricultural farming, natural reserve and sacred gardening, establishment of *ex situ* germplasm banks, development of micropropagation and agricultural methods for wild medicinal plants, and campaigns for public awareness, among others. Many laws and treaties that intend to protect the trade of biological resources also protect medicinal plants at the same time. Some widely used medicinal plants, such as *Dioscorea deltoidea* Wall., *Podophyllum hexandrum* Royle, *Rauvolfia serpentina* Baill., and *Taxus wallichiana* Zucc. are on the Appendix II list of the Convention on International Trade of Endangered Species (CITES), so that the trade of these species is strictly controlled by both exporting and importing nations. The Convention on Biological Diversity (CBD) also contributes to the retention of biodiversity of medicinal plants. In general, as both overall organism biodiversity decreases and populations of wild medicinal plants continue to diminish specifically, a more integrated and comprehensive cooperation among governments, various research institutes, relevant organizations, and the general public needs to be enacted in order to conserve these precious resources for the present and future health of humanity.[125]

2.7 BIOTECHNOLOGY / DIRECTED BIOSYNTHESIS ASPECTS

According to the United Nations Convention on Biological Diversity, biotechnology is defined as "any technological application that uses biological systems, living organisms, or derivatives thereof, to make or modify products or processes for specific use".[129] Perhaps the earliest application of plant biotechnology was the use of yeast to break down sugars into ethanol and carbon dioxide for the production of alcohol. In the 1830s, Cagniard de Latour, Schwann, and Kützing independently realized that alcoholic fermentation is a biological process involving the common yeast, and Payen and Persoz extracted diastase from malted barley, with this enzyme found to remove starch from grain husks and convert it into sugars.[130,131] These early findings allowed for the scientific understanding and the subsequent manipulation of the fermentation process. From a plant natural products perspective, biotechnology can refer to any of the following: the production of a plant natural product via cell culture of the source organism; the production of a plant natural product or a derivative thereof in a heterologous host via genetic engineering; the creation of novel metabolites by the introduction of "unnatural" precursors into a biosynthetic route through chemical or biological methods; and the metabolic transformation of a plant natural product by a heterologous organism. The examples provided in the following paragraphs correspond to these biotechnological classes and are presented sequentially. Biotechnology as pertaining to plant natural products is used generally as a means of solving sourcing problems, lowering production costs, generating libraries of "unnatural" natural products, and easing the strain on the environment by lowering ecosystem disruption and organic solvent consumption.

Paclitaxel (**21**) was originally obtained from the bark of the Pacific yew, *Taxus brevifolia* Nutt. However, as the tree takes 200 years to mature, stripping the bark to produce this compound for the market was not sustainable.[132] A better source of naturally occurring paclitaxel has not been found in other *Taxus* species and the total synthesis, on the other hand, has proven to be difficult and time consuming.[132] A major breakthrough in the sourcing of paclitaxel was its semisynthetic preparation from 10-deacetylbaccatin III found in the leaves of the European yew, *Taxus baccata* L., a renewable resource that can be easily cultivated in greenhouses.[26,132] This procedure, although sustainable and environmentally friendly, still requires an 11-step synthesis.[132] Since 2002, the major pharmaceutical company, Bristol-Myers Squibb, has produced paclitaxel solely using callus cell cultures of the Chinese yew, *Taxus chinensis* Rehder.[132] Elicitors such as methyl jasmonate, ethylene, and the endophytic fungus *Aspergillus niger* have been shown to increase paclitaxel production in *Taxus* cell cultures.[133,134]

Plant cell culture is useful in laboratory and in industry because it allows plant natural products to be produced in a relatively controlled manner, and provides a supply of plant material that is not affected by sourcing problems, such as environmental, seasonal, geographical, and political factors.[135] Also, plant cell culture allows for the "tweaking" and rearrangement of secondary metabolite biochemical pathways in order to produce novel metabolites, and to increase target compound yields, as well as allowing derivatives to be formed by introduction of analogs of natural intermediates.[135,136] Plant cell culture can be performed with callus and suspension cultures, as well as with shoot cultures and hairy root cultures. These latter two approaches are especially useful when a metabolite is found to be produced more readily in differentiated cells.[135,137]

Currently, artemisinin (**18**)-based antimalarial therapies remain too expensive to be afforded by those in the developing countries who need them most, due to low yields from the plant of origin, *Artemisia annua* L., and also due to the high cost of the specialized processing involved in the purification of this compound.[138,139] In order to solve this problem, the cloning of the genes responsible for the biosynthesis of artemisinic acid (**69**) into the yeast *Saccharomyces cerevisiae* has been proposed as a method of reducing the overall cost of production of artemisinin (**18**).[138,140] The basic transformations done were the upregulation of farnesyl pyrophosphate (FPP) production coupled with a downregulation of its use in steroid biosynthesis, and the introduction of two enzymes from *A. annua* to convert FPP into a precursor of artemisinin (**18**), artemisinic acid (**69**) (Fig. 4).[140] It was found that by using this method of production, which is not yet optimized for industrial processing, two to three orders of magnitude more artemisinin could be produced *per annum*, compared to isolation from cultivation of *Artemisia annua*.[140]

The *Catharanthus* bisindole alkaloids vinblastine (**4**) and vincristine (**5**), isolated from the Madagascar periwinkle, *Catharanthus roseus* (L.) G. Don., were discovered as anticancer agents during screening programs in the 1950s and 1960s, and remain in clinical use today.[81] Due to the range of clinical uses of these drugs, their unique mechanism of action, and their low yields in the producing organism, much research has been performed on these compounds including the application of biotechnology procedures.[81] It has been shown that *Catharanthus* alkaloid libraries can be produced in both cell culture and by seedlings through the introduction of different tryptamine precursors into the *Catharanthus* alkaloid biosynthetic pathway.[136]

Glycyrrhizin (**70**) (Fig. 4) from licorice (the roots and stolons of *Glycyrrhiza* spp.), is an oleanane-type triterpenoid diglucuronide occurring as a mixture of metallic salts. It is used principally in Japan as a sweetener and flavoring agent, as well as an antiallergy and antihepatitis agent.[141,142] Glycyrrhizin is 50 to 100 times sweeter than sucrose. However, β-D-monoglucuronide-glycyrrhizin (MGGR, **71**) (Fig. 4), glycyrrhizin with one glucuronide unit removed, has been rated as about 950 times sweeter than sucrose, and is more bioavailable than glycyrrhizin.[141,143] β-D-Monoglucuronide-glycyrrhizin is used in Japan for flavoring yogurt, chocolate

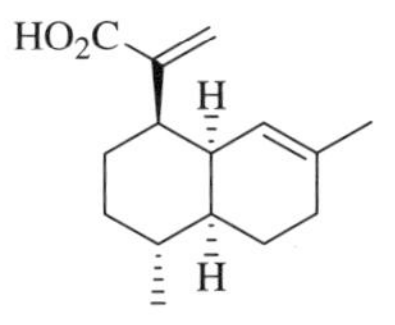

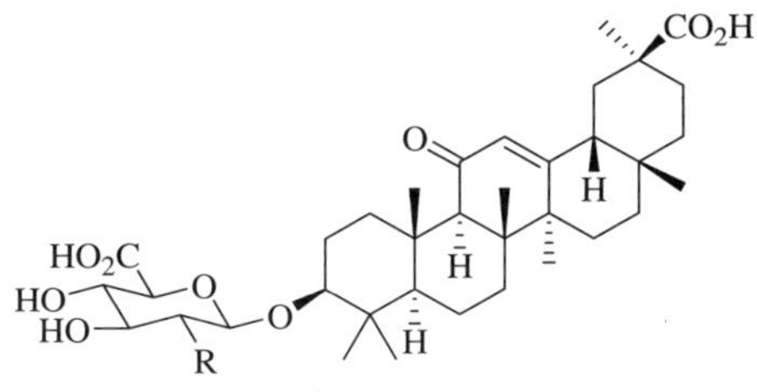

(69) Artemisinic acid

(70) R=

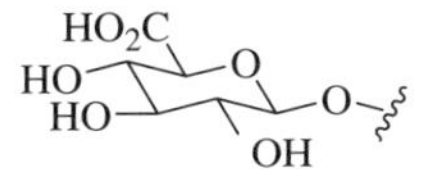

Glycyrrhizin

(71) R=OH

β-D-Mono-glucuronide-glycyrrhizin

Fig. 4. Structures of plant-derived compounds related to directed biosynthesis/biotechnology aspects.

milk, and fruit flavored soft drinks.[143] In order to produce enough of this compound for commercial use, biotechnology is currently being employed.[143] Although bacteria from the human gastrointestinal tract can carry out the conversion of glycyrrhizin to its monoglucuronide, it has been found that a *Penicillium* species from a field where *G. glabra* was growing is much more efficient at this transformation.[143] This process uses glycyrrhizin as the sole carbon source and elicitor for the fungus, as well as a low concentration of a surfactant to increase cell permeability.[143]

2.8 FUTURE PROSPECTS

Although many of the presently used plant drugs were developed initially as a result of ethnomedical observations on their plants of origin, there are surprisingly few modern examples of new plant-derived drugs that have been discovered from folkloric leads. There are several reasons for this, including the complexity of following up in the laboratory plants used traditionally with several ascribed medicinal attributes. Two other major factors are the erosion of tropical rainforests where much biodiversity is evident along with the concomitant loss of tribal ethnomedical knowledge.[144,145] In addition, current practices in biomedical research require higher ethical standards in the conduct of research using human subjects than previously, including the need to prepare written consent forms in the often obscure languages of "informants" who might provide salient information on ethnomedical practices for a given plant.[146] Following the Rio de Janeiro Convention (1992) on biodiversity, medicinal plants, among with other natural resources and indigenous knowledge, are regarded as the "intellectual property" of the source country, and collecting and transporting samples from one country to another require "benefit-sharing" agreements, which constitute another challenge in the plant-based drug discovery effort.[147]

While it is increasingly likely that in the future, new plant-derived drugs for an increasingly wide variety of human diseases will be discovered by high-throughput screening of plant extracts, there are certain problems inherent in the investigation of plants compared to other types of natural product extracts. The particular disadvantages that have emerged in plant-based natural product research stem from the complex nature of plant extracts. Some extracts might include compounds that are nonspecific protein binders (e.g. some plant polyphenols) or those that will interfere with spectrometric readings, which have a negative impact on bioassay studies. These problems have been duly addressed with relatively simple methods successfully developed to overcome them.[12,148]

Following a perceived lack of success of combinatorial chemistry to yield new drugs at the anticipated rate, some pharmaceutical and biotechnology companies are revisiting natural products as sources of new leads. New strategies have been developed to make plant-based natural product drug discovery more compatible with high throughput screening systems employed, the most significant approaches such as the need to obtain pure natural product libraries in a timely fashion.[29] Recent advances in hyphenated analytical and separatory techniques such as LC-SPE-NMR and LC-NMR-MS, which serve for concomitant rapid structure purification and elucidation purposes, and the development of specialized NMR probes that enable the structure elucidation of compounds at μg quantities, should have a significant impact in plant natural product research in the near future.[51–53,149]

It is pertinent to echo a sentiment expressed in our previous book chapter from 1993 which stated that "it is reasonable to expect that new plant sources of valuable and pharmaceutically interesting material remain to be discovered and developed".[25] Certainly, in the intervening period, interest in plant secondary metabolites with drug potential has not abated. Thus, it can be confidently expected that not only will new drug prototype molecules be afforded from plants in future, but that innovative methods for their more cost-effective procurement, including the use of biotechnology, will be developed.

REFERENCES

1. Kinghorn AD, Balandrin, MF (eds.), *Human Medicinal Agents from Plants*, in Symposium Series No. 534, American Chemical Society Books, Washington, DC, 1993.
2. Bruneton J, *Pharmacognosy, Phytochemistry, Medicinal Plants*, 2nd ed., Lavoisier Publishing, New York, 1995.
3. Dewick PM, *Medicinal Natural Products: A Biosynthetic Approach*, 2nd ed., John Wiley and Sons, Ltd., New York, 2002.
4. Evans WC, *Trease and Evans Pharmacognosy*, 15th ed., W.B. Saunders, New York, 2002.
5. Samuelsson G, *Drugs of Natural Origin: A Textbook of Pharmacognosy*, 5th Rev ed., Swedish Pharmaceutical Press, Stockholm, 2004.
6. Heinrich M, Barnes J, Gibbons S, Williamson EM, *Fundamentals of Pharmacognosy and Phytotherapy*, Churchill Livingstone, Edinburgh, 2004.
7. Bruhn JG, Bohlin L, Molecular pharmacognosy: An explanatory model, *Drug Discov Today* 2:243–246, 1997.
8. Rates SM, Plants as source of drugs, *Toxicon* **39**:603–613, 2001.
9. Kinghorn AD, Pharmacognosy in the 21st century, *J Pharm Pharmacol* **53**: 135–148, 2001.
10. Newman DJ, Cragg GM, Snader KM, Natural products as sources of new drugs over the period 1981–2002, *J Nat Prod* **66**:1022–1037, 2003.
11. Vuorelaa P, Leinonenb M, Saikkuc P, *et al.*, Natural products in the process of finding new drug candidates, *Curr Med Chem* **11**:1375–1389, 2004.
12. Butler MS, The role of natural product chemistry in drug discovery, *J Nat Prod* **67**:2141–2153, 2004.
13. Balunas MJ, Kinghorn AD, Drug discovery from medicinal plants, *Life Sci* **78**:431–441, 2005.
14. Butler MS, Natural products to drugs: Natural product derived compounds in clinical trials, *Nat Prod Rep* **22**:162–195, 2005.
15. Koehn FE, Carter GT, The evolving role of natural products in drug discovery, *Nat Rev Drug Discov* **4**:206–220, 2005.

16. Jones WP, Chin YW, Kinghorn AD, The role of pharmacognosy in modern medicine and pharmacy, *Curr Drug Targets* 7:247–264, 2006.
17. Chin YW, Balunas MJ, Chai HB, Kinghorn AD, Drug discovery from natural sources, *AAPS J* **8**: E239–253, 2006.
18. Evans FJ, Natural products as probes for new drug target identification, *J Ethnopharmacol* **32**:91–101, 1991.
19. Newman DJ, Cragg GM, Holbeck S, Sausville EA, Natural products and derivatives as leads to cell cycle pathway targets in cancer chemotherapy, *Curr Cancer Drug Targets* **2**:279–308, 2002.
20. Clement JA, Yoder BJ, Kingston DGI, Natural products as a source of CNS–active agents, *Mini-Rev Med Chem* **1**:183–208, 2004.
21. Tse WC, Boger DL, Sequence-selective DNA recognition: Natural products and nature's lessons, *Chem Biol* **11**:1607–1617, 2004.
22. Calixto JB, Kassuya CA, Andre E, Ferreira J, Contribution of natural products to the discovery of the transient receptor potential (TRP) channels family and their functions, *Pharmacol Ther* **106**:179–208, 2005.
23. El Sayed KA, Natural products as angiogenesis modulators, *Mini-Rev Med Chem* **5**:971–993, 2005.
24. Koehn FE, Therapeutic potential of natural product signal transduction agents, *Curr Opin Biotechnol* **17**:631–637, 2006.
25. Balandrin MF, Kinghorn AD, Farnsworth NR, Plant-derived natural products in drug discovery: An overview, in Kinghorn AD, Balandrin MF, eds., *Human Medicinal Agents from Plants*, Symposium Series No. 534, American Chemical Society Books, Washington, DC, pp. 2–12, 1993.
26. Sneader W, *Drug Discovery: A History*, John Wiley & Sons Ltd., Chichester, U.K., 2005.
27. Bohlin L, Bruhn JG, *Bioassay Methods in Natural Product Research and Drug Development*, Kluwer Academic, Boston, 1999.
28. Williamson EM, Okpako DT, Evans FJ, *Pharmacological Methods in Phytotherapy Research: Selection, Preparation and Pharmacological Evaluation of Plant Material.* Vol. 1, John Wiley & Sons, New York, 1996.
29. Bindseil KU, Jakupovic J, Wolf D, *et al.*, Pure compound libraries; a new perspective for natural product based drug discovery, *Drug Discov Today* **6**:840–847, 2001.
30. Firn RD, Jones CG, Natural products — A simple model to explain chemical diversity, *Nat Prod Rep* **20**:382–391, 2003.
31. Bajorath J, Chemoinformatics methods for systematic comparison of molecules from natural and synthetic sources and design of hybrid libraries, *J Comput-Aided Mol Des* **16**:431–439, 2002.

32. Feher M, Schmidt JM, Property distributions: Differences between drugs, natural products, molecules from combinatorial chemistry, *J Chem Inf Comput Sci* **43**:218–227, 2003.
33. Ganesan A, Natural products as a hunting ground for combinatorial chemistry, *Curr Opin Biotechnol* **15**:584–590, 2004.
34. Lee ML, Schneider G, Scaffold architecture and pharmacophoric properties of natural products and trade drugs: Application in the design of natural product-based combinatorial libraries, *J Comb Chem* **3**:284–289, 2001.
35. Ortholand JY, Ganesan A, Natural products and combinatorial chemistry: Back to the future, *Curr Opin Chem Biol* **8**:271–280, 2004.
36. Stahura FL, Godden JW, Xue L, Bajorath J, Distinguishing between natural products and synthetic molecules by descriptor Shannon entropy analysis and binary QSAR calculations, *J Chem Inf Comput Sci* **40**:1245–1252, 2000.
37. Lipinski CA, Lombardo F, Dominy BW, Feeney PJ, Experimental and computational approaches to estimate solubility and permeability in drug discovery and development settings, *Adv Drug Deliv Rev* **46**:3–26, 2001.
38. Williams DH, Why are secondary metabolites (natural products) biosynthesized? *J Nat Prod* **52**:1189–1208, 1989.
39. Mebs D, Toxicity in animals. Trends in evolution? *Toxicon* **39**:87–96, 2001.
40. Jarvis BB, Role of natural products in evolution, in Romeo JT, Ibrahim R, Varin L, De Luca V, (eds.), *Recent Advances in Phytochemistry: The Evolution of Metabolic Pathways*, Vol. 34, Pergamon Press, New York, pp. 1–24, 2000.
41. Field B, Jordan F, Osbourn A, First encounters — deployment of defence-related natural products by plants, *New Phytol* **172**:193–207, 2006.
42. Myles DC, Novel biologically active natural and unnatural products, *Curr Opin Biotechnol* **14**:627–633, 2003.
43. Dixon RA, Plant natural products: The molecular genetic basis of biosynthetic diversity, *Curr Opin Biotechnol* **10**:192–197, 1999.
44. Koch MA, Wittenberg LO, Basu S, Jeyaraj DA, Gourzoulidou E, Reinecke K, Odermatt A, Waldmann H, Compound library development guided by protein structure similarity clustering and natural product structure, *Proc Natl Acad Sci USA* **101**:16721–16726, 2004.
45. Allingham JS, Klenchin VA, Rayment I, Actin-targeting natural products: Structures, properties and mechanisms of action, *Cell Mol Life Sci* **63**: 2119–2134, 2006.
46. Chang TK, Waxman DJ, Synthetic drugs and natural products as modulators of constitutive androstane receptor (CAR) and pregnane X receptor (PXR), *Drug Metab Rev* **38**:51–73, 2006.

47. Hung DT, Jamison TF, Schreiber SL, Understanding and controlling the cell cycle with natural products, *Chem Biol* **3**:623–639, 1996.
48. Stromgaard K, Natural products as tools for studies of ligand-gated ion channels, *Chem Rec* **5**:229–239, 2005.
49. Suzuki T, Miyata N, Epigenetic control using natural products and synthetic molecules, *Curr Med Chem* **13**:935–958, 2006.
50. Tsukamoto S, Yokosawa H, Natural products inhibiting the ubiquitin-proteasome proteolytic pathway, a target for drug development, *Curr Med Chem* **13**:745–754, 2006.
51. Reynolds FR, Enriquez RG, Choosing the best pulse sequences, acquisition parameters, postacquisition processing strategies, probes for natural product structure by NMR spectroscopy, *J Nat Prod* **65**:221–244, 2002.
52. Jaroszewski JW, Hyphenated NMR methods in natural products research. Part 1: Direct hyphenation, *Planta Med* **71**:691–700, 2005.
53. Jaroszewski JW, Hyphenated NMR methods in natural products research. Part 2: HPLC-SPE-NMR and other new trends in NMR hyphenation, *Planta Med* **71**:795–802, 2005.
54. Gullo VP, Hughes DE, Exploiting new approaches for natural product drug discovery in the biotechnology industry, *Drug Discov Today* **2**:281–286, 2005.
55. Cordell GA, Biodiversity and drug discovery — a symbiotic relationship, *Phytochemistry* **55**:463–480, 2000.
56. Cordell GA, Quinn-Beattie ML, Farnsworth NR, The potential of alkaloids in drug discovery, *Phytother Res* **15**:183–205, 2001.
57. Kinghorn AD, Farnsworth NR, Soejarto DD, Cordell GA, Swanson SM, Pezzuto JM, Wani MC, Wall ME, Oberlies NR, Kroll DJ, Kramer RA, Rose WC, Vite GD, Fairchild CR, Peterson RW, Wild R, Novel strategies for the discovery of plant-derived anticancer agents, *Pharm Biol* **41**(Suppl.):53–67, 2003.
58. Exarchou V, Krucker M, van Beek TA, *et al.*, LC-NMR coupling technology: Recent advancements and applications in natural products analysis, *Magn Reson Chem* **43**:681–687, 2005.
59. Koch MA, Waldmann H, Protein structure similarity clustering and natural product structure as guiding principles in drug discovery, *Drug Discov Today* **10**:471–483, 2005.
60. O'Connor KA, Roth BL, Screening the receptorome for plant-based psychoactive compounds, *Life Sci* **78**:506–511, 2005.
61. Katz S, Harris R, Lau JT, Chau A, The use of gene expression analysis and proteomic databases in the development of a screening system to determine the value of natural medicinal products, *Evid Based Complement Alternat Med* **3**:65–70, 2006.

62. Rollinger JM, Langer T, Stuppner H, Integrated *in silico* tools for exploiting the natural products' bioactivity, *Planta Med* **72**:671–678, 2006.
63. Fabricant DS, Farnsworth NR, The value of plants used in traditional medicine for drug discovery, *Environ Health Perspect* **109**(Suppl. 1):69–75, 2001.
64. Golenser J, Waknine JH, Krugliak M, *et al.*, Current perspectives on the mechanism of action of artemisinins, *Int J Parasitol* **36**:1427–1441, 2006.
65. Li Y, Wu YL, How Chinese scientists discovered qinghaosu (artemisinin) and developed its derivatives. What are the future perspectives? *Med Trop (Mars)* **58**:9–12, 1998.
66. Kerr C, Companies agree to withdraw artemisinin monotherapy, *Lancet Infect Dis* **6**:397, 2006.
67. Barnes MP, Sativex®: Clinical efficacy and tolerability in the treatment of symptoms of multiple sclerosis and neuropathic pain, *Exp Opin Pharmacother* 7:607–615, 2006.
68. Buchtel L, Ventura HO, Lunar Society and the discovery of digitalis, *J La State Med Soc* **158**:26–30, 2006.
69. Warren B, Digitalis purpurea, *Am J Cardiol* **95**:544, 2005.
70. Hake AM, Use of cholinesterase inhibitors for treatment of Alzheimer disease, *Cleve Clin J Med* **68**:608–609, 613–614, 616, 2001.
71. Pearson VE, Galantamine: A new Alzheimer drug with a past life, *Ann Pharmacother* **35**:1406–1413, 2001.
72. Viegas C, Jr., da Bolzani VS, Barreiro EJ, Fraga CA, New anti-Alzheimer drugs from biodiversity: The role of the natural acetylcholinesterase inhibitors, *Mini-Rev Med Chem* **5**:915–926, 2005.
73. Snyder SH, Pasternak GW, Historical review: Opioid receptors, *Trends Pharmacol Sci* **24**:198–205, 2003.
74. Available from: http://www.rxlist.com/top200.htm. [cited February 2007]
75. Anekonda TS, Reddy PH, Can herbs provide a new generation of drugs for treating Alzheimer's disease? *Brain Res Brain Res Rev* **50**:361–376, 2005.
76. Rose JE, Behm FM, Westman EC, Kukovich P, Precessation treatment with nicotine skin patch facilitates smoking cessation, *Nicotine Tob Res* **8**:89–101, 2006.
77. Wani MC, Taylor HL, Wall ME, Coggon P, McPhail AT, Plant antitumor agents. VI. The isolation and structure of taxol, a novel antileukemic and antitumor agent from *Taxus brevifolia*, *J Am Chem Soc* **93**:2325–2327, 1971.
78. Cragg GM, Newman DJ, A tale of two tumor targets: Topoisomerase I and tubulin. The Wall and Wani contribution to cancer chemotherapy, *J Nat Prod* **67**:232–244, 2004.
79. Cook GC, History of tropical medicine: William Ernest Cooke, FRCSI (1879–1967) and his Asian tour of 1929–1930, *Acta Trop* **100**:1–10, 2006.

80. Lee JC, Harrison D, Timasheff SN, Interaction of vinblastine with calf brain microtubule protein, *J Biol Chem* **250**:9276–9282, 1975.
81. van Der Heijden R, Jacobs DI, Snoeijer W, Hallard D, Verpoorte R, The *Catharanthus* alkaloids: Pharmacognosy and biotechnology, *Curr Med Chem* **11**:607–628, 2004.
82. Cragg GM, Newman DJ, Plants as a source of anti-cancer agents, *J Ethnopharmacol* **100**:72–79, 2005.
83. Newman DJ, Cragg GM, Natural products as sources of new drugs over the last 25 years, *J Nat Prod* **70**:461–477, 2007.
84. van Agtmael MA, Eggelte TA, van Boxtel CJ, Artemisinin drugs in the treatment of malaria: From medicinal herb to registered medication, *Trends Pharmacol Sci* **20**:199–205, 1999.
85. Anonymous, *Guidelines for the Treatment of Malaria*, World Health Organization, Geneva, 2006.
86. Vennerstrom JL, Arbe-Barnes S, Brun R, *et al.*, Identification of an antimalarial synthetic trioxolane drug development candidate, *Nature* **430**:900–904, 2004.
87. Maerki S, Brun R, Charman SA, *et al.*, *In vitro* assessment of the pharmacodynamic properties and the partitioning of OZ277/RBx-11160 in cultures of *Plasmodium falciparum*, *J Antimicrob Chemother* **58**:52–58, 2006.
88. Rainsford KD, *Aspirin and the Salicylates*, Butterworths, London, 1984.
89. Patole J, Shingnapurkar D, Padhye S, Ratledge C, Schiff base conjugates of *p*-aminosalicylic acid as antimycobacterial agents, *Bioorg Med Chem Lett* **16**: 1514–1517, 2006.
90. Picco M, Krishna M, Cangemi J, Shelton D, Oral mesalamine and clinical remission are associated with a decrease in the extent of long-standing ulcerative colitis, *Inflamm Bowel Dis* **12**:537–542, 2006.
91. Rodrigo G, Rodrigo C, Burschtin O, A meta-analysis of the effects of ipratropium bromide in adults with acute asthma, *Am J Med* **107**:363–370, 1999.
92. Hansel TT, Barnes PJ, Tiotropium bromide: A novel once-daily anticholinergic bronchodilator for the treatment of COPD, *Drugs Today (Barc)* **38**:585–600, 2002.
93. Campone M, Cortes-Funes H, Vorobiof D, *et al.*, Vinflunine: A new active drug for second-line treatment of advanced breast cancer. Results of a phase II and pharmacokinetic study in patients progressing after first-line anthracycline/taxane-based chemotherapy, *Br J Cancer* **95**:1161–1166, 2006.
94. Foulds J, The neurobiological basis for partial agonist treatment of nicotine dependence: Varenicline, *Int J Clin Pract* **60**:571–576, 2006.
95. Barnes PJ, Drugs for asthma, *Br J Pharmacol* **147**:S297—S303, 2006.

96. McKiernan PJ, Nitisinone in the treatment of hereditary tyrosinaemia type 1, *Drugs* **66**:743–750, 2006.
97. Booth M, *Opium: A History*, St. Martin's Press, New York, 1984.
98. Blumberg H, Dayton HB, Wolf PS, Counteraction of narcotic antagonist analgesics by the narcotic antagonist naloxone, *Proc Soc Exp Biol Med* **123**: 755–758, 1966.
99. Volpicelli JR, Fenton M, Sustained-release naltrexone formulations for the treatment of alcohol and opioid dependence, *Future Neurol* **1**:389–398, 2006.
100. Paul IM, Shaffer ML, Yoder KE, *et al.*, Jr. Dose-response relationship with increasing doses of dextromethorphan for children with cough, *Clin Ther* **26**:1508–1514, 2004.
101. Bohlin L, Rosen B, Podophyllotoxin derivatives: Drug discovery and development, *Drug Discov Today* **1**:343–351, 1996.
102. Canel C, Moraes RM, Dayan FE, Ferreira D, Podophyllotoxin, *Phytochemistry* **54**:115–120, 2000.
103. Guénard D, Guéritte-Voegelein F, Potier P, Taxol and Taxotere: Discovery, chemistry and structure-activity relationships, *Accounts Chem Res* **26**:160–167, 1993.
104. Brownstein MJ, A brief history of opiates, opioid peptides, opioid receptors, *Proc Natl Acad Sci USA* **90**:5391–5393, 1993.
105. Alexander MD, Burkart MD, Leonard MS, *et al.*, A central strategy for converting natural products into fluorescent probes, *Chembiochem* 7:409–416, 2006.
106. Di Marzo V, A brief history of cannabinoid and endocannabinoid pharmacology as inspired by the work of British scientists, *Trends Pharmacol Sci* 27:134–140, 2006.
107. Mechoulam R, Hanus L, A historical overview of chemical research on cannabinoids, *Chem Phys Lipids* **108**:1–13, 2000.
108. Bellocchio L, Mancini G, Vicennati V, *et al.*, Cannabinoid receptors as therapeutic targets for obesity and metabolic diseases, *Curr Opin Pharmacol* **6**:586–591, 2006.
109. McCurdy CR, Scully SS, Analgesic substances derived from natural products (natureceuticals), *Life Sci* **78**:476–484, 2005.
110. Roth BL, Baner K, Westkaemper R, *et al.*, Salvinorin A: A potent naturally occurring nonnitrogenous κ opioid selective agonist, *Proc Natl Acad Sci USA* **99**:11934–11939, 2002.
111. Prisinzano TE, Psychopharmacology of the hallucinogenic sage *Salvia divinorum*, *Life Sci* **78**:527–531, 2005.
112. Rothman RB, Baumann MH, Targeted screening for biogenic amine transporters: Potential applications for natural products, *Life Sci* **78**:512–518, 2005.

113. Cortese F, Bhattacharyya B, Wolff J, Podophyllotoxin as a probe for the colchicine binding site of tubulin, *J Biol Chem* **252**:1134–1140, 1977.
114. Wall ME, Camptothecin and taxol: Discovery to clinic, *Med Res Rev* **18**:299–314, 1998.
115. Akiyama T, Ishida J, Nakagawa S, *et al.*, a specific inhibitor of tyrosine-specific protein kinases, *J Biol Chem* **262**:5592–5595, 1987.
116. Molokanova E, Kramer RH, Mechanism of inhibition of cyclic nucleotide-gated channel by protein tyrosine kinase probed with genistein, *J Gen Physiol* **117**: 219–234, 2001.
117. Kazanietz MG, Caloca MJ, Eroles P, *et al.*, Pharmacology of the receptors for the phorbol ester tumor promoters: Multiple receptors with different biochemical properties, *Biochem Pharmacol* **60**:1417–1424, 2000.
118. Caterina MJ, Schumacher MA, Tominaga M, *et al.*, The capsaicin receptor: A heat-activated ion channel in the pain pathway, *Nature* **389**:816–824, 1997.
119. Nagy I, Santha P, Jancso G, Urban L, The role of the vanilloid (capsaicin) receptor (TRPV1) in physiology and pathology, *Eur J Pharmacol* **500**:351–369, 2004.
120. Andrews AC, The silphium of the ancients: A lesson in crop control, *Isis* **33**: 232–236, 1941.
121. Rikhari HC, Palni, L.M.S., Sharma S, Nandi SK, Himalayan yew: Stand structure, canopy damage, regeneration and conservation strategy, *Environ Conserv* **25**:334–341, 1998.
122. Rikhari HC, Sharma S, Nadeem M, Palni, L.M.S. The effect of disturbance levels, forest types and associations on the regeneration of *Taxus wallachiana*: Lessons from the Central Himalaya, *Curr Sci* **79**:88–90, 2000.
123. Ma X, Tan C, Zhu D, Gang DR, A survey of potential huperzine A natural resources in China: The Huperziaceae, *J Ethnopharmacol* **104**:54–67, 2006.
124. Robbins CS, Comparative analysis of management regimes and medicinal plant trade monitoring mechanisms for American ginseng and goldenseal, *Conserv Biol* **14**:1422–1434, 2000.
125. Padulosi S, Leaman D, Quek P, Challenges and opportunities in enhancing the conservation and use of medicinal and aromatic plants, *J Herbs Spices Med Plants* **9**:243–267, 2002.
126. Leaman DJ, Conservation, trade, sustainability and exploitation of medicinal plant species, in Saxena PK, (ed.), *Development of Plant-Based Medicines: Conservation, Efficacy and Safety*, Kluwer Academic Publishers, Dordrecht, The Netherlands, pp. 1–15, 2001.
127. Anonymous, *Directory for Medicinal Plants Conservation: Networks, Organizations, Projects, Information Sources*, IUCN/SSC Medicinal Plant Specialist

Group and German Federal Agency for Nature Conservation, Bonn, Germany, 1996. Available from: http://www.dainet.de/genres/mpc-dir/ [cited January, 2007].
128. Anonymous. Conservation of medicinal plants in Central America and the Caribbean, *Indigenous Knowledge Notes* **93**:1–4, 2006.
129. Available from: http://www.biodiv.org/convention/convention.shtml. [cited January, 2007].
130. Manchester KL, Louis Pasteur (1822–1895) — Chance and the prepared mind, *Trends Biotechnol* **13**:511–515, 1995.
131. Turner MK, Biocatalysis in organic chemistry (Part II): Present and future, *Trends Biotechnol* **13**:253–258, 1995.
132. Ritter SK, Green innovations, *Chem Eng News*, pp. 25–30, 2004.
133. Roberts SC, Shuler ML, Large-scale plant cell culture, *Curr Opin Biotechnol* **8**:154–159, 1997.
134. Wang C, Wu J, Mei X, Enhancement of taxol production and excretion in *Taxus chinensis* cell culture by fungal elicitation and medium renewal, *Appl Microbiol Biotechnol* **55**:404–410, 2001.
135. Rao SR, Ravishankar GA, Plant cell cultures: Chemical factories of secondary metabolites, *Biotechnol Adv* **20**:101–153, 2002.
136. McCoy E, O'Connor SE, Directed biosynthesis of alkaloid analogs in the medicinal plant *Catharanthus roseus*, *J Am Chem Soc* **128**:14276–14277, 2006.
137. Uozumi N, Large-scale production of hairy root, *Adv Biochem Eng Biotechnol* **91**:75–103, 2004.
138. Fischbach MA, Walsh CT, Biochemistry. Directing biosynthesis, *Science* **314**:603–605, 2006.
139. Senior K, Shortfall in front-line antimalarial drug likely in 2005, *Lancet Infect Dis* **5**:75, 2005.
140. Ro DK, Paradise EM, Ouellet M, *et al.*, Production of the antimalarial drug precursor artemisinic acid in engineered yeast, *Nature* **440**:940–943, 2006.
141. Kinghorn AD, Compadre CM, Less common high potency sweeteners, in O'Brien Nabors L (ed.), *Alternative Sweeteners*, 3rd ed., Revised and Expanded, Marcel Dekker, New York, pp. 204–231, 2001.
142. Shibata S, A drug over the millennia: Pharmacognosy, chemistry, pharmacology of licorice, *Yakugaku Zasshi* **120**:849–862, 2000.
143. Feng S, Li C, Xu X, Wang X, Screening strains for directed biosynthesis of [β]-D-mono-glucuronide-glycyrrhizin and kinetics of enzyme production, *J Mol Catal B Enzymol* **43**:63–67, 2006.
144. Gentry AH, Tropical forest biodiversity and the potential for new medicinal plants, in Kinghorn AD, Balandrin MF (eds.), *Human Medicinal Agents from*

Plants, Symposium Series No. 534, American Chemical Society Books, Washington, DC, pp. 13–24, 1993.

145. Cox PA, Will tribal knowledge survive the millennium? *Science* **287**:44–45, 2000.
146. Rosenthal JP, Politics, culture, governance in the development of prior informed consent in indigenous communities, *Curr Anthropol* **47**:119–142, 2006.
147. Soejarto DD, Fong HHS, Tan GT, *et al.*, Ethnobotany/ethnopharmacology and mass bioprospecting: Issues on intellectual property and benefit-sharing, *J Ethnopharmacol* **100**:15–22, 2005.
148. Wall ME, Wani MC, Brown DM, *et al.*, Effects of tannins on screening of plant extracts for enzyme inhibitory activity and techniques for their removal, *Phytomedicine* **3**:281–285, 1996.
149. Wolters AM, Jayawickrama DA, Sweedler JV, Microscale NMR, *Curr Opin Chem Biol* **6**:711–716, 2002.

Chapter 3

PLANT AND BRAIN CANNABINOIDS: THE CHEMISTRY OF MAJOR NEW PLAYERS IN PHYSIOLOGY

Lumir Hanuš and Raphael Mechoulam

3.1 CANNABIS PLANT HISTORY

Cannabis sativa L. (hemp) is a dioecious annual flowering plant. Marihuana is the Spanish name for the dried leaves and female flowering tops of the hemp plant. Hashish is the resin which originates from these female flowering tops. The hemp's natural homeland is most likely in the regions north of Afghanistan and the Altai mountains of southern Siberia (Russia). It is not clear when and where cultivation of hemp (*Cannabis sativa* L.) started. It seems most likely that the cultivation of hemp may have originated in northeastern Asia (north and north-east China and southeastern Siberia).

Recent discoveries from Southern Moravia in the Czech Republic indicate that the inhabitants who lived at two of the most famous eastern Gravettian settlements, the upper paleolithic sites of Pavlov and Dolní Vìstonice, some 29 000 to 22 000 years ago were expert weavers. Excavations by American and Czech scientists revealed the evidence of plant fibres used in the manufacture of textiles, basketry, cordage and perhaps netting.[1] However the plant(s) used have not been identified and wild hemp is just one of the possibilities.

Cultivated hemp had many uses in ancient neolitic China. One of the earliests records of hemp use by man comes from a 12 000 year-old neolitic site at Yuan-shan (Taiwan island). Excavated items included coarse, sandy pottery with hemp cord marks covering the surface, and an incised, rod-shaped stone beater used to pound hemp.[2,3] A fragment of hemp cloth was discovered in 1972 in a grave from the Chou dynasty (1122–249 B.C.) in Shansi province (China). It is the oldest preserved specimen of cannabis.[4]

The emperor Chen Nung, who is the alleged discoverer of medicinal plants, taught his people how to cultivate grains as food. He is assumed to be the author of Shen-nung pen ts'ao ching, the earliest extant Chinese pharmacopoeia. According to this pharmacopoeia, ma-fen, the flowers of the female marijuana plant, contain the greatest amount of yin energy. Ma-fen was prescribed in cases of a loss of yin, such as in menstrual fatigue, rheumatism, malaria, beri-beri, constipation and absent-mindedness. The Pen Ts'ao warned that eating too many Ma seeds could cause one to see demons but, taken over a long period of time, marijuana seeds could enable one to communicate with spirits.[5]

The Assyrians who ruled over large parts of the Middle East for nearly a millennium, about 3 000 years ago, have left us a pharmaceutical legacy on hundreds of clay tablets. Cannabis was one of the major drugs of their pharmacopoeia. They named this plant according to its use. Campbell Thompson identified the Sumerian term as cannabis on the basis of their similarities to the Aramaic and Syriac *'azal*, meaning "to spin." Campbell Thompson then identified the Sumerian drug *gán-zi-gùn-nu* ("plant which takes away the mind") as hashish.[6] A letter written around 680 B.C. by an unknown woman to the mother of the Assyrian king, Esarhaddon, mentions a substance called *qu-nu-bu*, which also may have been cannabis.[7]

Cannabis was used by the ancient Egyptians in their medicine. The Ebers Papyrus,[8] found in the tomb at Thebes (about 1534 B.C.), is not the oldest known recorded mention of the medical uses of cannabis, but it is the oldest known "complete" medical textbook in existence. This document mentions medicinal cannabis under the name šmšmt.[9] There are two formulas mentioning medicinal use of cannabis: 1 remedy for a toe-nail (Formula No. 618) and 2a remedy to cool the uterus (Formula No. 821).

In the Indian scripture of the Atharva Veda, the fourth book of the Vedas, the ancient scriptures of the Brahman religion (ca 2,000–1,400 B.C.), bhang (hemp) was identified as one of five sacred plants of India.[10]

The Aryans who settled in Persia came from the same area in central Russia as their relatives who invaded India, hence it is hardly surprising that the Persian word bhanga is almost identical to the Indian term bhang. Zarathushtra (Zoroaster), the Persian prophet (around 1200 BC), who is said to have written the Zend-Avesta, was a user of bhanga (hemp).[11,12]

The ancient Aryans of India used cannabis in their worship of the deity Shiva. In one of the tantric scriptures, we find this revealing statement: "Intoxicating drink (containing bhang) is consumed in order to liberate oneself, and that those who do so, in dominating their mental faculties and following the law of Shiva (yoga) are to be likened to immortals on earth."[13]

Herodotus of Halicarnassus (430 and 424 B.C) described the tribal customs of the Scythians, nomads inhabiting what is now southern Russia. Herodotus relates how the Scythians inhaled hemp vapours.[14]

Benetowa,[15–17] claims that the term "cannabis" had its origin in Semitic languages, like Hebrew, and that it appears several times throughout the Old Testament. She identifies cannabis with *kneh bossem,*a not fully identified plant in the Old Testament. A different tentative identification is put forward by Rabin.[18] He suggests that *pannagh*, a material mentioned by the prophet, Ezekiel (Ezekiel 27:17), is in fact cannabis. In his view, *pannagh* was one of the original forms of the word (compare with Sanscrit *bhanga* and Persian *bhang*) which later underwent a linguistic metathesis in Semitic languages (compare Assyrian *qunnabu*, Syriac *qunnappa* and classical Arabic *kunnab*)

The French physician, Jacques Joseph Moreau, remains the most-cited connection between cannabis and the art community. Moreau first used hashish while traveling through the Middle East in the 1830s. He assumed that cannabis-induced sensations might model the hallucinations and delusions common in psychotic individuals. He had hoped that this research might help the treatment of the mentally ill. The outspoken hedonist and popular novelist, Theophile Gautier assisted Moreau in this research. He not only participated himself, but he also recruited other members of France's artistic community. This group of hedonists and experimenters met monthly in an old mansion in Paris which was known at the time as the Club Des Hachichins (Hashish Club).[19] For historical reviews on cannabis, see Abel[20] and Mechoulam.[21]

3.2 CANNABIS SATIVA (HEMP) AND ITS TYPICAL CANNABINOID COMPONENTS

3.2.1 Biogenesis

Until the mid 1960's the only plant cannabinoid whose structure was fully elucidated was cannabinol (CBN) — a constituent which actually may represent an oxidation artifact. However, on the basis of CBN, the main cannabinoid structure skeleton became known. Thus, cannabidiol (CBD), which had been independently isolated in pure form by Adams[22,23] and by Todd,[24] was correctly assumed to be, like CBN, a terpenoid derivative attached to olivetol. But its exact structure was not elucidated. The psychoactive components of cannabis were assumed to be related tricyclic derivatives. On the basis of the tentatively elucidated constituents, Todd[25] suggested that the cannabinoids may be formed initially in the plant by condensation of a menthatriene with olivetol.

Gaoni and Mechoulam[26] suggested that cannabigerol (CBG) originated in nature most probably by the condensation of geranyl pyrophosphate with olivetol. In their following papers,[27–29] they expanded the biogenetic scheme by suggesting subsequent conversions of CBG to CBD, tetrahydrocannabinol (THC) and finally to CBN.

Yamauchi *et al.*[30] demonstrated that the neutral cannabinoids and cannabinolic acid are artifacts produced from cannabinoid acids during the harvest and storage of cannabis.

Shoyama *et al.*[31] made use of labeled calcium malonate-[2-^{14}C]; sodium mevalonate-[2-^{14}C]; geraniol-[1-^{3}H]; nerol-[1-^{3}H] and labeled cannabigerolic acid (CBGA)-carboxyl-^{14}C; CBGA-[U-^{3}H] and CBDA-carboxyl-^{14}C to clarify the biogenesis. The plants were nourished with the labeled substances for the period of six days applied through a cotton wick or directly to the leaves and the production of radioactive cannabinoids was followed. On the evidence of the identified substances, a definitive biogenetic scheme could be formulated.[32] In a later publication, Shoyama *et al.*[33] reported that the cannabinoid acids with a n-propyl side chain, are biosynthesized along the same biogenetic pathway as the pentyl homologues.

In their next study, Shoyama and Nishioka[34] isolated new spirocompounds: cannabispirol and acetyl cannabispirol. This is in addition to the already known cannabispirone and cannabispirenone from a Japanese hemp variety. The two scientists included them in their biogenetic schema alongside the cannabinoid acids. In a further study, Shoyama *et al.*[35] dealt with the biosynthesis of propylcannabinoid acids by *in vitro* incubation with raw enzyme solution from three species of *Cannabis sativa KL*. A biogenetic schema is presented illustrating the relationship between methyl, propyl and pentyl cannabinoid acids.

Crombie *et al.*[36] isolated spiranes from Thai hemp and proposed the biogenesis of their origin. Turner and El-Sohly[37] isolated a series of polyoxidized cannabinoids as minor substances in cannabis. They modified the biogenetic scheme by assuming that Δ^9-THC is converted into CBN through epoxy and hydroxylated cannabinoids. More recently, a new cannabinoid, cannabinerolic acid was isolated from the leaves of a Mexican strain of *Cannabis sativa L.*[38] This cannabinoid acid is assumed to be involved in the biosynthesis of Δ^9-THCA. The same authors presented the first direct experimental evidence for the mechanism of Δ^9-THCA biosynthesis from CBGA through oxidocyclization by a new enzyme.[39] This Δ^9-THCA synthase was the first isolated and identified enzyme involved in cannabinoid biogenesis.

The absence of CBD and CBDA was reported in Cannabis of South African origin.[40,41] In samples of cannabis from three regions of Mexico,[42] likewise, no CBD was present. The absence of CBD and its acid, alongside the presence of THC and THC acid, indicated that the two types of compounds presumably originate by separate pathways. Indeed, Taura *et al.*[43] identified a unique enzyme, cannabidiolic acid synthase, that catalyzes the oxidocyclization of CBGA (predominantly) as well as cannabinerolic acid to CBDA. One year later, Morimoto *et al.*[44] identified in young leaves of *Cannabis sativa*, an enzyme, which is involved in the biosynthesis of cannabichromenic acid. Cannabichromenic acid synthase catalyze the oxidocyclization of CBGA and cannabinerolic acid to cannabichromenic acid (predominantly from CBGA). As THCA and CBCA originate in the plant by two independent different pathways from CBGA (and not from

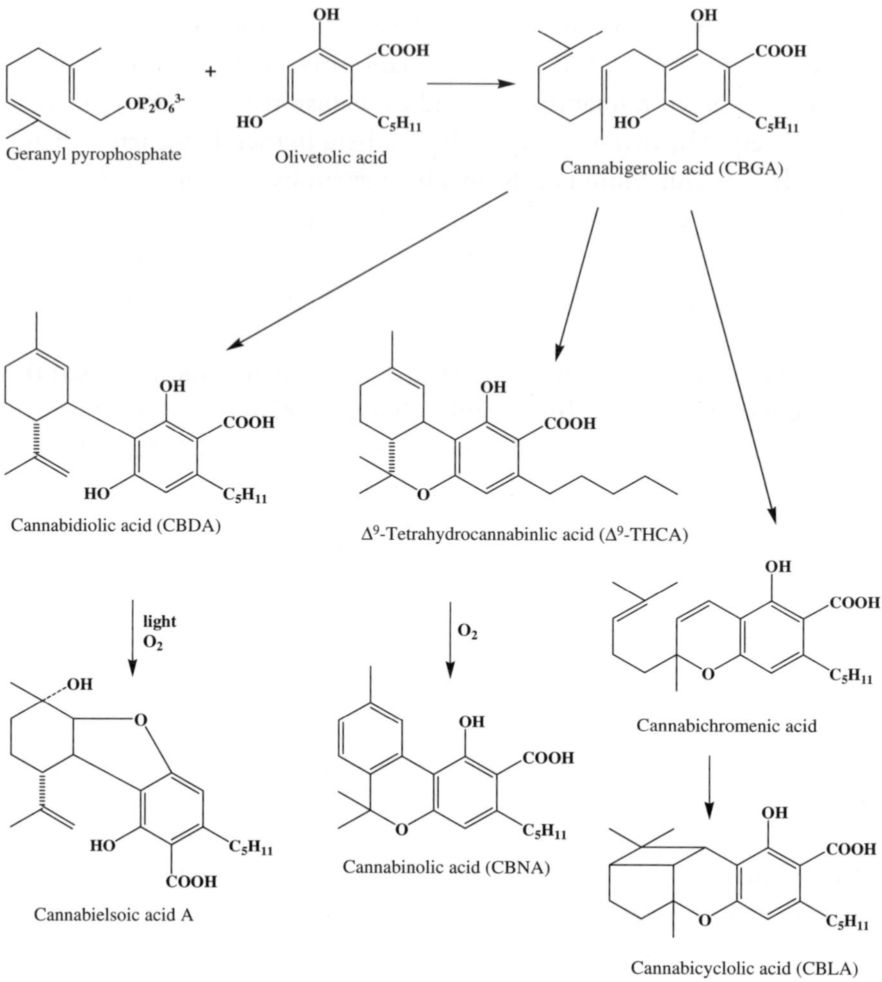

Fig. 1. Proposed biogenesis of cannabinoid acids in the plant *Cannabis sativa L.*

CBDA, as was theoretically assumed previously) with the help of appropriate enzymes[45] and because CBDA is synthesized in the plant by an independent pathway, it is easy to explain why some cannabis strains do not contain CBD and CBDA.

The biogenesis of cannabinoid acids in *Cannabis sativa L.* based on experimentally acquired knowledge is presented on Fig. 1.

3.2.2 Cannabinoid Substances from *Cannabis sativa L*

The first successful attempt to identify a typical cannabis constituent (today called cannabinoid) was achieved by Wood *et al.*[46], who isolated cannabinol (CBN) from the exuded resin of Indian hemp ("charas"), which analyzed for $C_{21}H_{26}O_2$. Another, big step was made by Cahn[47], who advanced the elucidation of the structure of CBN (Cahn, 1932). Several years later, Todd's group and Adam's group elucidated the correct structure of CBN by synthesizing and comparing various possible structures.[48–50]

A second cannabis constituent, cannabidiol (CBD) was also isolated, but its structure was only partially clarified.[51,52] Synthetic tetrahydrocannabinols, which showed cannabis-like activity in animal tests, were prepared, but they obviously differed from the active natural product, on the basis of their UV spectrum.[53–56]

In a systematic study of the antibacterial substances in hemp, Krejèí and Šantavý found that an extract containing carboxylic acids was effective against *Staphylococcus aureus* and other gram positive micro organisms. They isolated cannabidiololic acid (CBDA)[57,58] and partially elucidated the structure. The only problems remaining being the position of the double bond in monoterpene ring and the stereochemistry.

The advances in isolation methods made possible a clarification of the chemistry of cannabis. In 1963, our group reisolated CBD and reported its correct structure and stereochemistry.[59] A year later we finally succeeded in isolating pure Δ^9-tetrahydrocannabinol (Δ^9-THC), elucidated its structure, obtained a crystalline derivative and achieved a partial synthesis from CBD.[60] Several years later, a minor psychotomimetically active constituent, Δ^8-THC, was isolated from marijuana.[61] Whether this THC isomer is a natural compound, or an artifact formed during the drying of the plant, remains an open problem.

The absolute configuration of CBD, and hence also of THC, was established by the analysis of the shift of optical rotation values[62] and fully by correlation with known terpenoids.[63]

Several additional, non-psychotropic cannabinoids were also identified at that time. The best known are cannabigerol (CBG),[26] cannabichromene (CBC)[64,65] and cannabicyclol.[66] Cannabinolic acid (CBNA) and cannabigerolic acid (CBGA) were also identified,[63] followed by two

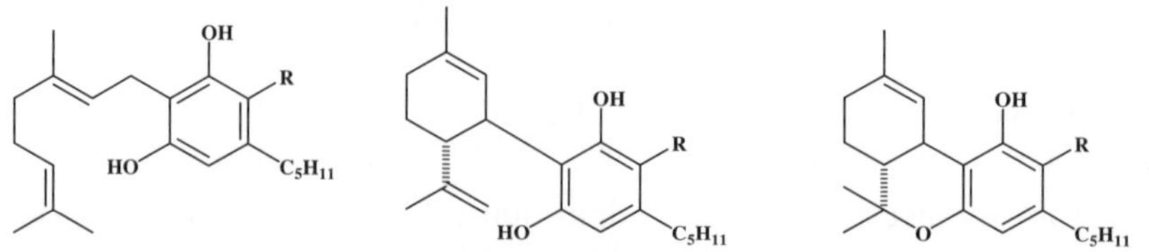

R = H, Cannabigerol (CBG)
R = COOH, Cannabigerolic acid (CBGA)

R = H, Cannabidiol(CBD)
R = COOH, Cannabidiolic acid (CBDA)

R = H, D^9-Tetrahydrocannabinol (D^9-THC)
R = COOH,D^9-Tetrahydrocannabinolic acid (D^9-THCA)

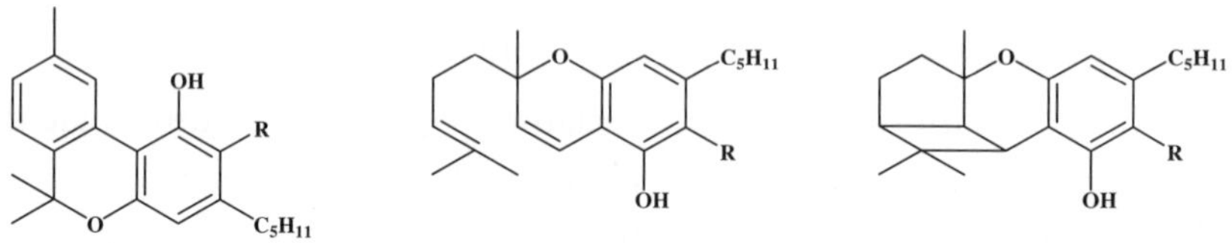

R = H, Cannabinol (CBN)
R = COOH, Cannabinolic acid (CBNA)

R = H, Cannabichromene (CBC)
R = COOH, Cannabichromenic acid (CBCA)

R = H, Cannabicyclol (CBL)
R = COOH Cannabicyclolic acid (CBLA)

Fig. 2. The main cannabinoids in *Cannabis sativa L.*

Δ^9-THC acids, A and B,[30,67,68] as well as Δ^8-THC acid[69] and cannabielsoic acid.[70] The decarboxylated product of cannabielsoic acid, cannabielsoin, is found in mammals as a metabolite of CBD.[71] The syntheses of some of the cannabinoid acids have been reported.[72]

It is possible, that some of the natural neutral cannabinoids (if not all of them) are artifacts formed by decarboxylation, photochemical cyclization (cannabicyclol), oxidation (cannabielsoic acid) or isomerization (Δ^8-THC and Δ^8-THC acid) of other constituents.

For the main representative cannabinoids see Fig. 2.

3.2.3 Non Cannabinoid Constituents of Cannabis sativa L

Like all plants, *Cannabis sativa L.* forms a huge number of chemicals. Many of these have been identified.[73–77] None of them, except the cannabinoids are specific for this plant and none of these non-cannabinoids has been found to contribute to the cannabinoid effects. However, very few have been directly tested together with cannabinoids for any biological effects. Recently, ElSohly and Slade,[78] published a comprehensive review. Up to date, 70 natural cannabinoid compounds were documented. Althogether, they record 419 constituents: nitrogenous compounds (27), amino acids

(**18**), proteins (**3**), enzymes (**6**), glycoproteins (**2**), sugars and related compounds (**34**), hydrocarbons (**50**), simple alcohols (**7**), simple aldehydes (**12**), simple ketones (**13**), simple acids (**20**), fatty acids (**23**), simple esters (**12**), lactones (**1**), steroids (**11**), terpenes (**120**), non cannabinoid phenols (**25**), flavonoids (**23**), vitamins (**1**), pigments (**2**) and elements (**9**).

3.3 CANNABINOID RECEPTORS IN LIVING ORGANISMS

3.3.1 Central (CB_1) and Peripheral (CB_2) Cannabinoid Receptors

The first creditable inkling, that Δ^9-THC acts through a specific receptor was demonstrated by Howlett, who reported that Δ^9-THC decreases cyclic AMP accumulation in neuronally derived cells.[79] Prostanoid-stimulated adenylate cyclase in a membrane preparation from these cells was inhibited by cannabinoid compounds. Howlett and Fleming[80] later showed that adenylate cyclase in plasma membranes was inhibited by Δ^8-THC and Δ^9-THC. Forskolin-activated adenylate cyclase was inhibited by all psychoactive cannabimimetic agents, but not by non-psychoactive cannabinoids. In membranes from N18TG2 neuroblastoma cells and the neuroblastoma X glioma hybrid cells, NG108-15, the psychoactive cannabinoid drugs and their nantradol analogs inhibited adenylate cyclase activity.[81] The cellular selectivity provided strong evidence for the existence of specific receptors for the cannabimimetic compounds. The requirement for a functional G_i in the action of cannabimimetic drugs was also indicated. Stereospecificity — a powerful indication of action on a biological target, be it an enzyme or a receptor — was also demonstrated with the help of the synthetic (-) HU-210 and (+) HU-211 enantiomers.[82]

In 1988, the same group reported the existence of a cannabinoid receptor in rat brain.[83] High affinity and stereoselectivity — essential criteria for a pharmacologically distinct cannabinoid receptor — in the brain was demonstrated. Several years later, the cDNA of the cannabinoid receptor was isolated from a rat cerebral cortex cDNA library and expressed in Chinese hamster ovary cells.[84] This G-protein-coupled receptor was isolated from a human brain stem cDNA library.[85] The deduced amino acid sequence encoded a protein of 472 residues, which shares 97.3% identity with the rat cannabinoid receptor cloned previously.[84] This receptor was named CB_1. Later, homologues of this central cannabinoid receptor were cloned from human[85,86] and mouse[87] tissues.

A peripheral cannabinoid receptor was cloned from HL-60 cells.[88] It is not expressed in the brain but in macrophages in the marginal zone of spleen. The peripheral cannabinoid receptor, mCB2, was cloned from a mouse splenocyte cDNA library.[89] This protein of 347 residues shares 82% overall identity with the only other known peripheral receptor, human CB_2 (hCB_2) and is shorter by 13 amino acids at the C-terminus.

Human CB_1 is present in the brain and throughout the central nervous system and has 472 amino acids; CB_2 is present in the spleen and the immune cells and has 360 amino acids. Both types are 7-helix transmembrane spanning receptors. There are three extra cellular and three intra cellular loops. A glycosylated extra cellular N-terminal domain, and an intra cellular C-terminal domain are involved in the interaction with the G-protein.

Originally, it was assumed that while the CB_1 receptor is found primarily in the brain and neuronal tissue, while the CB_2 receptor is found exclusively outside the central nervous system, primarily in immune tissue.[90] However, Van Sickle *et al.*[91] reported the expression of CB_2 receptor mRNA and protein localization on brainstem neurons. Recently, it was demonstrated that CB_2 receptor and their gene transcripts are also distributed in the mammalian brain.[92,93] The levels of CB2 receptors in brain are enhanced very significantly during numerous neurological conditions and may represent a protective reaction.[94]

Both receptor types are coupled through G-proteins to adenylyl cyclase and mitogen-activated protein (MAP) kinase. CB_1 receptors are also coupled through G-proteins to several types of calcium and potassium channels.

3.3.2 GPR55 Receptor

In 1999, Sawzdargo *et al.*[95] reported the identification and cloning of two novel human G-protein-coupled receptor genes, GPR52, GPR55 and a pseudogene Ψ GPR53. Human receptors show that the highest amino acid identity to GPR55 include P2Y5 (29%), GPR23 (30%), GPR35 (27%) and CCR4 (23%). A method for identification of an agent that modulates the activity of GPR 55 was patented several years later.[96]

Two years ago, Drmota *et al.*[97] discovered that the orphan G-protein-coupled receptor GPR55 binds and is activated by endogenous, natural and synthetic cannabinoid ligands. Thus, GPR55 represents a novel cannabinoid receptor that displays high affinity for several endogenous, natural and synthetic cannabinoid ligands.[98] This receptor is reported to be expressed in several tissues and may function in lipid or vascular biology. GPR55 apparently represents a new cannabinoid receptor. So far, no publications on the cannabinoid nature of this receptor have been published. The patent literature has however been reviewed[99].

Several tentative cannabinoid receptors have been proposed. The best one described is a receptor partly identified by Wagner *et al.*.[100] It mediates a mesenteric vasodilator response to anandamide and R-methanandamide and is distinct from the CB_1 receptor.

3.4 ENDOCANNABINOIDS

The discovery of central (CB_1) and later peripheral (CB_2) G-protein-coupled cannabinoid receptors suggested the existence of endogenous cannabinoid ligand(s), which bind(s) to activate these receptors. Several groups initiated programs to identify such endogenous materials. In 1992, our group reported the first isolation and identification of an endogenous cannabinoid ligand from porcine brain, which was named anandamide.[101] This endocannabinoid inhibited the specific binding of a radiolabelled cannabinoid probe to synaptosomal membranes in a manner typical of competitive ligands. It produced a concentration-dependent inhibition of the electrically evoked twitch responses of the mouse vas deferens, which is a characteristic effect of psychotropic cannabinoids. The structure of anandamide was established by mass spectrometry and nuclear magnetic resonance spectroscopy. Direct exposure chemical ionization (isobutene-DCI) mass spectrum indicated a molecular weight of 347. High resolution MS measurement suggested the elemental composition $C_{22}H_{37}NO_2$ (m/z347.2762), which is consistent with the presence of five double bond equivalents. Collision-induced dissociation (CID) measurement of the m/z348 MH^+ ion obtained under isobutene-DCI) gave rise to the following significant fragments: m/z287, 245, 203, 62 (highest abundance), and 44. The only reasonable composition of the most abundant m/z62 fragment ion is C_2H_8NO, which corresponds to protonated ethanolamine $HOCH_2CH_2NH_3^+$. The m/z44 ion may be formed by dehydration of the m/z62 fragment. The m/z287 {MH-61]$^+$ fragment ion corresponds to the loss of ethanolamine (C_2H_7NO) from MH^+. Additional data were obtained from the GC-MS and CID measurements of the trimethylsilyl (TMS) derivative of the purified material. Together, these results suggested that anandamide is an ethanolamide of arachidonic acid.

Supporting evidence for this general structure was found in the behavior of anandamide under GC/MS conditions. Thermal dehydration gave rise to m/z329 $M^{+\cdot}$ ion upon electron ionization (EI) and to $m/z$$MH^{+\cdot}$ ion under CI. Both self-CI m/z330$MH^{+\cdot}$ and m/z329 $M^{+\cdot}$ were formed under EI conditions in an ion-trap instrument. The fragmentation pattern of the dehydration products of anandamide and palmitoylethanolamide were similar in the low mass range of the EI mass spectra and included m/z85 (McLafferty rearrangement ion) and m/z98 (a product of a γ-cleavage). The EI mass spectrum of dehydrated palmitoylethanolamide exhibited an m/z112 ion that corresponded to a δ-cleavage fragment. The absence of this ion from the EI mass spectrum obtained in the GC/MS analysis of anandamide suggested the presence of the first double bond in the tetraenic acid at position 5 (as in arachidonoylethanolamide, which would not be expected to yield a δ-cleavage product).

Because of the small amount of natural anandamide available, we were able to record ^{1}H NMR spectra only. The peaks attributed to double-bond protons (δ 5.30 to 5.45, multiplet) were coupled with those of protons that have the chemical shifts of doubly allylic protons (δ 2.75 to 2.90, multiplet). Such doubly allylic protons are typically found in all-*cis*, nonconjugated polyunsaturated fatty acids such as linoleic and arachidonic acids. Three pairs of protons were observed between δ 2.01 and 2.27, which we attributed to two allylic methylene groups and one methylene group α to a carbonyl moiety. Only one methyl group was observed (0.99, triplet). The peaks observed for two protons at 3.42 (N-CH_2, triplet), two protons at 3.72 (O-CH_2, triplet), and two protons at 2.20 ($COCH_2$, triplet) were similar in chemical shifts and spin-coupling patterns to peak observed in the NMR spectrum of synthetic palmitoylethanolamide. The peaks for N-CH_2 and O-CH_2 were coupled.

We named the active constituent anandamide — based on the Sanskrit word *ananda* meaning delight, bliss, and on its chemical nature. A juxtaposition of the various analytical data led us to conclude that the structure of anandamide is that of arachidonoylethanolamide. This conclusion was confirmed by synthesis.

Anandamide inhibited the specific binding of [^{3}H]-HU-243 to synaptosomal membranes in a manner typical of competitive ligands, with an inhibition constant (K_i) of 39.0 $\pm$ 5.0 nM. In this system, the K_i of Δ^9-THC, a psychoactive compound of cannabis, was 46.0 $\pm$ 3.0 nM. These were exciting results — the psychoactive compound from a higher plant and a chemically completely different compound in the brain were found to bind to the same brain receptor at about the same level of activity. Soon after the identification of anandamide, this compound was tested for its pharmacological activity.[102] Anandamide administered i.p. in mice, caused lowering of activity in an immobility test and in an open field test, and produced hypothermia and analgesia. These effects parallel those caused by psychotropic cannabinoids.

Our research group expected that additional polyunsaturated fatty acid ethanolamides may be present in the brain. We also identified in porcine brain another two putative endocannabinoids, namely homo-γ-linoleoylethanolamide (K_i = 53.4 $\pm$ 5.5 nM) and 7,10,13,16-docosatetraenoylethanolamide (K_i = 34.4 $\pm$ 3.2 nM). The isolation of these two compounds[103] as constituents of porcine brain that bind to the cannabinoid receptor demonstrated that anandamide is not the sole representative of this class of potential mediators.

Later, we described the isolation of a second type of cannabinoid receptor ligand, 2-arachidonoyl glycerol (2-AG) (K_i = 5.85 $\pm$ 0.12 μM), an ester isolated from canine gut.[104] This was the first putative endogenous cannabinoid receptor ligand isolated from a peripheral tissue. Later, Sugiura *et al.*[105] isolated independently this compound from brain.

2-Arachidonyl glyceryl ether (noladin ether), isolated from porcine brain, is an example of a third, ether-type endocannabinoid.[106] The name is derived from the Hebrew word *nolad*, what means "to be born." The structure of noladin ether was determined by mass spectrometry and nuclear magnetic resonance spectroscopy. It was confirmed by the comparison with a synthetic sample. It binds to the CB_1 cannabinoid receptor ($K_i = 21.2 \pm 0.5$ nM) and causes sedation, hypothermia, intestinal immobility, and mild antinociception in mice. It binds weakly to the CB_2 receptor (Ki $> 3\,\mu$M).

However, two groups have been unsuccessful in identifying noladin ether in tissues. The first group did not find this compound in several mammalian brains (rat, mouse, hamster, guinea-pig and pig).[107] The second group did not identify, not only noladin ether, but also two additional endocannabinoids — arachidonoyl glycine and virodhamine in rod brain tissue.[108] We had originally identified and isolated noladin ether from porcine brain independently twice over two years. In view of the discrepancy of the results, it remains an open question whether noladin is an authentic endocannabinoid or an artifact, formed from some endogenous compound under the conditions of the isolation procedures.

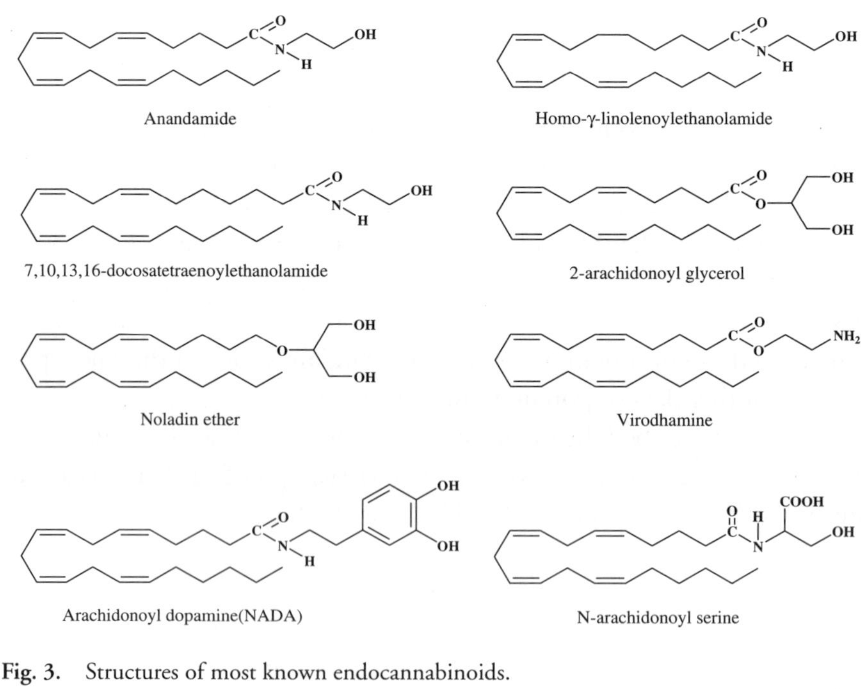

Fig. 3. Structures of most known endocannabinoids.

A compound with the same molecular weight as anandamide, but with a shorter retention time, was identified as O-arachidonoyl ethanolamine (arachidonic acid and ethanolamine joined by an ester linkage) (EC_{50} = 1906 nM). Based on this opposite orientation, the molecule was named virodhamine from the Sanskrit word, *virodha*, which means opposition.[109]

Huang *et al.*[110] assumed that N-arachidonoyl-dopamine (NADA) may exist as an endogenous "capsaicin-like" cannabinoid in mammalian nervous tissues and may possibly bind to the vanilloid receptor VR1. They found that NADA is indeed a natural endocannabinoid, in nervous tissues, with high concentrations found in the striatum, hippocampus, and cerebellum. They were also found in lower concentrations in the dorsal root ganglion. NADA binds to the cannabinoid receptors with a 40-fold selectivity for the CB_1 ($K_i = 250 \pm 130$ nM) over the CB_2 receptors.

3.5 THE ENDOCANNABINOID CONGENERS

The endocannabinoids are accompanied in the brain and other tissues by cannabinoid-like compounds, which however do not bind to the cannabinoid receptors. They are saturated and mono- or diunsaturated congeners, which may affect the metabolism and the function of the endocannabinoids (Fig. 4).

Palmitoylethanolamide, isolated from rat and quinea pig brains,[111] was shown to exhibit antiinflammatory and analgesic activity even though it does not activate central and peripheral cannabinoid receptors.[112,113] Palmitoylethanolamide binds with high potency ($EC_{50} = 3.2 \pm 1.3$ nM) to the orphan G-protein-coupled receptor GPR55, which as mentioned above, may represent a new cannabinoid receptor.[97]

2-AG is accompanied in the spleen, brain and gut by several 2-acyl glycerol esters, two major ones being 2-linoleoyl glycerol and 2-palmitoyl glycerol, which significantly potentiate the apparent binding of 2-AG and its apparent capacity to inhibit adenylyl cyclase. The data indicated that the biological activity of 2-AG can be increased by related, endogenous 2-acyl glycerols, which alone shows no significant activity in any of the tests employed. This effect, named by us "entourage effect", may represent a novel route for molecular regulation of endogenous cannabinoid activity.[114]

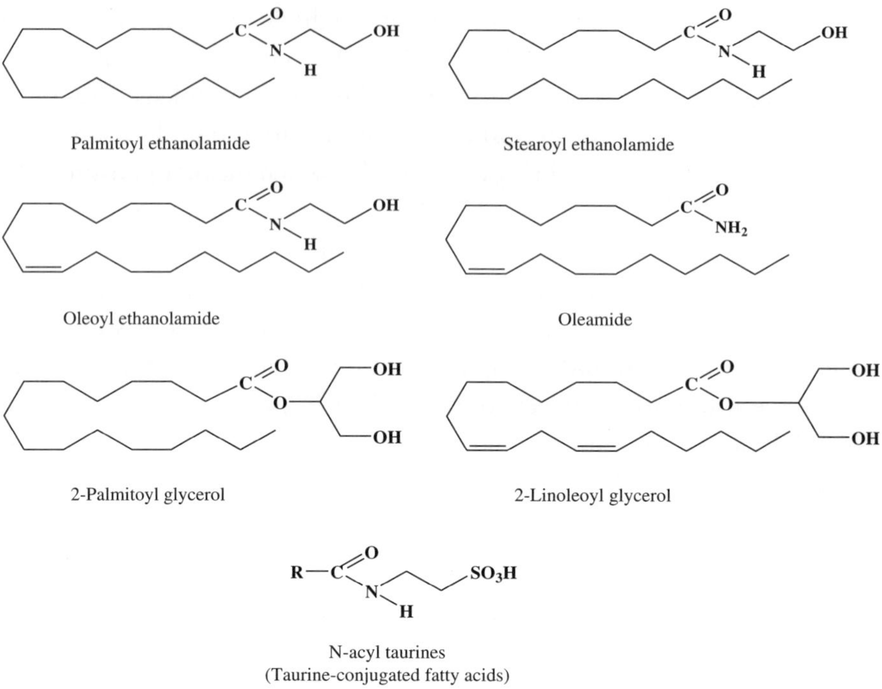

Fig. 4. Known endocannabinoids congeners.

Another saturated ethanolamide, stearoylethanolamide, exerts a marked dose-dependent anorexic effect. This congener reduces food intake in mice in a structurally selective manner.[115] Maccarone[116] reported that stearoylethanolamide binds to a specific site different from the known cannabinoid or vanilloid receptors. This site is not coupled to G-proteins, and apparently it is regulated by a different signaling pathway. Degradation and pro-apoptotic activity of stearoylethanolamide are regulated by NO in a way opposite to that reported for anandamide. Stearoylethanolamide potentiates the decrease of cAMP induced by AEA in mouse cortical slices, suggesting that SEA might also be an "entourage" compound.[117]

Oleoylethanolamide is an endogenous regulator of food intake, and intraperitoneal injection of this compound decreased food intake by rats starved for 24 hours.[118] This endogenous lipid mediator reduces food intake and thus represents a satiating factor. It decreases body weight in rodents by activating the nuclear receptor peroxisome proliferator-activated receptor-αg(gPAR-α). Oleoylethanolamide has central and peripheral anorexic effects.[119] A possible protective action of oleoylethanolamide against reactive oxygen species, could explain its beneficial effects on in *vitro* capacitated spermatozoa.[120]

Oleamide, an unsaturated fatty acid amide, which can modulate central nervous system functions, was isolated from the cerebrospinal fluid of sleep deprived cats.[121] Nanogram amounts of oleamide in biological fluids (using GC/MS) were measured. This quantitative assay may help determine the role of oleamide in additional physiological processes.[122] Its hypnotic properties were characterized.[123] Its mechanism of action is far from being understood. Although it does not bind with high affinity to CB_1 or CB_2 receptors, it exhibits some cannabimimetic actions, which could be explained at least in part by entourage effects. It is likely that oleamide and anandamide have common as well as distinct pathways of action. The 5-HT2A receptor appears to be a target for oleamide but the possibility of the existence of specific receptors for this compound is still open.[124] Legget *et al.*[125] have reported that oleamide is a full cannabinoid CB_1 receptor agonist. Therefore, in addition to allosteric modulation of other receptors and possible entourage effects due to fatty acid amide hydrolase (FAAH) inhibition, the effects of oleamide may be mediated directly via the CB_1 receptor. Oleamide also elicites vasorelaxation in rat small mesenteric arteries. This effect is partly dependent on the presence of endothelium, activation of Ca^{2+}- sensitive K^+ channels and involves capsaicin-sensitive sensory nerves.[126]

The anandamide precursor, phosphatidylethanol amine, is present in membranes almost always accompanied by phophatidylserine. It seemed reasonable to expect the formation of anandamide from its precursor will be paralleled by formation of N-arachidinoylserine from phosphatidyl serine. Indeed, *N*-arachidonoyl-L-serine (ARA-S) was found to be formed alongside anandamide (Fig. 4). This compound was isolated from bovine brain and its structure was elucidated by comparison with synthetic ARA-S.[127] Contrary to anandamide, ARA-S binds very weakly to the known cannabinoid CB_1 and CB_2 or vanilloid TRPV1 receptors. However, it produces endothelium-dependent vasodilation of rat isolated mesenteric arteries and abdominal aorta, and stimulates phosphorylation of p44/42 MAP kinase and protein kinase B/Akt in cultured endothelial cells. ARA-S also suppresses LPS-induced formation of tumor necrosis factor-α (TNF-α) in a murine macrophage cell line and in wild type mice, as well as in mice deficient in CB_1 or CB_2 receptors. Many of these effects parallel those reported for abnormal-cannabidiol (Abn-CBD), a synthetic agonist of a putative novel cannabinoid type receptor.[128,129] Hence, ARA-S may represent an endogenous agonist for this receptor.

A novel family of nervous system enriched natural products, the taurine-conjugated fatty acids was discovered recently.[130] N-acyl taurines were found to activate multiple members of the transient receptor potential family of calcium channels, including TRPV1 and TRPV4, which are both expressed in the kidney. The dramatic elevation in endogenous levels of N-acyl taurines following acute or chronic inactivation of FAAH, in conjunction with the pharmacological effects of these lipids on TRP channels, suggests the existence of a second major lipid signaling system regulated by FAAH *in vivo*.[131]

REFERENCES

1. Adovasio JM, Soffer O, Hyland DC, Klíma B, Svoboda J, Textil, košíkářství a sítí v mladém paleolitu Moravy. (Textiles, Basketry, and Nets in Upper Paleolithic Moravia.), *Archeologické rozhledy* **51**: 58–94, 1999.
2. Chang K, *The Archaeology of Ancient China*, New Haven, Yale University Press, pp. 111–112, 1968.
3. Kung CT, *Archeology in China*, Toronto, University of Toronto Press, Toronto, **1**:131, 1959.
4. Li Hui-Lin An archeological and historical account of Cannabis in China, *Econ Bot* **28**: 437–448, 1974.
5. Unschuld PU, *Medicine In China: A History Of Pharmaceutics*, University of California Press, p. 367, 1986.
6. Campbell Thompson R, A Dictionary of Assyrian Botany, London, The British Academy, 1949.
7. Waterman L, Royal Correspondence of the Assyrian Empire, University of Michigan Press, Ann Arbor, *letter* 368, 1930.
8. *The papyrus Ebers. The greatest Egyptian Medical document*, translated by Ebbell B, Copenhagen, Levin & Munksgaard, 1937.

9. Manniche L, *An Ancient Egyptian Herbal*, British Museum Publications Ltd, London, p. 176, 1989.
10. *Sacred Books of the East* Volume 42, *Hymns of the Atharva-Veda*, translated by Maurice Bloomfield, 1897.
11. The Vendidad (The Zend-Avesta, part I) Sacred books of the East, Volume 4., Oxford University Press, translated by James Darmesteter, 1880.
12. Khorda Avesta (Book of Common Prayer) in Sacred Books of the East, American Edition, translated by James Darmesteter, 1898.
13. Avalon Arthur *Tantra of the great liberation — Mahanirvana Tantra*, Dover publications, New York, 1972.
14. Herodotus, The History of Herodotus. Book 4: Melpomene [70], Macmillan, London and New York, translated by Macaulay GC, 1890.
15. Benetowa S, Konopie w wierzeniach i zwyczajach ludowych (Hemp in folk belief and customs), Instytutu Nauk Antropologicznych i Etnologicznych Towarzystwa Naukowego Warszawskiego, Warszawa, 1936.
16. Benetowa S, *Tracing One Word Through Different Languages*, in George Andrews, *The Book Of Grass: An Anthology on Indian Hemp*, Grove Press, 1936.
17. Benet S, Early Diffusion and Folk Uses of Hemp" in V. Rubin (ed.), *Cannabis and Culture*, The Hague, Mouton, pp. 39–49, 1975.
18. Rabin C. (1996), Rice in the Bible, *Journal of Semitic Studies* 11, 2–9.
19. Green J, *Cannabis*, Thunder's Mouth Press, New York, p. 256, 2002.
20. Abel EL, *Marijuana — The First Twelve Thousand Years*, Plenum Press, New York, 1980.
21. Mechoulam R, The Pharmacohistory of Cannabis sativa in Mechoulam R (ed.), Cannabinoids as Therapeutic Agents, CRC Press Inc., Boca Raton, Fla, pp. 1–19, 1986.
22. Adams R, Pease DC, Clark JH, Isolation of cannibinol and querbrachitol from red oil of minnesota wild hemp, *J Amer Chem Soc* **62**: 2194–2196, 1940.
23. Adams R (1942), Isolation of cannabidol from red oil obtained from hemp. *United States Patent* 2,304,669, December 8, 1942.
24. Jacob A, Todd AR (1940), Cannabis and Cannabol Constituents of Cannabis indica resin. *Nature* 145, 350.
25. Todd AR, The hemp drugs, *Endeavour* **2**: 69–72, 1943.
26. Gaoni Y, Mechoulam R, The structure and synthesis of cannabigerol, a new hashish constituent, *Proc Chem Soc* **82**, 1964.
27. Mechoulam R, Gaoni Y, The isolation and structure of cannabinolic, cannabidiolic and cannabigerolic acids, *Tetrahedron* **21**: 1223–1229, 1965.
28. Mechoulam R, Marihuana chemistry. Review, *Science* **168**: 1159- 1166, 1970.

29. Mechoulam R, Gaoni Y, Recent advances in the chemistry of hashish, Review article, *Fortschr Chem Org Naturstoffe* **25**: 175–213, 1967.
30. Yamauchi T, Shoyama Y, Aramaki H, Azuma T, Nishioka I, Tetrahydrocannabinolic acid a genuine substance of tetrahydrocannabinol, *Chem Pharm Bull* **15**: 1075–1076, 1967.
31. Shoyama Y, Yagi M, Nishioka I, Yamauchi T, Cannabis. 8. Biosynthesis of cannabinoid acids, *Phytochemistry* **14**: 2189–2192, 1975.
32. Shoyama Y, Yamauchi T, Nishioka I, Cannabis. 5. Cannabigerolic acid monomethyl ether and cannabinolic acid, *Chem Pharm Bull* **18**: 1327–1332, 1970.
33. Shoyama Y, Hirano H, Makino (née Tomita) H, Umekita N, Nishioka I, Cannabis .10. Isolation and structures of 4 new propyl cannabinoid acids, tetrahydrocannabivarinic acid, cannabidivarinic acid, cannabichromevarinic acid and cannabigerovarinic acid, from thai cannabis, Meao variant, *Chem Pharm Bull* **25**: 2306–2311, 1977.
34. Shoyama Y, Nishioka I, Cannabis. 13. 2 New spiro-compounds, cannabispirol and acetyl cannabispirol, *Chem Pharm Bull* **26**: 3641–3646, 1978.
35. Shoyama Y, Hirano H, Nishioka I, Cannabis. Part 16. Biosynthesis of propyl cannabinoid acid and its biosynthetic relationship with pentyl and methyl cannabinoid acids, *Phytochemistry* **23**: 1909–1912, 1984.
36. Crombie L, Crombie WML, Jamieson SV, Isolation of cannabispiradienone and cannabidihydrophenanthrene — Biosynthetic relationships between the spirans and dihydrostilbenes of Thailand cannabis, *Tetrahedron Lett* **7**: 661–664, 1979.
37. Turner CE, El-Sohly MA, Constituents of Cannabis sativa L. XVI. Possible decomposition pathway of Δ^9-tetrahydrocannabinol to cannabinol, *J Heterocycl Chem* **16**: 1667–1668, 1979.
38. Taura F, Morimoto S, Shoyama Y, Cannabis. 23. Cannabinerolic acid, a cannabinoid from Cannabis sativa, *Phytochemistry* **39**, 457–458, 1995.
39. Taura F, Morimoto S, Shoyama Y, Mechoulam R, First direct evidence for the mechanism of Δ^1-tetrahydrocannabinolic acid biosynthesis, *J Am Chem Soc* **117**: 9766–9767, 1995.
40. Turner CE, Hadley K, Constituents of Cannabis sativa. II. Absence of cannabidiol in an African variant, *J Pharm Sci* **62**: 251–254, 1973.
41. Krejčí Z, Hanuš L, Yoshida T, Braenden OJ, The effect of climatic and ecologic conditions upon the formation and the amount of cannabinoid substances in the cannabis of various provenance, *Acta Univ Olomuc, Fac Med* **74**: 147–160, 1975.
42. Guerrero Davalos S, Fournier G, Boucher F, Paris M, Contribution to the study of Mexican marihuana. Preliminary studies: Cannabinoids and essential oil, *Journal de Pharmacie de Belgique* **32**: 89–99, 1977.

43. Taura F, Morimoto S, Shoyama Y, Purification and characterization of cannabidiolic-acid synthase from Cannabis sativa L. Biochemical analysis of a novel enzyme that catalyzes the oxidocyclization of cannabigerolic acid to cannabidiolic acid, *J Biol Chem* **271**: 17411–17416, 1996.
44. Morimoto S, Komatsu K, Taura F, Shoyama Y, Enzymological evidence for cannabichromenic acid biosynthesis, *J Nat Prod* **60**: 854–857, 1997.
45. Taura F, Shoyama Y, Morimoto S, Biosynthetic study on THCA, the psychoactive component of marijuana, *Seibutsu Butsuri* **45**: 178–184, 2005.
46. Wood TB, Spivey WTN, Easterfield TH, Cannabinol. Part I, *J Chem Soc* **75**: 20–36, 1899.
47. Cahn RS, Cannabis indica resin, Part III. The constitution of Cannabinol, *J Chem Soc* 1342–1353, 1932.
48. Jacob A, Todd AR, Cannabis indica. Part II. Isolation of Cannabidiol from Egyptian Hashish. Observations on the Structure of Cannabinol, *J Chem Soc* 649–653, 1940.
49. Adams R, Baker BR, Wearn RB, Structure of Cannabinol. III. Synthesis of Cannabinol,1-Hydroxy-3-n-amyl-6,6,9-trimethyl-6-dibenzopyran, *J Amer Chem Soc* **62**: 2204–2207 1940.
50. Ghosh R, Todd AR, Wilkinson S, Cannabis indica, Part v, The synthesis of cannabinol, *J Chem Soc*, 1393–1396, 1940.
51. Adams R, Wolff H, Cain CK, Clark JH, Structure of Cannabidiol. V. Position of the Alicyclic Double Bonds, *J Amer Chem Soc* **62**: 2215–2219, 1940.
52. Adams R, Loewe S, Pease DC, Cain CK, Wearn RB, Baker BR, Wolff H, Structure of cannabidiol. VIII. Position of the double bonds in cannabidiol. Marihuana activity of tetrahydrocannabinols, *J Amer Chem Soc* **62**: 2566–2567, 1940.
53. Adams R, Baker BR, Structure of Cannabidiol. VII. A Method of Synthesis of a Tetrahydrocannabinol which Possesses Marihuana Activity, *J Amer Chem Soc* **62**: 2405–2408, 1940.
54. Adams R, Pease DC, Cain CK, Baker BR, Clark JH, Wolff H, Wearn RB, Conversion of cannabidiol to a product with marihuana activity. A type reaction for synthesis of analogous substances. Conversion of cannabidiol to cannabinol, *J Amer Chem Soc* **62**: 2245–2246, 1940.
55. Adams R,Pease DC, Cain CK, Clark JH, Structure of cannabinol. VI. Isomerization of cannabidiol to tetrahydrocannabinol, a physiologically active product. Conversion of cannabidiol to cannabinol, *J Amer Chem Soc* **62**: 2402–2405, 1940.
56. Ghosh R, Todd AR, Wilkinson S, *Cannabis indica,* Part IV. The synthesis of some tetrahydrodibenzopyran derivatives, *J Chem Soc* 1121–1125, 1940.

57. Krejčí Z, Šantavý F, Isolace dalších látek z listí indického konopí Cannabis sativa L, *Acta Univ Palacki Olomuc* **6**: 59–66, 1955.
58. Kabelík J, Krejčí Z, Šantavý F, *Cannabis* as a medicament, *Bull Narc* **12**: 5–23, 1960.
59. Mechoulam R, Shvo Y, The structure of cannabidiol, *Tetrahedron* **19**: 2073–2078, 1963.
60. Gaoni Y, Mechoulam R, Isolation, Structure, and Partial Synthesis of an Active Constituent of Hashish, *J Amer Chem Soc* **86**: 1646–1647, 1964.
61. Hively RL, Mosher WA, Hoffmann FW, Isolation of *trans*-Δ^6-tetrahydrocannabinol from marijuana, *J Am Chem Soc* **88**: 1832–1833, 1966.
62. Šantavý F, Notes on the structure of cannabidiol compounds, *Acta Univ Palacki Olomuc Fac Med* **35**: 5–9, 1964.
63. Mechoulam R, Gaoni Y, The absolute configuration of Δ^1-tetrahydrocannabinol, the major active constituent of hashish, *Tetrahedron Lett* **8**: 1109–1111, 1967.
64. Gaoni Y, Mechoulam R, Cannabichromene, a new active principle in hashish, *Chem Comm* 20–21, 1966.
65. Claussen U, v. Spulak F, Korte F, Zur chemischen klassifizierung von pflanzen XXXI. Haschisch X, Cannabichromen, ein neuer haschisch-inhalts-stoff, *Tetrahedron* **22**: 1477–1479, 1966.
66. Crombie L, Ponsford R, Hashish components. Photochemical production of cannabicyclol from cannabichromene, *Tetrahedron Lett* **9**: 5771–5772, 1968.
67. Korte F, Haag M, Claussen U, Tetrahydrocannabinol-carbonsäure, ein neuer Haschisch-Inhaltsstoff, *Angew Chem* **77**: 862, 1965.
68. Mechoulam R, Ben-Zvi Z, Yagnitinsky B, Shani A, A new tetrahydrocannabinolic acid, *Tetrahedron Lett* **10**: 2339–2341, 1969.
69. Hanuš L, Krejčí Z, Isolation of two new cannabinoid acids from Cannabis sativa L, of Czechoslovak origin, *Acta Univ Olomuc Fac Med* **74**: 161–166, 1975.
70. Shani A, Mechoulam R, Cannabielsoic acids. Isolation and synthesis by a novel oxidative cyclization, *Tetrahedron* **30**: 2437–2446, 1974.
71. Gohda H, Narimatsu S, Watanabe K, Yamamoto I, Yoshimura H, The formation mechanism of cannabielsoin from cannabidiol with guinea-pig hepatic-microsomal enzymes, *J Pharm Sci* **76**: S32, 1987.
72. Mechoulam R, Ben-Zvi Z, Carboxylation of resorcinols with methyl magnesium carbonate. Synthesis of cannabinoid acids, *Chem Commun* 343–344, 1969.
73. Hanuš L, The present state of knowledge in the chemistry of substances of Cannabis sativa L, VI. The other content substances, *Acta Univ Olomuc, Fac Med* **76**: 167–173, 1976.

74. Turner CE, Elsohly MA, Boeren EG, Constituents of Cannabis sativa L. XVII. A review of the natural constituents. *J Nat Prod* **43**: 169–234, 1980.
75. Mechoulam R, Alkaloids in Cannabis sativa L, in Brossi A (ed.), *The Alkaloids* Vol. 34, Academic Press, San Diego, pp. 77–93, 1988.
76. Ross SA, ElSohly MA, Constituents of Cannabis sativa L, XXVIII. A review of the natural constituents: 1980–1994. *Zagazig J Pharm Sci* **4**: 1–10, 1995.
77. Ross SA, ElSohly MA, Sultana GNN, Mehmedic Z, Hossain CF, Chandra S, Flavonoid glycosides and cannabinoids from the pollen of Cannabis sativa L, *Phytochem Anal* **16**: 45–48, 2005.
78. ElSohly MA, Slade D, Chemical constituents of marijuana: The complex mixture of natural cannabinoids. *Life Sci* **78**: 539–548, 2005.
79. Howlett AC, Inhibition of neuroblastoma adenylate cyclase by cannabinoid and nantradol compounds, *Life Sci* **35**: 1803–1810, 1984.
80. Howlett AC, Fleming RM, Cannabinoid inhibition of adenylate cyclase. Pharmacology of the response in neuroblastoma cell membranes, *Mol Pharmacol* **26**: 532–538, 1984.
81. Howlett AC, Qualy JM, Khachatrian LL, Involvement of Gi in the inhibition of adenylate cyclase by cannabimimetic drugs, *Mol Pharmacol* **29**: 307–313, 1986.
82. Howlett AC, Champion TM, Wilken GH, Mechoulam R, Stereochemical effects of 11-OH-Δ^6-tetrahydrocannabinol-dimethylheptyl to inhibit adenylate cyclase and bind to the cannabinoid receptor, Neuropharmacology **29**: 161–165, 1990.
83. Devane WA, Dysarz FA, Johnson MR, Melvin LS, Howlett AC, Determination and characterization of a cannabinoid receptor in rat brain, *Mol Pharmacol* **34**: 605–613, 1988.
84. Matsuda LA, Lolait SJ, Brownstein MJ, Young AC, Bonner TI, Structure of a cannabinoid receptor and functional expression of the cloned cDNA, *Nature* **346**: 561–564, 1990.
85. Gérard CM, Mollereau C, Vassart G, Parmentier M, Molecular cloning of a human cannabinoid receptor which is also expressed in testis, *Biochem J* **279**: 129–134, 1991.
86. Gérard C, Mollereau C, Vassart G, Parmentier M, Nucleotide sequence of a human cannabinoid receptor cDNA, *Nucleic Acids Res.* **18**: 7142, 1990.
87. Chakrabarti A, Onaivi ES, Chaudhuri G, Cloning and sequencing of a cDNA encoding the mouse brain-type cannabinoid receptor protein, *DNA sequence* **5**: 385–388, 1995.
88. Munro S, Thomas KL, Abu-Shaar M, Molecular characterization of a peripheral receptor for cannabinoids, *Nature* **365**: 61–65, 1993.

89. Shire D, Calandra B, Rinaldi-Carmona M, Oustric D, Pessegue B, Bonnin-Cabanne O, Le Fur G, Caput D, Ferrara P, Molecular cloning, expression and function of the murine CB2 peripheral cannabinoid receptor, *Biochim Biophys Acta* **1307**: 132–136, 1996.
90. Howlett AC, Barth F, Bonner TI, Cabral G, Casellas P, Devane WA, Felder CC, Herkenham M, Mackie K, Martin BR, Mechoulam R, Pertwee RG, International Union of Pharmacology. XXVII. Classification of cannabinoid receptors, *Pharmacol Rev* **54**: 161–202, 2002.
91. Van Sickle MD, Duncan M, Kingsley PJ, Mouihate A, Urbani P, Mackie K, Stella N, Makriyannis A, Piomelli D, Davison JS, Marnett LJ, Di Marzo V, Pittman QJ, Patel KD, Sharkey KA, Identification and functional characterization of brainstem cannabinoid CB2 receptors, *Science* **310**: 329–332, 2005.
92. Gong J-P, Onaivi ES, Ishiguro H, Liu Q-R, Tagliaferro PA, Brusco A, Uhl GR, Cannabinoid CB2 receptors: Immunohistochemical localization in rat brain, *Brain Res* **1071**: 10–23, 2006.
93. Onaivi ES, Ishiguro H, Gong J-P, Patel S, Perchuk A, Meozzi PA, Myers L, Mora Z, Tagliaferro P, Gardner E, Brusco A, Akinshola BE, Liu Q-R, Hope B, Iwasaki S, Arinami T, Teasenfitz L, Uhl GR, Discovery of the presence and functional expression of cannabinoid CB2 receptors in brain, *Ann N Y Acad Sci* **1074**: 514–536, 2006.
94. Fernandez-Ruiz J, Romero J, Velasco G, Tolon RM, Ramos JA, Guzman M, Cannabinoid CB2 receptor: A new target for controlling neural cell survival? *Trends Pharmacol Sci* **28**: 39–45, 2007.
95. Sawzdargo M, Nguyen T, Lee DK, Lynch KR, Cheng R, Heng HHQ, George SR, O'Dowd BF, Identification and cloning of three novel human G protein-coupled receptor genes GPR52, Ψ GPR53 and GPR55: GPR55 is extensively expressed in human brain, *Mol Brain Res* **64**: 193–198, 1999.
96. Wise A, Brown AJ, Screening for modulators of G-protein coupled receptor 55, WO0186305, p. 31, 2001.
97. Drmota T, Greasley P, Groblewski T, Screening assays for cannabinoid ligand-GPR55 receptor binding modulators, WO 2004/074844, p. 49, 2004.
98. Petitet F, Donlan M, Michel A, GPR55 as a new cannabinoid receptor: Still a long way to prove it, *Chem Biol Drug Design* **67**: 252–253, 2006.
99. Baker D, Pryce G, Davies WL, Hiley CR, *in silico* patent searching reveals a new cannabinoid receptor, *Trends Pharmacol Sci* **27**: 1–4, 2006.
100. Wagner JA, Varga, K, Jarai Z, Kunos G, Mesenteric Vasodilation Mediated by Endothelial Anandamide Receptors, *Hypertension* **33**: 429–434, 1999.
101. Devane WA, Hanuš L, Breuer A, Pertwee RG, Stevenson LA, Griffin G, Gibson D, Mandelbaum A, Etinger A, Mechoulam R, Isolation and structure of a brain

constituent that binds to the cannabinoid receptor, *Science* **258**: 1946–1949, 1992.

102. Fride E, Mechoulam R, Pharmacological activity of the cannabinoid receptor agonist, anandamide, a brain constituent. *Eur J Pharmacol* **231**: 313–314, 1993.
103. Hanuš L, Gopher A, Almog S, Mechoulam R, Two new unsaturated fatty acid ethanolamides in brain that bind to the cannabinoid receptor, *J Med Chem* **36**: 3032–3034, 1993.
104. Mechoulam R, Ben-Shabat S, Hanuš L, Ligumsky M, Kaminski NE, Schatz AR, Gopher A, Almog S, Martin BR, Compton DR, Pertwee RG, Griffin G, Bayewitch M, Barg J, Vogel Z, Identification of an endogenous 2-monoglyceride, present in canine gut, that binds to cannabinoid receptors, *Biochem Pharmacol* **50**: 83–90, 1995.
105. Sugiura T, Kondo S, Sukagawa A, Nakane S, Shinoda A, Itoh K, Yamashita A, Waku K, 2-Arachidonoylglycerol: A possible endogenous cannabinoid receptor ligand in brain, *Biochem Biophys Res Commun* **215**: 89–97, 1995.
106. Hanuš L, Abu-Lafi S, Fride E, Breuer A, Vogel Z, Shalev DE, Kustanovich I, Mechoulam R, 2-arachidonyl glyceryl ether, an endogenous agonist of the cannabinoid CB1 receptor, *PNAS* **98**: 3662–3665, 2001.
107. Oka S, Tsuchie A, Tokumura A., Muramatsu M, Suhara Y, Takayama H, Waku K, Sugiura T, Ether-linked analogue of 2-arachidonoylglycerol (noladin ether) was not detected in the brains of various mammalian species, *J Neurochem* **85**: 1374–1381, 2003.
108. Richardson, D, Ortori CA, Chapman V, Kendall DA, Barrett DA, Quantitative profiling of endocannabinoids and related compounds in rat brain using liquid chromatography-tandem electrospray ionization mass spectrometry, *Anal Biochem* **360**: 216–226, 2007.
109. Porter AC, Sauer JM, Knierman MD, Becker GW, Berna MJ, Bao JQ, Nomikos GG, Carter P, Bymaster FP, Leese AB, Felder CC, Characterization of a novel endocannabinoid, virodhamine, with antagonist activity at the CB1 receptor, *J Pharmacol Exp Ther* **301**: 1020–1024, 2002.
110. Huang SM, Bisogno T, Trevisani M, Al-Hayani A, De Petrocellis L, Fezza F, Tognetto M, Petros TJ, Krey JF, Chu CJ, Miller JD, Davies SN, Geppetti P, Walker JM, Di Marzo V, An endogenous capsaicin-like substance with high potency at recombinant and native vanilloid VR1 receptors, *PNAS* **99**: 8400–8405, 2002.
111. Bachur NR, Mašek K, Melmon KL, Fatty acid amides of ethanolamine in mammalian tissues, *J Biol Chem* **240**: 1019–1024, 1965.
112. Conti S, Costa B, Colleoni M, Parolaro D, Giagnoni G, Antiinflammatory action of endocannabinoid palmitoylethanolamide and the synthetic cannabinoid

nabilone in a model of acute inflammation in the rat, *Brit J Pharmacol* **135**: 181–187, 2002.

113. Lambert DM, Vandevoorde S, Jonsson KO, Fowler CJ, The palmitoylethanolamide family: A new class of anti-inflammatory agents?, *Curr Med Chem* **9**: 663–674, 2002.
114. Ben-Shabat S, Fride E, Sheskin T, Tamiri T, Rhee M,-H., Vogel Z, Bisogno T, De Petrocellis L, Di Marzo V, Mechoulam R, An entourage effect: Inactive endogenous fatty acid glycerol esters enhance 2-arachidonoyl-glycerol cannabinoid activity, *Eur J Pharmacol* **353**: 23–31, 1998.
115. Terrazzino S, Berto F, Carbonare MD, Fabris M, Guiotto A, Bernardini D, Leon A, Stearoylethanolamide exerts anorexic effects in mice via downregulation of liver stearoyl-coenzyme A desaturase-1 mRNA expression, *FASEB J* **18**: 1580–1582, 2004.
116. Maccarrone M, Cartoni A, Parolaro D, Margonelli A, Massi P, Bari M, Barrista N and Finazzi-Agro A, Cannabimimetic activity, binding and degradation of stearoylethanolamide within mouse antral nervous system, *Mol Cell Neurosci* **21**: 126–140, (2002a).
117. Maccarrone M, Pauselli R, di Rienzo M, Finazzi-agro A, Binding, degradation and apoptotic activity of stearoylethanolamide in rat C6 glioma cells, *Biochem J* **366**: 137–144, 2002.
118. Rodriguez de Fonseca F, Navarro M, Gómez R, Escuredo L, Nava F, Fu J, Murillo-Rodríguez E, Giuffrida A, LoVerme J, Gaetani S, Kathuria S, Gall C, Piomelli D, An anorexic lipid mediator regulated by feeding, *Nature* **414**: 209–212, 2001.
119. Brito B, Castro R, Cabrera de Leon A, Oleoylethanolamide: A molecular cross-talk with leptin in feeding behaviour regulation, *Lett Drug Des Disc* **3**: 741–746, 2006.
120. Ambrosini A, Zolese G, Ambrosi S, Ragni L, Tiano L, Littarru G, Bertoli E, Mantero F, Boscaro M, Balercia G, Oleoylethanolamide protects humans sperm cells from oxidation stress: Studies on cases of idiopathic infertility, *Biol Reprod* **74**: 659–665, 2006.
121. Cravatt BF, Prospero-Garcia O, Siuzdak G, Gilula NB, Henriksen SJ, Boger DL, Lerner RA, Chemical characterization of a family of brain lipids that induce sleep, *Science* **268**: 1506–1509, 1995.
122. Hanuš LO, Fales HM, Spande TF, Basile AS, A Gas Chromatographic-Mass Spectral Assay for the Quantitative Determination of Oleamide in Biological Fluids, *Anal Biochem* **270**: 159–166, 1999.
123. Basile AS, Hanuš L, Mendelson WB, Characterization of the hypnotic properties of oleamide, *NeuroReport* **10**: 947–951, 1999.

124. Lambert DM, Di Marzo V, The palmitoylethanolamide and oleamide enigmas: Are these two fatty acid amides cannabimimetic? *Curr Med Chem* **6**: 757–773, 1999.
125. Leggett JD, Aspley S, Beckett SRG, D'Antona AM, Kendall DA, Kendall DA, Oleamide is a selective endogenous agonist of rat and human CB1 cannabinoid receptors, *Brit J Pharmacol* **141**: 253–262, 2004.
126. Hoi PM, Hiley CR, Vasorelaxant effects of oleamide in rat small mesenteric artery indicate action at a novel cannabinoid receptor, *Brit J Pharmacol* **147**: 560–568, 2006.
127. Milman G, Maor Y, Abu-Lafi S, Horowitz M, Gallily R, Batkai S, Mo FM, Offertaler L, Pacher P, *PNAS* **103**: 2428–2433.
128. Offertaler L, Mo FM, Batkai S, Liu J, Begg M, Razdan RK, Martin BR, Bukoski RD, Kunos G, Selective ligands and cellular effectors of a G protein-coupled endothelial cannabinoid receptor, *Mol Pharmacol* **63**: 699–705, 2003.
129. Begg M, Pacher P, Batkai S, Osei-Hyiaman D, Offertaler L, Mo FM, Liu H, Kunos G, Evidence for novel cannabinoid receptors, *Pharmacol Therapeut* **106**: 133–145, 2005.
130. Saghatelian A, Trauger SA, Want EJ, Hawkins EG, Siuzdak G, Cravatt BF, Assignment of Endogenous Substrates to Enzymes by Global Metabolite Profiling, *Biochemistry* **43**: 14332–14339, 2004.
131. Saghatelian A, McKinney MK, Bandell M, Patapoutian A, Cravatt BF, A FAAH-Regulated Class of N-Acyl Taurines That Activates TRP Ion Channels, *Biochemistry* **45**: 9007–9015, 2006.

Chapter 4

NATURAL PRODUCTS AS BIOMARKER TRACERS IN ENVIRONMENTAL AND GEOLOGICAL PROCESSES

Bernd R. T. Simoneit

4.1 INTRODUCTION

This chapter provides the reader with an insight and the key literature for the potential of using natural products in biomarker determinations for research on global and environmental problems. It is not an exhaustive survey of the literature but a selection of some new applications.

4.1.1 Definitions

Natural products have been invoked by organic geochemists as the precursors of biomarker tracers for geological and environmental processes in numerous cases.[1–5] The carbon skeletons of the natural product precursor compounds as synthesized by biota were found either as such or altered directly or indirectly by diagenetic changes to their derivative products. The term "biomarker" evolved from early product-precursor relationships was proposed in the 1960s.[6] The relevance of those naturally derived saturated and aromatic hydrocarbons (and sometimes oxygenated analogs) to environmental geochemistry became evident when polynuclear aromatic hydrocarbons (PAHs) from anthropogenic or geogenic origins were found in recent and contemporary sediments.[7–9] In the geological record, the biomarkers were originally called chemical fossils,[10–12] and the concept has been extended to environmental chemistry.[13,14]

4.1.2 Glossary

Some of the terms and concepts used in this chapter need to be defined, because they are not necessarily used in the same way by natural product chemists. A brief selection follows where each term is printed in italics:

Aerosol is defined herein as an atmospheric suspension of solid particles including adsorbed water and absorbed volatile organic components.

Anthropogenic designates derivation from human activity.

Biomass is used here to describe any biosynthetic matter, such as all vegetation (flora), its recent detritus (e.g. litter, humus, peat to lignite), all fauna (e.g. rendering, cooking), and all products produced by industry utilizing biomass raw material (e.g. paper, rubber, etc.).

Bitumen is the geological equivalent of lipids, consisting in the widest sense of any sedimentary hydrocarbon ranging in state from tarry (asphalt) through viscous to liquid (petroleum).

Fossil fuels, such as coal, petroleum or natural gas, can be burned to produce heat or power.

Geosphere is used here analogous to lithosphere as the rocky outer crust of the earth (the sedimentary and igneous rock record).

Hydrothermal concerns magmatic circulation or emanations of hot water and the consequent processes and products.

Lipids is a broad term that includes all oil-soluble, water-insoluble organic substances such as fats, waxes, fatty acids, sterols, pigments and terpenoids synthesized by contemporary biota.

Particles can be solid or liquid in air (gas) and the mixture consitutes an aerosol. Their typical dimensions are >0.001 μm and they can be sampled by filtration.

Smoke is small gas-borne particles formed during incomplete combustion. It consists mainly of carbon and minor organic and inorganic compounds with typical dimensions of >0.01 μm.

Soil is the superficial, unconsolidated, and weathered part of the geosphere. This surface earth is a bioherm that can be tilled and a locale where vegetation grows.

Troposphere is used here to represent the lowest layer of the atmosphere, ranging from the ground to the base of the stratosphere at 10–15 km altitude. Essentially, all data on particulate organic matter is from near ground level in the lower troposphere.

4.1.3 Historical and New Horizons

Classical natural products chemistry involved the isolation and purification of major natural product compounds from specific representatives of flora or fauna for structure determination. The terpenoids are the compounds of interest in this chapter and their initial structure elucidations were carried out in the 1930s–1970s. There are numerous compilations and encyclopaedia on terpenoids listing the physicochemical properties and literature background of the pure natural products.[15,16] Many of the mass spectra of the underivatized terpenoid compounds are in the standard mass spectrum libraries of modern computerized data systems.

The novel horizons in natural product chemistry are a consequence of advances in mass spectrometry instrumentation.[5] Current applications comprise the elucidation of natural products as part of total extract mixtures in samples of interest to environmental chemists, archaelogists, paleobotanists, geologists, oceanographers, atmospheric chemists, forensic chemists and engineers. The list of applications is expected to expand and some examples are discussed in this chapter.

4.1.4 Natural Products and Derivatives

Many natural products, i.e. homologous aliphatic (lipids), polar (sugars, amino acids), and cyclic (terpenoids) compounds can be utilized as biomarkers. The following is a brief overview of the classical biomarkers commonly used, namely lipids and terpenoids.

4.1.4.1 *Lipids*

Lipids are important biomarkers because they carry a strong carbon-number predominance that is inherited from their biosynthesis (e.g. buildup from acetate), and their homolog distribution can reflect biogenic origin (i.e. marine vs. terrestrial origin).[8,17] They are generally derived from the epicuticular waxes and related lipids of higher plants. They consist primarily of *n*-alkanes (C_{27}, C_{29}, C_{31}, C_{33},), *n*-alkanols, and *n*-alkanoic acids (both homologous series have dominant C_{24}, C_{26}, C_{28}, or C_{30}), with lesser amounts of other oxygenated homologous species.

4.1.4.2 *Terpenoids*

The terpenoid biomarkers from higher plants and other flora are the natural and derivative products from the reductive and oxidative alteration of the precursors. That alteration can occur during transport, by diagenesis in sedimentary environments, or by thermal transformation processes.[3,18–21,83] Reductive alteration generally yields the parent compound skeleton with various isomerizations of chiral centers and, in some cases, loss of carbon due to decarboxylation and other reactions. Oxidative alteration occurs mainly by successive ring aromatizations that usually commence from a ring that has a functional group (e.g. OH, C=C, C=O, typically on ring-A), by direct dehydrogenation, dehydration, ring rearrangement, or ring opening and subsequent loss.

Mono-, sesqui-, and some diterpenoids are found in marine and terrestrial flora. They are, therefore, not always unambiguous tracers for higher plant sources. However, diterpenoids with the abietane and pimarane (Fig. 1), and less common phyllocladane and kaurane, skeletons are predominant constituents in resins and supportive tissue of coniferous vegetation (Coniferae), which evolved in the late Paleozoic (200–300 million years ago). Diterpenoid biomarkers have been characterized in

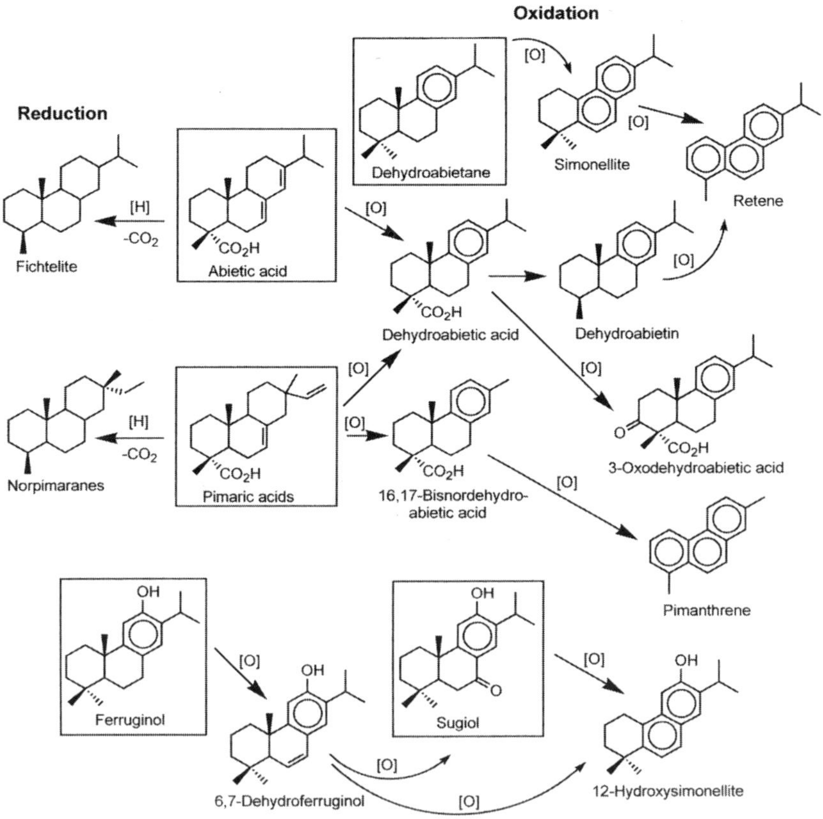

Fig. 1. Scheme for the alteration of diterpenoid natural product precursors to saturated and aromatic derivatives (natural product examples are boxed, stereochemistry is indicated where known).

ambers, coals, sediments, contemporary environments, and anthropogenic emissions.[3,6,11,23] The product-precursor relationship for diterpenoids has been presented by many authors[7–9,11,24–26] and is summarized in Fig. 1. Reductive preservation retains the C_{20} skeletons, which are the major diterpane biomarkers in the geological record. Decarboxylation of resin acids with subsequent reduction yields biomarkers with structures $\leq C_{19}$.

Triterpenoids are a major group of natural products in higher plants. The tetracyclic triterpenoids, based on the lanostane, euphane, onocerane, and dammarane skeletons, are found mostly in vascular plants and

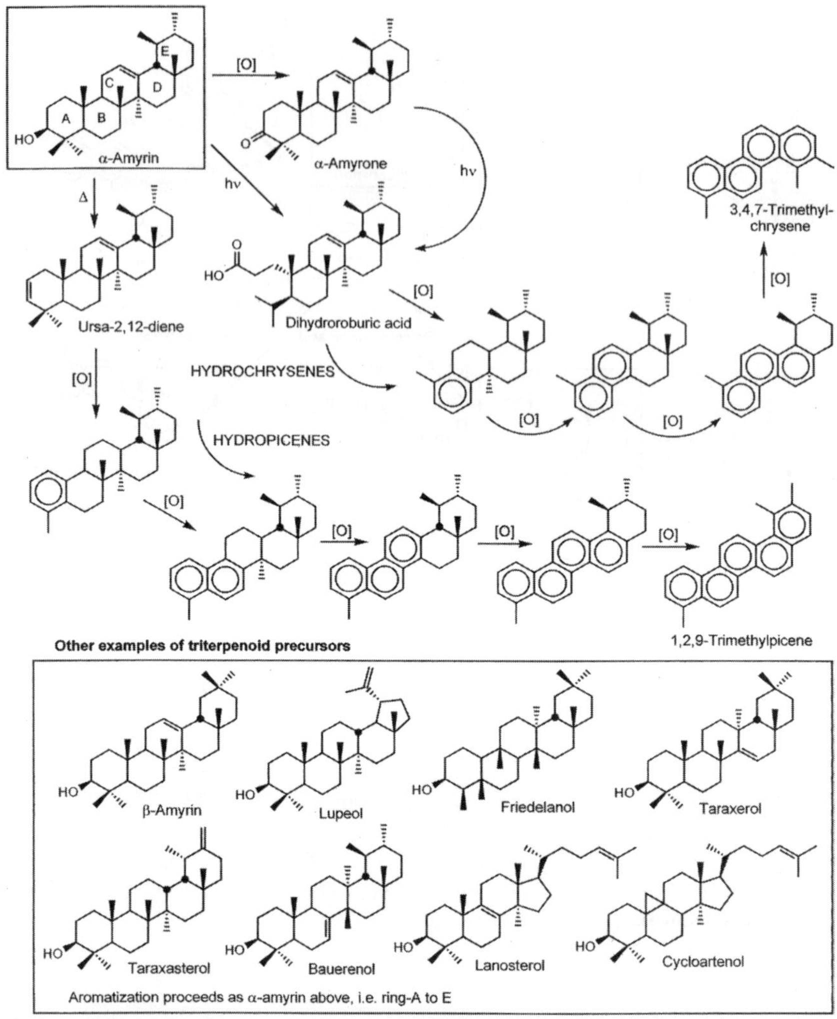

Fig. 2. Scheme for the alteration of triterpenoid precursors from higher plants to saturated and aromatic derivatives (natural product examples are boxed, stereochemistry is indicated where known).

have been reported in sedimentary rocks.[11,27] Many pentacyclic triterpenoids [e.g. the oleanane, ursane, taraxerane, lupane, friedelane, serratane, or bauerane skeletons (Fig. 2)] are characteristic natural product tracers for an origin from terrestrial higher plants. They occur as functionalized (e.g. alcohols, acids, ketones, esters) precursors but are not necessarily specific to individual classes of biota. The most frequently encountered compounds in sedimentary or contemporary environments are those derived from α- and β-amyrins (I, II, ursanes and oleananes, respectively, structures of key compounds are shown in Appendix I).[11,23] The reductive and oxidative alteration of the amyrins is illustrated in Fig. 2. It should be noted that the presence of a functionality at C-3 in those triterpenoids makes them more susceptible to microbial or photochemical degradation to yield ring-A opened products and ultimately compounds without ring-A (e.g. des-A-oleanane). Reductive alteration of triterpenoids yields mainly the parent skeleton with epimerization of key chiral centers. Aromatic triterpenoids were first isolated and characterized from brown coal extracts, and were inferred to be derived from triterpenoids based on the structural similarity and the origin of the coal.[6,28,29]

Steroids are natural product derivatives from triterpenoids and are ubiquitous in the geosphere and the ambient environment. The phytosterols (mainly C_{29}, minor C_{28} skeletons) have been used as indicators for higher plant sources, although many marine algae also biosynthesize C_{28} and C_{29} sterols with the same or different alkyl substituents on the side chain.[11,30] The reductive and oxidative alteration of steroids is illustrated in Fig. 3, and shows the geosteranes, biosteranes, aromatic steroid hydrocarbons, and thermal cracking derivatives. Geological reductive processes produce steranes and diasteranes (geosteranes) with typically the parent skeletons and various epimerizations of chiral centers. Products from dealkylation of the side chain are also encountered. The steroid aromatization reactions have been elucidated in the geological record.[31–33] Biological reductive processes produce *epi*-stanols and related derivatives (Fig. 3,[34]).

Tetraterpenoids and polyterpenoids are minor components of higher plants and are generally overwhelmed by the input of those compounds from microbial biomass in marine and lacustrine environments or sedimentary rocks. The natural cyclic tetraterpenoids have a maximum of two alicyclic rings, and thus the saturated and aromatic derivatives are limited. The common parent skeltons are lycopane, carotane, 1-(2',2',6'-trimethylcyclohexyl)-3,7,12,16,20,24-hexamethylpentacosane, and biphytane.[35,36]

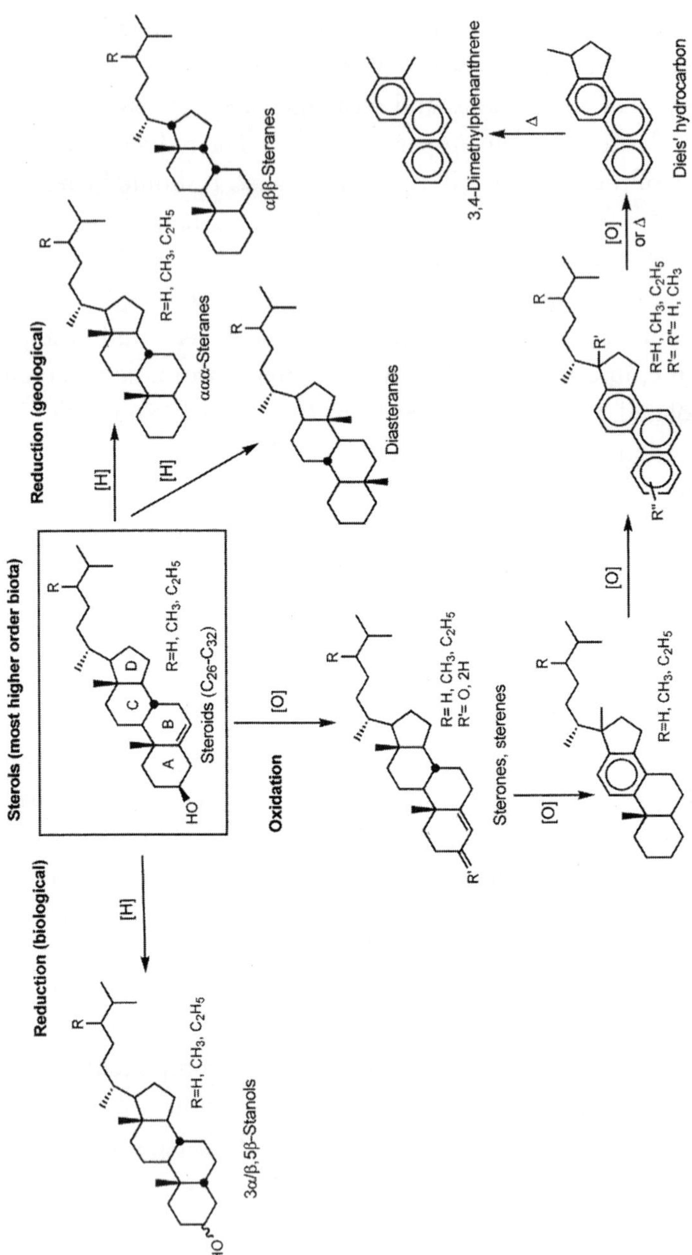

Fig. 3. Scheme for the alteration of sterols to saturated and aromatic hydrocarbons (natural product example is boxed.)

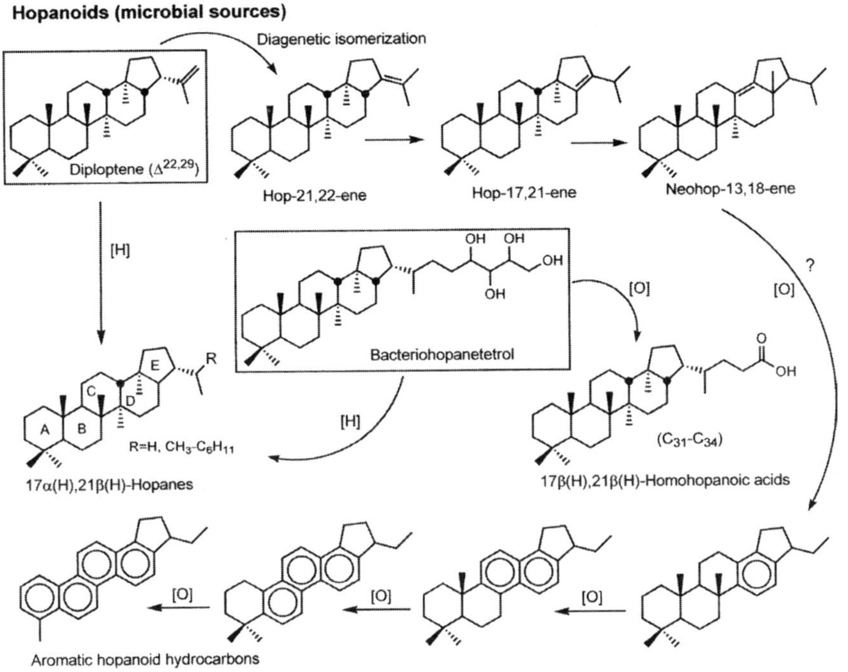

Fig. 4. Scheme for the alteration of hopenoids to aromatic hopanoids and hopane (natural product examples are boxed).

The hopane series are the natural product biomarkers elucidated initially as attributable to bacteria.[37–40] The 17α(H),21β(H)-hopanes ranging from C_{27} to C_{35} (no C_{28}) were encountered in numerous ancient sediments and petroleums, and diagenesis and maturation of the microbial precursors (e.g. bacteriohopanepolyol and diploptene, Fig. 4) were elucidated.[40,41] The diagenesis of diploptene in contemporary sediments proceeds by double bond migration from $\Delta^{22,29}$ via $\Delta^{21,22}$ to $\Delta^{17,21}$ and possibly to neohopene (Fig. 4). Oxidation of bacteriohopanetetrol yields mainly the homohopanoic acids ranging from C_{31} to C_{34} and minor homohopanols.

Aromatization of hopanoids, when derived from bacterial detritus in sediments, is inferred to proceed via diagenetic alteration or bonding of the precursors to organic matter (e.g. diploptene to neohopene, bacteriohopanepolyol), subsequent cracking and then dehydrogenation from ring-D to ring-A (Fig. 4.[42,43]).

4.2 METHODS

The experimental and analytical methods used in the numerous other reports cited in this chapter should be consulted directly if it is of use. Here, I provide a brief summary of the experimental and analytical methodology generally used for the studies discussed here as examples.

4.2.1 Extraction and Molecular Analyses

The extraction of environmental or geological particulate matter with solvents, solvent mixtures or supercritical carbon dioxide is the preferred method, because it results in minimum alteration of its polar components, avoids hydrolysis of anhydrides, esters, etc. and is highly efficient for most organic compounds (even sugars). Other workers have reported the direct vaporization by thermal desorption or flash pyrolysis of organic compounds from particulate matter into GC, MS, or GC-MS instruments. Those methods work fine for neutral compounds (e.g. hydrocarbons) but should be used with caution when analyzing polar or labile compounds.

4.2.1.1 *Extraction and Fractionation*

An example of a schematic for sample treatment, extraction, and separation procedures is given in Simoneit[44], and it follows the method first used by Simoneit and Mazurek[26] with minor modifications for data comparison purposes. The samples are powdered and dried, then typically extracted using ultrasonic agitation multiple times with a mixture of dichloromethane (CH_2Cl_2) and methanol (3:1 v/v). The solvent extracts are filtered through an annealed glass fiber filter for the removal of insoluble particles, concentrated by rotary evaporation and then by a stream of filtered nitrogen gas.

Aliquots are then taken for direct gas chromatography-mass spectrometry (GC-MS) analysis as total extract and as derivatized (silylated) total extract. Another aliquot is taken for some samples for derivatization (methylation) and subsequent fractionation. Alkanoic acid and phenolic moieties are methylated using diazomethane in diethyl ether prepared from the precursor N-methyl-N'-nitro-N-nitrosoguanidine. These methylated extracts are separated by preparative thin layer chromatography (TLC) on silica gel plates with a mobile phase eluent mixture of hexane:diethyl ether (9:1). This procedure allows for the determination of chemical information on single molecular groups or homologous compound series, which may not be detected due to coelution in the GC-MS analysis of the total extract mixture. It also provides additional chemical information on the molecular polarity or functional group constituents which aids in structure elucidation and identification. The four fractions removed from the TLC plates typically contain the following classes of compounds: (1) *n*-alkanes, *n*-alkenes and saturated and unsaturated cyclic di- and triterpenoid hydrocarbons; (2) *n*-alkanones and polycyclic aromatic hydrocarbons (PAH); (3) *n*-alkanoic acids (as methyl esters) and saturated and unsaturated di- and triterpenoid ketones; and (4) *n*-alkanols, sterols, terpenols and polar organics (saccharides remain at the origin in this procedure). The fourth fraction, as well as an aliquot of the total extract, is converted to trimethylsilyl derivatives prior to analysis by reaction with N, O-bis-(trimethylsilyl)trifluoroacetamide (BSTFA) plus 1% trimethylchlorosilane for approximately three hours at 70°C prior to GC-MS analysis.

4.2.1.2 *Instrumental Analyses*

Gaseous emissions from biomass or soil are determined directly in the field or in the laboratory on canister grab samples, and volatile organic compounds are analyzed using canister grab or cartridge samples by GC and GC-MS.[45–49,68]

Extracts and fractions from solid sample powders/particles are analyzed directly by capillary GC-MS, generally using a quadrupole mass spectrometer operated in the electron impact mode at 70 eV and coupled to a gas chromatograph. The GC is equipped with a capillary column usually coated with DB-5 (30 m × 0.25 mm i.d.) and operated using a temperature program as follows: hold at 65°C for 2 min, ramp to 300°C at 6°C/min, hold isothermal at 300°C for 20 min, with helium as carrier gas. Some samples can also be analyzed by high temperature (HT) GC and HTGC-MS using capillary columns coated with high temperature phases (e.g. OV-1701-OH) and GC oven temperatures to 400°C.[50]

Current detection limits are <1 ng for full scan mass spectra and <1 pg for multiple ion monitoring mass spectrometry. Compound identifications are based on the comparisons with authentic standards, GC retention time, literature mass spectra and the interpretation of mass spectrometric fragmentation patterns. The MS methods used for the various markers and studies are listed in Table 1.

Newer instrumental methods of potential utility in organic analysis of environmental and geological biomarkers are compound specific isotope analysis (CSIA) and carbon-14 dating with accelerator mass spectrometry (AMS). CSIA provides the carbon isotope composition of individual

Table 1. Natural Products Utilized as Molecular Tracers in Environmental and Geochemical Research

Natural Product Compound Class[a]	Common Source[b] Biotic/ Geologic/Synthetic	Analytical Detection Method
1. Labile compounds for assessment of contemporary or recent biological processes.		
Sugars/polysaccharides	Flora (some fauna)	GC, GC-MS, GC-MS/MS
Unsaturated lipid compounds	All biota	GC, GC-MS
L-Amino acids/peptides	All biota	GC, Py-GC, GC-MS
Nucleotides, bases, etc.	All biota	GC, GC-MS
Biopolymers	All biota	Py-GC-MS, specific methods
2. Lipid/bitumen compounds used as indicators for environmental and geological processes (distinguish contemporary natural products from fossil derivatives).		
Aliphatic hydrocarbons	Ubiquitous/specific if strong o/e	GC, GC-MS, GC-IRMS
Aliphatic acids, alcohols, ketones, aldehydes	Ubiquitous/some specific	GC, GC-MS, GC-IRMS
Isoprenoids	Biogenic	GC, GC-MS, GC-IRMS
Steroids	Flora/fauna	GC-MS, GC-IRMS
Triterpenoids	Flora/microbes	GC-MS, GC-IRMS
Diterpenoids	Flora/microbes	GC-MS, GC-IRMS
Pigments	Flora/microbes/fauna	HPLC-MS, HPLC-IRMS
Biopolymers	Flora/microbes	Py-GC-MS
Novel and unknown natural products (biomarkers)	–	GC-MS, Py-GC-MS, GC-IRMS, NMR, X-ray

Table 1. (Continued)

Natural Product Compound Class[a]	Common Source[b] Biotic/ Geologic/Synthetic	Analytical Detection Method
3. Synthetic organic compounds (distinguish from natural products and fossil derivatives).		
Halogenated compounds	Anthropogenic	GC-MS, ECGC, GC-CIMS, specific methods
Drugs, hormones, supplements	Anthropogenic/natural	GC-MS, HRMS, specific methods
Plasticizers, other industrial chemicals	Anthropogenic	GC-MS, specific methods
4. Other organic compounds (distinguish abiotic and thermal resynthesis from natural products).		
D/L-Amino acids	Abiotic	GC, Py-GC, GC-MS
Aliphatic alcohols, acids, hydrocarbons	Abiotic	GC, GC-MS, GC-IRMS
PAH	Abiotic (thermogenic)	GC, GC-MS, GC-IRMS

[a] Labile markers are stable for shorter geological periods after diagenetic preservation. Group 2 compounds are stable for longer geologic periods.
[b] Slash indicates that a distinction is possible.
GC = gas chromatography, Py = pyrolysis (flash or hydrous heating), MS = mass spectrometry (HR = high resolution), IRMS = isotope ratio mass spectrometry, HPLC = high-pressure liquid chromatography, EC = electron capture, CI = chemical ionization, NMR = nuclear magnetic resonance spectrometry, X-ray = X-ray crystallography.

compounds as they elute off a GC column[51,52] (Table 1). Both carbon and nitrogen isotope compositions for bulk (total) organic matter fractions have also been applied.[51,53–55] The AMS method for determining the ^{14}C content of organic matter was initially used to distinguish the component from contemporary biomass versus fossil fuel utilization in urban aerosols.[56–58] However, AMS determinations of ^{14}C content cannot distinguish the direct natural background emissions from vegetation and the organic matter from burning of vegetation.

4.3 NATURAL PRODUCTS AS TRACERS

This section aims to illustrate the application of natural products as tracers for biotic sources in environmental and geological processes. It will concentrate on polar and functionalized examples.

4.3.1 Environmental Applications

Environmental chemistry is the study of the sources, reactions, transport effects and fates of chemical species, both organic and inorganic, in the water, soil, and air environments.[13,44,59–61] The topic has considerable overlap with environmental biogeochemistry, i.e. the study of effects of environmental chemical species on ecosystems, and with toxicological chemistry, i.e. the study of toxic substance effects on living organisms. It is important in all those fields to be able to distinguish natural products and geological biomarker compounds from the superimposed synthetic (anthropogenic) compounds. This is illustrated here with the following examples.

4.3.1.1 *Tocopherols*

Vitamin E, a natural antioxidant, is essential for growth, disease prevention, tissue integrity and reproduction in all fauna. Natural vitamin E, as it occurs in plants, consists mainly of α-tocopherol (III, R = H) with minor amounts of γ-tocopherol (IV), although this ratio can vary as for example in the vegetable oil composition shown in Fig. 5a.[62] In animal husbandry, such as dairy farms or cattle feed lots, the stock diets are commonly supplemented with vitamin E, because processed grain-based cereal fodder, hay, and silage are deficient in vitamin E.[63–65] This results in higher levels of α-tocopherol in animal tissues, which in turn increases tissue stability towards lipid oxidation. Thus, cattle fodder in such operations is normally supplemented with α-tocopherol acetate (III, R = Ac), which is stable to *in vitro* oxidation and readily hydrolyzes in the animal gut.[63–65]

Consequently, soil samples from feedlots have a dual origin of tocopherols. This is illustrated with the GC-MS data of a study done in the San Joaquin Valley of California (Fig. 5b,[34]). The γ-tocopherol is derived exclusively from the vegetation fodder, while α-tocopherol is derived from both vegetation and the hydrolysis of the tocopheryl acetate feed supplement. Excess α-tocopherol acetate, not hydrolyzed in the animal gut, is also present. Because α-tocopherol acetate is relatively stable, its presence in soil dust samples indicates the level of its usage as a feed supplement. A significant metabolite of α-tocopherol elutes after the acetate and the

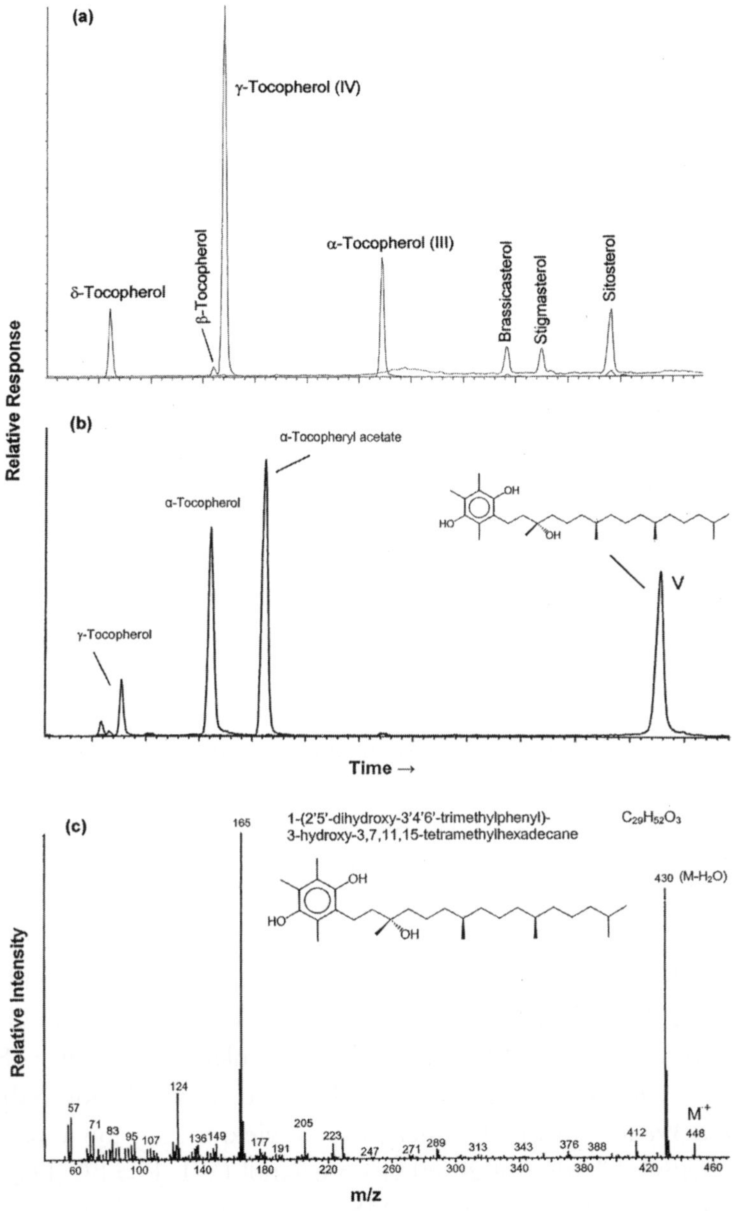

Fig. 5. Examples of GC-MS data for vitamin E in a vegetable oil and cattle feedlot soil: (a) vegetable oil, sum of key ions *m/z* 129 (TMS for sterols), 474, 488 and 502 ($M^{\cdot+}$ of tocopherols as TMS), (b) feedlot soil, sum of key ions *m/z* 416 and 430 for tocopherols, α-tocopheryl acetate, and the metabolite (as free phenols), and (c) mass spectrum of α-tocopherol hydroquinone (V).

mass spectrum is shown in Fig. 5c. The molecular weight of 448 dalton fits with the composition $C_{29}H_{52}O_3$ and the fragmentation pattern, as the underivatized compound, is interpreted to be the hydration product of the chroman ring of α-tocopherol, namely 1-(2',5'-dihydroxy-3',4',6'-trimethylphenyl)-3-hydroxy-3,7,11,15-tetramethylhexadecane (α-tocoph -erol hydroquinone, V).

In order to further confirm such an input of feedlot soil particles to the atmosphere, the steroid signature is utilized.[34] Sterols and stanols are environmentally quite stable and thus are useful markers for differentiating sewage from human sources (mainly coprostanol) and herbivores (mainly 5β-stigmastanol and *epi*-5β-stigmastanol). There is no known major direct natural product source for these compounds.[66–68] The sterol distribution in vegetable oil is shown in Fig. 5a. Only the primary C_{28} and $C_{29}\Delta^5$-steroids are present. During digestion in the gut, Δ^5-steroids are biohydrogenated by microflora to their stanol derivatives with configurational changes at the C-3 and C-5 positions in three isomers: stanols (5αH, 3βOH, VI), coprostanols (5βH, 3βOH, VII) and *epi*-coprostanols (5βH, 3αOH, VIII).[69–72] When excreted, stanols in the soil are rather stable? and have been used to identify ancient animal farming.[73]

One example of soil from cattle feedlots is shown in Fig. 6 as the GC-MS key ion traces and mass spectra of the major tracer both as the free and silylated compound.[34] The coprostanols and *epi*-coprostanols are poorly separated by GC as the underivatized compounds but do resolve as the TMS ethers. The major compound is the C_{29} *epi*-stanol (VIII, $R = \beta C_2H_5$, 24-ethyl-*epi*-coprostanol), with its molecular weight of m/z 416 and fragmentation by loss of CH_3 to m/z 401, loss of H_2O to m/z 398 and a base peak at m/z 215. The mass spectrum of *epi*-5β-stigmastanol trimethylsilyl ether (VIII, $R = \beta C_2H_5$, OTMS) has the molecular weight of m/z 488 and fragments by loss of CH_3 to m/z 473, loss of trimethylsilanol to the base peak m/z 398 and further fragmentation to yield the key ion at m/z 215. The trimethylsilyl ether of 24-ethylcoprostanol (VII, $R = \beta C_2H_5$) elutes just prior to *epi*-5β-stigmastanol-TMS and also has a key ion at m/z 215.

Thus, the most promising marker compounds for fugitive dust from dairy farms and cattle feedlots are the biohydrogenated phytosterols from fodder, i.e. 5β-stigmastanol and *epi*-5β-stigmastanol. Furthermore,

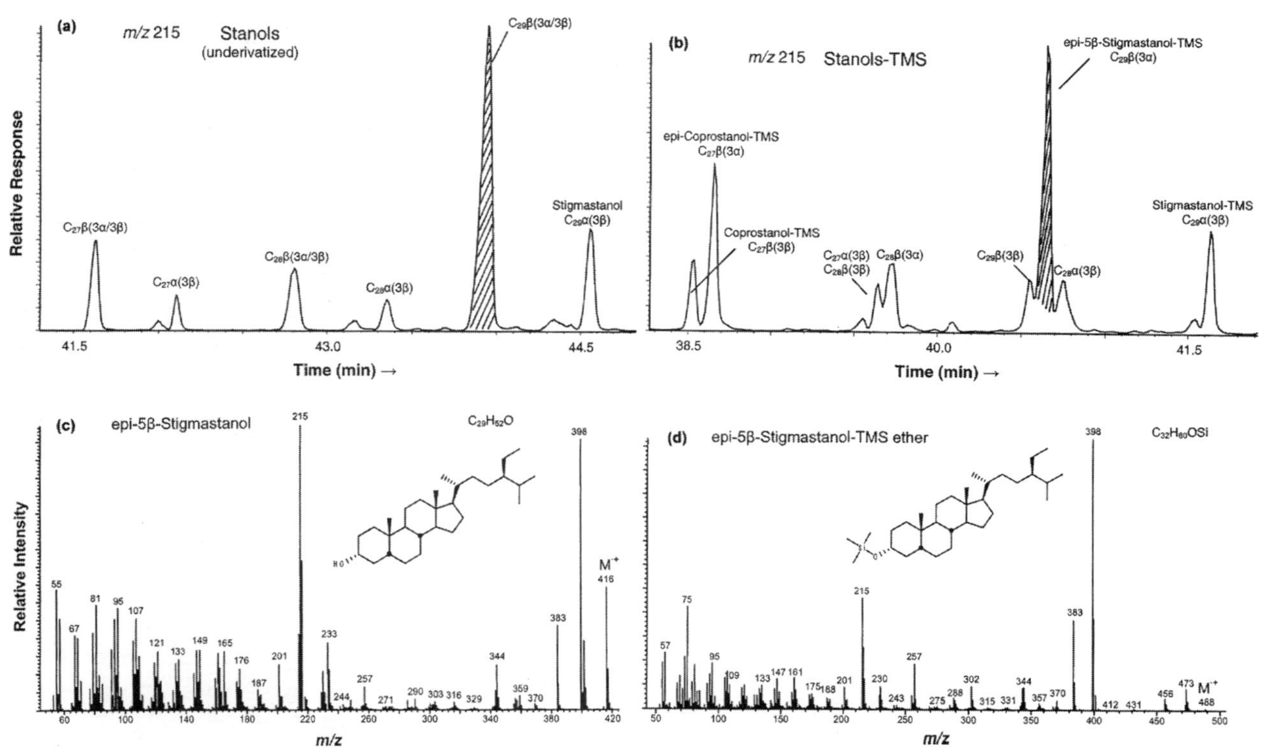

Fig. 6. GC-MS key ion fragmentograms of m/z 215 and associated mass spectra: (a) the stanol elution range for a feedlot dust extract, (b) the same range for the silylated extract, (c) *epi*-5β-stigmastanol, and (d) *epi*-5β-stigmastanol-TMS ether. C_i are the carbon skeletons of the other isomers.

antioxidant feed supplements (vitamin E products) are also detectable in such surface soils and the more stable synthetic α-tocopheryl acetate is a supporting source tracer when vitamin supplements are fed to cattle.

4.3.1.2 *Biomass Burning*

Biomass fuels burned by natural or anthropogenic processes are all extant forms of vegetation and its detritus (e.g. deadfall, duff, composites, etc.). Emissions from fossil fuels (originally of a biogenic origin) are excluded, although lignite (brown coal) and peat can emit the same compounds as contemporary biomass upon burning, except in different amounts and ratios.[21] The major source emission profiles of organic compounds in smoke from burning major plant taxa and anthropogenic activities such as food grilling/frying, tobacco use, garbage burning, etc. have been reviewed.[5,21,83] In urban environments, there are numerous fugitive emission sources from biomass burning, with many untested, as for example from building fires, dung chip burning, crematoria, rendering, rubber tire burning, etc. In rural and remote environments, the major untested emissions are from agricultural burning and wildfires of vegetation and plant detritus from the countless species diversity in agricultural and ecosystem environments.

Biomass burning smoke tracers

The numbers of identifiable organic compounds emitted from burning of biomass are generally large,[50,74–82,84–86] and certain major compounds have a longer atmospheric stability than others. The natural products from biomass lipids, both floral and faunal, are vaporized primarily unchanged or with limited thermal alteration into smoke by steam stripping. The biopolymers, e.g. cellulose, lignin, hemicellulose, chitin, are in part degraded during burning and the cracking/breakdown products are also vaporized into smoke.[14,87,153] This is illustrated with the degradation products from cellulose shown in Fig. 7, where levoglucosan is the major anhydro product in equilibrium with the 1,4 derivative. Mannosan and galactosan, as well as 1,6-anhydro-β-D-glucofuranose are minor products, where the former are derived from hemicellulose. Thus, the dominant organic components of biomass smoke particulate matter are monosaccharide derivatives from the

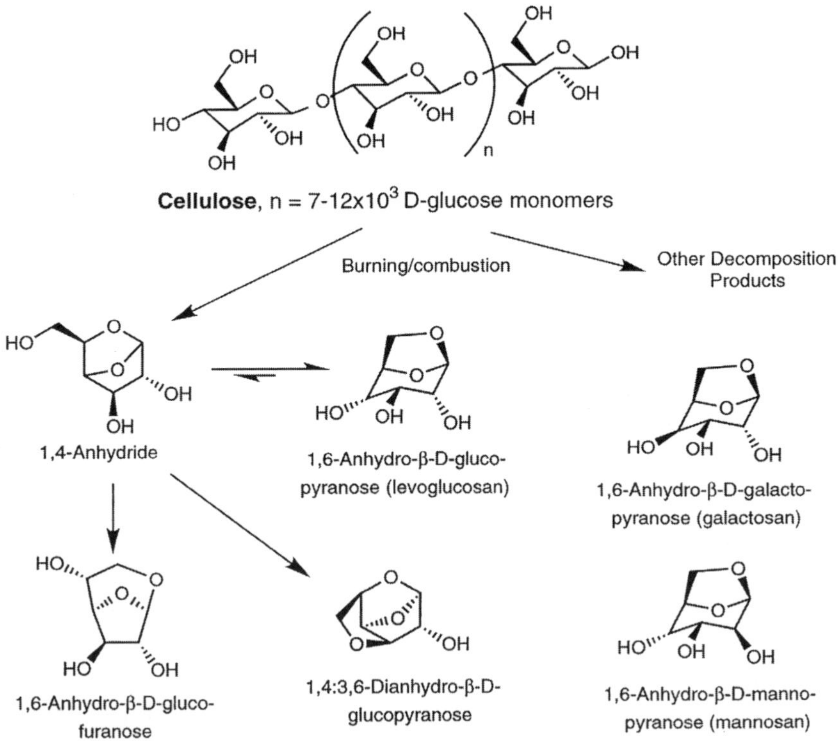

Fig. 7. The major decomposition products from burning of cellulose. Levoglucosan is the dominant compound, and mannosan and galactosan are formed as well as levoglucosan from burning of hemicelluloses.

thermal breakdown of cellulose, phenolics from the thermal degradation of lignins, accompanied by lesser amounts of terpenoids and straight-chain, aliphatic and oxygenated lipids from vegetation waxes, resins, gums and other biopolymers.[21] The major molecular tracers identified to be specific for angiosperms, gymnosperms and gramineae are summarized in Table 2 and illustrated in the GC-MS TIC traces for total aerosol extracts from a small town and a large city during winter (Fig. 8). Levoglucosan is the dominant tracer, complemented with pimaric acid (IX), dehydroabietic acid (X) and vanillic acid as secondary indicators for conifer wood fuel and lignin in biomass fuels, respectively. Thus, various ratios of the major molecular groups in smoke (e.g. vanillyl-, syringyl- and

Table 2. Major Molecular Tracers Identified in Smoke Particles from Biomass Burning

Compound Group	Molecular Tracers	Biomass Source
Monosaccharide derivatives	Levoglucosan (mannosan, galactosan)	All with cellulose
Methoxyphenols	Vanillin, vanillic acid	Conifers (Gymnosperms)
	Syringaldehyde, syringic acid	Angiosperms
	p-Hydroxybenzaldehyde, p-hydroxybenzoic acid	Gramineae
Diterpenoids	Abietic, primaric, iso-pimaric, sandaracopimaric acids	Conifers
	Dehydroabietic acid	Conifers
Triterpenoids	α-Amyrin, β-amyrin, lupeol	Angiosperm
Phytosterols	Sitosterol, stigmasterol	All
	Campesterol	Gramineae
Sterols	Cholesterol	Meat cooking/algae
Monoacyl glycerols	1- & 2-Palmityl glycerol	Food frying
Chitin derivatives	Anhydroacetamido-deoxyglucose	Seafood cooking

coumaryl-type compounds from lignin) and other biomarker tracers are useful for identifying the vegetation source that was burned.[14] The *n*-alkanes, *n*-alk-1-enes, *n*-alkanoic acids, *n*-alkan-2-ones,*n*-alkanols, PAH, levoglucosan and phytosterols are not fuel source specific, because they are generally found in all biomass combustion emissions.[74,88,89] However, the major compounds are specific for smoke from biomass burning, which when coupled with the directly emitted and thermally altered molecular markers, can be used as key tracers for assessing and tracing emissions from burning of specific biomass fuels.[21,78–80]

Source apportionment and global transport

Organic compounds, natural, fossil or anthropogenic, can be used to provide a chemical mass balance for atmospheric particles and a receptor model was developed that relates source contributions to mass concentrations in airborne fine particles.[90] The approach uses organic compound distributions in both source and ambient samples to determine source contributions to the airborne particulate matter. This method was validated for southern California and is being applied in numerous other airsheds.[90,91]

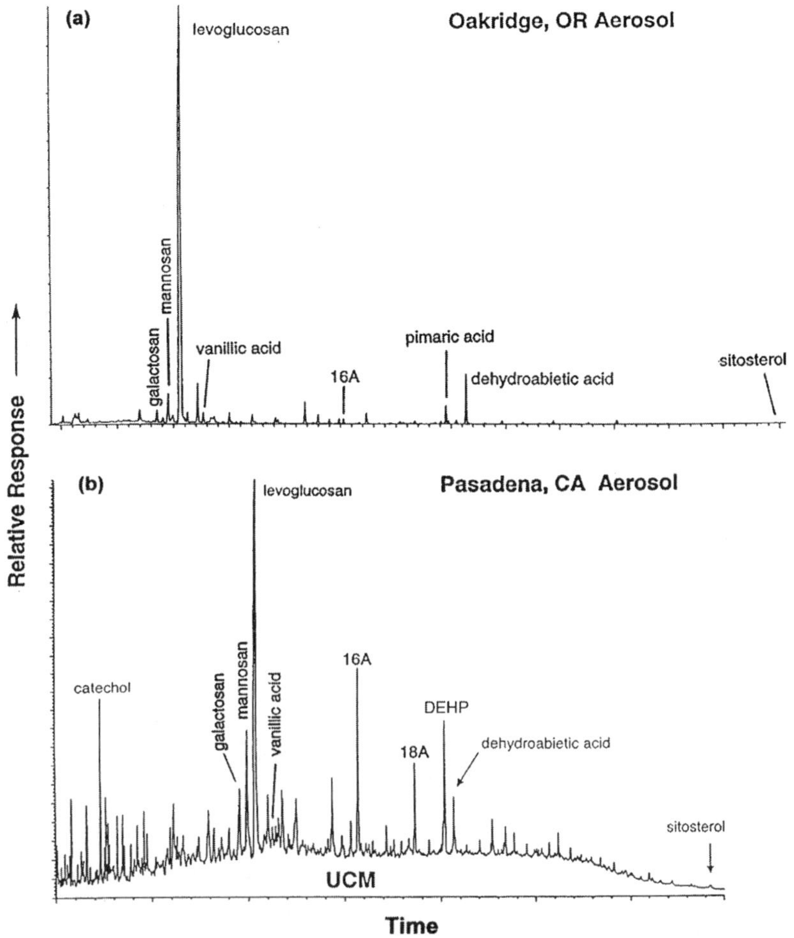

Fig. 8. Examples of GC-MS data for typical total extracts from rural and urban aerosol particle samples (analyzed as silylated derivatives): (a) total ion current (TIC) trace for Oakridge, OR, rural aerosol (Jan/Feb. 1985) and (b) TIC trace for Pasadena, CA, aerosol (Feb. 1989), DEHP = diethylhexyl phthalate, iA = carbon chain length of *n*-alkanoic acids.

In urban areas, the typical dominant sources of fine organic aerosol particles are diesel exhaust, gasoline-powered vehicle exhaust, meat cooking operations, smoke from wood combustion, and paved road dust; followed by four smaller sources of particles: tire wear, vegetative detritus, natural gas combustion, and cigarette smoke.[90]

The long range transport of continentally derived particulate matter is an active area of research.[92–94] The major finding of the research on organic matter composition of long range transported marine aerosols is their terrestrial higher plant wax content.[89,95–101] Since those initial results, there have been other organic compound analyses of long range transported particles from over the oceans taken from ships and aircraft.[102–104] Biomass combustion is an important primary source of many trace substances that are reactants in atmospheric chemistry and of carbonaceous soot particulate matter. These smoke particles are transported over long geographic distances and thus become part of the total particle burden in the global atmosphere.[105,106]

4.3.1.3 *Saccharides*

Saccharides (sugars) are ubiquitous and abundant compounds in nature and thus potentially powerful tools in elucidating sources and processes of biologically important organic materials in natural environments. They are the most common class of polar and water-soluble compounds found in environmental samples such as plants, soils, aerosols and sediments. The most important saccharide observed from the energy cycle of viable biota is glucose (found as α- and β-glucopyranose, XI) and the dominant energy storage sugars are sucrose (XII) and the fungal metabolite mycose (XIII), both of which are disaccharides.[107] These are the common saccharides observed in environmental samples (e.g. Fig. 9b,d), although other saccharides, e.g. inositols, arabinose, rhamnose, galactose, fructose, xylitol, sorbitol, etc. are often also present (cf., Fig. 9a,c).[107–109]

The example of a total extract composition of a tropical soil from the Amazon, Brazil, shows mycose as the major compound, numerous other monosaccharides, lipid components such as fatty acids and fatty alcohols, and natural product biomarkers (Fig. 9a). The mycose and elevated levels of the other saccharides reflect the efficient fungal/microbial degradation of plant detritus in the tropics. This can be compared to the saccharides in the soil from an almond orchard in California, where glucose and mycose are the main sugars with lipids, sterols and triterpenoids (Fig. 9b,[110]).

Sugars are also found in river and marine sediments which are probably associated with microbiota and fungi.[111] Natural product biomarkers

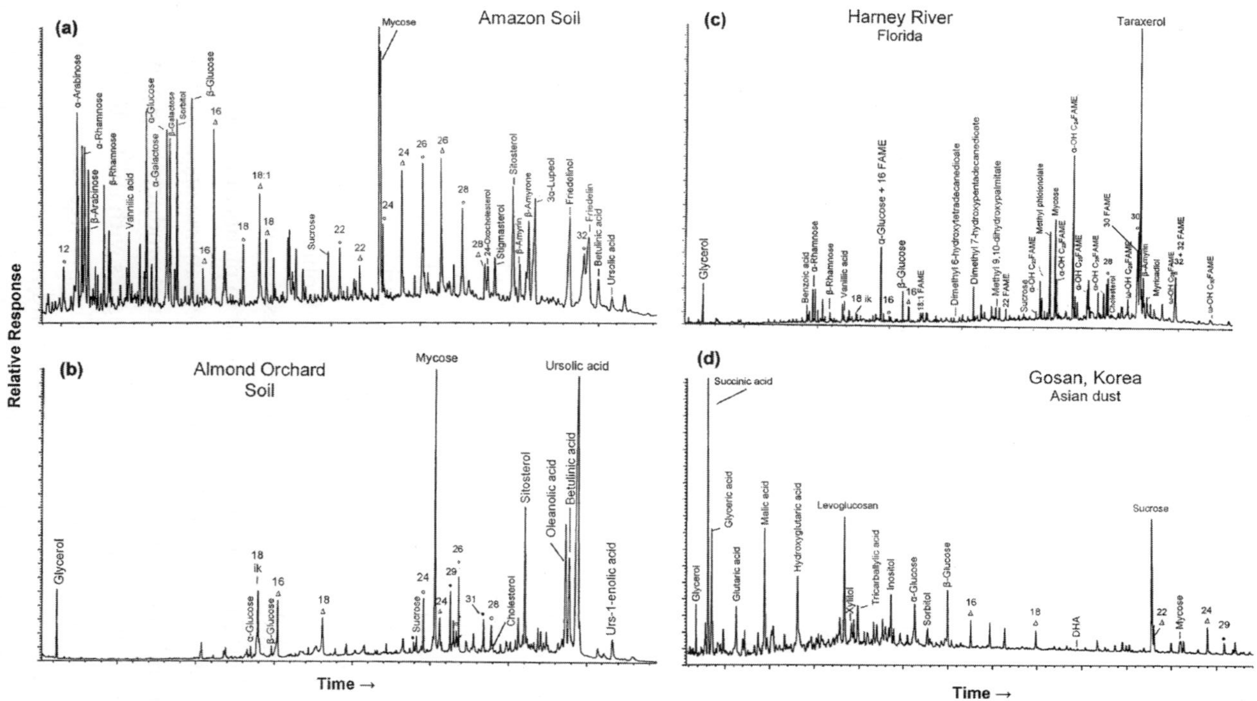

Fig. 9. GC-MS TIC traces for silylated total extracts of soil, river sediment and aerosol samples: (a) Amazon Forest soil (Manaus, Brazil); (b) almond orchard agricultural field soil (CA, USA); (c) Harney River sediment in Everglades National Park (FL, USA), and (d) Gosan Island (Korea) aerosol during Asian dust event (April 27–28, 2001). Numbers refer to carbon chain length of homologous series (● = *n*-alkane, ○ = *n*-alkanol, Δ = *n*-alkanoic acid, DHA = dehydroabietic acid, ik = isoprenoid ketone, S = sitosterol).

can be used to assess the dominant organic compounds in river/estuarine systems, as well as their redox alteration processes during transit from land to sea.[112,113] The total extract of river mud from Everglades National Park, Florida, has major compounds from lipids and taraxerol with intermediate levels of glucose, rhamnose and mycose (Fig. 9c).

Soil can be advected by wind and transported in aerosols over regional and long distances.[103,104,107,109,113] An example of a total extract of Asian dust aerosol collected from Gosan Island, South Korea, is shown in Fig. 9d. The glucose, sucrose, mycose, inositol, xylitol and sorbitol are the tracers for soil input, the lipids are natural background from terrestrial vegetation, and levoglucosan and dehydroabietic acid are indicators from biomass burning emissions.[103,104]

Thus, a multibiomarker tracer analysis of both water-soluble and hydrophobic organic compounds in various environmental samples using the GC-MS analytical method is useful. The saccharides, i.e. monosaccharides, disaccharides, anhydrosaccharides, and polyols (reduced sugars) are an important class of water-soluble compounds to be considered as environmental tracers.[14,107,112]

4.3.1.4 *Photochemistry*

The environmental photochemistry of natural products, specifically triterpenoids, is ready to be reexamined. Many 3,4-*seco*-triterpenoidal acids have been characterized as natural or secondary products from epicuticular waxes of higher plants, e.g. canaric acid,[114,115] nyctanthic acid,[116,117] roburic acid,[118] and methyl dihydronyctanthate.[119–121] The current list of 3,4-*seco*-triterpenoids comprises acids, aldehydes, alcohols and hydrocarbons, with some compounds glycosylated, and also less commonly, some 2,3-*seco*-triterpenes.[122] These compounds are beneficial to plants due to their bitter taste and metabolic effects, which are defense mechanisms against predation and pathogenic microorganisms.[122] These natural products have generally not been reported in sediments, but their analogues with the C-4(23) double bond hydrogenated are common.[123,124] The 3,4-*seco*-3-acids form by photolysis in solution under UV as the acids and methyl esters.[125] Here, I illustrate the photochemistry of friedelin with an example in a tropical lake sediment (Fig. 10a). The total extract is comprised mainly of lipids from higher plant wax, quite similar to the distribution of the epicuticular wax of *Pandanus helicorpus*, a dominant reed around the lake. The major biomarkers in *P. helicorpus* wax are 24-methyl- and 24-ethylcycloart-22-enol

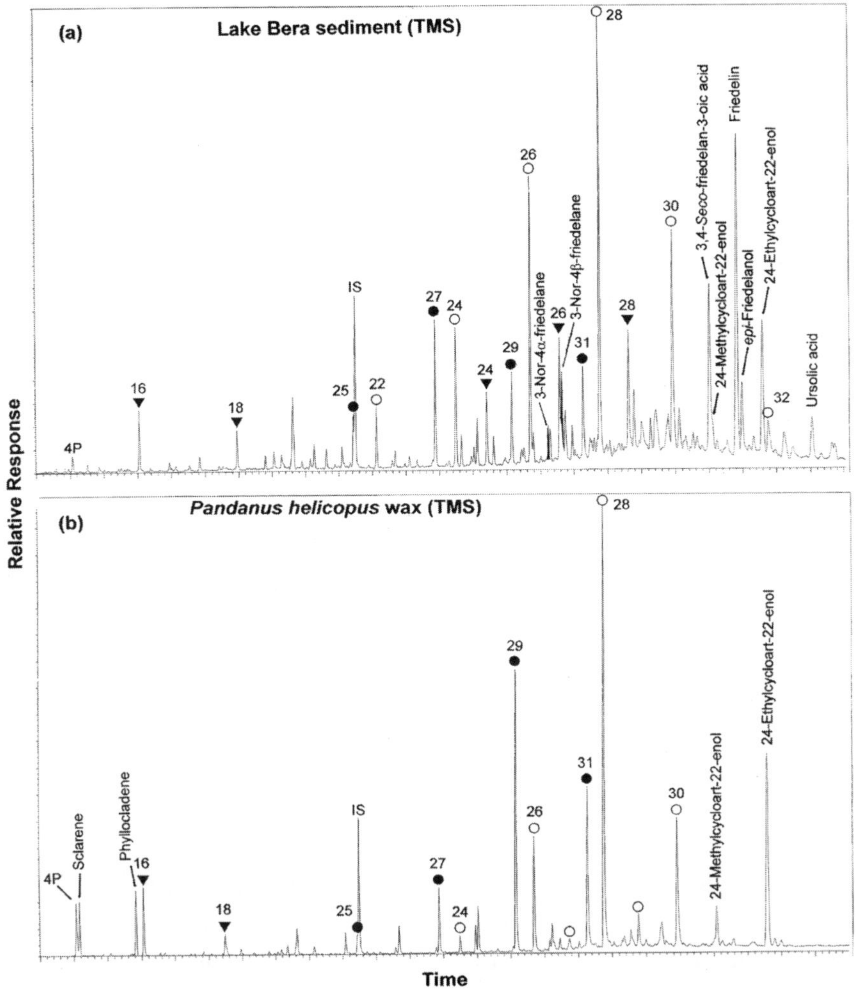

Fig. 10. GC-MS TIC traces for total extracts of: (a) a tropical lake sediment, and (b) epicuticular wax from a dominant reed growing in the shallow areas of the lake (analyzed as TMS). Numbers and symbols as in Fig. 9, 4P = dibutyl phthalate contaminant, IS = internal standard.

(XIV, R = CH_3 and C_2H_5, respectively, Fig. 10b), whereas the sediment contains additional friedelin, *epi*-friedelanol and ursolic acid. Furthermore, the friedelin photoproducts, i.e. dihydroputranjivic acid (3,4-*seco*-friedelan-3-oic acid, XV) and 4α- and 4β-3-norfriedelane (XVI) are dominant sedimentary components. Thus, characterizing photoproducts from terpenoids may provide additional biomarker tracer compounds for environmental studies.

4.3.1.5 *Geological applications*

The application of biomarker research in the geologic record has dealt with the derivative hydrocarbons as found in petroleum, coals, and sedimentary rocks. Reports on biomarkers in discrete fossils compared to the host rocks are sparse because (1) previous studies focused on the highly degraded "geoterpenoids," i.e. saturated and aromatic hydrocarbons,[2,31,126,127] and (2) the preservation potential of polar compounds (natural product "bioterpenoids") was believed to be low.[6] However, recent investigations of conifer fossils demonstrated that unaltered natural product terpenoids can be preserved in resin material.[24,25,128] This will be illustrated here with an example.

Another aspect to be described by example is hydrothermal petroleum and the related high temperature alteration of natural products in aqueous medium. In such cases immature organic detritus with natural products is altered mainly to hydrocarbons by rapid reductive hydrous pyrolysis.

4.3.1.6 *Natural Products in Fossils and Amber*

Most of the terpenoids reported from plant fossils are degradation products from the natural precursor compounds synthesized by the living plant. Despite various biological and physicochemical degradation processes during the decay and diagenesis of the plant material, the geoterpenoids retain their basic natural product skeleton and can be assigned to their respective structural classes.[2,127,129] The composition of geoterpenoids in fossil plants can thus be compared directly to the distribution of certain bioterpenoid classes in extant plants, i.e. markers for (paleo)chemosystematics.[130–132] However, it should be emphasized that the degradation of some functionalized natural product skeletons may yield the same series of geoterpenoids. Therefore, the bioterpenoids are more specific biomarkers than the geoterpenoids, this is because their occurrence in extant plants may be unique compared to the distribution of structural classes of terpenoids.[129] The analyses of terpenoids and lipids from fossil plant remains started early with fossil wood and resins[133–135] and with sediments rich in fossil plants.[136–138] Numerous terpenoids and aliphatic lipids have been reported from fossil conifer shoots and cones,[24,25,128,139–147] fossil angiosperm leaves[148–150] and fossil resins.[144,151–153] Extracts of the fossil materials and extant plants are analyzed with the same analytical methods, i.e. GC-MS.

This procedure is illustrated here with the analysis of the extractable organic matter from the seed cone of a fossil Cupressaceae from the Miocene Clarkia Flora of Idaho and of a related extant species to evaluate the preservation of the characteristic biomarkers in the fossils (Fig. 11,[25]). The major compounds in the fossil and extant samples are oxygenated terpenoids. They are dehydroabietane (XVII), 6,7-dehydroferruginol (XVIII), ferruginol (XIX), taxodione acetate (XX), 7-acetoxy-6,7-dehydroroyleanone (XXI), sugiol (XXII), *iso*-chamaecydin (XXIII), chamaecydin (XXIV), 18- or 19-hydroxyferruginol (XXV), *trans*-communic acid (XXVI); the minor geoterpenoids simonellite (XXVII) and 12-hydroxysimonellite (XXVIII) for the fossil cone. The extant *Taxodium distichum* contained additional natural products, such as royleanone, taxoquinone, 7-hydroxytaxodione acetate, 6-hydroxytaxoquinone, and taxodone; some of these were described previously.[154]

4.3.1.7 *Hydrothermal Petroleum/Bitumen*

Hydrothermal petroleums/bitumens, which are products of rapid diagenesis, catagenesis and metagenesis in aqueous systems (e.g. marine rift systems), have alkane and biomarker distributions analogous to those of conventional crude oils.[20,155–160] The carbon number distributions, biomarker compositions, and other geochemical parameters of marine hydrothermal petroleums generally reflect the source organic matter and the degree of thermal alteration or maturity.[20,32,155,156,160–162] An example of the alkane and biomarker patterns representative of

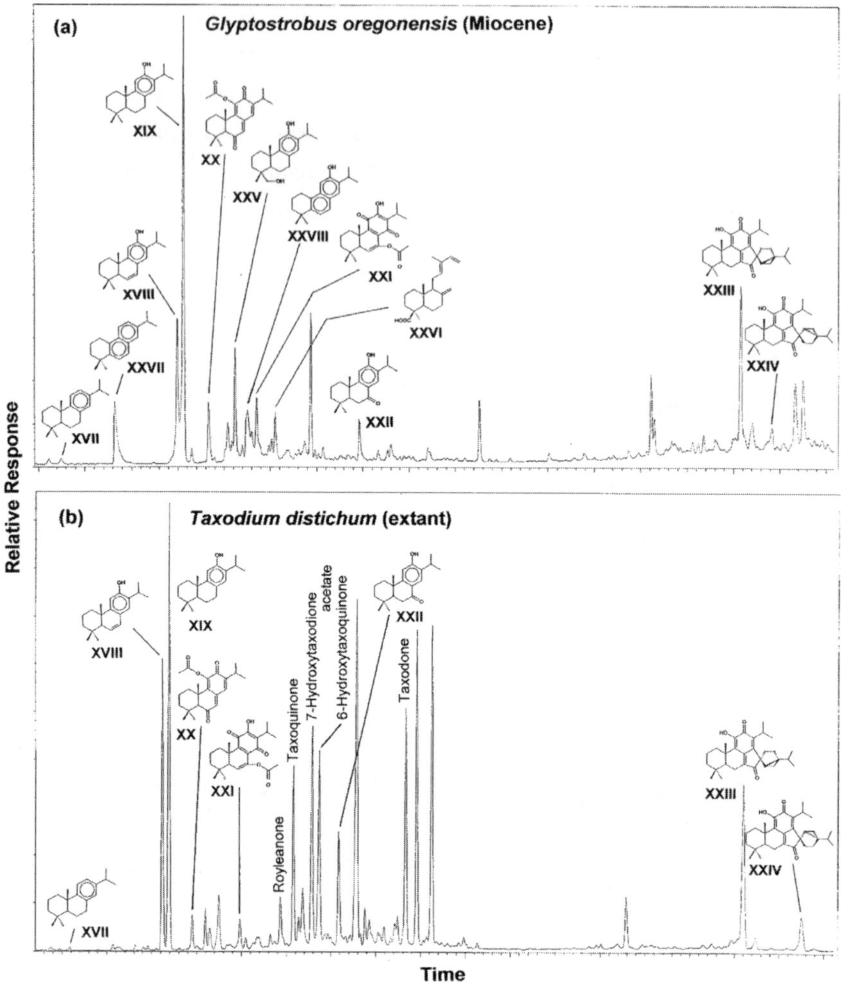

Fig. 11. Annotated GC-MS TIC traces of the total extracts (TMS derivatives) of seed cones of: (a) *Glyptostrobus oregonensis* (Miocene), and (b) *Taxodium distichum* (extant).

hydrothermal petroleums is shown in Fig. 12, where alteration has proceeded from immature natural product precursors to the fully mature petroleum biomarkers. Immature biomarkers occur in low-temperature regions of sedimentary hydrothermal systems, and maturation is observed with increasing temperature and depth below the seafloor (sub-bottom depth).[159,163–167] For example, sterols are major components in unaltered

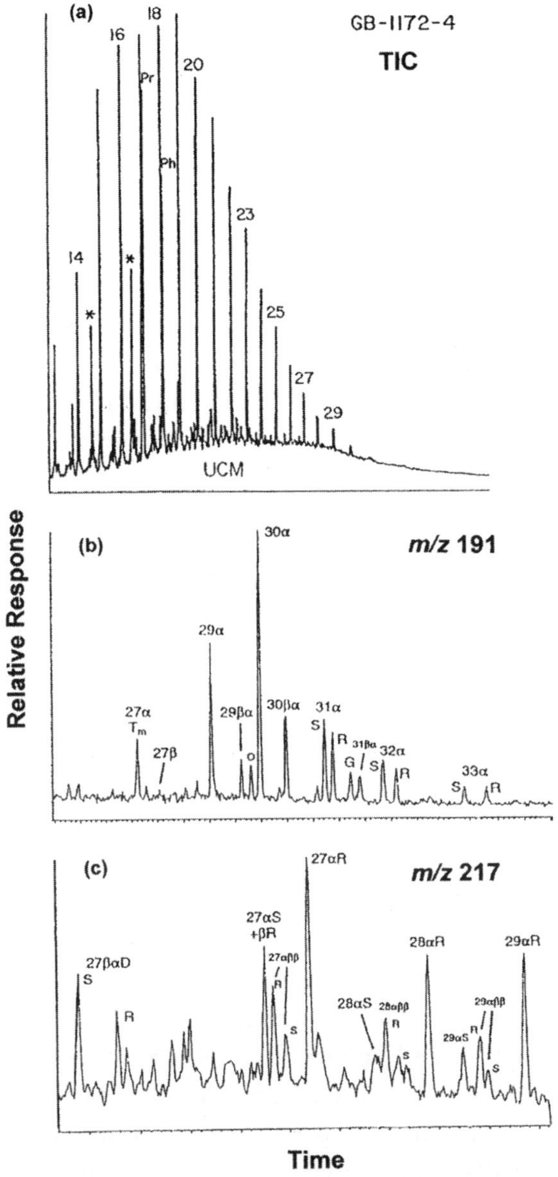

Fig. 12. Salient features of the GC-MS data for a hydrothermal petroleum (Guaymas Basin, Gulf of California, Mexico): (a) TIC trace of total oil, (b) m/z 191 key ion for hopanes, and (c) m/z 217 key ion for steranes. Numbers refer to carbon chain (n-alkanes) or skeleton, UCM = unresolved complex mixture, Pr = pristane, Ph = phytane, asterisks = other isoprenoids, α, βα, αββ, R, S = configurations of biomarkers.

sedimentary sections and range from C_{27} to C_{29}, with cholesterol dominant or equal to sitosterol. Diagenetic alteration of sterols accelerated by thermal stress yields stenones and stanones with the same range from C_{27} to C_{29} and ultimately steranes and diasteranes (Fig. 12c). The hopanes (Fig. 12b) are derived from the reductive alteration of hopanoid precursors primarily in bacterial detritus.

Novel biomarkers, i.e. tracer derivatives from unknown natural products, are sometimes encountered in geological or environmental samples, typically as hydrocarbons. The detection and determination of these compounds are usually based on the interpretation of mass spectra in GC-MS analyses. The proofs of chemical structures are based on the proposed interpretation of the MS data, separation and purification of the unknown compounds, exact structure determination by NMR methods or X-ray crystallography (if the compound is a solid that can be crystallized), and finally, comparison with a synthetic standard.[168–173] The next question concerns the biological source of the biomarker precursor compound. Many biomarkers still have no proven natural product precursors nor known biological sources (e.g. perylene, tricyclic terpanes).[174–176]

The characterization of a novel series of biomarkers is illustrated with the *gem*-dialkylalkanes in bitumen from a hydrothermal system on the Mid-Atlantic Ridge.[177] The total bitumen consists of hydrocarbons, a major UCM (unresolved complex mixture of branched and cyclic compounds) and mature biomarkers (e.g. hopanes) (Fig. 13a). The bitumen contains a series of cyclopentylalkanes (C_nH_{2n}) that range from $n = 14$ to 34, with only even-chained pseudohomologs and a concentration maximum (C_{max}) at $n = 18$. Their source is biogenic, based on the presence of only even-carbon number homologs, but the precursors are unknown.[177]

Four other significant homologous series present are *gem*-diethyl substituted *n*-alkanes. The *gem* prefix designates geminal-substituted compounds, i.e. two substituents on the same atom of a disubstituted compound. Several research groups have reported the presence of 5,5-diethylalkanes, as well as lesser amounts of other *gem*-dialkylalkanes, in sedimentary rocks back to the Precambrian and in hydrothermal fluids.[178–181] The correct structural assignments of the 3,3-diethylalkanes (XXIX) and the 5,5-diethylalkanes (XXX) were described recently by comparison with

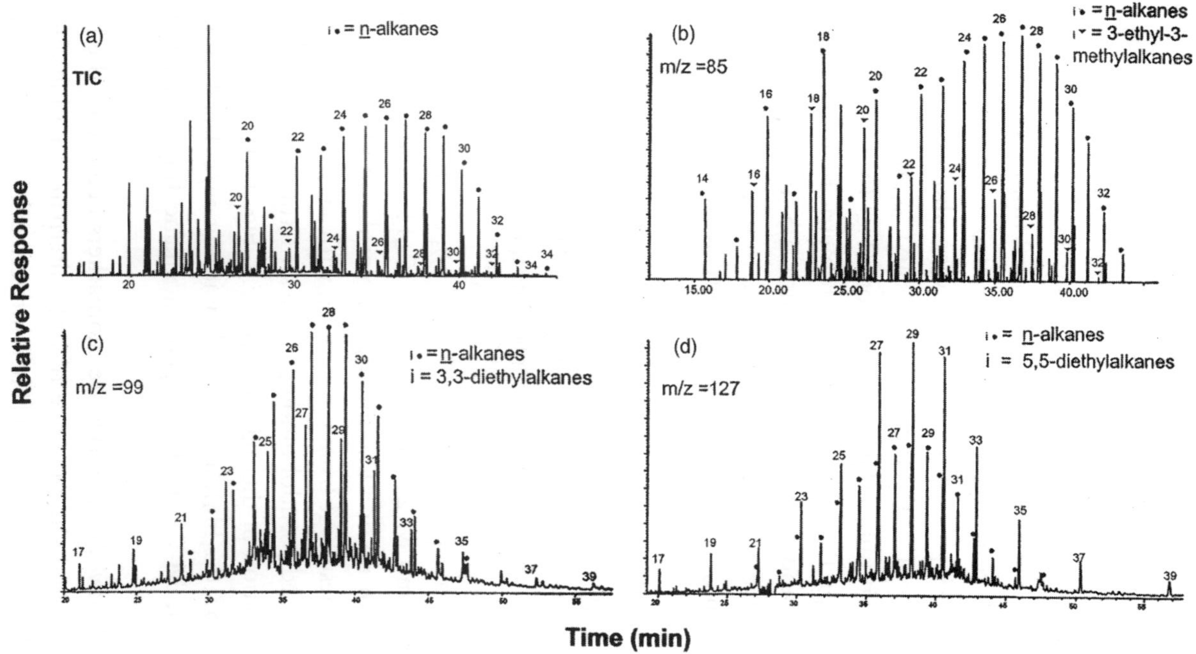

Fig. 13. Salient features of the GC-MS data for the saturated fraction of a hydrothermal bitumen sample from the Mid-Atlantic Ridge: (a) TIC - background trace, (b) m/z 85 fragmentogram (key ion for 3-ethyl-3-methylalkanes and *n*-alkanes), (c) m/z 99, key ion for 3,3-diethylalkanes with *n*-alkanes, and (d) m/z 127, key ion for 5,5-diethylalkanes with *n*-alkanes. (Numbers refer to total carbon number, dots over peaks are *n*-alkanes.)

synthetic standards.[179] The other *gem*-dialkylalkane series were tentatively identified based on the interpretation of their characteristic mass spectrometric fragmentation patterns and gas chromatographic retention factors.[178,179] The major homologous series can be visualized by key ion plots. The 3-ethyl-3-methylalkanes (i.e. 2,2-diethyl substitution) range from C_{14} to C_{34}, with only even-chained pseudohomologs detectable and a C_{max} at 18 (e.g. Fig. 13b). Their structures are interpreted from the mass spectra (e.g. Fig. 14a), which consist of a cleavage of C2-C3 to yield the base peak (C_6H_{13}, *m/z* 85), M-C_2H_5, minor alkane cleavage, and no molecular ion. The 3,3-diethylalkanes (XXIX) range from C_{15} to C_{38}, with a C_{max} at 27 and only odd-carbon numbered pseudohomologs (Fig. 13c). The structures are based on an interpretation of the mass spectrometric fragmentation patterns (e.g. Fig. 14b) and coinjection of authentic 3,3-diethylpentadecane. The mass spectra have a base peak at *m/z* 57, an intense ion at *m/z* 99 (C_7H_{15}, key ion) from C3-C4 cleavage, M-29 (C_2H_5), minor M-57, typical alkane fragments, and no molecular ion.

The 5,5-diethylalkanes (XXX) also range from C_{15} to C_{39} with odd-carbon numbered pseudohomologs and a C_{max} at 29 (Fig. 13d). The structures are based on the interpretation of the mass spectrometric fragmentation patterns, and on prior reports of the occurrence of those compounds in sedimentary sulfides and rocks.[178–180] The mass spectra generally exhibit a base peak at *m/z* 57 (C_4H_9), an intense key ion at *m/z* 127 (C_9H_{19}), and M-29 (C_2H_5), M-57 (C_4H_9) and other general alkane fragments, but no or a low-intensity molecular ion (e.g. Fig. 14c). The 6,6-diethylalkanes range from C_{16} to C_{38}, with a C_{max} at 30 and only even-carbon numbered pseudohomologs. The structures are based on the analogous interpretation of the mass spectra (e.g. Fig. 14d), which have a base peak at *m/z* 57 (C_4H_9), an intense key ion at *m/z* 141 ($C_{10}H_{21}$), and M-29 (C_2H_5), M-71 (C_5H_{11}), general alkane fragments, but no molecular ion.

The branched *gem*-alkane series are biomarkers because of the locale of their occurrence and the presence of only alternate pseudohomologs (even- or odd-carbon numbers only). Their inferred origin is from probable microbial precursors of unknown species, where methylation and ethylation, diethylation, butylation and ethylation occurred during biosynthesis

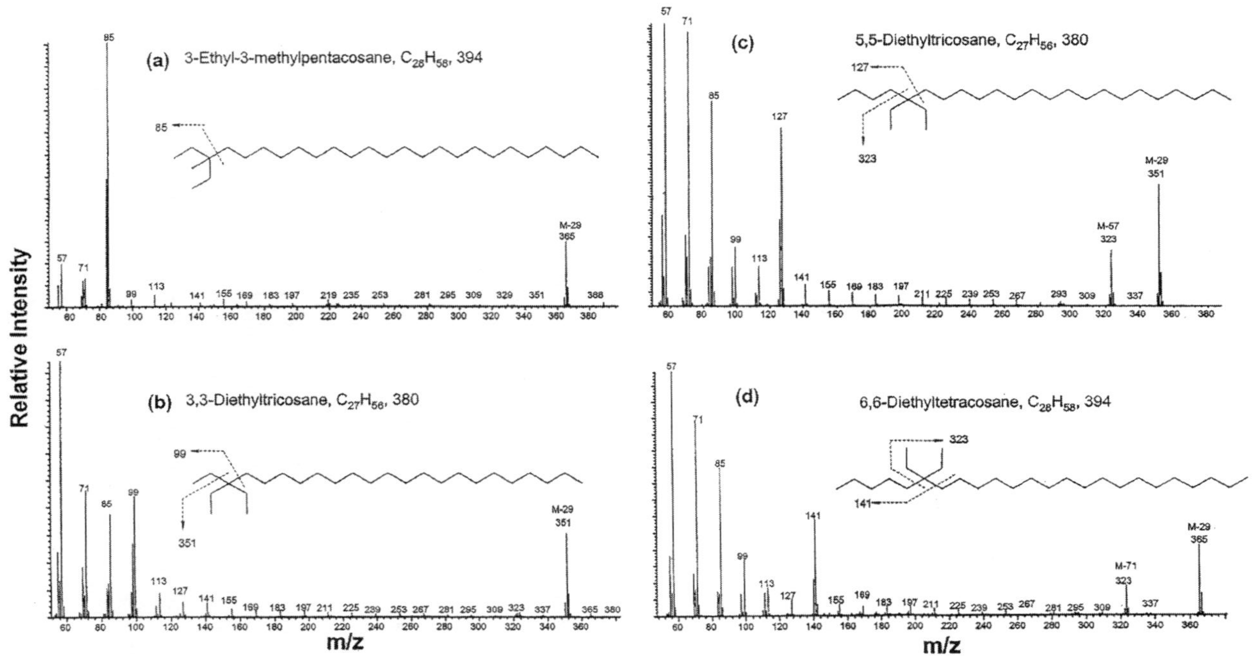

Fig. 14. Representative mass spectra of *gem*-dialkylalkanes in a hydrothermal bitumen sample: (a) 3-ethyl-3-methylpentacosane, $C_{28}H_{58}$, (b) 3,3-diethyltricosane, $C_{27}H_{56}$, (c) 5,5-diethyltricosane, $C_{27}H_{56}$, and (d) 6,6-diethyltetracosane, $C_{28}H_{58}$.

at the C-2, C-3, C-5, or C-6 positions of odd- or even-carbon chained substates (i.e. C_{11}-C_{33}). These compound series occur in samples of relict or weathered hydrothermal talus at the base of active vent systems, consistent with a lower input of detritus from archea and with an enrichment of lipid residues from microbial mats and/or sulfide-oxidizing bacteria as proposed for the ancient examples[178,180,181] and hydrothermal fluids on the Juan de Fuca ridge flank.[179]

4.4 PROGNOSIS

This chapter has presented an insight with some key literature about the elucidation of natural products for applications as biomarkers and molecular tracers for numerous environmental and geological processes. Natural product biomarker elucidation and analysis has been illustrated with various examples to clarify the concepts, applications, and procedures. The number of novel compounds, as well as the application of natural products as biomarker tracers, is expected to increase, especially with the use of mass spectrometry methods in the contemporary interdisciplinary sciences.

Most analytical methods for current natural product/biomarker applications usually deal with individual compounds in the concentration range of pg to $\mu g\,g^{-1}$, small sized samples (5–1000 μl solution), and in complex mixtures. At those low concentrations, the contamination by extraneous organic matter during sample acquisition, preparation, extraction, and analysis can be a major problem and analysts need to take precautions. The interpretation by an organic mass spectrometrist/chemist should be sought when mass spectra of unknown compounds are encountered. This may lead to the characterization of novel biomarkers with the ultimate structure proof by other chemical methods (e.g. NMR). Finally, I encourage the classical natural product chemists to continue to collaborate with environmental and geological organic chemists by providing their expertise and invaluable standards.

Appendix I: Chemical Structures Cited in the Text

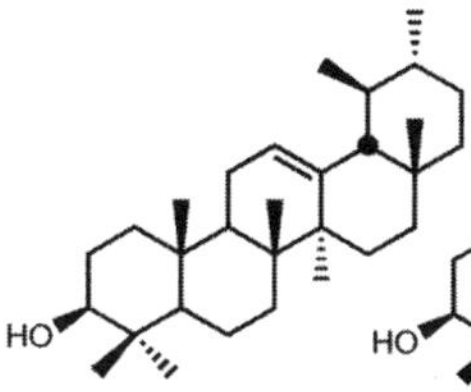
I. α-Amyrin

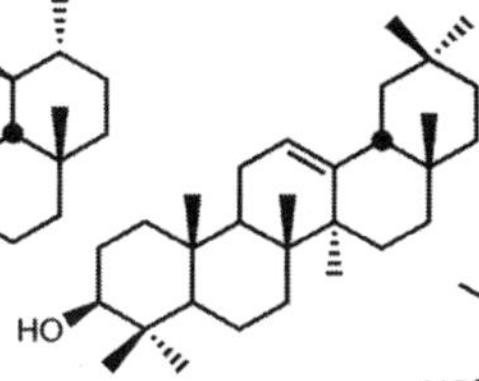
II. β-Amyrin

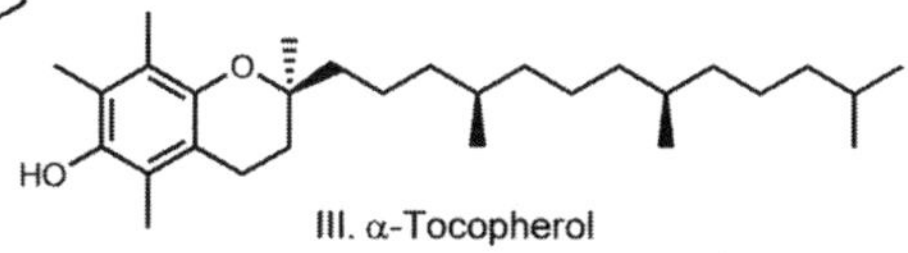
III. α-Tocopherol

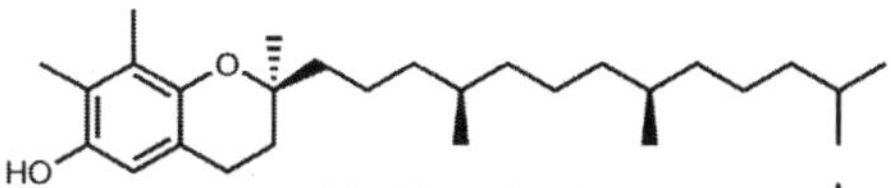
IV. γ-Tocopherol

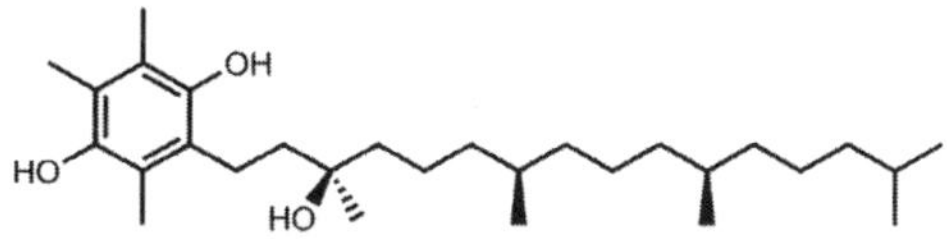
V. α-Tocopherol hydroquinone

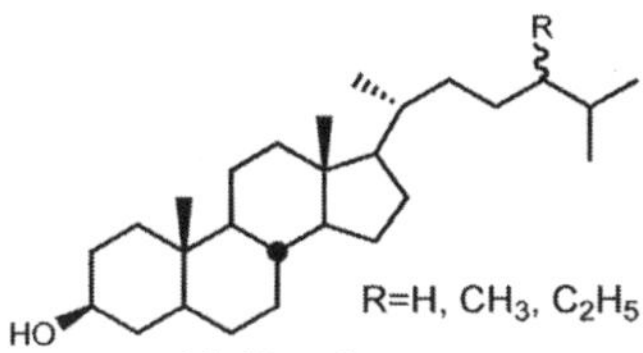

VI. Stanols

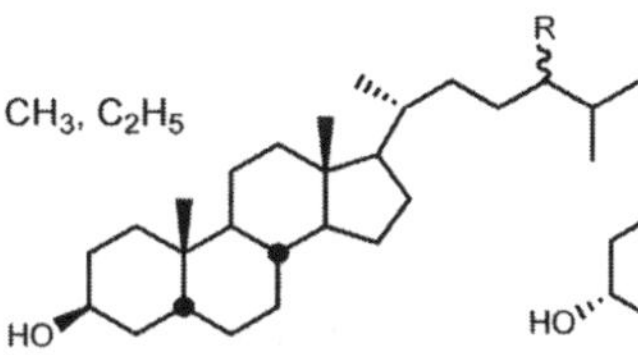
VII. Coprostanols

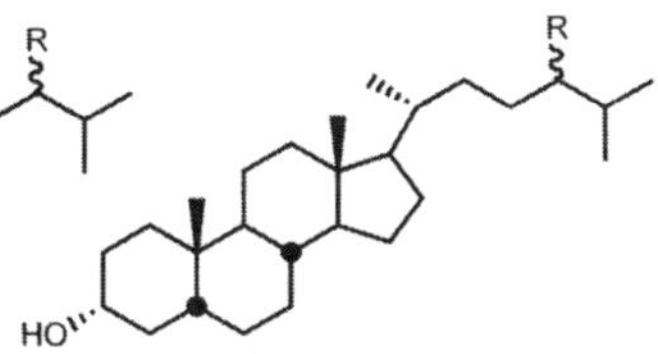
VIII. *epi*-Coprostanols

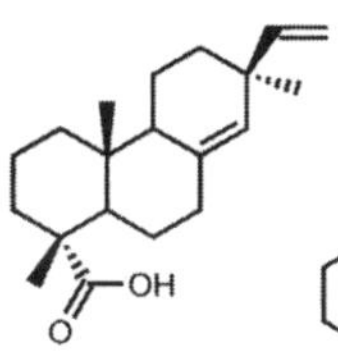
IX. Pimaric acid

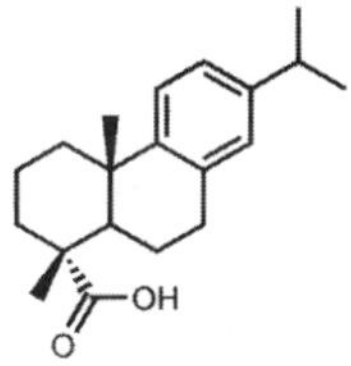
X. Dehydroabietic acid

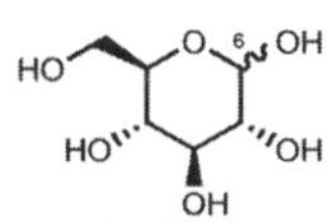
XI. Glucose (α=6αOH, β=6βOH)

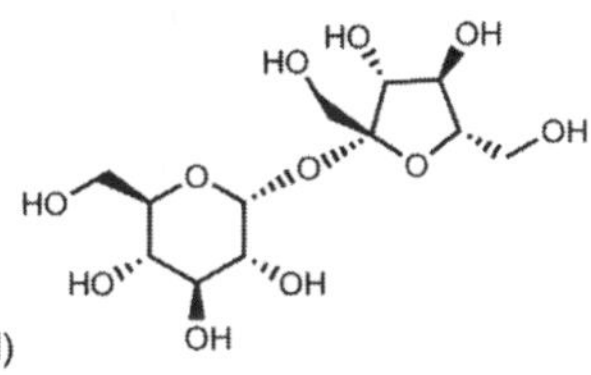
XII. Sucrose

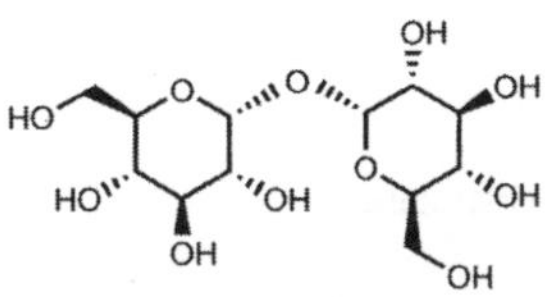
XIII. Mycose (trehalose)

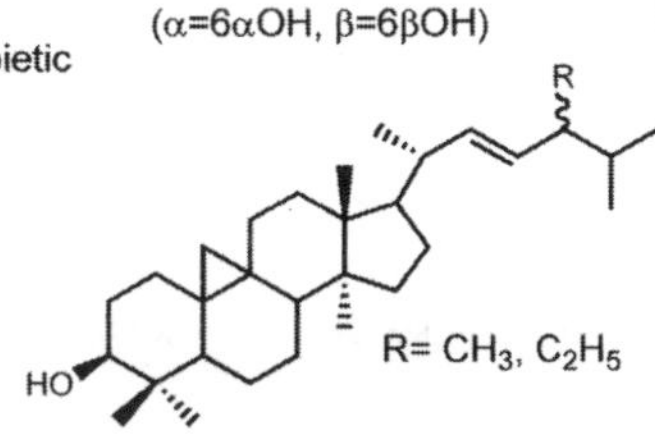

XIV. 24-Alkylcycloart-22-enols

Appendix I: (continued)

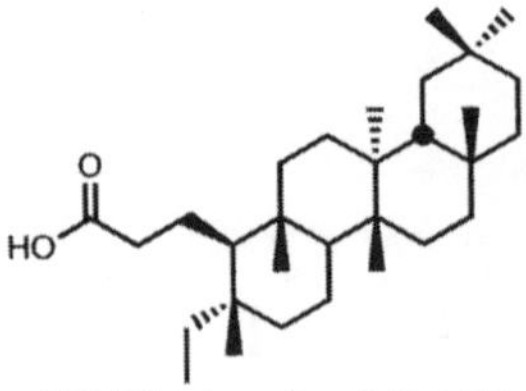

XV. Dihydroputranjivic acid

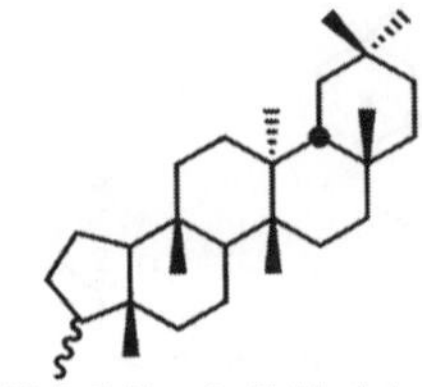

XVI. 3-Nor-4α/β-friedelane

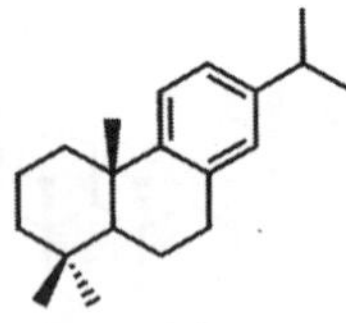

XVII. Dehydro-abietane

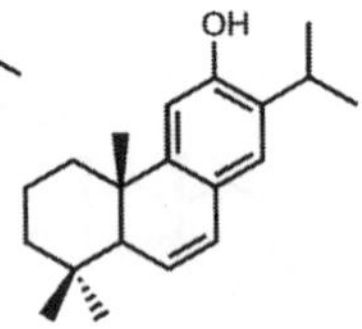

XVIII. 6,7-Dehydro-ferruginol

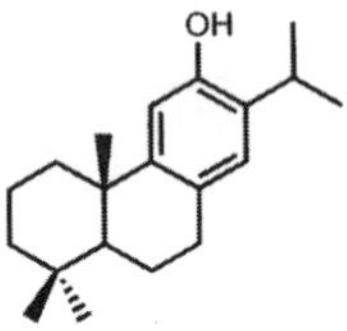

XIX. Ferruginol

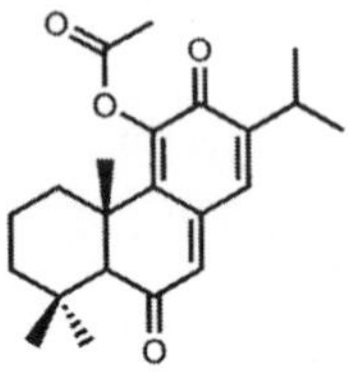

XX. Taxodione acetate

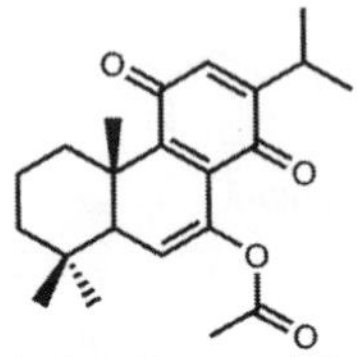

XXI. 7-Acetoxy-6,7-dehydroroyleanone

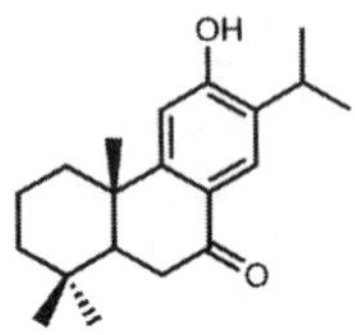

XXII. Sugiol

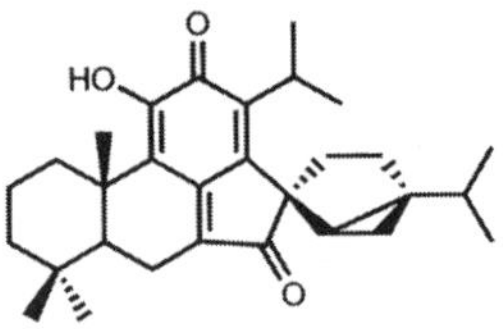

XXIII. *iso*-Chamaecydin

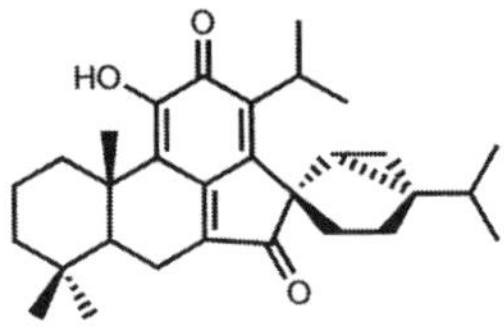

XXIV. Chamaecydin

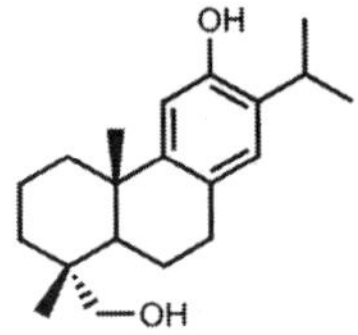

XXV. 18/19-Hydroxy-ferruginol

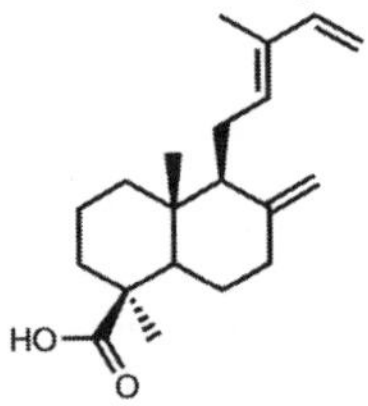

XXVI. *trans*-Communic acid

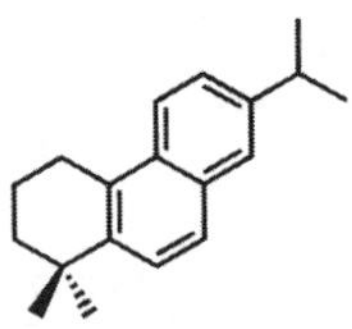

XXVII. Simonellite

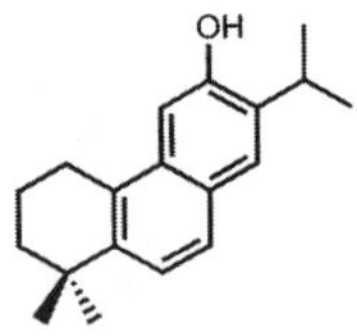

XXVIII. 12-Hydroxy-simonellite

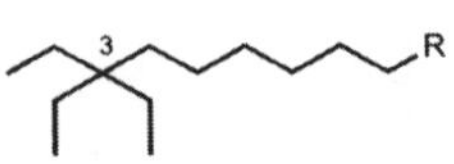

XXIX. 3,3-Diethylalkanes

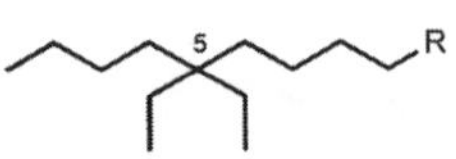

XXX. 5,5-Diethylalkanes

REFERENCES

1. Simoneit BRT, The organic chemistry of marine sediments, in Riley JP, Chester R (eds.), *Chemical Oceanography*, 2nd ed., Academic Press, New York, Vol. 7, Chap. 39, pp. 233–311, 1978.

2. Simoneit BRT, Cyclic terpenoids of the geosphere, in Johns RB. (ed.), *Biological Markers in the Sedimentary Record*, Elsevier, Amsterdam, pp. 43–99, 1986.
3. Simoneit BRT, Biomarker PAHs in the environment, in Neilson AH (ed.), *The Handbook of Environmental Chemistry. Hutzinger O, (ed.-in-chief), PAHs and Related Compounds*, Vol. 3, Part I, Springer-Verlag, Berlin, pp. 175–221, 1998.
4. Simoneit BRT, Molecular indicators (biomarkers) of past life, *The Anatomical Record* **268**:186–195, 2002.
5. Simoneit BRT, A review of current applications of mass spectrometry for biomarker/molecular tracer elucidations, *Mass Spectrometry Reviews* **24**:719–765, 2005.
6. Streibl M, Herout V, Terpenoids-especially oxygenated mono-, sesqui-, di-, and triterpenoids, in Eglinton G, Murphy MTJ (eds.), *Organic Geochemistry: Methods and Results*, Springer-Verlag, Berlin, pp. 401–424, 1969.
7. LaFlamme RE, Hites RA, Tetra-and pentacyclic, naturally-occurring, aromatic hydrocarbons in recent sediments, *Geochim Cosmochim Acta* **43**:1687–1691, 1979.
8. Simoneit BRT, Diterpenoid compounds and other lipids in deep-sea sediments and their geochemical significance, *Geochim Cosmochim Acta* **41**:463–476, 1977.
9. Wakeham SG, Schaffner C, Giger W, Polycyclic aromatic hydrocarbons in Recent lake sediments-II. Compounds derived from biogenic precursors during early diagenesis, *Geochim Cosmochim Acta* **44**:415–429, 1980.
10. Eglinton G, Calvin M, Chemical fossils, *Scientific American* **216**:32–43, 1967.
11. Johns RB (ed.) Biological Markers in the Sedimentary Record, Elsevier, Amsterdam, 1986.
12. Mackenzie AS, Brassell SC, Eglinton G, Maxwell JR, Chemical fossils—the geological fate of steroids, *Science* **217**:491–504, 1982.
13. Eganhouse RP (ed.), Molecular Markers in Environmental Geochemistry, *Amer Chem Soc Symp Ser 671*, Washington, DC, 1997.
14. Simoneit BRT, Schauer JJ, Nolte CG, Oros DR, Elias VO, Fraser MP, Rogge WF, Cass GR, Levoglucosan, a tracer for cellulose in biomass burning and atmospheric particles, *Atmosph Environ* **33**:173–182, 1999.
15. Devon TK, Scott AI, Handbook of Naturally Occurring Compounds, Vol. II, Terpenes, Academic Press, New York, pp. 576, 1972.
16. Glasby JS, Encyclopaedia of the Terpenoids, John Wiley & Sons, New York, pp. 2020, 1982.
17. Simoneit BRT, The Black Sea, a sink for terrigenous lipids, *Deep-Sea Research* **24**:813–830, 1977.

18. Currie BR, Johns RB, An organic geochemical analysis of terrestrial biomarkers in a transect of the Great Barrier Reef Lagoon, *Australian J Marine and Freshwater Research* **40**:275–284, 1989.
19. Poinsot J, Adam P, Trendel JM, Connan J, Albrecht P, Diagenesis of higher plant triterpenes in evaporitic sediments, *Geochim Cosmochim Acta* **59**:4653–4661, 1995.
20. Simoneit BRT, Hydrothermal petroleum: Genesis, migration and deposition in Guaymas Basin, Gulf of California, *Canadian Journal of Earth Sciences* **22**:1919–1929, 1985.
21. Simoneit BRT, Biomass burning — A review of organic tracers for smoke from incomplete combustion, *Applied Geochem* **17**:129–162, 2002.
22. Versteegh GJM, Schefu E, Dupont L, Marret F, Sinninghe Damsté JS, Jansen JHF, Taraxerol and *Rhizophora* pollen as proxies for tracking past mangrove ecosystems, *Geochim Cosmochim Acta* **68**:411–422, 2004.
23. Wang TG, Simoneit BRT, Organic geochemistry and coal petrology of Tertiary brown coal in Zhoujing Mine, Baise Basin, South China: 2. Biomarker assemblage and significance, *Fuel* **69**:12–20, 1990.
24. Otto A, White JD, Simoneit BRT, Natural product terpenoids in Eocene and Miocene conifer fossils, *Science* **297**:1543–1544, 2002.
25. Otto A, Simoneit BRT, Rember WC, Resin compounds preserved in the cones of three fossil conifer species from the Miocene Clarkia flora, Emerald Creek, Idaho, USA and related extant species, *Rev Palaeobot Palynol* **126**:225–241, 2003.
26. Simoneit BRT, Mazurek MA, Organic matter of the troposphere — II. Natural background of biogenic lipid matter in aerosols over the rural western United States, *Atmosph Environ* **16**:2139–2159, 1982.
27. Kimble BJ, Maxwell JR, Philp RP, Eglinton G, Albrecht P, Ensminger A, Arpino P, Ourisson G, Tri- and tetraterpenoid hydrocarbons in the Messel oil shale, *Geochim Cosmochim Acta* **38**:1165–1181, 1974.
28. Münch W, Über die Harze der Braunkohle. Die Sterine des Harzbitumens, *Öl und Kohle* **2**:564–566, 1934.
29. Ruhemann S, Raud H, Über die Harze der Braunkohle. I. Die Sterine des Harzbitumens. *Brennstoff-Chemie* **13**:341–345, 1932.
30. Volkman JK, A review of sterol markers for marine and terrigenous organic matter, *Org Geochem* **9**:83–99, 1986.
31. Brassell SC, Eglinton G, Maxwell JR, The geochemistry of terpenoids and steroids, *Biochemical Society Transactions* **11**:575–586, 1983.
32. Kawka OE, Simoneit BRT, Polycyclic aromatic hydrocarbons in hydrothermal petroleums from the Guaymas Basin spreading center, in Simoneit BRT (ed.),

Organic Matter in Hydrothermal Systems — Petroleum Generation, Migration and Biogeochemistry, *Applied Geochem* **5**:17–27, 1990.

33. Riolo J, Ludwig B, Albrecht P, Synthesis of ring C monoaromatic steroid hydrocarbons occurring in geological samples, *Tetrahedron Letters* **26**:2697–2700, 1985.
34. Rogge WF, Medeiros PM, Simoneit BRT, Organic marker compounds for soil and fugitive dust from open lot dairies and cattle feedlots, *Atmospheric Environment* **40**:27–49, 2006.
35. Hartgers WA, de Leeuw JW, Sinninghe Damsté JS, Diagenetic and catagenetic products of isorenieratene: Molecular indicators for photic zone anoxia, *Geochim Cosmochim Acta* **60**:4467–4496, 1996.
36. Ocampo R, Repeta DJ, Structural determination of purpurin-18 (as methyl ester) from sedimentary organic matter, *Org Geochem* **30**:189–193, 1999.
37. Ensminger A, van Dorsselaer A, Spyckerelle C, Albrecht P, Ourisson G, Pentacyclic triterpenes of the hopane type as ubiquitous geochemical markers: Origin and significance, in Tissot B, Bienner F (eds.), *Advances in Organic Geochemistry* Editions Technip, Paris, pp. 245–260, 1974.
38. Kajukova GI, Pustulnikova CD, Abruitina NN, Golovkina LS, Meshtueva VL, Petrov AA, Structural variations in hopane type hydrocarbons, *Neftechimiya* **21**:803–811, 1981.
39. Ourisson G, Albrecht P, Rohmer M, The hopanoids, paleochemistry and biochemistry of a group of natural products, *Pure and Applied Chemistry* **51**:709–729, 1979.
40. Ourisson G, Albrecht P, Rohmer M, Predictive microbial biochemistry — from molecular fossils to prokaryotic membranes, *Trends in Biological Sciences* 236–239, 1982.
41. Rohmer M, The biosynthesis of triterpenoids of the hopane series in the eubacteria: A mine of new enzyme reactions, *Pure and Applied Chemistry* **65**:1293–1298, 1993.
42. Spyckerelle C, Greiner AC, Albrecht P, Ourisson G, Aromatic hydrocarbons from geological sources. Part III. A tetrahydrochrysene derived from triterpenes, in recent and old sediments: 3,3,7-trimethyl-1,2,3,4-tetrahydrochrysene, *J Chemical Research* **12**:330–331, 1977.
43. Spyckerelle C, Greiner AC, Albrecht P, Ourisson G, Aromatic hydrocarbons from geological sources. Part IV. An octahydrochrysene derived from triterpenes, in oil shale: 3,3,7,12a-tetramethyl-1,2,3,4,4a,11,12,12a-octahydrochrysene, *J Chemical Research* **12**:332–333, 1977.
44. Simoneit BRT, A review of biomarker compounds as source indicators and tracers for air pollution, *Environ Sci Pollut Res* **6**:153–163, 1999.

45. Andreae MO, Atlas E, Cachier H, Cofer WR III, Harris GW, Helas G, Koppmann R, Lacaux JP, Ward DE, Trace gas and aerosol emissions from savanna fires, in Levine JS (ed.), *Biomass Burning and Global Change*, Vol. 1, MIT Press, Cambridge, MA, pp. 278–295, 1996.
46. Blake NJ, Blake DR, Collins JE Jr, Sachse GW, Anderson BE, Brass JA, Riggan PJ, Rowland FS, Biomass burning emissions of atmospheric methyl halide and hydrocarbon gases in the South Atlantic region, in Levine JS (ed.), *Biomass Burning and Global Change*, Vol. 2, MIT Press, Cambridge, MA, pp. 575–594, 1996.
47. Fraser MP, Cass GR, Simoneit BRT, Rasmussen RA, Air quality model evaluation data for organics: 4. C_2 to C_{36} non-aromatic hydrocarbons, *Environ Sci Technol* **31**:2356–2367, 1997.
48. Fraser MP, Cass GR, Simoneit BRT, Rasmussen RA, Air quality model evaluation data for organics: 5. C_6-C_{22} nonpolar and semipolar aromatic compounds, *Environ Sci Technol* **32**:1760–1770, 1998.
49. Fraser MP, Cass GR, Simoneit BRT, Measurement of gas-phase and particle-phase compounds in a roadway tunnel, *Environ Sci Technol* **32**:2051–2060, 1998.
50. Elias VO, Simoneit BRT, Pereira AS, Cabral JA, Cardoso JN, Detection of high molecular weight organic tracers in vegetation smoke samples by high temperature gas chromatography–mass spectrometry, *Environ Sci Technol* **33**:2369–2376, 1999.
51. Ballentine DC, Macko SA, Turekian VC, Gilhooly WP, Martincigh B, Chemical and isotopic characterization of aerosols collected during sugar cane burning in South Africa, in Levine JS (ed.), *Biomass Burning and Global Change*, Vol. 1, MIT Press, Cambridge, MA, pp. 460–465, 1996.
52. Simoneit BRT, Compound-specific carbon isotope analyses of individual long-chain alkanes and alkanoic acids in Harmattan aerosols, *Atmosph Environ* **31**: 2225–2233, 1997.
53. Cachier H, Buat-Menard P, Fontagne M, Chesselet R, Long-range transport of continentally derived particulate carbon in the marine atmosphere: Evidence from stable isotope studies, *Tellus* **38B**:161–177, 1986.
54. Cachier H, Liousse C, Pertuisot MH, Gaudichet A, Echalar F, Lacaux JP, African fire particulate emissions and atmospheric influence, in Levine JS (ed.) *Biomass Burning and Global Change*, Vol. 1, MIT Press, Cambridge, MA, pp. 428–440, 1996.
55. Gilhooly WP, Macko SA, Turekian VC, Swap RJ, Garstang M, Ruddiman WF, Stable carbon isotopic analysis of charcoal from single plant sources, in Levine JS

(ed.), *Biomass Burning and Global Change*, Vol. 1, MIT Press, Cambridge, MA, pp. 466–471, 1996.

56. Berger R, McJunkin D, Johnson R, Radiocarbon concentration of California aerosols, *Radiocarbon* **28**:661–667, 1986.
57. Currie LA, Klouda GA, Continetti RE, Kaplan IR, Wong WW, Dzubay TG, Stevens RK, On the origin of carbonaceous particles in American cities: Results of radiocarbon dating and chemical characterization, *Radiocarbon* **25**:603–614, 1983.
58. Klinedinst DB, Currie LA, Direct quantitation of $PM_{2.5}$ fossil and biomass carbon within the Northern Front Range Air Quality Study's domain, *Environ Sci Technol* **33**:4146–4154, 1999.
59. Evershed RP, Combined gas chromatography-mass spectrometry, *Gas Chromatography* 359–391, 1993.
60. Lee ML, Novotny MV, Bartle KD, Analytical Chemistry of Polycyclic Aromatic Compounds, Academic Press, London, 1981.
61. Schwarzenbach RP, Gschwend PM, Imboden DM, *Environmental Organic Chemistry*, 2nd ed., Wiley-Interscience, Hoboken, NJ, 2003.
62. Wyatt CJ, Carballido SP, Mendez RO, Alpha- and gamma-tocopherol content of selected foods in the Mexican diet: Effect of cooking losses, *J Agricultural and Food Chemistry* **46**:4657–4661, 1998.
63. Arnold RN, Scheller KK, Arp SC, Williams SN, Buege DR, Schaefer DM, Effect of long- or short-term feeding of α-tocopheryl acetate to Holstein and crossbred beef steers on performance, carcass characteristics and beef color stability, *J Animal Science* **70**:3055–3067, 1992.
64. Lynch MP, Kerry JP, Buckley DJ, Faustman C, Morrissey PA, Effect of dietary vitamin E supplementation on the colour and lipid stability of fresh, frozen and vacuum-packed beef, *Meat Science* **52**:95–99, 1999.
65. McDowell LR, Willikams SN, Hidiroglou N, Njeru CA, Hill GM, Ochoa L, Wilkinson NS, Vitamin E supplementation for the ruminant, *Animal Feed Science Technology* **60**:273–296, 1996.
66. Isobe KO, Tarao M, Zakaria MP, Chiem NH, Minh LY, Takada H, Quantitative application of fecal sterols using gas chromatography-mass spectrometry to investigate fecal pollution in tropical waters: Western Malaysia and Mekong Delta, Vietnam, *Environ Sci Technol* **36**:4497–4507, 2002.
67. Leeming R, Ball A, Ashbolt N, Nichols P, Using faecal steroids from humans and animals to distinguish faecal pollution in receiving waters, *Water Research* **30**:2893–2900, 1996.
68. Standley LJ, Kaplan LA, Smith D, Molecular tracers of organic matter sources to surface water resources, *Environ Sci Technol* **34**:3124–3130, 2000.

69. Grimalt JO, Fenandez P, Bayona JM, Albaiges J, Assessment of fecal sterols and ketones as indicators of urban sewage inputs to coastal waters, *Environ Sci Technol* **24**:357–363, 1990.
70. Skurikhin VN, Identification of neutral steroids in the diet, chyme and feces of heifers by chromatography and mass spectrometry, *Doklady Vsesoyuznoi Akademii Sel'skokhozyaistvennykh Nauk* 7:42–44, 1984.
71. Smith LL, Gouron RE, Sterol metabolism. VI. Detection of 5β-cholestan-3β-ol in polluted waters, *Water Research* **3**:141–148, 1969.
72. Snog-Kjaer A, Prange I, Dam H, Conversion of cholesterol into coprosterol by bacteria *in vitro*, *J General Microbiology* **14**:256–260, 1956.
73. Evershed RP, Bethell PH, Reynolds PJ, Welch NJ, 5β-Stigmastanol and related 5β-stanols as biomarkers of manuring: Analysis of modern experimental material and assessment of the archaeological potential, *J Archaeological Science* **24**:485–495, 1997.
74. Abas MRB, Simoneit BRT, Elias VO, Cabral JA, Cardoso JN, Composition of higher molecular weight organic matter in smoke aerosol from biomass combustion in Amazonia, *Chemosphere* **30**:995–1015, 1995.
75. Hawthorne SB, Miller DJ, Barkley RM, Krieger MS, Identification of methoxylated phenols as candidate tracers for atmospheric wood smoke pollution, *Environ Sci Technol*, **22**:1191–1196, 1988.
76. McDonald JD, Zielinska B, Fujita EM, Sagebiel JC, Chow JC, Watson JG, Fine particle and gaseous emission rates from residential wood combustion, *Environ Sci Technol* **34**:2080–2091, 2000.
77. Oros DR, Simoneit BRT, Identification of molecular tracers in organic aerosols from temperate climate vegetation subjected to biomass burning, *Aerosol Sci Tech* **31**:433–445, 1999.
78. Oros DR, Simoneit BRT, Identification and emission factors of molecular tracers in organic aerosols from biomass burning: Part 1. Temperate climate conifers, *Applied Geochem* **16**:1513–1544, 2001.
79. Oros DR, Simoneit BRT, Identification and emission factors of molecular tracers in organic aerosols from biomass burning: Part 2. Deciduous trees, *Applied Geochem* **16**:1545–1565, 2001.
80. Oros DR, Radzi bin Abas M, Omar NYMJ, Rahman NA, Simoneit BRT, Identification and emission factors of molecular tracers in organic aerosols from biomass burning: Part 3. Grasses, *Appl Geochemistry* **21**:919–940, 2006.
81. Rogge WF, Hildemann LM, Mazurek MA, Cass GR, Simoneit BRT, Sources of fine organic aerosol: 6. Cigarette smoke in the urban atmosphere, *Environ Sci Technol* **28**:1375–1388, 1994.

82. Rogge WF, Hildemann LM, Mazurek MA, Cass GR, Simoneit BRT, Sources of fine organic aerosol: 9. Pine, oak, and synthetic log combustion in residential fireplaces, *Environ Sci Technol* **32**:13–22, 1998.
83. Simoneit BRT, Schauer JJ, Nolte CG, Oros DR, Elias VO, Fraser MP, Rogge WF, Cass GR, Levoglucosan, a tracer for biomass burning and atmospheric particles, *Atmos Environ* **33**:173–182, 1999.
84. Standley LJ, Simoneit BRT, Composition of extractable organic matter in smoke particles from prescribed burns, *Environ Sci Technol* **21**:163–169, 1987.
85. Standley LJ, Simoneit BRT, Preliminary correlation of organic molecular tracers in residential wood smoke with the source of fuel, *Atmosph Environ* **24B**:67–73, 1990.
86. Standley LJ, Simoneit BRT, Resin diterpenoids as tracers for biomass combustion aerosols, *J Atmos Chem* **18**:1–15, 1994.
87. Simoneit BRT, Rogge WF, Lang Q, Jaffé R, Molecular characterization of smoke from campfire burning of pine wood (*Pinus elliottii*), *Chemosphere: Global Change Science* **2**:107–122, 2000.
88. Simoneit BRT, Organic matter of the troposphere: III – Characterization and sources of petroleum and pyrogenic residues in aerosols over the Western United States, *Atmos Environ* **18**:51–67, 1984.
89. Simoneit BRT, Organic matter of the troposphere—V: Application of molecular marker analysis to biogenic emissions into the troposphere for source reconciliations, *J Atmos Chem* **8**:251–275, 1989.
90. Schauer JJ, Rogge WF, Hildemann LM, Mazurek MA, Cass GR, Simoneit BRT, Source apportionment for airborne particulate matter using organic compounds as tracers, *Atmos Environ* **30**:3837–3855, 1996.
91. Schauer JJ, Cass GR, Source apportionment of wintertime gas-phase air pollutants using organic compounds as tracers, *Environ Sci Technol* **32**:1821–1832, 2000.
92. Andreae MO, Raising dust in the greenhouse, *Nature* **380**:389–390, 1996.
93. Prospero JM, The atmospheric transport of particles to the Ocean, in Ittekot V, Schäfer P, Honjo S, Depetris PJ, (eds.), *Particle Flux in the Ocean*, J Wiley and Sons, New York, pp. 19–56.
94. Prospero JM, Saharan dust transport over the North Atlantic Ocean and Mediterranean: An overview, in Guerzoni S, Chester R. (eds.), *The Impact of Desert Dust Across the Mediterranean*, Kluwer Academic Publishers, Amsterdam, pp. 133–151, 1996.
95. Gagosian RB, Peltzer ET, The importance of atmospheric input of terrestrial organic material to deep sea sediments, in Advances in Organic Geochemistry 1985, *Org Geochem* **10**:661–669, 1986.

96. Gagosian RB, Peltzer ET, Zafiriou OC, Atmospheric transport of continentally derived lipids to the tropical North Pacific, *Nature* **291**:321–324, 1981.
97. Marty JC, Saliot A, Aerosols in equatorial Atlantic air: *n*-alkanes as a function of particle size, *Nature* **298**:144–147, 1982.
98. Schneider JK, Gagosian RB, Cochran JK, Trull TW, Particle size distributions of *n*-alkanes and ^{210}Pb in aerosols off the coast of Peru, *Nature* **304**:429–432, 1983.
99. Sicre MA, Marty JC, Saliot A, Aparicio X, Grimalt J, Albaiges J, Aliphatic and aromatic hydrocarbons in different sized aerosols over the Mediterranean Sea: Occurrence and origin, *Atmos Environ* **21**:2247–2259, 1987.
100. Simoneit BRT, Organic matter in eolian dusts over the Atlantic Ocean, *Mar Chem* **5**:443–464, 1977.
101. Simoneit BRT, Biogenic lipids in eolian particulates collected over the ocean, in Novakov T (ed.), *Proceedings Carbonaceous Particles in the Atmosphere*, NSF-LBL, pp. 233–244, 1979.
102. Simoneit BRT, Cardoso JN, Robinson N, An assessment of terrestrial higher molecular weight lipid compounds in air particulate matter over the South Atlantic from about 30–70°S, *Chemosphere* **23**:447–465, 1991.
103. Simoneit BRT, Kobayashi M, Mochida M, Kawamura K, Huebert BJ, Aerosol particles collected on airplane flights over the northwestern Pacific region during the ACE-Asia campaign: Composition and major sources of the organic compounds, *J Geophys Res — Atmos* **109**, doi:10.1029/2004JD004565, D19S09/1–13, 2004.
104. Simoneit BRT, Kobayashi M, Mochida M, Kawamura K, Lee M, Lim HJ, Turpin BJ, Komazaki Y, Composition and major sources of organic compounds of aerosol particulate matter sampled during the ACE-Asia campaign, *J Geophys Res — Atmos* **109**, doi:10.1029/2004JD004598, D19S10/1–22, 2004.
105. Simoneit BRT, Elias VO, Organic tracers from biomass burning in atmospheric particulate matter over the Ocean, *Mar Chem* **69**:301–312, 2000.
106. Swap RJ, Garstang M, Macko SA, Kållberg P, Comparison of biomass burning emissions and biogenic emissions to the tropical South Atlantic, in Levine JS (ed.), *Biomass Burning and Global Change*, Vol. 1, MIT Press, Cambridge, MA, pp. 396–402, 1996.
107. Simoneit BRT, Elias VO, Kobayashi M, Kawamura K, Rushdi AI, Medeiros PM, Rogge WF, Didyk BM, Sugars — dominant water-soluble organic compounds in soils and characterization as tracers in atmospheric particulate matter, *Environ Sci Technol* **38**:5939–5949, 2004.
108. Graham B, Mayol-Bracero OL, Guyon P, Roberts GC, Decesari S, Facchini MC, Artaxo P, Maenhaut W, Köll P, Andreae MO, Water-soluble organic compounds in biomass burning aerosols over Amazonia. 1. Characterization

by NMR and GC-MS, *J Geophysical Research* **107**:8047, 2002, doi:10.1029/2001JD000336

109. Medeiros PM, Conte MH, Weber JC, Simoneit BRT, Sugars and other compounds as source indicators of biogenic organic carbon in aerosols collected above the Howland Experimental Forest, Maine, *Atmospheric Environment* **40**:1694–1705, 2006.
110. Rogge WF, Medeiros PM, Simoneit BRT, Organic tracers in soil samples of crop fields from the Fugitive Dust Characterization Study, *Atmospheric Environment*, 2007.
111. Jaffé R, Rushdi AI, Medeiros PM, Simoneit BRT, Natural product biomarkers as indicators of sources and transport of sedimentary lipids in a subtropical river, *Chemosphere* **64**:1870–1884, 2006.
112. Medeiros PM, Simoneit BRT, Analysis of sugars in environmental samples by gas chromatography–mass spectrometry, *J Chromatography A* **1141**:271–278, 2007.
113. Simoneit BRT, Atmospheric transport of terrestrial organic matter to the sea, in Volkman JK (ed.), *The Handbook of Environmental Chemistry*, Vol. 2, Part N, Marine Organic Matter: Biomarkers, Isotopes and DNA, Springer Verlag, Berlin, 165–208, 2006.
114. Carman RM, Cowley DE, Canaric acid — a 3,4-secotriterpene acid from *Canarium muelleri, Tetrahedron Letters* **12**:627–629, 1964.
115. Carman RM, Cowley D, The structure and partial synthesis of canaric acid, *Australian J Chemistry* **18**:213–217, 1965.
116. Arigoni D, Barton DHR, Bernasconi R, Djerassi C, Mills JS, Wolff RE, The constituents of dammarenolic and nyctanthic acid, *J Chem Soc* (London), 1900–1905, 1960.
117. Witham GH, The structure of nyctanthic acid, *J Chem Soc* 2016–2020, 1960.
118. Mangoni L, Belardini M, Roburic acid, a new triterpene 3,4-*seco*-acid, *Tetrahedron Letters* **14**:921–924, 1963.
119. Baas WJ, Investigations on leaf waxes III. Pentacyclic triterpenes, *seco*-triterpenes and non-volatile aliphatics of four Hoya species and *Ficus benjamina* in relation to leaf age, *Acta Bot Neerlande* **31**:449–476, 1982.
120. Baas WJ, Dihydronyctanthic acid methyl ester and other 3,4- *seco*-pentacyclic triterpenoids from *Hoya lacunosa, Phytochemistry* **22**:2809–2812, 1983.
121. Baas WJ, van Berkel IEM, Versluis C, Heerma W, Kreyenbroek MN, Ring-A fissioned 3,4- *seco*-3-nor-triterpene-2-aldehydes and related pentacyclic triterpenoids from the leaf wax of *Hoya australis, Phytochemistry* **31**:2073–2078, 1992.
122. Baas WJ, Naturally occurring *seco*-ring-A-triterpenoids and their possible biological significance, *Phytochemistry* **24**:1875–1889, 1985.

123. Huang YS, Lockheart MJ, Logan GA, Eglinton G, Isotope and molecular evidence for the diverse origins of carboxylic acids in leaf fossils and sediments from the Miocene Lake Clarkia deposit, Idaho, U.S.A, *Org Geochem* **24**:289–299, 1996.
124. Trendel JM, Lohmann F, Kintzinger JP, Albrecht P, Identification of the des-A-triterpenoid hydrocarbons occurring in surface sediments, *Tetrahedron* **45**:4457–4470, 1989.
125. Corbet B, Albrecht P, Ourisson G, Photochemical or photomimetic fossil triterpenoids in sediments and petroleum, *J Amer Chem Soc* **102**:1171–1173, 1980.
126. Chaffee AL, Hoover DS, Johns RB, Schweighardt FK, Biological markers extractable from coal, in Johns RB (ed.), *Biological Markers in the Sedimentary Record*, Elsevier, Amsterdam, pp. 311–345, 1986.
127. Peters KE, Moldowan JM, The Biomarker Guide – Interpreting Molecular Fossils in Petroleum and Ancient Sediments, Prentice Hall, Englewood Cliffs, NJ, 1993.
128. Otto A, Simoneit BRT, Chemosystematics and diagenesis of terpenoids in fossil conifer species and sediment from the Eocene Zeitz Formation, Saxony, Germany, *Geochim Cosmochim Acta* **65**:3505–3527, 2001.
129. Otto A, Wilde V, Sesqui-, di- and triterpenoids as chemosystematic markers in extant conifers, *Botanical Review* **67**:141–238, 2001.
130. Chaloner WG, Allen K, Palaeobotany and phytochemical phylogeny, in Harbourne JB (ed.), *Phytochemical Phylogeny*, Academic Press, London, pp. 21–31, 1969.
131. Thomas BA, The biochemical analysis of fossil plants and its use in taxonomy and systematics, in Spicer RA, Thomas BA (eds.), *Systematic and Taxonomic Approaches in Palaeobotany*, Systematic Association Special Vol. 31. Clarendon Press, Oxford, pp. 39–51, 1986.
132. Thomas BR, Modern and fossil plant resins, in Harborne JB (ed.), *Phytochemical Phylogeny*, Academic Press, London, pp. 59–79, 1970.
133. Gough LJ, Mills JS, The composition of succinite (Baltic amber), *Nature* **239**:527–528, 1972.
134. Langenheim JH, Amber: A botanical inquiry. *Science* **163**:1157–1169, 1969.
135. Sterling EC, Bogert MT, The synthesis of 12-methylperhydroretene (abietane) and its non-identity with fichtelite, *J Organic Chemistry* **4**:20–29, 1939.
136. Anderson KB, Crelling JC. (eds.), Amber, Resinite and Fossil Resins, *Amer Chem Soc Symposium Series 617*, Washington, DC, 1995.
137. Knoche H, Ourisson G, Organic compounds in fossil plants (Equisetum; horsetails), *Angewandte Chemie*, International Edition **6**:1085, 1967.
138. Knoche H, Albrecht P, Ourisson G, Organic compounds in fossil plants (*Voltzia brogniarti*, Coniferales), *Angewandte Chemie*, International Edition 7:631, 1968.

139. Anderson KB, LePage BA, Analysis of fossil resins from Axel Heiberg Island, Canadian Arctic, in Anderson KB, Crelling, JC (eds.), Amber, Resinite, and Fossil Resins, *Amer Chem Soc Symposium Series 617*, Washington, DC, pp. 171–192, 1995.
140. Otto A, Walther H, Püttmann W. Sesqui- and diterpenoid biomarkers preserved in Taxodium-rich Oligocene oxbow lake clays, Weisselster basin, Germany, *Org Geochem* **26**:105–115, 1997.
141. Otto A, Kvacek J, Goth K, Biomarkers from the taxodiaceous conifer *Sphenolepis pecinovensis* and isolated resin from Bohemian Cenomanian, *Acta Palaeobotanic Supplementum* **2**:153–157, 1999.
142. Otto A, Simoneit BRT, Wilde V, Kunzmann L, Püttmann W, Terpenoid composition of three fossil resins from Tertiary and Cretaceous conifers, *Review of Palaeobotany and Palynology* **2452**:1–13, 2001.
143. Otto A, Simoneit BRT, Rember WC, Conifer and angiosperm biomarkers in clay sediments and fossil plants from the Miocene Clarkia formation, Idaho, U.S.A., *Org Geochem* **36**:907–922, 2005.
144. Simoneit BRT, Grimalt JO, Wang TG, Cox RE, Hatcher PG, Nissenbaum A, Cyclic terpenoids of contemporary resinous plant detritus and of fossil woods, ambers and coals, *Org Geochem* **10**:877–889, 1986.
145. Staccioli G, Mellerio G, Alberti MB, Investigation on terpene-related hydrocarbons from a Pliocenic fossil wood, *Holzforschung* **47**:339–342, 1993.
146. Stout SA, Aliphatic and aromatic hydrocarbons in a Tertiary angiospermous lignite, *Org Geochem* **18**:51–66, 1992.
147. Vávra N, Walther H, Chemofossilien aus dem Harz von *Cunninghamia miocenica* ETTINGSHAUSEN (Taxodiaceae; Oligo/Miozän), *Neues Jahrbuch für Geologie und Paläontologie, Monatshefte* **11**:693–704, 1993.
148. Huang Y, Lockheart MJ, Collister JW, Eglinton G, Molecular and isotopic biogeochemistry of the Miocene Clarkia Formation: Hydrocarbons and alcohols, *Org Geochem* **23**:785–801, 1995.
149. Lockheart MJ, van Bergen PF, Evershed RP, Chemotaxonomic classification of fossil leaves from the Miocene Clarkia lake deposit, Idaho, USA based on n-alkyl lipid distributions and principal component analyses, *Org Geochem* **31**:1223–1246, 2000.
150. Logan GA, Smiley CJ, Eglinton G, Preservation of fossil leaf waxes in association with their source tissues, Clarkia, N. Idaho, U.S.A., *Geochim Cosmochim Acta* **59**:751–763, 1995.
151. Czechowski F, Simoneit BRT, Sachanbinski M, Wolowiec S, Physicochemical structural characterization of ambers from deposits in Poland, *Applied Geochem* **11**:811–834, 1996.

152. Grantham PJ, Douglas AG, The nature and origin of sesquiterpenoids in some Tertiary fossil resins, *Geochim Cosmochim Acta* **44**:1801–1810, 1980.
153. Grimalt JO, Simoneit BRT, Hatcher PG, Nissenbaum A, The molecular composition of ambers, *Org Geochem* **13**:677–690, 1988.
154. Kupchan SM, Karim A, Marcks C, Tumor inhibitors. XLVIII. Taxodione and taxodone, two novel diterpenoid quinine methide tumor inhibitors from *Taxodium distichum, J Organic Chemistry* **34**:3912–3918, 1969.
155. Simoneit BRT (ed.), Organic matter in hydrothermal systems-maturation, migration and biogeochemistry, *Applied Geochem* **5**:1–248, 1990.
156. Simoneit BRT, Petroleum generation, an easy and widespread process in hydrothermal systems: An overview, *Applied Geochem* **5**:3–15, 1990.
157. Simoneit BRT, Natural hydrous pyrolysis — Petroleum generation in submarine hydrothermal systems, in Whelan JK, Farrington JW (eds.), *Productivity, Accumulation and Preservation of Organic Matter in Recent and Ancient Sediments*, Columbia University Press, New York, 368–402, 1992.
158. Simoneit BRT, Aqueous organic geochemistry at high temperature/high pressure, *Origins of Life and Evolution of the Biosphere* **22**:43–45, 1992.
159. Simoneit BRT, Lipid/bitumen maturation by hydrothermal activity in sediments of Middle Valley, Leg 139, in Mottl M, Davis E, Fisher A, Slack J (eds.), *Proceedings of the Ocean Drilling Program, Scientific Results*, Vol. 139, College Station, TX: Ocean Drilling Program, pp. 447–465, 1994.
160. Simoneit BRT, Lonsdale PF, Hydrothermal petroleum in mineralized mounds at the seabed of Guaymas Basin, *Nature* **295**:198–202, 1982.
161. Kawka OE, Simoneit BRT, Survey of hydrothermally-generated petroleums from the Guaymas Basin spreading center, *Org Geochem* **11**:311–328, 1987.
162. Kvenvolden KA, Rapp JB, Hostettler FD, Morton JL, King JD, Claypool GE, Petroleum associated with polymetallic sulfide in sediment from Gorda Ridge, *Science* **234**:1231–1234, 1986.
163. Brault M, Simoneit BRT, Steroid and triterpenoid distributions in Bransfield Strait sediments: Hydrothermally-enhanced diagenetic transformations, *Org Geochem* **13**:697–705, 1988.
164. Rushdi AI, Simoneit BRT, Hydrothermal alteration of organic matter in sediments of the Northeastern Pacific Ocean: Part 1. Middle Valley, Juan de Fuca Ridge, *Appl Geochem* **17**:1401–1428, 2002.
165. Rushdi AI, Simoneit BRT, Hydrothermal alteration of organic matter in sediments of the Northeastern Pacific Ocean: Part 2. Escanaba Trough, Gorda Ridge, *Appl Geochem* **17**:1467–1494, 2002.
166. Simoneit BRT, Mazurek MA, Brenner S, Crisp PT, Kaplan IR, Organic geochemistry of Recent sediments from Guaymas Basin, Gulf of California, *Deep-Sea Research* **26A**:879–891, 1979.

167. Simoneit BRT, Philp RP, Jenden PD, Galimov EM, Organic geochemistry of Deep Sea Drilling Project sediments from the Gulf of California — hydrothermal effects on unconsolidated diatom ooze, *Org Geochem* 7:173–205, 1984.
168. Chaffee AL, Fookes CJR, Polycyclic aromatic hydrocarbons in Australian coals—III. Structural elucidation by proton nuclear magnetic resonance spectroscopy, *Org Geochem* **12**:261–271, 1988.
169. Hauke V, Graff R, Wehrung P, Trendel JM, Albrecht P, Schwark L, Keely BJ, Peakman TM, Novel triterpene-derived hydrocarbons of arborane/fernane series in sediments. Part I, *Tetrahedron* **48**:3915–3924, 1992.
170. Hauke V, Graff R, Wehrung P, Trendel JM, Albrecht P, Riva A, Hopfgartner G, Gulaçar FO, Buchs A, Eakin PA, Novel triterpene-derived hydrocarbons of the arborane/fernane series in sediments: Part II, *Geochim Cosmochim Acta* **56**:3595–3602, 1992.
171. Hussler G, Albrecht P, Ourisson G, Cesario M, Giulhem J, Pascard C, Benzohopanes, a novel family of hexacyclic geomarkers in sediments and petroleums, *Tetrahedron Letters* **25**:1179–1182, 1984.
172. Moldowan JM, Fago FJ, Carlson RMK, Young DC, van Duyne G, Clardy J, Schoell M, Pillinger CT, Watt DS, Rearranged hopanes in sediments and petroleum, *Geochim Cosmochim Acta* **55**:3333–3353, 1991.
173. Smith GW, Fowell DT, Melsom BG, Crystal structure of 18α(H)-oleanane, *Nature* **228**:355–356, 1970.
174. Aquino Neto FR, Restle A, Connan J, Albrecht P, Ourisson G, Novel tricyclic terpanes (C_{19}, C_{20}) in sediments and petroleums, *Tetrahedron Letters* **23**:2027–2030, 1982.
175. Simoneit BRT, Leif RN, Radler de Aquino Neto F, Almeida Azevedo D, Pinto AC, Albrecht P, On the presence of tricyclic terpane hydrocarbons in Permian tasmanite algae, *Naturwissenschaften* 77:380–383, 1990.
176. Venkatesan I, Occurrence and posible sources of perylene in marine Sadiments — A review. *Marine Chemistry* **25**:1–27, 1988.
177. Simoneit BRT, Lein AY, Peresypkin VI, Osipov GA, Composition and origin of hydrothermal petroleum and associated lipids in the sulfide deposits of the Rainbow Field (Mid-Atlantic Ridge at 36°N), *Geochim Cosmochim Acta* **68**:2275–2294, 2004.
178. Greenwood PF, Arouri KR, Logan GA, Summons RE, Abundance and geochemical significance of C_{2n} dialkylalkanes and highly branched C_{3n} alkanes in diverse Meso- and Neoproterozoic sediments, *Org Geochem* **35**:331–346, 2004.
179. Kenig F, Simons DJH, Crich D, Cowen JP, Ventura GT, Rehbein-Khalily T, Brown TC, Anderson KB, Branched aliphatic alkanes with quaternary substituted carbon atoms in modern and ancient geologic samples, *Proceedings of the National Academy of Sciences* (U.S.) **100**:12554–12558, 2004.

180. Logan GA, Hinman MC, Walter MR, Summons RE, Biogeochemistry of the 1640 Ma McArthur River (HYC) lead-zinc ore and host sediments, Northern Territory, Australia, *Geochim Cosmochim Acta* **65**:2317–2336, 2001.
181. Mycke B, Michaelis W, Degens ET, Biomarkers in sedimentary sulfides of Precambrian age, *Org Geochem* **13**:619–625, 1988.

Chapter 5

TOXINS OF MARINE INVERTEBRATES AND MICROORGANISMS

Yoel Kashman and Yehuda Benayahu

5.1 INTRODUCTION

Many marine organisms and plants, including unicellular protists, invertebrates (e.g. jellyfish, sponges, shells, starfish and worms), fish and some sea snakes have evolved venom as either a defense mechanism or as a primary means for the capture of their prey. The plethora of biologically active molecules that constitute venom is used to disrupt essential organ systems in the envenomed animal. Toxic marine organisms have attracted the attention of biologists, chemists, nature lovers, aquarium hobbyists over the years. Knowledge of the biological features of the specific marine animals is crucial for any research that deals with their toxins and their activity. These include their structure, function, natural history taxonomic status (classification), placement in the evolutionary tree, and the possible role played by the toxins in the biology and ecology of the various species. In the following sections, we will present some of these aspects in marine several groups of animals.

5.1.1 Dinoflagellates

Dinoflagellates are unicellular, flagellated, often photosynthetic protists, commonly regarded as "algae" of the Dinoflagellata.[1] The majority of dinoflagellate species are marine, and they make up the most important primary producers in the oceans. Dinoflagellates are also common in freshwater lakes and rivers.[2] They can occur in blooms of sufficient concentration to discolor the water, producing what are known as "red tides".[3–5] Their cells are characterized by a transverse flagellum that encircles the body and a longitudinal flagellum oriented perpendicular to the transverse flagellum. This imparts a distinctive spiral to their swimming motion.[6] Dinoflagellates possess a unique nuclear structure at some stage of their life cycle—a dinokaryotic nucleus, as opposed to eukaryotic or prokaryotic, in which the chromosomes are permanently condensed.[7] The cell wall of many dinoflagellates is divided into plates of cellulose ("armor") with distinctive geometry and morphology, which is the main means for their classification. Both heterotrophic (feeding) and autotrophic (photosynthetic) dinoflagellates are known, some may exhibit both modes of nutrition.[8] Most dinoflagellates go through moderately complex life cycles involving both sexual and asexual stages which are motile and nonmotile. During their annual reproduction, some species which produce large populations appear as algal bloom ("red tide") commonly occurring in warm, shallow water and tend to become discolored by the sheer concentration of algae seeking the sunlight. This discoloration is a result of the various photosynthetic pigments depending on the species of algae present in the water that may reflect a variety of colors.[9] Since red is the most common pigment, the phenomenon has come to be called red tide. Dinoflagellates are also known for producing toxins, particularly when they occur in large numbers.[10] Besides being bad for a large range of marine life, red tides can also introduce nonfatal or fatal amounts of toxins into filter feeding animals such as shellfish (bivalves), crabs and fish that may be eaten by humans, who are also affected by the toxins.[11,12] Many of these toxins are quite potent, and if not fatal, can still cause neurological and other nasty effects. It is believed that due to anthropogenic impacts on the marine environment, mainly inputs of nutrients, and sea water temperatures rise due to global change, events of red tides will become more spread and frequent.[13]

5.1.2 Porifera

Poriferans (sponges) are considered as the simplest multicellular animals.[14,15] The vast majority of the sponge species are marine. Unlike most other animals, sponges lack a nervous system and have no true musculature. They are benthic and filter food particles suspended in the water. They have no specialized organ systems, often they are amorphous and asymmetrical animals. Only a few different cell types are encountered within sponges which are functionally independent to the extent that an entire sponge can be dissociated into its constituent cells. Special flagellated cells called choanocytes generate currents that help maintain water circulation within the sponge and capture food particles.[16]

Sponges produce a large variety of toxins[17] which are present either on the surface of the sponge or released into the water. The intended function of these toxins is to deter predators that would otherwise feed on the sponges,[18] but humans can become the unintended targets by handling sponges or by abrading against sponges. Small particles from the sponge surface including their tiny skeletal elements (spicules) may also dislodge and remain adherent to the skin surface of the unfortunate traveler causing local inflammation. This may cause pain, itching, tingling at the site of injury, swelling, etc.

Since space and especially hard substrata is limited in marine ecosystem interactions along borders between pairs of species, mainly of sponges, are of ecological interest. Several different forms of overgrowth are common but only in a few cases, death of the underlying species take place due to overgrowth of the underlying sponges. Overgrowth in these types of sponges may be a defensive strategy aimed at maintaining their occupied space.[19] Only one species, *Aplysilla rosea*, gained space along borders and this occurred without any overgrowth of the space-losing species. This sponge was found to be very toxic toward a variety of indicator organisms and it is possible that these toxins form the basis of the mechanism behind its success in maintaining and obtaining space. The encrusting sponges were damaged experimentally to investigate the influence of their different rates of regeneration on the outcome of competitive interactions. Again, *A. rosea* gained the most space but some species which did not gain space along borders gained space from neighboring sponges in this experiment.

Most carnivorous animals avoid sponges as they are of poor nutritional value and are protected by spicules and toxins. Many sea sponges use toxins to repel would-be predators. Some nudibranchs, however, such as *Platydoris scabra*, have evolved immunity against the toxins of specific sponges[20,21] (in this case, Microcionidae). This adaptation benefits the slugs in two ways. First, they don't have to compete with many other organisms for the sponges. The sea slugs can also concentrate the sponge toxins to foil their own predators, at least until the slugs' predators also evolve immunity to the toxins. Many sponges have escaped predation and also evolved chemical weapons for use against fouling organisms.[22]

5.1.3 Cnidaria

The Phylum Cnidaria are incredibly diverse in form and contains some 10 000 species, as evidenced by sea anemones, corals, jellyfish, fire corals, fresh water *Hydra* and the Portuguese-man-of-war. The name Cnidaria comes from the Greek word "cnidos", which means stinging nettle. These animals are united because they are all armed with stinging cells called cnidocytes and contain a structure called a nematocyst.[23] The stinging cells are used to capture and subdue prey. All cnidarians have a basic radial symmetry, are simple in structure and posses only two cell layers: the epidermis and gastrodermis which are separated by a gelatinous layer, the mesoglea. All cnidarians posses stinging cells that are unique to the species of this phylum. These small subcellular organelles are formed within specialized cells termed cnidoblasts. When touched or irritated, they can discharge a barbed thread that is connected to a venom sac.[23,24] Cnidarians use their stinging cells to incapacitate their prey which include mainly tiny planktonic organisms. Large cnidarians jellyfish and sea anemones are predators that can attack and capture conspicuous prey items such as fish, crabs, worms etc.[25] Stinging cells can be used for defense against predators, and among benthic species such as sea anemones and corals, for exclusion of neighboring organisms.[26] Most stinging cells are of insignificant strength, being able to cause only discomfort to man. In most cases, stinging cnidarians such as of fire coral, stony corals, some sea anemones or most jellyfish, can cause discomfort. Often pieces of tentacles cling to the skin after contact, and they can continue to release poison into the skin for several hours. There are exceptions, where cnidarians can be very venomous. One of the most dangerous species is the "box jelly" (or "sea wasp"), found only in the

Pacific waters near Australia.[27] These jellyfish have been known to cause death within hours or even minutes of a sting.[28,29] Along the beaches of Australia, warning signs are often posted to remind swimmers to beware of the box jellyfish.

The Portuguese-man-of-war is a colony of four kinds of highly modified polyps.[30] The float (pneumatophore) is a single individual and supports the rest of the colony. The tentacles (dactylozooids) are polyps in the colony concerned with the detection and capture of food and convey their prey to the digestive polyps (gastrozooids). Sexual reproduction is carried out by the gonozooids, another type of polyp. The float is a bottle or pear-shaped sac that can exceed 15 cm and the tentacles may be more than 50 cm in length. The float is a living, muscular bag that secretes its own gas, which is similar to air and has aerodynamic properties. It seems likely that sailing characteristics may be modified by muscular contraction of the crest. The Portuguese-man-of war that is found in the warmer regions of the Atlantic and Indo Pacific Ocean and its stinging cells can cause very painful injuries to humans and even fatalities.[28,31]

Many soft corals and some other benthic cnidarians such as zooanthids, hydrozoans and sea anemones produce a range of chemical substances which effectively deter predators, prevent overgrowth by fouling organisms or algae or screen UV radiation.[32] These substances have been termed secondary metabolites, but a change in terminology to fundamental and complementary metabolites has been proposed to replace the previously used term. Many of these substances are toxic and act as feeding deterrent, and are effective against predators such as fish, egg cowries, nudibranch snails or even predatory star fish such as the crown of thorns star fish.[33,34] Other compounds produced by some soft corals are slowly released into the seawater and act as chemical defense against space competitors, causing tissue necrosis in surrounding benthic organisms such as stony corals and sponges etc.[35]

5.1.4 Mollusca

The snails, members of the Prosobranchia, have a few members who are both venomous and poisonous. Whelks, for example, have poison salivary glands but no venom apparatus.[36] This is used in penetrating their prey shellfish for consumption. The best known snails for toxicity are the cone shell species, which are all exclusively carnivorous.[29,37] Most of them are predatory and feed on worms, crabs and other molluscs, including their smaller relatives and conspecifics. About 70 of the larger cone shell species on the other hand are piscivorous, which means that they feed on small fish. Cone shells, being gastropods, are much slower than their prey and they are only able to catch and eat it, because they evolved a suitable hunting method — venom. This venom paralyses and kills the prey in seconds. Cone shell venom is applied by the radula teeth which is comonly found among snails such as the rasp tongue snails. But the cone shell's radula has changed during evolution and in contrary to other gastropods, and instead of having tens of thousands of toothlets on their radula, the cone shell only has one at a time. That, however, has evolved to a chitinous blade rolled together to form a barb-hooked hollow harpoon. At its inner end, the harpoon tooth is connected to a venom gland.[38] Additional replacement teeth are stacked in a sack-like organ at the radula's end. From here, they are replaced when one tooth has been expended. Cone shell toxins are composed of oligopeptides and to avoid self-poisoning, a cone shell's venom is generated in various nonpoisonous components that are stored in a safe way. Only shortly before application, the components merge and build the effective nerve poison, the conotoxin. Through the proboscis, a cone shell's elongated snout, the harpoon is aimed and then stung into the prey's body.[39] Through the hollow of the harpoon, the venom is then injected into the victim's organism.

Nudibranchs, members of the Opisthobranchia, or sea slugs, "steal" venom from other animals and use it for self-defense.[19] For example, when they hydroid, any unfired nematocysts pass through the nudibranch's body and accumulate in the cerata, which are appendages arranged along its back.[40] This "second-hand" venom protects the nudibranch from potential predators.[41] Many nudibranchs are predators with very particular food preferences, and each species may consume only one or two different prey items. However, considered as a group, they feed on a wide variety of benthic animals, including sponges, bryozoans, ascidians, soft corals and hydroids.[42] Most of these prey organisms contain complex chemical compounds that are toxic to fish and other predators. However, nudibranchs have evolved physiological processes that enable them to consume their prey without suffering adversely from the toxins.[43] Many species store the toxins, in the cerata or their skin. Thus, they in turn become unpalatable to potential predators,[44] and many species advertise this unpalatability with bright warning color patterns.

The octopuses members of the Cephalopoda have toxins as well.[45] Most bites from octopuses are rather mild and only cause a burning/itching sensation, although some severe cases have been reported that include feeling of detachment and paralysis.

5.1.5 Echinodermata

The phylum Echinodermata includes a diverse group of marine animals that are slow moving and nonaggressive, including brittle stars, sea lilies, starfish, sea urchins, and sea cucumbers. Nearly all of the 6 000 species are marine, and none live in freshwater. Although their free living larval stages are bilateral, most adult animals have a basic five-point, pentamerous, radial symmetry.[46] They lack a head structure and do not have anterior and posterior ends. Instead, body surfaces are designated as being oral (bearing the mouth opening) and aboral (not bearing the mouth). Echonoderms have calcareous skeletons that form thick plates and protective spines in some; hence, they are named Echinodermata, which means spiny skin.[39] The skeletal elements are internal, commonly embedded in the body wall. The major unifying characteristics of echinoderm species is the presence of water vascular system of a unique structural feature which involves both movement of the animals and transport of dissolved gases and nutrients. Injury and envenomation by echonoderms occur almost exclusively from accidental contact or careless handling by[47] bathers, divers, and fishermen. While most echinoderms are poisonous, and many have sharp spines capable of causing injury, only a few members of the starfish, echinoids and sea cucumbers classes are capable of causing venomous injuries in humans. Among echinoderms, envenomation usually refers to the application of toxins produced in specialized glands and tissues with modified application structures (spines, pedicellaria, tentacles).[48] This definition is in contrast to poisoning or intoxication, which refers to the oral ingestion of toxins produced or accumulated in nonspecialized glands or tissues.

Brittle stars (class Ophiuroidea) are not generally considered capable of causing venomous injuries in humans. However, some brittle stars (*Ophiomastix annulosa*) do possess toxins and are capable of causing paralysis and death in small animals. Starfish (Asteroidea) envenomation in humans is well described, with the crown-of-thorns starfish (*Acanthaster planci*) as the main culprit. *Acanthaster* species possess long (5–6 cm), extremely sharp spines projecting from the aboral surfaces of their bodies and numerous arms (7–23, a notable exception to the usual 5 arms found in most other species). These spines are covered with an integument that, in turn, is associated with glandular cells that produce a variety of toxins.[49] Rupture of the overlying integument during spine penetration results in the release of a range of bioactive substances capable of causing local and generalized toxicity.[50] Other starfish potentially capable of envenomation include members of the genus *Echinaster*, which possess thorny spines and small pits from which toxins are secreted, and *Plectaster* and *Solaster* species, which are reported to cause contact dermatitis. Sea urchins (Echinoidea) capable of causing venomous injuries in humans use specialized spines (long or short) and pedicellaria (delicate seizing organs equipped with pincerlike jaws) to deliver their venom.[51] Although both structures are present, generally only one is venomous in a given species. Thus, grouping the venomous urchins into the one of the following three categories is convenient:

(1) Long-spined species may inject venom during a puncture with the rupture of the overlying integument (*Diadema* species) or with fracture and release of venom from hollow-lumen spines (*Echinothrix* species).
(2) Short-spined species similarly may envenom during puncture when downward pressure ruptures the surrounding integument (*Phormosoma* species), or they may deliver a severe sting without puncture via venom glands located at the spine tips (*Asthenosoma* and*Araeosoma* species).
(3) Species with pedicellaria include those reputed to be the most venomous of all sea urchins, the flower urchins (*Toxopneustes* species). Pedicellaria are small, delicate, tripled-jawed seizing organs that are supported by basal calcareous plate or embedded in the body stiff integument.

Threatened or wounded sea cucumber (*Holothuroidea*) will contract its body exposing the small skeletal bones that make up the body wall, which can act as hooks to the mouths of predators. For most species, the connective tissue that makes up the greater part of the body wall is the primary deterrent to predators.[52,53] The contracted body not only makes a more compact body to bite, but increases the stiffness of the body wall. The sea cucumber also has a very unusual defense mechanism. Many species use the Cuvierian tubules, which are located in the digestive system of the animal to confuse would-be predators.[54,55] The sea cucumber, when stimulated, will raise the anus, point it in the direction of the stimulation, and contract the body wall to expel the long tubules through the anal opening. The Cuvierian tubules are sticky and toxic, and are used to simply confuse the predator. These digestive tubules are lost when the sea cucumber takes this action, and can be regenerated later. Many species contain toxins in their body wall, thus they become unpalatable to potential predators such as fish, crabs and worms.

The knowledge of marine toxins comes mainly from behavioral-biological observations of marine invertebrates in the sea, for example, ichthiotoxicity, antibacterial activity, allelopathy (a chemical produced by one organism affecting a neighbouring one) and toxicity to humans. Toxins from one category can of course also be active in the second category. Often the role of human-toxins in the natural marine environment is unknown. Below are examples of toxins from both groups starting with human-toxins.[56]

5.2 POLYETHER TOXINS

5.2.1 Tetrodotoxin (TTX)

Among the marine toxins relevant for human intoxication, tetrodotoxin (TTX) has been known as one of the most potent low-molecular weight neurotoxins.[57–61] Puffer fish (family Tetraodontidae — the source of the TTX name) were originally thought to be the only animal from which TTX could be isolated. Subsequent works determined that TTX also exists in a large number of other marine organisms of different phylogenic classes e.g.

echinoderms, mollusks, a very few amphibians, as well as microorganisms like dinoflagellate.

The production of TTX is generally accepted to be associated with bacteria that are transmitted to marine organisms through the food chain. Yasumoto's group[62] succeeded in obtaining TTX and congeners from cultures of several bacteria. Cultures of the bacteria *Shewanella alga* afforded TTX and 4,9-anhydro TTX. Obtaining TTX congeners from different organisms supports the idea of symbiosis between the host animal and the guest bacteria. Indirect evidence for a symbiont comes from studies on hatchery-raised puffer fish, which are not toxic, but become so when fed tetrodotoxin-containing liver of wild-caught puffer fish and after being fed tetrodotoxin-producing bacteria.

The absolute structure of TTX was approved in 1964[62] and later synthesized by many groups.[63–66] TTX has a unique tetracyclic structure presented in Fig. 1.

Tetrodotoxin is a virulent poison; the LD_{50} for the mouse is 10 nanograms. It acts by blocking the conduction of nerve impulses along nerve fibers and axons. The victim eventually dies from respiratory paralysis. Tetrodotoxin is quite specific in blocking the Na^+ ion channel and therefore the flow of Na^+ ions while having no effect on K^+ ions. Binding to the channel is relatively tight. Whereas the hydrated sodium ion binds reversibly on a nanosecond time-scale, tetrodotoxin is bound for tens of seconds. Tetrodotoxin, much larger than the sodium ion, acts like a cork in a bottle, preventing the flow of sodium until it slowly diffuses off. A mortal dose of tetrodotoxin is but a single milligram. Tetrodotoxin competes with the hydrated sodium cation and enters the Na^+-channel where it binds. It is proposed that this binding results from the interaction of the positively charged guanidino group on the tetrodotoxin and negatively charged carboxylate groups on side chains in the mouth of the channel.

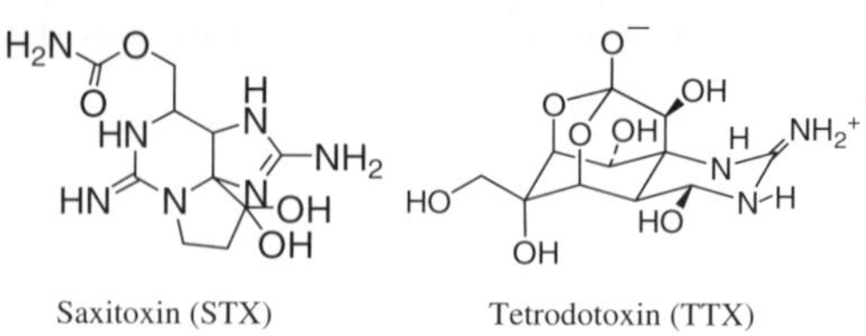

Fig. 1. Structure of tetrodotoxin and saxitoxin.

5.2.2 Saxitoxin (STX)

In 1987, marine scientists were presented with a series of very disturbing mass die-offs involving humpback whales and other sea creatures. In the case of the whales, the number of deaths represented 50 years of natural mortality. The cause of these puzzling deaths was eventually blamed on pollution, since none of the animals showed any signs of disease following post-mortem examination. By the end of the same year, there had been two more major incidents involving mass poisoning of humans. A whole array of symptoms was reported, including dizziness, diarrhea and vomiting, disorientation, respiratory distress, and eye irritation. By the end of these poisoning episodes, three people were dead and over a hundred had been severely incapacitated by illness.[67] The cause of these incidents was eventually traced to a natural source, in fact, several sources were identified: a series of algal. The species each producing their own potent biotoxins. One of these naturally occurring poisons is called saxitoxin, reported to be one of the most toxic non-protein substances known. LD_{50} in mice is approximately 8 μ/kg which means that a single dose of 0.2 mg would prove fatal for the average weight human. The causative algal species are actually dinoflagellates, which include Alexandrium tamarense, Gymnodinium catenatum, and Pyrodinium bahamense, and although they themselves do not usually affect humans, filter-feeding mollusks may ingest them, thereby concentrating the toxins which they produce.[68] The algal species itself can grow incredibly fast when conditions are right, leading to spectacular phenomena like that shown above, which are called red tides (or harmful algal blooms). The red tide problem is becoming increasingly prevalent, and has been blamed on pollution in coastal areas, and also on farmland run-off into rivers and lakes. The structure of STX was determined in 1975[69,70] (Fig. 1), since then, several syntheses were reported.[71,72] STX, like many other naturally occurring toxins, has been an indispensable tool in medical research. It is a potent and extremely selective sodium-channel blocker. It has no effect on potassium or calcium channels, or the flux of chloride ions or on acetylcholine release (whereas other marine toxins have similarly selective effects on these aspects of nerve function). Saxitoxin has been used for the labeling, characterization and isolation of various sodium channel components, and this has opened up new avenues for the study of various nervous disorders.[67]

5.2.3 Ciguatoxin (CTX)

More than 20 000 people in subtropical and tropical regions suffer annually from ciguatera, which is the most widespread human poisoning caused by the consumption of seafood.[61,73] The disease is characterized by gastrointestinal, neurological, and cardiovascular disturbances, which often persist for months or years, and in severe cases, paralysis, coma, and death may occur. More than 400 species of fish can be vectors of the ciguatera toxins, which are produced by a marine dinoflagellate, *Gambierdiscus toxicus,* living on macro algae. These neurotoxins, designated as ciguatoxins, are far more dangerous [acute toxicity on mice median lethal dose ($LD_{50}0.2 \sim 4\,\mu/kg$] than the structurally related red-tide toxins, brevetoxins (LD_{50} 100 μg/kg). Reef fish are increasingly exported to other areas and because ciguateric fish look, taste, and smell normal, ciguatera may become a world wide health problem. Currently, there are no rapid and reliable methods of detecting ciguatoxins at fisheries. Furthermore, the content of ciguatoxins in fish is extremely low, which has hampered the isolation, detailed biological studies, and most importantly, the preparation of the anti ciguatoxin antibodies for detecting these toxins. Yasumoto's group in Tohoku University[13] succeeded in the isolation of 0.35 mg toxin from four tons (!) of moray eels. Tests in mice have shown that one of the toxins, CTX3C, is 300 times as toxic as TTX isolated from puffer fish. There are more then 20 related ciguatoxins that share similar chemical structures. The structure elucidation of CTX3C ($C_{60}H_{86}O_{19}$) was accomplished by extensive use of NMR spectroscopy and FAB MS/MS measurements of the natural compound and degradation products.[11,74] Recently, the toxin was synthesized by Hirama *et al.* from Tuhoku University[75] (Fig. 2), a synthesis, which will provide a practical supply of ciguatoxins for further studies. Ciguatoxin (CTX) and CTX3C, along with brevetoxins, are structurally classified as ladder-like polyethers. From pharmacological studies, ciguatoxins and brevetoxins exert their toxicities by binding to a common site on the voltage-sensitive sodium channels (VSSCs), resulting in the persistent activation and/or prolonged open time of VSSCs. The significantly higher VSSC affinity of

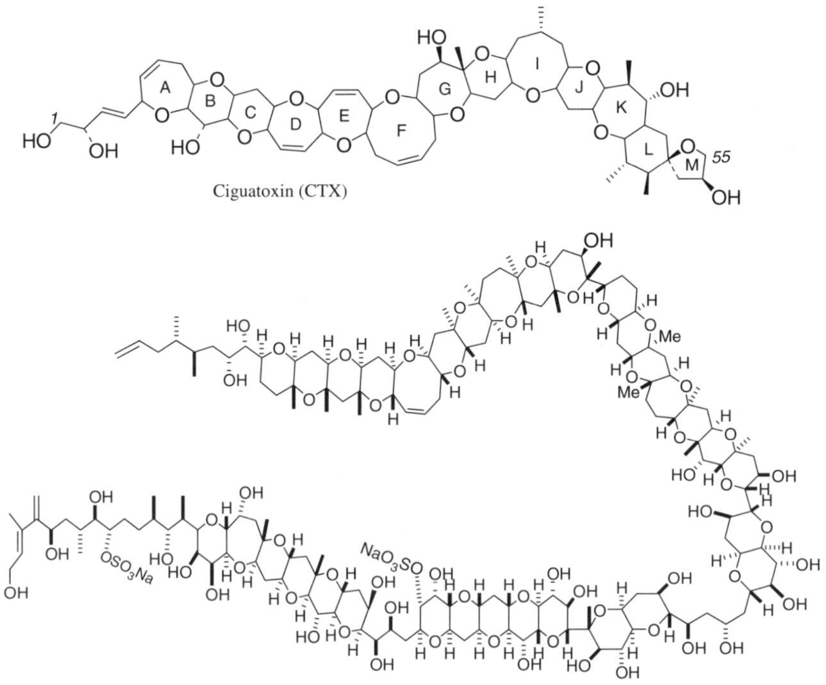

Ciguatoxin (CTX)

Maitotoxin (MTX)

Fig. 2. The structure of ciguatoxin and maitotoxin.

ciguatoxins than brevetoxins can be useful in the investigation of the VSSC function itself.[76]

The total synthesis of CTX3C[75] was achieved via the convergent assembly of two comparably complex fragments, which were synthesized by coupling two simple cyclic ethers. The latter versatile synthetic strategy should be applicable for synthesizing congeners and should help accelerate the preparation of anti-ciguatoxin antibodies for detecting intoxicated ciguateric fish and create VSSC probes that may provide valuable insight into the VSSC-ligand interaction at the molecular level, as well as the activation and gating mechanism of VSSCs.

5.2.4 Maitotoxin

Maitotoxin (MTX) is one of the most potent marine toxins known, it is found in the "red tide" dinoflagellate *Gambierdiscus toxius* and is responsible in part for ciguatera seafood poisoning.[61,77] MTX, having the molecular weight of 3422, is the largest natural product hitherto elucidated, except for biopolymers. MTX possesses 32 rings (Fig. 2). Using primarily NMR spectroscopy on the ^{13}C enriched MTX, prepared by adding $NaH^{13}CO_3$ to the culture of *G. toxius*, Yasumoto and coworkers have successfully determined the relative structure of MTX,[25,78] a structure that was confirmed by the synthesis of the compound by Kishi's group.[79] MTX is the most toxic compound, except for a few proteinous bacterial toxins. Its lethality (500 ng/kg, mice, ip) is 200 times more potent then TTX. MTX causes Ca^{2+} ion influx into cells in all cell lines tested. Though potent, MTX may play little role in ciguateria poisoning because of its low concentration in fish and a poor absorption rate from digestive tracts.

5.2.5 Okadaic acid (OA)

Okadaic acid (OA) is another polyether toxin, it is a C_{38} fatty acid derivative isolated independently from two sponges, *Halichondria* (syn *Reniera*) *okadai* Kadota, a black sponge, commonly found along the Pacific coast of Japan, and from *H. melanodocia*, a Caribbean sponge collected in the Florida Keys. Mammalian toxicity of a crude extract of *H. okadai* guided the isolation of a the pure compound. Okadaic acid is toxic (LC_{50} 192 μg/kg; ip mice) and inhibited the growth of KB cells by more than 30% at 2.5 ng/mL and more than 80% at 5 ng/mL. OA's structural features suggest that it belongs to the class of compounds known as ionophores (compounds that intercelate metal ions) which hitherto had been known only from terrestrial microorganisms. It is likely that okadaic acid is a metabolite of an epiphytic microorganism, rather than of *Halichondria*. Quite often, compounds isolated from sponges are produced by guest microorganisms living within the sponge — it is difficult to determine unambiguously the real source of the compounds. The structure of OA was determined by NMR and X-ray diffraction analysis[61,77,80] (Fig. 3). The compound was also synthesized by Urbenek.[81]

The cytotoxicities of okadaic acid as EC_{50} against the P388 and L1210 cell lines are 1.7 nano-molar and 17 nano-molar, respectively. Additionally, okadaic acid strongly inhibits protein serine/threonine phosphatase 1, 2A,

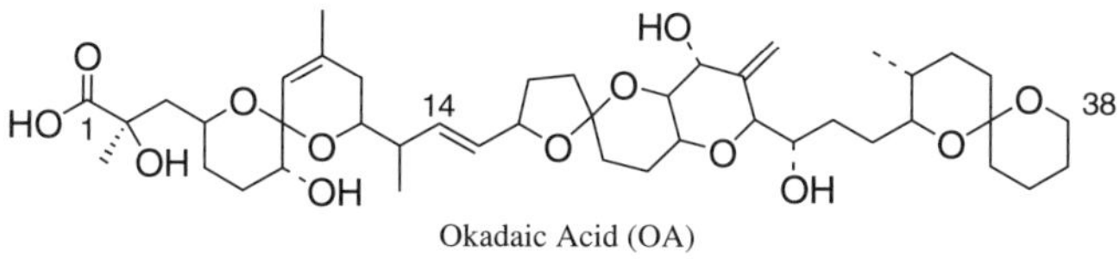

Fig. 3. Structure of okadaic acid.

and 2B. The inhibitory effect of okadaic acid is strongest for 2A, followed by 1, and then 2B. The dissociation constant of the inhibition on protein serine/threonine phosphatase 2A is 30 pico-molar. Okadaic acid is also a powerful tumor promoter even without protein kinase C. There are several other closely related diarrhetic shellfish poisoning (DSP) congeners.[61,82]

5.2.6 Metabolites of Cyanobacteria

Freshwater cyanobacteria blooms implicated in human and livestock intoxications have been extensively studied. And they have become a serious health problem in recent years. Blooms of marine cyanobacteria are also becoming an increasingly familiar occurrence within the tropical and subtropical regions of the world. Several systematic surveys within Europe and the USA have concluded that the two most commonly isolated groups of cyanotoxins are the alkaloids-neurotoxins and the cyclic peptide hepatotoxins, both of which are destructive to liver cells.[56,83–85] Several genera are responsible for the production of the neurotoxins which include the anatoxin-a (Fig. 4) — the first identified cyanotoxin isolated from several cyanobacteria.

Hepatotoxins are the most commonly encountered toxins involving cyanobacteria, that include the cyclic peptides microcystin and nodularin (Fig. 4). Microcystins are cyclic seven amino acid peptides, and nodularins are cyclic five amino acid peptides. Over 50 different variants of microcystin have been isolated.

Marine cyanobacteria continue to be an exceptionally rich source of metabolites with unusual features as well as interesting biological activities.[85]

One of the most toxic genera of marine cyanobacteria is *Lyngbya* — filamentous cyanobacteria within tropical and subtropical waters. *Lyngbya*

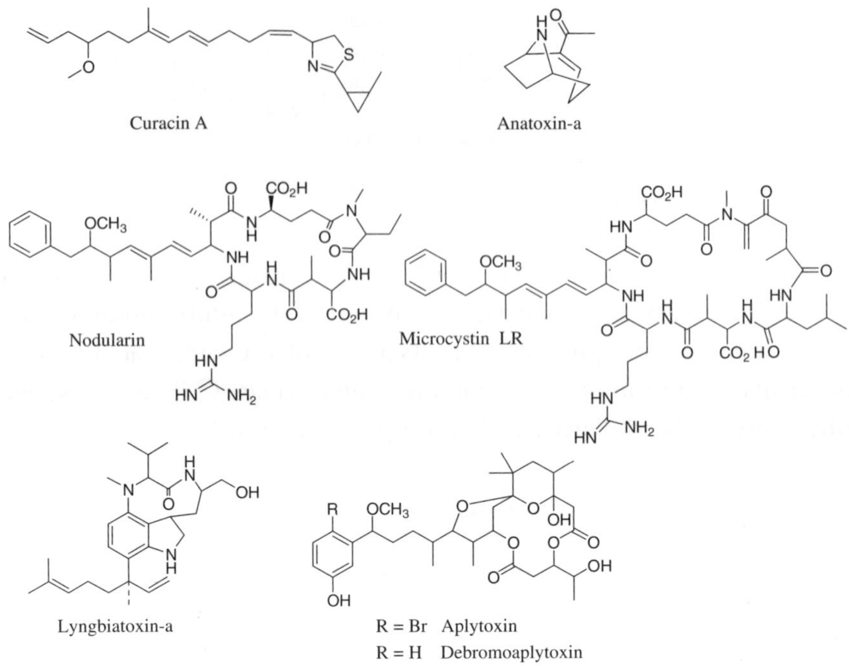

Fig. 4. Structure of several cyanobacteria toxins.

spp are amid the more common ones causing health problems, responsible for such cytotoxic compounds as antillatoxin, aplysiatoxin, debromoaplysiatoxin and lyngbyatoxin (Fig. 4).

Of particular interest is *Lyngbya majusculata* which has been the subject of extensive studies.[86] Interest in this species was first spurred by its ability to induce a dermatitis-like condition in swimmers ("swimmers itch") which was traced to potent inflammatory, blister-producing and tumor promoting compounds such as aplysiatoxin, debromoaplysiatoxin and lyngbiatoxin (Fig. 4).

Gerwick's group in Oregon has intensively investigated on the latter cyanobacteria, finding many additional interesting compounds. Among the more exciting ones are the curacins.[86] Curacin A (Fig. 4), a potent brine shrimp toxin, has shown promising antiproliferative activity due to inhibition of tubulin polymerization. The curacins are excellent potentially anticancer drugs.[62,87]

It is likely that cyanobactria synthesize poisons to ward off attack by other planktonic species. Carmichael *et al.* have found that cyanobacterial neurotoxins and hepatotoxins can be extremely harmful to the zooplankton living in the marine environment. This is a good example of the natural task of the toxins in protecting the organism that produces them.[88]

5.2.7 Palytoxin (PTX)

Palytoxin is a complex marine natural product containing 71 stereochemical elements (Fig. 5). The structure of PTX was elucidated by Moore.[89] PTX is isolated from a zoanthid (order Zoanthidea) a type of soft coral commonly found in coral reefs all around the world. These animals come in a variety of different colonizing formations and in numerous colors. They can be found as individual polyps, attached by a fleshy stolon or a mat that can be created from pieces of sediment, sand and rock (soft coral). PTX is considered to be one of the most toxic nonpeptide substances known, second only to Maitotoxin. Typical symptoms of palytoxin poisoning are angina-like chest pains, asthma-like breathing difficulties, tachycardia, unstable blood pressure, hemolysis (destruction of red blood

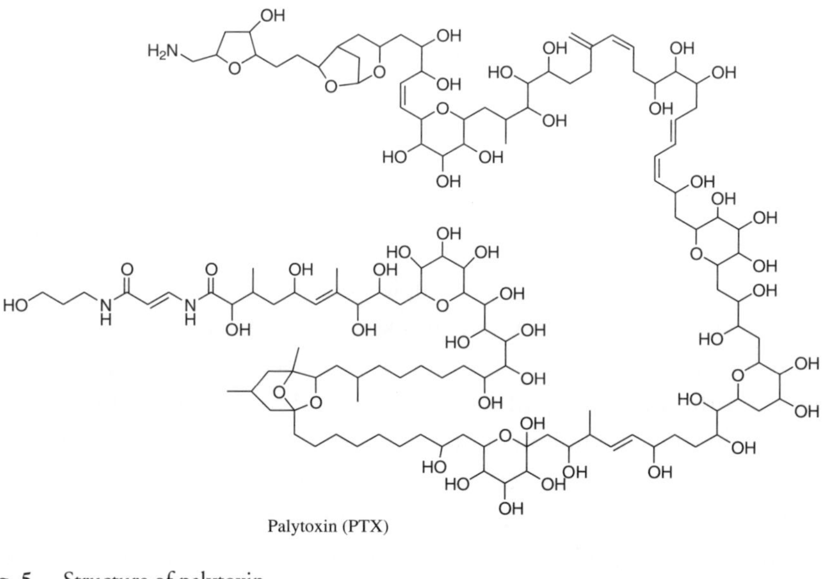

Palytoxin (PTX)

Fig. 5. Structure of palytoxin.

cells), and an electrocardiogram showing an exaggerated T wave. The onset of symptoms is rapid, and death usually follows within minutes. Kishi and coworkers first synthesized palytoxin in 1994.[90] To date, this feat is still considered by many to be the greatest synthetic accomplishment ever.

Palytoxin targets the sodium-potassium pump protein by binding to the molecule in such a way that the molecule is locked in a position where it allows passive transport of both the sodium and potassium ions, thereby destroying the ion gradient that is essential for most cells.[91]

5.2.8 Conotoxins

Conotoxins represent one of the more inspiring examples of marine toxins because of their outstanding biological activities in their habitat, in the sea, as well as their being drug leads.[92–96] Ziconotide, (Prialt) a 25 amino acid peptide[97] isolated from cones, became the first FDA approved drug for the treatment of neuropathic pain and severe chronic pain that do not respond to other forms of treatment.

The genus Conus is a large and successful group of 500 species of carnivorous predators found in all tropical marine habitats. Although individual *Conus* species can be highly specialized, as a whole, the genus shows a remarkably broad phylogenetic range of prey. At least five different phyla of animals are envenomated by cone snails. All members of this species-rich group use venom as the major weapon for prey capture and have a delivery system consisting of a venom duct, where the venom is synthesized and stored; a venom bulb, believed to transfer venom from the duct, and most remarkably, hollow, harpoon like teeth which serve as a hypodermic needle for injecting the venom. Most cone snails have a long distensible proboscis, and when they forage for prey, a single harpoon tooth is transferred into the lumen of the proboscis. Once the extended proboscis, which serves as a fishing line, touches the prey, the tooth is propelled out, grasped by circular muscles at the anterior tip of the proboscis and the venom is injected into the prey through the hollow tooth. The fish typically jerks suddenly after being struck but remains tethered through the proboscis. This efficient venom delivery system is characteristic of all cone snails, as well as certain other toxoglossate mollusks. This fish venom is toxic to humans and often, depending on the cone, causes death.

The increasing interest in cone snail venom components arises, in part, from their pharmacological potential. Not only are conopeptides used as basic science tools for neurobiologists, but several cone snail toxins are being directly developed for therapeutic applications. At present, a few conopeptides have reached human clinical trials as drug candidates. As mentioned above, Prialt (ziconotide), *vide infra*, is a pain relieving drug. Several other drug candidates are being actively explored.

Although small in size, the conotoxins contain many of the structural elements present in larger proteins, including α-helices, β-sheets and β-turns, hence, they are often referred to as mini-proteins. Their relative ease of synthesis allows accurate three-dimensional structures to be obtained using techniques such as X-ray crystallography and NMR spectroscopy.

5.2.9 Ziconotide

The first conotoxin approved as a drug was ziconotide.[97] In 2004, the Food and Drug Administration approved two new drugs intended to treat a form of pain that often proves resistant to anti-inflammatories and opiates. These two predominant classes of pharmaceuticals for analgesia are Lyrica and Prialt (ziconotide) from Elan. The latter requires that a pump be implanted, or used externally, to send the drug by catheter into the spinal fluid, a technology often reserved to deliver morphine to critically ill AIDS and cancer patients. Prialt, from the perspective of neuroscientists and pharmacologists, is by far the more interesting of the two compounds.

Prialt [Fig. 6] is a synthetic copy of a toxin from the Magician's cone snail, *Conus magus*, a mollusk from the Indo-Pacific region. This is also one of the first pharmaceuticals that demonstrate the promise that marine life, particularly invertebrates, holds for drug developers.

Because the conotoxins are small enough for direct chemical synthesis and are sufficiently constrained for three-dimensional conformation determination, conotoxins bridge protein chemistry and molecular genetics. Furthermore, the strategy that the cone snails have evolved over millions of years for the generation and design of an enormous array of small peptide ligands, each with high affinity and specificity for a particular receptor protein target, may be adaptable for use *in vitro*.

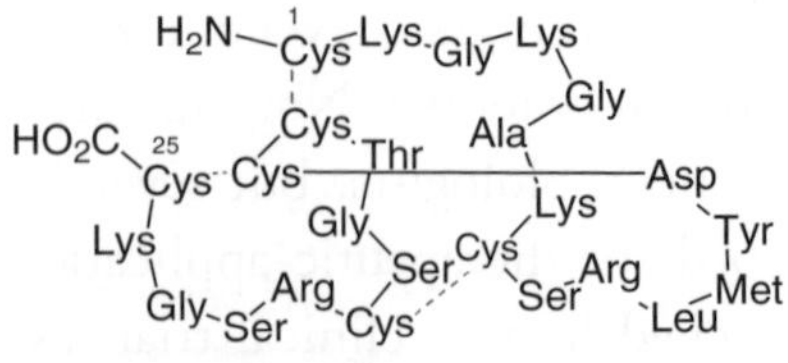

Fig. 6. Structure of conotoxin.

5.2.10 Sponge, Soft Coral and Tunicate Toxins

In the search for bioactive marine secondary metabolites, drugs or drug leads with new modes of action, many invertebrates with new modes of action have been explored and tested in a variety of assays. Many marine natural products were found to be cytotoxic to an assortment of tumor cells.[98] Some parts of the marine natural products were also found to be ichthyotoxic so as to defend themselves. Two examples, out of many, are the latrunculins[1] and sarcophine.[99,100]

While working on the ecology of coral reefs in the Red Sea, it was observed that many of the soft corals growing exposed on rocky surfaces are not attacked by predators, whereas others, which were exposed experimentally, were immediately devoured. One of these, common nonattacked species is the alcyonarian *Sarcophyton glaucum* which forms large patches on the reefs.[100] Experiments in the sea and in aquaria showed that these soft-bodied animals are avoided by predators, despite their high glyceride content which can go up to 20–30%. It was also observed that fish could not live together with *S.glaucum* in a common container, due to the soft coral toxic effect. Bioguided chromatography (lethality of *Gambusia affinis* fish) of the hexane extract of the soft coral led to the isolation of the ichthiotoxic, allelochemical compound responsible for the toxicity of the soft coral. The structure of this toxin, sarcophine, was determined to be a 14-membered macrocyclic diterpene — a cembranoid (Fig. 7). Sarcophine was one of the first identified cembranoids, since then, hundreds of additional ones were isolated, mainly from soft corals, many of them possessing interesting biological activities. Sarcophine by itself is also toxic to mice. It showed strong anticholinergic activity as well as cytototoxcity.[101,102] Many

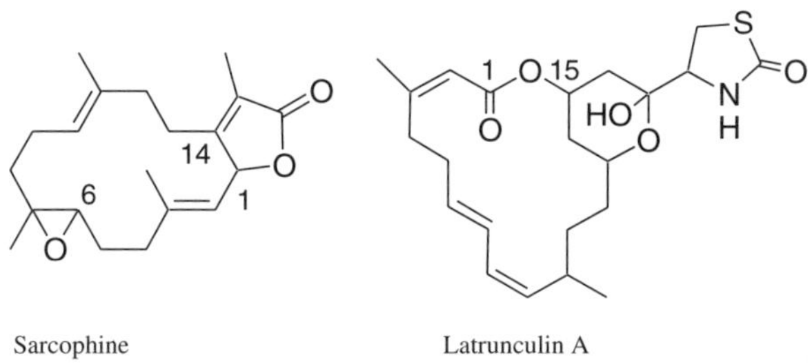

Fig. 7. Structure of Sarcophine and Latrunculin A.

other activities were reported for sarcophins, e.g. potent cancer chemopreventive[103] and insecticidal activity. Sarcophine and analogues were also synthesized for structure-activity studies so as to find compounds less toxic but with similar activities.

Among the shallow-water coral reefs of the northern Red Sea, most sponge species grow below or beneath coral plates and rocks and, if artificially exposed, are immediately devoured by various fish, especially wrasses, parrot fish and angel fish. Only a few sponge species grow exposed and, among them, the most prominent are colonies of the branching, red-coloured *Latrunculia magnifica* (*Negombata magnifica*).[49] Colonies of the latter were never observed to be damaged or eaten by fish. When manually squeezed, these sponges exude a reddish "juice" which causes fish to immediately escape from the vicinity of the sponge. In the aquarium, the latter "juice" causes poisoning (loss of balance) and death of the fish within 10 minutes. In addition, it was found that pieces of the sponge placed in the aquarium with sea water, remained intact for two years; they are not degraded by bacteria in full contrast to other sponges, most likely due to toxic antifouling compound(s). Bioguided chromatography (lethality of *Gambusia affinis*) of an extract of the sponge resulted in the isolation of the toxin designated latrunculin A or B (according to the collection place, either A or B was isolated).[98]

The structure of the toxins was elucidated by NMR and approved by X-ray diffraction analysis (Fig. 7). The two compounds were found to contain a new class of 16- and 14-membered macrolides attached via 6-membered lactol to the rare 2-thiazolidinone moiety. Both toxins are potent mammalian toxins.[1] It has been shown that the latrunculins alter the actin-monomer subunit interface to prevent polymerization. [*Nature Cell Biology*, 2000]. Because of the interesting activity of the latrunculins, they became an important tool in cell biology studies of processes in which actin is involved, the latrunculins and analogues have been synthesized.

Notable, from the ecological point of view, is the transfer of latrunculins in the food chain from the pray sponge to a predator nudibranch.[65]

Similarly to the latter two examples, there are many other defense toxins isolated from a variety of invertebrates, sponges, tunicates, sea urchins, sea cucumbers, and others. Many of the latter compounds are bioactive.[98]

In summary, it has been shown that many marine invertebrates and microorganisms produce toxins which influence the environment (are allomones) and numerous, which are toxic to humans, are drug leads which are indispensable research tools for cell biology studies.

REFERENCES

1. Sournia A, *Cryptogamie Algologie* **14**(2–3):133–144, 1993.
2. Canion AK, Ochs C, *Journal of Fresh Water Ecology* **4**:617–626, 2005.
3. Tyler MA, Seliger HH, *Limnology and Oceanography* **23**(2):227–246, 1978.
4. Hallegraeff GM, *Phycologia* **32**(2):79–99, 1993.
5. Horner RA, Garrison DL, Plumley FG, *Limnology and Oceanography* **42**(5):1076–1088, 1997.
6. Fenchel T, *Protist* **152**(4):329–338, 2001.
7. Daugbjerg N, Hansen G, Larsen J, *Phycologia* **39** (4):302–317, 2000.
8. Schnepf E, Elbrachter M, *European Journal of Protitology* **28**(1):3–24, 1992.
9. Millie DF, Schofield OM, Kirkpatrick GJ, *Limnology and Oceanography* **42**(5):1240–1251, 1997.
10. Landsberg JH, *Reviews in Fisheries Science* **10**(2):113–390, 2002.
11. Murata M, Shimatani M, Sugitani H, *Bulletin of the Japanese Society of Scientific Fisheries* **48**(4):549–552, 1982.
12. Todd ECD, *Journal of Food Protection* **56**(1):69–83, 1993.
13. Mudie PJ, Rochon A, Levac E, *Palaeogeography Palaeoclimatology Palaeoecology* **180**(1–3):159–186, 2002.
14. Cavalier-Smith T, A revised six-kingdom system of life, *Biol Rev* **73**:203–266, 1998.

15. Dohrmann M, Voigt O, Erpenbeck D, Worheide G, Non-monophyly of most supraspecific taxa of calcareous sponges (porifers, Calcarea) revealed by increased taxon sampling and partitioned Bayesian analysis of ribonomal DNA, *Molecular Phylogenetics and Evolution* **40**:830–843, 2006.
16. Maldonado M, Choanoflagellates, choanocytes, and animal multicellularity, *Invertebrate Biology* **123**:1–22, 2004.
17. Ianora A, Boersma M, Casotti R, Fontana A, Harder J, Hoffmann F, Pavia H, Potin P, Poulet SA, Toth G, New trends in marine chemical ecology, *Estuaries and Coasts* **29**:531–551, 2006.
18. Hill MS, Lopez NA, Young KA, *Marine Ecology-progress Series* **291**:93–102, 2005.
19. Becerro MA, Uriz MJ, Turon X, Chemically mediated interactions in benthic organisms: The chemical ecology of Crambe crambe (Porifera, Poecilosclerida), *Hydrobiologia* **356**:77–89, 1997.
20. Proksch P, Defensive roles for secondary metabolites from marine sponges and sponge-feeding nudibranchs, *Taxicon* **32**:639–655, 1994.
21. Gavagnin M, Mollo E, Docimo T, Guo YW, Cimino G, Scalarane metabolites of the nudibranch Glossodoris rufomarginata and its dietary sponge from the South China Sea, *Journal of Natural Product* **67**:2104–2107, 2004.
22. Qian PY, Dobretsov S, Dahms HU, *Marine Ecology-progress Series* **324**:151–165, 2006.
23. Tardenr P, *Bioessays* **17**(4):351–362, 1995.
24. Kass-Simon G, Scappaticci AA, *Canadian Journal of Zoology-revue Canadienne de Zoologie* **80**(10):1772–1794, 2002.
25. Rifkin J, Endean R, *Cell and tissue research* **233**(3):563–577, 1983.
26. Langmead O, Chadwick-Furman NE, *Marine Biology* **134**(3):479–489, 1999.
27. Tibballs J, *Toxicon* **48**(7):830–859 Sp. Iss. SI, 2006.
28. Burnett JW, Gable WD, *Toxicon* **27**(7):823–824, 1989.
29. Currie BJ, Jacups SP, *Medical Journal of Australia* **183**(11–12):631–636, 2005.
30. Dunn CW, Wagner GP, *Development Genes and Evolution* **216**(12):743–754, 2001.
31. Burnett JW, Medical aspects of jellyfish envenomation: Pathogenesis, case reporting and therapy, *Hydrobiologia* **451**:1–9, 2001.
32. Coll JC, *Chemical Reviews* **92**(4):613–631, 1992.
33. Sammarco PW, *Journal of Experimental Marine Biology* **200**(1–2):135–168, 1996.
34. Sammarco PW, Coll JC, *Marine Ecology-progress Series* **88**(1):93–104, 1992.
35. Aceret TL, Sammarco PW, Coll JC, *Marine Biology* **122** (2):317–323, 1995.

36. Ito K, Asakawa M, Beppu R, *Marine Pollution Bulletin* **48**(11–12):1116–1121, 2004.
37. Cruz LJ, Gray WR, Olivera BM, *Journal of Biological Chemistry* **260**(16):9280–9288, 1985.
38. Olivera BM, *Science* **230**:1338–1343, 1985.
39. Olivera BM, Rivier J, Scott JK, Hillyard DR, Conotoxins LJ, *J of Biolog Chem* **266**(33):22067–22070, 1991.
40. Olivera BM, *Annual Review of Ecology and Systematics* **33**:25–47, 2002.
41. Le Gall F, Favreau P, Richard G, *Toxicon* **37**(7):985–998, 1999.
42. Stewart J, Gilly WF, *Biological Bulletin* **209**(2):146–153, 2005.
43. Becerro MA, Starmer JA, Paul VJ, Chemical defenses of cryptic and aposematic gastropterid molluscs feeding on their host sponge dysidea granulosa, *Journal of Chemical Ecology* **32** (7):1491–1500, 2006.
44. Frick KE, *Marine Biology* **147**(6):1313–1321, 2005.
45. Greenwood PG, Garry K, Hunter A, *Biological Bulletin* **206**(2):113–120, 2004.
46. Cimino G, Fontana A, Gavagnin M, *Current Organic Chemistry* **3**(4):327–372, 1999.
47. Rogers SD, Paul VJ, *Marine Ecology-progress Series* 77(2–3):221–232, 1991.
48. Mcclintock JB, Baker BJ, Slattery M, *Journal of Chemical Ecology* **20**(12):3361–3372, 1994.
49. Grisley MS, Boyle PR, Key LN, *Journal of Experimental Marine Biology and Ecology* **202**(2):225–237, 1996.
50. Smith AB, Pisani D, Mackenzie-Dodds JA, *Molecular Biology and Evolution* **23**(10):1832–1851, 2006.
51. Stauber M, Markel K, *Zoomorphology* **108**(3):137–148, 1988.
52. Nakagawa H, Yamaguchi C, Hayashi H, *Journal of Natural Toxins* **6**(2):193–202, 1997.
53. Nakagawa H, Tanigawa T, Tomita K, *Journal of Toxicology-toxin Reviews* **22**(4):633–649, 2003.
54. Shiomi K, Midorikawa S, Ishida M, *Toxicon* **44**(5):499–506, 2004.
55. Komori T, *Toxicon* **35**(10):1537–1548, 1997.
56. Nakagawa H, Kimura A, *Folia Pharmacologica Japonica* **75**(7):P194–P194, 1979.
57. Stonik VA, Kalinin VI, Avilov SA, *Journal of Natural Toxins* **8**(2):235–248, 1999.
58. Iyengar EV, Harvell CD, *Journal of Experimental Marine Biology and Ecology* **264**(2):171–188, 2001.
59. Habermeh G, Volkwein G, *Toxicon* **4**:319, 1971.
60. Hamel JF, Mercier A, *Marine and Freshwater Behaviour and Physiology* **33**(2):115–139, 2000.
61. Yasumoto T, Murata M, Marine toxins, *Chem Rev* **93**:1897–1909, 1993.

62. Yasumoto T, Chemistry, etiology, and food chain dynamics of marine toxins, *Proc Japan Acad* **81**:Ser B 81, 43–50, 2005.
63. Mosher HS, Fuhrman FA, in Ragelis EP (ed.), ACS Symposium Series No 262 Seafood Toxins, *Am Chem Soc* **28**:333–344, 1984.
64. Walker MJA, Masuda VL, in Hall S, Strichartz G (eds.), ACS Symposium Series No 418 Marine Toxins, *Am Chem Soc* **24**:313–332, 1990.
65. Miyazawa K, Noguchi T, Distribution and origin of Tetrodotoxin, *J Toxicol Rev* **20**:11–33, 2001.
66. Daly JW, *J Nat Prod* **67**:1211–1215, 2004.
67. Woodward RB, Gougoutas JZ, Structure of tetrodotoxin, *J Am Chem Soc* **86**:5030–5033, 1964.
68. Ohyabu N, Nishikawa T, Isobe M, First asymmetric total synthesis of tetrodotoxin, *J Am Chem Soc* **125**:8798–8805, 2003.
69. Kishi Y, Synthesis of tetrodotoxin and its derivatives, *Farumashia* **8**:284–285, 1972.
70. Hinman A, Du Bois J, A Stereoselective Synthesis of (-)-Tetrodotoxin, *J Am Chem Soc* **125**:11510–11511, 2003.
71. Sato K, Akai S, Sugita N, Ohsawa T, Kogure T, Shoji H, Yoshimura J, Novel and stereo controlled synthesis of (□)-tetrodotoxin from myo-inositol, *J Org Chem* **70**:7496–7504, 2005.
72. Watters MR, Organic neurotoxins in seafoods, *Clinical Neurology and Neurosurgery* **97**:119–124, 1995.
73. Sevcik C, Noriega G, D'Suze G, Identification of enterobacter bacteria as saxitoxin producers in cattle's rumen and surface water from Venezuelan Savannahs, *Toxicon* **42**:359–366, 2003.
74. Bordner J, Thiessen WE, Bates HA, Rapoport H, The structure of a crystalline derivative of saxitoxin. The structure of saxitoxin, *J Am Chem Soc* **97**:6008–6012, 1975.
75. Schantz EJ, Ghazarossian VE, Schnoes HK, Strong FM, Springer JP, Pezzanite JO, Clardy J, Structure of saxitoxin, *J Am Chem Soc* **97**:1238–1239, 1975.
76. Tanino H, Nakata T, Kaneko T, Kishi Y, A stereospecific total synthesis of *d,l*-saxitoxin, *J Am Chem Soc* **99**:2818, 1977.
77. Fleming James J, Du Bois J, A Synthesis of (+)-Saxitoxin, *J Am Chem Soc* **128**:3926–3927, 2006.
78. Lewis RJ, The changing face of ciguatera, *Toxicon* **39**:97–106, 2001.
79. Murata M, Legrand AM, Ishibashi Y, Yasumoto T, Structures of ciguatoxin and its congener, *J Am Chem Soc* **111**:8929–8931, 1989.
80. Satake M, Murata M, Yasumoto T, The structure of CTX3C, a ciguatoxin congener isolated from cultured *Gambierdiscus toxicus*, *Tetrahedron Lett* **34**:1975–1978, 1993.

81. Hirama M, Oishi T, Uehara H, Inoue M, Maruyama M, Oguri H, Satake M, Total synthesis of ciguatoxin CTX3C, *Science* **294**(5548):1904–1907, 2001.
82. Anger T, Madge DJ, Mulla M, Riddall D, Medicinal Chemistry of Neuronal Voltage-Gated Sodium Channel Blockers, *J Med Chem* **44**:115–137, 2001.
83. Murata M, Naoki H, Iwashita T, Matsunga S, Sasaki M, Yokoyama A, Yasumoto T, Structure of maitotoxin, *J Am Chem Soc* **115**:2060–2062, 1993.
84. Satake M, Ishida S, Yasumoto T, Murata M, Utsumi H, Hinomoto T, Structural confirmation of maitotoxin based on complete ^{13}C NMR assignments and the three-dimensional PFG NOESY-HMQC spectrum, *J Am Chem Soc* **117**:7019–7020, 1995.
85. Kishi Y, Synthesis of maitotoxin, *Pure and Appl Chem* **70**:339–344, 1998.
86. Tachibana K, Scheuer PJ, Tsukitani Y, Kikuchi H, Van Engen D, Clardy J, Gopichand Y, Schmitz FJ, Okadaic acid a cytotoxic polyether from two marine sponges of the genus *Halichondria*, *J Am Chem Soc* **103**:2469–2471, 1981.
87. Urbanek RA, Sabes SF, Forsyth CJ, Efficient synthesis of okadaic acid 1. Convergent assembly of the C15-C38 domain, *J Am Chem Soc* **120**:2523–2533, 1998.
88. Britton R, Roberge M, Brown C, Van Soest R, Andersen RJ, New okadaic analogues from the marine sponge *Merriamum oxeato* and their effect on mitosis, *J Nat Prod*, **66**:838–8433, 2003.
89. Burja AM, Bernard B, Abou-Mansour E, Burgess JG, Wright PC, Marine cyanobacteia — a prolific source of natural products, *Tetrahedron* **57**:9347–9377, 2001.
90. Borner T, New bioactive compounds. Toxins of cyanobacteria, *Biologie in Unserer Zeit* **31**:108–115, 2001.
91. Carmichael WW, Mahmood NA, Edward GH, Natural toxins from cyanobacteria (blue-green algae), in Marine toxins, *ACS symposium series 418*, Halls S, strihartz G (eds.), *Am Chem Soc* Washington DC 1990, 1987.
92. Carmichael WW, Status report on planktonic cyanobacteria (blue-green algae) and their toxins, *Report* 149, 1992.
93. Arif JM, Kunhi M, Siddiqui YM, El Sayed KA, Orabi KY, Al-Hazzani A, Al-Ahdal MN, Al-Khodairy FM, Role of intermediary biomarkers in determining the anticancer efficacy of marine compounds, *Medicinal Chemistry Research* **13**:553–562, 2004.
94. Han B, McPhail KL, Gross H, Goeger D, Mooberry SL, Gerwick WH, Isolation and structure of five lyngbyabellin derivatives from a Papua New Guinea collection of the marine cyanobacterium *Lyngbya majuscule, Tetrahedron* **61**:11723–11729, 2005.
95. Wipf P, Reeves JT, Day BW, Chemistry and biology of curacin A, *Current Pharmaceutical Design* **10**:1417–1437, 2004.

96. Muir JC, Pattenden G, Ye T, Total synthesis of (+)-curacin A, a novel antimitotic metabolite from a cyanobacterium, *J Chem Soc* 2243–2250, 2002.
97. Carmichael WW, Mahmood NA, Hyde EG, in Hall S. Strichartz G (eds.), ACS Symposium Series No 418, Marine Toxins, *Am Chem Soc* **6**:87–106, 1990.
98. Moore RE, Bartolini G, Structure of palytoxin. *J Am Chem Soc* **103**:2491–2494, 1981.
99. Kishi Y, Palytoxin: An inexhaustible source of inspiration-personal perspective, *Tetrahedron* **58**:6239–6258, 2002.
100. Li S, Wattenberg EV, Differential activation of mitogen-activated protein kinases by palytoxin and ouabain, two ligands for the Na^+, K^+ -ATPase, *Toxicology and Applied Pharmacology* **151**:377–384, 1998.
101. Terlau H, Olivera BM, Conus venoms: A rich source of novel ion channel-targeted peptides, *Phisiol Rev* **84**:41–68, 2004.
102. Grant MA, Morelli XJ, Rigby AC, Conotoxins and structural biology: A prospective paradigm for drug discovery, *Curr Protein and Peptide Sci* **5**:235–248, 2004.
103. Armishaw CJ, Alewood PF, Conotoxins as research and drug leads, *Curr Protein and Peptide Sci* **6**:221–224, 2005.
104. Gary S, A toxin against pain, *Sci Am* **292**(4):88–93, 2005.
105. Newman DJ, Gordon CM, Marine natural products and related compounds in clinical and advanced preclinical trials, *J Nat Prod* **67**:1216–1238, 2004.
106. Spector I, Shochet N, Kashman Y, Groweiss A, Latrunculins: Novel marine toxins that disrupt microfilament organization in cultured cells, *Science* **219**(4584):493–495, 1983.
107. Ne'eman I, Fishelson L, Kashman Y, Sarcophine. New toxin from the soft coral *Sarcophyton glaucum* (Alcyonaria) *Toxicon* **12**:593–598, 1974.
108. Bernstein J, Shmeuli U, Zadock E, Kashman Y, Neeman I, Sarcophine, a new epoxy cembranolide from marine origin, *Tetrahedron* **30**:2817–2824, 1974.
109. Chang-Yih D, Rei-Sheu H, Cytotoxic cembranoids from the soft corals Sinularia gibberosa and *Sarcophyton trocheliophorum, J Nat Prod* **59**:595–598, 1996.
110. Fahmy H, Khalifa S, Konoshima T, Zjawiony JK, Potent cancer chemopreventive activity, *Marine Drugs* **2**:1–7, 2002.
111. Ne'eman I, Fishelson L, Kashman Y, *Marine Biology* **30**:293–296, 1975.
112. Okuda RK, Scheuer PJ, Latrunculin-A, ichthyotoxic constituent of the nudibranch Chromodoris elisabethina, *Experientia* **41**:1355–1356, 1985.

Chapter 6

ENANTIOMERIC DISTRIBUTION OF ODOROUS OXYGENATED MONOTERPENES IN AROMATIC PLANTS

Uzi Ravid

6.1 INTRODUCTION

For millions of years, nature has made use of enantiomerically pure molecules in living organisms. In recent decades, there has been an increasing interest in selectively building chiral molecules. Naves[1] concluded that enantiomers can differ in odor quality as well as in odor strength. According to the stereochemical theory of Amoore,[2,3] an odorous molecule must possess a stereostructure complementary to the sites of the receptors. Two enantiomers should be different in their odors because they are not superimposable.[4] The mechanisms of recognition of an odorant by one or several receptors have not been fully elucidated until the end of the 20th century.[5] The 2004 Nobel laureates, Dr Linda Buck and Dr Richard Axel, became the first to identify a family of genes that control the olfactory system, a complex network that governs our sense of smell. The genes are blueprints for a family of smell-receptor proteins in the nose that work in different combinations so that the brain can identify a nearly infinite array of odors.[6,7]

Although the role of chirality in odor perception is still a rather new area of interest, it was noted that more than 285 enantiomeric pairs are known to exhibit either differing odors or odor intensities.[8–10]

Interesting examples of single stereoisomers of natural and synthetic odorants, prepared via bioorganic routes, support the statement that our sense of smell is enantioselective.[11] Nowadays, we have enough good quality data to know that sometimes absolute configuration affects odor perception.[12]

Recently, "chiral economic" techniques have been developed which allow the complete transformation of a starting material into the desired enantiomer. According to a study by the market research firm of Frost and Sullivan,[13] worldwide revenues due to chiral technology, which amounted to US$4.8 billion in 1999, will have reached more than triple the sum by 2009 — US$14.9 billion. One valuable approach is using the "chiral pool" as a large reservoir of optically pure building blocks, mainly derived from natural sources.[14–16]

These chiral building blocks are incorporated into the target molecules in such a way that the configuration at the stereo centers remains unchanged. Since the relative configuration of newly produced centers of chirality can be controlled, virtually any enantiomerically pure product can be built around the chiral starting molecule. In the case of pheromones, chirality has a similar influence on their biological activity: while one enantiomer attracts the insect species, the other may act as a repellent.[17]

There is currently great interest in the flavor and fragrance industry in developing natural chiral chemicals from plant materials. The interest in natural chiral chemicals continues to grow for several reasons:

(1) It is tempting to utilize natural substances presumed to be safe, since they occur in foods which have been used for hundreds of years. Thus, one could avoid the problems of proving the safety of synthetic compounds.
(2) Many industrial "nature identical" compounds are racemates, or they are not optically pure. In many cases, this leads to differences in the organoleptic properties and biological activity of the natural and "nature-identical" compounds.
(3) Optically pure natural compounds are the building blocks of optically pure synthetic flavors, fragrances, pharmaceuticals, pheromones and others.[14,18,19]

Natural chiral essential oil components generally have a characteristic enantiomeric distribution that is attributable to stereoselectivity — controlled biogenetic formation mechanisms. An excess of one enantiomer or the other occurs and can be detected, in a variety of essential oils and oleoresins.[20] The authentic enantiomeric ratio of some essential oil components can be modified by the addition of synthetic racemic or natural ingredients (adulteration).[21–25]

The use of enantiomers as additional markers for taxonomic characterization of aromatic plants may be very helpful. The differences in the enantiomeric distribution of *trans*- and *cis*- sabinene hydrates in two *Origanum* species enable the species to be distinguished in spite of their similar essential oil compositions.[26] A further study on the natural variation of the enantiomeric composition within a wild population may be carried out in order to examine the stability in a taxon (including the possible presence of chemotypes). The enantiomeric ratio of essential oil components is a reliable parameter for assessing quality because it may be indicative of adulteration, contamination, aging, shelf life, technological process and the botanical source of a specific chiral compound.[24]

Most of the herbs which contain chiral volatile components are collected from the wild or grown as primitive crops. They have not been subjected to systematic breeding efforts or to modern agronomic development. Consequently, yields are still relatively low, cultivation procedures are old fashioned, and the quality of the essential oil is frequently variable.

6.2 OLFACTORIAL PROPERTIES OF CHIRAL OXYGENATED MONOTERPENES

Enantiomeric compounds differ from one another only in two aspects: The chiroptical characteristics (optical rotation) and the speed of their reaction with chiral molecules. The human olfactory organ is also capable of distinguishing chiral molecules. The odor quality and potency of enantiomeric compounds may show considerable differences. Thus, distinct differentiation in odor perception could be observed in the pairs of enantiomeric oxygenated monoterpenoid odorants.[8,10,27–34] However, the enantiomers of camphor cannot be distinguished at all.[35] The intensity of the odor of enantiomeric compounds may exhibit as great a variability as their quality does. For example, the odor threshold value of (*R*) (−) – carvone (0.043 ppm) is about ten times lower than that of its (*S*) (+) –enantiomer (0.60 ppm).[36]

Emberger and Hopp[37] and Werkhoff and Hopp[38] reported that there are significant sensory differences between the eight menthol and the four menthone enantiomers. (*S*)(-)-7-Hydroxy-6, 7-dihydrocitronellal has a lily-of-the-valley odor, while the odor of its enantiomer is weaker and has green, leaf like and minty notes.[39] The enantiomers of *cis*- and *trans*-rose oxide have closely similar odors with slight but detectable quality differences.[32]

6.3 CHIRAL- GC DETERMINATION OF ENANTIOMERIC COMPOSITION OF OXYGENATED MONOTERPENES

The classical procedure for determining enantiomeric composition is polarimetry. The basic requirements are a relatively large sample, a relatively high specific rotation and a knowledge of the specific rotation of the pure enantiomer under similar conditions.[17] High performance methods of enantiomer analysis and chromatographic and electroseparation techniques for chiral separation, have become an urgent necessity in order to keep pace with the development of highly enantioselective reactions and to cope with the requirements of government laboratories. The guidelines for the development of flavoring agents should be issued by regulating authorities to require efficient methods for enantiomeric purity control, as in the legislative control of new drugs.[40–42]

The methods for determination of enantiomeric composition are:

(1) Chiroptic analysis: polarimetry; circular dichroism; optical rotatory dispersion.
(2) X-ray crystallography of a diastereomeric derivative.
(3) Nuclear magnetic resonance (^{1}H and ^{13}C): chiral shift reagent; chiral derivatizing agent; chiral solvating agent.

(4) Achiral (separation of diastereomeric derivatives) and chiral (separation of enantiomers) chromatography: gas chromatography; high-performance liquid chromatography; super- and sub-critical fluid chromatography; thin-layer chromatography.
(5) Electromigration techniques: capillary electrophoresis; capillary electrochromatography.

One of the most sophisticated methods is the use of chiral gas chromatographic capillary columns for the direct separation of volatile enantiomers.[43] Complexation gas chromatography with enantioselective transition metal β-ketoenolates permits the stereochemical analysis of volatile oxygenated compounds in the nanogram range with high accuracy.[44,45]

Two other direct methods of GC analysis on chiral stationary phases of enantiomeric mixtures are:

(1) Enantiomer separation on optically active amino acid, dipeptide, diamide and amide phases by association via hydrogen bonding.[43,46]
(2) Enantiomer separation on modified macrocyclic cyclodextrin phases via inclusion into chiral cavities.[47,48]

Capillary gas chromatography on optically active modified cyclodextrin phases is a simple, fast, accurate and highly sensitive method for the enantiomeric analysis of chiral volatile compounds.

Cycledextrins are cyclic oligosaccharaides derived from six (α-cyclodextrin), seven (β-cyclodextrin) or eight (γ-cyclodextrin) $\alpha - 1 - > 4$ bonded glucose units. Cyclodextrins can be converted into tailor-made GC phases by total or partial alkylation or arylation of their numerous hydroxy groups. In addition to inclusion effects, the separation mechanism of modified cyclodextrins involves dipole-dipole and hydrogen bonding interactions. Separation can be achieved if the volatile enantiomers pack the cavity of the macrocyclic cyclodextrin molecule to different degrees.

Nowadays, derivatized cyclodextrins are the most common chiral selector in the direct GC enantiomer separation of flavors, fragrances, essential oils, pheromones and other natural, nature identical and synthetic volatile compounds.[20,22,23,47,49–57]

A lot of published data on the separation of enantiomers of flavors and fragrances by GC is reviewed by Chirbase/Flavor[58] database.[59] Table 1. summarizes the enantiomer separation of oxygenated monoterpenes on chiral stationary phases of cyclodextrin derivatives by high resolution gas chromatography.

Table 1. Enantiomer Separation of Oxygenated Monoterpenes on Chiral Stationary Phases of Cyclodextrin Derivatives by High Resolution Gas Chromatography

Monoterpene	Chiral phase	Order of elution	References
Piperitone	I	(4*R*)(−), (4*S*)(+)	König[47] Ravid *et al.*[73]
	IV	(4*R*)(−), (4*S*)(+)	König[47]
	V	(4*R*)(−), (4*S*)(+)	Larkov *et al.*[75]
Pulegone	II	(1*R*)(+), (1*S*)(−)	Köpke and Mosandl[80]; Ravid *et al.*[78]
	VIII	(1*R*)(+)(1*S*)(−)	Köpke *et al.*[81]
Menthone	I	(1*S*, 4*R*)(+), (1*R*, 4*S*)(−)	König[47] Ravid *et al.*[90]
	IV	(1*S*, 4*R*)(+), (1*R*, 4*S*)(−)	König[47]
	VI	(1*R*, 4*S*)(−), (1*S*, 4*R*)(+)	Kreis and Mosandl[87]
Isomenthone	I	(1*R*, 4*R*)(+), (1*S*, 4*S*)(−)	König[47]; Ravid *et al.*[90]
	IV	(1*R*, 4*R*)(+), (1*S*, 4*S*)(−)	König[47]
	VI	(1*R*, 4*R*)(+), (1*S*, 4*S*)(−)	Kreis and Mosandl[87]
Verbenone	I	(1*R*)(+), (1*S*)(−)	König[47]; Ravid *et al.*[97]
	II	(1*S*)(−), (1*R*)(+)	Schmalzing *et al.*[60]; Ravid *et al.*[97]
Carvone	I	(4*S*)(+), (4*R*)(−)	Ravid *et al.*[104]
	II	—	Bicchi *et al.*[61]
	III	(4*R*)(−), (4*S*)(+)	König *et al.*[74] Blank and Grosch[107]; König[47]
	IV	(4*S*)(+), (4*R*)(−)	König *et al.*[110]
	IX	—	Bicchi *et al.*[62]

Table 1. (Continued)

Monoterpene	Chiral phase	Order of elution	References
Fenchone	I	(1*R*)(−), (1*S*)(+)	König *et al.*[74]; König[47]; Ravid *et al.*[121]
	II	(1*R*)(−), (1*S*)(+)	Spilkin, 1991[63]
Camphor	I	(1*R*, 4*R*)(+), (1*S*, 4*S*)(−)	König[47]; Ravid *et al.*[122]
	VI	(1*S*, 4*S*)(−), (1*R*, 4*R*)(+)	Kreis *et al.*[123]; Mosandl[54]
	X	(1*S*, 4*S*)(−), (1*R*, 4*R*)(+)	Tateo *et al.*[124]
Piperitone oxide	V	(1*R*, 2*R*, 4*R*)(+)-*trans*, (1*R*, 2*R*, 4*S*)(+)-*cis*, (1*S*, 2*S*, 4*R*)(−)-*cis*, (1*S*, 2*S*, 4*S*)(−)-*trans*	Larkov *et al.*[75]
cis- and *trans*- Sabinene hydrate	II	(1*R*, 4*S*, 5*S*)-*trans*, (1*S*, 4*R*, 5*R*)-*trans*, (1*R*, 4*R*, 5*S*)-*cis* (1*S*, 4*S*, 5*R*)-*cis*	Cornwell *et al.*[153]; Marriott *et al.*[52] Larkov *et al.*[26]
	V	(1*R*, 4*S*, 5*S*)-*trans*, (1*S*, 4*R*, 5*R*)-*trans*, (1*R*, 4*R*, 5*S*)-*cis*, (1*S*, 4*S*, 5*R*)-*cis*	Larkov *et al.*[26]
cis- and *trans*- Sabinene hydrate acetate	II	(1*R*, 4*S*, 5*S*)-*trans*, (1*S*, 4*R*, 5*R*)-*trans*, (1*R*, 4*R*, 5*S*)-*cis*, (1*S*, 4*S*, 5*R*)-*cis*	Larkov *et al.*[26]
	V	(1*R*, 4*S*, 5*S*)-*trans*, (1*S*, 4*R*, 5*R*)-*trans*, (1*R*, 4*R*, 5*S*)-*cis*, (1*S*, 4*S*, 5*R*)-*cis*	Larkov *et al.*[26]
Linalool	I	(3*S*)(+), (3*R*)(−)	König and Hochmuth[23]
	II	(3*R*)(−), (3*S*)(+)	Werkhoff *et al.*[57]; Ravid *et al.*[160]
	V + VI	(3*R*)(−), (3*S*)(+)	Bayer and Mosandl[183]

Table 1. (Continued)

Monoterpene	Chiral phase	Order of elution	References
	VI	(3*R*)(−), (3*S*)(+)	Kreis and Mosandl[87]; Mosandl[54]
	VII	(3*R*)(−), (3*S*)(+)	Tamogami *et al.*[163]
	IX	(3*R*)(−), (3*S*)(+)	König et al.[120]; Bicchi *et al.*[62]
	X	(3*R*)(−), (3*S*)(+)	Dugo *et al.*[157]
	XI	(3*R*)(−), (3*S*)(+)	Bernreuther and Schreier[154]
Linalyl acetate	I	(3*S*)(+), (3*R*)(−)	Weinreich and Nitz[172]; König and Hochmuth[23]; Ravid *et al.*[171]
	V + VI	(3*R*)(−), (3*S*)(+)	Bayer and Mosandl[183]
	VI	(3*R*)(−), (3*S*)(+)	Mosandl[54]
	VIII	(3*R*)(−), (3*S*)(+)	Maas *et al.*[170]
	X	(3*R*)(−), (3*S*)(+)	Dugo *et al.*[157]
Citronellol	VI	(3*S*)(−), (3*R*)(+)	Kreis and Mosandl[87]; Mosandl[54]
Citronellol - trifluoroacetyl derivative (TFA)	I	(3*R*)(+), (3*S*)(−)	König[47]; Ravid *et al.*[77]
Borneol	II	(1*S*, 2*R*, 4*S*)(−), (1*R*, 2*S*, 4*R*)(+)	Kreis *et al.*[140] Ravid *et al.*[178]
	V	(1*S*, 2*R*, 4*S*)(−), (1*R*, 2*S*, 4*R*)(+)	Kreis *et al.*[123]
	IX	—	Bicchi *et al.*[62]
Borneol - TFA	I	(1*S*, 2*R*, 4*S*)(−), (1*R*, 2*S*, 4*R*)(+)	König et al.[64]
α-Terpineol	II	(4*S*)(−), (4*R*)(+)	Weinreich and Nitz[172]; Leach *et al.*[181]; Werkhoff *et al.*[20]; Ravid *et al.*[180]
	V + VI	(4*S*)(−), (4*R*)(+)	Bayer and Mosandl[183]
	VI	(4*S*)(−), (4*R*)(+)	Kreis and Mosandl[87]

Table 1. (Continued)

Monoterpene	Chiral Phase	Order of elution	References
	IX	(4*S*)(−), (4*R*)(+)	Bicchi *et al.*[61,65]; König *et al.*[120]
	X	(4*S*)(−), (4*R*)(+)	Dugo *et al.*[157] Shellie *et al.*[182]
	XII	(4*S*)(−), (4*R*)(+)	Askari *et al.*[6666]
Terpinen-4-ol	II	(4*S*)(+), (4*R*)(−)	Ravid *et al.*[184]; Leach *et al.*[181]
	V + VI	(4*R*)(−), (4*S*)(+)	Bayer and Mosandl[183]
	VI	(4*R*)(−), (4*S*)(+)	Kreis *et al.*[123]; Mosandl[54]
	IX	(4*S*)(+), (4*R*)(−)	König *et al.*[120]; Bicchi *et al.*[62]
	X	(4*S*)(+), (4*R*)(−)	Dugo *et al.*[157]; Shellie *et al.*[182]

Legend of chiral phases:
I = Octakis (3-O-butyryl-2,6-di-O-pentyl)-γ-cyclodextrin
II = Heptakis (2,3,6-tri-O-methyl)-β-cyclodextrin
III = Hexakis (2,3,6-tri-O-pentyl or hexyl)-α-cyclodextrin
IV = Octakis (6-O-methyl-2,3-di-O-pentyl)-γ-cyclodextrin
V = Heptakis (2,3-di-O-methyl-6-O-tert-butyldimethylsilyl)-β-cyclodextrin
VI = Heptakis (2,3-di-O-acetyl-6-O-tert- butyldimethylsilyl)-β-cyclodextrin
VII = Heptakis (2,6-di-O-methyl-3-O-pentyl)-β-cyclodextrin + Octakis (2,6-di-O-methyl-3-O-trifluoroacetyl)-γ-cyclodextrin
VIII = Octakis (2,3-di-O-acetyl-6-O-tert-butyldimethylsilyl)-γ-cyclodextrin
IX = Heptakis (2,6-di-O-methyl-3-O-pentyl)-β-cyclodextrin
X = Heptakis (6-O-tert-butyldimethylsilyl-2,3-di-O-ethyl)-β-cyclodextrin
XI = Heptakis (2,3,6-tri-O-pentyl)-β-cyclodextrin
XII = Heptakis (2,3,6-tri-O-ethyl)-β-cyclodextrin
Table 1 includes many, but not all, chiral stationary phases used to separate the oxygenated monoterpenes.

6.4 SELECTED CHIRAL OXYGENATED MONOTERPENES

6.4.1 Piperitone

Piperitone, with its fresh minty camphor-like odor, is present in various leaf essential oils. Among *Mentha* and *Eucalyptus* genera, there are piperitone-rich species which are considered common sources of this natural monoterpene ketone.[67] The (4*S*)(+)- enantiomer was reported to be present in the essential oils of *Mentha arvensis* var. *piperascens*, *M. silvestris*, *Cymbopogon sennaavensis* and *Andropogon jwarancusa*.[68–70] (4*R*)(−)-Piperitone is present in approximately 30 varieties of *Eucalyptus* species such as *E. dives* and *E. radiata*, and also in *Orthodon perforatum*.[68,69,71] The racemate has been reported in the oils of *A. jwarancusa*, *M. pulegium* var. *hirsuta* and *E. dives*.[70,71] Piperitone is used for scenting flavoring preparations, and has been used as a starting material in the synthesis of menthone and menthol. Piperitenone is the precursor of both piperitone enantiomers in the biosynthetic reduction of monoterpenes in *Mentha*.[72] High enantiomeric purities of (4*R*)(−) piperitone were detected in *M. longifolia* (Negev type) oil.[73] The (4*S*)(+) enantiomer was detected in peppermint oils[74] and in *M. longifolia* (Yizre'el Valley) oils in a high enantiomeric ratio. Relatively high enantiomeric purities of (4*S*)(+) piperitone were detected in *M. arvensis* and in *Artemisia judaica* oils.[73] Highly variable enantiomeric compositions have been observed within *Micromeria fruticosa* populations. In *M. longifolia*, the difference was observed between populations.[75]

6.4.2 Pulegone

(1*R*)(+)-Pulegone, with an odor reminiscent of peppermint and camphor, is the most abundant natural enantiomer.[68] It is an important "chiral pool" compound and has been used as a chiral starting material for many chiral auxiliaries in the enantioselective synthesis of natural products.[76] (+)-*cis*-Isopulegone is the only precursor of (1*R*)(+)-pulegone in the biosynthesis of monoterpenes in the *Mentha* genus.[72,77] (1*R*)(+)- pulegone is a major component in the oils of *Mentha pulegium*,[69] *M. rotundifolia*[71] and *M. longifolia*.[70] It has also been detected in the oils of *M. arvensis* var. *piperascens*, *Pycanthemum lanceolatum*, *Calamintha nepeta* and others.[69] High enantiomeric purity of (1*R*)(+)-pulegone was detected in the essential oils from fresh leaves of *M. piperita*, *M. longifolia*, *M. pulegium*, *M. sylvestris*, *Calamintha incāna* and *Micromeria fruticosa*.[78]

The (1*S*)(−)- enantiomer is a major constituent of the oil of *Agastache formosana*,[71] and has been reported in the oils of Buchu leaf (*Agathosma* *betulina*), *A. crenulata*,[79,80] Israeli orange[68] and black currant (*Ribes nigrum*) flavor concentrate.[81] (1*S*)-Configured pulegone, menthone and isomenthone may serve as appropriate indicators in the authenticity control of buchu leaf oil as well as "cassis" type fruit aromas and fragrances.[82]

Toxicological observations indicate that pulegone is an insect neurotoxin,[83] as well as an abortifacient in women and in animals.[84] (1*S*)-(−)-Pulegone is far less toxic in mice than (1*R*)-(+) pulegone (both hepatotoxic and pneumotoxic).[85,86]

6.4.3 Menthone and Isomenthone

The two diastereomeric forms of menthone and isomenthone occur in many essential oils as a single enantiomer species. Menthone, possessing an odor reminiscent of peppermint but harsher,[69] and isomenthone, with its slightly musty odor,[79] are significant distinguishing marks in quality control of *Mentha* and *Pelargonium* oils.[87] The two diastereomers are easily chemically interconvertible; the isomerization of (1*R*, 4*S*)(−)-menthone results in (1*R*, 4*R*)(+)- isomenthone as the two materials are structurally related despite their opposing signs of rotation. (1*S*, 4*R*)(+)-Menthone is similarly converted into (1*S*, 4*S*)(−)-isomenthone.[88]

Enantiomerically pure (1*R*, 4*S*)(−)-menthone and (1*R*, 4*R*)(+)-isomenthone were detected in many oils of *Mentha*, such as *M. piperita*,[37,89] *M. longifolia*, *M. pulegium*, *M. sylvestris*, *M. requienii*, *Micromeria fruticosa* and *Calamintha incãna*.[90] (1*R*, 4*S*)(−)-Menthone is also present in essential oils of *M. arvensis*, *M. gattefosei*, *M. timija*, *Calamintha nepeta*, *Andropogon fragrans*, *Pinus palturis* and *Pelargonium odoratissimum*. (1*R*, 4*R*)(+)-Isomenthone occurs in the oils of *Micromeria abissinica* and *Pelargonium tomentosum*.[68–71,91]

In freshly distilled geranium oils, (1*S*, 4*R*)(+)-menthone and (1*S*, 4*S*)(−)-isomenthone were detected in high enantiomeric purities (>99%).[90] (1*S*, 4*R*)(+)-Menthone is a constituent of the oils of buchu leaf,[82] *Micromeria biflora*, *Nepeta japonica* and *Barosma pulchella*. (1*S*, 4*S*)(−)-Isomenthone has been identified in Réunion and Algerian geranium oils, *Pelargonium capitatum* and in buchu leaf oil.[68–71,82,91]

$(1R)(+)$-Pulegone is the precursor of $(1R,4S)(-)$-menthone and $(1R,4R)(+)$-isomenthone in the biosynthesis of monoterpenes in the *Mentha* genus.[92] (−)-Menthone is reduced to the epimeric alcohols (−)-menthol and (+)-neomenthol,[93] while (+)-isomenthone is reduced to (+)-isomenthol and (+)-neoisomenthol.[94] Both enantiomers of labeled pulegone were converted into the corresponding labeled (−)-menthone and (+)-isomenthone by *Mentha piperita*, indicating an unspecific reduction process.[95] Genuine mint and peppermint oils contain enantiopure $(1R)$-configured monoterpenoids.[54]

6.4.4 Verbenone

Verbenone possesses an odor reminiscent of camphor, menthol and celery[69]; $(1R)(+)$-verbenone is recognized as a character impact constituent in rosemary oil.[96] Very high enantiomeric purity of $(1R)(+)$-verbenone was detected in rosemary oil from various origins or cultivars.[97] $(1R)(+)$-Verbenone also occurs in Spanish verbena (*Verbena triphylla*), Spanish *Eucalyptus globulus* and Spanish rosemary oils.[69,96] Natural $(1S)(-)$-verbenone has not yet been isolated from plant material. However, it has been identified in old oxidized laevorotatory turpentine oils or in other oils containing pinene.[69] Enantiomeric ratios of verbenone have been reported as important factors in inhibiting the attraction responses of the male southern pine beetle (*Dendroctonus frontalis*).[98] It was found that in another bark beetle , *Ips paraconfusus*, (−)-α−pinene serves as the precursor of (+)-*cis*-verbenol and (−)-verbenone, whereas (+)-α−pinene leads to (+)-*trans*-verbenol and (+)-verbenone.[99] Enantiomeric blends of verbenone significantly reduced the percentage of females of *Ips paraconfusus* reaching the attractant source. Verbenone has been actively pursued as a potential candidate for the protection of individual trees and forest stands because it decreases the response to pheromone in many bark beetle species.[100]

Biotransformation of (+)- and (−)-*cis/trans*-verbenols to (+)-verbenone has been achieved by a cell-free system of *Cannabis sativa* Callus.[101] The (−)- and (+)-verbenone (via verbenol) were the starting materials for the first synthesis of the (−)- and (+)- enantiomers of Δ^1-tetrahydrocannabinol.[102]

6.4.5 Carvone

Carvone occurs in nature in the (4*S*)(+), (4*R*)(−) and (*RS*) forms. Optically pure (4*R*) (−)-carvone is a major component of native spearmint (*Mentha spicata*), *M. longifolia*,[47,74,103,104] and kuromoji oils.[68,70] (4*R*)(+)-Limonene obtained from orange oil is the preferred starting material for the chemical synthesis and biotransformation of (4*R*)(−)-carvone.[105,106] It has also been found in *M. viridis* var. *crispa*, *Eucalyptus globulus* and several other mint species.[70] (4*R*)(−)-Carvone has a warm-herbal-spicy-pungent odor with a slightly sweeter tone than that of (*S*)(+)-carvone. Optically pure (4*S*)(+)-carvone has a warm-herbal-spicy-pungent impression, more dry and hay-like than (4*R*)(−)-carvone, and is the character impact and major component of caraway (*Carum carvi*) and dill (*Anethum graveolens*) seed aroma.[47,103,104,107–110] It has also been detected in *Anethum sowa*, *Lippia carviodora*, *M. arvensis* and others.[68,70] (−)-*trans*-Carveol is the biosynthetic precursor of (−)-carvone in *Mentha*.[72] (4*S*)(+)-carvone is the stating material for the synthesis of (*R*)-Z-3-methyl-6-isopropenyl-3, 9-decadien-1-yl acetate, a pheromone component of the female California red scale.[111]

(4*R*)(−)-Carvone is detectable at a lower threshold than its enantiomer.[28,112,113] (4*R*)(−)-Carvone is used as a starting material in the preparation of picrotoxinin.[114] The (*RS*) form is present in gingergrass, *Listea guatemaleusis*, lavender and *Artemisia ferganensis*.[68,70]

Humans,[115–117] mole rats[118] and squirrel monkeys[116] can distinguish between suprathreshold concentrations of the two enantiomers of carvone. The (4*R*)(−)-carvone is detected at lower concentrations than the (4*S*)(+)-carvone by humans. At equivalent suprathreshold concentrations some individuals perceive (4*S*)(+)-carvone to be more intense than its enantiomer.[119]

6.4.6 Fenchone

The two enantiomers of fenchone occur in a number of essential oils. Optically pure (1*S*)(+)-fenchone has been detected in bitter fennel oil (*Foeniculum vulgare* var. *vulgare*) and in sweet fennel oil (*F. vulgare* var. *dulce*) from various sources.[120,121] It has also been reported to exist in the oil of *Lavndula stoechas*.[69] (1*S*)(+)-Fenchone has a camphor-like odor, powerful and sweet, with a warm, somewhat bitter, burning taste.[68] Fennel oil gets its typical flavor from enantiomerically pure (1*S*)(+)-fenchone.[74]

Enantiomerically pure (1*R*)(−)-fenchone has been detected in the oils of wormwood (*Artemisia absinthum*), tansy (*Tanacetum vulgare*) and cedar leaf (*Thuja occidentalis*).[121] Fenchone has also been reported in the oils of *Thuja plicata*, *T. standishii*, Russian anise, a few *Artemisia* varieties and *Lavandula burmanuii*.[68] However, its enantiomeric composition has not been reported. Dehydrogenation of (−)-fenchol produces (+)-fenchone with minute amounts of the (−)-enantiomer.[79] Fenchone enantiomers were subjected to a series of olfactory discrimination tests on humans, squirrel monkeys and on mole rats. It was found that humans cannot discern the enantimers of fenchone.[31,119] Mole rat females were attracted to the odor of (−)-fenchone, while males responded ambiguously.[118]

6.4.7 Camphor

Camphor, with its small, quasispherical, rigid molecule, has a characteristic musty, penetrating, slightly minty odor, and a burning bitter fresh taste.[68,79] Both enantiomers of camphor have the same odor[35] and are widespread in herbs and spices. Camphor that was isolated from commercial camphor tree oil[122] and in authentic rosemary oils[123] was racemic. The main component in oil obtained from various parts of the camphor tree, *Cinnamomum camphora*, is (1*R*, 4*R*)(+)-camphor[68,69,91] (1*R*, 4*R*)(+)-camphor with high enantiomeric purity was detected in *Salvia officinalis*, *S. sclarea*, *S. glutinosa* and *Ocimum basilicum*.[122,124] The two enantiomers were detected in *Lavandula* oil.[54] The (1*R*, 4*R*)(+)-enantiomer has also been reported in the oils of sassafras, spike lavender, Reunion basil, *Ocimum canum* and Dalmatian sage, and in other Lamiaceae.[68,69]

Chrysanthemum parthenium oil [100%(1*S*, 4*S*)(−)] and the two chemotypes of *Artemisia judaica* oils contained (1*S*, 4*S*)(−)-camphor as the dominant enantiomer. The (1*S*, 4*S*)(−)-enantiomer, with high enantiomeric purity, was detected in two types of *S. officinalis*[122] and in *Coriandrum sativum*[124] (1*S*, 4*S*)(−)-Camphor has been reported in the oils of *S. grandiflora*, *S. fruticosa*, *Matricaria parthenium* and in other oils.[68,69,71] Natural camphor is prepared commercially by fractional distillation and crystallization of camphor tree oil. The two enantiomers of camphor are important as chiral starting materials in the enantiospecific syntheses of natural products with high enantiomeric excesses.[125–127] (1*S*, 4*S*)(−)-Camphor is the chiral starting material for the fungal sesquiterpenoid analogues of *cis*-sativenediol and helminthosporal,[128] (−)-laurolenic acid,[129] steroid intermediates[130] and vitamin B_{12}.[131] (1*S*, 4*S*)(−)-Camphor and its enzyme precursor (−)-bornyl pyrophosphate cyclase were identified in the extracts of *Tanacetum vulgare*.[132] (1*R*, 4*R*)(+)-Camphor is a starting material for the synthesis of (−)-oestrene,[133] carotenoid and terpenoid intermediates.[134–136]

6.4.8 *cis*- and *trans*-Piperitone Oxide

The two diastereomers of *cis*- and *trans*-piperitone oxide are 1,2-epoxy-3-keto-p-menthanes, which have been found in various plant species. The most abundant natural sources of these compounds are plants belonging to the Lamiaceae family, especially *Mentha* and *Calamintha*.[75] In many publications, the authors gave no details on the diastereomeric composition of piperitone oxides, probably because of their very similar retention times on a non-polar column.[137]

(1*S*, 2*S*, 4*S*)(−)-*trans*-Piperitone oxide (dihydro rotundifolone) has been isolated for the first time from *Mentha longifolia* by Reitsema and Varnis.[138] The presence of the (1*S*, 2*S*, 4*R*)(−)-*cis*-isomer was reported in a *M. spicata* mutant.[139] Only enantiomerically pure levorotatory piperitone oxides, (1*S*, 2*S*, 4*S*)-*trans* piperitone oxide and (1*S*, 2*S*, 4*R*)-*cis* piperitone oxide, were detected by chiral GC analysis of *Micromeria fruticosa* and *Mentha longifolia*.[75] The occurrence of the *cis*- and *trans*- piperitone oxides was dependent on the population of the species. In most cases, the enantiomeric composition of *trans*- and *cis*- piperitone oxides detected in various essential oils is unknown.

The antimicrobial activity of piperitone oxide rich *Mentha rotundifolia* (*M. suaveolens*) oil was investigated by Oumzil *et al.*,[191] who found it less significant than that of pulegone or piperitone rich oils. The biosynthesis of the two levorotatory diastereomers from (1*S*, 2*S*)(+)-piperitenone oxide was described in *M. spicata* mutant[139] and in *M. piperita* low menthofuran mutant.[141]

6.4.9 *trans*- and *cis*-Sabinene Hydrates and their Acetates

The bicyclic monoterpenes cis-sabinene hydrate and *cis*-sabinene hydrate acetate are known as the major and original flavor compounds of marjoram (*Origanum majorana*).[142–146] Relatively large quantities of *cis*- and *trans*-sabinene hydrate were detected in the genus *Origanum*.[26] *cis*-Sabinene hydrate acetate was found in relatively large amounts in solvent and CO_2 extracts of *O. majorana*[142,143,145–148] and as a minor component in hydro-distilled and in steam-distilled essential oils of marjoram. The enantiomers of *trans*- and *cis*- sabinene hydrate and their acetates were detected in the extracts of *Origanum ramonense*, *O. dayi*, *O. majorana*, *O. vulgare* ssp. *vulgare* and *O. syriacum* ssp.*syriacum*. An unusual enantiomeric distribution of sabinene hydrates was observed in *O. dayi* extracts.[26] (1*R*, 4*S*, 5*S*)-*trans*-Sabinene hydrate was isolated from the oil of *Mentha candicans* in very large quantities.[149] The enantiomers of *trans*-sabinene hydrate and of *cis*-sabinene hydrate were found in *Melaleuca alternifolia* (tea tree)[52,150,151] and in other *Melaleuca* ssp.[152,153] (+)-*trans*-Sabinene hydrate, which is present in real peppermint oil (~1%), is close to zero in *Mentha arvensis*.[23]

6.4.10 Linalool

The enantiomeric differentiation of linalool is useful in the quality control of essential oils and oleoresins, as it was found to provide an important indication of the authenticity of many herbs and spices. The enantiomeric composition of linalool has been determined in many essential oils, including basil, bergamot, rosemary, lavandin, lavender, balm, coriander, mace, *Pelargonium*, rose, *Cymbopogon*, lemon, mandarin, *Osmanthus*, davana, jasmine, *Lippia alba* and orange, as well as in many fruit extracts.[20,23,49,53,57,154–162] (3*S*)(+)-Linalool is a major component in essential oils such as those of coriander, palmarosa, mace, petitigrain, *Lippia alba* and sweet orange flowers. (3*R*)(−)-Linalool is a main component in the oils of *Ocimum* species, including sweet basil, and in the oils of neroli, linaloe, bergamot, lavender and others. Linalool enantiomers in

Gardenia, *Osmanthus* and other flowers were detected using a mixture of two chiral stationary phases.[163] As to their olfactory qualities, the linalool enantiomers differ considerably. (3*R*)(−)-Linalool was found to have an odor reminiscent of lavender and a woody note. The (3*S*)(+) enantiomer has a petitgrain odor with more of a citrus and fruity note, and its threshold value is higher than that of the (3*R*)(−)-enantiomer.[27] The sedative effect of inhaling linalool enantiomers on humans was studied.[164] Linalool and linalyl acetate are present exclusively as (*R*)-enantiomers in genuine bergamot oil.[22,55]

6.4.11 Linalyl Acetate

Linalyl acetate is the principal constituent of the essential oils of bergamot, clary sage, lavender and lavandin. It is also present in the essential oils of *Salvia officinalis*, petitgrain, sassafras, neroli, lemon, lime, a few *Mentha* species and others.[68,69]

Optically pure (3*R*)(−)-linalyl acetate was detected in the oils of clary sage (*Salvia sclarea*), *Salvia dominica*, lavender and lavandin using ^{1}H-NMR spectroscopy with a chiral lanthanide shift reagent, $Eu(hfc)_3$.[165] This enantiomer was also detected in the oils of lavender, lavandin and bergamot using complexation gas chromatography on $Ni(hfc)_2$,[166–168] and a capillary gas chromatography on chiral phases of β-[155,157,161,169,190] and γ-cyclodextrin derivatives.[23,170–172] The (3*R*)(−)-enantiomers of linalyl acetate and linalool have been introduced as indicators of genuineness of *Lavandula* oils.[54,173]

6.4.12 Citronellol

The odor quality and strength of the two citronellol enantiomers were found to be different.[174] The (3*S*)(−)-enantiomer (Rhodinol) possesses a much finer rose odor than the (3*R*)(+)-enantiomer.[69] (3*S*)(−)-Citronellol has a sweat, peach-like flavor, while (3*R*)(+)-citronellol has a bitter taste.[68]

(3*S*)(−)-Citronellol has been found in a number of geranium and rose oils.[70,91] (3*R*)(+)-Citronellol occurs in the oils of Ceylon and Java citronella,[70] *Cymbopogon winterianus*,[71] *Boronia citriodora*, *Eucalyptus citriodora*,[79] geranium, Spanish verbena[69] and others. Because of the characteristic rose-like odor of (3*S*)(−)-citronellol, its enantiomeric excess is important for the estimation of the odor quality of geranium oils and of synthetic rose compounds. Commercially prepared (3*S*)(−)-citronellol is obtained from geranium Reunion oil by fractional distillation[79] or by enantioselective reduction of the aldehyde group of racemic citronellal with particular yeast strains.[175,176] Adulteration of geranium and rose oils by addition of racemic or (3*R*)(+)-citronellol or natural enantiomeric mixture of citronellol can be detected easily by a lowering of the enantiomeric excess values of (3*S*)(−)-citronellol.[23,34,54,87,177]

6.4.13 Borneol

(1*R*, 2*S*, 4*R*)(+)-Borneol has a pungent camphor-like odor with a slightly sharp earthy-peppery tang.[79] High enantiomeric purities of (1*R*, 2*S*, 4*R*)(+)-borneol were detected in lavandin essence and in lavandin oil. It has also been found in the oils of rosemary, lavender, camphor, olibanum, nutmeg and in many other plants. (1*S*, 2*R*, 4*S*)(−)-Borneol with high enantiomeric purities was detected in the oils of *Coridothymus capitatus*, *Artemisia herba alba*, *Origanum vulgare*, *Ocimum canum* and feverfew.[178] It has been reported in the oils from the Pinaceae, Compositae and Graminaceae species, citronella, thuja, coriander, valerian root and others.[68–71,79]

The different enantiomeric distributions of borneol in lavender and lavandin may aid in distinguishing between the two oils or essences. The wide variation of borneol enantiomers among rosemary cultivars indicates genetic variability in the genus *Rosmarinus*.[178] The enantiomeric composition of borneol from Spanish and Tunisian oils differs significantly.[22] Mosandl[54] claimed that (1*S*, 2*R*, 4*S*)(−)-borneol of high enantiomeric purity has been detected as a reliable indicator of genuine rosemary oils. (1*R*, 2*S*, 4*R*)(+)-Borneol is the biosynthetic precursor of (1*R*, 4*R*)(+)-camphor in *Salvia officinalis*.[179]

6.4.14 α-Terpineol

(4*R*)(+)-α-Terpineol has a heavy floral, typically lilac odor. The (4*S*)(−)-enantiomer has a coniferous, tarry odor character.[27] (4*R*)(+)-α-Terpineol is the predominant enantiomer in many essential oils. In one case, that of the oil of *Micromeria fruticosa*, it was detected as a pure enantiomer.[180] High enantiomeric purities of (4*R*)(+)-α-terpineol were detected in the oils of *Origanum vulgare*, cardamom, *O. syriacum*, *Mentha citrata*, tea tree, lavender, coriander, rosemary, eucalyptus and others.[55,123,172,180,181]

Relatively high enantiomeric purities of (4*S*)(−)- α-terpineol were detected in the essential oils of cinnamon and *Laurus nobilis*.[180] In litchi flavor, the (4*S*)(−)- enantiomer was predominant, while in mango flavor, the (4*R*)(+)-enantiomer was predominant.[20]

Natural racemic mixtures of α-terpineol were detected in geranium oils[87,180] and in Morio Muscat-wine aroma.[66] α-Terpineol enantiomers, found in the Myrtaceae family and in lavender and citrus oils, were separated using a two-columned coupled system[157,182] and a mixture of two chiral phases,[183] respectively. When a homogenate of bergamot fruit was treated by steam distillation, linalool was completely racemized and partially rearranged to racemic α-terpineol.[22]

6.4.15 Terpinen-4-ol

Terpinen-4-ol occurs in the essential oils of lavender, hyssop, pine, *Origanum*, nutmeg, thyme, marjoram, tarragon, cubeb, neroli, rosemary, juniper berry and passion fruit and raspberry flavor as a mixture of enantiomers.[20,120,123,157,184] In many cases, the (4*S*)(+)-enantiomer is predominant. (4*S*)(+)-terpinen-4-ol with high enantiomeric purity was detected in lavender oils of different origin.[184] Terpinen-4-ol enantiomers were detected in several essential oils of the Myrtaceae family using a two-column coupled system.[182] The enantiomers were also detected in lavender and citrus oils by using mixed chiral stationary phases.[183]

Separation of terpinen-4-ol enantiomers performed by a chiral GC column[184] and a chiral lanthanide shift reagent $Eu(hfc)_3$,[185] showed that the enantiomeric composition of an isolated compound from sweet marjoram oil was 73% (*S*)(+) : 27%(*R*)(−). The (4*R*)(−)-enantiomer was found in the oil of *Eucalyptus dives*.[186] Terpinen-4-ol was also found in several bark beetle species[187] and is the main component in the aggregation pheromone of *Polygraphus poligraphus*.[188]

Enantiomeric purities of terpinen-4-ol are significantly different when *Melaleuca* and *Eucalyptus* oils are compared in view of authenticity assessment.[55] The absolute configuration of terpinen-4-ol was determined by Ohloff and Uhde.[189]

REFERENCES

1. Naves YR, The relationship between the stereochemistry and odorous properties of organic substances, in *Molecular Structure and Organoleptic Quality, S.C.I.* Monograph No. 1, 1957.
2. Amoore JE, *The Molecular Basis of Odour* C.C. Thomas, Illinois, USA, 1970.
3. Amoore JE, Odor theory and odor classification, in: Theimer ET (ed.), *Fragrance Chemistry, The Science of the Sense of Smell*, Academic Press, New York, USA, pp. 27–76, 1982.
4. Beets MGJ, Odor and stimulant structure, in Theimer ET (ed.), *Fragrance Chemistry, The Science of the Sense of Smell*, Academic Press, New York, USA, pp. 77–122, 1982.
5. Breer H, Raming K, Krieger J, Signal recognition and transduction in olfactory neurons, *Biochem Biophys Acta* **1224**:277–287, 1994.
6. Buck LB, The search for odorant receptors, *Cell* **116**:117–119, 2004.
7. Buck L, Axel R, A novel multigene family may encode odorant receptors – a molecular basis for odor recognition, *Cell* **65**:175–187, 1991.
8. Bernreuther A, Epperlein U, Koppenhoefer B, Enantiomers: Why they are important and how to resolve them, in Marsili, R (ed.) *Techniques for Analyzing Food Aroma*. Marcel Dekker, New York, USA, pp. 143–207, 1997.
9. Brenna E, Fuganti C, Serra S, Enantioselective perception of chiral odorants, *Tetrahedron: Assymetry* **14**:1–42, 2003.
10. Leffingwell JC, Chirality and Odour Perception, www.leffingwell.com/chirality, 2003.
11. Abate A, Brenna E, Fuganti C, Gatti FG, Serra S, Odor and (bio)diversity: Single enantiomers of chiral fragrant substances, *Chemistry and Biodiversity* **1**:1888–1898, 2004.
12. Sell CS, Scent through the looking glass, *Chemistry & Biodiversity* **1**:1899–1920, 2004.
13. Stinson SC, *Chiral Chemistry* C&EN, **79**(20):45–57, 2001.
14. Challener CA (ed.), Chiral Intermediates. Ashgate: Brookfield, Vermont, USA, 2001.
15. Fluka Chemika, *Chiral Compounds Chemistry*, Fluka Chemie AG, Switzerland, 1994.

16. Merck-Schuchardt, Chiralica — Reagents, catalysts and building blocks for enantioselective synthesis; *Resolving Agents*, E Merck Co., Germany, 1985.
17. Silverstein RM, Chiral semiochemicals, in Acree TE, Soderlund DM (eds.), *Semiochemistry: Flavors and Pheromones*, Walter de Gruyter, Berlin, Germany, pp. 121–140, 1985.
18. Ho TL, Enantioselective synthesis: Natural products from chiral terpenes, Wiley-Interscience, New York, USA, 1992.
19. Mori K, Semiochemicals – Synthesis, stereochemistry and bioactivity. *Eur J Org Chem* 1479–1489, 1998.
20. Werkhoff P, Brennecke S, Bretschneider W, Güntert M, Hopp R, Surburg H, Chirospecific analysis in essential oil, fragrance and flavor research, *Z Lebensm Unters Forsch* **196**:307–328, 1993.
21. Formâĉek V, Kubeczka KH, *Essential Oils Analysis by Capillary Gas Chromatography and Carbon-13 NMR Spectroscopy*, John Wiley & Sons, Chichester, Great Britain, 1982.
22. König WA, Fricke C, Saritas Y, Momeni B, Hohenfeld G, Adulteration or natural variability? Enantioselective gas chromatography in purity control of essential oils, *J High Resolut Chromatogr* **20**:55–61, 1997.
23. König WA, Hochmuth DH, Enantioselective gas chromatography in flavor and fragrance analysis: Strategies for the identification of known and unknown plant volatiles *J Chromatogr Sci.* **42**:423–439, 2004.
24. Marchelli R, Dossena A, Pala G, The potential of enantioselective analysis as a quality control tool, *Trends Food Sci Technol* **7**:113–119, 1996.
25. Mosandl A, Juchelka D, Advances in the authenticity assessment of citrus oils. *J Essent Oil Res* **9**:5–12, 1997.
26. Larkov O, Dunkelblum E, Zada A, Lewinsohn E, Freiman L, Dudai N, Ravid U, Enantiomeric composition of *trans*- and *cis*- sabinene hydrate and their acetates in five *Origanum* spp., *Flavour Fragr J* **22**:109–114, 2005.
27. Boelens M, Boelens H, van Gemert LJ, Sensory properties of optical isomers, *Perfum Flavor* **18**(6), 1–16, 1993.
28. Friedman L, Miller JG, Odor incongruity and chirality, *Science* **172**:1044–1046, 1971.
29. Jones FN, Elliot D, Individual and substance differences in the discriminability of optical isomers, *Chem Senses and Flavor* **1**:317–321, 1975.
30. Koppenhoefer B, Behnisch R, Epperlein U, Holzschuh H, Bernreuther A, Piras P, Roussel C, Enantiomeric odor differences and gas chromatographic properties of flavors and fragrances, *Perfum Flavorist* **19**(5):1–14, 1994.
31. Laska M, Teubner P, Olfactory discrimination ability of human subjects for ten pairs of enantiomers, *Chem. Senses* **24**:161–170, 1999.

32. Ohloff G, in Schneider D (ed.), Odorous properties of enantiomeric compounds: Olfaction and Taste IV, Wissenschaft Verlagsgesellschaft, Stuttgart, Germany, pp. 156–160, 1972.
33. Theimer ET, McDaniel MR, Odor and optical activity, *J Soc Cosmet Chem* **22**:15–26, 1970.
34. Zawirska-Wojtasiak R, Chirality and the nature of food authenticity of aroma, *Acta Sci Pol Technol Aliment*, **5**:21–36, 2006.
35. Theimer ET, Yoshida T, Klaiber EM, Olfaction and molecular shape. Chirality as a requisite for odor, *J Agric Food Chem* **25**:1168–1177, 1977.
36. Pelosi P, Viti R, Specific anosmia to l-carvone: The minty primary odour, *Chem Senses* **3**:331–337, 1978.
37. Emberger R, Hopp R, Synthesis and sensory characterization of menthol enantiomers and their derivatives for the use in nature identical peppermint oils, in Berger RG, Nitz N, Schreier P, (eds.), *Topics in Flavour Research*, H. Eichhorn, Marzling-Hangenham, Germany, pp. 201–218, 1985.
38. Werkhoff P, Hopp R, Isolation and gas chromatographic separation of menthol and menthone enantiomers from natural peppermint oils, in Brunke EJ (ed.), *Progress in Essential Oil Research*, Walter de Gruyter, Berlin, Germany, pp. 529–549, 1986.
39. Skorianetz W, Giger H, Ohloff G, Darstellung von (+) und (−)-7-hydroxy-dihydrocitronellal aus (+)-pulegon. Ein Beitrag zur kenntnis olfactorischer eigenschaften von enantiomern, *Helv Chim Acta* **54**:1797–1801, 1971.
40. Grob K, Neukom HP, Etter R, Romann E, Food control by government laboratories: Innovation, flexibility and no restriction by reglementation, *Chimia* **46**:420–424, 1992.
41. Gübitz G, Schmid G, Chiral separation by choromatographic and electromigration techniques, A review, *Biopharm. Drug Dispos* **22**:291–336, 2001.
42. Siegwart Y, Some critical comments on the legislative control of flavouring in food, *Flavour Fragr J* **8**:87–90, 1993.
43. Schurig V, Separation of enantiomers by gas chromatograpy, *J Chromatogr* **906**:275–299, 2001.
44. Schurig V. Enantiomer separation by complexation gas chromatography — Applications in chiral analysis of pheromones and flavours, in Schreier P (ed.), Bioflavour'87, Walter de Gruyter, Berlin, Germany, pp. 35–54, 1988.
45. Weber R, Schurig V, Complexation gas chromatography — a valuable tool for the stereochemical analysis of pheromones, *Naturwissenschaften*, **71**:408–413, 1984.
46. König WA, *The Practice of Enantiomer Separation by Capillary Gas Chromatography*, Hüthig, Heidelberg, Germany, 1987.

47. König WA, *Gas Chromatographic Enantiomer Separation with Modified Cyclodextrins*, Hüthig, Heidelberg, Germany, 1992.
48. Schurig V, Nowotny HP, Gas chromatographic separation of enantiomers on cyclodextrin derivatives, *Angew Chem* **29**, 939–957, 1990.
49. Bicchi C, Manzin V, D'Amato A, Rubiolo P, Cyclodextrin derivatives in GC separation of enantiomers of essential oil, aroma and flavour compounds, *Flavour Fragr J* **10**:127–137, 1995.
50. Bicchi C, D'Amato A, Rubiolo P, Cyclodextrin derivatives as chiral selectors for direct gas chromatographic separation of enantiomers in the essential oil, aroma and flavor fields, *J Chromatogr A* **843**:99–121, 1999.
51. König WA, Enantioselective capillary gas chromatography in the investigation of stereochemical correlations of terpenoids, *Chirality* **10**:499–504, 1998.
52. Marriott PJ, Shellie R, Cornwell C, Gas chromatographic technologies for the analysis of essential oils, *J Chromatogr A* **936**:1–22, 2001.
53. Mosandl A, Capillary gas chromatography in quality assessment of flavours and fragrances *J Chromatogr* **624**:267–292, 1992.
54. Mosandl A, Enantioselective capillary gas chromatography and stable isotope ratio mass spectrometry in the authenticity control of flavors and essential oils, *Food Rev Int* **11**:597–664, 1995.
55. Mosandl A, Authenticity assessment: A permanent challenge in food flavor and essential oil analysis, *J Chromatogr Sci* **42**:440–449, 2004.
56. Ravid U, Enantiomeric distribution of oxygenated monoterpenes in some *Mentha* essential oils, *Perfum Flavor* **23**(4):25–30, 1998.
57. Werkhoff P, Brennecke S, Bretschneider W, Progress in the chirospecific analysis of naturally occurring flavour and fragrance compounds, *Chem Mikrobiol Technol Lebebsm* **13**:129–152, 1991.
58. Chirbase/Flavor; http://www.uni-tubingen.de/AK_Koppenhoefer/chirgc1.html (accessed on Nov. 28, 2006)
59. Koppenhoefer B, Epperlin U, Schlunk R, Information technology for enantiomer separation of flavors and fragrances, Chimica OGGI/ Chemistry Today, Jan–Feb, 1996.
60. Schmalzing D, Jung M, Mayer S, Rickert J, Schurig V, Extending the scope of enantiomer separation on Chirasil-Dex by GLC: Comparison with permethylated β-cyclodextrin dissolved in 0V-1701, *J High Resolut Chromatogr* **15**:723–729, 1992.
61. Bicchi C, Artuffo G, D'Amato A, Nano GM, Galli A, Galli M, Permethylated cyclodextrins in the GC separation of racemic mixtures of volatiles: Part 1, *J High Resolut Chromatogr* **14**:301–305, 1991.

62. Bicchi C, Artuffo G, D'Amato A, Manzin V, Galli A, Galli M, Cyclodextrin derivatives for the GC separation of racemic mixtures of volatile compounds, Part VI: The influence of the diluting phase on the enantioselectivity of 2,6-di-O-methyl-3-O-pentyl-β-cyclodextrin, *J High Resolut Chromatogr* **16**:209–214, 1993.
63. Spilkin A, GC analysis using Cyclodex-B^{TM}, *The J & W Separation Times*, **5**(2): 6–7, 1991.
64. König WA, Krebber R, Mischnick P, Cyclodextrins as chiral stationary phases in capillary gas chromatography; Part V: Octakis(3-O-butyryl-2,6-di-O-pentyl)-γ-cyclodextrin, *HRC* **12**:732–738, 1989.
65. Bicchi C, Artuffo G, D'Amato A, Manzin V, Galli A, Galli M, Cyclodextrin derivatives in the GC separation of racemic mixtures of volatile compounds, Part V: Heptakis 2,6-dimethyl-3-pentyl-β-cyclodextrins, *J High Resolut Chromatogr* **15**:710–714, 1992.
66. Askari C, Hener U, Schmarr HG, Rapp A, Mosandl A, Stereodifferentiation of some chiral monoterpenes using multidimensional gas chromatography, *Fresenius J Anal Chem*, **340**:768–772, 1991.
67. Lawrence BM, Essential oils as sources of natural aroma chemicals, *Perfum. Flavor* **17**(5):15–28, 1992.
68. Furia TE, Bellanca N (eds.), *Fenaroli's Handbook of Flavor Ingredients*, CRC Press, Press, Boca Raton, Florida, USA, 1975.
69. Guenther E, *The Essential Oils*, Van Nostrand, New York, USA, 1949.
70. O'Neil MJ (ed.), The Merck Index, 13th ed., Merck & Co., Whitehouse Station, New Jersey, USA, 2001.
71. Dev S (ed.), *Handbook of Terpenoids*, CRC Press, Boca Raton, Florida, USA, 1982.
72. Dewick PM, The biosynthesis of C_5-C_{25} terpenoid compounds, *Nat Prod Rep* **14**:111–114, 1997.
73. Ravid U, Putievsky E, Katzir I, Enantiomeric distribution of piperitone in essential oils of some *Mentha* spp., *Calamintha incāna* (Sm.) Heldr. and *Artemisia judaica* L, *Flavour Fragr J*, **9**:85–87, 1994.
74. König WA, Krebber R, Evers P, Bruhn G, Stereochemical analysis of constituents of essential oils and flavor compounds by enantioselective capillary gas chromatography *HRC* **13**:328–332, 1990.
75. Larkov O, Matasyoh JC, Dudai N, Lewinsohn E, Mayer AA, Ravid U, Distribution of piperitone oxide stereoisomers in *Mentha* and *Micromeria* species and their chemical syntheses, *Flavour Fragr J* **22**:328–333, 2007.
76. Merck-Schuchardt, MS-INFO leaflet 84–107.

77. Croteau R, Venkatachalam KV, Metabolism of monoterpenes: Demonstration that (+)-*cis*-isopulegone, not piperitenone, is the key intermediate in the conversion of (−)- isopiperitenone to (+)-pulegone in peppermint (*Mentha piperita*), *Arch Biochem Biophys* **249**:306–315, 1986.
78. Ravid U, Putievsky E, Katzir I, Chiral GC analysis of (1*R*)(+)- pulegone with high enantiomeric purity in essential oils of some Lamiaceae aromatic plants, *Flavour Fragr J* **9**:205–207, 1994.
79. Bauer K, Garbe D, Surburg H, Common Fragrance and Flavor Materials, VCH, Weinheim, Germany, 1990.
80. Köpke T, Mosandl A, Stereoisomeric flavor compounds. Part LIV: 8-Mercapto-p-menthane-3-one -optically pure stereoisomers and chirospecific analysis, *Z Lebensm Unters Forsch* **194**:372–376, 1992.
81. Köpke T, Schmarr HG, Mosandl A, Stereoisomeric flavour compounds. Part LVII: The stereoisomers of 3-oxo-p-menthane-8-thiol acetate, simultaneously stereoanalyzed with their corresponding thiols, *Flavour Fragr J* 7:205–211, 1992.
82. Köpke T, Dietrich A, Mosandl A, Chiral compounds in essential oils XIV: Simultaneous stereoanalysis of buchu leaf oil, *Phytochem Anal* **5**(19):61–67, 1994.
83. Grundy DL, Still CC, Inhibition of acetylcholinesterases by pulegone — 1,2-epoxide, *Pest Biochem Physiol* **23**:383–388, 1985.
84. Riddle JM, Estes JW, Oral contraceptives in ancient and medieval times, *Am Sci* **80**:226–233, 1992.
85. Madyastha KM, Thulasiram HV, Transformation of a monoterpene ketone, (*R*)-(+) pulegone, a potent hepatotoxin, in *Mucor piriformis*, *J Agric Food Chem*, 47:1203–1207, 1999.
86. Nelson SD, McClanahan RH, Knebel N, Thomassen D, Gordon WP, Doshi S, The metabolism of (*R*)-(+)-pulegone, a toxic monoterpene, in Petrosky RJ, McCormick SP (eds.), *Secondary-Metabolite Biosynthesis and Metabolism*, Plenum Press, New York, USA, 1992.
87. Kreis P, Mosandl A, Chiral compounds of essential oils, Part XIII, Simultaneous chirality evaluation of geranium oil constituents, *Flavour Fragr J* **8**:161–168, 1993.
88. Clark GS, An aroma chemical profile: Menthone, *Perfum Flavor* **19**(3):41–45, 1994.
89. Askari C, Kreis P, Mosandl A, Schmarr HG, Chiral compounds of essential oils VII: Quality evaluation of *Mentha* oils, using enantioselective CGC-analysis of monoterpenoid constituents, *Arch Pharm* (Weinheim) **325**:35–39, 1992.
90. Ravid U, Putievsky E, Katzir I, Chiral GC analysis of menthone and isomenthone with high enantiomeric purities in laboratory-made and commercial essential oils, *Flavour Fragr J* **9**:139–142, 1994.

91. Glasby JS, *Encyclopedia of the Terpenoids*, John Wiley, Chichester, GB, 1982.
92. Lawrence BM, Monoterpene interrelationships in the *Mentha* genus: A biosynthetic discussion, in Mookherjee BD, Mussinan CJ (eds.), *Essential Oils*, Allured, Wheaton, Illinois, USA, pp. 1–81, 1981.
93. Croteau R, Sood VK, Renstrom B, Bhushan R, Metabolism of monoterpenes: Early steps in the metabolism of d-neomenthyl-beta-d-glucoside in peppermint (*Mentha piperita*) rhizomes, *Plant Physiol* **76**:647–653, 1984.
94. Karp F, Croteau R, Role of hydroxylases in monoterpene biosynthesis, in Schreier P (ed.), Bioflavor '87, Walter de Gruyter, Berlin, Germany, pp. 173–198, 1988.
95. Fuchs S, Beck T, Sandvoss M, Mosandl A, Biogenetic studies in *Mentha* x *piperita*. 2. Stereoselectivity in the bioconversion of pulegone into menthone and isomenthone, *J Agric Food Chem* **47**:3058–3062, 1999.
96. Klein E, Rojahn W, (+)-Verbenone — A newly discovered component of Spanish rosemary oil, *Dragoco Report* **14**:75–76, 1967.
97. Ravid U, Putievsky E, Katzir I, Lewinsohn E, Dudai N, Identification of (1*R*)(+)-verbenone in essential oils of *Rosmarinus officinalis* L., *Flavour Fragr J* **12**:109–112, 1997.
98. Salom SM, Billings RF, Upton WW, Dalusky MJ, Grosman DM, Payne TI, Berisford CW, Shaver TN, Effect of verbenone enantiomers and racemic endo-brevicomin on response of *Dendroctonus frontalis* (Coleoptera: Scolitidae) to attractant-baited traps, *Can J For Res* **22**:925–931, 1992.
99. Renwick JAA, Hughes PR, Krull IS, Selective production of *cis*- and *trans*- verbenol from (−)-and (+)-α-pinene by a bark beetle, *Science* **191**:199–201, 1976.
100. McPheron LJ, Seybold SJ, Storer AJ, Wood DL, Ohtsuka T, Kubo I, Effects of enantiomeric blend of verbenone on response of *Ips paraconfusus* to naturally produced aggregation pheromone in the laboratory, *J Chem Ecol* **23**:2825–2839, 1997.
101. Takeya K, Itokawa H, Stereochemistry in oxidation of allylic alcohols by cell-free system of callus induced from *Canabis sativa* L., *Chem Pharm Bull* (Tokyo), **25**:1947–1951, 1977.
102. Mechoulam R, Braun P, Gaoni Y, A stereospecific synthesis of (−)Δ^1 and (−)Δ^6-tetrahydrocannabinols, *J Am Chem Soc* **89**:4552–4554, 1967.
103. Ravid U, Bassat M, Putievsky E, Weinstein V, Ikan R, Isolation and determination of optically pure carvone enantiomers from caraway (*Carum carvi* L.), dill (*Anethum graveolens* L.), spearmint (*Mentha spicata* L.) and *Mentha longifolia* (L.) Huds, *Flavour Fragr J* **2**:95-97, 1987.
104. Ravid U, Putievsky E, Katzir I, Weinstein V, Ikan R, Chiral GC analysis of (*S*)(+)- and (*R*)(−)-carvone with high enantiomeric purity in caraway, dill and spearmint oils, *Flavour Fragr J* **7**:289–292, 1992.

105. Kieslich K, Abraham WR, Stumpf B, Thede B, Washausen P, Transformation of terpenoids, in Brunke EJ, *Progress in Essential Oil Research*, Walter de Gruyter, Berlin, Germany, pp. 367–394, 1986.
106. Singaram B, Verghese J, Synthetic routes to carvone, *Perfum Flavor* **2**(3):47–51, 1977.
107. Blank I, Grosch W, Evaluation of potent odorants in dill seed and dill herb (*Anethum graveolens* L.) by aroma extract dilution analysis, *J Food Sci* **56**:63–67, 1991.
108. Bouwmeester HJ, Davies JAR, Toxopeus H, Enantiomeric composition of carvone, limonene, and carveols in seeds of dill and annual and biennial caraway varieties, *J Agric Food Chem* **43**:3057–3064, 1995.
109. Clark GS, Carvone, *Perfum Flavor* **14**(3):35–40, 1989.
110. König WA, Icheln D, Runge T, Pforr I, Krebs A, Cyclodextrins as chiral stationary phases in capillary gas chromatography, VII, Cyclodextrins with an inverse substitution pattern — Synthesis and enantioselectivity, *J High Resolut Chromatogr* **13**:702–707, 1990.
111. Roelofs W, Gieselmann M, Cardé A, Tashiro H, Moreno DS, Henrick CA, Identification of the California red scale sex pheromone, *J Chem Ecol* **4**:211–224, 1978.
112. Leitereg TJ, Guadagni DG, Harris J, Mon TR, Teranishi R, Evidence for the difference between the odor of the optical isomers (+)- and (−)-carvone, *Nature* **230**:455–456, 1971.
113. Russel GF, Hills JI, Odor differences between enantiomeric isomers, *Science* **172**:1043–1044, 1971.
114. Corey EJ, Pearce HL, Total synthesis of picrotoxinin, *J Am Chem Soc* **101**:5841–5843, 1979.
115. Hormann CA, Cowart BJ, Olfactory discrimination of carvone enantiomers, *Chem. Senses* **18**:573, 1993.
116. Laska M, Liesen A, Teubner P, Enantioselectivity of odor perception in squirrel monkeys and humans, *Am J Physiol Regul Integr Comp Physiol* **277**:(R)1098–(R)1103, 1999.
117. Pike LM, Enns MP, Hornung DE, Quality and intensity differences of carvone enantiomers when tested separately and in mixtures, *Chem Senses* **13**:307–309, 1988.
118. Heth G, Nevo E, Ikan R, Weinstein V, Ravid U, Duncan H, Differential olfactory perception of enantiomeric compounds by blind subterranean mole rats (*Spalax ehrenbergi*), *Experientia* **48**:897–902, 1992.
119. Cowart BJ, Olfactory responses to enantiomers, *Chem Senses* **15**:562–563, 1990.

120. König WA, Gehrcke B, Icheln D, Evers P, Dönnecke J, Wang W, New, selectively substituted cyclodextrins as stationary phases for the analysis of chiral constituents of essential oils, *J High Resolut Chromatogr* **15**:367–372, 1992.
121. Ravid U, Putievsky E, Katzir I, Ikan R, Chiral GC analysis of enantiomerically pure fenchone in essential oils, *Flavour Fragr J* 7:169–172, 1992.
122. Ravid U, Putievsky E, Katzir I, Determination of the enantiomeric composition of (1*R*)(+)- and (1*S*)(−)-camphor in essential oils of som Lamiaceae and Compositae herbs, *Flavour Fragr J* **8**:225–228, 1993.
123. Kreis P, Dietrich A, Mosandl A, Chiral compounds in essential oils, Part 18: On the authenticity assessment of the essential oil of *Rosmarinus officinalis* L., *Pharmazie* **49**:761–765, 1994.
124. Tateo F, Bononi M, De Dominicis E, Fumagalli V, Update on enantiomeric composition of (1*R*)(+)- and (1*S*)(−)-camphor in essential oils by enantioselective gas chromatography, *Anal Commun* **36**:149–151, 1999.
125. Money T, Camphor: A chiral starting material in natural product synthesis, *Nat Prod Rep* **2**:253–289, 1985.
126. Oppolzer W, Camphor derivatives as chiral auxiliaries in asymmetric synthesis, *Tetrahedron* **43**:1969–2004, 1987.
127. Szabo WA, Lee HT, Chiral starting materials and reagents. An outline of recent synthetic applications, *Aldrichimica Acta* **13**:13–20, 1980.
128. Allen MS, Lamb N, Money T, Salisbury P, Synthesis and biological activity of monoterpenoid analogues of *cis*- sativenenediol and helminthosporal, *JCS Chem Commun*: 112–117, 1979.
129. Meyer WL, Capshew CE, Johnson JH, Klusener AR, Lobo AP, McCarthy RN, rearrangement of α-bromocamphoric anhydride. 2. Competitive mechanism in the formation of laurolenic acid, *J Org Chem* **42**:527–534, 1977.
130. Stevens RV, Gaeta FCA, Lawrence DS, Camphorae: Chiral intermediates for the enantiospecific total synthesis of steroids, *J Am Chem Soc* **105**:7713–7719, 1983.
131. Stevens RV, Beaulieu N, Chan WH, Daniewsky AR, Takeda T, Waldner A, Williard PG, Zuther U, Studies on the synthesis of vitamin B_{12}, *J Am Chem Soc* **108**:1039–1049, 1986.
132. Croteau R, Biosynthesis of cyclic monoterpenes, in Parliment TH, Croteau R (eds.), *Biogeneration of Aromas*, ACS Symposium series 317, Washington DC, USA pp. 134–156, 1986.
133. Hutchinson JH, Money T, Enantiospecifc synthesis of estrone, *Can J Chem* **65**: 1–6, 1987.
134. Kitagawa I, Tsujii S, Fujioka H, Kajiwara A, Yamamoto Y, Shibuya H, Chemical transformation of terpenoids, VI, Syntheses of chiral segments, key building-blocks for the left half of taxane-type diterpenoids, *Chem Pharm Bull* **32**:1294–1302, 1984.

135. Shibuya H, Fujioka H, Yamamoto Y, Suzuki K, Kitagawa I, Chemical transformation of terpenoids, IV, Acid treatment of (3*R*)-1-vinyl-, (3*R*)-1-hydroxypropenyl and (3*R*)-1-epoxyethyl-5-methoxy-1,2,2-trimethylcyclopentane derivatives: Ring enlargement relations and successive migrations of methyl residues, *Chem Pharm Bull* **30**:1280–1288, 1982.
136. Weedon BC, Some recent studies on carotenoids and related compounds, *Pure Appl Chem* **35**:113–130, 1973.
137. Baldovini N, Ristorcelli D, Tomi F, Casanova J, Infraspecific variability of the essential oil of *Calamintha nepeta* from Corsica (France), *Flavour Fragr J*, **15**: 50–54, 2000.
138. Reitsema RN, Varnis JV, The isolation of piperitone oxide from *Mentha sylvestris, J Am Chem Soc* **78**:3792–3794, 1956.
139. Croteau R, Karp F, Wagschal KC, Satterwhite DM, Hyatt DC, Skotland KB, Biochemical characterization of a spearmint mutant that resembles peppermint in monoterpene content, *Plant Physiol* **96**:744–752, 1991.
140. Kreis P, Juchelka D, Motz C, Mosandl A, Chiral compounds in essential oils, IX: Stereodifferentiation of borneol, iso-borneol and bornyl acetate, *Dtsch Apoth Ztg*, **131**:1984–1987, 1991.
141. Mahmoud SS, Croteau R, Menthofuran regulates essential oil biosynthesis in peppermint by controlling a downstream monoterpene reductase, *Proc Natl Acad Sci USA* **100**:14481–14486, 2003.
142. Fischer N, Nitz S, Drawert F, Original flavour compounds and the essential oil composition of marjoram (*Majorana hortensis* Moench), *Flavour Fragr J* **2**:55–61, 1987.
143. Fischer N, Nitz S, Drawert F, Original composition of marjoram flavor and its changes during processing, *J Agric Food Chem* **36**:996–1003, 1988.
144. Novak J, Bitsch C, Langbehn J, Pank F, Skoula M, Gotsiou Y, Franz CM, Ratios of *cis*- and *trans*-sabinene hydrate in *Origanum majorana* L. and *Origanum microphyllum* (Bentham) Vogel, *Biochem Syst Ecol* **28**:697–704, 2000.
145. Novak J, Bitsch C, Pank F, Langbehn J, Franz CM, Distribution of the *cis*-sabinene hydrate acetate-chemotype in accessions of marjoram (*Origanum majorana* L.), *Euphytica* **127**:69–74, 2002.
146. Novak J, Langbehn J, Pank F, Franz CM, Essential oil compounds in a historical sample of marjoram (*Origanum majorana* L., Lamiaceae), *Flavour Fragr J* **17**:175–180, 2002.
147. Nitz S, Fischer N, Drawert F, Flavour compounds in plants I. Commun.: Volatile terpenoid compounds from bound precursors in "*Majorana hortensis* Moench," *Chem Mikrobiol Technol Lebensm* **9**:87–94, 1985.
148. Rodriguez MRA, Caramão EB, dos Santos JG, Dariva C, Oliveira JV, The effect of temperature and pressure on the characteristics of the extracts from

high-pressure CO_2 extraction of *Majorana hortensis* Moench, *J Agric Food Chem* **51**:453–456, 2003.

149. Karasawa D, Shimizu S, *Mentha candicans*, a new chemical strain of section Spicatae, containing *trans*-sabinene hydrate as the principal component of essential oil, *Agric Biol Chem* **42**:433–437, 1978.
150. Cornwell CP, Leach DN, Wyllie SG, Incorporation of oxygen-18 into terpinen-4-ol from the $H_2^{18}O$ steam distillates of *Melaleuca alternifolia* (tea tree), *J Essen Oil Res* 7:613–620, 1995.
151. Shellie R, Marriott P, Cornwell C, Application of comprehensive two-dimensional gas chromatography (GC'GC) to the enantioselective analysis of essential oils, *J Sep Sci* **24**:823–830, 2001.
152. Cornwell CP, Ph.D. Thesis, University of Western Sydney, Hawkesbury, Australia, 1999.
153. Cornwell CP, Leach DN, Wyllie SG, The origin of terpinen-4-ol in the steam distillates of *Melaleuca argenta, M. dissitiflora* and *M. linariifolia, J Essen Oil Res* **11**:49–53, 1999.
154. Bernreuther A, Schreier P, Multidimensional gas chromatography-mass spectrometry: A powerful tool for the direct chiral evaluation of aroma compounds in plant tissues, II, Linalool in essential oils and fruits *Phytochem Anal* **2**:167–170, 1991.
155. Casabianca H, Graff JB, Faugier V, Fleig F, Grenier C, Enantiomeric distribution studies of linalool and linalyl acetate; A powerful tool for authenticity control of essential oils, *J High Resol Chromatogr* **21**:107–112, 1998.
156. Cotroneo A, d'Alcontres IS, Trozzi A, On the genuineness of citrus essential oils. Part XXXIV. Detection of added reconstituted bergamot oil in genuine bergamot essential oil by high resolution gas chromatography with chiral capillary columns, *Flavour Fragr J* 7:15–17, 1992.
157. Dugo G, Bartle KD, Bonaccorsi I, Catalfamo M, Cotroneo A, Dugo P, Lamonica G, McNair H, Mondello L, Previti P, d'Alcontres IS, Trozzi A, Verzera A, Advanced analytical techniques for the analysis of citrus essential oils., Part 2. Volatile fraction: LC-HRGC and MDGC, *Essenz Deriv Agrum* **69**:159–217, 1999.
158. Oliver JE, (*S*)(+)-Linalool from oil of coriander, *J Essen Oil Res* **15**:31–33, 2003.
159. Ravid U, Putievsky E, Weinstein V, Ikan R, Determination of the enantiomeric composition of natural flavouring agents by ^{1}H-NMR spectroscopy, in Baerheim Svendsen A, Scheffer JJC (eds.), *Essential Oils and Aromatic Plants*, Martinus Nijhoff/Dr. W. Junk Publishers, Dordrecht, The Netherlands, pp. 135–138, 1985.

160. Ravid U, Putievsky E, Katzir I, Lewinsohn E, Enantiomeric composition of linalool in the essential oils of *Ocimum* species and in commercial basil oils, *Flavour Fragr J* **12**:293–296, 1997.
161. Schubert V, Mosandl A, Chiral compounds of essential oils, VIII: Stereodifferentiation of linalool using multidimensional gas chromatography, *Phytochem Anal* **2**:171–174, 1991.
162. Siani AC, Tappin MRR, Ramos MFS, Mazzei JI, Ramos MCKV, de Aquino Neto FR, Frighetto N, Linalool from *Lippia alba*: Study of the reproducibility of the essential oil profile and the enantiomeric purity, *J Agr Food Chem* **50**:3518–3521, 2002.
163. Tamogami S, Awano K, Amaike M, Takagi Y, Kitahara T, Analysis of enantiomeric ratios of aroma components in several flowers using a Chiramix column, *Flavour Fragr J* **19**:1–5, 2004.
164. Sugawara Y, Hara C, Tamura K, Fujii T, Nakamura K, Masujima T, Aoki T, Sedative effect on humans of inhalation of essential oil of linalool: Sensory evaluation and physiological measurements using optically active linalools, *Anal Chim Acta* **365**:293–299, 1998.
165. Ravid U, Putievsky E, Bassat M, Ikan R, Weinstein V, Isolation of optically pure (−)-linalyl acetate from clary sage, *Salvia dominica* L., lavender and lavandin, *Flavour Fragr J* **1**:121–124, 1986.
166. Bicchi C, Pisciotta A, Use of two-dimensional gas chromatography in the direct enantiomer separation of chiral essential oil components, *J Chromatogr* **508**: 341–348, 1990.
167. Mosandl A, Schubert V, Stereoisomeric flavor compounds XXXVII: Enantiomer separation of 1-alken-3-yl esters and their chirality evaluation from essential oils using multidimensional gas chromatography (MDGC), *J Essen Oil Res* **2**:121–132, 1990.
168. Mosandl A, Schubert V, Stereoisomeric flavour compounds XXXIX: Chiral constituents of essential oils (I), stereodifferentiation of linalyl acetate – A new possibility for quality evaluation of lavender oil, *Z Lebensm Unters Frosch* **190**:506–510, 1990.
169. Casabianca H, Graff JB, Separation of linalyl acetate enantiomers: Application to the authentication of bergamot food products, *J High Resolut Chromatogr* **17**:184–186, 1994.
170. Maas B, Dietrich A, Mosandl A, Enantioselective capillary gas chromatography — Olfactometry in essential oil analysis, *Naturwissenschaften* **80**:470–472, 1993.
171. Ravid U, Putievsky E, Katzir I, Chiral GC analysis of enantiomerically pure (*R*)(−)-linalyl acetate in some Lamiaceae, myrtle and petitgrain essential oils, *Flavour Fragr J* **9**:275–276, 1994.

172. Weinreich B, Nitz S, Influences of processing on the enantiomeric distribution of chiral flavour compounds, Part A : Linalyl acetate and terpene alcohols, *Chem Mikrobiol Technol Lebensm* **14**:117–124, 1992.
173. Mosandl A, On the assessment of natural and nature-identical flavour and essential oil compounds, *Chem Mikrobiol Technol Lebensm* **14**:187–188, 1992.
174. Rienäcker R, Ohloff G, Opticsch actives β-citronellol aus (+)-oder (−)-pinan, *Angew Chem* **73**:240, 1961.
175. Takasago Perfumery Co Ltd, Jap Pat 7316191, 1970.
176. Yamaguchi Y, Komatsu A, Moroe T, Microbial transformation of odorous compounds I: Microbial reduction of d, l-citronellal, *J Agric Chem Soc* (Japan) **50**:443–445, 1976.
177. Ravid U, Putievsky E, Katzir I, Ikan R, Weinstein V, Determination of the enantiomeric composition of citronellol in essential oils by chiral GC analysis on a modified γ-cycledextrin phase, *Flavour Fragr J* 7:235–238, 1992.
178. Ravid U, Putievsky E, Katzir I, Stereochemical analysis of borneol in essential oils using permethylated β-cyclodextrin as a chiral stationary phase, *Flavour Fragr J* **11**:191–195, 1996.
179. Dehal SS, Croteau R, Metabolism of monoterpenes: Specificity of the dehydrogenases responsible for the biosynthesis of camphor, 3-thujone and 3-isothujone, *Arch Biochem Biophys* **258**:287–291, 1987.
180. Ravid U, Putievsky E, Katzir I, Determination of the enantiomeric composition of α−terpineol in essential oils. *Flavour Fragr J* **10**:281–284, 1995.
181. Leach DN, Wyllie SG, Hall JG, Kyratzis I, Enantiomeric composition of the principal components in the oil of *Melaleuca alternifolia*, *J Agric Food Chem* **41**:1627–1632, 1993.
182. Shellie R, Mondello L, Dugo G, Marriott P, Enantioselective gas chromatographic analysis of monoterpenes in essential oils of the family Myrtaceae, *Flavour Fragr J* **19**:582–585, 2004.
183. Bayer M, Mosandl A, Improved gas chromatographic stereodifferentiation of chiral main constituents from different essential oils using a mixture of chiral stationary phases, *Flavour Fragr J* **19**:515–517, 2004.
184. Ravid U, Putievsky E, Katzir I, Ikan R, Determination of the enantiomeric composition of terpinen-4-ol in essential oils using a permethylated β-cyclodextrin coated chiral capillary column, *Flavour Fragr J* 7:49–52, 1992.
185. Ravid U, Bassat M, Putievsky E, Ikan R, Weinstein V, Determination of the enantiomeric composition of (+)-terpinen-4-ol from sweet marjoram *Origanum majorana* L. using a chiral lanthanide shift reagent, *Flavour Fragr J* **2**:17–19, 1987.

186. Naves YR, Tullen P, Études sur les matières végétales volatiles. CLXX(I). Présence de β-myrcène, de Δ^3-carène et de (+)-terpinène-1 ol-4 dans l'huile essentielle de lavande *Bull Soc Chim* Fr, 2123–2124, 1960.
187. Francke W, Sauerwein P, Vité JP, Klimetzek D, The pheromone bouquet of *Ips amitinus*, *Naturwissenschaften* **67**:147–148, 1980.
188. Kohnle U, Francke W, Bakke A, *Polygraphus poligraphus* (L.): Response to enantiomers of beetle specific terpene alcohols and a bicyclic ketal, *Z Angew Entomol* **100**:5–8, 1985.
189. Ohloff G, Uhde G, Absolute konfiguration von terpinenol-(4), *Helv Chim Acta* **48**:10–28, 1965.
190. Kreis P, Mosandl LA, Chiral compounds of essential oils. Part XI. Simultaneous stereoanalysis of *Lavandula* oil constituents, *Flavour Fragr J* 7:187–193, 1992.
191. Oumzil H, Ghoulami S, Rhajaoui M, Ilidrissi A, Fkih-Tetouani S, Faid M, Benjouad A, Antibacterial and antifungal activity of essential oils of *Mentha suaveolens*, *Phytother Res* **16**:727–731, 2002.

Chapter 7

RECENT TRENDS OF SOME NATURAL SWEET SUBSTANCES FROM PLANTS

Bernard Crammer

7.1 INTRODUCTION

Man's craving for sweet food is natural, considering that the newly born infant first tastes his mother's milk which contains lactose, also known as milk sugar. The infant's smile implies he likes the taste of his mother's sweet milk. Man's addiction to sweet foods is presenting ever increasing health problems such as obesity and diabetes. Obesity is becoming a global epidemic. It is so common now that it is ranked with infectious diseases and malnutrition as one of the most significant contributors to ill health. Obesity is associated with diabetes mellitus, certain forms of cancer, sleep breathing disorders, and coronary heart disease. There are several commercial natural and synthetic sweeteners available but all of them do not satisfy the ideal sweetener. The sweetener should have a taste profile like sugar, highly sweet, odorless, colorless, stable to cooking foods, noncarcinogenic, noncaloric, tooth friendly and cheap to market the product.

The worldwide demand for high potency sweeteners is increasing and, with blending of different sweeteners becoming a standard practice, the demand for the search of alternative natural sweeteners is also increasing. Due to many adverse effects of artificial sweeteners such as, for example, aspartame, sucralose, acesulphame K and the natural sweetener sucrose found in sugar cane and bee's honey, the search for natural sweeteners derived from plants has been intensified during the past decade.[1,2]

The continuing search for sweeteners that will reduce obesity, is now focused on natural sweeteners derived from plants and most of those discovered are many times sweeter than sugar. Two classes of such sweeteners are certain sweet glycosides and sweet proteins.

7.2 DITERPENE GLYCOSIDES

7.2.1 Ent-kaurene Diterpene Glycosides from the Leaves of Stevia rebaudiana

The Guarani Indians had known for centuries about the unique advantages of *kaa he-he* (a native term which translates as "sweet herb"). These Indians living in Paraguay knew that the leaves of the wild stevia shrub (a perennial indigenous to the Amambay Mountain region) have a sweetening power unlike any other natural sweetener. The stevia leaves were used to enhance the taste of the bitter mate (a tea-like beverage) and medicinal potions, or simply chewed for their sweet taste. An Italian Botanist Dr Moises Santiago Bertoni discovered the pleasant sweet properties of this shrub more than 150 years ago.[3]

The plant Stevia rebaudiana *bertoni* has been studied in depth because it was discovered that this plant is the source of six sweet-tasting diterpene glycosides. They are stevioside, rebaudiosides A, C, D, E and dulcoside A. (Table 1). These sweet diterpene glycosides, as well as a complex mixture of organic compounds of which more than a hundred compounds have been identified, are found mainly in the leaves of the plant. The leaves contain a complex mixture of labdane diterpenes, triterpenes, stigmasterol, tannins and volatile oils. There is an abundance of reviews and patents relating to these sweet diterpene glycosides.[4–6]

7.2.2 Stevia rebaudiana bertoni

Stevioside is the most abundant sweet-tasting compound in the leaves. Bridel and Lavielle isolated the crystalline glycoside, stevioside from an alcoholic extract of *S. rebaudiana* and found it to be 300 times sweeter than

Table 1. **Diterpene Glycosides Isolated from Stevia Leaves**

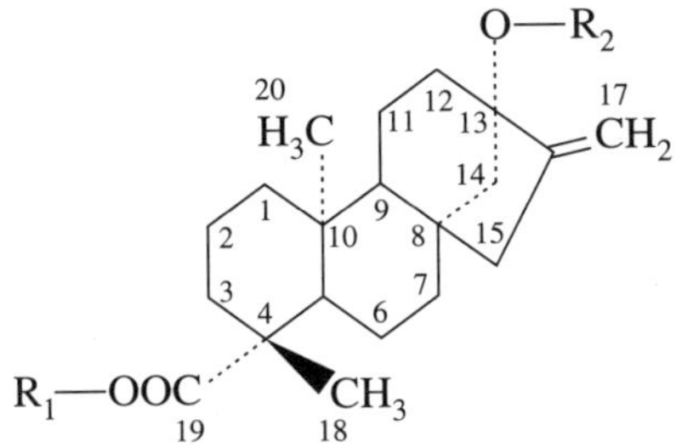

Diterpene glycoside	R_1[a]	R_2[a]	*Sweetening potency (sucrose = 1)*
Steviobioside	H	glc^{2}—^{1}glc	100–125
Rubusoside	glc	glc	100–120
Stevioside	glc	glc^{2}—^{1}glc	150–300
Rebaudioside A	glc	glc$^{2}_{3}$—^{1}glc ╲^{1}glc	250–450
Rebaudioside B	H	glc$^{2}_{3}$—^{1}glc ╲^{1}glc	300–350
Rebaudioside C (dulcoside B)	glc	glc$^{2}_{3}$—^{1}rham ╲^{1}glc	50–120
Rebaudioside D	glc^{2}—^{1}glc	glc$^{2}_{3}$—^{1}glc ╲^{1}glc	250–450
Rebaudioside E	glc^{2}—^{1}glc	glc^{2}—^{1}glc	150–300
Dulcoside A	glc	glc^{2}—^{1}rham	50–120

[a] glc, *β*-D-glucopyranosly; rham, α-L-rhamnopyranosyl.

sucrose.[7] From their hydrolysis experiments, Bridel and Lavielle showed that an enzyme catalyzed the hydrolysis of stevioside (Table 1) using a gut extract of the vineyard snail, *Halix pomatia*, yielded the aglycone steviol (Table 1). Acid–catalyzed hydrolysis of stevioside afforded steviol and isosteviol. Bridel and Lavielle were unable to determine fully the chemical structures of stevioside, steviol and isosteviol.

Mosettig and Ness determined that steviol and isosteviol were diterpenoid acids with the former containing a 2,11-cyclopentanoperhydrophenanthrene skeleton, a hydroxyl acid and a terminal methylene group.[8] It was in 1963 that Mosettig and his colleagues finally showed the unambiguous structure of stevioside.

Saponification of stevioside with a strong alkaline base yielded steviolbioside. Although steviolbioside has been identified in some *S. rebaudiana* extracts, it is generally thought to be an artifact of extraction or isolation procedures rather than a naturally occurring glycoside.[9]

In the 1970s, further investigations of the sweet and tasteless constituents of *S. rebaudiana* were conducted by several Japanese laboratories especially by Tanaka and his coworkers at Hiroshima University. Eight sweet tasting glycosides from the leaves of *S. rebaudiana* contain a common aglycone steviol (13-hydroxy-*ent*-kaur-16-en-19-oic acid), and differ only in the glycosidic constituents attached at C-13 and/or C-19. Tanaka and coworkers reported the isolation of the *ent*-kaurenes rebaudioside A and rebaudioside B.[10]

There are many patents and journal articles that describe processes to produce the principal sweet diterpene glycosides stevioside and in particular, rebaudioside A. Many of the reported methods of production require the use of ion exchange columns or gases and are not satisfactory if the scale-up to commercial quantities are required.[11] Most methods for extraction and purification of the sweet diterpene glycosides from Stevia use complicated processing of the crude extracts. Generally, the recovery processes of the principal two diterpene glycosides stevioside and rebaudioside A involve:

1. Extraction of the leaves of *S. rebaudiana* with a highly polar solvent such as water or methanol to afford the glycoside extract.
2. Applying liquid chromatography by eluting with a polar solvent to separate two of the principal glycosides.
3. Removing the eluting solvent from the two diterpene glycosides.[12]

7.3 SWEET PROTEINS

Most sweet compounds including the commercial sweeteners, are small molecular weight compounds but there are also sweet macromolecules both synthetic[13] and natural.[14] It was thought that compounds with molecular masses over 2 500 would generally be tasteless. It was assumed that macromolecules such as proteins could elicit a sweet taste similar to small molecules such as sucrose and stevioside until the discovery of miraculin, a taste modifying glycoprotein which has the unusual property of modifying the sour taste of substances into sweet taste. Miraculin was shown to have a molecular weight of 24 600.[15,16] Sweet proteins have the potential to replace the commercial artificial sweeteners, by acting as natural low calorie sweeteners because it is known that proteins do not trigger a demand for insulin whereas sugars such as sucrose and fructose do.

There are seven known sweet and taste-modifying proteins, namely (1) monellin and (2) thaumatin;[17,18] (3) mabinlin.[19] and (4) curculin;[20] (5) pentadin,[21] (6) brazzein[22] and (7) miraculin.[23] The properties and characteristics of these proteins are illustrated in Table 2. There are several recent reviews relating to sweet proteins.[1,24] Apart from curculin and Mabinlin, the other known sweet proteins were discovered in West Africa (Table 2).

The key group on the protein surface responsible for biological activity has not yet been identified with certainty for any of the known sweet proteins.[25] The sweet taste in man is mainly due to recently discovered T1R2-T1R3 receptor.[26–28] The human T1R2-T1R3 receptor recognizes natural and synthetic sweeteners and T1R1-T1R3 recognizes umami taste.[29,30]

7.3.1 Brazzein

Brazzein is the smallest, most heat-stable and pH-stable member of the set of proteins known to have intrinsic sweetness. The protein is about 6 473 Daltons in size and consists of 52 amino acid residues. It is reported to be between 2 000 times sweeter than sucrose in a 2% solution than a 2% sucrose solution[31,32] and represents an excellent alternative to commercially available low calorie sweeteners. Brazzein was originally isolated from the fruit of an African plant, *Pentadiplandra brazzeana* Baillon. The *Pentadiplandra-brazzeana* plant was discovered in 1984 by Marcel and Anette Hladik, both working at the Paris National Nature Museum, who were studying the eating habits of apes in Gabon.[33] Heat and pH stability of the protein make it an ideal system for investigating the chemical and structural requirements of a sweet-tasting protein. Based on the wild type brazzein, 25 mutants were produced to identify critical regions important for sweetness. To assess their sweetness, psychophysical experiments were carried out with 14 human subjects. Firstly, the results suggest that

Table 2. Comparison of Thaumatin, Monellin, Mabinlin, Pentadin, Brazzein, Curculin and Miraculin

	Thaumatin	Monellin	Mabinlin	Pentadin	Brazzein	Curculin	Miraculin
Source	Thaumatococcus danielli Benth	Dioscoreophyllum cumminsii Diels	Capparis masakai Levl	Pentadiplandra brazzeana Baillon	Pentadiplandra brazzeana Baillon	Curculingo latifolia	Richadella dulcifica
Geographic distribution	West Africa	West Africa	China	West Africa	West Africa	Malaysia	West Africa
Variants	I, II, a, b, c[a]	–	I-a, II, III, IV[a]	–	–	–	–
Sweetness factor (weight basis)	3000	3000	100	500	2000	550	–
Molecular mass (active form, kDa)	22.2	10.7	12.4	12.0[b]	6.5	24.9	98.4
Amino acids	207	45 (A chain) 50 (B chain)	33 (A chain) 72 (B chain)	?	54	114	191
Active form	Monomer	Dimer (A + B)	Dimer (A + B)	?	Monomer	Dimer (A + A)	Tetramer (A + A+A+A+A)

Source: Adapted from Kurihara.[34] [a]At least five different forms of thaumatin[35] and four different forms of mabinlin[36] have been identified. [b]A chromatographic fraction containing a 12 kDa protein was sweet. This same fraction, when subjected to electrophoresis under nonreducing conditions showed bands in the region between 22 kDa and 41 kDa, suggesting the presence of subunits.

residues 29–33 and 39–43, plus residue 36 between these stretches, as well as the C-terminus are involved in the sweetness. Secondly, the charge of these residues plays an important role in its interaction with the sweet taste receptor.

Unlike most other sugar substitutes, brazzein is a protein, not a carbohydrate. In addition, brazzein has fewer calories than sugar. Also, unlike sugar, it can be eaten by diabetics without any problems. It is more sugar-like than most other sugar substitutes and it can withstand heat, which is a big bonus for industrial food manufacture. Brazzein can be commercially extracted from genetically modified maize through ordinary milling. Approximately one ton of maize yields 1 kg–2 kg of brazzein. It can also be engineered into plants like wheat to make presweetened grains, e.g. cereals.

The amino acid sequence of brazzein was determined by peptide sequencing[37] (Fig. 1) and the three-dimensional structure of brazzein was solved by homonuclear ^{3}H NMR spectroscopy.[25] The protein has a highly compact structure consisting of one short α-helix and three anti-parallel β-strands held together by four disulfide bridges. No significant sequence or structural similarity was found between brazzein and the two other sweet tasting proteins of known three-dimensional structure: monellin[38] and thaumatin[39,40] and the sweetness profile was shown to be undiminished after incubation at 100°C for 4 hours.[41] Hellekant and Ming succeeded to produce large quantities of brazzein by means of a recombinant host.[32]

The original NMR studies of fruit brazzein indicated that the protein adopts a cysteine-stabilized αβ (CSαβ) fold in which the α-helix and β-strands are stabilized by the presence of four disulfide bridges.[25] Other proteins with this fold include members of the rapeseed family of serine

Asp Lys Cys Lys Lys Val Tyr Glu Asn Tyr Pro Val Ser Lys Cys Gln
1 5 10 15
Leu Ala Asn Gln Cys Asn Tyr Asp Cys Lys Leu Asp Lys His Ala Arg
20 25 30
Ser Gly Glu Cys Phe Tyr Asp Glu Lys Arg Asn Leu Gln Cys Ile Gly
35 40 45
Asp Tyr Cys Gly
50

Fig. 1. Brazzein comprising 52 amino acids.

proteinase inhibitors, scorpion toxins, insect defensins and plant-derived γ-thionins. Apart from the conserved cysteines, little sequence indentity is found between members of the different families, Brazzein is the only CSαβ protein known to be sweet. The modification of the Brazzein protein produced even sweeter characteristics by altering and lengthening the amino acids in the brazzein peptide chain. Some of these modified proteins were eight times sweeter than brazzein.[42,43]

7.3.2 Thaumatin

Thaumatin is a low-calorie (virtually calorie-free) protein sweetener and flavor modifier. The substance is often used primarily for its flavor modifying properties and not exclusively as a sweetener. Totally natural thaumatin is metabolized by the body like any other dietary protein. The thaumatins were first found as a mixture of proteins isolated from the katemfe fruit (*Thaumatococcus daniellii* Benth), a plant native to tropical West Africa. Some of the proteins in the thaumatin family are natural sweeteners roughly 2 000 times more potent than sugar. Although it tastes very sweet, thaumatin's taste is markedly different from sugar's. The sweetness of thaumatin builds very slowly. The perception of its sweetness lasts a long time leaving a liquorice-like aftertaste at high usage levels.

Naturally occurring thaumatin consists of six closely related proteins (I, II, III, a, b, and c), all with a molecular mass of 22 kDa (207 amino acids).[18] The protein crystallizes in a hexagonal lattice after a temperature shift from 293 to 277 K. The structure has been solved at 1.6 Å resolution. Its fold was found to be identical to that found in three other crystal forms grown in the presence of crystallizing agents of differing chemical natures. It consists of 207 amino acid residues with eight intramolecular disulfide bonds and contains no free cysteine residues (Figs. 2a, 2b).[35]

It aggregates upon heating at pH 7.0 above 70°C, whereupon its sweetness disappears. The protein is approximately 10 000 times sweeter than

1 MATFEIVNRC SYTVWAAASK GDAALDAGGR QLNSGESWTI NVEPGTNGGK IWARTDCYFD
61 DSGSGICKTG DCGGLLRCKR FGRPPTTLAE FSLNQYGKDY IDISNIKGFN VPMNFSPTTR
121 GCRGVRCAAD IVGQCPAKLK APGGGCNDAC TVFQTSEYCC TTGKCGPTEY SRFFKRLCPD
181 AFSYVLDKPT TVTCPGSSNY RVTFCPTA

Fig. 2a. Amino acid sequence of Thaumatin I.

1 MAATTCFFFL FPFLLLLTLS RAATFEIVNR CSYTVWAAAS KGDAALDAGG RQLNSGESWT
61 INVEPGTKGG KIWARTDCYF DDSGRGICRT GDCGGLLQCK RFGRPPTTLA EFSLNQYGKD
121 YIDISNIKGF NVPMDFSPTT RGCRGVRCAA DIVGQCPAKL KAPGGGCNDA CTVFQTSEYC
181 CTTGKCGPTE YSRFFKRLCP DAFSYVLDKP TTVTCPGSSN YRVTFCPTAL ELEDE

Fig. 2b. Amino acid sequence of Thaumatin II.

Table 3. Table of the 20 Amino Acid Names, Three- and One-letter Standard Abbreviations, and Linear Structures

Name	Abbreviation	Linear Structure
Alanine	ala A	CH3–CH(NH2)–COOH
Arginine	arg R	HN=C(NH2)–NH–(CH2)3–CH(NH2)–COOH
Asparagine	asn N	H2N–CO–CH2–CH(NH2)–COOH
Aspartic Acid	asp D	HOOC–CH2–CH(NH2)–COOH
Cysteine	cys C	HS–CH2–CH(NH2)–COOH
Glutamic Acid	glu E	HOOC–(CH2)2–CH(NH2)–COOH
Glutamine	gln Q	H2N–CO–(CH2)2–CH(NH2)–COOH
Glycine	gly G	NH2–CH2–COOH
Histidine	his H	NH–CH=N–CH=C–CH2–CH(NH2)–COOH
Isoleucine	ile I	CH3–CH2–CH(CH3)–CH(NH2)–COOH
Leucine	leu L	(CH3)2–CH–CH2–CH(NH2)–COOH
Lysine	lys K	H2N–(CH2)4–CH(NH2)–COOH
Methionine	met M	CH3–S–(CH2)2–CH(NH2)–COOH
Phenylalanine	phe F	Ph–CH2–CH(NH2)–COOH
Proline	pro P	NH-(CH2)3–CH–COOH
Serine	ser S	HO–CH2–CH(NH2)–COOH
Threonine	thr T	CH3–CH(OH)–CH(NH2)–COOH
Tryptophan	trp W	Ph-NH–CH=C–CH2–CH(NH2)–COOH
Tyrosine	tyr Y	HO–Ph–CH2–CH(NH2)–COOH
Valine	val V	(CH3)2–CH–CH(NH2)–COOH

sugar on a molar basis. It is a protein that tastes intensely sweet only to Old World monkeys and to higher primates, including man, as it has been found that the protein binds to certain elements in the taste pores of Rhesus monkey foliate papillae. Thaumatin has been approved for use in many countries as both a flavor enhancer and a high-intensity sweetener.

Thaumatin is highly water-soluble, and stable to heating under acidic conditions. Thaumatin production is induced in katemfe in a response to an attack on the plant by viroid pathogens. Several members of the thaumatin protein family display significant *in vitro* inhibition of hyphal growth and sporulation by various fungi. The thaumatin protein is considered a prototype for a pathogen response protein domain. This thaumatin domain has been found in species as diverse as rice and *Caenorhabditis elegans*.

Within West Africa, the katemfe fruit has been locally cultivated and used to flavor foods and beverages for some time. The fruit's seeds are encased in a membranous sac, or aril, that is the source of thaumatin. In the 1970s, the Talin Food Company of Merseyside, in the United Kingdom, began extracting thaumatin from the fruit and selling it under the trade name Talin. In 1990, researchers at Unilever reported the isolation and sequencing of the two principal proteins found in thaumatin, which they dubbed *thaumatin I* and *thaumatin II*. These researchers were also able to express thaumatin in genetically engineered bacteria.

Thaumatin has been approved as a sweetener in the European Union (E957), Israel, and Japan. In the USA, it is a "Generally Recognized as Safe" flavoring agent. (FEMA GRAS 3732).

7.3.3 Safety evaluation of thaumatin (Talin protein)

Thaumatin, the sweet proteinaceous extract of the arils of Thaumatococcus daniellii, (Benth.) has been studied for its subacute toxicity in rats and dogs and its ability to produce anaphylactic antibodies following oral administration to rats and normal human subjects. Thaumatin was readily digested prior to absorption in rats and no adverse effects resulted from its continuous administration to rats and dogs at the dietary concentrations of 0, 0.3, 1.0 and 3.0% for 13 weeks. It was not teratogenic when administered orally to rats at 0, 200, 600 and 2 000 mg/kg body weight/day from day 6 to day 15 of gestation. It was without effect on the incidence of dominant lethal mutations when administered on five consecutive days to male mice at 200 and 2 000 mg/kg/day. The lack of mutagenic potential was confirmed in bacterial mutagenic assays with *Salmonella typhimurium* (strains TA1535, TA1537, TA1538, TA98 and TA100) and *Escherichia coli* WP2, at levels of addition of 0.05–50 mg/plate. In rats, thaumatin was found to be a weak sensitizer, comparable with egg albumen, when administered systemically but to be inactive when administered orally. The prick testing of laboratory personnel who had been intermittently exposed by inhalation to thaumatin for periods up to seven years showed that 9.3% (13/140) responded positively to commercial thaumatin, while 30.7% were positive to *Dermatophagoides pteronyssinus* (house dust mite). None of the subjects who gave a positive skin reaction to commercial thaumatin responded to the plant components remaining after the removal of the specific sweet Thaumatin proteins. The challenge tests in man did not demonstrate any oral sensitization. The results indicated that thaumatin, when used as a flavor modifier and extender, and partial sweetener, is unlikely to be hazardous at the anticipated level of consumption.

7.3.4 Monellin

Monellin is a two chain sweet protein from the West African serendipity plant which is estimated to be about 70 000 times sweeter than sugar on a molecule for molecule basis (and probably about 1 000 times sweeter on a gram for gram basis). It is currently being investigated for commercial use as a noncarbohydrate sweetener by the Kirin Brewery company in Japan. While useful in sweetening some materials, it may have limited use because it may denature under high temperature conditions, although, as a protein it may become popular owing to the popularity of the Atkins Nutritional Approach (Atkins diet). Monellin, a sweet protein, consists of two noncovalently associated polypeptide chains, an A chain of 44 amino acid residues and a B chain of 50 amino acid residues (Figs. 3a, 3b).[44]

Monellin can be purified from the fruit of *Dioscoreophyllum cumminsii* grown in West Africa and is approximately 100 000 times sweeter than sugar on a molar basis and several thousand times sweeter on a weight basis.[26] Single-chain monellin (SCM), which is an engineered 94-residue polypeptide, has proven to be as the sweet as native two-chain monellin. It is more stable than the native monellin at high temperature and in acidic

FREIKGYEYQ LYVYASDKLF RADISEDYKT RGRKLLRFNG PVPPP

Fig. 3a. Chain A 44 amino acid residues.

GEWEIIDIGP FTQNLGKFAV DEENKIGQYG RLTFNKVIRP CMKKTIYEEN

Fig. 3b. Chain B 50 amino acid residues.

environments. Native monellin is relatively sensitive to heat or acid treatment, which may cause the separation of the subunits and the denaturation of the protein. Despite misgivings about the stability of the protein to heat and acid, downstream processes have been established. Its D-enantiomer has been crystallized and analyzed by X-ray crystallography at 1.8 Å resolution. Two crystal forms (I and II) were found under crystallization conditions similar, but not identical, to the crystallization conditions of natural L-monellin. One NMR study of a nonsweet analog in which the Asp^{B7} of protein was replaced by Abu^{B7} (L-2-Amino-butylic acid), showed similar 3-dimensional structures of these two proteins, indicating that the lack of the betacarboxyl group in the Abu^{B7} analog is responsible for the loss of its sweetness. Recent research on identifying binding sites on the receptor by means of structure-taste relationships, found that four monellin analogues, [AsnA16]-, [AsnA22]-, [GlnA25]-, and [AsnA26]-monellin were 7 500, 750, 2 500, and 5 500 times as sweet as sucrose on a weight basis, respectively. Thus, among them, [AsnA22]-monellin and [GlnA25]-monellin were less sweet than the native monellin.

7.3.5 Curculin

Curculin which is extracted with 0.5 M sodium chloride from the fruits of *Curculigo* latifolia and purified by ammonium sulphate fractionation, CM-sepharose ion-exchange chromatography and gel filtration.[45] The protein acts as a low calorie sweetener and has a maximum sweetness equal to 0.35 M of sucrose. In addition to its sweetness, curculin also has taste modifying abilities since water and sour substances elicit a sweet taste after consumption of curculin. Currently, there is no other protein that has both sweet taste and taste modifying abilities.

The taste modifying activity of the protein remains unchanged when it is incubated at 50°C for 1 hour between pH 3 and 11. Curculin elicits a sweet taste. It is 20 000 times sweeter than sucrose on a molar basis and 550

1 KFLLTILVTF AAVASLGMAD NVLLSGQTLH ADHSLQAGAY TLTIQNKCNL VKYQNGRQIW
61 ASNTDRRGSG CRLTLLSDGN LVIYDHNNND VWGSACWGDN GKYALVLQKD GRFVIYGPVL
121 WSLGPNGCRR VNGGITVAKD STEPQHEDIK MVINN

Fig. 4. Amino acid sequence of curculin.

times sweeter than sucrose on a weight basis. The sweet taste of curculin disappeared a few minutes after holding the sweet protein in the mouth. Then, the application of water to the mouth elicited a sweet taste. The action of making water sweet lasted for about five minutes. After curculin was held in the mouth for three minutes and its sweet taste had disappeared, it was found that 0.02 M citric acid after 10 M curculin held in the mouth was equivalent to that of 0.35 μM sucrose. This taste modifying action of curculin lasted for about ten minutes.

The molecular weight of Curculin was determined by low angle laser light scattering and was found to be 27 800.[45] Its three-dimensional model has been built from the X-ray coordinates of GNA, a mannose-binding lectin from snowdrop (*Galanthus nivalis*). The three mannose-binding sites present in GNA were found in curculin but were not functional. Some well exposed regions on the surface of the three-dimensional model of the said protein could act as epitopes responsible for the sweet tasting properties of the protein. The protein can be crystallized by the vapor diffusion method using polyethylene glycol 400 as a precipitant. The crystals belong to orthorhombic space group P2(1)2(1)2(1) with unit cell dimensions: a = 105 Å, b = 271 Å, c = 48.7 Å. The crystals diffract X-rays to a resolution of 3.0 Å and are suitable for X-ray crystallographic studies.[46]

7.3.6 Mabinlin

Mabinlin is a sweet protein with the highest known thermostability. It is extracted from the mature seeds of the Chinese plant, *Capparis masaikai Levi* (local name mabinlang). It grows in a subtropical region of the Yunnan province of China. It is suspected that at least four sweet tasting polypeptides exist in the *C. masaikai* plant. The most studied sweet protein has been named mabinlin II. Its sweetness was estimated to be around 100 times that of sucrose on weight basis. It consists of two polypeptide chains, an A chain with 33 amino acid residues, and a B chain composed of 72 amino acid residues. The B chain contains two intramolecular disulfide bonds and is connected to the A chain through two intermolecular disulfide bridges (Figs 5a, 5b). Its heat stability is due to the presence of these four disulfide bridges.

E L W R C Q R Q F L Q H Q R L R A C Q R
1 5 10 15 20
F I H R R A Q F G G Q P D
21 25 30 33

Fig. 5a. The amino acid sequence of the A chain of mabinlin II.

E P R R P A L R Q C C N Q L R Q V D R P
1 5 10 15 20
C V C P V L R Q A A Q Q V L Q R Q I I Q
21 25 30 35 40
G P Q Q L R R L F D A A R N L P N I C N
41 45 50 55 60
I P N I G A C P F R A W
61 65 70 72

Fig. 5b. The amino acid sequence of the B chain of mabinlin II.

E P L C R R Q F Q Q H Q H L R A C Q R Y I R R R A
1 10 20

Q R G G L V D
30 32

Fig. 6a. The amino acid sequence of the A chain of mabinlin I-1.

E Q R G P A L R L C C N Q L R Q V N K P C V C P V
1 10 20
L R Q A A H
31
Q Q L Y Q G Q I E G P R Q V R Q L F R A A R N L P
32 40 50
N I C K I P A V G R C Q F T R W
60 63 64 70 72

Fig. 6b. The amino acid sequence of the B chain of mabinlin I-1.

Other variants of mabinlin have been isolated.[36] They have been named mabinlin I-1, III and IV (Figs. 6a, 6b, 7a, 7b, 8a, 8b). The sweet activity of mabinlin I-1 disappears after 1 hour incubation at 80°C while Mabinlin II, III and IV remain unchanged after 1 hour at 80°C.

7.3.7 Pentadin

Pentadin is a sweet protein extracted from the plant, *Pentadiplandra brazzeana* Baillon, a shrub found in tropical forests of a few African

E P L C R R Q F Q Q H Q H L R A C Q R Y L R R R A Q R G G
1 10 20

L A D
30 32

Fig. 7a. The amino acid sequence of the A chain of mabinlin III.

E Q R G P A L R L C C N Q L R Q V N K P C V C P V L R Q A
1 10 20

A H Q Q
30 33

L Y Q G Q I E G P R Q V R R L F R A A R N L P N I C K I P A
34 40 50 60

V G R C Q F T R W
67 68 70 72

Fig. 7b. The amino acid sequence of the B chain of mabinlin III.

E P L C R R Q F Q Q H Q H L R A C Q R Y L R R R A Q R G
1 10 20 28

Fig. 8a. The amino acid sequence of the A chain of mabinlin IV.

E Q R G P A L R L C C N Q L R Q V N K P C V C P V L R Q A
1 10 20

A H Q Q
30 33

L Y Q G Q I E G P R Q V R R L F R A A R N L P N I C K I P A
34 40 50 60

V G R C Q F T R W
67 68 70 72

Fig. 8b. The amino acid sequence of the B chain of mabinlin IV.

countries. The protein was reported to be around 500 times sweeter then sucrose on a weight basis. Electrophoretic studies in the presence and absence of 2-mercaptoethanol suggested that the mature protein consists of subunits coupled by disulphide bonds.

An aqueous extract from the pulp of this plant, Baillon (Pentadiplandraceae), yielded a strong sweet tasting material. This sweet principle was first isolated by van der Wel *et al.* in 1989[21] basis by water extraction, ultra and gel filtration. The conclusion that this substance must be of a proteinaceous nature was based on amino acid analysis, characteristic UV-absorption spectrum and a positive color reaction with Coomassie brilliant blue. The molecular weight of the subunit of the sweet protein was estimated to be (small tilde) 12 000 Daltons. The sweetness intensity of the whole protein was (small tilde) 500 times that of sucrose on a weight basis. The taste response in a Rhesus monkey to a 0.1% solution was comparable to the response to a 0.02% monellin solution. The name "pentadin" was proposed by H van der Wel, and his colleagues for this sweet-tasting protein.[21]

7.3.8 Miraculin

Richadella dulcifica or *Synsepalum dulcificum* is a native shrub of tropical West Africa. The local names for the plant include *taami, asaa*, and *ledidi*. It yields red berries which have the unusual property in modifying sour taste into sweet taste. The red fruit bears a single olive-shaped seed it is about 2 cm to 3 cm long and quite tasteless. If one's tongue is coated with the pulp of this fruit, one can swallow a litre of vinegar followed by a kilogram of lemons without a wince. And the effect can last up to two hours. The issue is that the consumption of a miracle berry will modify a sour taste into a sweet taste. It is something that the West Africans have known for a long time and for years, they have been using it in food and beverages to suppress sourness.

Because of this unusual property, the berry has been called "miracle fruit." Kurihara and Beidler first isolated the active principle of the miracle fruit and showed that it is a basic glycoprotein.[16] Brouwer and his colleagues also isolated the active principle and named it "miraculin".[15]

Miraculin is a glycoprotein which not only shields a sour taste, it can also make you believe that what you are eating or drinking is actually sweet! It is a 190 amino acid glycoprotein and its amino acid sequence was determined completely by Theerasilip and his Japanese colleagues.[47] Miraculin is extracted with 0.5 M NaCl solution. The extracted solution is colorless and shows the strong sweet-inducing activity. Miraculin was purified from the extracted solution by ammonium sulfate fractionation, CM-Sepharose ion-exchange chromatography, and concanavalin A-Sepharose affinity chromatography. The purified miraculin thus obtained gave a single sharp peak in reverse phase high performance liquid chromatography, indicating that it is highly pure. The sample also gave a single band having a molecular weight of 28 000 in sodium dodecyl sulfate-polyacrylamide gel electrophoresis. This value was much lower than the values reported previously (40 000–48 000). The amino acid composition of the purified miraculin was determined. The sequence analysis of the purified miraculin indicated that it is composed of a pure single polypeptide and identified 20 amino-terminal amino acids. The purified miraculin contained as much as 13.9% of sugars, which consisted of glucosamine, mannose, galactose, xylose, and fucose in a molar ratio of 3.03:3.00:0.69:0.96:2.12. The purified miraculin contains almost 14% sugar: glucosamine, mannose, galactose, xylose and fucose.

A structural model of the sweet-inducing protein and its binding to the sweet receptor has been suggested. Miraculin may bind to sweet receptors although, in the absence of sourness, its active site does not. As a result, there is no sweet taste. However, when sour substances are presented to the tongue, the sweet receptors undergo a conformational change and, in doing so, give miraculin the opportunity to reposition its active site within the sweet receptor. The net result is a strong sweet taste in the mouth. And since miraculin binds particularly firmly to the receptor, the sweet taste can last a maximum of two hours. It is said that the actual taste of the food is kept yet the sourness is warded off. Hence, a slice of lemon would taste like lemon candy.

The detailed mechanism of its taste-inducing behavior is still unknown. It has been suggested that the miraculin molecule can change the structure of taste cells on the tongue. As a result, the sweet receptors are activated by acids, which are sour in general. This effect remains until the taste buds return to normal.

While attempts to express it in *E. coli* bacteria have failed, Japanese researchers have succeeded in preparing genetically modified plants, such as lettuce, that express miraculin. This efficient method to produce miraculin might be applied to create a new sugar-free sweetener.

7.4 CONCLUSIONS

There may be a bright future for the high molecular weight sweet proteins for commercial application as food additives. They are much sweeter than sugar, and in general, are more stable to heat and lighter than the small molecular weight natural sweeteners such as stevioside and glycyrrhizin. There is still some controversy about the safety of some of the natural sweeteners such a stevioside and rebudioside A. The FDA has allowed the sweet proteins to be available for the consumers market. They would be extremely useful as natural low calorie sweeteners for people suffering from diseases linked to consumption of sugar such as obesity, diabetes and hyperlipemia.

REFERENCES

1. Kim NC, Kinghorn AD, Highly Sweet Compounds of Plant Origin, *Arch Pharm Res* **25**(6):725–746, 2002.
2. Kinghorn AD, Soejarto DJ, Discovery of terpenoid and phenolic sweeteners from plants, *Pure Appl Chem* 74(7):1169–1179, 2002.
3. Bertoni MS, *Revista de Agronomia de l'Assomption*, **1**:35, 1899.
4. Crammer B, Ikan R, in Grenby TH (ed.), *Developments in Sweeteners-3*, Elsevier Applied Science, London and New York, 1987.
5. Crammer B, Ikan R, Weinstein V, *Process for the extraction of Diterpene Glycosides From certain Perennial Plants of the Compositae Family*, Israel Patent No. 81351, 1990.
6. Kinghorn AD (ed.) *Stevia*, Taylor & Francis Inc, New York, USA, 2002.
7. Bridel M, Lavieille R, Le Principe a saveur sucree du Kha-he-e(Stevia Rebaudiana Bertoni) Proprietes du stevioside, *Journal de Pharmacie et de Chimie*, **14**:99–113; 154–163, (1931a, b, c, d and e).
8. Mosettig E, Ness WR, Stevioside. II. The structure of the aglucon. *J Org Chem* **20**:884–899, 1955.
9. Kim SH, Dubois GE, Natural High Potency Sweeteners in Marie S, Piggott JR (eds.), *Handbook of Sweeteness*, Avi, New York, 116–185, 1991.

10. Kohda H, Kasai R, Yamasaki K, Murakami K, Tanaka O, New Sweet diterpene glycosides from stevia rebaudiana, *Phytochemistry* **15**:981–983, 1976.
11. Brandle J, *Stevia Rebaudiana with altered steviol glycoside composition*, US Patent No. 6,255,557, 2001.
12. Dobberstein RH, Ahmed MS, *Extraction, separation and recovery of diterpene glycosides from Stevia rebaudiana plants.* US Patent No. 4,361,697, 1982.
13. Zaffaroni A, *Nonabsorbable, Nonnutrative Sweeteners*, US Patent No. 3,876,816, 1975.
14. Morris JA, *Lloydia* **39**:25–38, 1976. [See reference 2 in FEBS 526 1–4, 2002]
15. Brouwer JN, van der Wel H, Francke A, Henning GJ, Miraculin, the sweetness-inducing protein from miracle fruit, *Nature* **220**:373–374, 1968.
16. Kurihara K, Beidler LM, Taste modifying protein from miracle fruit, *Science* **161**:1241–1243, 1968.
17. Morris JA, Cagan RH, *Biochim Biophys Acta* **261**:112–122, 1972.
18. Van der Wel H, Loeve K, Isolation and characterization of Thaumatin I and II, the sweet tasting proteins from Thaumatococcus daniellii Benth, *Eur J Biochem* **31**:221–225, 1972.
19. Liu X, Maeda S, Hu Z, Aiuchi T, Nakaya K, Kurihara Y, Purification, complete amino acid sequence and structural characterization of the heat stable sweet protein, marbinlin II, *Eur J Biochem* **211**:281–287, 1993.
20. Hu Z, He M, Studies on mabinlin, a sweet protein from the seeds of Capparis masaikai levl. L. extraction, purification and certain characteristics, *Acta Botan Yunnan*, **5**:207–212, 1983.
21. Van der Wel H, Larson G, Hladik A, Hladik CM, Hellekant G, Glaser D, Isolation and characterization of pentadin, the sweet principle of Pentadiplandra brazzeana Baillon, *Chem Senses* **14**:75–79, 1989.
22. Ming D, Hellekant G, Brazzein, a new high potency thermostable sweet protein from Pentadiplandra Brazzeana B, *FEBS Lett*, **355**:106–108, 1994.
23. Takahashi N, Hitotsuya H, Hanzawa H, Arata Y, Kurihara Y, Structural study of asparagine-linked oligosaccharide moiety of taste-modifying protein, miraculin, *J Biol Chem* **265**:7793–7798, 1990.
24. Kant R, Sweet proteins — Potential replacement for artificial low calorie sweetners, *Nutrition Journal* **4**:5, 2005.
25. Caldwell JE, Abildgaard F, Dzakula Z, Ming D, Hellekant G, Solution Structure Of The Thermo Stable Brazzein, *Nature Structure Biology* **5**:427–431, 1998.
26. Margolskee RF, Molecular mechanisms of bitter and sweet taste transduction, *J Biol Chem* **277**:1–4, 2002.
27. Nelson G, Hoon MA, Chandrashekar J, Zhang Y, Ryba NJ, Zuker CS, Mammalian sweet taste receptors, *Cell* **106**:381–390, 2001.

28. Straszewski L, Li X, Xu H, Durick K, Zoller M, Adler E, Human receptors for sweet and umami taste, *Proc Natl Acd Sci USA* **99**:4692–4696, 2002.
29. Conte C, Ebeling M, Marcuz A, Nef P, Andres-Barquin PJ, **98**:45–53, 2002.
30. Nelson G, Hoon MA, Chandrashekar J, Feng G, Ryba NJ, Zhao G, An amino-acid taste receptor, *Nature* **416**:199–202, 2002.
31. Hellekant BG, Ming D, *Brazzein Sweetener*, US Patent No. 5,346,998, 1994.
32. Hellekant BG, Ming D, *Brazzein Sweetener*, US Patent No. 5,741,537, 1998.
33. Hladik A, Bahuchet S, Ducatillion C, Hladik CM, Les plantes à tubercules de la foret dense d'afrique centrale, *Rev Ecol* (Terre Vie), **39**:249–290, 1984.
34. Kurihara Y, Harada S, Maeda S, Kai Y, Kasai N, Crystallization and preliminary X-ray diffraction studies of curculin: A new type of sweet protein having taste-modifying action, *J Mol Biol* **238**:286–287, 1994.
35. Lee JH, Weickmann JL, Koduri RK, Ghosh-Dastidar P, Saito K, Blair LC, Date T, Lai JS, Hollenberg SM, Kendall RL, Expression of synthetic thaumatin genes in yeast, *Biochemistry* **27**(14):5101–5107, 1988.
36. Nirasawa S, Nishino T, Katahira M, Uesugi S, Hu Z, Kurihara Y, Structures of heat-stable and unstable homologues of the sweet protein mabinlin, *Eur J Biochem* **223**:989–995, 1994.
37. Ming D, Hellekant, US Patent No. 5,346,998, 1994.
38. Somoza JR, Jiang F, Tong L, Kang CH, Cho JM, Kim SH, *J Mol Biol* **234**:390–404, 1993.
39. Ko TP, Day J, Greenwood A, McPherson A, *Acta Crystallogr D* **50**:813–825, 1994.
40. Ogata CM, Gorden PF, de Vos AM, Kim SH, *J Mol Biol* **228**:893–908, 1992.
41. Ming D, Hellekant G, Zhong H, *Acta Botanica Yunnanica* **18**:123–133, 1996.
42. Jin Z, Markley JL, Assadi-Porter FM, Hellekant BG, *Protein Sweetener*, US Published Patent Application No. 2004/0018290, 2004.
43. Markley JL, Assadi-Porter FM, *Protein Sweetener*, US Patent No. 6,274,707, 2001.
44. Kohmura M, Nio N, Ariypshi Y, *Agric Biol Chem* **54**:2219–2224, 1990.
45. Yamashita H, Theerasilp S, Aiuchi T, Nakaya K, Nakamura Y, Kurihara Y, Purification and complete amino acid sequence of a new type of sweet protein with taste-modifying activity, curculin, *J Biol Chem* **265**(26):15770–15775, 1990.
46. Harada S, Otani H, Maeda S, Kai Y, Kasai N, Kurihara Y, Crystallization and preliminary X-ray diffraction studies of curculin. A new type of sweet protein having taste modifying action, J Mol Biol **238**:286–287, 1994.
47. Theerasilp S, Hitotsuya H, Nakajo S, Nakaya K, Nakamura Y, Kurihara Y, Complete amino acid sequence and structure characterization of the taste-modifying protein, miraculin, *J Biol Chem* **264**(12):6655–6659, 1989.

Chapter 8

NATURAL PRODUCTS FOR PEST MANAGEMENT

Stephen O. Duke, Agnes M. Rimando, Kevin K. Schrader, Charles Cantrell, Kumudini M. Meepagala, David E. Wedge, Nurhayat Tabanca, and Franck E. Dayan

8.1 INTRODUCTION

Most organisms synthesize secondary products with biological activity that is useful in their defense. Defense can be against vertebrates, arthropods, molluscs, plants (both algal and higher plants), and microbes. Many of these compounds have been used from ancient times to the present as pharmaceuticals. Others have had utility as pesticides or as the molecular building blocks of pesticides. This brief chapter will highlight some of the more important natural product-based pesticides, as well as some of those that our group has discovered. More detailed reviews and books on this general subject are available,[1,2] as are reviews on specific types of natural product-based pesticides.[3]

8.2 ALGICIDES

Throughout the world, there are many different types of innocuous phytoplankton that can inhabit freshwater ecosystems. However, in some of these ecosystems, certain species of cyanobacteria (blue-green algae) are highly undesirable due to their production of toxins or odorous compounds.

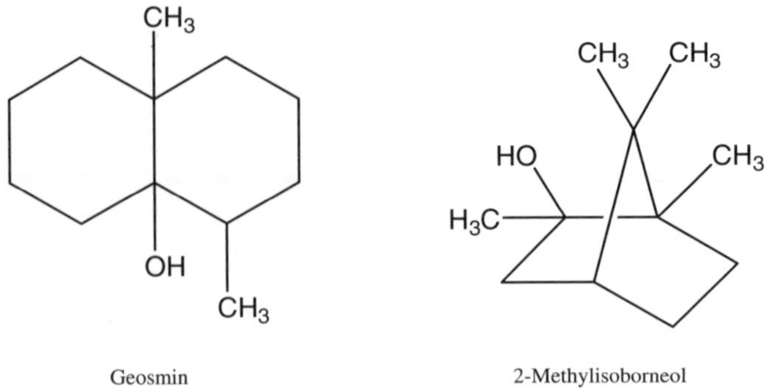

Fig. 1. Structures of geosmin and 2-methylisoborneol, the two most common causes of off-flavor in farm-raised channel catfish.

For example, in channel catfish (*Ictalurus punctatus*) aquaculture ponds in the southeastern USA, some species of cyanobacteria produce "earthy" and "musty" metabolites such as geosmin and 2-methylisoborneol (MIB) (Fig. 1). These compounds are rapidly absorbed across the gills of the catfish and accumulate in the flesh of the catfish, thereby rendering them unpalatable and unmarketable.

In the USA, the MIB-producing cyanobacterium *Planktothrix perornata* (Skuja) Anagnostidis & Komárek (syn. *Oscillatoria perornata*) has been attributed as the major contributor to musty off-flavor problems in catfish ponds in West Mississippi.[4] It is also a contributor to off-flavor problems in East Mississippi and Alabama.[5] Algicides have been the management approach used by most catfish producers in the southeastern USA. The two most common types of algicides currently used are copper-based products, such as chelated-copper compounds and copper sulfate, and the synthetic herbicide diuron [*N′*-(3,4-dichlorophenyl)-*N*,*N*-dimethylurea]. Copper-based products and diuron have several negative attributes including high environmental persistence and broad-spectrum toxicity towards phytoplankton. The misapplication of these products can lead to poisoning of the entire phytoplankton community and subsequent water quality problems such as low dissolved oxygen, thereby leading to stressed catfish and potential die-offs of the entire crop of catfish in the pond.

Until recently, there has been limited research to discover natural and natural product-based algicides as alternatives to the compounds currently used for the management of noxious cyanobacteria. One of the first concerted attempts to provide a natural algicide found ricinoleate to be selectively toxic towards cyanobacteria in the laboratory.[6] Ricinoleate is in the same class of compounds as the oxygenated fatty acids isolated from the aquatic plant *Eleocharis microcarpa*. It is found to inhibit the growth of cyanobacteria.[7] This discovery lead to the development of the commercial algicide, Solricin 135®, containing the active constituent potassium ricinoleate. However, efficacy testing determined that applications of potassium ricinoleate to catfish ponds did not reduce the abundance of cyanobacteria and did not prevent severe musty off-flavor episodes in catfish from the treated ponds.[8]

Another natural-based approach that has been investigated for managing odor-causing cyanobacterial blooms in catfish ponds is the use of decomposing barley straw. Decomposing barley straw can be used to control algal blooms in freshwater canals.[9] Cyanobacteria appear to be very sensitive to the antialgal compounds released by the decomposing barley straw.[10] One hypothesis provided by Barrett[11] proposed that lignin components in the straw could be oxidized under aerobic conditions to form quinones and then humic acids. Cooper and Zika[12] found that sunlight catalyzes the production of singlet oxygen or hydrogen peroxide from humic acids in surface water. Hydrogen peroxide and compounds that break down in water to form hydrogen peroxide directly such as sodium carbonate peroxyhydrate (SCP) have been demonstrated to inhibit the growth of certain species of cyanobacteria.[13–15] However, research by Wills[16] found that decomposing barley straw in catfish ponds did not improve the flavor quality of the catfish from the treated ponds compared to those from control ponds.

Several commercial products for cyanobacteria control have been developed from decomposing barley straw constituents. Schrader[17] recently tested two of these commercial products in the laboratory for their effect on the growth of *P. perornata* and found that neither commercial product was effective in inhibiting the growth of *P. perornata*, even at test concentrations greater than 100 times the label-recommended application rates. Commercial products containing SCP have also been developed recently and approved by the USEPA for use in aquaculture. However, recent efficacy testing of one of these commercial products did not consistently reduce the abundance of *P. perornata* in catfish ponds (Schrader and Tucker, unpublished observations). Additional efficacy studies are needed to determine the potential of the use of SCP-formulated products in managing musty off-flavor episodes in pond-cultured catfish.

Recent research by Schrader *et al.*[19] has focused on using a rapid bioassay to discover natural compounds from plants and marine organisms to identify novel selective algicides for managing odor-producing cyanobacteria in aquaculture ponds. Among the thousands of crude extracts and natural compounds evaluated so far, one of the most promising compounds that has been discovered is 9,10-anthraquinone.[19] The efficacy testing of 9,10-anthraquinone in limnocorrals (fiberglass enclosures) placed in catfish ponds found that the abundance of *P. perornata* and MIB levels in pond water were not reduced by the applications of 9,10-anthraquinone dissolved in ethanol.[20] Due to the lack of solubility in water, different formulations of 9,10-anthraquinone (e.g. Tween 80 and canola oil emulsion) were subsequently tested to attempt to maintain 9,10-anthraquinone in the water column of the pond. None of the formulations of 9,10-anthraquinone were effective in reducing the abundance of *P. perornata*.[20] Ultimately, a different approach was taken in which the chemical structure of 9,10-anthraquinone was modified to impart water solubility.

The efficacy testing of two of these water-soluble derivatives of 9,10-anthraquinone found them to be effective in reducing the abundance of *P. perornata* and MIB levels in catfish ponds. They were also much less persistent than diuron in the pond water.[20] These novel derivatives of 9,10-anthraquinone have recently been patented.[21] One of the more promising anthraquinone analogs, anthraquinone-59 (Fig. 2), has undergone additional testing, including toxicological evaluation towards channel catfish that determined a safety margin of at least one order of magnitude between the concentration effective in reducing the abundance of *P. perornata* in catfish ponds and the 96-h LC50 for channel catfish.[22]

Another quinone-related natural compound evaluated by Schrader *et al.*[23] is SeaKleen®, a commercial biocide used to treat the ballast water of ships to help prevent the spread of pest organisms such as algae.[24] SeaKleen consists of menadione sodium bisulfite, a water-soluble form of menadione

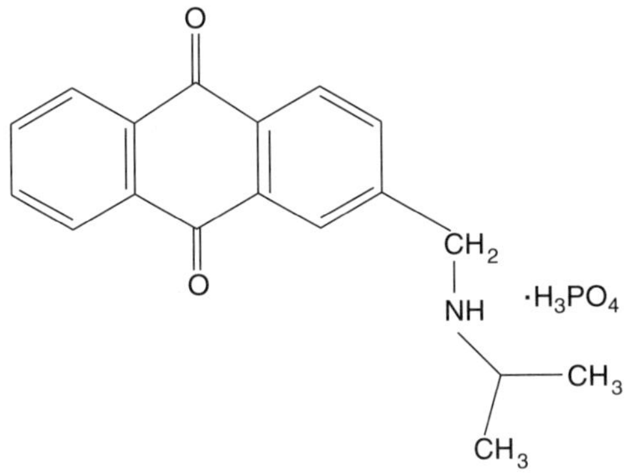

Fig. 2. Water-soluble derivative of anthraquinone referred to as anthraquinone-59.

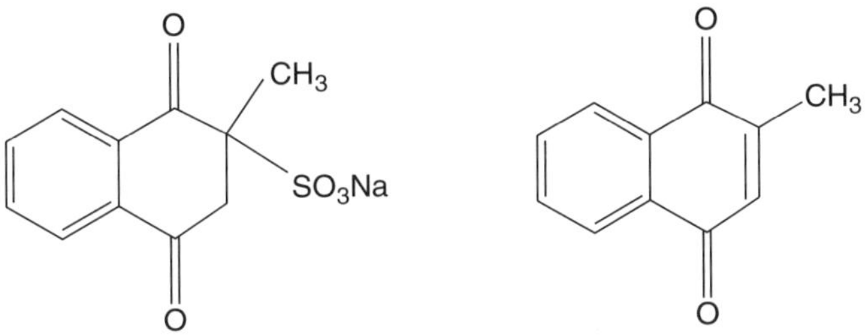

(A) (B)

Fig. 3. Structures of (A) menadione sodium bisulfite and (B) menadione.

that is also known as vitamin K_3 or 2-methyl-1,4-naphthoquinone (Fig. 3). In limnocorrals, SeaKleen was effective in reducing the abundance of *P. perornata* at the application rates of 1.3 mg/L.[23] In addition, the half-life of SeaKleen in pond water was determined to be less than one day compared to two weeks for diuron.

Further evaluation of SeaKleen and other water-soluble derivatives of 9,10-anthraquinone for use as selective algicides in catfish aquaculture will require additional research such as determining the environmental fate of these quinone-based algicides and whether there is any potential accumulation in the flesh of channel catfish. In addition, studies required by the USEPA for the consideration of the approval of quinone-based algicides in food-fish production ponds will require at least several more years.

8.3 INSECTS

8.3.1 Insect Repellents

The United States Centers for Disease Control and Prevention (CDC) recommends the use of products containing active ingredients which have been registered with the U.S. Environmental Protection Agency (EPA) for use as repellents applied to skin and clothing. EPA registration of repellent active ingredients indicates that the materials have been reviewed and approved for efficacy and human safety when applied according to the instructions on the label (http://www.cdc.gov/ncidod/dvbid/westnile/RepellentUpdates.htm). Two of the active ingredients registered with the EPA, DEET (*N*,*N*-diethyl-*m*-toluamide) and picaridin [2-(2-hydroxyethyl)-1-piperidinecarboxylic acid 1-methylpropyl ester] (Fig. 4), have arguably demonstrated higher degrees of efficacy in the peer-reviewed, scientific literature.[25] Also recognized by the CDC as effective insect repellents are those containing oil of lemon eucalyptus [primarily *p*-menthane 3,8-diol (PMD)] (Fig. 4), a plant-derived active ingredient. This recognition by the CDC is a testament to the changing perceptions in the USA and the desire by consumers for effective and natural alternatives to conventional synthetic based repellents.

In addition to the repellent constituents mentioned above, numerous commercial botanical repellents have recently been evaluated in the popular

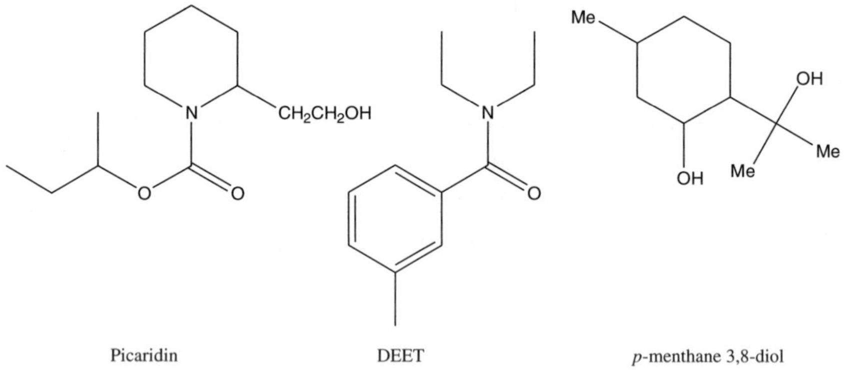

Fig. 4. Structures of three insect repellent compounds, picardin, DEET, and *p*-menthane 3,8-diol.

press[26,27] as well as the peer-reviewed scientific literature.[28] Barnard and Xue evaluated the effectiveness of four synthetic mosquito repellents (Autan [10% KBR3023], IR3535 [7.5%], Off! [15% deet], Skinsations [7% deet]) and eight natural (primarily plant extracts and/or essential oils) product-based repellents (Bite Blocker [2% soybean oil], ByGone, GonE!, Natrapel [10% citronella], Neem Aura, Sunswat, MosquitoSafe [25% geraniol], and Repel [26% *p*-menthane-3,8-diol]) against *Aedes albopictus*, *Culex nigripalpus*, and *Ochlerotatus triseriatus*. When the estimated mean protection time (eMPT) responses for each repellent were averaged for all three mosquito species, Autan, Bite Blocker, Off!, and Repel prevented biting for ≥7.2 hours; IR3535, MosquitoSafe, and Skinsations for 3.2 hours–4.8 hours; and ByGone, Natrapel, GonE, NeemAura, and SunSwat for 0.9 hours–2.3 hours.

Repellent plant essential oils, like some of those mentioned above, are generally composed of complex mixtures of terpenoids and phenolics. Recently, DEET, plant-based repellent Repel Care (*Eucalyptus citriodora* and turmeric oils), and essential oils from ginger root rhizomes (*Boesenbergia rotunda*), guava leaves (*Psidium guajava*), and turmeric rhizomes (*Curcuma longa*) steam distilled and formulated as insect repellents were evaluated in the field on human volunteers against hematophagous mosquitoes (mainly *Aedes albopictus*) and black flies (*Simulium nigrogilvum* and *S. chumpornense*) in Thailand. All five agents provided complete protection from mosquito landing and biting for up to 9 hours (experiment duration). Similar results were obtained with the five products against black flies, providing 100% protection for 9 hours, but 82%–96% protection at 10 hours and 11 hours post-treatment.[29]

Numerous additional examples of the evaluation of plant essential oils as insect repellents exist in the literature;[30,31] however, rarely are bioassay-guided fractionations performed to determine the constituents responsible for the activity of these plant essential oils. When attempts are made to identify active constituents, they are often of a non-systematic manner and usually focus on those constituents that are commercially available or the major constituents. Such approaches have the potential of missing unusual or novel structural types which could be highly active compounds with the potential for commercial development.

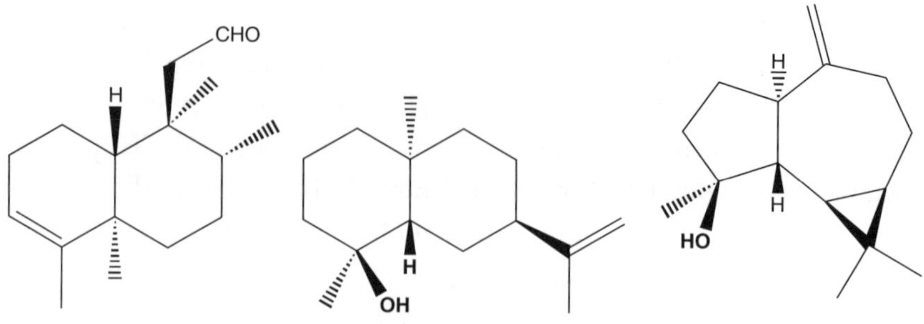

Fig. 5. Three insect-repelling compounds isolated from *Callicarpa americana*.

Cantrell *et al.*[32] investigated the leaf essential oil extract of a folk remedy from the USA, the American beautyberry (*Callicarpa americana*), using a bioassay-guided approach to identify the active constituents. *C. americana* had been used as recently as the 1980s in Mississippi by crushing fresh leaves attached to a stem and placing this stem under the harness of stock animals to repel arthropod pests (ticks, mosquitoes, flies). Researchers isolated two common terpenoids, intermedeol and spathulenol, as well as a newly isolated compound, callicarpenal, with a unique truncated transclerodane skeleton (Fig. 5). All three compounds possessed significant repellent activity against *Aedes aegypti* and *Anopheles stephensi* in laboratory bioassays.

Additional examples of isolated natural repellents include the compounds sandaracopimarinol and (1*S*,6*R*)-2,7(14),10-bisabolatrien-1-ol-4-one isolated and identified from *Cryptomeria japonica* as repellents against *Armadillidium vulgare* which is well known as an unpleasant household pest and as a vegetable pest in Japan (Fig. 6).[33] These compounds strongly repelled *A. vulgare* when they were combined, although each compound alone did not show any activity.

Okunade and Wiemer[34] reported isolation of a potent ant-repellent compound, (-)-loliolide (Fig. 7) from *Xanthoxyllum setulosum*. The plant is one of the many native Costa Rican plants found to escape attack of the highly polyphagous leafcutter ants *Atta cephalotes*. Additional leafcutter ant repellent constituents have been isolated from *Piper tuberculatum*.[35]

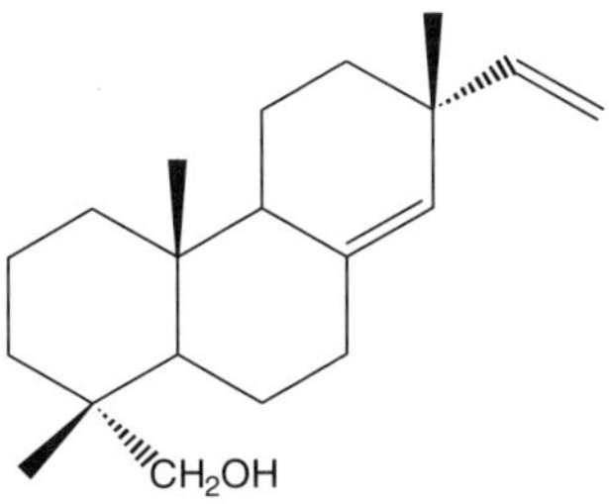

Sandaracopimarinol

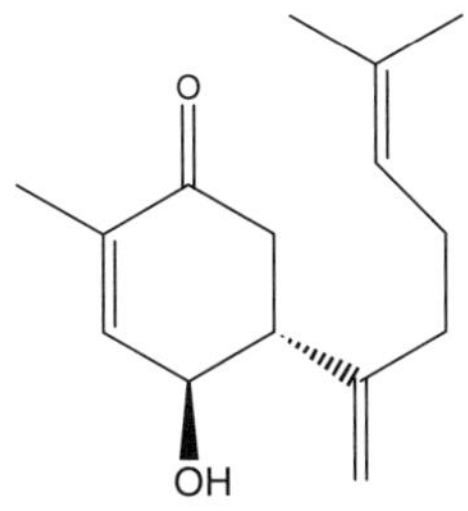

(1S,6R)-2,7(14),10-bisabolatrien-1-ol-4-one

Fig. 6. Insect repellents identified from *Cryptomeria japonica.*

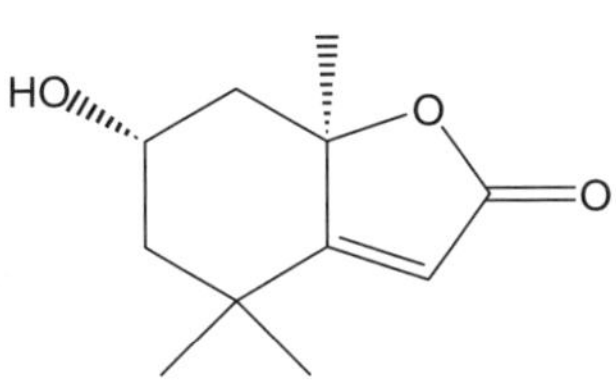

(-)-loliolide

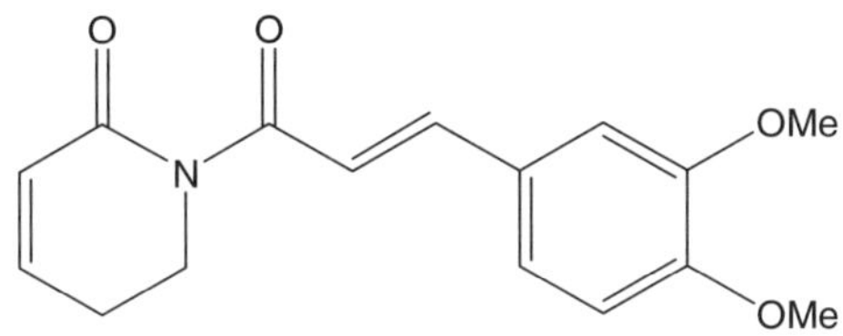

Piplaroxide

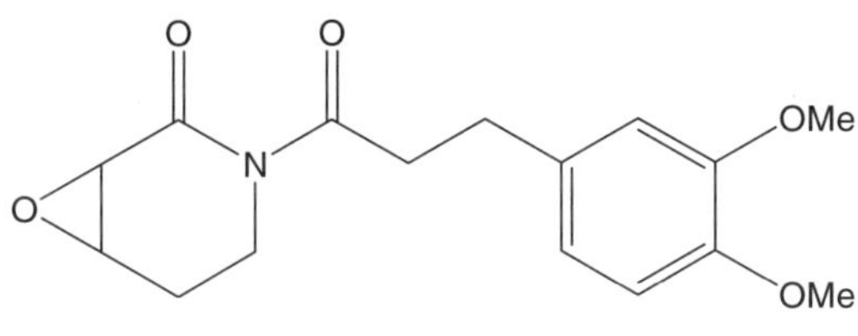

Demethoxypiplartine

Fig. 7. Insect repellents isolated from *Xanthoxyllum setulosum* and *Piper tuberculatum.*

Of the three compounds isolated, piplaroxide and demethoxypiplartine (Fig. 7) demonstrated significant activity in a laboratory bioassay measuring repellency to the leafcutter ant *Atta cephalotes.*

Evidence of repellent properties in catnip, *Nepeta cataria,* to flies and cockroaches was observed in preliminary studies.[36] This study compared catnip essential oil obtained by steam distillation and elemol (Fig. 8), a major constituent of osage orange essential oil, to current commercial repellents. These comparative studies found both the catnip steam distillate and elemol to be as good, and in some cases better, at repelling house

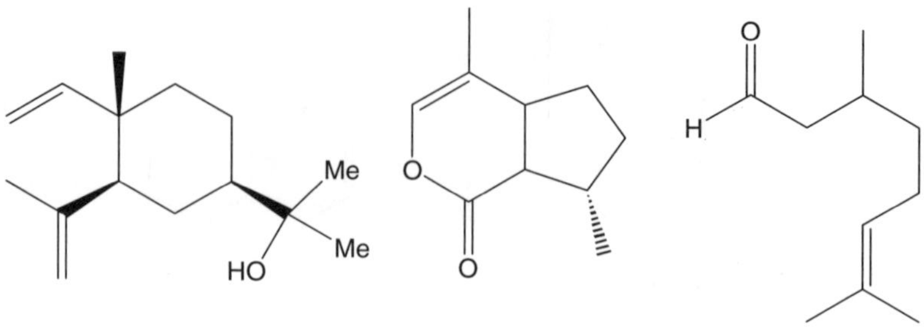

Elemol Nepetalactone Citronellal

Fig. 8. Insect repellent compounds from catnip (elemol and nepetalactone) and osage orange (citronellal).

flies, *Musca domestica*, and American cockroaches, *Periplaneta americana*, as DEET or citronellal (Fig. 8). Nepetalactone (Fig. 8), the major constituent of catnip essential oil, is present as two isomers, and previous studies have shown that the E,Z-nepetalactone isomer is even more repellent to cockroaches than the dominant isomer, Z,E-nepetalactone.[37]

8.3.2 Insecticidal Natural Products

The peer-reviewed literature is full of reports on natural products possessing insecticidal activity [[1,38] and references therein]; therefore, the focus of this section will be on both recent and well investigated natural products. Of those well investigated natural products, at least three natural product derived products (neem, spinosyn, and pyrethrum) seem to be used more than all the others.

The seeds from the Indian neem tree, *Azadirachta indica*, are the source of two types of neem-derived botanical insecticides; neem oil and medium polarity extracts. Neem seeds contain numerous azadirachtin (Fig. 9) analogs, but the major form is azadirachtin and the remaining minor analogs are likely to contribute little to the overall efficacy of the extracts.[38] Typically, solvent partitions or other chemical processes are required to concentrate this active ingredient to the level of 10% to 50% seen in the technical grade material used to produce commercial products.[38]

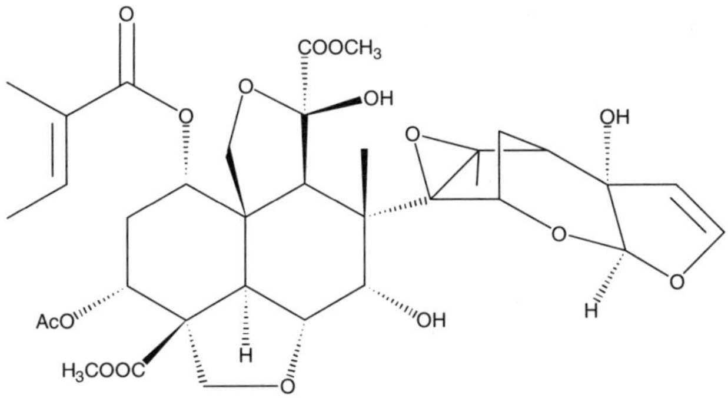

Azadirachtin

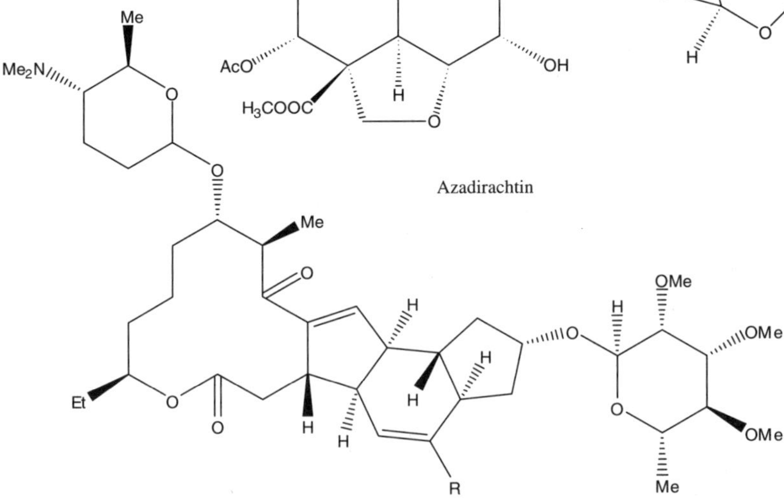

Spinosyn AR = H
Spinosyn DR = Me

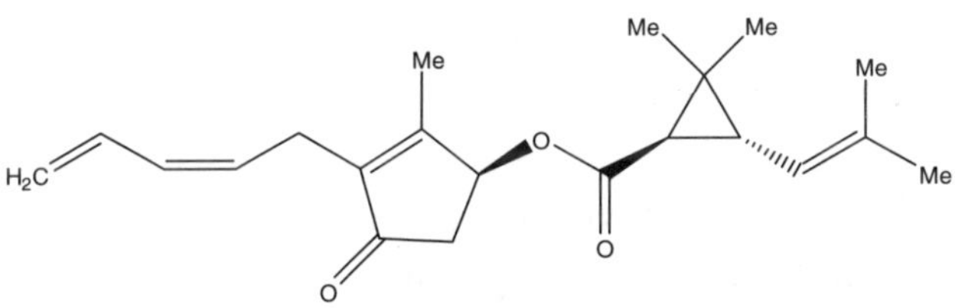

Pyrethrin I

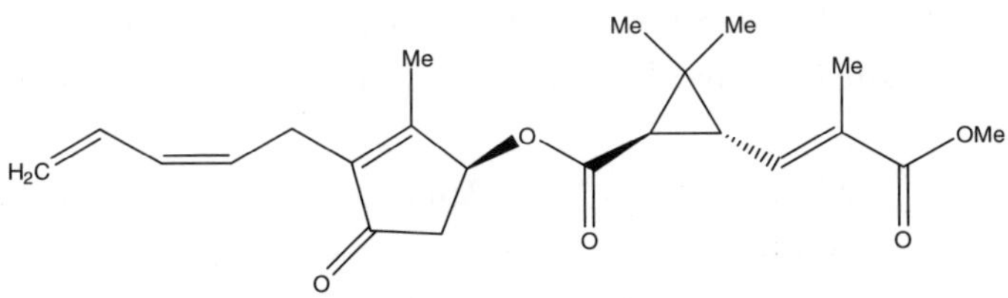

Pyrethrin II

Fig. 9. Natural products that are used commercially as insecticides.

Spinosad (Fig. 9) is a mixture of spinosyn A and spinosyn D, originally isolated from the soil Actinomycete, *Saccharopolyspora spinosa*. Spinosad is recommended for the control of a very wide range of caterpillars, leaf miners, thrips and foliage-feeding beetles. Spinosad is sold as an aqueous based suspension concentrate formulation under several trade names.[2,39]

Pyrethrum refers to the oleoresin extracted from the dried flowers of *Tanacetum cinerariaefolium* (Asteraceae) and is the source of the pyrethrins, chrysanthemates and pyrethrates.[2,38] Among the natural pyrethrins, those incorporating the alcohol pyrethrolone, namely pyrethrins I and II (Fig. 9), are the most abundant and account for most of the insecticidal activity.[38] The pyrethrins are recommended for control of a wide range of insects and mites on fruit, vegetables, field crops, ornamentals, glasshouse crops and house plants, as well as in public health, stored products, animal houses and on domestic and farm animals. Pyrethrins are sold in a wide variety of formulations, under many different trade names by a large number of different manufacturers.

The Formosan subterranean termite (*Coptotermes formosanus*) has become a devastating pest throughout the world. Over the past two decades, organo-chlorines and organo-phosphates, the two prominent classes of termite control agents have been banned due to environmental and human health concerns.[40] At the present time, phenylpyrazoles (e.g. fipronil), pyrethroids (permethrin and cypermethrin), chloronicotinyls (imidacloprid), and pyrroles (chlorfenapyr) are marketed as termite control agents (Fig. 10). The use of creosote and chromated copper aresenates as wood preservatives have also been limited because of environmental/health concerns. Termiticides have a history of being banned by EPA due to mammalian toxicity and environmental problems, resulting in the need for "greener" chemistries for this use. Plants are a good potential source of such compounds, since many plant species must cope with termites and related insects.

Asteraceae,[41,42] Milliaceae[43–45] and Apiaceae,[46,47] are among some plant families that are known to have insecticidal constituents. Vulgarone B (isolated from *Artemisia douglasiana;* Asteraceae), apiol (isolated from *Ligusticum hultenii;* Apiaceae), and cnicin (isolated from *Centaurea maculosa;* Asteraceae) cause significant mortality to Formosan subterranean

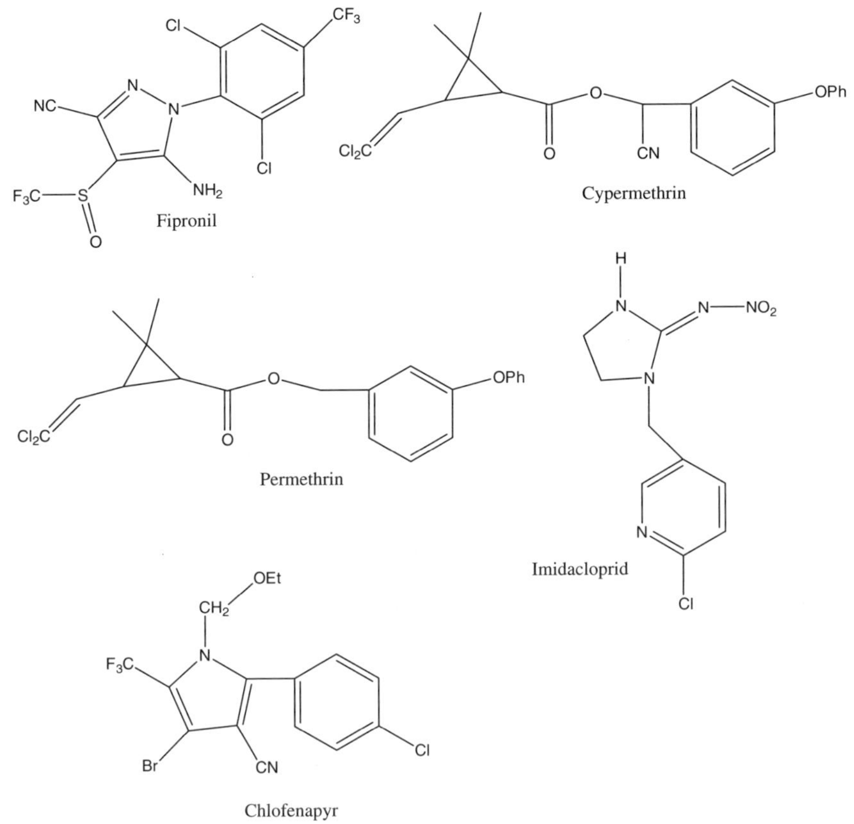

Fig. 10. Structures of some of the currently marketed termiticides.

termites in laboratory bioassays[48] (Fig. 11). These compounds are present in high levels in the respective plant sources, suggesting some ecological importance of these compounds. Nootkatone (Fig. 11), a sesquiterpene ketone, isolated from vetiver (*Vetiveria zizanioides*) oil is another example of a natural product that acts as a strong termiticide repellent and toxicant to Formosan subterranean termites.[49] The lowest effective concentration tested was 10 μg/g substrate.

The *Aedes aegypti* (Diptera: Culicidae) mosquito is the primary vector in transmitting dengue and yellow fever. Insecticide use has been the primary method of control of this and other mosquitoe species. Piperine [(E, E)-1-piperoyl-piperidine], is the major constituent in *Piper nigrum*

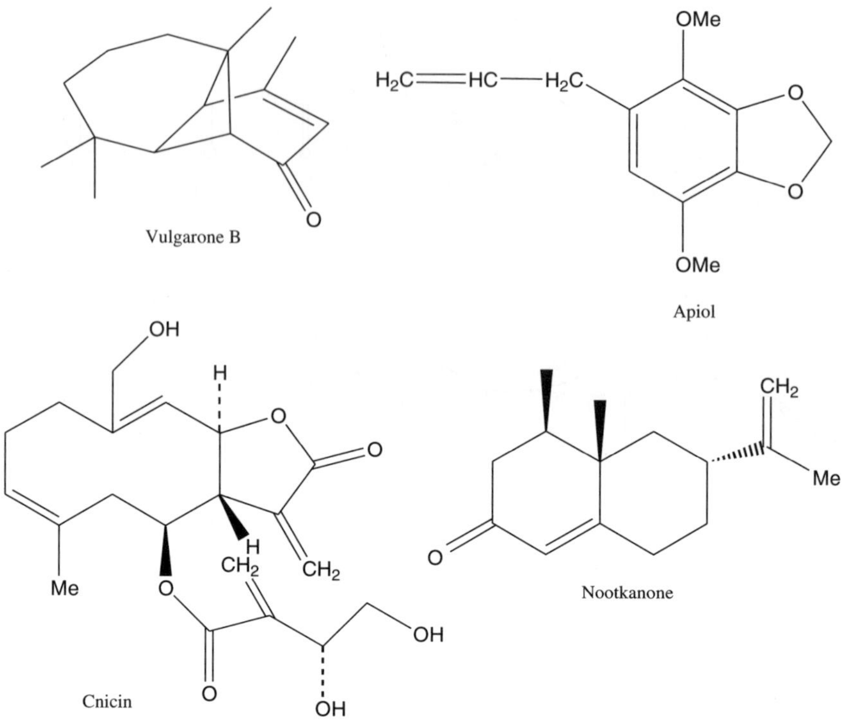

Fig. 11. Some insecticidal compounds from Asteraceae and Apiaceae.

fruit. Amides of some piperidines containing the 1-undec-10-enoyl moiety have also been synthesized and tested against female adults of *A. aegypti* (L.). According to the 24 hours LD_{50} values after topical application, the most toxic compound was found to be 2-ethyl-1-undec-10-enoyl-piperidine (LD_{50} = 0.80 μg per mosquito.[50] Piperine was less toxic (LD_{50} = 8.13 μg per mosquito) than any of the tested ethyl- or methyl-derivatives of 1-undec-10-enoyl-piperidine. When a benzyl moiety was added to the piperidine ring, regardless of the carbon position, the toxicities of 1-undec-10-enoyl-piperidines were significantly decreased. Thus, substituted piperidine derivatives could be further developed as insecticides for adult mosquitoes.

The phthalide derivatives cnidilide, (*Z*)-ligustilide, (3*S*)-butylphthalide, neocnidilide, and Z-butyldienepthalide and the furanocoumarins xanthotoxin and isopimpinellin have been reported as insecticides and

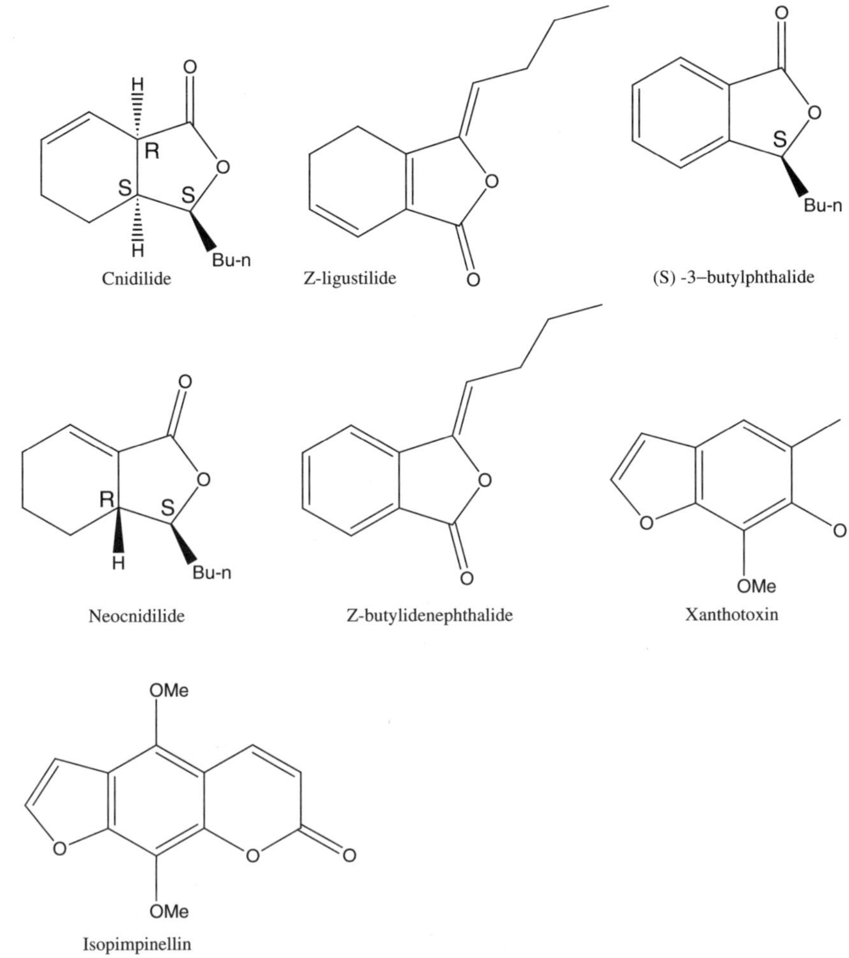

Fig. 12. Some naturally occurring insecticides against the fruit fly.

larvicides against fruit flies (*Drosophila melanogaster*)[51,52] (Fig. 12). *Z*-ligustilide has been reported to be the major constituent in *Lomatuium hultenii* seeds.[53] It is also present in *Angelica sinensis* (Danggui in Chinese) and considered to be the pharmacologically active constituent.[54]

The members of the plant family Piperaceae are well known in traditional pharmacology in Asia and Africa as medicines and insecticides.[55] *Piper nigrum* L. (Piperaceae, black pepper) has traditionally been used as a

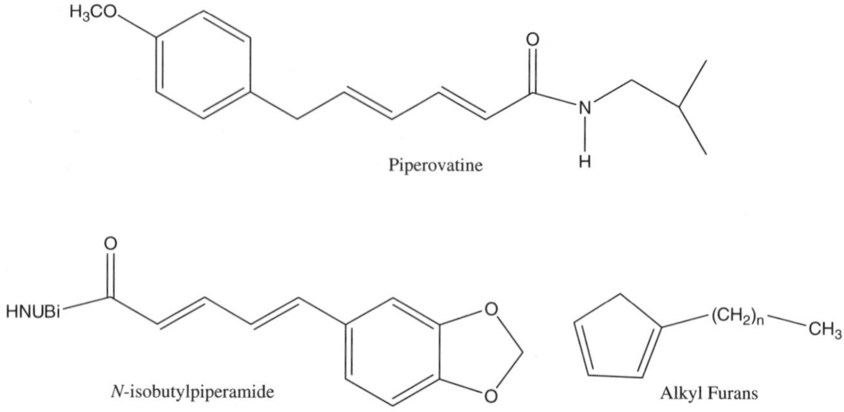

Fig. 13. Insecticidal compounds from the Piperaceae and the general structure of natural and synthetic insecticidal alkyl furans.

spice and as a medicine. Piperamides from *P. nigrum* have exhibited potential insecticide activity.[56–60] Piperovatine (Fig. 13) isolated from a number of plants in the Piperaceae family has been shown to possess sodium channel agonistic activity.[59] This activity also explains the ethnopharmacological use of the plant extracts containing piperovatine as toothache remedies, fish poisons, and local anesthetics. Other alkylamide-containing plants such as "toothache tree" a *Zanthoxylum* spp. in the Rutaceae family[61] and the "purple coneflower," an *Echinacea* spp. in the Asteraceae family,[62] are also being used in traditional medicine as local anesthetics.

Extracts of *P. nigrum* have insecticidal activity towards a variety of pests.[58] *Piper* species are rich sources of amides particularly piperamides.[59,60] The piperamides have two functional groups that are responsible for the insecticidal activity: the isobutylamide functional group and methylenedioxyphenyl group[59,63] (Fig. 13). The isobutylamide moiety is a sodium channel agonist, and the methylenedioxyphenyl moiety is an inhibitor of polysubstrate monooxygenase,[59,64] thereby slowing the metabolism of the insecticide.[64]

Alkyl furans are another class of naturally occurring compounds with insecticidal activity[65,66] (Fig. 13). This group of compounds, commonly referred to as avocado furans is present in avocados, *Persea americana* (Lauraceae) and related plants. When tested against beet armyworm, *Spodoptera*

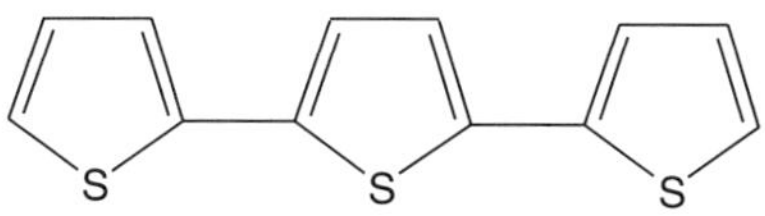

Alpha-terthienyl

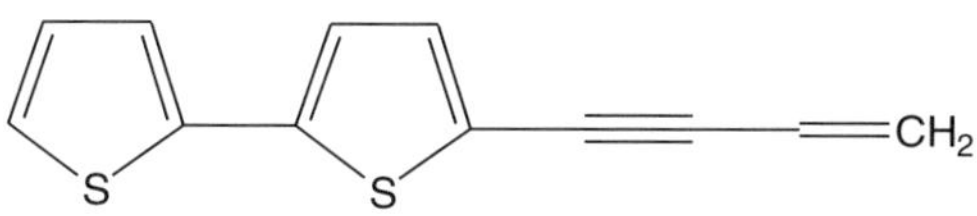

5′-(3-buten-1-ynyl)-2,2′-bithiophene

Fig. 14. Insecticidal compounds from *Echinops* species.

exigua, the highest mortality was observed with the two naturally occurring furans: 2-(pentadecyl)furan and 2-(heptadecyl)furan, previously isolated from avocado idioblast oil cells. These compounds and their synthetic analogs ($n = 11$, and $n = 15$) were inactive against Formosan termites.

Recently, Fokialakis *et al.*[67] evaluated over two-hundred and twenty crude extracts from repositories generated from plants native to Greece and Kazakhstan for termiticidal activity against the Formosan subterranean termite. Emerging from this screening effort were bioactive extracts from four species of *Echinops (E. ritro, E. spinosissimus subsp. spinosissimus, E. albicaulis* and *E. transiliensis*). Bioassay-guided fractionation of the most active extracts from each of the four species led to the isolation of eight thiophenes possessing varying degrees of termiticidal activity. α-Terthienyl and 5′-(3-buten-1-ynyl)-2,2′-bithiophene (Fig. 14) caused 100% mortality against *C. formosanus* within 9 days at 1 and 2 percent (wt/wt) concentrations, respectively.

8.3.3 Molluscicides

Members of the phylum mullusca range from snails, clams, abalone, squid, cuttlefish to octopi. Although some molluscs (sometimes mollusks) are ornamental animals and food for humans, some of them are significant agricultural pests and disease vectors of humans, livestock, fish, wild game and pets. The most troublesome molluscs are snails and slugs. Snails without shells are called slugs.

Schistosomiasis is second to malaria as a major parasitic disease. Snails, mainly of the genera *Biomphalaria*, *Bulinus*, and *Oncomelania*, serve as intermediate hosts to schistosomiasis trematodes, and humans are the final hosts. Thus, there is a need to have cost effective, environmentally friendly molluscicides. Copper sulphate has been used as a molluscicide in standing bodies of water, but its use has been limited due to toxicities to non-target species such as fish and algae. Other synthetic molluscicides such as N-trityl morpholine and pentachlorophenol (Fig. 15) are toxic to fish.[68] Currently, niclosamide (Fig. 15) has replaced copper sulphate and other organo metallic molluscicides such as organo-tin molluscicides, as it is relatively safer to the environment.[69] Despite its biodegradability, niclosamide is toxic to fish. Metaldehyde (Fig. 15) is also a commonly used mulluscicide. It is classified as a Class II (moderately hazardous) pesticide by the World Health Organization.

A primary factor in the search for natural product-based molluscicides is the cost of the product, as many of the mollusc-borne diseases occur in developing countries where the most affected people are impoverished. Synthetic molluscicides are expensive, and many of them are non-biodegradable and persist in the environment, causing damage to non-target species. One approach to finding inexpensive natural mulluscicides is to look into traditional folk medicine and ethno-pharmacological information in order to search for natural product-based molluscicides. Also important is the availability of the plant material in the affected area, the ability to extract active material (preferably with water), the ease of application, and low toxicities to non-target species. The use of the berries of endod (*Phytolaca dodecandra*) as a mollusicide to control schistosomiasis has been successful in Ethiopia.[70] The active constituents have been isolated and identified as saponins.[71] *Lonicera nigra*, *Hedera helix*, *Cornus florida*, *Asparagus curillus* are among some other plant species that have been investigated for molluscicidal saponins.[71] *Hedera helix* contains a hederagenine glycoside (Fig. 15) with LC_{100} at 3 ppm against *B. glabrata* snails.

Efforts are underway in our laboratory to discover natural product-based molluscicides to control snails that are harmful for agricultural commodities such as catfish, rice, tarro, and orchids. Catfish (*Ictalurus punctatus*) is one of the major farm-raised fish in the U.S. The ram's

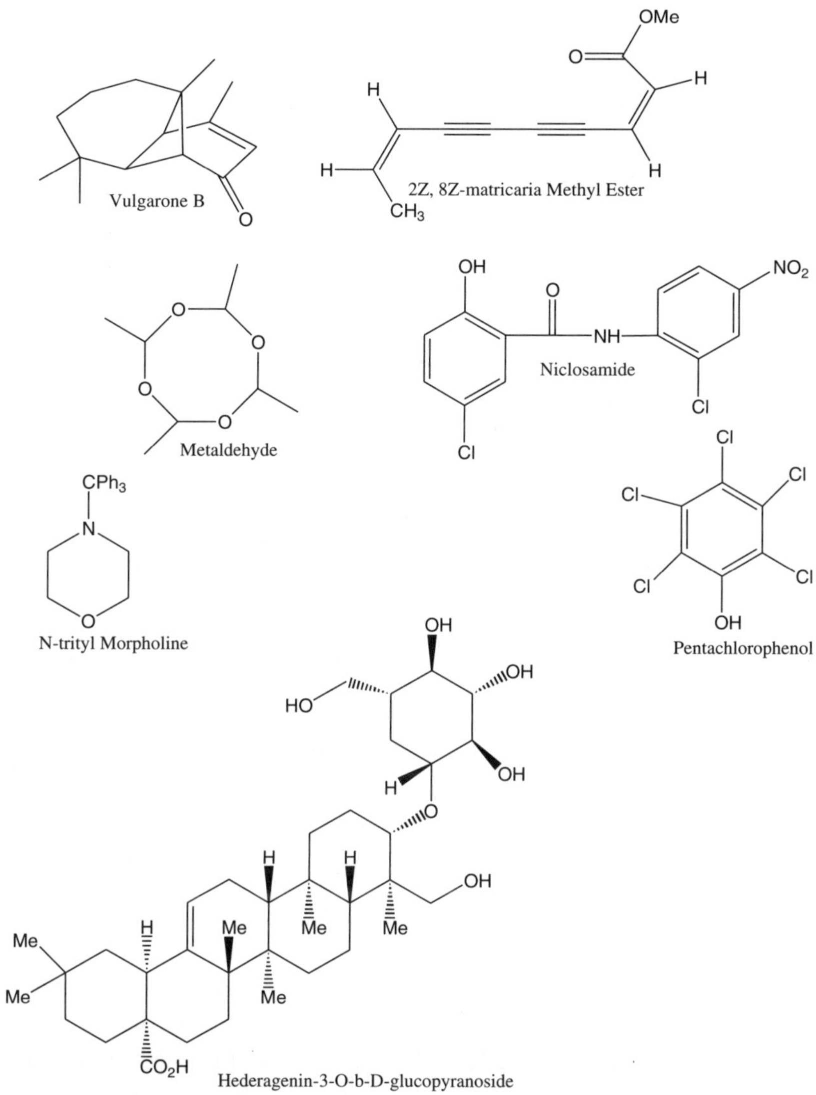

Fig. 15. Natural and synthetic molluscicides.

horn snail (*Planobdella trivolvis*) is an intermediate host for the digenetic trematode (*Bolbophorus confusus*) that has recently been discovered to be a significant problem in commercial channel catfish production ponds in the Mississippi Delta region.[72] The life cycle of this parasitic trematode involves the snail, channel catfish, and the American white pelican (*Pelecanus erythrorhynchos*). Catfish infested by the trematode develop cysts, have impaired growth, and are prone to other diseases that can weaken and kill the catfish. At present, there is no cure or treatment for the infected fish. One of the possible approaches to eradicate or control this problem is to interrupt the life cycle of the parasitic trematodes by eliminating the snails, which are an essential component of the life cycle.

Vulgarone B (Fig. 15), isolated from the steam distillate of the aerial parts of the plant *Artemisia douglasiana* (Asteraceae), was found to be active towards *P. trivolvis* with an LC_{50} of *ca* 24 μM.[73] The treated snails suffered severe hemolysis associated with lethal activity. Channel catfish toxicity studies indicated a LC_{50} of *ca* 207 μM. Thus, Vulgarone B may be an environmentally acceptable alternative for snail control in aquaculture. 2Z, 8Z-matricaria methyl ester (Fig. 15) isolated from *Erigeron speciosus*, a plant in the Asteraceae family also has molluscicidal activity against *P. trivolvis*, with a LC_{50} of 50 μM associated with marked hemolysis.[74] In laboratory experiments, yucca extract at 10 ppm caused 100% mortality of *P. trivolvis,* and ethanol extracts of *Phytolacca americana* berries (American poke weed) and *Lonicera nigra* (black-berried honeysuckle) had no significant activity (unpublished data). On the other hand, hederagenin-3-O-β-D-glucopyranoside (Fig. 15), isolated from English ivy (*Hedera helix*), an invasive plant in the northwestern states was active against *P. trivolvis* with an LC_{50} of 30 μM in laboratory studies (unpublished data). This compound also has activity against *B. glabrata*.[71]

The golden apple snail (GAS, Pomacea *canaliculata* (Lamarck)) is a major pest of rice in all rice-growing countries where it was either intentionally or accidentally introduced.[75] The economic losses due to this pest are estimated to be up to US$1.2 billion in the Phillippines alone.[76] GAS is also a problem in taro plantations in Hawaii where $CuSO_4$ is used to control the snail population.[77]

GAS is a highly productive nocturnal herbivore. A single female GAS can lay up to 500 eggs at a time. There are no predators for the eggs. The hatched young GAS grows fast, feeding on rice seedlings in water-rich rice paddies that provide the perfect habitat. GAS causes much greater damage to direct-seeded rice and young seedlings transplanted for 18–21 days.[78]

It causes 100% destruction of rice seedlings in the germinating stage and at least 20% destruction of rice seedlings in transplanted seedlings.

Integrated management methods are recommended for GAS, but many farmers depend on commercially available synthetic molluscicides (niclosamide and metaldehyde). There are numerous cases of poisoning caused by metaldehyde.[79] Therefore cost effective, target specific, and environmentally friendly molluscicides are needed due to the economic burden and undesirable effects of currently available commercial molluscicides. The LC_{50} value for vulgarone B is about 30 μM at 24 hours for GAS.[80] In the same bioassay, the standard commercial molluscicide, metaldehyde, also had an LC_{50} of about 30 μM. This corresponds to about 6.5 mg/L and 4.4 mg/L of vulgarone B and metaldehyde, respectively. The concentrations needed for 100% mortality at 24 hours were about 75 and 200 μM, respectively, for vulgarone B and metaldehyde. In practical terms, a rice farmer who consumes about 250 liters of water for spraying one hectare will require 4.8 grams of pure vulgarone B for GAS control. Vulgarone B did not cause further mortality for 48 hours–96 hours after treatment, unlike the observed increased mortality with time associated with metaldehyde. This indicates that vulgarone B is fast-acting, unlike metaldehyde.

There was no toxicity to 14-day-old rice plants ten days after being sprayed with concentrated vulgarone B solutions that cause complete or nearly complete mortality of GAS.[80] However, when incorporated into agar growth medium, pronounced chlorosis occurred after 14 days of growth. Therefore vulgarone B should be used after the germination of rice seeds. Vugarone B has also been shown in field and laboratory experiments to have potential as a molluscicide in taro paddies in Hawaii (unpublished data).

8.4 PLANT PATHOGENS

8.4.1 Compounds that Induce Resistance to Pathogens in Plants

Plant resistance to pathogens can be systemically induced by an endogenous signal molecule produced at the infection site that is then translocated to other parts of the plant. This phenomenon is called systemic acquired resistance (SAR).[81] The discovery of the putative signal is of great interest to many plant scientists because such molecules have possible uses as "natural product" disease control agents. Salicylate accumulation is one of the first steps in SAR; however, salicylate itself is too rapidly biodegraded to be used as an exogenous resistance inducer. Synthetic compounds have been used to induce SAR-like resistance.

In response to pathogen attack, fatty acids of the jasmonate cascade are formed from membrane-bound α-linolenic acid by lipoxygenase-mediated peroxidation.[82] Analogous to the prostaglandin cascade in mammals, α-linolenic acid is thought to participate in a lipid-based signaling system where jasmonates induce the synthesis of a family of wound-inducible defensive proteinase inhibitors[83] and low and high molecular weight phytoalexins (natural fungicides) such as certain flavonoids, alkaloids, and terpenoids.[84,85] This type of resistance induction is termed as induced systemic resistance (ISR).[81]

Several plant or bacterial natural products that act through the induction of SAR or ISR processes are used as plant protectants. Commercial products that induce SAR include a harpin protein which switches on natural plant defenses in plant response to bacterial leaf spot, wilt, and blight and fungal diseases such as powdery mildew and those casued by *Botrytis* spp. A partially refined extract of the giant knotweed plant (*Reynoutria sachalinensis*) is sold for use against a wide range of fungal pathogens, including *Botrytis* spp. and powdery mildews in ornamental and glasshouse crops.[2] It acts through induction of SAR. However, elicitors with no direct antifungal activity will not appear active in most current high throughput screening systems.

8.4.2 Compounds that Act Directly on Pathogens

Plants, fungi, and other marine and terrestrial organisms produce hundreds of thousands of compounds that often suppress the invasion and growth of pathogenic microorganisms. One can hypothesize that plants produce chemicals such as these to defend themselves directly by inhibiting pathogen proliferation, or indirectly by disrupting chemical signal processes related to growth and development of pathogens.[86]

Chemical control of plant diseases has become more difficult as several effective fungicides have lost their registration for use, and some pathogens have evolved to be resistant to the more commonly used fungicides.[86] Fungicides with new modes of action are needed to combat these resistant pathogens and in strategies to delay evolution of resistance to fungicides in general.

Strobilurins are a class of fungitoxic compounds (e.g. strobilurin A, Fig. 16) derived from basidiomyces growing on decaying wood[87] that have become highly successful during the past decade as templates for highly successful synthetic fungicides. The relatively recent development of the strobilurin class of fungicides, a class with a new mode of action, has increased the interest in natural products as crop protectants. This class includes, synthetic compounds such as azoxystrobin, trifoxystrobin, and fluoxastrobin (Fig. 16), based on natural strobilurins. Azoxystrobin is

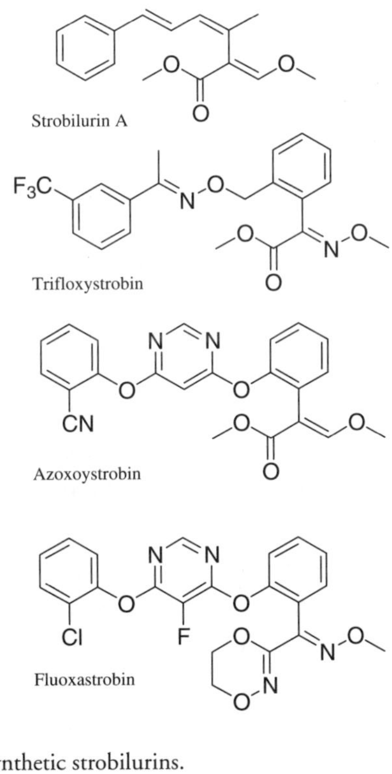

Fig. 16. Natural and synthetic strobilurins.

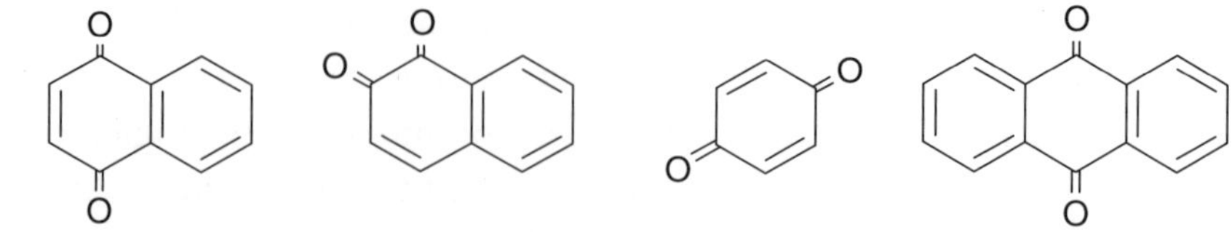

1,4-naphthoquinone 1,2-naphthoquinone 1,4-benzoquinone Anthraquinone

Fig. 17. Quinone classes upon which fungicides can be based.

registered for the control of anthracnose and powdery mildew on strawberries, and pyraclostrobin is registered for the control of anthracnose, leaf spot, and powdery mildew and is known to suppress Botrytis fruit rot. Trifloxystrobin, in combination with the systemic fungicide propiconazole, is labeled for wheat leaf disease. Fluoxastrobin provides plant disease control in crops and turf. The evolution of resistance to strobilurin fungicides in the North American and the EU has raised concerns over the use and long term viability of these products.

Quinone metabolites serve vital roles in electron transport and pathogen defense and they are also widespread throughout the plant.[88] Quinone fungicides are in a general class of "organic fungicides" that are considered as multisite inhibitors.[89] Thirty-three naturally occurring quinones of four major classes (Fig. 17), 1,4-naphthoquinones, 1,2-naphthoquinones, 1,4-benzoquinones, anthraquinones, and other miscellaneous compounds from our natural products collection were tested for antifungal activity using bioautography. Quinones demonstrated good to moderate antifungal activity against *Colletotrichum* species. In this study, *Colletotrichum fragariae* appeared to be the most sensitive *Colletotrichum* species to quinone-based chemistry, and *C. gloeopsorioides* had intermediate sensitivity. *Colletotrichum acutatum* was the most insensitive species to quinones.

Sampangine (Fig. 18), benzo-[4,5]-sampangine, and eupolauridine are alkaloids isolated from the root bark of *Cleistopholis patens*. These novel broad-spectrum compounds have promising antifungal activity against several serious pathogenic fungi of plants including *B. cinerea*, *C. fragariae*, *C. acutatum*, *C. gloeosporioides*, and *F. oxysporum*.[90] In 96-well microbioassays, sampangine and its analogs are active against synthetic

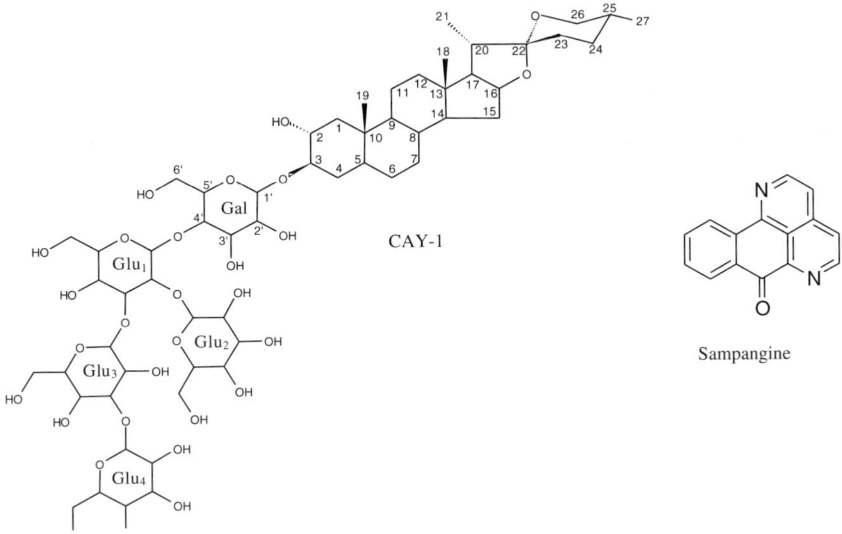

Fig. 18. Structures of CAY 1 and sampangine.

fungicide-resistant strains of *B. cinerea* ($IC_{50} < 3.0\,\mu M$), *C. gloeosporioides* ($IC_{50} < 3.0\,\mu M$), *C. fragariae* ($IC_{50} < 3.0\,\mu M$), *C. acutatum* ($LC_{50} < 3.0\,\mu M$), and *F. oxysporum* ($IC_{50} < 30.0\,\mu M$).

CAY-1 (Fig. 18) is a fungicidal steroidal saponin recently isolated and identified[91] from the fruit of cayenne pepper (*Capsicum frutescens*). CAY-1 was lethal to germinating conidia of *Aspergillus flavus*, *A. fumigatus*, *A. parasiticus* and *A. niger*. It was also active against agricultural and medicinally important fungi and yeast. *In vitro* dose-response assays with CAY-1 against plant pathogenic fungi showed that $3\mu M$ CAY-1 inhibited the growth of *C. gloeosporioides* and *C. acutatum* by 100% and *C. fragariae*, *P. obscurans*, and *P. viticola* by 90%. Detached leaf assays using strawberry leaves from anthracnose susceptible cv. Chandler demonstrated that CAY-1 decreased *C. fragariae* induced lesion number by 95% compared to the untreated control.

The detached leaf assays indicated that 625 ppm CAY-1 and sampangine provided effective "protectant" activity for disease control of anthracnose on the leaf surface (Table 1). These compounds were as effective as the commercial fungicide, azoxystrobin.

Table 1. Number of Leaf Spots on Detached Leaves Following Inoculation with *Colletotrichum fragariae* Isolate CF-75 and Post Treatment with Commercial and Experimental Fungicides (unpublished data by Wedge, Smith, and DeLucca)

Concentration	Azoxystrobin	CAY-1	Sampangine
ppm	*Mean number of leaf spots/leaflet*		
0	22.34 a[z]	22.34 a	22.34a
625	4.46 b	0.77 b	8.31 b
1250	1.58 b	0.35 b	1.26 c
2500	3.54 b	0.08 b	0.80 c
LSD[y]	5.97	4.99	5.54

[z] Means followed by the same letter are not significantly different.
[y] Least significant difference = LSD at $p < 0.05$.

Thiophenes from *Echinops* spp. possess many biological activities, including fungicidal activity. Bioassay-directed isolation of antifungal compounds from the dichloromethane extract of the radix of *E. ritro* (Asteraceae) led to the isolation of eight fungicidal thiophenes. The antifungal activities of isolated compounds, together with a previously isolated thiophene from *E. transiliensis*, were first evaluated by bioautography and microbioassay. 5'-(3-buten-1-ynyl)-2,2′-bithiophene, α-terthienyl, and 2-[pent-1,3-diynyl]-5-[4-hydroxybut-1-ynyl]thiophene at 3 μM were active against three *Colletotrichum* species, *Fusarium oxysporum*, *Phomopsis viticola*, and *P. obscurans*.[92]

Furanocoumarins from *Ruta graveolens* possess antifungal activity to several plant pathogenic fungi. 7-Hydroxycoumarin, 4-hydroxycoumarin, and 7-methoxycoumarin were active against *Colletotrichum* species that cause Anthracnose disease of strawberry.[93]

Bioassay-directed isolation of antifungal compounds from the bark of *Macaranga monandra* yielded two clerodane-type diterpenes, kolavenic acid and 2-oxo-kolavenic acid, with moderate activity against *Phomopsis viticola* and *Botrytis cinerea*.[94] Bioassay-guided isolation of antifungal compounds from *Ligusticum hulteni* and *Lomatium californicum* seeds led to the isolation of two active compounds, (*Z*)-ligustilide and apiol.[53] Both compounds were active against *C. fragariae*. (*Z*)-Ligustilide and apiol showed the highest activity against *B. cinerea*.

8.5 VERTEBRATE PESTS

As reviewed by Ujváry,[95] some of the earliest natural product-based pesticides were those for the elimination of vertebrate pests. For example, strychnine (Fig. 19), obtained from seeds of *Strychnos nux-vomica*, is a rodenticide that is an antagonist to the neurotransmitter glycine and is used against a few mammal species, as well as pest birds and fish. The first generation of anticoagulant rodenticides were based on dicoumarin,

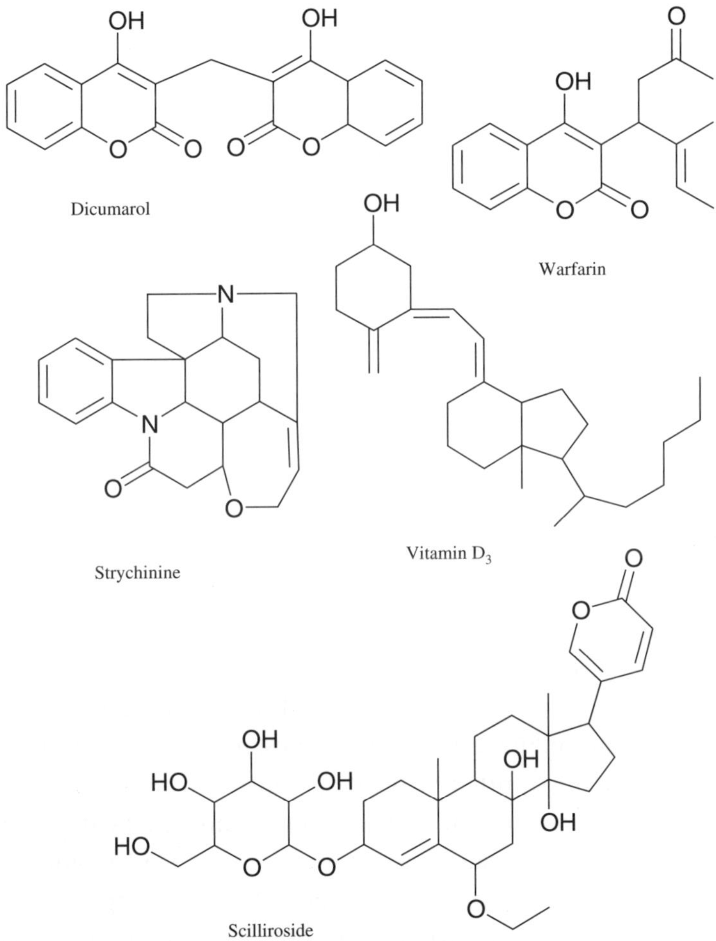

Fig. 19. Structures of some natural product rodenticides.

a contaminant of spoiled hay that causes hemorrhagic disease of cattle. Dicoumarol (Fig. 19) itself was used as a rodenticide in the UK until more potent synthetic analogues such as warfarin (Fig. 19) were introduced. Red squill (*Urginea maritima*) bulbs contain the toxin scilliroside (Fig. 19). The products containing this compound were used as a rodenticides from the 13th century in Europe until production was discontinued in 1980. Vitamins D_2 (ergocalciferol) and D_3 (chlolecalciferol) (Fig. 19) have been used as rodenticides.

8.6 WEEDS

8.6.1 The Use of Natural Products in Weed Management

Modern agriculture has relied heavily on synthetic pesticides for managing weeds, which amount to more than half of all agricultural pesticides utilized in the developed world. Synthetic herbicides are highly effective (with some compounds active at rates as low as a gram or less per hectare) and relatively inexpensive to manufacture. Most of them, but not all, have low impact on the environment and wildlife. Nevertheless, concern over the potential impact of pesticides on the environment has led many nations to implement stricter pesticide regulatory policies, which has resulted in the removal of many pesticides from the market. The increased popularity of organic agriculture also introduced a new dimension in the agrochemical market, where the use of natural products may be the only permitted form of pesticides.

No herbicide with new modes of action has been introduced since the triketones in 1986. At the same time, weeds have evolved resistance to most of the commercial herbicide families. Meanwhile, the introduction of glyphosate and its companion glyphosate-resistant crops has greatly reduced the use of many other herbicides, resulting in diminished economic incentive for herbicide discovery programs in most pesticide companies. The emergence of glyphosate-resistant weeds is increasing rapidly and may provide the impetus necessary to revive those discovery programs. The convergence of the new stricter regulations, the desire of consumers for safer and environmentally friendly pest management tools, and the need for new chemical classes may provide the conditions to develop more natural product-based herbicides.

As illustrated in the other sections of this chapter, nature has been a valuable source of compounds and chemistries with advantageous properties for disease and insect pest management. This section illustrates that natural products have also been useful for the development of new weed management either in their original forms or as templates for natural products derived compounds.

8.6.2 Natural Product Herbicides

Organic agriculture is a specialty area where synthetic herbicides are not allowed. While weed management under organic agriculture practices relies primarily on cultivation and biocontrol, the use of some natural products is permitted (Table 2). Maize gluten meal has been tested as a pre-emergence herbicide on lawns and high-value crops.[96] However, the control of grasses and other weeds requires the extremely high rate of nearly 2 tons per hectare. Herbicidal soaps, composed of various fatty acid preparations (e.g. pelargonic acid), and vinegar are common active ingredients in many natural burndown (contact) herbicides.[2] They all act relatively rapidly and have broad-spectrum (no selectivity) weed control. However, most weeds tend to recover because there is no residual activity after the initial burndown

Table 2. Examples of Natural Products used for Weed Management in Organic Agriculture

Compound	Method	Commercial Products
Corn gluten	Pre-emergent	WeedBan™
		Corn Weed Blocker™
Pelargonic acid	Post-emergence	Scythe™
Acetic acid	Post-emergence	Burnout™
		Bioganic™
Clove oil	Post-emergence	Matran II
Cinnamon oil	Post-emergence	

effect on the shoot which takes place soon after application.[97] Essential oils have also shown some potential as herbicides. For example, clove oil or cinnamon oil at the concentrations of 1 to 5% controlled most small weeds.[98] Surfactants, which are also limited in organic agriculture, are often required to assist in the spreading of the material.

The natural herbicidal compounds permitted in organic agriculture are not very active, so they must be applied in relatively large quantities. Ironically, this may lead to undesirable environmental effects.

The literature is filled with reports of the isolation and characterization of natural phytotoxins, but very few of them have been successfully used as herbicides when applied in their native form. Discovery efforts by the industry have focused primarily on microbes as sources of herbicidal natural products, and many compounds have been patented.[99] However, only bialaphos (isolated from *Streptomyces*) and phosphinothricin (Fig. 20) have been successfully commercialized. Glufosinate, the synthetic form of phosphinothricin (a racemic mixture of the natural L form and the unnatural D isomer), acts directly on the plant's target site, whereas bialaphos must be metabolically bioactivated into phosphinothricin by plants before having herbicidal activity.[100] Bialaphos has been produced by industrial fermentation and marketed in eastern Asia whereas glufosinate is sold around the world. Both glufosinate and bialaphos are broad-spectrum post-emergence herbicides that can be used to control a wide range of weeds in agricultural settings, for total vegetation control on non-cultivated areas, or to desiccate crops before harvest.

Both compounds inhibit glutamine synthetase, which is necessary for the production of glutamine and for ammonia detoxification. Plants that are exposed to glufosinate have reduced glutamine and increased ammonia levels in their tissues, which stops photosynthesis and results in death within a few days. Other natural products (e.g. tabtoxine-β-lactam, oxetin, and methionine sufoximine) are known to have this mode of action, but they have not been developed as commercial products.

Glufosinate has a broad weed spectrum (little to no selectivity), and is therefore sometimes marketed along with genetically engineered glufosinate-resistant crops (cotton, canola, maize, and rice). While this raises some concern in some parts of the world, transgenic technology has been widely accepted in the western hemisphere. Nevertheless, neither

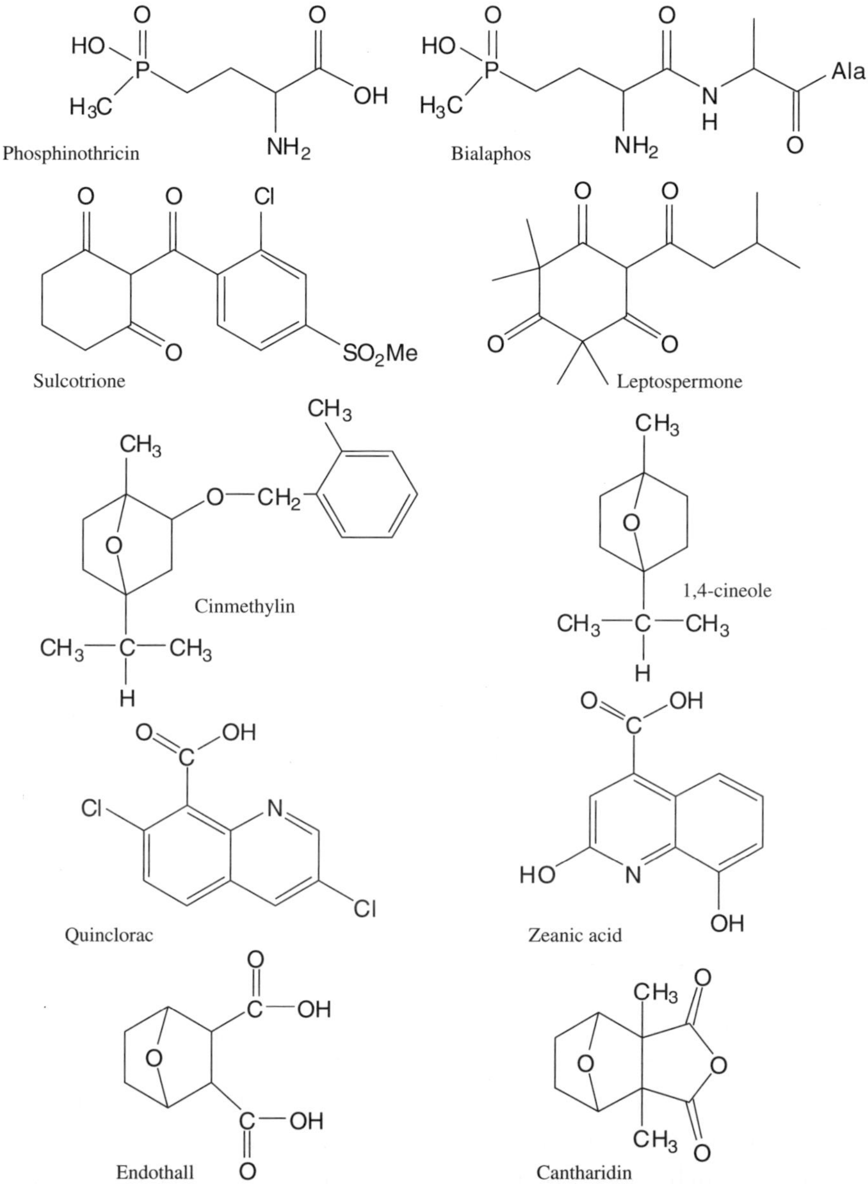

Fig. 20. Structures of commercial herbicides and their natural product analogues.

glufosinate nor glufosinate-resistant crops have been accepted by organic farmers. Although the L form of glufosinate is exactly the same molecule as phosphinothricin, we assume that it is rejected because of the presence of the D form in the racemic mixture.

8.6.3 Natural Products as Lead Compounds for Conventional Herbicides

Several herbicide classes are similar to natural products. However, there is some debate whether many or any of these were actually derived from natural products or were discovered as part of large herbicide screening programs. Regardless of how these compounds were discovered, the commercial products are the results of intensive structure optimization because most natural products do not have the physicochemical properties required to be good herbicides. In general, the structures of bioactive natural products are too complex to be manufactured at reasonable costs. Furthermore, natural products tend to have relatively short environmental half-life which prevents them from having great efficacy.

The triketones (Fig. 20), the most recent class of herbicides, were discovered in 1986 and introduced commercially (e.g. sulcotrione) in 1991. While several papers report that the triketones herbicides were derived from the plant secondary metabolite leptospermone (Fig. 20) that was isolated from the bottlebrush plant (*Callistemon* spp.) [Ref. 101], there is a report that suggests that the discovery may have been more serendipitous.[102]

The molecular target site of triketone herbicides is the enzyme *p*-hydroxyphenylpyruvate dioxygenase (HPPD).[101] Inhibition of this enzyme disrupts the biosynthesis of carotenoids and causes a bleaching (loss of chlorophyll) effect on the foliage similar to that observed with inhibitors of phytoene desaturase (e.g. norflurazon). However, the mechanism of action of HPPD inhibitors is different. Inhibtion of HPPD stops the synthesis of homogentisate (HGA), which is a key precursor of the 8 different tocochromanols (tocopherols and tocotrienols) and prenyl quinones. In the absence of prenylquinone plastoquinone, phytoene desaturase activity is interrupted. The bleaching of the green tissues ensues as if these compounds inhibited phytoene desaturase.[103]

Natural triketones and other natural products also inhibit HPPD.[104–106] The natural products tend to be competitive inhibitors of HPPD, whereas the synthetic herbicides have been optimized and bind irreversibly to the enzyme.

Cinmethylin is a structural analogue of the monoterpene 1,4-cineole (Fig. 20). However, cinmethylin was not derived from the natural product. Instead, it was discovered by Shell Chemical during a biorational insecticide design program that targeted the glycerol-3-phosphate shuttle in 1968. Some of the bioactive bicyclic intermediates had the basic backbone of the monoterpene cineoles. An important aspect in the final development of cinmethylin involved the optimization of the physical properties of the compound to lower the volatility through structural modification of the benzyl ether side chain.[107] Cinmethylin (Fig. 20) was commercialized outside the United States in 1982 under the trade names of Cinch® and Argold®.

Cinmethylin is active on several important grasses in rice (*Echinochloa* spp., *Cyperus* spp., and *Monochoria vaginalis*) at rates from 25 to 100 g ai/ha.[108] Its chemical and toxicological features (low persistence in environment and low mammalian toxicity) generated a significant level of interest in its use as a rice herbicide. Seedlings of *Echinochloa* species cannot easily be distinguished from rice seedlings and are often transplanted along with rice. Cinmethylin was tested as a potential tool to control *Echinochloa* in rice nurseries. While this herbicide was able to control the weed, it also had a negative effect on rice.

Quinclorac (Fig. 20) is an auxin-type herbicide related to the natural quinolinecarboxylic acid phytohormones oryzanin and zeanic acid (Fig. 20) that arose from a herbicide discovery program involving the synthesis of more than 3000 compounds done in the early 1980s (reviewed by Grossmann[109]). Quinclorac has been commercially available since 1988 and it is used primarily for pre-emergence weed control in rice. It is most effective against barnyardgrass and foxtail but also has activity against some annual and perennial broadleaf weeds. Quinclorac is a systemic herbicide and its mode of action appears to differ between dicot and monocot plants. In dicots, it appears to behave as a regular auxin herbicide, whereas, in monocots, it causes increases in ethylene and cyanide production.[110,111]

Rice appears to be tolerant to quinclorac by having an ACCase that is not as affected by the herbicide as well as having a high β-cyano-alanine synthase activity.[111]

Endothall is a structural analogue of cantharidin (Fig. 20). Its mode of action is apparently inhibition of protein phosphatase(s).[112] A structure/activity relationship study demonstrated that the presence of the oxygen bridge and the location of the two carboxylic groups play important roles in the activity of the molecules.[113] The unusual 7-oxabicyclo[2.2.1]heptane ring of endothal is similar to that of cinmethylin, another natural product-like herbicide for which the mode of action has also eluded scientists for many years.

8.6.4 Summary

Biobased herbicides such as acetic acid, fatty acids and oils are commonly used as alternatives to synthetic compounds in organic agriculture. However, these compounds have not transferred successfully in the more conventional crop production systems. The only natural herbicide available for large scale cropping system is glufosinate (a metabolite of bialaphos). The relative success of glufosinate, especially when used in conjunction with genetically engineered glufosinate-resistant crops, supports the premises that natural products can serve as herbicides. As well, the commercialization of the triketones herbicides should also encourage industry to renew its discovery efforts based on natural products backbones.

While allelopathy is not covered in this chapter, excellent weed control can be achieved at reduced herbicide rate when used in conjunction with allelopathic rice varieties. Therefore, selection of highly allelopathic crop varieties, either through traditional breeding or using genetic engineering techniques, may also provide novel and low input approaches to weed control.[114]

8.7 CONCLUSIONS

Natural compounds offer a vast array of chemical structural diversity associated with evolved biological activities attacking many different molecular targets. In some cases, this activity evolved in defense against pests of the same type that humans desire to control. In other cases, although the natural compound has limitations, it has been the basis for the design of highly effective synthetic pesticides. We have only provided a brief sampling of natural products that have been used as pesticides or modified for such use. The strategy of using natural products for pesticide discovery is still highly valid.

REFERENCES

1. Rimando AM, Duke SO (eds.), *Natural Products for Pest Management*, Amer Chem Soc Symp Ser No 927 American Chemical Society, Washington, D.C., 2006.
2. Copping L, Duke SO, Natural products that have been used commercially as crop protection agents, *Pest Manag Sci* **63**: 524–554, 2007.
3. Duke SO, Dayan FE, Rimando AM, Schrader KK, Aliotta G, Oliva A, Romagni JG, Chemicals from nature for weed management, *Weed Sci* **50**:138–151, 2002.
4. van der Ploeg M, Tucker CS, Boyd CE, Geosmin and 2-methylisoborneol production by cyanobacteria in fish ponds in the southeastern United States, *Water Sci Technol* **25**:283–290, 1992.
5. Schrader KK, Dennis ME, Cyanobacteria and earthy/musty compounds found in commercial catfish (*Ictalurus punctatus*) ponds in the Mississippi Delta and Mississippi-Alabama Blackland Prairie, *Water Res* **39**:2807–2814, 2005.
6. van Aller RT, Pessoney GF, USM algal research team makes major off-flavor/water quality discovery, *Aquacult Mag* **8**:18–22, 1982.
7. van Aller RT, Pessoney GF, Rogers VA, Watkins EG, Leggett HG, Oxygenated fatty acids: A class of allelochemicals from aquatic plants, *ACS Symp Ser* **268**:387–400, 1985.
8. Tucker CS, Lloyd SW, Evaluation of potassium ricinoleate as a selective blue-green algicide in channel catfish ponds, *Aquaculture* **65**:141–148, 1987.
9. Welch IM, Barrett PRF, Gibson MT, Ridge I, Barley straw as an inhibitor of algal growth. I. Studies in the Chesterfield canal, *J Appl Phycol* **2**:231–239, 1990.
10. Newman JR, Barrett PRF, Control of *Microcystis aeruginosa* by decomposing barley straw, *J Aquat Plant Manage* **31**:203–206, 1993.
11. Barrett PRF, Field and laboratory experiments on the effects of barley straw on algae, Proceedings, BCPC Monograph 59, Comparing field and glasshouse pesticide performance II, 191–200, 1994.

12. Cooper WJ, Zika RG, Photochemical formation of hydrogen peroxide in surface and ground water exposed to sunlight, *Science* **220**:711–712, 1983.
13. Barrion G, Feuillade M, Hydrogen peroxide as a potential algicide for *Oscillatoria rubescens* D.C., *Water Res* **20**:619–623, 1986.
14. Barrett PRF, Newman JR, Annual Progress Report, Aquatic Weeds Research Unit, Broadmoor Lane, Sonning, Reading, UK, 1992.
15. Schrader KK, de Regt MQ, Tidwell PD, Tucker CS, Duke SO, Compounds with selective toxicity towards the off-flavor metabolite-producing cyanobacterium *Oscillatoria* cf. *chalybea Aquaculture* **163**:85–99, 1998.
16. Wills GD, Tucker CS, Jones EJ, Effect of barley straw for the control of off-flavor in pond-raised catfish, *Proc Southern Weed Sci Soc* **52**:227–230, 1999.
17. Schrader KK, Evaluation of several commercial algicides for control of odor-producing cyanobacteria, *J Aquat Plant Manage* **43**:100–102, 2005.
18. Schrader KK, de Regt MQ, Tucker CS, Duke SO, A rapid bioassay for selective algicides, *Weed Technol* **11**:767–774, 1997.
19. Schrader KK, de Regt MQ, Tidwell PR, Tucker CS, Duke SO, Selective growth inhibition of the musty-odor producing cyanobacterium *Oscillatoria* cf. *chalybea* by natural compounds, *Bull Environ Contam Toxicol* **60**:651–658, 1998.
20. Schrader KK, Nanayakkara NPD, Tucker CS, Rimando AM, Ganzera M, Schaneberg BT, Novel derivatives of 9,10-anthraquinone are selective algicides against the musty-odor cyanobacterium *Oscillatoria perornata Appl Environ Microbiol* **69**:5319–5327, 2003.
21. Schrader KK, Nanayakkara NPD, Selective algaecides for control of cyanochloronta, U.S. Patent No. 6,949,250, 2005.
22. Schrader KK, Foran CM, Holmes BD, Schlenk DK, Nanayakkara NPD, Schaneberg BT, Toxicological evaluation of two anthraquinone-based cyanobactericides to channel catfish, *N Am J Aquacult* **66**:119–124, 2004.
23. Schrader KK, Rimando AM, Tucker CS, Glinski J, Cutler SJ, Cutler HG, Evaluation of the natural product SeaKleen® for controlling the musty-odor-producing cyanobacterium *Oscillatoria perornata* in catfish ponds, *N Am J Aquacult* **66**:20–28, 2004.
24. Cutler HG, Culter SJ, Wright D, Dawson R, Methods of controlling zoological and aquatic plant growth, U.S. Patent No. 6,340,468, 2002.
25. Fradin MS, Day JF, Comparative efficacy of insect repellents against mosquito bites, *N Engl J Med* **347**:13–18, 2002.
26. Knight W, Natural Bug Sprays: To Swat or Not?, *The New York Times*, E8, July 27, 2006.
27. Consumer Reports, Insect Repellents: Which keep bugs at bay?, June, 2006.

28. Barnard DR, Xue RD, Laboratory evaluation of mosquito repellents against *Aedes albopictus, Culex nigripalpus*, and *Ochlerotatus triseriatus* (Diptera: Culicidae), *J Med Entomol* **41**:726–730, 2004.
29. Tawatsin A, Thavara U, Chansang U, Chavalittumrong P, Boonruad T, Wongsinkongman P, Bansidhi J, Mulla MS, Field evaluation of Deet, Repel Care, and three plant-based essential oil repellents against mosquitoes, black flies (Diptera: Simuliidae), and land leeches (Arhynchobdellida: Haemadipsidae) in Thailand, *J Amer Mosquito Control Assoc* **22**:306–313, 2006.
30. Zhu J, Zeng XY, Liu T, Qian K, Han Y, Xue S, Tucker B, Schultz G, Coats J, Rowley W, Zhang A, Adult repellency and larvicidal activity of five plant essential oils against mosquitoes, *J Amer Mosquito Control Assoc* **22**:515–522, 2006.
31. Thorsell W, Mikiver A, Tunon H, Repelling properties of some plant materials on the tick *Ixodes ricinus* L, *Phytomedicine* **13**:132–134, 2006.
32. Cantrell CL, Klun JA, Bryson C, Kobaisy M, Duke SO, Isolation and Identification of Mosquito-Bite-deterrent Terpenoids from Leaves of American (*Callicarpa americana*) and Japanese (*Callicarpa japonica*) Beautyberry, *J Agric Food Chem* **53**:5948–5953, 2005.
33. Morisawa J, Kim C, Kashiwagi T, Tebayashi S, Horike M, Repellents in the Japanese Cedar, *Cryptomeria japonica,* against the Pill-bug, *Armadillidium vulgare, Biosci Biotechnol Biochem* **66**:2424–2428, 2002.
34. Okunade AL, Wiemer DF, (-)-Loliolide, an ant-repellent compound from *Xanthoxyllum setulosum. J Nat Prod* **48**:472–473, 1985.
35. Capron MA, Wiemer DF, Piplaroxide, an ant-repellent piperidine epoxide from *Piper tuberculatum. J Nat Prod* **59**:794–795, 1996.
36. Schultz G, Simbro E, Belden J, Zhu J, Coats J, Catnip, *Nepeta cataria* (Lamiales: Lamiaceae) — a closer look: Seasonal occurrence of nepetalactone isomers and comparative repellency of three terpenoids to insects. *Environ Entomol* **33**:1562–1569, 2004.
37. Peterson CJ, Nemetz LT, Jones LM, Coats JR, Behavioral activity of catnip (Lamiaceae) essential oil components to the German cockroach (*Blattodea: Blattellidae*). *J Econ Entomol* **95**:377–380, 2002.
38. Isman MB, Botanical insecticides, deterrents, and repellents in modern agriculture and an increasing regulated world, *Annu Rev Entomol* **51**:45–66, 2006.
39. Sarfraz M, Dosdall LM, Keddie BA, Spinosad: A promising tool for integrated pest management. *Outlooks Pest Manag* **16**:78–84, 2005.
40. Potter MF, Termites, in Hedges SA (ed.), *Handbook of Pest Control,* Franzak & Foster, Cleveland, OH, pp. 233–333, 1997.
41. Chiasson H, Belanger A, Bostanian N, Vincent C, Poliquin A, Acaricidal properties of *Artemisia absinthium* and *Tanacetum vulgare* (Asteraceae) essential oils obtained by three methods of extraction, *J Econ Entomol* **94**:167–171, 2001.

42. Gonzalez-Coloma A, Guadano A, Tonn CE, Sosa ME, Antifeedant/insecticidal terpenes from Asteraceae and Labiatae species native to Argentinean semi-arid lands, *Z Naturforsch C: J Biosci* **60**:855–861, 2005.
43. Gupta SC, Misra AK, Management of okra shoot of fruit borer, *Earias vittella fabr.* through bio-rational insecticides, *Pestic Res J* **18**:33–34, 2006.
44. Prasad SM, Prasad N, Barnwal MK, Evaluation of insecticides against aphid vectors of virus diseases of French Bean, *J Plant Protect Environ* **3**:87–90, 2006.
45. Sharma DC, Badiyala A, Choudhary A, Bioefficacy and persistent toxicity of biopesticides and insecticides against potato tuber moth, *Phthorimaea opercullela* Zell. on spring potato, *Pestic Res J* **18**:43–46, 2006.
46. Hadacek F, Mueller C, Werner A, Greger H, Proksch P, Analysis, isolation and insecticidal activity of linear furanocoumarins and other coumarin derivatives from *Peucedanum* (Apiaceae : Apioideae), *J Chem Ecol* **20**:2035–2054, 1994.
47. Kang S, Kim H, Lee W, Ahn Y, Toxicity of bisabolangelone from *Ostericum koreanum* roots to *Dermatophagoides farinae* and *Dermatophagoides pteronyssinus* (Acari: Pyroglyphidae), *J Agric Food Chem* **54**:3547–3550, 2006.
48. Meepagala K, Osbrink W, Sturtz G, Lax A, Plant-derived natural products exhibiting activity against Formosan subterranean termites (*Coptotermes formosanus*), *Pest Manag Sci* **62**:565–570, 2006.
49. Zhu BCR, Henderson G, Chen F, Maistrello L, Laine RA, Nootkatone is a repellent for Formosan subterranean termite (*Coptotermes formosanus*), *J Chem Ecol* **27**:523–531, 2001.
50. Pridgeon JW, Becnel JJ, Wang Z, Katritzky AR, Meepagala KM, Clark GG, Structure-activity relationships of 33 piperidines as toxicants against female adults of *Aedes aegypti* (Diptera: Culicidae), *J Med Entomol* **44**:263–269, 2007.
51. Miyazawa M, Tsukamoto T, Anzai J, Ishikawa Y, Insecticidal effect of phthalides and furanocoumarins from *Angelica acutiloba* against *Drosophila melanogaster*, *J Agric Food Chem* **52**:4401–4405, 2004.
52. Tsukamoto T, Ishikawa Y, Miyazawa M, Larvicidal and adulticidal activity of alkylphthalide derivatives from rhizome of Cnidium officinale against *Drosophila melanogaster*, *J Agric Food Chem* **53**:5549–5553, 2005.
53. Meepagala KM, Sturtz G, Wedge DE, Schrader KK, Duke SO, Phytotoxic and antifungal compounds from two Apiaceae species, *Lomatium californicum* and *Ligusticum hultenii*, rich sources of Z-ligustilide and apiol, respectively, *J Chem Ecol* **31**:1567–1578, 2005.
54. Lao SC, Li SP, Kan KKW, Li P, Wan JB, Wang YT, Dong TTX, Tsim KWK, Identification and quantification of 13 components in *Angelica sinensis* (Danggui) by gas chromatography-mass spectrometry coupled with pressurized liquid extraction, *Analytica Chimica Acta* **526**:131–137, 2004.

55. Parmar VS, Jain SC, Bisht KS, Jain R, Taneja P, Jha A, Tyagi OD, Prasad AK, Wengel J, Olsen CE, Boll PM, Phytochemistry of the genus *Piper Phytochemistry* **46**:597–673, 1997.
56. Su HCF, Horvat R, Isolation, identification, and insecticidal properties of the *Piper nigrum* amides, *J Agric Food Chem* **29**:115–118, 1981.
57. Bernard CB, Krishnamurty DC, Durst T, Philogène BJR, Sanchez-Vindas P, Hasbun C, Poveda L, San Roman L, Arnason JT, Insecticidal defenses of Piperaceae of the Neotropics, *J Chem Ecol* **21**:801–814, 1995.
58. Scott IM, Jensen H, Arnason JT, Philogéne BJR, Lesage L, Efficacy of *Piper* (Piperaceae) extracts on various insect pests of the home and garden, *J Econ Entomol* **97**:1390–1403, 2004.
59. McFerren MA, Cordova D, Rodriguez E, Rauh JJ, *In vitro* neuropharmacological evaluation of piperovatine, an isobutylamide from *Piper piscatorum* (Piperaceae), *J Ethnopharmacol* **83**:201–207, 2002.
60. Zlotkin E, The insect voltage-gated sodium channel as target of insecticides, *Annu Rev Entomol* **44**:429–455, 1999.
61. Rao KV, Davies R, The ichthyotoxic principles of Zanthoxylum clava-herculis, *J Nat Prod* **49**:340–342, 1986.
62. Wolski T, Ludwiczuk AKF, Wydz Farm A, Med LP, Naturally occurring pesticides and their plant protection applications, *Przemysl Chemiczny* **81**:370–373, 2002.
63. Hodgson E, Philpot RM, Interaction of methylenedioxyphenyl (1,3-benzodioxole) compounds with enzymes and their effect on mammals, *Drug Metab Rev* **3**:231–301, 1974.
64. Scott IM, Jensen H, Scott JG, Isman MB, Arnason JT, Philogéne BJR, Botanical insecticides for controlling agricultural pests: Piperamides and the Colorado potato beetle *Leptinotarsa decemlineata* Say (Coleoptera: Chrysomelidae), *Arch Insect Biochem Physiol* **54**:212–225, 2003.
65. Rodriguez-Saona C, Millar JG, Maynard DF, Trumble JT, Novel antifeedant and insecticidal compounds from avocado idioblast cell oil, *J Chem Ecol* **24**:867–889, 1998.
66. Rodriguez-Saona CR, Maynard DF, Phillips S, Trumble JT, Alkylfurans: Effects of alkyl side-chain length on insecticidal activity, *J Nat Prod* **62**:191–193, 1999.
67. Fokialakis N, Osbrink WLA, Mamonov LK, Gemejieva NG, Mims AB, Skaltsounis AL, Lax A, Cantrell CL, Antifeedant and toxicity effect of thiophenes from four *Echinops* species against the Formosan subterranean termite, *Coptotermes formosanus*, *Pest Manag Sci* **62**:832–838, 2006.
68. Lemma A, Yau P, Various molluscicides to fish and snails, *Ethiopian Med J* **12**:109–114, 1974.

69. Perrett S, Whitefield PJ, Currently available molluscicides, *Parasitology Today* **12**:156–159, 1996.
70. Goll PH, Lemma A, Duncan J, Mazengia B, Control of schistosomiasis in Adwa, Ethiopia, using the plant molluscicide endod (*Phytolacca dodecandra*), *Tropenmedizin Parasitologie* **34**:177–183, 1983.
71. Marston A, Hostetmann K, Plant molluscicides, *Phytochemistry* **24**:639–652, 1985.
72. Mitchell AJ, Update and impact of a trematode that infects cultured channel catfish, *The Catfish J* **16**:17–27, 2001.
73. Meepagala KM, Sturtz G, Mischke CC, Wise D, Duke SO, Molluscicidal activity of vulgarone B against ram's horn snail (*Planorbella trivolvis*), *Pest Manag Sci* **60**:479–482, 2004.
74. Meepagala KM, Sturtz G, Wise D, Wedge DE, Molluscicidal and antifungal activity of *Erigeron speciosus* steam distillate, *Pest Manag Sci* **58**:1043–1047, 2002.
75. Joshi RC, Baucas NS, Joshi EE, Verzola EA, CD-ROM: Information database on the golden apple snail (golden kuhol), *Pomacea canaliculata* (Lamarck): A terminal report submitted to the Department of Agriculture-Cordillera Highland Agricultural Resource Management Project, pp. 18, 2003.
76. Sin TS, Damage potential and control of *Pomacea canaliculata* (Lamarck) in irrigated rice and its control by cultural approaches, *Int J Pest Manag* **49**:49–55, 2003.
77. Hill SA, Miyasaka SC, Taro responses to excess copper in solution culture, *Hort Science* **35**:863–867, 2000.
78. Litsinger JA, Estano DB, Management of the GAS (*Pomacea canaliculata* Lamarck) in rice, *Crop Prot* **12**:363–370, 1993.
79. Shih CC, Chang SS, Chan Y-L, Chen JC, Chang MW, Tung MS, Deng JF, Yang CC, Acute metaldehyde poisoning in Taiwan, *Vet Human Toxicol* **46**:140–143, 2004.
80. Joshi RC, Meepagala KM, Sturtz G, Cagauan AG, Mendoza CO, Dayan FE, Duke SO, Molluscicidal activity of vulgarone B from *Artemisia douglasiana* (Besser) against the invasive, alien, mollusc pest, *Pomacea canaliculata* (Lamarck), *Internat J Pest Manag* **51**:175–180, 2005.
81. Vallad GE, Goodman RM, Systemic acquired resistance and induced systemic resistance in conventional agriculture, *Crop Sci* **44**:1920–1934, 2004.
82. Vick BA, Zimmerman DC, Biosynthesis of jasmonic acid by several plant species, *Plant Physiol* **75**:458–461, 1984.
83. Farmer EE, Ryan CA, Octadecanoid precursors of jasmonic acid activate the synthesis of wound-inducible proteinase inhibitors, *Plant Cell* **4**:129–134, 1992.

84. Gundlach H, Muller MJ, Kutchan TM, Zenk MH, Jasmonic acid is a signal transducer in elicitor-induced plant cell cultures, *Proc Natl Acad Sci USA* **89**:2389–2393, 1992.
85. Mueller MJ, Brodschelm W, Spannagl E, Zenk MH, Signaling in the elicitation process is mediated through the octadecanoid pathway leading to jasmonic acid, *Proc Natl Acad Sci USA* **90**:7490–7494, 1993.
86. Wedge DE, Camper ND, Connections between agrochemicals and pharmaceuticals, in Cutler HG, Cutler SJ, (eds.), *Biologically Active Natural Products: Pharmaceuticals*, CRC Press, Boca Raton, FL. pp. 1–15, 2000.
87. Sauter H, Ammermann E, Roehl F, Strobilurins —- From natural products to a new class of fungicides, in *Crop Protection Agents from Nature*, Royal Soc Chem Cambridge UK, pp. 50–81, 1996.
88. Cape JL, Bowman MK, Kramer DM, Computation of the redox and protonation properties of quinones: Towards the prediction of redox cycling natural products, *Phytochemistry* **67**:1781–1788, 2006.
89. Meazza G, Dayan FE, Wedge DE, Activity of quinones on *Colletotrichum* species, *J. Agric Food Chem* **51**:3824–3828, 2003.
90. Wedge DE, Nagle DG, Preparation of sampangine and its analogs as fungicides, U.S. Pat. Appl. Publ. 12 pp. CODEN: USXXCO US 2004192721 A1 20040930 CAN 141:273006 AN 2004:803936, 2004.
91. De Lucca AJ, Bland JM, Vigo CB, Cushion M, Selitrennikoff CP, Peter J, Walsh TJ, CAY-1, a fungicidal saponin from *Capsicum sp fruit Med Mycol* **40**:131–137, 2002.
92. Fokialakis N, Cantrell CL, Duke SO, Skaltsounis AL, Wedge DE, Antifungal activity of thiophenes from *Echinops ritro.* L, *J Agric Food Chem* **54**:1651–1655, 2006.
93. Oliva A, Meepagala KM, Wedge DE, Hale AL, Harries D, Aliotta G, Duke SO, Natural fungicides from *Ruta graveolens* L. leaves, including a new quinolone alkaloid, *J Agric Food Chem* **51**:890–896, 2003.
94. Salah AM, Bedir E, Ngeh TJ, Kahn IA, Wedge DE, Antifungal clerodane diterpenes from *Macaranga monandra* (L) Muell. et Arg. (Euphorbiaceae), *J Agric Food Chem* **51**:7607–7610, 2003.
95. Ujváy I, Natural product pesticides, in Plimmer JR (ed.) *Encyclopedia of Agrochemicals, Vol. 3*, Wiley-Interscience. Hoboken, NJ, pp. 1090–1103, 2003.
96. Quarles W, Non-toxic weed control in the lawn and garden, *Common Sense Pest Cont Quarter* Summer, 4–14, 1999.
97. Anonymous, Vinegar wipes out thistles organically, *Stockman Grass Farmer*, July, 1, 2002
98. Tworkoski T, Herbicide effects of essential oils, *Weed Sci* **50**:425–431, 2002.

99. Duke SO, Abbas HK, Amagasa T, Tanaka T, Phytotoxins of microbial origin with potential for use as herbicides, in Copping LG (ed.), *Crop Protection Agents from Nature: Natural Products and Analogues, Critical Reviews on Applied Chemistry, Vol. 35*. Society for Chemical Industries, Cambridge, UK, pp. 82–113, 1996.
100. Lydon J, Duke SO, Inhibitors of glutamine synthesis, in Singh BK (ed.), *Plant Amino Acids: Biochemistry and Biotechnology*, Marcel Dekker, New York, pp. 445–464, 1999.
101. Lee DL, Prisbylla MP, Cromartie TH, Dagarin DP, Howard SW, Provan WM, Ellis MK, Fraser T, Mutter LC, The discovery and structural requirements of inhibitors of *p*-hydroxyphenylpyruvate dioxygenase, *Weed Sci* **45**:601–609, 1997.
102. Lee DL, Knudsen CG, Michaely WJ, Chin H-L, Nguyen NH, Carter CG, Cromartie TH, Lake BH, Shribbs JM, Fraser T, The structure-activity relationships of the triketone class of HPPD herbicides, *Pestic Sci* **54**:377–384, 1998.
103. Pallett KE, Little JP, Sheekey M, Veerasekaran P, The mode of action of isoxaflutole I. Physiological effects, metabolism, and selectivity, *Pestic Biochem Physiol* **62**:113–124, 1998.
104. Meazza G, Scheffler BE, Tellez MR, Rimando AM, Nanayakkara NPD, Khan IA, Abourashed EA, Romagni JG, Duke SO, Dayan FE, The inhibitory activity of natural products on plant *p*-hydroxyphenylpyruvate dioxygenase, *Phytochemistry* **59**:281–288, 2002.
105. Romagni JG, Meazza G, Nanayakkara NPD, Dayan FE, The phytotoxic lichen metabolite usnic acid is a potent inhibitor of plant *p*-hydroxyphenylpyruvate dioxygenase, *FEBS Lett* **480**:301–305, 2000.
106. Dayan FE, Duke SO, Sauldubois A, Singh N, McCurdy C, Cantrell CL, *p*-Hydroxyphenylpyruvate dioxygenase is a target site for β-triketones from *Leptospermum scoparium Phytochemistry*, 2007.
107. Grayson BT, Williams KS, Freehauf PA, Pease RR, Ziesel WT, Sereno RL, Reinsfelder RE, The physical and chemical properties of the herbicide cinmethylin, *Pestic Sci* **21**:143–153, 1987.
108. Grayson BT, Webb JD, Factors offecting the performance and crop phytotoxicity of a new rice herbicide, cinmethylin. I. Effects of water depth and soil type on the distribution and uptake of anmethylin by transplanted and direct-seeded rice, *Pestic Sci* **32**:207–218, 1991.
109. Grossmann K, The mode of action of quinclorac: A case study of a new auxin-type herbicide, in Cobb AH, Kirkwood RC (eds.), *Herbicides and their Mechanisms of Action*, Sheffield Academic Press, UK, pp. 181–214, 2000.

110. Grossmann K, Kwiatkowski J, Evidence for a causative role of cyanide, derived from ethylene biosynthesis, in the herbicidal mode of action of quinclorac in barnyard grass, *Pestic Biochem Physiol* **51**:150–160, 1995.
111. Grossmann K, Kwiatkowski J, The mechanism of quinclorac selectivity in grasses, *Pestic Biochem Physiol* **66**:83–91, 2000.
112. Agrawal GK, Jwa NS, Rakwal R, A novel rice (*Oryza sativa* L.) acidic PR1 gene highly responsive to cut, phytohormones, and protein phosphatase inhibitors, *Biochem Biophys Res Commun* **274**:157–165, 2000.
113. Matsuzawa M, Graziano MJ, Casida JE, Endothal and cantharidin analogues: Relation of structure to herbicidal activity and mammalian toxicity, *J Agric Food Chem* **35**:823–829, 1987.
114. Duke SO, Baerson SR, Rimando AM, Pan FE, Dayan FE, Belz RG, Biocontrol of weeds with allelopathy. in Gressel J, Vurro M (eds.), *Novel Biotechnologies for Biocontrol Agent Enhancement and Management*, Springer, pp. 75–85, 2007.

Chapter 9

NATURAL PRODUCTS IN MYCELIAL MICROORGANISMS: IMPACT OF MORPHOLOGY

Sergei Braun

9.1 NATURAL PRODUCTS FROM MICROORGANISMS: CURRENT STATUS

The industrial production of penicillin by Pfizer in 1943 transformed the chemistry of natural products. Microbiology became its field and pharmaceutical research became its leading trend. The main driving force behind the research of the chemistry of natural products was and still remains to be human health care. Since the advent of antibiotics, many sources of natural chemicals, such as plants, fungi and bacteria, have been searched for antibacterials, anticancer products, immuno-suppressors and chemicals that affect the human psyche. This research was so successful that, even today, more than half of all pharmaceuticals are either natural products or derived from, or inspired by them. The market of antibiotics alone was ca. US$30 billion in 2005. The majority of antibiotics is produced by submerged fermentation of filamentous microorganisms such as soil prokaryotic actinomycetes and eukaryotic moulds, ascomycetes.

Since the late 1940s, large international pharmaceutical companies provided the bulk of research incentives, although government and the public also contributed their share. In the 1950s and 1960s, about 80%

of the reviews in the fields of microbiology and natural products' chemistry have been devoted to antibiotics. However, in the last 25 years, there was a dramatic decrease in the output of antibiotic research. The rate of FDA approval of new antibiotics has dropped drastically. Only three new antimicrobials were approved between 2004 and 2006.[1]

Although commercial antibiotics resulting from the research boom of the mid-20th century provided for the control of infections, pathogens were never eradicated or even controlled completely. The losses for global economics from infectious diseases are staggering. They affect the population as a whole both in the developed countries and developing countries. Infections are the scourge of the Third World, but they are not a trifle in the richer countries. Microbial infection is the third leading cause of death in the developed countries![2] According to the Infectious Disease Society of America (IDSA), 90 000 people die of infectious diseases in the US yearly; 70% of these are due to antibiotic-resistant pathogens such as methicillin-resistant *Staphylococcus aureus* (MRSA) and *Mycobacterium tuberculosis*.[3]

Due to the saturation of the market with existing antibiotics and the high costs needed to get FDA approval, there is a current lull in the research of antibiotics. However, the main reason is the relative size of different sections of the healthcare industry. Medical statistics show that money is spent on drugs to treat age-related chronic conditions such as heart diseases, cancer, obesity, low sexual performance etc. These conditions are the main concern of the aging population, and the medically well insured societies of Europe, Japan and North America. This simple fact dictates preference for treating chronic *versus* acute patients. Nevertheless, the danger of uncontrolled spread of drug-resistant infections exists and serves as a reason to seek new remedies. As in other types of enterprises, the restriction of spending may have a positive effect on the research, making it lean and efficient. The current revolution in life sciences carries with it many new tools for the discovery and the effective production of new types of antimicrobial agents from microorganisms. The potential of natural products from mycelial organisms is far from being exhausted. In the genus *Streptomyces*, ca. 10^5 antibiotics are waiting to be discovered![4] The known antibiotics represent only 3% of this potential.

Pharmaceuticals are only a part of natural products derived from filamentous organisms. Filamentous fungi are also grown industrially for the production of natural products other than pharmaceuticals. Organic acids such as citric, itaconic and gluconic acids, carotenoids, taste and aroma compounds, and a variety of industrial enzymes such as amylases, lipases, dextranases and proteases are also extracted from these fungi.

Modern technologies offer new efficient methods for improved industrial exploitation of the microbial world. Intelligent screening replaces brute force in the search for valuable metabolites. Genetic tools for screening of rare bacteria from unusual niches (deep sea mud, lichens, algae), and hard-to-culture soil organisms, as well as metagenomics — the screening of metagenomic gene clusters directly from soil, speed up the discovery of new types of natural products. Artificial evolution of key genes, combinatorial biosynthesis and metabolic pathway engineering allow for new efficient fermentation processes and efficient use of available research resources. These new tools provide ways of acquiring new knowledge of the metabolism, improving the understanding of metabolic chains resulting in biologically active metabolites. This review focuses on one important, though less known, aspect of filamentous microorganisms — the connection between form and productivity, the common genetic regulation of morphogenesis and secondary metabolism.

9.2 COMPLEXITY OF MYCELIAL MICROORGANISMS

Filamentous microorganisms, both pro- and eukaryotic, provide an evolutionary link between a single cell and a truly multicellular organism. The latter contain numerous nuclei in an extended single body. Their survival and reproductive success require coordination and specialization of cells and cellular complexes, to which the term "organ" may be justifiably applied. The filamentous body extends by branching and vegetative growth at specialized hyphal tips. Some cells differentiate to form a variety of reproductive organs such as spore-forming aerial mycelium and spores in actinomycetes, and in ascomycetes, aerial mycelium, conidiophores, fruiting bodies and a variety of dispersive propagules,[5] both sexual and asexual. Mycelial organisms traffic water and nutrients across large filamentous networks. They are able to bring water to the dry nutrient substrate which is otherwise inaccessible due to low water activity. This ability allowed them to become the ultimate scavengers, being able to colonize niches where the others cannot exist.

Filamentous organisms colonize the humid nutrient substrate, producing, in general, highly branched and densely interwoven vegetative or feeding mycelium. When the nutrients are becoming scarce, the branching rate decreases, and linear hyphal growth ensues. The daughter colonies of feeding mycelium are established when the outgrowing hyphae reach unexhausted substrate. The primary colony differentiates in response to nutrient deprivation, thus forming aerial mycelium. The surface of aerial mycelia is less smooth and much more hydrophobic than that of the feeding mycelium, allowing its separation from the moist substrate phase by lowering water tension and initiating the growth of hyphae in the hydrophobic air phase. The cessation of vegetative growth and the formation of aerial mycelium are concomitant with fundamental changes in metabolism. Secondary metabolites of various nature are synthesized with many of them being protective pigments. Stress-resistant dispersive dormant seed cells such as spores and conidia are at last produced to be dispersed in the air. The daughter colonies, which are more or less equidistant from the primary colony, undergo the same cycle of vegetative growth, nutrient stress and formation of dispersive seed cells, giving rise to the ring phenomena abundant in the world of fungi. Fruiting bodies of mushrooms form so called witches' rings. Colonies of many moulds such as *Trichoderma* (Fig. 1) and *Alternaria*, grown from a single spore also form concentric rings on the surface of the substrate.

The complexity of the architecture and the lifestyle of mycelial organisms require intricate intercellular signaling system responsive to environmental conditions, both biotic and abiotic, to the ecosystem as the whole. It is of no coincidence that mycelial organisms are rich in secondary metabolites, which are potent biological regulators. These metabolites function not only as antibiotics, regulators of foreign species, but also as essential signals for cellular differentiation in the very organisms, which produce them.[6] The richness of natural product chemistry is, thus, the reflection of the elaborate biology responsible for the regulation of mycelial organisms

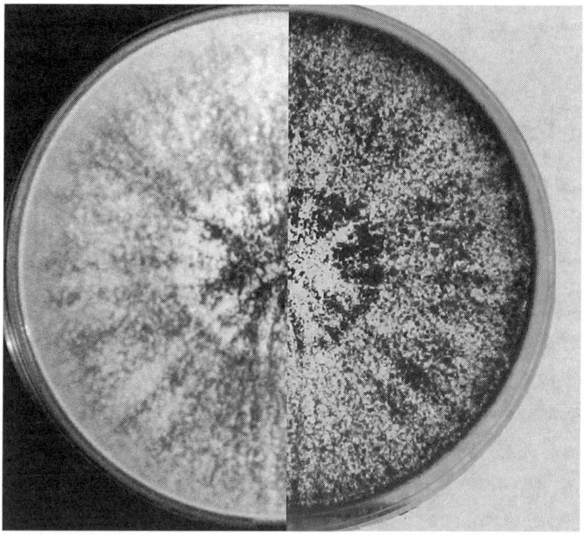

Fig. 1. *Trichoderma harzianum* colony forms ring structures on potato-dextrose agar: white feeding mycelium (right) and green sporulating mycelium (left).

in their natural state. The complex and elaborate biology is, in turn, the reflection of environmental complexity and the essential heterogeneity of natural habitat. Once we employ mycelial cultures for our industrial goals, this complexity is lost. In the pursuit of our industrial goals, we have to recur to essentially homogeneous submerged fermentation of filamentous microorganisms in large reactors, engineered as a bubble column, airlift fermenter or stirred tank. Submerged fermentation eliminates the essential prerequisite of secondary metabolism — the emergence of aerial mycelia. Thus, the preservation of the ability of these organisms to produce the desired metabolite, especially secondary metabolite, and the optimization of its yield demand either global redesign of their metabolic pathways (and of their regulation) or mimicking of their natural habitat. The complexity of this task undertaken with limited knowledge of the metabolites, boggles the mind. How could anyone achieve this goal?

The proof that this task is possible is the very existence of successful industrial processes with working concentrations of secondary metabolites overreaching by many orders of magnitude their accumulation in the natural habitat. It was mostly done by brute force selection, by trial and error, although recent progress in the techniques of molecular biology, genomics and metabolomics[7] increasingly provides for some rationality. Yet, in spite of continuous progress in technology, and the accumulation of knowledge and experience, the process of achieving productive growth of filamentous microorganisms and then exploiting their ability to overproduce certain natural products has never been a simple task. It requires years of genetic (strain improvement) and morphological (optimization of medium and process parameters) engineering.

9.3 MORPHOLOGY OF FILAMENTOUS MICROORGANISMS IN INDUSTRIAL HABITAT

Depending on process parameters, culture age and genetic make-up, filamentous microorganisms in an industrial fermenter can either grow as free mycelia or as mycelial aggregates — pellets or flocks (Fig. 2). Of course, this polar definition is far from reality; all kinds of intermediate morphologies exist between free unassociated mycelia, loose flocks and dense pellets. The

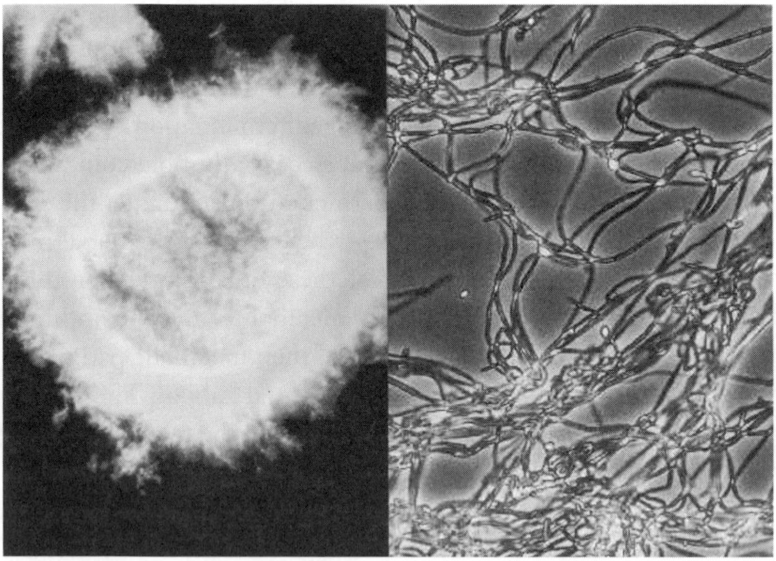

Fig. 2. Cross-section of the pellet (Ø = 2 mm, left) and filaments (right) of *Streptomyces tendae.*

use of automatic image analysis during the last decade has introduced the necessary degree of quantization in the description and the characterization of mycelial morphology. These advances, factors impacting fungal morphology and their impact on metabolism in fungi were recently summarized in an excellent review, both exhaustive and concise, by Papagianni.[8] Culture morphology can greatly affect fermentation productivity. For example, the optimal production of citric acid by *Aspergillus niger*[9], itaconic acid by *A terreus*[10] and heterologous proteins by *A niger*[11] requires pellet morphology. Filamentous growth, on the other hand, is preferable for penicillin production by *Penicillium chrysogenum*[12] and fumaric acid by *Rhizopus arrhizus*.[13] The ability to obtain and maintain a particular morphology is one of the key parameters in the development of a productive filamentous fermentation. Process conditions such as mechanical shear, oxygen concentration, substrate concentration, nitrogen-carbon ratio, pH, ionic strength, and inoculum concentration[14] have effects on morphology. Some of these parameters such as oxygen transfer and shear rates do not coscale. The preservation of the desired morphology, while scaling up the production, requires the development of process-specific tools to control it.[15]

Both modes of growth present significant technical difficulties for the efficient production of metabolites. Dense filamentous growth increases the apparent viscosity of fermentation broth, reduces homogenation times, and creates stagnant zones in the fermentation vessel. Filaments are more sensitive to mechanical shear than dense smooth pellets. Pellets decrease the culture viscosity, which, in turn, increases the efficiency of mixing and, thus, mass transfer. However, the nonhomogeneity is merely scaled down to microscale. Diffusion is slow within the pellets. Oxygen concentration falls rapidly with depth. Below the surface of pellets of *Phanerochaete chrysosporium*[16], oxygen was completely depleted 0.8 mm below the surface. The product output also declined precipitously at pellet diameters greater than 1 mm. The central zone of pellets with diameters above 1 mm is usually completely or at least partially lysed (Fig. 2, left panel). We have shown that the deposition of the bikaverin pigment, the water insoluble secondary metabolite of the mould *Gibberella fujikuroi*, occurs neither in the vegetative outer surface of the pellet, nor in the oxygen starved central core, but only in the narrow (less than 0.2 mm) layer lying about 0.2 mm below the surface.[17] All these indicate that there is a practical upper limit for pellet size associated with high product output in aerobic bioprocesses — less than 1 mm in diameter.

There are, however, more intricate interrelationships between the mycelial architecture of the filamentous organism and the productivity of these organisms. We have observed, for example, that bikaverin is deposited only at the branching points within the hyphae of *G fujikuroi* (Fig. 3). Below, we shall show that there are correlations between the global translational regulators responsible for both morphogenesis and secondary metabolism.

The issue of the relation between the form and the productivity in filamentous microorganisms is of utmost importance for a variety of reasons ranging from purely fundamental to purely economic. This importance resulted in the multiplicity of research timely published and reviewed. However, the field remains descriptive but not predictive. It is not yet possible to make a molecular model or algorithm, where one uses morphological parameters to come up with the output and the assortment of metabolites.

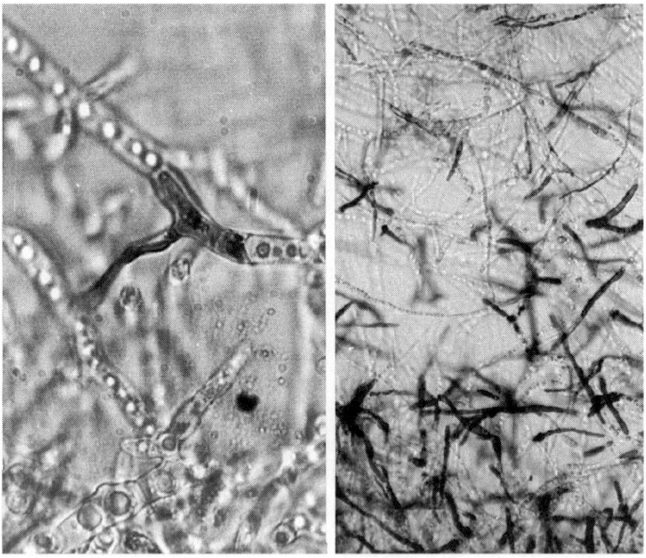

Fig. 3. Preferential deposition of water insoluble pigment bikaverin (dark) at the filament branching points in the fungus *Gibberella fujikuroi*.

The reason for this failure is in the complexity of regulatory pathways, mostly unresolved in any details, as well as in the fragmentation of research in this field. The economic interest has driven the research into specific processes and targets and not systematic coherent basic studies. True, complete genomes of *Streptomyces coelicolor* and *Aspergillus nidulans* are available, and genomic analysis has shed a considerable light on the development of these and other filamentous microorganisms. Despite this, the molecular biology of filamentous growth has not reached the detail of resolution comparable to that of other model organisms such as Baker's yeast, *Saccharomyces cerevisiae*, *E coli* and the nematode *Caenorhabditis elegans*.

9.4 THE UNITY OF LIVING MATTER

The impact of growth morphology on the productivity of mycelial organisms cannot yet be described in the terms of the chemistry of natural products. After reflection, I arrived at the conclusion that my review must be significantly different from the normal pattern of "reports and recipes." It should drift into a more synthetic and speculative mode. It should point out to a variety of areas seemingly unrelated, yet holding the promise of breaching the walls of ignorance separating filamentous architecture from the formation of certain metabolites. In this mode, I will draw analogies from fields as distant as bacterial cross talk, flocculation of *S cerevisiae*, filamentous growth and invasiveness of pathogenic *Candida* species quorum sensing, adhesion and bacterial biofilm formation. I shall mix regulatory circuits in *Mixobacteria* with cell wall proteins in fungi. I shall strive to show the underlying similarity in the general principles tying certain morphemes with discrete metabolic activities.

The justification for such an approach is the basic unity of living matter. It has been estimated that the chance to stumble upon a physiologically active compound among natural products is higher by two orders of magnitude than in chemical compound libraries.[18] This is hardly surprising. Recent genomic studies suggest broad evolutionary links between bacteria and eukaryotes both of which share similarity in the domain architecture allowing for the combinatorial engineering of functional chimerical signaling and differentiation systems. I quote from the recent opinion: "Spurred on in part by comparative genomics, recent efforts in many microbial systems have begun to divulge the high level operating principles that underlie biology, some of which have been enshrouded by the mists of evolutionary time. The ever closer relationships between microbial and metazoan cellular functions have now surpassed even the wildest expectations. . . . these similarities now span subsystems once perceived either as being in the exclusive domain of multicellular organisms, such as apoptosis, now firmly documented in yeast, or in the eukaryotic domain, such as cytoskeletal substructures, now evident as microtubule-, actin- and intermediate filament-like forms in bacterial systems."[19]

The new trends in biology underscore the unity of nature. Specific is better understood in the context of general. It is my intention to show that the general regulatory principles involved in the regulation of morphogenesis and secondary metabolism in prokaryotes and eukaryotes are similar. Both prokaryotes and eukaryotes use similar evolutionary adaptations in similar circumstances. I would like the reader to see this similarity and, perhaps, to take one step further into devising strategies for the identification of genetic targets for strain improvement. The stress of this review is not on fine details, which are, sometimes, not available, but rather on general phenomena. Thus, the description of pellet formation in actinomycetes can be enhanced by the analysis of adhesive behavior of unicellular bacteria in microbial mats and in yeast flocks.

9.5 MICROBIAL ADHESION: FLOCCULATION AND BIOFILM FORMATION

The formation of microbial pellets is only one example of aggregative behavior in microorganisms; other examples include flocculation and biofilm formation. Flocculation plays an important role in brewing. Many industrial strains of the yeast, *Saccharomyces cerevisiae* flocculate following the exhaustion of fermentable sugars. The flocks containing thousands of cells either sink or float, making the separation of the biomass and the beer a simple procedure. Flocculation phenomenon is well dissected by molecular tools because of its industrial interest and the well-studied genetics of the Brewer's yeast.

Yeast flocculation mechanism can be described as a phenomenon of adhesion to certain surfaces. The ability to adhere to surfaces and to form biofilm is the basis of the pathogenicity of *Candida* species. Pathogens adhere to mucous membranes and wounds, they stick to medical instruments and prosthesis, and thus contaminate surfaces in food processing facilities. The high mortality rate in disseminated fungal infections caused an increase in the amount of research on the molecular basis of the adhesive phenomena in *Candida*. This research discovered a considerable overlap in the molecular regulation of all forms of adhesive behavior.[20]

In *S cerevisiae*, cell to cell adhesion is mediated by a family of cell surface proteins called flocculins. Gene products of *flo* 1, 5, 9 and 10 confer flocculative behavior (cell to cell adhesion), while *flo*11 is responsible for adhesion to hydrophobic surfaces and to biofilm formation.[21] Similar specialization of adhesin genes for different surfaces within the adhesion protein families is characteristic of other fungi as well as *als* genes' family of *C albicans* and *epa* in the human pathogen *C glabrata*. The expression of various adhesins is the result of environmental challenges and it displays plasticity and complexity reflected in multitiered regulation systems. In *S cerevisiae*, at least five different systems are involved: (1) the Ras-cAMP (PKA)-dependent pathway; (2) the mitogen activated protein kinase (MAPK)-dependent filamentous growth pathway; (3) the main glucose repression pathway; (4) the TOR nitrogen starvation pathway; and (5) the poorly understood pathway involving transcription regulators *sok2*, *phd1* and *ash1* (for the review see[20]). Whatever is the precise molecular architecture of these systems, they have a feature in common: they mediate responses to stress and are mediated by intercellular signaling. Thus, MAP kinase pathway, which is, among other things, involved in the responses for nitrogen starvation is triggered by alcohols such as butanol.

Flocculins and especially Flo11 are responsible for morphogenic phenomena such as pseudohyphal growth and biofilm formation in yeast.[21] Biofilm formation is an adhesive phenomenon akin to flocculation. The majority of bacteria exist in highly organized natural biofilm populations rather than in free floating cultures. Within the biofilm, bacteria display coordinated behavior. They form structures, release toxins or emit light. They, sometimes, differentiate to form physiologically defined subpopulations, fruiting bodies and spores. Mature biofilm consists of regular microcolonies encased in an extracellular polymeric substance (EPS), which is made of typical bacterial capsid materials such as polysaccharides, lipids, proteins and DNA, they also contain a regular system of water and nutrient channels. It is worth noting that *algC* promoter responsible for the biosynthesis of bacterial alginate in *P aeruginosa*, one of the best characterized EPS components, is activated within 15 min of its attachment to the biofilm.[22] Under other environmental conditions, the activation of *algC* promoter is triggered by a variety of stresses including high osmolarity and nitrogen limitation.

Filamentous aggregates display many traits remindful of biofilms. Thus, microbial pellets of *S coelicolor* dispersed into filament fragments by treatment using enzymes, DNase and hyaluronidase.[23] The flocculin gene *flo11*, which is necessary for biofilm formation and pseudohyphal growth in diploid yeast, promotes the invasive growth of yeast colonies in agar. Both phenomena are mediated through cAMP/PKA and MAPK pathways[24] by extracellular cAMP. Invasive growth on agar is also typical of many actinomycetes, although this phenomenon is not well investigated. Pellet formation in actinomycetes also involves the modification of cell surface properties — growth morphology is influenced by the addition of surfactants.[25] Individual hyphae can display flocculative behavior. Thus, homogenized filaments produced from pellets of *S. tendae* tend to flocculate, while linear mycelium does not.[14] Aerial mycelia of *Streptomycetes* are coated with hydrophobic cell wall proteins (see below in this review). Perhaps, pelletal growth mimics the physiological conditions, causing the outgrowth of aerial mycelium.

It is possible that extracellular cAMP is the unifying regulatory molecule behind aggregative growth and the invasiveness of both ascomycetes and actinomycetes. In our old review,[17] we have proposed to view "mycelial aggregates not merely as mechanical conglomerates, but rather as complex differentiated tissues." We compared the transition from dispersed to aggregated growth in filamentous organisms with its differentiation in slime moulds. Under starvation, free living cells of the slime mold, *Dictyostelium discoideum*, aggregate to form a multicellular body. Cyclic AMP serves as the signal for this metamorphosis.[26]

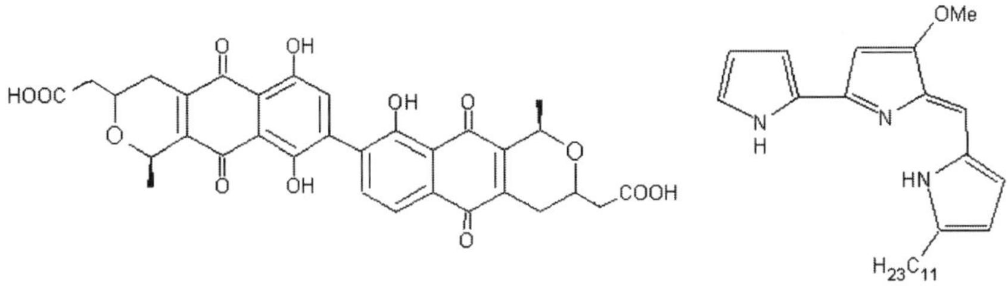

Fig. 4. Actinorhodin (ACT) and undecylprodigiosin.

In *Streptomyces coelicolor*, cAMP production peaks, when aerial mycelium is formed and the polyketyde antibiotic pigment, actinorhodin (ACT), is produced (Fig. 4).[27] In the adenylate cyclase-disrupted mutant of *S. coelicolor* lacking ACT, the external cAMP induced the pigment production in μM concentrations in a dose-dependent manner. In contrast, cAMP, albeit at higher concentrations, inhibited the production of another pigmented antibiotic, tripyrrol undecylprodegiosin (Fig. 4). It was suggested that cAMP may serve as a diffusible signaling molecule to coordinate antibiotic biosynthesis, probably through the phosphorylated products of the regulatory genes *afsR* and *afsR2*. In related *S. lividans*, glucose prevents ACT production by repressing the synthesis of *afsR2* mRNA. A shift to glycerol as the sole carbon source leads to extensive ACT synthesis and eventually obliterates this phenotypic distinction between *S. lividans* and *S. coelicolor*. Transcription from the *afsR2* promoter during growth in glycerol was dependent on *afsR* gene function and was developmentally regulated, occurring specifically at the time of aerial mycelium formation and coinciding temporally with the onset of ACT production.[28]

9.6 AERIAL MYCELIUM

One of the morphogenic activities common to all filamentous microorganisms is the formation of aerial mycelium.[29] Fungi and filamentous bacteria normally colonize moist environment, in which they grow by hyphal extension and branching, forming extensive, interwoven networks termed as the feeding mycelium. Upon stress or starvation, aerial hyphal growth ensues followed by further differentiation into reproductive structures such as spores and fruiting bodies, which can be very diverse and complex. Thus, aerial mycelium is one of the key events in the differentiation of mycelial organisms, inseparably linked to the formation of secondary metabolites. In actinomycetes, the majority of mutants lacking the ability to form aerial hyphae ("bold" *bld* mutants) are devoid of antibiotic production capacity.

The aerial growth is achieved through the modification of the surface of the hyphae. It is mediated, among other things, by the excretion of specialized small proteins, which in fungi are referred to as hydrophobins. In the vicinity of hydrophobic surfaces, hydrophobins undergo a spontaneous aggregation into rodlet-like structures. They share a limited amino acid sequence homology with a conserved arrangement of eight cystein residues and, probably, have structural similarity.[30] The genome analysis of fungi revealed the existence of multiple hydrophobins with specialized function: some are located on aerial hyphae and conidia; others are expressed during the invasive growth of several plant pathogens.[31,32]

Nevertheless, many mutants lacking hydrophobins produce aerial mycelium. Other hydrophobic proteins have been implicated as possible candidates for the hydrophobin-independent formation of aerial mycelium such as repellents of *Ustillago maydis*[33] or Sc15 protein[34] of *Schizophyllum commune*.

A similar function in the aerial mycelium of filamentous bacteria is played by three distinct groups of proteins such as rodlins, chaplins and SapB of *S. coelicolor*. The surface of aerial hyphae in many *Streptomyces* species is coated by protein-made "rodlet layer." The genes of the two rodlin proteins (*rdlA* and *rdlB*) are only expressed during the formation of aerial hyphae. Although *Streptomyces* strains and mutants, lacking both rodlin genes show reduced ability to adhere to hydrophobic surfaces, their ability to form aerial mycelium and sporulation are not impaired. The surface of hyphae reveals a pattern of smaller rodlets made of chaplins, a family of eight hydrophobic proteins in *S. coelicolor*. All share a highly conserved hydrophobic domain of ca. 40 amino acids, the chaplin domain. The deletion of all chaplin genes results in *bld* ("bold", unable to form aerial mycelium) phenotype[35] — loss of hyphal hydrophobicity and severe defects in the formation of aerial mycelium. Rodlins and chaplins are assumed to interact between them and with an additional small hydrophobic "spore-associated peptide," SapB.[36] SapB is a lantibiotic-like peptide

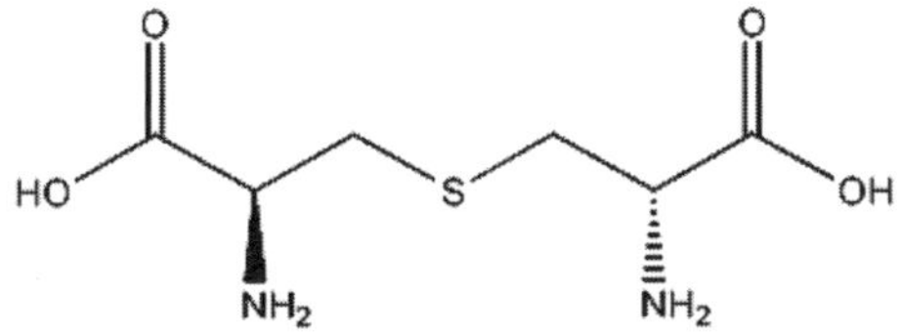

Fig. 5. Lanthionine bridge between didehydroalanine (Dha) and cysteine (Cis).

containing 21 amino-acids with two hydrophobic loops connected by lanthionine bridges[37] (Fig. 5). The gene for SapB was mapped to *ram*, rapid aerial mycelium, gene cluster.[38]

The majority of *bld* mutants are unable to produce and secrete SapB.[36] Sap B is able to reduce water surface tension considerably.[39] Purified Sap B, SapT,[40] lanthionine-containing peptide from *S. tendae* and SC3 hydrophobin[41] of *Schizophylum communae*, added to the medium, restore the ability of *bld* mutants to raise aerial hyphae.

There are indications that an additional regulatory pathway, *sky*, is involved in the full differentiation of aerial mycelium following its release from the substrate.[42]

9.7 QUORUM SENSING

The question of morphogenesis and the productivity in filamentous microorganisms could be reduced, in general terms, to the differential gene regulation during the development of the culture in response to environmental conditions. This implies physicochemical perception of the environment, processing of the signals and global regulation of gene expression bringing out the adaptive response. We, industrial microbiologists, tend to perceive a microbial culture as a monoculture, and each cell as an atomic particle, perfect and independent, because that is how we grow them in our industrial fermentation vessels. In its natural state, a microorganism is only a part of a diverse microbial community, responsive to others in the environment, to its kin, its friend and it's foe. In the industrial fermentation vessels, the abiotic environment is contained within the pre-set levels. Within monoculture, the biotic environment is uniform with regard to biological variety; only the cell density changes. Thus, the cell to cell communication involves only auto-inductive signals and density-dependent regulation.

Many, if not all, populations of unicellular organisms behave as a community. They sense both the foreign environment and their own population density. Gene regulation in response to population density is termed quorum sensing. It is mediated by extracellular chemical signals.[43,44]

Aggregative behavior produces high biomass densities. Quorum sensing signals are involved in the formation of multicellular microbial mats;[45] therefore one would expect that they may regulate morphogenesis and differentiation of mycelial aggregates and, thus, the production of secondary metabolites. The best studied signal molecules are known from industrially important actinomycetes *Streptomyces* species, ascomycetes *Saccharomyces* and pathogenic *Candida* species. In *C. albicans*, tyrosol and farnesol form a part of a complex quorum sensing and regulation system. In *S. cerevisiae*, aromatic alcohols[46] (tyrosol,[47] tryptophol and phenylethanol) (Fig. 6), which accumulate in the medium at high cell densities and/or following nitrogen starvation, enhance *flo11* expression and induce diploid pseudohyphal growth. Aromatic alcohol-mediated signaling involves cAMP/PKA but not MAPK pathway.[48]

It is interesting to note that, while *Saccharomyces* and *Candida* quorum sensing signals share similarity, their effects seem different. In *Saccharomyces*, the signals lead to adhesion and invasive filamentous growth; in *Candida*, filamentation is suppressed. Farnesol seems to promote the dispersion of cells from the *Candida* biofilm.[49]

Antibiotic production and morphological differentiation (formation of aerial mycelium and sporulation) in *Streptomyces* species and in some other actinomycetes is regulated by small signaling molecules called γ-butyrolactones or butanolids.[50] These quorum-sensing signals bind to

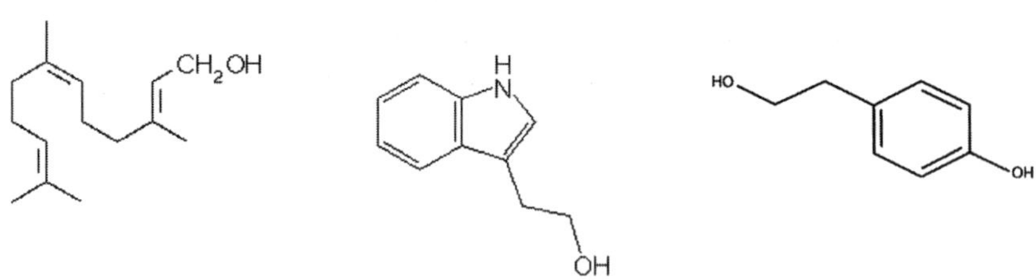

Fig. 6. Quorum sensing molecules in yeast: farnesol, tryptophol and tyrosol.

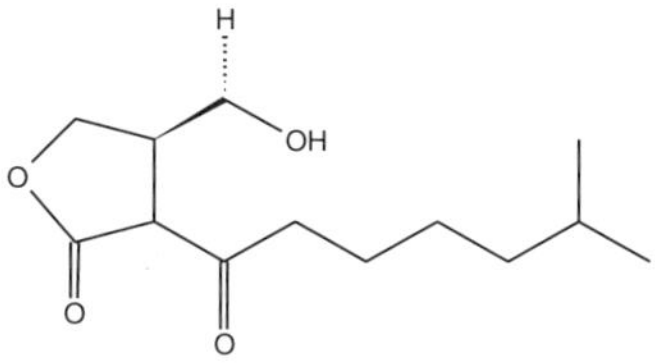

Fig. 7. A-factor from *S. griseus.*

receptors, many of which are involved in the regulation of antibiotic biosynthesis clusters. The first γ-butyrolactone regulating antibiotic production and the formation of aerial mycelium in *Streptomyces* was discovered in *S. griseus*[56]. The antibiotic became known as A-factor (2-S-isocaprolyl-3R-hydroxymethyl-γ-butyrolactone; Fig. 7). A-factor is effective in nM concentrations. The binding of A-factor to its receptor, ArpA, results in the release of this repressor from the *adp* promoter, and in the transcription of Adp, the transcriptional activator of gene clusters responsible for streptomycin biosynthesis, resistance, and the formation of aerial mycelia. More than ten species-specific butinolide molecules have been identified.[52]

Many γ-butyrolactone receptor homologues have been identified in the genome sequences of *Streptomyces.* At least 22 out of 33 genes encoding homologues of γ-butyrolactone receptors are located close to antibiotic biosynthesis gene clusters or have been shown to regulate antibiotic production.[53]

Bacteria lacking quorum sensing genes are not able to form organized biofilm[54] including EPS formation. Quorum sensing accounts for only a part of biofilm-specific gene expression patterns. Some are, probably related to the "surface sensing."[55]

9.8 APOPTOSIS

Although the precise molecular mechanisms of programmed cell death (PCD) in fungi are not known, fungal genomes contain the complement of genes involved in PCD in *S. cerevisiae*, and also have homologs of genes, involved in PCD in metazoans, which are not present in *S. cerevisiae* or *S. pombe.*[56] Apoptosis in filamentous fungi occurs in response to treatment with hydrogen peroxide, amphotericin B and farnesol.[57] Farnesol is involved in quorum sensing in *Candida* indicating that quorum sensing signals may serve antagonistic interactions between species in microbial communities.

In *C. albicans*, mutations that block cAMP/PKA pathway suppress or delay the apoptotic response to H_2O_2 and amphotericin B. By contrast, mutations that result in constitutive activation of the RAS pathway accelerate entry into the apoptotic pathway.[58] Apoptosis and quorum sensing in filamentous fungi are phenomena associated with stress responses, a recurring motif in morphogenesis and secondary metabolism. The treatment of *Colletotrichum trifolii* with proline, a known stress relief chemical, suppressed apoptosis associated with Ras as well as apoptosis associated with a variety of other stresses.[59]

As the fungal culture progresses from the stage of colonization and feeding mycelium to nutrient limitation and the onset of stresses, it enters the stage of secondary phenomena or idiophase. This stage is characterized by changes in cell wall architecture, vacuolization and cell death. One would expect that apoptotic systems control idiophase. Indeed, the heterologous expression of mammalian mitochondrial PCD genes, the pro-apoptotic *bax* and the anti-apoptotic *bcl-2* result in modified morphogenesis and the production of conidia in *Colletotrichum gloeosporioides*. Of particular interest is the observation that *bcl-2* does not only extend the life span of the culture but also causes activation of processes which do not normally occur at the same time. "Similar to many other fungi, when *C. gloeosporioides* grows in low nutrient media or under stress conditions, mycelium production is arrested and conidia production is activated. Instead, the *bcl-2* isolates retained conidia and mycelium production highly active at the same time. Conidia production was high in rich medium without affecting mycelium growth, while mycelium growth continued under stress conditions. As a result, the logarithmic growth phase of fungal liquid cultures was extended significantly and biomass production was increased by 2- to 2.5-fold per culture volume, while conidia production on solid media in young, actively growing cultures was greatly enhanced."[60] Sporulation in *Agaricus bisporus*[61] and *Aspergillus nidulans*[62] was accompanied by morphogenic cell death and the activation of PCD-associated caspase-like activity. We speculate that the preferential production of bikaverin in senescent area of *G. fujikuroi* pellet, as well as its location at septated branching points (Fig. 2), may well be related to PCD phenomena.

Although no gene homologues to PCD genes in eukaryotes have been found in filamentous bacteria, there are phenomenological analogues. A broad review that considers the evolutionary origins, mechanisms and functions of programmed cell death in various cell communities including bacterial multicellular aggregates and symbiotic systems has been published by Ameisen.[63]

9.9 STRINGENT RESPONSE

Antibiotic production in ascomycetes is a secondary phenomenon temporarily tied to nutrient limitation and reduction in growth rate. It occurs at the same time, slightly precedes or follows the formation of aerial mycelium, and is followed by sporulation. A seeming involvement of growth control factors in secondary metabolism attracted considerable attention to (p)ppGpp, the guanyl nucleotide involved in the growth rate-controlled expression of genes in unicellular bacteria. Gene regulation mediated by (p)ppGpp is conserved in many bacteria and even in plants. This nucleotide acts as a global regulator during the physiological adaptation of the organism to a variety of environmental conditions.[64]

E. coli possesses two enzymes, RelA and SpoT (with 32% amino acid sequence identity), which can synthesize (p)ppGpp from GTP and ATP. On nitrogen and amino acid starvation, uncharged tRNAs bind to the ribosomal A site and activate the ribosome-bound RelA, while carbon or phosphate starvation promotes SpoT-dependent (p)ppGpp synthesis in a ribosome-independent manner.[65] In *S. coelicolor*, there are two homologues of *relA: relA* and *rshA*.[66] RelA shows a somewhat greater level of amino acid sequence, identity to the SpoT homologue of *E. coli* (42.6%), than to its RelA counterpart (38.5%).[67]

The ribosome-associated ppGpp synthetase (RelA) is required for antibiotic production under the conditions of nitrogen limitation in *S. coelicolor* A3[68] and for cephamycin C production in *S. clavuligerus*.[69] Deletion of the *relA* in *S. coelicolor* A3 results in the loss of production of the antibiotics actinorhodin (Act) and undecylprodigiosin (Red) and delayed morphological differentiation when the mutant is grown under the conditions of nitrogen limitation. However, the deletion of *rshA* has no effect on the ability of the *relA* mutant to make Act and Red when grown under the conditions of phosphate limitation. The high-level induction of *tipAp*::*rshA* in the *relA* mutant results in growth inhibition, while low-level induction restores antibiotic production and sporulation. In neither case, (p)ppGpp synthesis could be detected. Thus, a ppGpp-independent mechanism exists to activate antibiotic production under the conditions of phosphate limitation, which can be mimicked by the over-expression of *rshA*.[70]

9.10 CELL DIVISION GENES

Experienced industrial microbiologists can frequently predict the productivity of the culture by just observing it under the microscope. One of the most interesting investigations pointing to a direct link between the filament morphology and the productivity in actinomycetes belongs to Bushell and his collaborators. In-depth studies by Bushell and his colleagues on erythromycin production by *Saccharopolyspora erythraea* showed a striking correlation between the diameter of the mycelium fragments and productivity.[71] They observed that there is a critical size (80 μm–90 μm), below which mycelial fragments are not productive, despite having the same growth rate as larger particles. Bushell *et al.* proposed that this observation may account for frequently occurring phenomena such as the loss of biosynthesis in liquid cultures in species that are able to produce antibiotic on agar, the differences in productivity in different small-scale culture vessels and the morphology-related effects on productivity in bioreactor cultures.[72]

Interestingly, the selection of variants with reduced branching rates and enhanced strength of the cell wall results in a significantly enhanced production of erythromycin. This led Wardell *et al.* to search for the correlation between resistances to cell wall inhibitors: bacitracin, tunicamycin and penicillin and production of the antibiotic. Significantly greater erythromycin titers were observed in the cultures of the tunicamycin- and penicillin-resistant mutants, whereas bacitracin-resistant mutant produced less antibiotic than the original strain.[73]

The molecular mechanisms underlying the morphology of actinomycetes in liquid-grown cultures are poorly understood. However, at least one important morphogenic factor is known: SsgA, an acidic 15 kDa protein, a member of a novel family of actinomycetes-specific proteins with six or seven members in streptomycetes that relate functionally to cell division and morphogenesis.[74] The SsgA-like proteins appear to be primarily involved in the control of peptidoglycan maintenance; *ssgA* is essential for the sporulation of *S. coelicolor* and stimulates septum formation making the mycelium more sensitive to shear stress. Enhanced SsgA protein levels result in mycelial fragmentation. In *S. coelicolor* strain over-expressing *ssgA*, the morphology of the hyphae was characterized by irregular septa with a significantly wider diameter, thereby forming spore-like compartments. This suggests that *ssgA* can induce a process similar to submerged sporulation in *Streptomyces* strains, which otherwise fails to do so. Recently, several *S. griseus* mutants that produced submerged spores in rich media were identified. This phenotype could be suppressed by the expression of *ssgA* gene.[75]

The enhanced expression of *ssgA* in *S. coelicolor*, *S. lividans*, and *S. roseosporus* results in improved growth rates with a reduced lag phase and a smaller average mycelium size. In the cultures of *S. coelicolor* GSA2, over-expressed *ssgA* stimulated the production of undecylprodigiosin by the order of magnitude; actinorhodin production was somewhat reduced. *S. lividans* is an important commercial strain; it is used as the *Streptomyces* host for enzyme production. In cultures over-expressing *ssgA*, tyrosinase productivity increased more than two-fold.[76]

The tubulin-like protein FtsZ is a highly abundant 43 kDa protein, which can polymerize under GTP hydrolysis. It plays a key role in bacterial cell division. FtsZ is recruited to the division site, where it forms a dynamic ring structure (the so-called Z ring), leading to septum formation and cell division.[77] While in most eubacteria, *ftsZ* is an essential gene, null mutants could be obtained in *Streptomyces coelicolor*.[78] *S. coelicolor ftsZ* mutants form long vegetative and aerial hyphae without any cross walls and fail to sporulate. The over-expression of *ftsZ* in *S. coelicolor* M145 inhibited aerial mycelium formation leading to so-called bald (*bld*) phenotype and also blocked sporulation. The timing and the level of FtsZ expression are crucial, as increased and constitutive expression of FtsZ results in the inhibition of morphological differentiation, a phenotype essentially similar to that of the *ftsZ* null-mutant. Over-expression of *ftsZ* also inhibited morphological differentiation in *S. lividans* 1326, although aerial mycelium formation was less reduced. Antibiotic production was increased in both strains, and, in particular, in the otherwise dormant actinorhodin biosynthesis cluster of *S. lividans*, which was activated in liquid and solid cultures. In *S. coelicolor*, large amounts of prodigiosin were produced as soon as colonies could be discerned. Actinorhodin production was also enhanced, but it appears much later than Red on plates. Act production was especially high in submerged culture, resulting in almost black cultures. The colonies were much more solid than those produced by the wild type strain, indicating the altered consistency of the mycelium.[79]

Streptomycetes harbor an additional putative cell division gene that seems to be unique for these organisms, *facC* (Factor C) that, like *ssgA,* is involved in the regulation of sporulation in liquid culture. Factor C was identified as a 34 kDa protein, which restores submerged sporulation for an *S. griseus bld* mutant.[80] Factor C is induced by nutrient limitation, which is a very important signal for sporulation. Submerged sporulation is typically elicited after nutritional shift down from rich to minimal medium, suggesting that (p)ppGpp plays a key role in this process.[81] Therefore, it is likely that growth cessation — signaled *via* the stringent response — is a prerequisite for the onset of sporulation not only in solid, but also in submerged cultures. Schematically, mycelial growth can be seen as a combination of hyphal elongation and branching. Interestingly, antibiotic production is also often induced by the stringent response (see above), suggesting a temporal correlation between sporulation and antibiotic production in submerged culture.

9.11 GENES OF FUNGAL MORPHOLOGY

Knowledge of the genes involved in the complex process of fungal morphology and the understanding of their function is a prerequisite for the application of genetic engineering for the control of morphology in fungi. Unfortunately, very little is known of the genes related to morphology. In a recent work, subtractive hybridization was applied for the identification of genes associated with morphology in *Aspergillus niger*.[82] The morphology of citric acid production in strains of *Aspergillus niger* is sensitive to the concentration of manganese (Mn^{2+}). Upon increasing the Mn^{2+} concentration in *A. niger* cultures to 14 ppb or higher, the morphology switches from pelleted to filamentous, accompanied by a rapid decline in citric acid production. The use of suppression subtractive hybridization has identified 22 genes that are responsive to Mn^{2+}. Fifteen genes were differentially expressed, when *A. niger* was grown in the filamentous form, and seven genes were expressed in the pelleted form. Of the 15 filament-associated genes, seven were novel and eight shared 47 to 100% identity with genes from other organisms. Five of the pellet-associated genes were novel, and the other two genes encoded a pepsin-type protease and polyubiquitin. All 10 genes with deduced functions are either involved in amino acid metabolism, protein catabolism or cell regulatory processes. Northern blot analysis showed that the transcripts of all 22 genes were rapidly enhanced or suppressed by Mn^{2+}. Steady state mRNA levels of six selected filament-associated genes remained high during five days of culture in a filamentous state and remained low under pelleted growth conditions. The opposite behavior was observed for four selected pellet-associated genes. The filament-associated clone, *Brsa-25*, was isolated. Antisense expression of *Brsa-25* permitted pelleted growth and increased citrate production at concentrations of Mn^{2+}, which were higher ones that the parent strain could tolerate. These results suggest the involvement of the newly isolated genes in the regulation of *A. niger* morphology.

The gene products with homology to known proteins fell into one of the two groups: those involved in signal transduction or those involved in amino acid synthesis or in protein catabolism. There are two genes associated with the signaling group. The translated sequence of *Balu-4* is 96% identical to the *A. nidulans* G-protein β-subunit gene, *sfaD*. This heterotrimeric G-protein component is known to be required for normal growth and repression of sporulation in *A. nidulans*.[83,84] The translated sequence of *Brsa-47* is 60% identical to the *Pichia pastoris* inositol-1-phosphate synthase (*ino1*) gene. This is the first enzyme on the biosynthetic pathway to inositol phosphates involved in intracellular signaling and morphogenesis. For example, inositol-1,4,5-triphosphate induces Ca^{2+} release, which stimulates hyphal tip growth in *N. crassa*,[85] thus affecting the branching rate.

9.12 RECIPES FOR THE FUTURE

The material presented in this review shows how intimately intertwined are morphogenetic and biosynthetic activities in filamentous organisms. It also shows the convergence of regulatory mechanisms related to the differentiation of actinomycetes and ascomycetes in response to external stimuli such as nutrient deprivation, stress, quorum and surface sensing. This differentiation is related to both — the change in form and the change in metabolism. In the wild, the growth of filamentous microorganisms occurs in the moist hydrophilic substrate, while seed production and dispersion occurs in the dry hydrophobic air. Secondary metabolism coincides with the transfer of main biological activities from the substrate layer to the air. Many secondary metabolites are related to the survival of aerial structures and of the seed outside the substrate; they prevent drying, radiation and oxidative damage. Others, perhaps, are useful in the preservation of the colony and the seed in repeated cycles of desiccation and wetting — the antimicrobial metabolites.

In the industrial milieu, the essential water-air heterogeneity of mycelial habitat is abolished, especially in the filamentous mode of growth. Therefore, the production of many secondary metabolites in submerged fermentation requires redesign of regulatory systems by mutagenesis necessary to develop an industrial strain. In the pellet mode of growth, the heterogeneity of water-air is substituted by the inside-outside heterogeneity mimicking the former. The inside of the pellet is poorer in nutrients, more hydrophobic, amenable to biofilm and quorum sensing regulations. This is, probably, the reason that the majority of antibiotics are produced in mycelial aggregates. They require less drastic reshuffling of regulatory systems and, therefore, crop up more frequently in strain development programs.

Strain development is the prerequisite for the development of efficient processes for the production of natural products in filamentous microorganisms. Although the morphology of filamentous growth can be handled by process engineering, this ability is limited. The breakthrough in efficiency can only come from the manipulation of the genetics. Classical methods of strain improvement have long ago reached the point of diminishing returns. Random mutagenesis in highly efficient strains damages the fine balance between good fermentative behavior and good yields. In the development of new strains and new products, the long way of incremental strain improvement is not supported by the financial reality.

The time has come to achieve better understanding of regulatory processes controlling the output and the nomenclature of secondary metabolites. Nowhere is such understanding more needed than in the understanding of morphogenesis. Secondary metabolism in filamentous microorganisms is inseparably tied with the functional changes in the form of mycelia and of colonies. The engineering of productivity is tied with the engineering of morphology.

Our short review, if anything, shows fragmentation of the research. Although the new knowledge is steadily accumulating, there is no structural understanding of interactions between various phenomena. One knows that the formation of aerial mycelia is triggered by nutrient limitation, but nobody knows how. What are the signals? How are they related to quorum sensing, to stringent response, to surface recognition? Do aerial mycelia formation and secondary metabolite biosynthesis share common regulatory steps or is the former a prerequisite, a signal for the latter? In this maze of unknowns, one must, for the sake of efficiency, target the upper tier of regulatory chain — global transcription regulators.

The very observation that morphogenesis and productivity of industrial filamentous organisms are linked is indicative of regulation on the basis of global transcriptional regulators in a mode resembling that of butanolide autoinducers. This makes the global regulators attractive targets for rational strain improvement. It seems that they were the unidentified target of classical random mutagenesis-based techniques. Indeed, production cultures of the wild source strain of *Streptomyces fradiae* and its derived, classically improved, industrial strain, expressed the 85-kb polyketide antibiotic tylosin biosynthetic gene cluster similarly, while they expressed several tens of other *S. fradiae* genes and their *S. coelicolor* homologues differently[86] indicating regulatory override of biosynthetic capacity.

I can envision a highly efficient research program, in which transcription regulators of an industrial filamentous organism are identified by the tools of genomic analysis, cloned, if necessary, and either over-produced or nullified by antisense expression. The genes related to morphology could then be identified by observation, and the genes related to productivity by differential expression of antibiotic gene clusters. Of course, there are many possible strategies aimed at the same goal, which is to identify targets for rational industrial strain development.

One way or another, the new techniques of genetic manipulation and analysis will find their way into industrial production of more efficient biologically active novel natural products from filamentous microorganisms. This process is already under way. The major pharmaceutical companies are at the forefront of progress in applying state-of-the-art techniques to strain and product development.

ACKNOWLEDGMENT

The author wants to express his thanks to Professor David Engelberg from the Hebrew University for his valuable comments on this review.

REFERENCES

1. Peláez F, The historical delivery of antibiotics from microbial natural products—Can history repeat? *Biochem Pharmacol* 71:981–990, 2006.
2. Fauci AS, Infectious diseases: Considerations for the 21st century, *Clin Infec Dis* **32**:675–685, 2001.
3. von Bubnoff A, Seeking new antibiotics in Nature's backyard, *Cell* **127**:867–869, 2006.
4. Watve MG, Tickoo R, Jog MM, Bhole BD, How many antibiotics are produced by the genus *Streptomyces*? *Arch Microbiol* **176**:386–390, 2001.
5. Jennings DH, Lysek G, Fungal Biology, BIOS Sci Publ, Oxford, UK, pp. 117–124, 1996.
6. Beppu T, Secondary metabolites as chemical signals for cellular differentiation, *Gene* **115**:159–165, 1992.
7. Kutchan T, Dixon RA, Physiology and Metabolism. Secondary Metabolism: Nature's chemical reservoir under deconvolution, *Curr Opinion Plant Biol* **8**: 227–229, 2005.

8. Papagianni M, Fungal morphology and metabolite production in submerged mycelial processes, *Biotechnol Adv* **22**:189–259, 2004.
9. Clark DS, Submerged citric acid fermentation of ferrocyanide-treated beet molasses: Morphology of pellets of *Aspergillus niger, Can J Microbiol* **8**:133–136, 1962.
10. Metz B, Kossen NWF, van Suizdum JC, The rheology of mold suspensions, *Adv Biochem Eng* **11**:103, 1979.
11. Xu J, Wang L, Ridgway D, Gu T, Moo-Young M, Increased heterologous protein production in *Aspergillus niger* fermentation through extracellular proteases inhibition by pelleted growth, *Biotechnol Prog* **16**:222–227, 2000.
12. Smith JJ, Lilly MD, Fox RI, The effect of agitation on the morphology and penicillin production of *Penicillium chrysogenum, Biotechnol Bioeng* **35**:1011–1023, 1990.
13. Byrne GS, Ward OP, Effect of nutrition on pellet formation by *Rhizopus arrhizus, Biotechnol Bioeng* **33**:912–914, 1989.
14. Vecht-Lifshitz SE, Magdassi S, Braun S, Pellet formation and cellular aggregation in *Streptomyces tendae, Biotechnol Bioeng* **35**:890–896, 1990.
15. Junker BH, Hesse M, Burgess B, Masurekar P, Connors N, Seeley A, Early phase process scale-up challenges for fungal and filamentous bacterial cultures, *Appl Biochem Biotechnol* **119**:241–277, 2004.
16. Michel FC, Jr, Grulke EA, Reddy CA, Determination of the respiration kinetics for mycelial pellets of *Phanerochaete chrysosporium, Appl Environ Microbiol* **58**:1740–1745, 1992.
17. Braun S, Vecht-Lifshitz SE, Mycelial morphology and metabolite production, *Trends Biotechnol* **9**:63–68, 1991.
18. Berdy J, Bioactive microbial metabolites: A personal view, *J Antibiot* **58**:1–26, 2005.
19. Aravind L, Anantharaman V, Iyer LM, Evolutionary connections between bacterial and eukaryotic signaling systems: A genomic perspective, *Curr Opinion Microbiol* **6**:490–497, 2003.
20. Verstrepen KJ, Klis FM, Flocculation, adhesion and biofilm formation in yeasts, *Mol Microbiol* **60**:5–15, 2006.
21. Reynolds TB, Fink GR, Baker's yeast, a model for fungal biofilm formation, *Science* **291**:878–881, 2001.
22. Davies DG, Geesey GG, Regulation of the alginate biosynthesis gene *algC* in *Pseudomonas aeruginosa* during biofilm development in continuous culture, *Appl Environ Microbiol* **61**:860–867, 1995.
23. Kim Y, Kim J, Formation and dispersion of mycelial pellets of *Streptomyces coelicolor* A3(2), *J Microbiol* **42**:64–67, 2004.

24. Stanhill A, Schick N, Engelberg D, The Yeast Ras/Cyclic AMP pathway induces invasive growth by suppressing the cellular stress response, *Mol Cell Biol* **19**:64–67, 1999.
25. Vecht-Lifshitz SE, Magdassi S, Braun S, Effect of addition of surface active agents on pellet formation in submerged fermentation of *Streptomyces tendae*, *J Disp Sci Technol* **10**:265–275, 1989.
26. Newel PC, Aggregation and cell surface receptors in cellular slime molds. In Cuatrecasas P, Greaves MF (eds.), *Receptors and Recognition*, pp. 1–58, 1977.
27. Süsstrunk U, Pidoux J, Taubert S, Ullmann A, Thompson CJ, Pleiotropic effects of cAMP on germination, antibiotic production and morphological development in *Streptomyces coelicolor*, *Mol Microbiol* **30**:33–46, 1998.
28. Kim E-S, Hong H-J, Choi C-Y, Cohen SN, Modulation of actinorhodin biosynthesis in *Streptomyces lividans* by glucose repression of *afsR2* gene transcription, *J Bacteriol* **183**:2198–2203, 2001.
29. Elliot MA, Talbot NJ, Building filaments in the air: Aerial morphogenesis in bacteria and fungi, *Current Opinion in Microbiology* 7:594–601, 2004.
30. Hakanpää J, Paananen A, Askolin S, Nakari-Setälä T, Parkkinen T, Penttilä M, Linder MB, Rouvinen J, Atomic resolution structure of the HFBII hydrophobin, a self-assembling amphiphile, *J Biol Chem* **279**:534–539, 2004.
31. Whiteford JR, Lacroix H, Talbot NJ, Spanu PD, Stage-specific cellular localization of two hydrophobins during plant infection by the pathogenic fungus *Cladosporium fulvum*, *Fungal Gen Biol* **41**:624–634, 2004.
32. Fuchs U, Czymmek KJ, Sweigard JA, Five hydrophobin genes in *Fusarium verticillioides* include two required for microconidial chain formation, *Fungal Gen Biol* **41**:852–864, 2004.
33. Wösten HAB, Bohlmann R, Eckerskom C, Lottspeich F, Bolker M, Kahmann R, A novel class of small amphipathic peptides affect aerial growth and surface hydrophobicity in *Ustilago maydis*, *EMBO J* **15**:4274–4281, 1996.
34. Lugones LG, de Jon JF, de Vries OMH, Jaiving R, Dijkterhuis J, Wösten HAB, The SC15 protein of *Schizophyllum commune* mediates formation of aerial hyphae and attachment in the absence of the SC3 hydrophobin, *Mol Microbiol* **53**:707–716, 2004.
35. Elliot MA, Karoonuthaisiri N, Huang J, Bibb MJ, Cohen SN, Kao CM, Buttner MJ, The chaplins: A family of hydrophobic cell-surface proteins involved in aerial mycelium formation in *S. coelicolor*, *Genes Dev* **17**:1727–1740, 2003.
36. Willey J, Santamaria R, Guijarro J, Geistlich M, Losick R, Extracellular complementation of a developmental mutation implicates a small sporulation protein in aerial mycelium formation by *S. coelicolor*, *Cell* **65**:641–650, 1991.
37. Kodani S, Hudson ME, Durrant MC, Buttner MJ, Nodwell JR, Willey JM, The SapB morphogen is a lantibiotic-like peptide derived from the product of the

developmental gene rams in *S. coelicolor*, *Proc Natl Acad Sci USA* **101**:11448–11453, 2004.

38. Ma H, Kendall K, Cloning and analysis of a gene cluster from *S. coelicolor* that causes accelerated aerial mycelium formation in *S. lividans*, *J Bacteriol* **176**:3800–3811, 1994.
39. Willey JM, Schwedock J, Losick R, Multiple extracellular signals govern the production of a morphogenetic protein involved in aerial mycelium formation by *Streptomyces coelicolor*, *Genes Dev* 7:895–903, 1993.
40. Kodani S, SapT, a lanthionine-containing peptide involved in aerial hyphae formation in the streptomycetes, *Mol Microbiol* **58**:1368–1380, 2005.
41. Tillotson RD, A surface active protein involved in aerial hyphae formation in the filamentous fungus *Schizophyllum commune* restores the capacity of a bald mutant of the filamentous bacterium *Streptomyces coelicolor* to erect aerial structures, *Mol Microbiol* **30**:595–602, 1998.
42. Claessen D, de Jong W, Dijkhuizen L, Woesten HAB, Regulation of Streptomyces development: Reach for the sky! *Trends Microbiol* **14**:313–319, 2006.
43. Visick KL, Fuqua C, Decoding microbial chatter: Cell-cell communication in bacteria, *J Bacteriol* **187**:5507–5519, 2005.
44. Hogan DA, Talking to themselves: Autoregulation and quorum sensing in fungi, *Eukaryot Cell* **5**:613–619, 2006.
45. Wagner-Dobler I, Biebl H, Environmental biology of the marine *Roseobacter lineage*, *Annu Rev Microbiol* **60**:255–280, 2006.
46. Chen H, Fink GR, Feedback control of morphogenesis in fungi by aromatic alcohols, *Genes Dev* **20**:1150–1161, 2006.
47. Chen H, Fujita M, Feng Q, Clardy J, Fink GR, Tyrosol is a quorum-sensing molecule in *Candida albicans*, *Proc Natl Acad Sci USA* **101**:5048–5052, 2004.
48. Hornby JM, Jensen EC, Lisec AD, Tasto JJ, Jahnke B, Shoemaker R, Dussault P, Nickerson KW, Quorum sensing in the dimorphic fungus *Candida albicans* is mediated by farnesol, *Appl Environ Microbiol* **67**:2982–2992, 2001.
49. Ramage G, Saville SP, Wickes BL, and Lopez-Ribot JL, Inhibition of *Candida albicans* biofilm formation by farnesol, a quorum-sensing molecule, *Appl Environ Microbiol* **68**:5459–5463, 2002.
50. Takano E, g-Butyrolactones: Streptomyces signalling molecules regulating antibiotic production and differentiation, *Curr Opin Microbiol* **9**:287–294, 2006.
51. Khokhlov AS, Tovarova II, Borisova LN, Pilner SA, Schevchenko LA, Kornitskaya EI, Ivkina NS, Rappoport IA, A-factor responsible for the biosynthesis of streptomycin by a mutant strain *Actinomyces streptomycini*, *Dokl Akad Nauk SSSR* **177**:232–235, 1967.

52. Piepersberg W, Streptomycin and related aminoglycosides In Vining LC, Stuttard C (eds.), *Biochemistry and Genetics of Antibiotic Biosynthesis*, Butterworths-Heinemann, Boston, MA, pp. 531–570, 1999.
53. Takano E, Kinoshita H, Mersinias V, Bucca G, Hotchkiss G, Nihira T, Smith C, Bibb M, Wohlleben W, Chater KF, A bacterial hormone (the SCB1 extracellular signalling system) directly controls an antibiotic pathway-specific regulator in the cryptic type I polyketide biosynthetic cluster of *Streptomyces coelicolor* A3(2), *Mol Microbiol* **56**:465–479, 2005.
54. Davies DG, Parsek MR, Pearson JP, Iglewski BH, Costerton JW, Greenberg EP, The involvement of cell-to-cell signals in the development of a bacterial biofilm, *Science* **280**:295–298, 1998.
55. Stoodley P, Sauer K, Davies DG, Costeron JW, Biofilms as complex differentiated communities, *Annu Rev Microbiol* **56**:187–209, 2002.
56. Fedorova N, Badger J, Robson G, Wortman J, Nierman W, Comparative analysis of programmed cell death pathways in filamentous fungi, *BMC Genomics* **6**:177, 2005.
57. Semighini CP, Hornby JM, Dumitru R, Nickerson KW, Harris SD, Farnesol-induced apoptosis in *Aspergillus nidulans* reveals a possible mechanism for antagonistic interactions between fungi, *Mol Microbiol* **59**:753–764, 2006.
58. Phillips AJ, Crowe JD, Ramsdale M, Ras pathway signaling accelerates programmed cell death in the pathogenic fungus *Candida albicans*, *Proc Natl Acad Sci USA* **103**:726–731, 2006.
59. Chen C, Dickman MB, Proline suppresses apoptosis in the fungal pathogen *Colletotrichum trifolii*, *Proc Natl Acad Sci USA* **102**:3459–3464, 2005.
60. Barhoom S, Sharon A, Bcl-2 proteins link programmed cell death with growth and morphogenetic adaptations in the fungal plant pathogen *Colletotrichum gloeosporioides*, *Fung Gen Biol* **44**:32–43, 2007.
61. Umar MH, Van Griensven LJLD, Morphogenetic cell death in developing primordia of *Agaricus bisporus*, *Mycologia* **89**:274–277, 1997.
62. Thrane C, Kaufmann U, Stummann BM, Olsson S, Activation of caspase-like activity and poly (ADP-ribose) polymerase degradation during sporulation in *Aspergillus nidulans*, *Fungal Genet Biol* **41**:361–368, 2004.
63. Ameisen JC, On the origin, evolution, and nature of programmed cell death: A timeline of four billion years, *Cell Death Differ* **9**:367–393, 2002.
64. Gralla JD, *Escherichia coli* ribosomal RNA transcription: Regulatory roles for ppGpp, NTPs, architectural proteins and a polymerase-binding protein, *Mol Microbiol* **55**:973–977, 2005.
65. Cashel M, Gentry DR, Hernandez VJ, Vinella D, The stringent reponse, In Neidhardt FC (ed.), *Escherichia coli* and *Salmonella typhimurium*, 2nd edn., Vol. 1, American Society for Microbiology, Washington DC, pp. 1458–1496, 1995.

66. Jin W, Ryu YG, Kang SG, Kim SK, Saito N, Ochi K, Lee SH, Lee KJ, Two *relA/spoT* homologous genes are involved in the morphological and physiological differentiation of *Streptomyces clavuligerus*, *Microbiology* **150**:1485–1493, 2004.
67. Chakraburtty R, White J, Takano E, Bibb MJ, Cloning, characterization and disruption of a (p)ppGpp synthetase gene (*relA*) of *Streptomyces coelicolor* A3(2), *Mol Microbiol* **19**:357–368, 1996.
68. Chakraburtty R, Bibb MJ, The ppGpp synthetase gene (*relA*) of *Streptomyces coelicolor* A3(2) plays a conditional role in antibiotic production and morphological differentiation, *J Bacteriol* **179**:5854–5861, 1997.
69. Jin W, Kim HK, Kim JY, Kang SG, Lee SH, Lee KJ, Cephamycin C production is regulated by *relA* and *rsh* genes in *Streptomyces clavuligerus* ATCC27064, *J Biotechnol* **114**:81–87, 2004.
70. Sun J, Hesketh A, Bibb M, Functional analysis of *relA* and *rshA*, two *relA/spoT* homologues of *Streptomyces coelicolor* A3(2), *J Bacteriol* **183**:3488–3498, 2001.
71. Bushell ME, Dunstan GL, Wilson GC, Effect of small scale culture vessel type on hyphal fragment size and erythromycin production in *Saccharopolyspora erythreae*, *Biotechnol Lett* **19**:849–852, 1997.
72. Martin SM, Bushell ME, Effect of hyphal morphology on bioreactor performance of antibiotic-producing *Saccharopolyspora erythreae* cultures, *Microbiology* **142**:1783–1788, 1996.
73. Wardell JN, Stocks SM, Thomas CR, Bushell ME, Decreasing the hyphal branching rate of *Saccharopolyspora erythraea* NRRL 2338 leads to increased resistance to breakage and increased antibiotic production, *Biotechnol Bioeng* **78**:141–146, 2002.
74. Noens EE, Mersinias V, Traag BA, Smith CP, Koerten HK, van Wezel GP, SsgA-like proteins determine the fate of peptidoglycan during sporulation of *Streptomyces coelicolor*, *Mol Microbiol* **58**:929–944, 2005.
75. Kawamoto S, Ensign JC, Cloning and characterization of a gene involved in regulation and sporulation and cell division in *Streptomyces griseus*, *Actinomycetologica* **9**:136–151, 1995.
76. Gilles P, van Wezel GP, Krabben P, Traag BA, Keijser BJF, Kerste R, Vijgenboom E, Heijnen JJ, Kraal B, Unlocking *Streptomyces* spp. for use as sustainable industrial production platforms by morphological engineering, *Appl Env Microbiol* **72**:5283–5288, 2006.
77. Bramhill D, Bacterial cell division, *Annu Rev Cell Dev Biol* **13**:395–424, 1997.
78. McCormick JR, Losick R, Cell division gene *ftsQ* is required for efficient sporulation but not growth and viability in *Streptomyces coelicolor* A3(2), *J Bacteriol* **178**:5295–5301, 1996.

79. van Wezel GP, van der Meulen J, Taal E, Koerten H, Kraal B, Effects of increased and deregulated expression of cell division genes on the morphology and on antibiotic production of streptomycetes, *Antonie Leeuwenhoek* **78**:269–276, 2000.
80. Birko Z, Sumegi A, Vinnai A, van Wezel GP, Szeszak F, Vitalis S, Szabo PT, Kele Z, Janaki T, Biro S, Characterization of the gene for Factor C, an extracellular signal protein involved in morphological differentiation of *Streptomyces griseus*, *Microbiology* **145**:2245–2253, 1999.
81. Daza A, Martin JF, Dominguez A, Gil JA, Sporulation of several species of *Streptomyces* in submerged cultures after nutritional downshift, *J Gen Microbiol* **135**:2483–2491, 1989.
82. Dai Z, Mao X, Magnuson JK, Lasure LL, Identification of genes associated with morphology in *Aspergillus niger* by using suppression subtractive hybridization, *Appl Environ Microbiol* **70**:1740–1745, 2004.
83. Rosen S, Yu JH, Adams TH, The *Aspergillus nidulans sfaD* gene encodes a G protein beta subunit that is required for normal growth and repression of sporulation, *EMBO J* **18**:5592–5600, 1999.
84. Yang Q, Poole SI, Borkovich KA, A G-protein beta subunit required for sexual and vegetative development and maintenance of normal G alpha protein levels in *Neurospora crassa*, *Eukaryot Cell* **1**:378–390, 2002.
85. Silverman-Gavrila LB, Lew RR, An IP(3)-activated Ca^{2+} channel regulates fungal tip growth, *J Cell Sci* **115**:5013–5025, 2002.
86. Lum AM, Huang J, Hutchinson CR, Cao CM, Reverse engineering of industrial pharmaceutical-producing actinomycete strains using DNA microarrays, *Metab Eng* **6**:186–196, 2004.

Chapter 10

RECENT ADVANCES IN THE CHEMISTRY OF INSECT PHEROMONES

Mangesh J. Goundalkar and Francis X. Webster

10.1 INTRODUCTION

Pheromones, especially insect pheromones, have become common news stories in the popular press and hence are well known to most people. For instance, most elementary schools in the USA now include coverage of pheromones in general science and biology courses. Concomitant with this widespread coverage and inclusion in elementary school curricula is ongoing basic and applied research, which leads to important practical uses and beneficial applications. Since Butenandt's initial report on the pheromone of the silk worm moth,[1] there have been many reviews of pheromones and recent ones are cited here.[2] This review of the chemistry of insect pheromones will cover the isolation and identification of new pheromones and the synthesis of these compounds as well as other recently reported syntheses of important pheromones.

While pheromones for many organisms have been elucidated and compilations are available,[2] there remains considerable research activity on the isolation and identification of new pheromones. This work remains important since, if for no other reason, there are many important pests for which we still have not yet elucidated their pheromones. Even though there are well established methods for isolating tiny quantities of volatile organic compounds and identifying them, there have been variations and improvements in methodology, which we note. Compound names used in this review are the same as those reported in the original publications.

The synthesis of pheromones remains an important cornerstone of pheromone research and development. The initial synthesis of an insect pheromone serves not only to confirm the structural identification, but also provides authentic material for either laboratory or field bioassay. Oftentimes, larger quantities of the pheromones are desired for the development of monitoring and pest control methods. The scale up of the synthesis of complex pheromone compounds is often a major bottleneck in the practical application of pheromones. Also, the unusual structure of some pheromone compounds beckons synthetic chemists to apply new methodology in order to "show off" their methods.

Work continues on the application of pheromones for the monitoring and control of pest insects. Unlike the general application of pesticides, which indiscriminately kills both target and beneficial insects alike, the application of pheromones in controlling pests is species specific. This specificity is desirable, time consuming, and expensive. The desirability of species specificity is obvious; only the target organism is affected. Since each pest is unique in terms of its pheromone, the application thereof means that each application of pheromones in pest control will require considerable time and money in order to be successful. The economic hurdles in applying pheromones have slowed their adoption.

Most practical applications of insect pheromones involve one of four popular modalities: monitoring, mass trapping, mating disruption, or attract and kill. One of the first applications of insect pheromones was monitoring, which helped to create the idea of integrated pest management (IPM). Integrated pest management programs now encompass many ideas and are beyond the scope of this review. Mass trapping of pest insects can be an effective method when aggregation pheromones are used such as with bark beetles. Mating disruption is not well established for many agricultural pests. These methods require no pesticides. Attract and kill programs have not been studied or employed as widely as mating disruption but it may offer promise under conditions where mating disruption is ineffective.

10.2 ISOLATION AND IDENTIFICATION OF NEW PHEROMONES

This section of the chapter summarizes recent identifications of pheromone compounds. Coverage includes from approximately the year 2000 to present. The methods are tersely described along with the compounds. The nomenclature used throughout takes the name from the original publication. The coverage is arranged based on the "order" of the species.

10.2.1 Coleoptera

Field bioassays with adult cerambycid beetles, *Neoclytus acuminatus acuminatus* (F.) (Coleoptera: Cerambycidae), revealed that males produce a pheromone that attracts both sexes. Male extracts revealed a single major male-specific compound (2*S*, 3*S*)-hexanediol. Field trials showed that a racemic blend of (2*S*, 3*S*) and (2*R*, 3*R*)-hexanediols attracted both sexes and that activity was similar to enantiomerically enriched (2*S*, 3*S*)-hexanediol (e.e. 80.2%). However, a blend of all four stereoisomers attracted only a few beetles.[3]

An ether extract of the abdominal glands of the female black chafers, *Holotrichia loochooana loochooana* (Coleoptera: Scarabaeidae), elicited responses from males. Anthranilic acid was identified as the major compound by GC-MS and the amount extracted from each female chafer was estimated to be approximately 1.3 μg/female.[4]

Male locust borers, *Megacyllene robiniae* (F.) (Coleoptera: Cerambycidae), respond to females only after contacting them with their antennae, suggesting the presence of a contact pheromone. This contact pheromone was identified as (*Z*)-9-pentacosene, the major component present on the surface of the cuticular wax layers of the female locust borers.[5] The major sex pheromone components of the click beetle, *Agriotes brevis* Candeze (Coleoptera: Elateridae), were identified as geranyl butyrate and (*E*, *E*)-farnesyl butyrate.[6]

Analyses of volatiles from the male coffee white stemborer of Indian origin, *Xylotrechus quadripes* Chevrolat (Coleoptera: Cerambycidae), showed three male specific components comprising more than 90% of the volatile compounds, two of which elicited EAG responses. The major EAG-active compound was identified as (*S*)-2-hydroxy-3-decanone by comparing data with synthetic samples using GC-MS, IR, and NMR. The second component was identified as 3-hydroxy-2-decanone and the third minor component was identified as (*S*, *S*)-2,3-dihydroxyoctane. Less importantly, 2-hydroxy-3-octanone, 2,3-decanedione, 2-phenyl-ethanol and octanoic acid were also identified in the volatiles of the male beetles.[7]

The female sex pheromone of the cranberry root grub, *Lichnanthe vulpina* Hentz (Coleoptera: Glaphyridae), was identified as (*Z*)-7-hexadecenol and (*Z*)-7-hexadecenal. Both compounds captured males although it was observed that the alcohol was slightly more effective than the aldehyde.[8]

The male aggregation pheromone of the date palm fruit stalk borer, *Oryctes elegans*, is a blend of 4-methyloctanoic acid (major) and ethyl-4-methyloctanoate, which were occasionally mixed with minor components: 4-methyloctanyl acetate, methyl-4-methyloctanoate, 4-methyloctanol and nonanyl acetate. EAG and field trapping experiments showed that 4-methyloctanoic acid is an essential component of the male aggregation pheromone. It was barely attractive by itself but was synergistic with the odor of fresh date palms. Females did not seem to emit sex specific volatiles, when adults feeding on sugarcane were sampled.[9]

Six male specific compounds were isolated from the crucifer flea beetle, *Phyllotreta cruciferae*, and the same compounds plus two additional compounds were isolated from males of *Aphthona flava*, *A. czwalinae*, and *A. cyparissiae*. Three of the compounds were identified as (+)-*ar*-himachalene, (+)-*trans*-α-himachalene; (+)-γ-cadinene. Two other compounds were identified as new enantiomers of himachalene hydrocarbons that were previously identified from the fir trees, *Abies alba* and *Abies nordmanniana*. Finally, there were two himachalene alcohols and one nonsesquiterpene ketone that is a himachalene analog that were identified. The chemical and electrophysiological patterns are consistent with, but do not prove, a pheromonal function.

The aggregation pheromone of the leaf beetle *Diorhabda elongata* Brulle (Coleoptera: Chrysomelidae) have been identified as two compounds namely, (2*E*, 4*Z*)-2,4-heptadienal and (2*E*, 4*Z*)-2,4-heptadien-1-ol produced exclusively by males. They were also detected in trace quantities from females but the levels in males were 8–40 times higher.[10]

Males from *Galerucella calmariensis* and *Galerucella pusilla* (Coleoptera: Chrysomelidae) emit an aggregation pheromone while feeding on host foliage. The compound was identified by spectrometric and other microchemical tests as the novel dimethylfuran lactone, 12,13-dimethyl-5,14-dioxabicyclo[9.2.1]-tetradeca-1(13),11-dien-4-one. The structure was confirmed by synthesis and the synthetic compound attracted both males and females in field tests.[11]

Male *Megacyllene caryae* Gahan (Coleoptera: Cerambycidae) respond to females only after touching them with their antennae, indicating the presence of a contact sex pheromone. The hydrocarbon, (*Z*)-9-nonacosene, was identified as the major component of the contact sex pheromone of the beetle.[12]

A terpenoid, (−)-β-caryophyllene, was identified as a gender specific sesquiterpene produced by the female multicolored Asian lady beetle, *Harmonia axyridis* Pallas (Coleoptera: Coccinellidae).[13]

The male-produced volatiles eliciting responses from male and female antennae of the sap beetle, *Colopterus truncatus* Randall (Coleoptera: Nitidulidae), were identified as (2*E*,4*E*,6*E*)-3,5-dimethyl-2,4,6-octatriene, (2*E*, 4*E*, 6*E*)-4,6-dimethyl-2,4,6-nonatriene and (2*E*, 4*E*,6*E*,8*E*)-3,5,7-trimethyl-2,4,6,8-decatetraene. A fourth male-specific compound (2*E*,4*E*,6*E*,8*E*)-4,6,8-trimethyl-2,4,6,8-undecatetraene was not EAG active.[14]

Field trapping experiments showed that (*R*)-(+)-γ-decalactone was the male emitted pheromone attracting females in the endangered scarab beetle, *Osmoderma eremita*. This fruity, peachlike or plumlike compound is the first evidence of a lactone. The female sex pheromone of the scarab beetle, *Hoplia equina* LeConte (Coleoptera: Scarabaeidae: Melolonthinae) has been identified as 2-tetradecanone. It was the only compound of the eight other compounds found that attracted males.[15] (*R*, *Z*)-5-(−)-(oct-1-enyl)oxacyclopentan-2-one was found to attract male scarab beetle, *Anomala solida* Er. (Coleoptera: Scarabaeidae: Rutelinae).[16]

The female-produced sex pheromone of the scarab beetle, *Phyllophaga lanceolata* was identified as the methyl ester of an essential amino acid, L-leucine.[17] Three amino acid derivatives were identified in the extracts of pheromone glands of the scarab beetle, *Phyllophaga elenans*. These were

L-isoleucine methyl ester, N-formyl L-isoleucine methyl ester and N-acetyl L-isoleucine methyl ester.[18] The major component of the male attractant pheromone of the scarab beetle, *Holotrichia reynaudi*, was identified as anisole.[19]

Three male specific compounds were identified from the strawberry blossom weevil, *Anthonomus rubi*, namely (*Z*)-2-(3,3-dimethylcyclohexylidene)ethanol, (*cis*)-1-methyl-2-(1-methylethenyl) cyclobutaneethanol (grandisol), and 2-(1-methylethenyl)-5-methyl-4-hexen-1-ol (lavandulol). The first two being the aggregation pheromone of the boll weevil, *Anthonomous grandis*, grandlure II and grandlure I. Grandlure I was the (1*R*, 2*S*)-(+) enantiomer and lavandulol was a single enantiomer, whose absolute configuration was not determined. Trace amounts of other two grandlure components (*Z*)-(3,3-dimethylcyclohexylidene)acetaldehyde (grandlure III) and (*E*)-(3,3-dimethylcyclohexylidene) acetaldehyde (grandlure IV) were also detected.[20]

The chiral alcohol, 2-methyl-4-octanol, was identified as an aggregation pheromone from airborne volatiles produced by male and females of the sugarcane weevil *Sphenophorus levis*. Enantiomeric resolution by GC with a chiral column showed that the natural alcohol possessed the *S* configuration. The alcohol elicited aggregation behavior among adults.[21]

Aggregation pheromones were studied from two geographical isolates of the New Guinea sugarcane weevil, *Rhabdoscelus obscurus*. Extracts from the males and females of the Hawaiian *R. obscurus* revealed a single EAD-active compound, 2-methyl-4-octanol. Corresponding volatile analyses from the Australian *R. obscurus* consistently revealed three EAD-active male-specific candidate pheromone components, identified as, 2-methyl-4-octanol, (*E*2)-6-methyl-2-hepten-4-ol (rhynchophorol) and 2-methyl-4-heptanol.[22]

Extracts of the summer chafer, *Amphimallon solstitiale* L. (Coleoptera: Scarabaeidae), a common European scarab beetle were analyzed by GC-MS and GC-EAD. Both male and female extracts were shown to contain Acetoin — $(R):(S) > 9 : 1$, as well as 2,3-butanediol — $(2R, 3R) : (2S, 3S) : meso = 1 : 1 : 9$. Although (2*S*, 3*S*)-butanediol did not show any activity, the other compounds elicited strong responses exclusively with male antennae.[23]

The well known aldehyde, 4,8-dimethyldecanal, was shown to be a common pheromone of five *Tribolium* flour beetle species, *T. castaneum, T. confusum, T. freemani, and T. madens*. Two other volatiles were found; 1-pentadecene was shown to be a common semiochemical of flour beetles and 1,6-pentadecadiene was detected in five species, *T.audax, T. brevicornis, T. destructor, T.freemani and T. madens.*[24]

The sex pheromone emitted by the female yellow mealworm beetles, *Tenebrio molitor* L. (Coleoptera: Tenebrionidae), 4-methyl-1-nonanal is well known. However, it was observed that males emit a pheromone that attracts females, which was identified as (*Z*)-3-dodecenyl acetate.[25]

10.2.2 Dictyoptera

In the Madera cockroach, *Leucophaea maderae*, male calling behavior involves the release of a sex pheromone from the abdominal sternal glands. The extracts of sternal glands attracted conspecific females over a distance and the compounds responsible were identified as hydroxy-3-butan-2-one, (2*R*, 3*R*)-butanediol, senecioic acid and (*E*)-2-octenoic acid. The same components are also present in the male tergal glands.[26]

Volatiles collected from emissions of the female mantid, *Sphodromantis lineola* Burmeister, revealed a mixture of pentadecenal and tetradecenal. This was the first instance of a putative sex pheromone in a praying mantid.[27]

10.2.3 Diptera

(*Z*,*Z*)-4,7-tridecadienyl-(*S*)-2-yl acetate was identified as the sex pheromone of the female Douglas-Fir cone gall midge, *Contarinia oregonesis* Foote (Diptera: Cecidomyiidae). The *R* isomer was behaviorally benign.[28]

Pharmacophagy of methyl eugenol (ME), a highly potent male attractant, by the fruit fly *Bactrocera papayae*, results in the hydroxylation of ME to sex pheromonal components, 2-allyl-4,5-dimethoxyphenol (DMP) and (*E*)-coniferyl alcohol (CF).[29]

The sex pheromone of the female aphidophagous midge *Aphidoletes aphidimyza* Rondi (Diptera: Cecidomyiidae), was identified as (2*R*, 7*S*)-diacetoxytridecane, which elicited significant responses from males. The stereoisomeric mixture failed to attract males indicating the need for a single enantiomer.[30]

The sex pheromone of the pea midge, *Contarinia pisi* Winn. (Diptera: Cecidomyiidae), consists of 2-acetoxytridecane, (2*S*, 11*S*)-diacetoxytridecane and (2*S*, 12*S*)-diacetoxytridecane.[31]

(*S*, *S*)-2,12-,(*S*, *S*)-2,13- and (*S*, *S*)-2,14-diacetoxyheptadecanes were identified as the sex pheromone components of the female red cedar cone midge, *Mayetiola thujae* Hedlin (Diptera: Cecidomyiidae), the combination of which elicited response from males.[32]

Ovipositor extracts of calling females of the swede midge, *Contarinia nasturtii* Kieffer (Diptera: Cecidomyiidae) showed two compounds that elicited responses from males. These were identified as (2*S*, 9*S*)-diacetoxyundecane and (2*S*, 10*S*)-diacetoxyundecane. In addition, a trace amount of 2-acetoxyundecane was also found in the extracts. In wind tunnels experiments, males were attracted only when a blend of these three was used.[33]

10.2.4 Hemiptera

Extract of odors collected from sexually mature males of the green stink bug, *Acrosternum hilare* Say (Hemiptera: Pentatomidae), showed one major sex-specific component, (4*S*)-*cis*-(*Z*)-bisabolene epoxide [(4*S*)-cis-*Z*-BAE]. The crude extract was attractive to females, but when separated into four fractions, only the portion containing (4*S*)-*cis*-*Z*-BAE and the minor component (4*S*)-*trans*-*Z*-BAE was attractive to females. The fraction was as attractive as the crude extract suggesting that the former contained all the components. Neither the *cis*- nor the *trans*- BAE alone was attractive to females, but a 95:5 *cis*:*trans* blend, mimicking the ratio naturally produced by males was attractive to females in Y-tube bioassays.[34]

The male produced sex pheromone of the red-shouldered stink bug, *Thyanta pallidovirens* Stal (Hemiptera: Pentatomidae), was shown to consist of a blend of methyl (*E*2, *Z*4, *Z*6)-decatrienoate and the sesquiterpenes (+)-α-curcumene, (−)-zingiberene and (−)-β-sesquiphellandrene. In laboratory bioassays, sexually mature males attracted sexually mature females but not males, and females did not attract either sex. Pheromone components were not attractive to females individually, but combinations with any of the male-produced sesquiterpenes were attractive.[35]

Two male specific compounds were identified as the sex pheromones of the Brazilian rice stalk stink bug, *Tibraca limbativentris* Stal (Hemiptera: Pentatomidade), as isomers of (1′*S*)-zingiberenol.[36]

Two pheromonal components were detected in airborne collections from the vine mealybug, *Planococcus ficus* Signoret (Hemiptera: Pseudococcidae) in Israel. They were identified as (*S*)-lavandulyl senecioate (I) and (*S*)-lavandulyl isovalerate (II). Compound I has been identified as the sex pheromone of *P. ficus* in California. The report shows that feral P. ficus mealybugs produce and respond only to (I) whereas mealybugs reared in the laboratory on potato sprouts produce and respond to both (I) and (II).[37]

10.2.5 Heteroptera

(*E*)-2-octenyl acetate was identified from the volatiles produced by the adult female predatory bug, *Geocoris punctipes* that attracted male bugs. This compound comprised a significant portion of female volatiles. Other components in the extracts from adults of both sexes included (*E*)-2-hexenyl acetate, (*E*)-2-octenal and several saturated hydrocarbons but were not part of the attractant. Homogenized adults also yielded (*E*)-2-hexenal and (*E*)-4-oxo-2-hexenal whereas cast skins from late instar nymphs contained (*E*)-2-octenal, (*E*)-4-oxo-2-octenal, (*E*)-2-octenoic acid, and other saturated hydrocarbons.[38]

Volatiles collected from the male red-shouldered stink bug, *Thyanta perditor*, attracted females. The major component was identified as a male-specific compound, methyl (2*E*, 4*Z*, 6*Z*)-decatrienoate.[39]

Field trapping studies indicated that *n*-hexyl *n*-hexanoate, (*E*)-2-hexenyl *n*-hexanoate, and *n*-octyl *n*-butyrate were pheromone components of the rice leaf bug, *Trigonotylus caelestialium*, and mixtures in ratios of 1000:400:500:10–100 were attractive to males.[40]

10.2.6 Homoptera

The sex pheromone component of the citrus mealybug, *Pseudococcus cryptus* Hempel (Homoptera: Pseudococcidae), was identified as

3-isopropenyl-2,2-dimethylcyclobutylmethyl-3-methyl-3-butanoate. The pheromone component is an ester whose alcohol portion is identical to the known alcohol moiety of the pheromone of *Planococcus citri*. The absolute configuration of the pheromone was shown to be (1*R*, 3*R*) and the structure was confirmed by synthesis. Bioassay of the synthetic material indicated good activity in a glasshouse.[41]

The sex pheromone of the obscure mealybug, *Pseudococcus viburni* Signoret (Homoptera: Coccidae), consists of (1*R**, 2*R**, 3*S**)-(2,3,4,4-tetramethylcyclopentyl)methyl acetate. This compound is interesting because it is the first example of a new monoterpenoid structural motif in which the two isoprene units forming the carbon skeleton are joined by the 2′-2 and 3′-4 connections rather than the usual 1′-4 head-to-tail connections.[42]

Analyses of the volatiles released by females of the peach aphid *Tuberocephalus momonis* showed two sex pheromone components that were identified as (4a*S*,7*S*,7a*R*)-nepetalactone and (4a*S*,7*S*,7a*R*)-nepetalactol, in a ratio of 4:1. The most effective blend for trapping males was found to be 85:15 nepetalactone to nepetalactol respectively. In addition to *T. momonis*, over 20 other species of aphids were also caught.[43]

Females of the potato aphid *Macrosiphum euphorbiae* produced (1*R*, 4*aS*, 7*S*, 7*aR*)-nepetalactol and (4a*S*,7*S*,7a*R*)-nepetalactone in ratios that varied with age from 4:1 to 2:1. Males did not respond to either of the components when presented alone. However, the 5:1 ratio, which proved quite attractive in lab tests, was not as effective during three years of field trials.[44]

10.2.7 Hymenoptera

The pheromone for *Eurytoma amygladi* Enderlein (Hymenoptera: Eurytomidae), the almond seed wasp, was recently reported. Bioassays suggested that two alkadienes, (*Z*, *Z*)-6,9-tricosadiene [(*Z*, *Z*)-6,9-$C_{23:2}$] and (*Z*, *Z*)-6,9-pentacosadiene [(*Z*, *Z*)-6,9-$C_{25:2}$] and to a lesser extent alkenes identified in the extracts of virgin female *E. amygladi* were male attractants. Identification was based on GC, MS, and gas phase IR data.[45]

The trail pheromone taken from the hindguts of the formicine species, *Camponotus castaneus, C. balzani* and *C. sericeiventris* were identified. The trail pheromone in *C. castaneus* is 3,5-dimethyl-6-(1′-methylpropyl)-tetrahydro-2H-pyran-2-one and that of the other two species is 3,4-dihyrdo-8-hydroxy-3,5,7-trimethylisocoumarin. Although both the compounds induce precise trail following behavior, the major recruitment signal however, seems to be from formic acid released from the poison glands.[46] The major component of the trail pheromone of the myrmicine ant,*Crematogaster castanea*, was identified as (*R*)-2-dodecanol from the tibial glands of the hind legs.[47]

A lactone, (*Z*)-9-octadecen-4-olide, has been identified as a female specific antennally active compound from the currant stem girdler, *Janus integer* Norton (Hymenoptera: Cephidae). Comparison of the γ-lactone and a synthesized racemic mixture of (*Z*)-9-octadecen-4-olide on a chiral GC column showed the presence of a single component in the natural material.[48]

The extracts of three species of male North American decorator wasps, *Eucerceris rubripes, E. conata* and *E. tricolor*, were analyzed to reveal the presence of one major component in large quantities. This component detected in the head extract of males was identified as (*Z*)-3-hexenyl-3-hydroxybutanoate. The structure was confirmed by synthesis and the absolute configuration of the chiral center was determined to be *R* for the three species. In addition, 2- and 3-hexenoic acid and a few other aromatic compounds were also detected in varying quantities in males and females, along with hydrocarbons and fatty acids.[49]

A new alarm component discovered in the sting apparatus of the Africanized honeybees [*Apis mellifera* L.] was identified as 3-methyl-2-buten-1-yl acetate.[50]

Enantiomerically pure (*S*)-(+)-linalool was the main constituent in the extracts of the solitary bee *Colletes cunicularius* L. (Hymenoptera: Colletidae). Field tests using pure enantiomers and the racemate of linalool showed that the highest activity for male bees was for (*S*)-(+)-linalool.[51]

Besides a variety of alkanes and alkenes, four major compounds were identified from the cephalic glands of the male European beewolves, *Philanthus triangulum* F. (Hymenoptera: Sphecidae). Two of these, (*Z*)-11-eicosen-1-ol and (*Z*)-10-nonadecen-2-one were previously described as constituents of the cephalic gland. The two new components identified were 1-octadecanol and enantiopure (*S*)-2,3-dihydrofarnesoic acid, the latter being identified for the first time in nature.[52]

The Dufour's gland is the origin of the trail pheromone of the ponerine ant, *Gnamptogenys striatula.* Analyses of the glandular extracts showed a series of new natural products, especially esters of (2*E*)-3,4,7-trimethyl-2,6-octadien-1-ol (4-methylgeraniol) and (2*E*)-3,4,7-trimethyl-2,6-nonadien-1-ol (a bishomogeraniol isomer) with medium chain fatty acids. The authors concluded that the trail pheromone constituted a mixture of at least the 4-methylgeranyl esters identified in the gland after doing bioassays with synthetic racemates of the esters.[53]

A total of 21 compounds present in the surface and cephalic extracts of the queen bumblebee, *Bombus terrestris*, evoked responses in male antennae. These included saturated and unsaturated fatty acids, ethyl and methyl esters of the fatty acids, heptacosene, 2-nonanone, and geranyl geraniol. A blend of synthetic versions of these compounds elicited typical male mating behavior.[54]

The sex pheromone of the slave-making ant, *Polyergus breviceps*, has been identified as a blend of methyl-6-methylsalicylate and 3-ethyl-4-methylpentanol. Each compound alone was completely unattractive to males whereas a blend of the two compounds attracted hundreds of males in a couple of hours.[55]

Hexyl decanoate has been identified as the first trail pheromone in a stingless bee, *Trigona recursa.*[56]

The sex attractant of the male ectoparasitoid, *Melittobia digitata*, was identified as a mixture of α- and β-*trans*-bergamotene. This is the first instance of a male-produced sex attractant pheromone in the parasitic Hymenoptera.[57]

Volatiles and cuticular extracts of the wheat stem sawfly, *Cephus cinctus* Norton (Hymenoptera: Cephidae), were analyzed. The bifunctional 9-acetyloxynonanal gave the strongest EAD response. Additional EAD-active compounds included 13-acetyloxytridecanal, aldehydes with 9–16 carbon chain lengths, acids with 8–10 carbon chain lengths, and phenylacetic acid. The odor of phenylacetic acid was detected instantly by human nose when several males were allowed to interact, but not from isolated insects.[58]

10.2.8 Isoptera

The diterpene, Neocembrene A, or (1*E*, 5*E*, 9*E*, 12*R*)-1,5,9-trimethyl-12-(1-methyl-ethenyl)-1,5,9-cyclotetradecatriene has been identified after SPME and GC-MS analysis as the major component of the trail-following pheromone of the Rhinotermitidae Prorhinotermitinae, *Prorhinotermes canalifrons* and *P. simplex*. In all other Rhinotermitidae studied so far, the major component of the trail pheromone has been dodecatrienol.[59] Hydroquinone was identified for the first time as a phagostimulating pheromone in the Australian termite species, *Mastotermes darwiniensis*.[60]

10.2.9 Lepidoptera

Pheromone gland extracts of the Australian guava moth, *Coscinoptycha improbana* (Lepidoptera: Carposinidae), contained four compounds that elicited responses from male moth antennae in GC-EAD analyses. These compounds, identified on the basis of GC-MS, were found to be (*Z*)-7-tricosene along with three monounsaturated ketones, namely (*Z*)-7-ocatadecen-11-one, (*Z*)-7-nonadecen-11-one and (*Z*)-7-tricosen-11-one; they were found in a ratio of 65:23.5:1.5:10 respectively.[61]

The ester, 1-methylethyloctanoate, a new Lepidopteran sex pheromone of the bagworm moth, *Megalophanes viciella*, was discovered and identified. This identification was further confirmed when conspecific males were trapped in baits treated with this compound. In addition to the ester, octanal and dodecanal were also identified by GC-MS. The two aldehydes did not significantly affect the catches.[62]

Coupled GC-EAD of both gland extracts and effluvial collections from female blueberry leafminer, *Caloptilia porphyrectica* Braun (Lepidoptera: Gracillariidae), showed that females produce a single EAD-active compound. (*E*)-11-hexadecenal was determined to be the sex pheromone based on comparison of the retention time of an authentic standard on both polar and non-polar capillary columns. Microreaction-GC-EAD analyses and field trapping results confirmed the identification.[63]

Analysis of the pheromone-gland extracts of the female brinjal fruit and shoot borer, *Leucinodes orbonalis* Guenee (Lepidoptera: Pyralidae: Pyraustinae), of Indian and Taiwanese origin confirmed (*E*)-11-hexadecenyl acetate as the major pheromone component with 0.8% to 2.8% of the related (*E*)-11-hexadecen-1-ol. In this report, it was shown that the traps baited with a blend of the ester and the alcohol increased the catches. However, the data suggested that the optimal ratio maybe dose-dependant and further work is required to confirm this observation.[64]

Simultaneous GC-EAD analyses of the extracts of the sex pheromone gland of the female bronzed cutworm, *Nephelodes minians*, indicated two compounds which elicited strong EAD responses from conspecific male antennae, (*Z*)-11-hexadecenal and (*Z*)-11-hexadecenyl acetate. Double bond positions were confirmed by the dimethyl disulfide derivatives of the pheromone component. In a field test, a 5:1 blend of aldehyde and ester attracted male *N. minians*.[65]

The sex pheromone of the buck moth, *Hemileuca maia* (Lepidoptera: Saturniidae) has been identified as a blend, with the major component of this blend being (*E*10, *Z*12)-hexadeca-10,12-dienal, in combination with (*E*10, *Z*12)-hexadeca-10,12-dien-1-ol and (*E*10, *Z*12)-hexadeca-10,12-dien-1-yl acetate. Ratios of the compounds in extracts of the female pheromone glands varied around a mean of 100:7.4:6.3. None of the three components however, were attractive to male moths when tested as single components.[66]

A combination of (*Z*)-8-dodecenyl acetate and (*Z*)-8-dodecenol was identified from the sex pheromones glands of the carambola fruit borer, *Eucosma notanthes* Meyrick (Lepidoptera: Tortricidae). The ratio of the alcohol to the ester in the extracts was 2:7. GC, GC-MS, chemical derivatization, and comparison of retention times with authentic standards were used for identification.[67]

Collections of volatiles released by female carpenterworm moths, *Cossus insularis* (Lepidoptera: Cossidae), showed that two compounds elicited EAG responses from the antennae of male moths. These compounds were identified as (*E*)-3-tetradecenyl acetate and (*Z*)-3-tetradecenyl acetate in a ratio of 95:5 respectively.[68]

Three components extracted from the female glands of the large aspen tortrix, *Choristoneura conflictana* (Lepidoptera: Tortricidae), elicited response from antennae of conspecific males using coupled GC-EAD system. The main component extracted from the glands was

(*Z*)-11-tetradecenal and the other two minor components were (*E*)-11-tetradecenal and (*Z*)-11-tetradecen-1-ol.[69]

Extracts of the male oil palm bunch moths, *Tirathaba mundella* Walker (Lepidoptera: Pyralidae), were shown to contain four compounds namely, (3*S*, 6*S*)-2,2,6-trimethyl-6-vinyl-tetrahydro-pyran-3-ol, 4-hydroxy-3-methoxy benzaldehyde (vanillin), 6,10,14-trimethyl-2-pentadecanone, and the corresponding alcohol 6,10,14-trimethyl-2-pentadecanol, which elicited responses from the females. However, when synthetic compounds were used in lures, the ketone and alcohol by themselves did not attract females, suggesting that they might not be an essential part of the blend. This can only be confirmed once their absolute configuration has been determined and the proper stereoisomers tested in the field.[70]

Three compounds extracted from the female northern (beech) winter moth, *Operophtera fagata* Scharf (Lepidoptera: Geometridae), elicited response from male antennae, and were identified as the female sex pheromone. The two major compounds found were (9*Z*)-nonadecene and (6*Z*, 9*Z*)-nonadecediene and the third compound found in very small quantities was identified as (1, 3*Z*, 6*Z*, 9*Z*)-nonadecatetraene. The third component has previously been identified as the sex pheromone of the common winter moth, *O. brumata*.[71]

A mixture of (*Z*, *Z*)-11,13-hexadecadienyl acetate and (*Z*, *E*)-11,13,15-hexadecatrienyl acetate was identified as the female (synergistic) sex pheromone of the oak processionary moth, *Thaumetopoea processionea* (Lepidoptera: Thaumetopoeidae). In field experiments, a combination of both esters attracted a significant number of male moths, but not when used individually in traps.[72]

Extracts from the sex pheromone glands of the female *Ostrinia latipennis* (Lepidoptera: Crambidae) revealed one EAD active compound novel to *Ostrinia*. It was identified as (*E*)-11-tetradecenol. Field trapping experiments confirmed the attractiveness of this alcohol to males.[73]

Male satin moths, *Leucoma (Stilpnotia) salicis* L. (Lepidoptera: Lymantriidae), were attracted only to the (3*Z*, 6*R*, 7*S*, 9*R*, 10*S*)-isomer out of the four (3*Z*)-cis-6,7-cis-9,10-diepoxy-3-henicosenes (leucomalure). This finding was confirmed by field tests conducted in Hungary, which conclusively showed that males were attracted to the (3*Z*, 6*R*, 7*S*, 9*R*, 10*S*)-leucomalure. Interestingly, the British Columbia population of *L. salicis* responded to the (3*Z*, 6*S*, 7*R*, 9*S*, 10*R*)-leucomalure.[74]

Analyses of methanolic extracts of the male webbing clothes moth (WCM), *Tineola bisselliella* (Hum.) (Lepidoptera: Tineidae) showed three candidate pheromone components namely, hexadecanoic acid methyl ester, (*Z*)-9-hexadecenoic acid methyl ester, and octadec-anoic acid methyl ester. In bioassay experiments, the 16 carbon esters were attractive to both males and virgin females but the 18 carbon ester was inactive. The extracts of female WCM showed two compounds as candidate sex pheromone components, namely (*E*, *Z*)-2,13-octadecadienal and (*E*, *Z*)-2,13-octadecadienol. The synthetic samples of the aldehyde and alcohol were attracting WCM males in bioassay experiments successfully.[75]

GC-EAD analysis of the sex pheromone gland of the citrus flower moth, *Prays nephelomima* (Lepidoptera: Yponomeutidae) revealed a single active component, (*Z*)-7-tetradecenal. Field trials with this monounsaturated aldehyde in citrus orchards on the north island of New Zealand captured males successfully.[76]

Four EAG-active compounds were detected from the extracts of the female clear-winged tussock moth, *Perina nuda* Fabricius (Lepidoptera: Lymantriidae). The most abundant component was identified as (3*Z*,6*S*,7*R*,9*Z*)-6,7-epoxyhenicosa-3,9-diene. Other minor components identified were, (3*Z*, 9*Z*)-*cis*-6,7-epoxyicosa-3,9-diene (A), (3*R*,4*S*,6*S*,7*R*,9*Z*)-3,4,6,7-diepoxyhenicos-9-ene (B) and its 3*S*, 4*R*, 6*R*, 7*S*-isomer (C). Identification of the major component was done by GC-MS analyses of the extracts along with chemical derivatization and chiral HPLC. The major component showed weak attractiveness to males; however, the effect was enhanced by the addition of components (B) and (C). In the field test that was conducted males of *Hyopcala rostrata* were also attracted by synthetic *P. nuda* pheromone.[77]

The larval aggregation pheromone of the codling moth, *Cydia pomonella* L. (Lepidoptera: Olethreutidae), was shown to be a blend of (*E*)-2-octenal, (*E*)-2-nonenal, sulcatone and geranyl acetone in combination with either 3-carene and/or three saturated aldehydes (octanal, nonanal, decanal).[78]

The sex pheromone of a New Zealand geometrid moth, the common forest looper, *Pseudocoremia suavis* Butler (Lepidoptera: Geometridae), was identified as a mixture of four compounds. The two major active compounds were identified as (6*Z*)-*cis*-9,10-epoxynonadec-6-ene and (3*Z*, 6*Z*)-*cis*-9,10-epoxynonadeca-3,6-diene. Of the other two active compounds, one was identified tentatively as (3*Z*, 6*Z*)-*cis*-9,10-epoxyhenicosa-3,6-diene, the fourth however could not be identified due to insufficient amounts in the extract.[79]

The female sex pheromone of the cranberry blossom worm, *Epiglaea apiata* Grote (Lepidoptera: Noctuidae), was identified as a blend of (*Z*)-9-hexadecenyl acetate, (*Z*)-9-tetradecenyl acetate, and tetradeceyl acetate. They were present in the ratio of 65:2:33 respectively in the female pheromone gland.[80]

The female sex pheromone of the currant shoot borer, *Lampronia capitella* Clerck (Lepidoptera: Prodoxidae), was identified as a mixture of (*Z*, *Z*)-9,11-tetradecadienol and its corresponding acetate and aldehyde. The EAG-activity of all the four geometric isomers was studied, and the respective (*Z*, *Z*)-isomers were found to be the most active in all cases. Aldehydes elicited the highest response followed by the alcohol; the ester was the least active compound.[81]

The sex pheromone of the female dogwood borer, *Synanthedon scitula* Harris (Lepidoptera: Sesiidae), was identified as a blend of (*Z*, *Z*)-3,13-octadecadienyl acetate(a), (*E*, *Z*)-2,13-octadecadienyl acetate (b), and (*E*, *Z*)-3,13-octadecadienyl acetate (c) in a ratio of 88:6:6 respectively. Further analysis indicated that (a) was the major component and was attractive by itself. A mixture of (a) and (b) enhanced the activity. Attraction was strongly antagonized by the addition of (*E*, *Z*)-3,13-octadecadienyl acetate.[82]

The major component of the female sex pheromone produced by the herald moth, *Scoliopteryx libatrix* (L.) (Lepidoptera: Noctuidae), has been identified as (6*Z*)-13-methylheneicosene. This is the first example of a branched chain alkene as a sex pheromone in the Noctuidae family and is markedly different from the pheromones of the other members in the group. The authors have not conclusively stated, but they suggest that it is probably the (13*S*)-isomer.[83]

Two unsaturated alcohols, (6Z, 9Z, 11S)-6,9-heneicosadien-11-ol and (6Z, 9Z, 11R)-6,9-heneicosadien-11-ol, were identified as the female sex pheromone components of the tussock moth, *Orgyia detrita* Guerin-Meneville (Lepidoptera: Lymantriidae). Both of the alcohols in combination, but not singly, attracted male moths.[84]

The sex pheromones of four Plusiinae species in the family Noctuidae namely, *Ctenoplusia albostriata* (CA), *Macdunnoughia purissima* (MP), *Syngrapha ain* (SA) and *Diachrysia stenochrysis* (DS), were identified. CA females produced (Z)-5-decenyl acetate (I), (Z)-7-dodecenyl acetate (II), and (Z)-7-dodecen-1-ol (III) in a ratio of 2:100:13. The MP females produce II, III, and (Z)-5-dodecenyl acetate (IV) in a ratio of 100:80:20. Compounds II and III were also identified from the SA females, whereas the DS females produced just one active compound, (Z)-7-decenyl acetate. DS species were the first Plusiinae species identified using only the ω3-compound and none of the ω5-compounds, such as II and III that are common components of Plusiinae pheromones.[85]

The sex pheromone of the fir coneworm moth, *Dioryctria abietivorella* (Pyralidae: Phycitinae) has been identified as a blend of (3Z, 6Z, 9Z, 12Z, 15Z)-pentacosapentaene and (9Z, 11E)-tetradecadienyl acetate.[86] A single component in the extracts of virgin female geometrid moths *Miliona basalis pryeri* Druce (Lepidoptera: Geometridae) elicited responses from male moth antennae and in field tests. This compound was identified as (Z, Z)-(3S, 4R)-3,4-epoxynonadeca-6,9-diene by GC-MS and NMR. The opposite enantiomers, the racemic mixture, and nonvirgin females attracted no more moths than the solvent controls.[87]

Chemical analysis of pheromonal gland extracts of the female giant geometrid moth *Biston robustum* Butler (Geometridae: Ennominae), showed the presence of (Z, Z)-6,9-nonadecadiene (I), (Z, Z, Z)-3,6,9-nonadecatriene (II), *cis*-(Z)-6,7-epoxy-9-nonadecene (III) and *cis*-(Z, Z)-6,7-epoxy-3,9-nonadecadiene (IV) in a ratio of 13:2:70:15. Field tests showed that (III) and (IV) with the 6S, 7R configurations were essential components and that (I) and (II) showed a synergistic effect on male attraction.[88]

Analysis of the sex pheromone glands of the female grapevine moths *Lobesia botrana* Denis and Schiffermuller (Lepidoptera: Tortricidae), showed three previously unidentified compounds, (E)-7-dodecenyl acetate and the (E,E) and (Z,E)-isomers of 7,9,11-dodecatrienyl acetate. This is the first account of a triply unsaturated component in a tortricid moth. The monoenic acetate (E)-7-dodecenyl acetate and the trieneic acetate ($7Z,9E,11$)-dodecatrienyl acetate significantly enhanced the male response to the main pheromone compound, ($7E,9Z$)-7,9-dodecadienyl acetate in wind tunnel experiments.[89] ($7R$,$8S$)-*cis*-7,8-epoxy-2-methyloctadec-17-ene has been identified as a novel trace component from the sex pheromone gland of the gypsy moth, *Lymantria dispar* (Lepidoptera: Lymantriidae).[90]

Two components of the female produced sex pheromone of the hornet moth, *Sesia apiformis* Cl. (Lepidoptera: Sesiidae), were identified as ($3Z,13Z$)-octadeca-3,13-dien-1-ol and ($2E,13Z$)-octadeca-2,13-dienal. Although single components were not attractive, a 2:3 mixture proved to be highly active towards males in field tests.[91]

EAG experiments on the composition of the sex pheromone of the horse-chestnut leafminer, *Cameraria ohridella* Deschka et Dimic (Lepidoptera, Gracillariidae, Lithocolletinae) showed that ($9E$)-tetradecenal and the stereoisomers of ($8E,10Z$)-tetradeca-8,10-dienal exhibited significant electrophysiological activity and could, therefore, be considered as possible minor pheromone components along with ($8E,10Z$)-tetradeca-8,10-dienal which is the major component.[92]

Coupled GC-EAD detection analyses of pheromone gland extracts revealed one EAD active compound produced by female *Lymentra lucescens* and by female *L. serva* (Lepidoptera: Lymantriidae). This compound was identified as 2-methyl-(Z)-7-octadecene by the usual analytical techniques and by comparison with an authentic synthetic compound.[93]

Two compounds collected from calling females of the leafminer moth, *Phyllonorycter emberizaepenella* (Lepidoptera: Gracillariidae) were identified as potential sex pheromone components: ($8E,10E$)-8,10-tetradecadienol and ($8E,10E$)-8,10-tetradecadienyl acetate, the former occurring in trace amounts. In field tests, no males were attracted to traps baited with either the potential pheromone or virgin females. (It has however been shown that the moth reproduces by parthenogenesis of the thelytoky type).[94]

The female sex pheromone of the leaf-miner *Phyllonorycter platini* Staudinger (Lepidoptera: Gracillariidae) has been identified as (*Z*10)-tetradecenyl acetate. Activity of the synthetic compound was confirmed by field tests.[95]

A three-component blend consisting of the major sex pheromone component of the legume podborer, *Maruca vitrata* (syn. M. testulalis) (F.) (Lepidoptera: Pyralidae), (*E*, *E*)-10,12-hexadecadienal along with the minor components, (*E*, *E*)-10,12-hexadecadienol and (*E*)-10-hexadecenal in a ratio of 100:5:5 caught more males than traps baited with the major component alone, either two-component blend or virgin female moths.[96]

(7*Z*, 9*E*)-2-methyl-7,9-octadecadiene has been identified as the single sex pheromone component of *Lymantria bantaizana* also known as *Lymantria (Spinotria) grisescens bantaizana* (Lepidoptera: Lymantriidae). Field tests substantiated that it is the sex pheromone of *L. bantaizana*, a compound previously unknown in the Lepidoptera.[97]

The components of the sex pheromone from female nettle caterpillars *Darna trima* and *Darna bradleyi* (Lepidoptera: Limacodidae) were identified. The compounds from *D. trima* were 2-methylbutyl-(*E*)-7,9-decadienoate (**A**) and (*E*)-2-hexenyl (*E*)-7,9-decadienoate (**B**) and from *D. bradleyi* were identified as methyl (*E*)-7,9-decadienoate (**C**) and isobutyl (*E*)-7,9-decadienoate (**D**). In Malaysia, (*S*)-2-methylbutyl- (*E*)-7,9-decadienoate in combination with **B** proved to be essential and synergistic pheromone components for attraction of males in field tests. (*R*)-2-methylbutyl- (*E*)-7,9-decadienoate had no behavioral activity. Compound **D** singly attracted male *D. bradleyi*, but the addition of **C** and **D** in a ratio of 1:10 significantly enhanced attractiveness of the bait.[98]

The major sex pheromone components of the painted apple moth, *Teia anartoides* Walker (Lepidoptera: Lymantriidae), were identified as (6*Z*, 9*Z*)-henicosa-6,9-dien-11-one and (6*Z*,9*Z*)-henicosa-6,9-diene on the basis of GC-EAD, GC-MS, HPLC-MS and HPLC-UV/vis spectroscopy of pheromone gland extracts and authentic standards. Other minor components included (6*Z*)-9*R*,10*S*-epoxyeicos-6-ene, (6*Z*)-9*R*, 10*S*-epoxyhenicos-6-ene, (6*Z*, 9*Z*)-henicosa-6,9-dien-1-ol, (6*Z*)-henicos-6-en-11-one and (6*Z*, 8*E*)-henicosa-6,8-dien-11-one; however, the roles of these minor components remain unknown. (6*Z*,9*Z*)-henicosa-6,9-dien-11-one is thermally labile and rearranges to (6*Z*,

8*E*)-henicosa-6,8-dien-11-one and other products at ambient temperature, thereby rendering the synthetic pheromone lure inactive after two days of field exposure. In field cage and field experiments, a blend of all seven identified compounds attracted males as well as calling females.[99,100]

The female Paulownia bagworm, *Clania variegata* Snell (Lepidoptera: Psychidae), produces two major compounds that elicit strong responses from male antennae. The more abundant of the two components was identified as 1-ethyl-2-methylpropyl 3,13-dimethylpentadecanoate by transesterification, GC-MS, and comparing spectral and GC retention with synthetic compounds. The absolute configuration of this compound and the structure of the other component remain to be determined.[101]

Three EAG-active compounds were identified from the extracts of pheromone glands of the persimmon fruit moth, *Stathmopoda masinissa* Meyrick (Lepidoptera: Oecophoridae). These were (4*E*, 6*Z*)-4,6-hexadecadienal and the corresponding acetate (*E*4, *Z*6-16:OAc) and alcohol (*E*4, *Z*6-16:OH). A preliminary field trail confirmed that the acetate as a single component attracted male moths, the possible roles of the aldehydes and the alcohol as components of lures remains to be determined.[102]

The sex pheromone of the female pistachio twig borer, *Kermania pistaciella* (Lepidoptera: Oinophilidae), from Turkey was identified as (2*S*, 12*Z*)-2-acetoxy-12-heptadecene. In field tests in Turkey, lures containing synthetic (2*S*, 12*Z*)-2-acetoxy-12-heptadecene attracted a large number of male moths. However, the attractiveness was reduced significantly by the presence of the (*R*)-enantiomer or of either enantiomer of the corresponding alcohol. It is the first pheromone component identified in the Oinophilidae and the first secondary acetate identified in Lepidoptera.[103]

Analyses of the pheromone glands of the female *Anadevidia peponis* and *Maddunnoughia confusa*, showed that *A. peponis* produces six monoene acetates and two monoene alcohols and that *M. confusa* produces five monoene acetates. These components include (*Z*)-7-dodecenyl acetate as a major common constituent and three other acetates as minor common constituents. The minor constituents are quite different in blend composition. An indispensable component for male attraction is (*Z*)-5-decenyl acetate for *A. peponis* and (*Z*)-9-tetradecenyl acetate is essential for *M. confusa.* Field tests showed synergistic effects of some other minor components and male attraction of three other Plusiinae species, *Macdunnoughia purissima, Ctenoplusia albostriata* and *Chrysodeixis eriosoma.*[104]

The female sex pheromone component of the New Zealand raspberry budmoth, *Heterocrossa rubphaga,* has been identified as (*Z*)-12-nonadecen-9-one.[105]

The extracts of the female sex pheromone gland of the sandthorn carpenterworm moth, *Holcocerus hippophaecolus* Hua (Lepidoptera: Cossidae), were found to contain (*E*)-3-tetradecenyl acetate, (*Z*)-3-tetradecenyl acetate, (*Z*)-7-tetradecenyl acetate, the corresponding alcohols and (*E*)-9-tetradecenyl acetate. A combination of field trials and experiments were done to conclude that the sex pheromone of *H. hippophaecolus* is composed of (*Z*)-7-tetradecenyl acetate and (*E*)-3-tetradecenyl acetate. Optimal ratios and doses of these two and the possible role of other minor components remain to be determined.[106]

The sex pheromone of the South American tortricid moth, *Argyrotaenia sphaleropa,* was identified as a mixture of (*Z*)-11-tetradecenal, (*Z*)-11,13-tetradecadienal, (*Z*)-11-tetradecenyl acetate and (*Z*)-11,13-tetradecadienyl acetate in the ratio of 1:4:10:40. Best results for trapping were obtained with mixtures of (*Z*)-11-tetradecenal and (*Z*)-11,13-tetradecadienal in the ratio of 1:4 to 1:9.[107]

A sex pheromone component was identified for three North American moth species of the lepidopteran genus *Saturnia.* (*E*4, *Z*9)-tetradecadienal was identified as a component for *S. walterorum, S. Mendocino* and *S. albofasciata.*[108]

Five candidate pheromone components were identified from pheromone gland extracts of the tomato fruit borer, *Neoleucinodes elegantalis* (Lepidoptera: Crambidae): (*E*)-11- hexadecenol, (*Z*)-11-hexadecenol, (*E*)-11-hexadecenal, (*E*)-11-hexadecenyl acetate and (*Z*3,*Z*6,*Z*9)-tricosatriene. A two-component blend of synthetic (*E*)-11- hexadecenol and the hydrocarbon was observed to be the most effective.[109]

The unstaturated ketone, (*Z*)-6-heneicosen-11-one, was identified as the only essential sex pheromone component of the whitemarked tussock moth, *Orgyia leucostigma* J.E. Smith (Lepidoptera: Lymantriidae).[110]

Comparative GC, GC-EAD and GC-MS analyses of extracted *Setora nitens* (nettle caterpillars) compounds and authentic standards showed that the candidate pheromone components were (*Z*)-9-dodecenal and (*Z*)-9,11-dodecadienal. The other two EAD-active compounds were the corresponding alcohols of these aldehydes.[111]

Sex pheromones of five olethreutine species (Lepidoptera: Tortricidae), were shown to be associated with seedlings and fruits of mangrove plants in the Ryuku Islands in Japan, where these compounds were identified and tested in fields.[112]

10.2.10 Neuroptera

GC-EAD analysis of thoracic extracts of the male green lacewing, *Chrysopa nigricornis* Burmeister, showed that two compounds elicited response from conspecific male antennae: 1-tridecene and (1*R*, 2*S*, 5*R*, 8*R*)-iridodial. Iridodial also attracted males of the goldeneyed lacewing, *C. oculata* Say, and to a lesser extent, *C. coloradensis* Banks males.[113]

Enantiomerically pure diastereoisomers (1*R*, 4*S*, 4a*R*, 7*S*, 7a*R*)-(**1**) and (1*R*, 4*R*, 4a*R*, 7*S*, 7a*R*)-dihydronepetalactol (**2**) were synthesized diastereoselectively from a renewable natural source (4a*S*,7*S*,7a*R*)-nepetalactone (**3**), isolated from the catmint plant. Compound (**1**) was found to catch significant numbers of three species of lacewing in the field: in Korea, *Chrysopa cognata* and in the United Kingdom, *Nineta vittata* and most notably *Peyerimhoffina gracilis*. All species caught were found more frequently in traps releasing (**1**) than (**2**), while more *C. cognata, C. formosa* and *C. phyllochroma* were found in traps releasing (1*R*, 4a*S*,7*S*, 7a*R*)-nepetalactol (**4**).[114]

10.2.11 Thysanoptera

Defensive secretions of adult and larval thrips, *Suocerathrips linguis* (Phlaeothripidae, Thysanoptera), were found to contain a long chain acetate, (11*Z*)-11,19-eicosadienyl acetate. This acetate, which was not known to occur naturally, was found along with octadecyl acetate and other long chain acetates. The ecosadienyl acetate repels ants and spreads on the surface of potential predators.[115]

Two major components have been detected from adult male western flower thrips, *Frankliniella occidentalis* Pergande (Thysanoptera: Thripidae) that are not present in females. They were identified as (*R*)-lavandulyl acetate and neryl (*S*)-2-methylbutanoate. The catch of males and females on traps was significant by the butanoate alone or by a 1:1 blend of the acetate and the butanoate, but the acetate was not active alone.[116]

10.2.12 Acarina

A new major aggregation pheromone emitted by the storage mite, *Chortoglyphus arcuatus*, was identified as (4*R*, 6*R*, 8*R*)-4,6,8-trimethyldecan-2-one. To identify this new pheromone from mites, headspace of undisturbed colonies of storage mites was analyzed by GC-MS by the use of a closed-loop stripping apparatus or solid phase microextraction. The structure was then elucidated by MS, synthesis of authentic samples, and GC on a chiral phase.[117]

The female sex pheromone of a mite, *Rhizoglyphus robini* Claparede (Astigmata: Acaridae), was identified as α-acaridial [2(*E*)-(4-methyl-3-pentenyl)-butenedial], which stimulated males and enhanced male mounting behavior. This was the first time that two pheromones (the alarm pheromone neryl formate and the sex pheromone α-acaridial) have been demonstrated to be components of the same opisthonotal gland secretion in astigmatid mites.[118]

10.2.13 Aranea

Cupilure or (*S*)-1,1′-dimethyl citrate was identified as the contact sex pheromone of the female spider, *Cupiennius salei*. Chemical analysis, electrophysiology and behavioral assay demonstrated that it was indeed the contact sex pheromone.[119]

10.2.14 Some Other Interesting Examples

An example of chemical mimicry of pheromones involves the female bolas spider, *Mastophora hutchinsoni*: it feeds exclusively on male moths of a few species, of which, one is the male brisly cutworm, *Lacinipolia renigera*. Female brisly cutworms produce a pheromone blend consisting of (*Z*)-9-tetradecenyl acetate and (*Z*, *E*)-9,12-tetradecenyl acetate. On analyzing volatiles collected from a hunting female spider, a similar blend was identified which elicited response from the antennae of the male brisly cutworm.[120] Although not a pheromone, butyl hexanoate was identified from apple fruit volatiles as a female specific attractant for the codling moth, *Cydia pomonella*.[121] Another interesting discovery was reported wherein buibuilactone, a pheromone of various scarab beetles and ant pyrazines were produced by marine bacteria.[122]

10.3 SYNTHESIS OF PHEROMONES

Below is a tabular list of pheromone components that have been synthesized since 2000. This list contains the organism and the structure of the individual compounds. Also included is the name of the compound as given in the original publication. An outline of each individual synthesis is beyond the scope of this review.

Table 1. List of Pheromone Components

Synthetic Methodology/Species	Structure of Synthesized Pheromone/ Precursor
Almond bark beetle, *Scolytus amygladi* (Coleoptera: Scolytidae)[123]	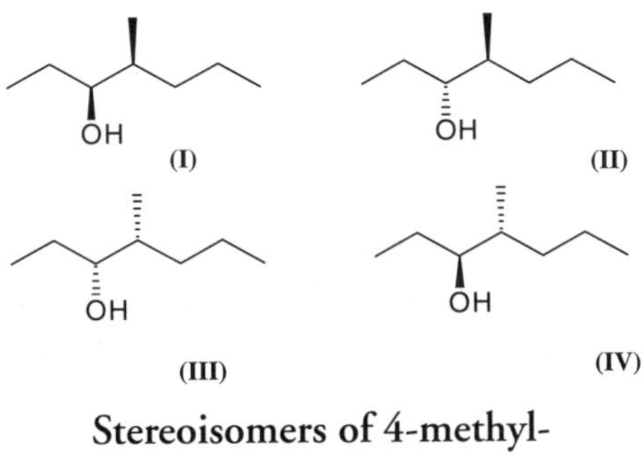 **Stereoisomers of 4-methyl-3-heptanol**

Citrus fruit borer, *Ecdytolopha aurantiama* Lima (Lepidoptera: Tortricidae)[124]

(1)

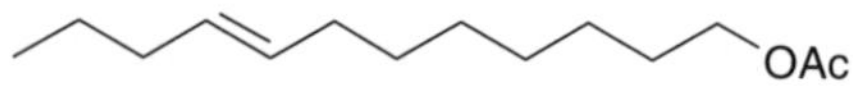

(2)

(1) (*E*)-8-dodecenol
(2) (*E*)-8-dodecenyl acetate

Currant stem girdler, *Janus integer* Norton (Hymenoptera: Cephidae)[125–127]

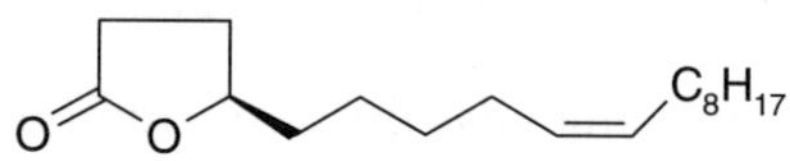

(*R*, *Z*)-9-octadecen-4-olide

Citrus leafminer, *Phyllocnitis citrella* Stainton (Lepidoptera: Gracillariidae)[128,129]

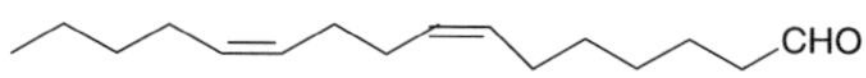

(A)

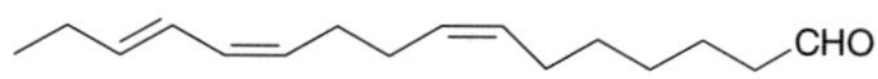

(B)

A = (*Z*, *Z*)-7,11-hexadecadienal
B = (*Z*, *Z*, *E*)-7,11,13-hexadecatrienal

Autumn gum moth, *Mnesampela privata* Guenee (Lepidoptera: Geometridae)[130]

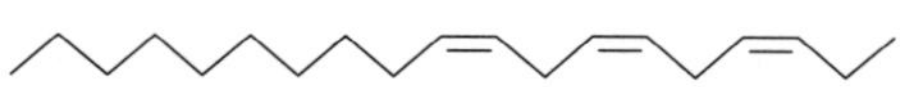

(3*Z*, 6*Z*, 9*Z*)-3,6,9-nonadecatriene

Brazilian fall armyworm, *Spodoptera frugiperada* Smith (Lepidoptera: Noctuidae)[131]

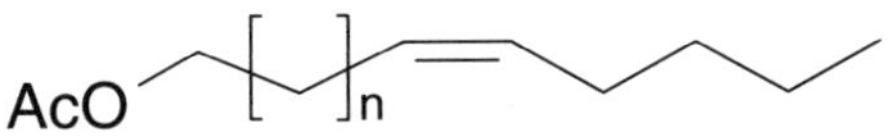

n = 5, *Z*7-12:Ac [(*Z*)-7-dodecenyl acetate]
n = 7, *Z*9–14:Ac [(*Z*)-9-tetradecenyl acetate]
n = 9, *Z*11–16:Ac [(*Z*)-hexadecenyl acetate]

Simple synthesis of (*E*)-4-oxo-2-decenal and homologues, which are common components of the defensive secretions of true bugs (Hemiptera)[132]

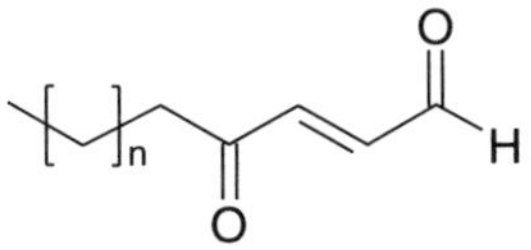

n = 4, (*E*)-4-oxo-2-decenal

Satin moth, *Leucoma salicis*. All four isomers of (*Z*)-3-*cis*-6,7-*cis*-9,10-diepoxyhenicosenes were synthesized using D-xylose as the chirally pure starting material.[133]

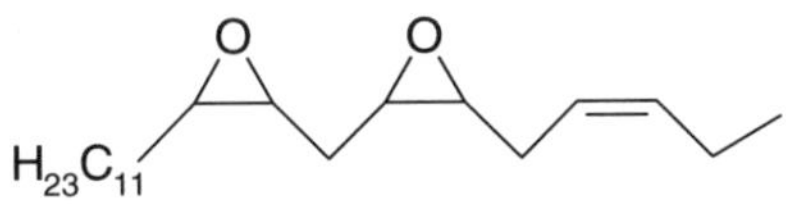

(*Z*)-3-*cis*-6,7-*cis*-6,7-*cis*-9,10-diepoxyhenicosene

Diepoxyalkene derivatives (all stereoisomers) were synthesized from (3*Z*, 6*Z*, 9*Z*) trienes with a C_{21}, C_{19} or C_{18} straight chain, lymantriid sex pheromones, and their candidates were synthesized by MCPBA oxidation of optically active epoxyalkadienes.[134]

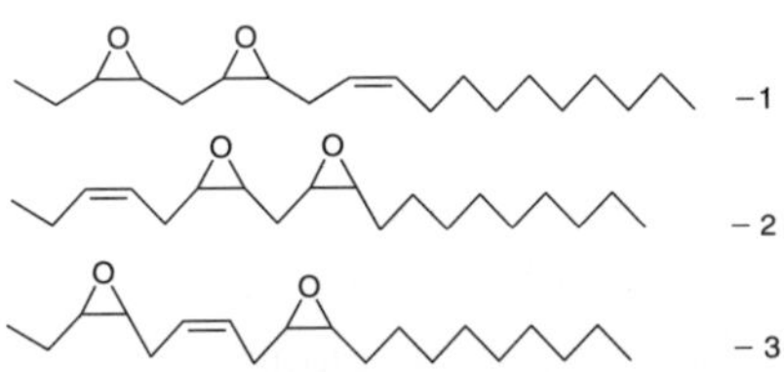

(1) (*Z*)-*cis*-3,4-*cis*-6,7-diepoxy-9-nonadecene
(2) (*Z*)-*cis*-6,7-*cis*-9,10-diepoxy-3-nonadecene
(3) (*Z*)-*cis*-3,4-*cis*-9,10-diepoxy-6-nonadecene

3,7-dimethyl-2-undecanol, 3, 7,9-trimethyl-2-tridecanol, and 3,7,11-trimethyl-2-tridecanol, which are known precursors of the female sex pheromones of the pine sawfly species *Diprion nipponica, Macrodiprion nemoralis* and *Microdiprion pallipies* respectively were synthesized in moderate yields.[135]

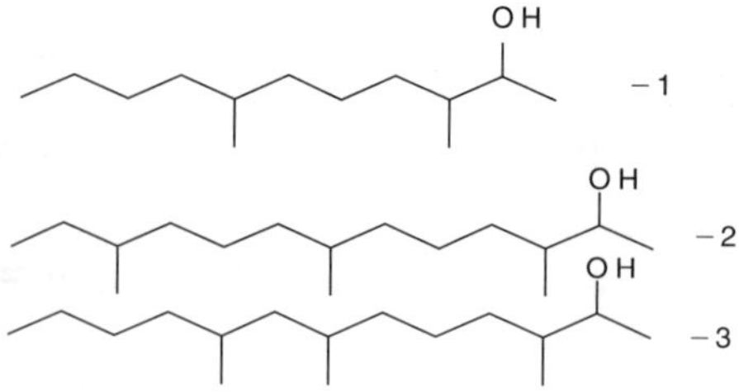

(1) 3,7-dimethyl-2-undecanol
(2) 3,7,11-trimethyl-2-tridecanol
(3) 3,7,9-trimethyl-2-tridecanol

Male-produced sex pheromones of the Brazilian soybean stink bugs, *Euschistus heros* and *Piezodorous guildinli*, are methyl-2,6,10-trimethyltridecanoate and methyl-2,6,10-trimethyldodeacanoate respectively. These two compounds were synthesized as a mixture of stereoisomers.[136] Another synthesis for the sex pheromone of the male stink bug *Euschistus heros*, along with the synthesis of the pheromone of the rice moth, *Corcyra cephalonica*, has been reported.[137]

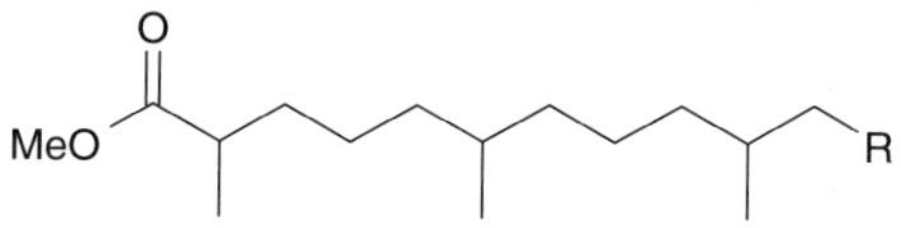

R = Me, Methyl-2,6,10-trimethyldodeacanoate
R = Et, Methyl-2,6,10-trimethyltridecanoate

This chapter reviews literature regarding the use of 10-undecenoic acid in the synthesis of insect pheromones.[138]

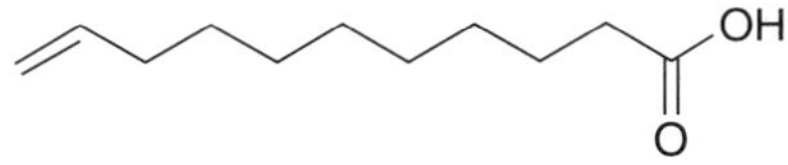

10-undecenoic acid

Absolute configuration of Quercivorol, (1*S*, 4*R*)-*p*-Menth-2-en-1-ol, an aggregation pheromone of the Ambrosia beetle *Platypus quercivorus* (Coleoptera: Platypodidae) was determined.[139,140]

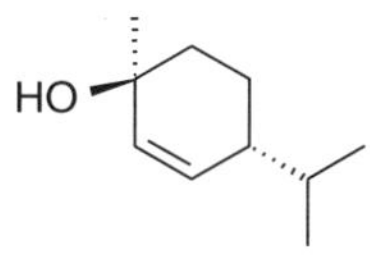

1

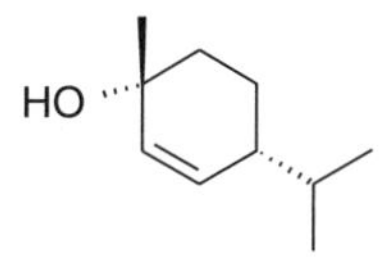

2

(1) (1*S*,4*R*)-quercivorol
(2) (1*R*,4*R*)-quercivorol

Three linear methyl-branched pheromones were synthesized using the chiral starting material (+)-aromadendrene.[141]

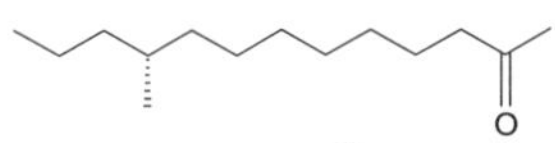

(1)

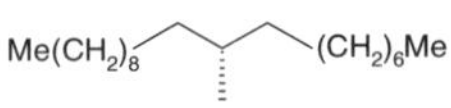

(2)

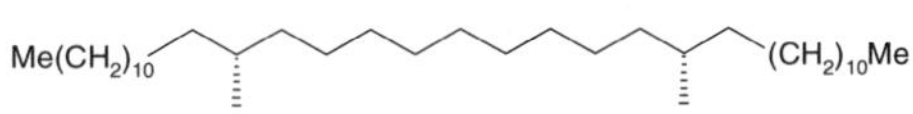

(3)

(1) (10*R*)-10-methyl-2-tridecanone
(2) (9*S*)-9methyl-nonadecane
(3) (13*R*,23*S*)-13,23-dimethylhex-atriacontane

One-pot stereoselective synthesis of trisubstituted (*E*)-2-methylalk-2-enoic acids was performed by treatment of unactivated Baylis-Hillman adducts. This method was used to synthesize three insect pheromones namely, (4*S*, 2*E*)-2,4-dimethyl-2-hexenoic acid, (+)-(*S*)-manicone and (+)-(*S*)-normanicone.[142]

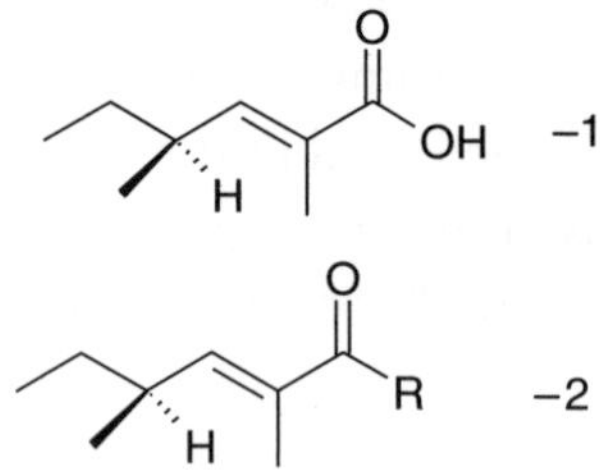

(1) (4*S*,2*E*)-2,4-dimethyl-2-hexenoic acid
(2) R = Me, (+)-(*S*)-normanicone
(3) R = Et, (+)-(*S*)-manicone

The Baylis-Hillman reaction has also been used to synthesize (+)-dominicalure I and (+)-dominicalure II along with (4*S*, 2*E*)-2,4-dimethyl-2-hexenoic acid, (+)-(*S*)-manicone and (+)-(*S*)-normanicone.[143]

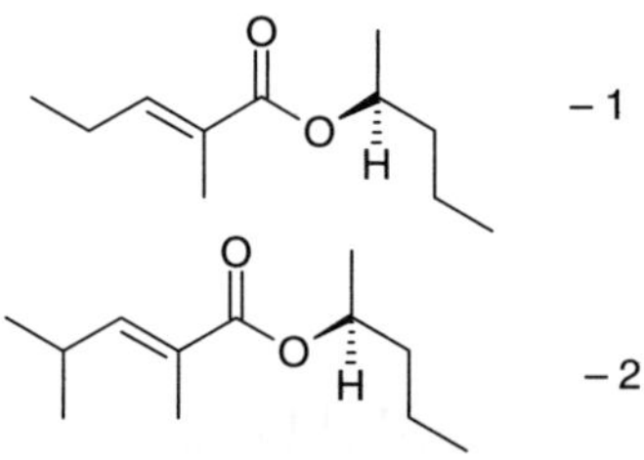

(1) (+)-(*S*)-dominicalure I
(2) (+)-(*S*)-dominicalure II

Acetyl derivatives of the Baylis-Hillman adducts on treatment with Zn in saturated aq. NH_4Cl solution under reflux has also been utilized as a method for the preparation of chiral insect pheromones dominicalure I and dominicalure II of the lesser grain borer *Rhyzopertha dominica*.[144]

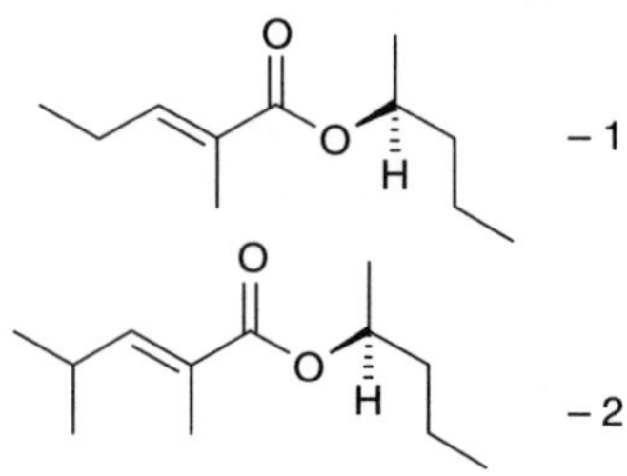

(1) (+)-(*S*)-dominicalure I
(2) (+)-(*S*)-dominicalure II

A polyketide/fatty acid-type metabolic route using three propionate units has been shown by stable isotope-labeled probes and MS to the biosynthetic route towards the production of the insect pheromone (*S*)-4-methyl-3-heptanone in the ant *Harpegnathos saltator*.[145]

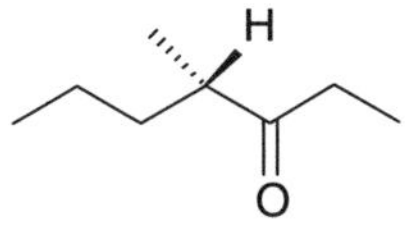

(*S*)-4-methyl-3-heptanone

The sex attractant of the black carpet beetle, *Attagenus megatoma* (Fabricius), (3*E*, 5*Z*)-3,5-tetradecadienoic acid also known as megatomic acid has been synthesized. The key step in this synthesis is the Cadiot-Chodkiewicz cross coupling of 1-decyne and 4-bromo-3-butyn-1-ol in the presence of CuCl.[146]

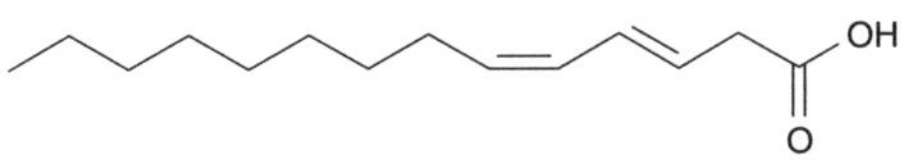

(3*E*,5*Z*)-3,5-tetradecadienoic acid (Megatomic acid)

A rapid and low cost preparation of the (3*E*, 5*Z*)-alkadienyl system encountered in several insect pheromones has been developed. Knoevenagel condensation of (*E*)-2-alkenals with ethyl hydrogen malonate in DMSO, in the presence

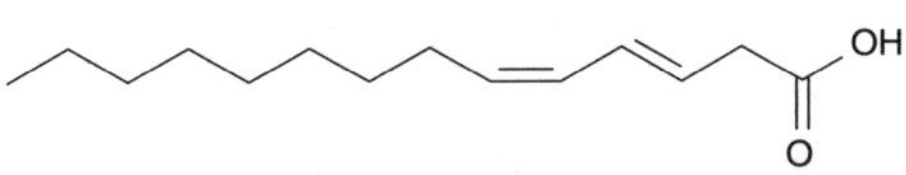

(3*E*,5*Z*)-3,5-tetradecadienoic acid (Megatomic acid)

of piperidinium acetate as a catalyst gave a mixture of geometrical isomers of ethyl 3,5- and 2,4-alkadienoates from which the (3*E*, 5*Z*)-3,5-alkadienoate was separated by urea inclusion complex formation.[147]

Diprionyl acetate, which is an attractant of the *Neodiprion* species of sawflies, has been synthesized by diastereoselective protonation.[148]

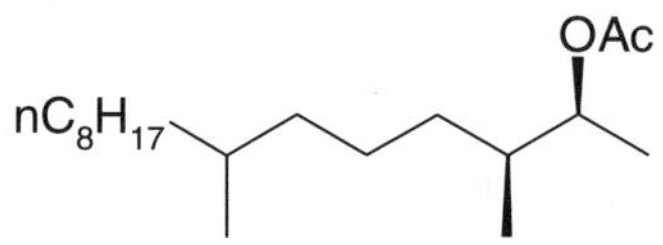

(2*S*,3*S*,7*RS*)-diprionyl acetate

An enantioselective approach to both enantiomers of α-alkyl-α-methoxyarylacetic acid derivatives has been described from L-(+)-tartaric acid. Key steps include stereoselective addition of Grignard reagents to 1,4-diketones derived from tartaric acid. This methodology has been applied in synthesizing the pine beetle pheromone frontalin.[149]

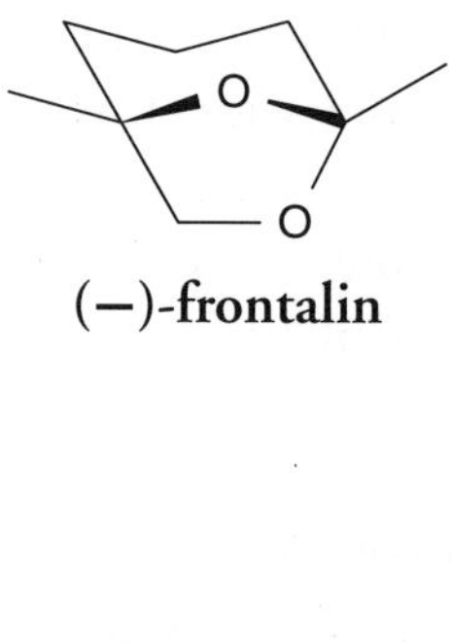

(−)-frontalin

The importance of hydrolytic kinetic resolution (HKR) in providing a wide range of highly enantiomerically enriched terminal mono- and bis epoxides has been showed by the conversion of such epoxides efficiently to some important insect pheromones.[150,151]

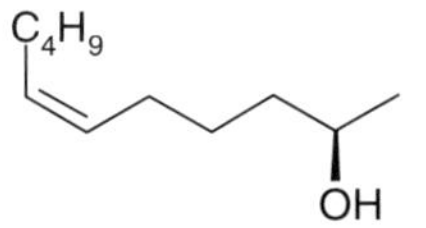

–1

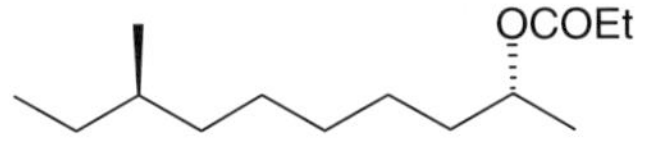

–2

(1) (−)-(*R*)-(*Z*)-undec-6-en-2-ol (Nostrenol)
(2) (−)-(1*R*,7*R*)-1,7-dimethylnonyl propanoate

The male bean weevil sex attractant has been made by a new method of allene synthesis from aldehydes and alkenyl aryl sulfoxides by sulfoxide-metal exchange as the key reaction.[152]

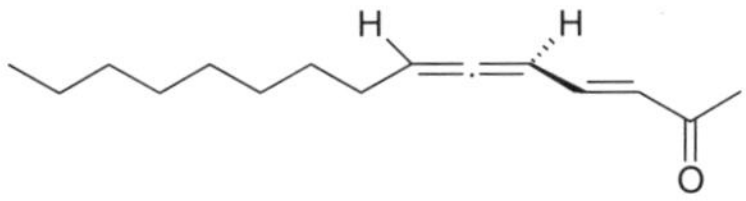

– 1

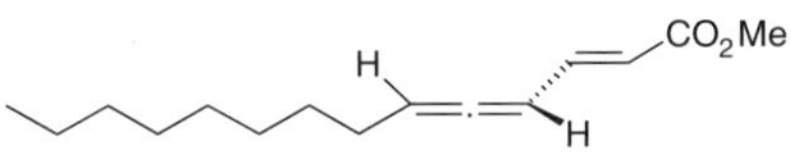

– 2

1 & 2 = (*R*) and (*S*) methyl 2,4,5-tetradecatrienoate

A procedure was developed for the synthesis of (*Z*)-5 and (*Z*)-7-monoene components of sex pheromones of the Lepidoptera insects based on cometathesis of ethylene and cycloocta-1,5-diene.[153]

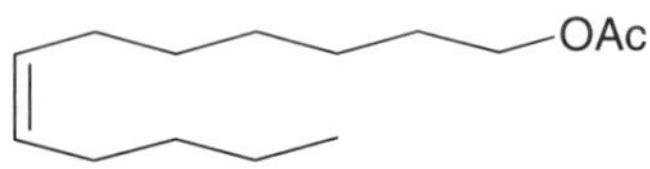

(*Z*)-dodec-7-en-1-yl acetate

Enantioselective syntheses of four different stereoisomers of the major component of the sex pheromone of the pink hibiscus mealybug, *Maconellicoccus hirsutus*, maconelliyl 2-methylbutanoate from α-pinene have been reported. Absolute configuration of both the major and the minor component, lavandulyl 2-methylbutanoate was established.[154]

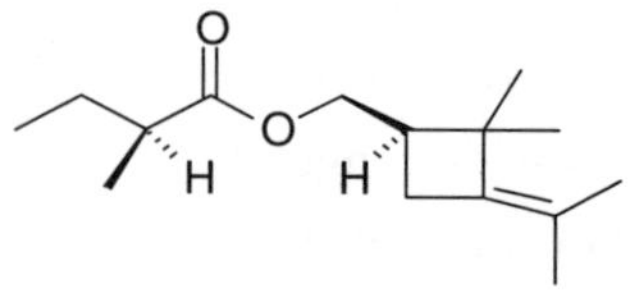
– 1

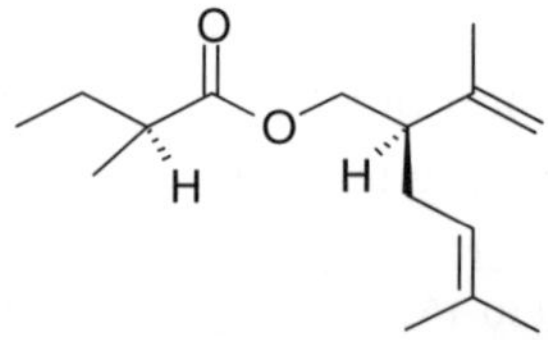
– 2

(1) Maconelliyl 2-methylbutanoate
(2) Lavandulyl 2-methylbutanoate

4-substituted butenolides and the corresponding γ-lactones have been synthesized by hydrolysis of 1-methoxy-8-oxabicyclo-[3.2.1]-oct-6-en-3-one. These butenolides are precursors of insect pheromones.[155]

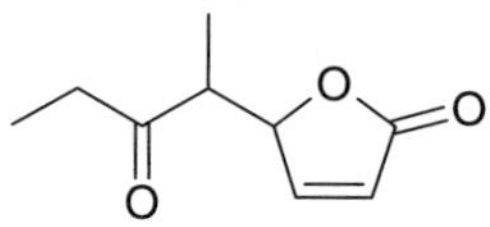
– 1

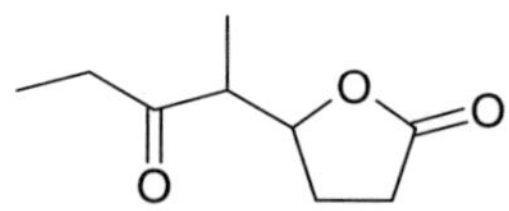
– 2

(1) 4-(-1-methyl-2-oxo-butyl)-2-butenolide
(2) 4-(-1-methyl-2-oxo-butyl)-γ-lactone

Two insect pheromones, (2*S*, 5*R*)-2-methylhexanolide and (*R*)-hexadecanolide were synthesized as examples of a new protocol for asymmetric reduction of long chain aliphatic ketones.[156]

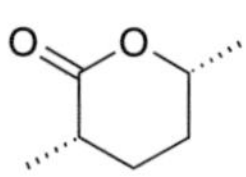
– 1

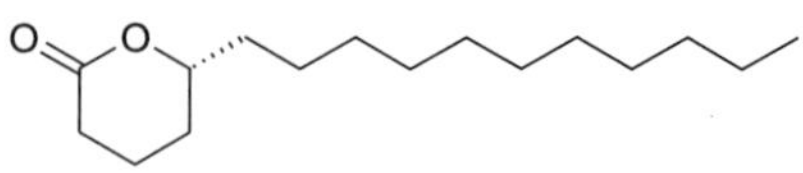
– 2

(1) (2*S*,5*R*)-2-methylhexanolide
(2) (*R*)-hexadecanolide

(*Z*, *E*)-5,7-dodecadienol, a component of the Siberian moth, *Dendrolimus superans sibiricus*, and (*E*, *Z*)-10,12-hexadecadienol, a pheromone component of various Lepidoptera pheromones were synthesized by *cis* reduction of the corresponding enynols with activated Zn.[157]

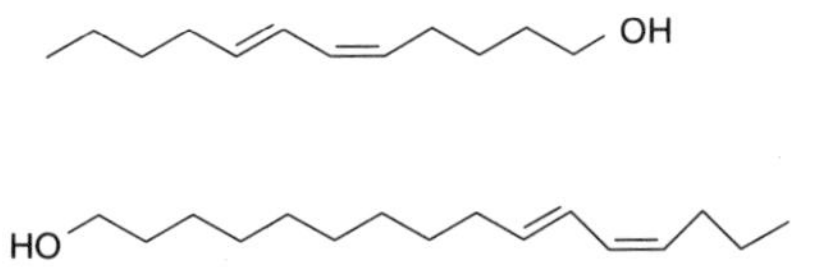

– 1
– 2

(1) (*Z,E*)-5,7-dodecadienol
(2) (*E,Z*)-10,12-hexadecadienol

2*S*-tridecylacetate and 2*S*-tridec-10*E*-enylacetate, the sex pheromones of the fruit fly (*Drosophilia mulleri*) and the Hessian fly (*Mayetiola destructor*) have been synthesized using ethyl-3*S*-hydroxybutanoate, a product from the enzymatic reduction of ethyl acetoacetate by the soil yeast strain "80–1".[158]

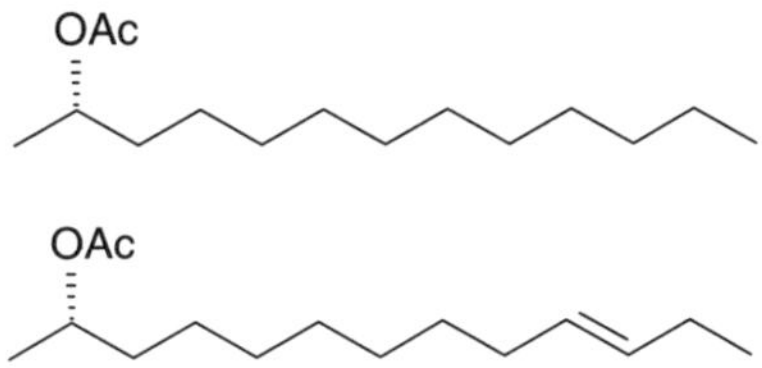

– 1
– 2

(1) 2*S*-tridecylacetate
(2) 2*S*-tridec-10*E*-enylacetate

Trifluoro and monofluoro-analogues of Frontalin were synthesized and evaluated for their biological activity.[159]

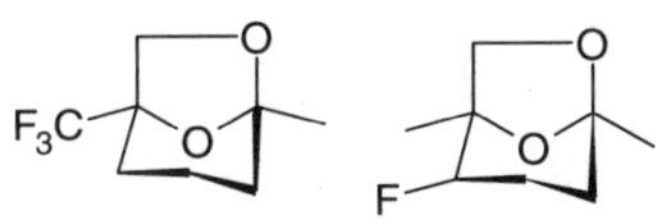

Fluorinated frontalins

A highly stereoselective Claisen rearrangement was used to synthesize 4*E*, 10*Z*-tetradecadien-1-yl acetate, principal component

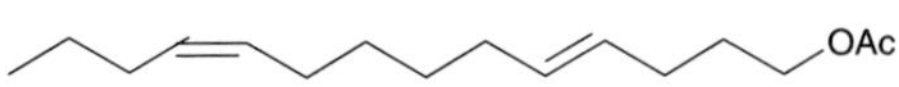

4*E*,10*Z*-tetradecadien-1-yl acetate

of the apple leaf miner (*Lithocolettis ringoniella*) sex pheromone.[160] (incorrect structure in the chapter)

β- and δ-lactones present in the pheromone system of the giant white butterfly, *Idea leuconoe*, were synthesized.[161]

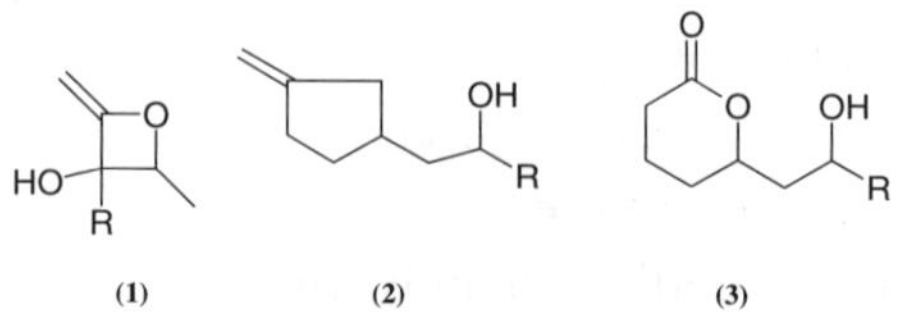

(1) R = $CH(CH_3)_2$ [Viridifloric β-lactone], CH_2CH_3 [2-ethyl-2-hydroxy-3-butanolide]
(2) R = C_4H_9, C_5H_{11}, C_6H_{13}, C_7H_{15} [6-hydroxy-4-alkanolides]
(3) R = C_4H_9, C_5H_{11}, C_6H_{13} [7-hydroxy-4-alkanolides]

Four pheromone components of the female painted apple moth (*Teia anartoides*), (*Z*)-6-henicosen-11-one, (6*Z*, 8*E*)-6,8-henicosadien-11-one, and the enantiomers of (*Z*)-*cis*-9,10-epoxy-6-henicosene and (*Z*)-*cis*-9,10-epoxy-6-icosene, were synthesized.[162]

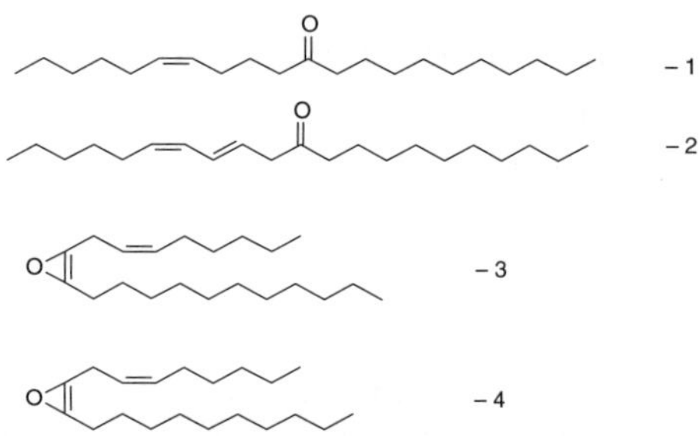

(1) (*Z*)-6-henicosen-11-one
(2) (6*Z*,8*E*)-6,8-henicosadien-11-one
(3) (*Z*)-*cis*-9,10-epoxy-6-henicosene
(4) (*Z*)-*cis*-9,10-epoxy-6-icosene

The application of fluorous mixture synthesis (FMS) for deriving natural products and their stereoisomers was demonstrated by the total synthesis of all 16 stereoisomers of the pine sawfly sex pheromone.[163]

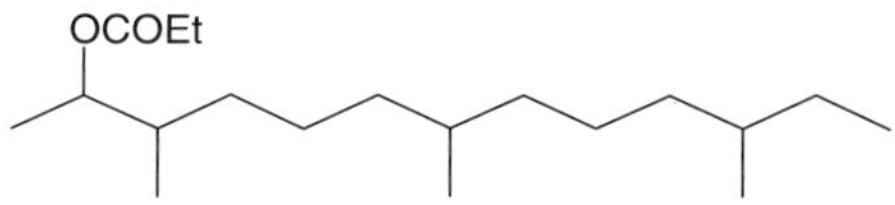

Propionic acid 1,2,6,10-tetramethyldodecyl ester

Two alternative syntheses of (1*R*)-(+)-*cis*-2,2-dimethyl-3-isopropenyl-cyclobutane methanol acetate, the female sex pheromone of the citrus mealybug (*Planococcus citri*) have been reported.[164]

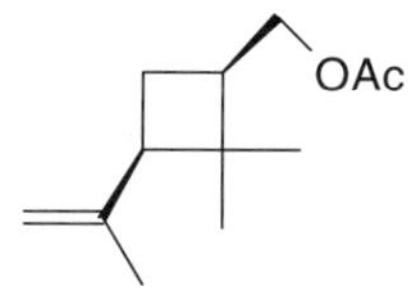

(1*R*)-(+)-*cis*-2,2-dimethyl-3-isopropenyl-cyclobutane methanol acetate

An alternative synthetic method for preparing chiral 1,2-epoxy-3-alkanol tosylates using Sharpless AD reaction as the key step has been reported and was used in the total synthesis of two insect pheromones namely, (6*Z*,9*S*,10*R*)-9,10-epoxy-6-henicosene and (3*Z*,6*Z*,9*S*,10*R*)-9,10-epoxy-3,6-henicosadiene of *Phragmatobia fuliginosa*.[165]

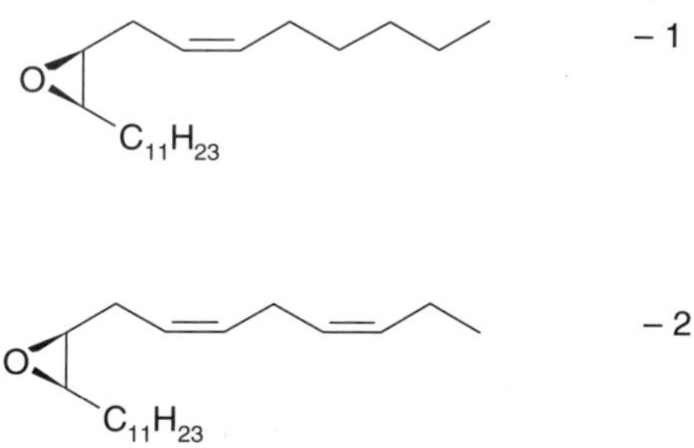

(1) (6*Z*,9*S*,10*R*)-9,10-epoxy-6-henicosene
(2) (3*Z*,6*Z*,9*S*,10*R*)-9,10-epoxy-3,6-henicosadiene

Double [2 + 2] photocycloaddition of ethylene to a bis(α, β-butenolide), which was readily available from D-mannitol, was a key step reported in the enantioselective synthesis of the main component of the sex pheromone of the cotton boll weevil, (+)-grandisol.[166] Asymmetric synthesis of grandisol via kinetic resolution has also been reported.[167]

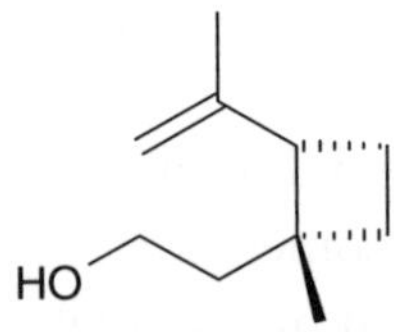

(+)-grandisol

Synthesis of Haptens and the development of an enzyme-linked immunoassay (ELISA) for the olive fruit fly pheromone, *Bactrocera oleae* has been reported.[168]

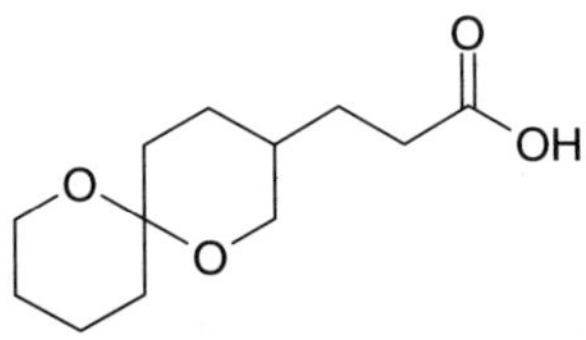

Hapten I

(*R*)- and (*S*)-3-octanol, (*R*)-2-dodecanol, (*R*)-2-methyl-4-heptanol and (*R*)-2-methyl-4-octanol, the pheromones of *Myrmica scabrinodis, Crematogaster castanea, C. liengmei, C. auberti* and *Metamasius hemipterus* were synthesized starting from nonracemic β-hydroxy sulfides.[169]

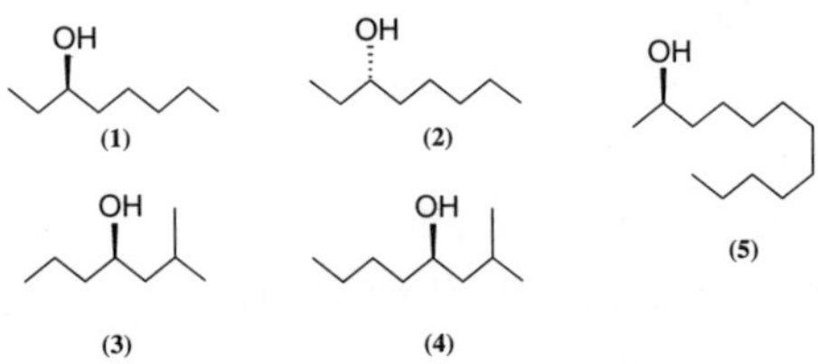

(1) & (2) (*R*)- and (*S*)-3-octanol
(2) (*R*)-2-methyl-4-heptanol
(3) (*R*)-2-methyl-4-octanol
(4) (*R*)-2-dodecanol

Two alternative syntheses of (9*E*, 11*Z*)-hexadeca-9,11-dienal, the sex pheromone of the pecan nut casebearer, *Acrobasis nuxvorella*, have been reported, the key step in each sequence is an ortho-ester Claisen rearrangement.[170]

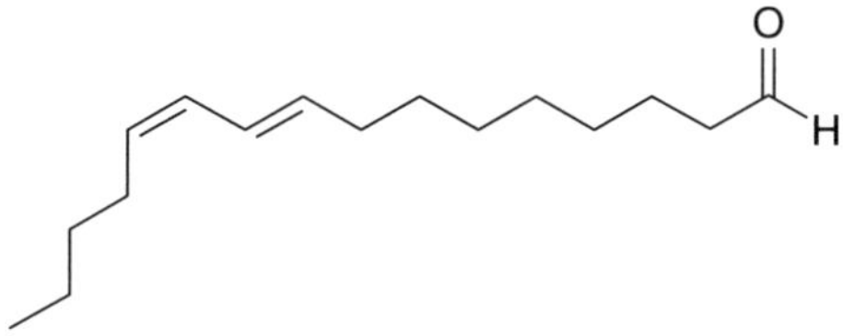

(9*E*,11*Z*)-hexadeca-9,11-dienal

A new approach to synthesize racemic analogues of acylic 1,5-dimethyl-branched insect pheromones based on monoalkylation of ethylacetoacetate with 1-acetoxy-5-bromo-3-methyl-pentane, which is a decyclization product of 4-methyltetrahydropyran has been reported.[171]

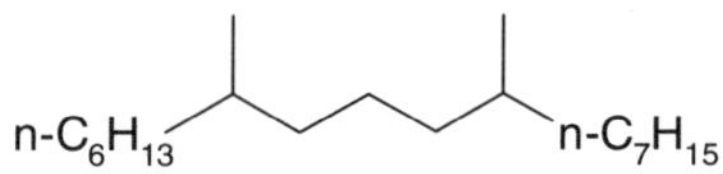

7,11-dimethyloctadecane

1,8-cineole, a natural and abundant monoterpene was converted to suedenone, a sex pheromone of *Dendroctonus pseudotsugae* using a highly efficient and simple route.[172]

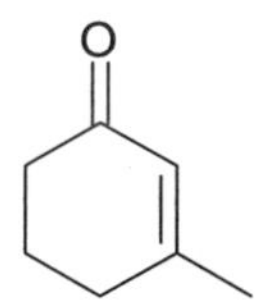

Suedenone
(3-methyl-2-cyclohexenone)

The role of tellurium in the synthesis of (*Z*, *E*)-dienic precursors of insect pheromones is discussed.[173]

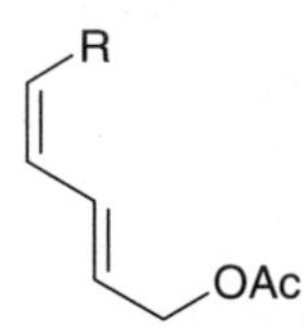

R = Et, *n*-Pr, *n*-Bu

Starting from (±)-citronellol, chiron antipodes were developed via a sequential acetylation protocol using two lipases. The different stereoisomers of the chiron were then functionalized by simple routes to furnish (4*R*,8*R*)-dimethy ldecanal, an insect pheromone commonly known as tribolure, along with a marine phospholipid fatty acid.[174]

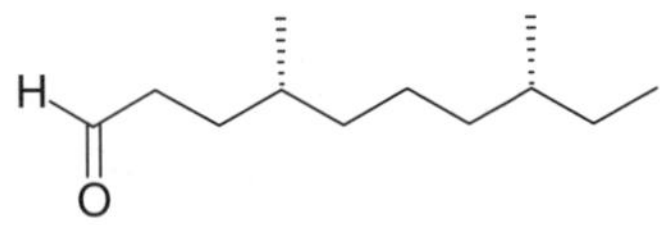

(4*R*,8*R*)-dimethyldecanal (Tribolure)

The sex pheromone of the German cockroach, *Blattella germanica*, has been characterized and its structure confirmed by its synthesis to be gentisyl quinone isovalerate (blattellaquinone).[175,176]

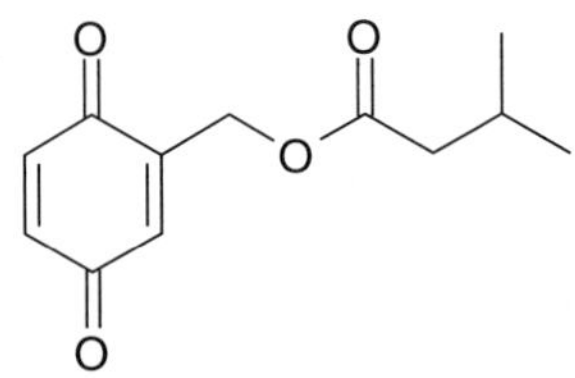

Blattellaquinone

A three-step synthesis of a mixture of stereoisomers of 5,9-dimethylpentadecane, the sex pheromone of the coffee leaf miner, *Leucoptera coffeella*, has been described. The key step being the unsymmetrical Wittig olefination to build the carbon skeleton of the molecule.[177]

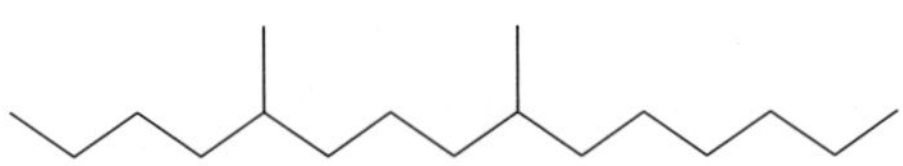

5,9-dimethylpentadecane

The sex pheromone of the pine sawfly (*Diprion pini*) is (2*S*, 3*R*, 7*R*)-3,7-dimethyltridec-2-yl acetate. This acetate was prepared from diastereomerically enriched (2*S**, 3*R**, 7*R*)-3,7-dimethyltridecan-2-ol.[178]

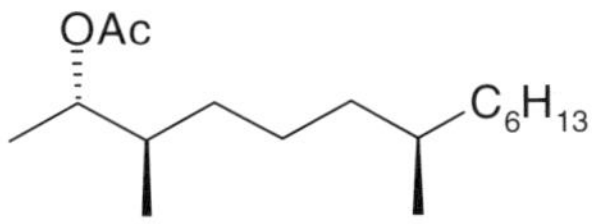

(2*S*,3*R*,7*R*)-3,7-dimethyltridec-2-yl acetate

Zinc promoted reduction of Baylis-Hillman adduct derived allylic bromides gave access to 2-methylalk-2-enoates. This methodology was used in the synthesis of (*E*)-2,4-dimethyl-2-hexenoic acid, a male ant pheromone in the genus *Camponotus*.[179]

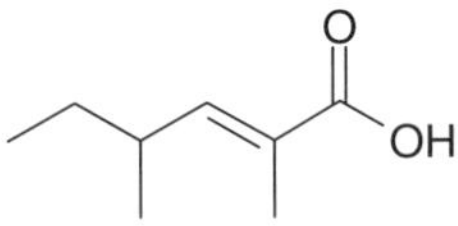

(*E*)-2,4-dimethyl-2-hexenoic acid

Organocatalytic transfer hydrogenation of enals has been discussed as a route for synthesis of some lepidopteran pheromones.[180]

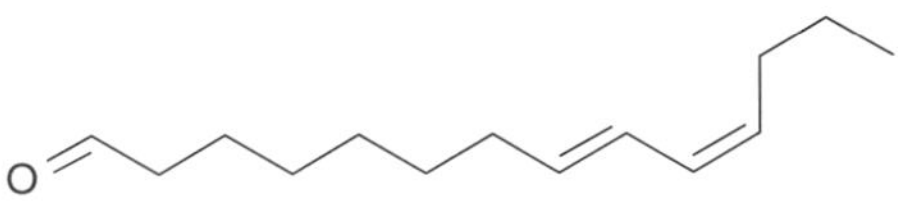

(8*E*,10*Z*)-8,10-tetradecadienal

Japonilure (*R*)-, the sex pheromone of the female Japanese beetle, *Popilla japonica* and its enantiomer, (*S*)-(+)-(5*Z*)-tetradecen-4-olide, which is the pheromone of the Osaka beetle, *Anomala osakana* have been synthesized using a highly convergent procedure and in satisfactory overall yields.[181]

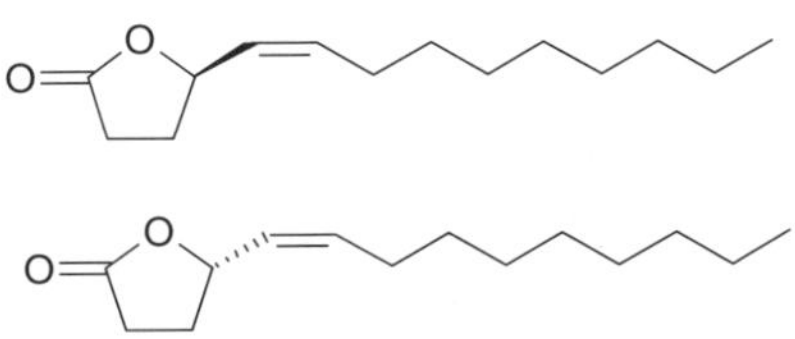

– 1

– 2

(1) (*R*)-japonilure
(2) (*S*)-(+)-(5*Z*)-tetradecen-4-olide

The sex pheromone of the female cigarette beetle (*Lasioderma serricorne*), commonly known as serricornin has been synthesized in seven steps from readily available racemic 1,4-diaoxa-8-thiaspiro-[4.5]decane-6-carboxyaldehyde.[182–185]

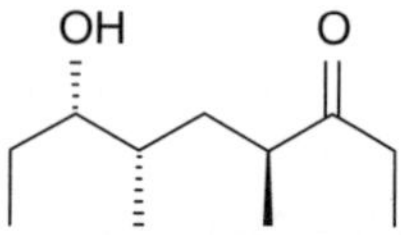

Serricornin- [(4*S*,6*S*,7*S*]-7-hydroxy-4,6-dimethylnonan-3-one

(4*R*, 8*R*)-4,8-dimethyldecanal, an aggregation pheromone of Tribolium flour beetles was produced using radical mediated stereoselective synthesis.[186]

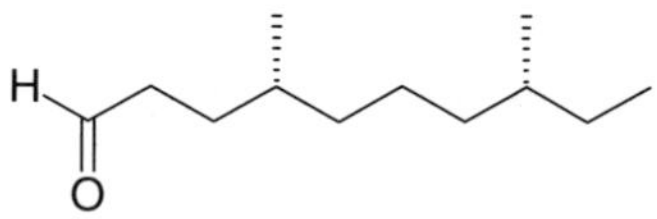

(4*R*,8*R*)-4,8-dimethyldecanal

Sequential treatment of ω-bromoalkyl triflates with an alkyllithium at 0°C followed by addition of a second alkynyllithium and NaI and heating the reaction mixture gave unsymmetrical diynes in one-pot in good yields. This methodology was used to stereoselectively transform these diynes to diene pheromones such as (*Z*, *Z*)- and (*E*, *Z*)-3,13-octadecadienyl acetate.[187]

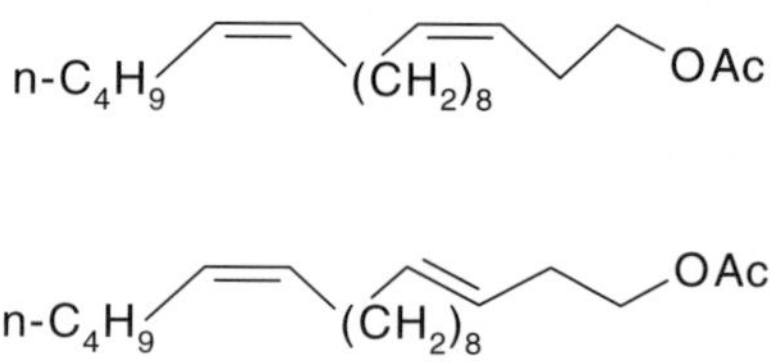

(*Z*,*Z*)- and (*E*,*Z*)-3,13-octadecadienyl acetate

(2*S*, 3*S*, 7*S*)-3,7-dimethylpentadecan-2-yl acetate and its propionate analogue, the sex pheromones of the pine sawflies were synthesized by a concise asymmetric pathway.[188] (2*R*, 3*R*, 7*S*)-3,7-dimethylpentadecan-2-ol, sex pheromone component of the pine sawfly *Neodiprion sertifer*, has also been synthesized enantioselectively.[189]

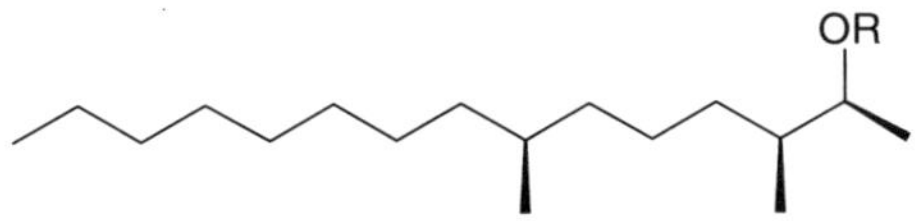

R = COMe [(2*S*,3*S*,7*S*)-3,7-dimethylpentadecan-2-yl acetate], COEt [(2*S*,3*S*,7*S*)-3,7-dimethylpentadecan- 2-yl propionate]

Sex pheromone of the citrus mealybug (*Pseudococcus cryptus*), [(1*R*, 3*R*)-3-isopropenyl-2,2-dimethylcyclobutyl]methyl 3-methyl-3-butenoate, was synthesized from (+)-α-pinene in five steps.[190]

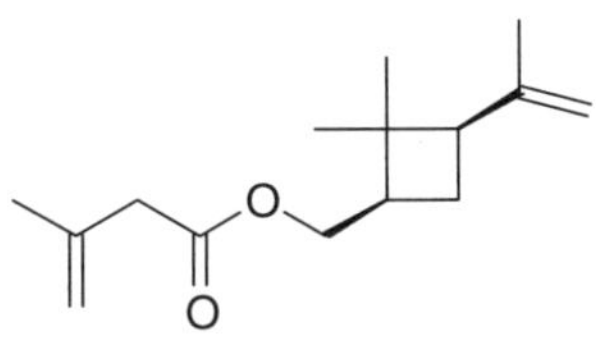

[(1*R*,3*R*)-3-isopropenyl-2,2-dimethylcyclobutyl]methyl 3-methyl-3-butenoate

Efficient asymmetric total syntheses of naturally occurring insect pheromones, (*R*)-4-dodecanolide and (*S*)-5-hexadecanolide have been reported.[191]

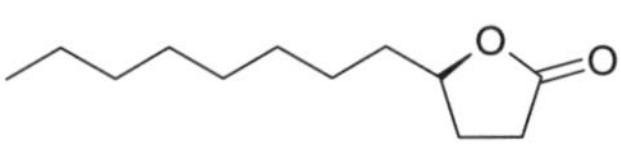

– 1

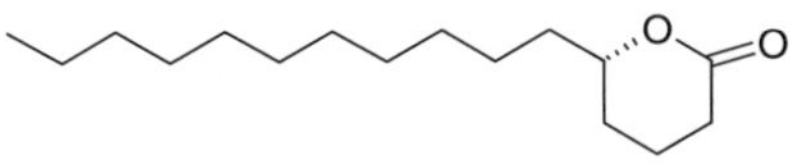

– 2

(1) (*R*)-4-dodecanolide
(2) (*S*)-5-hexadecanolide

Synthesis of (−)-(5*R*, 6*S*)-6-acetoxyhexadecanolide, a mosquito oviposition pheromone of *Culex pipiens fatigans*, was also prepared by asymmetric synthesis.[192]

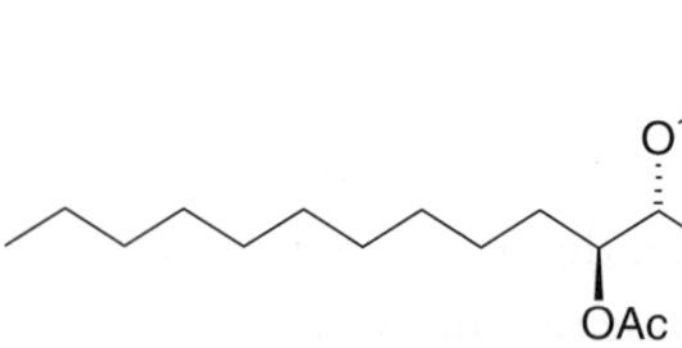

(−)-(5*R*,6*S*)-6-Acetoxyhexadecanolide

Synthesis of (*R*)-ar-himachalene, a pheromone component of the flea beetle, has been reported.[193]

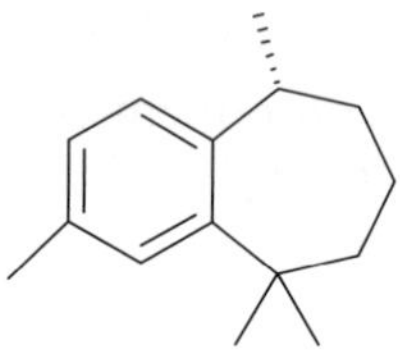

(*R*)-ar-himachalene

(1*S*, 3*S*, 7*R*)-3-methyl-α-himachalene, the sex pheromone of the Brazilian (from Jacobina, Brazil) male sandfly, *Lutzomyia longipalpis*, has been synthesized. The ring junction of this pheromone has the absolute configuration opposite to that of the known (1*S*, 3*S*, 7*R*)-3-methyl-α-himachalene (1*R*, 7*R*)-α-himachalene of plant origin.[194] The synthesis of the pheromone of the sandfly from Lapinha, Brazil has also been reported, the active pheromone compound being (*S*)-9-methylgermacrene-B.[195]

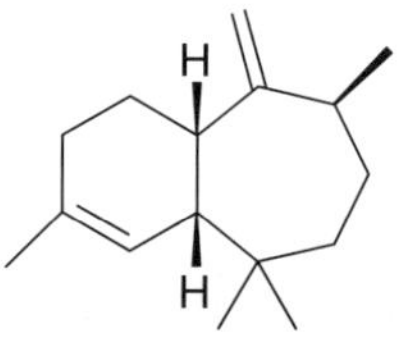

(1*S*,3*S*,7*R*)-3-methyl-α-himachalene

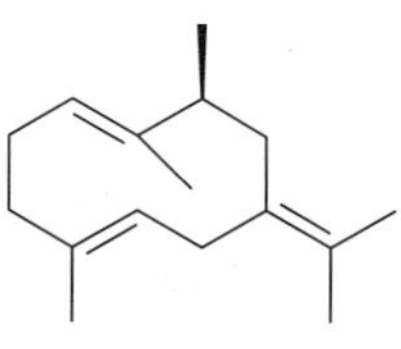

(*S*)-9-methylgermacrene-B

Synthesis of candidate structures for putative sex pheromone of the sandfly *Lutzomyia lichyi* has also been reported.[196]

Male produced pheromone candidates of the flea beetle *Aphthona Flava*, which were four himachalene type sesquiterpenes were synthesized from (*R*)-(+)-citronellal and their stereochemistry discussed. However, there is some ambiguity regarding the stereochemistry.[197]

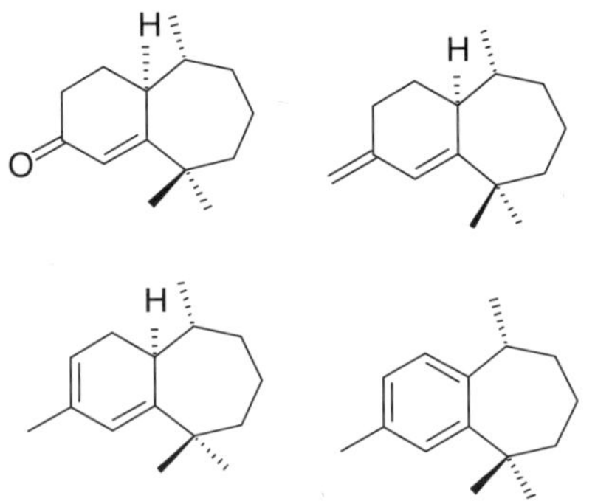

Himachalene type sesquiterpenes

Racemic and diastereomeric mixtures of 3,7,11,15-tetramethylhentriacontane and 4,8,12,16-tetramethyldotriacontane were synthesized as the possible contact sex pheromones of the tsetse fly, *Glossina brevipalpis*.[198] The synthesis of all the stereoisomers of the contact sex pheromones of the tsetse fly, *Glossina austeni*, has also been reported.[199]

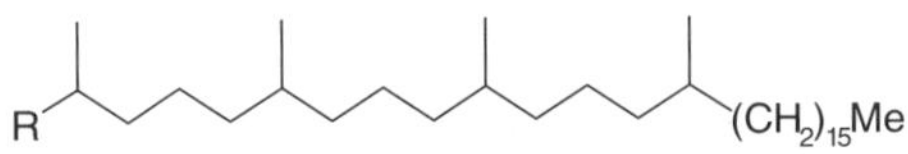

R = Et, 3,7,11,15-tetramethylhentriacontane
R = n-Pr, 4,8,12,16-tetramethyldotriacontane

Possible candidates for the female pheromone of the screwworm fly, *Cochliomyia hominivorax*, 6-Acetoxy-19-methylnonacosane (**1**), 7-acetoxy-19-methylnonacosane (**2**), 8-acetoxy-19-methylnonacosane (**3**), 7-acetoxy-15-methylnonacosane (**4**), and 21-methyl-7-hentriacontanone (**5**) were synthesized as racemic and diastereomeric mixtures.[200–203] The structures (**1–5**) are shown.

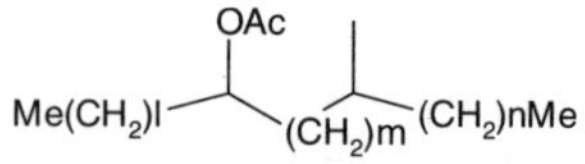

–1 to 4

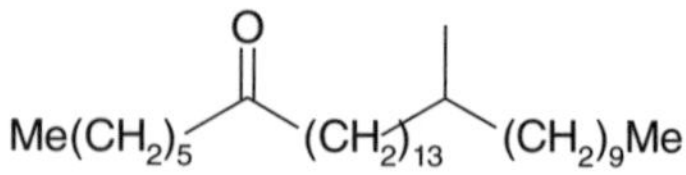

– 5

(1) l = 4, m = 12, n = 9
(2) l = 5, m = 11, n = 9
(3) l = 6, m = 10, n = 9
(4) l = 5, m = 7, n = 13

(4*R*, 6*S*, 7*R*)-7-Hydroxy-4,6-dimethyl-3-nonanone and (3*R*,5*S*,6*R*)-6-hydroxy-3,5-dimethyl-2-octanone, the pheromone components of the bostrychid beetle, *Dinoderus bifoveolatus*, as well as their (4*R*, 6*S*, 7*S*)- and (3*R*, 5*S*, 6*S*)-isomers were synthesized from (2*R*, 4*S*, 5*R*)- and (2*R*, 4*S*, 5*S*)-2,4-dimethyl-5-heptanolide, respectively.[204]

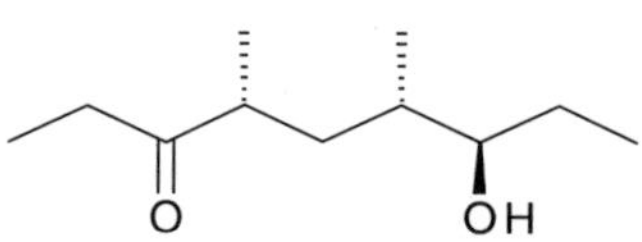

– 1

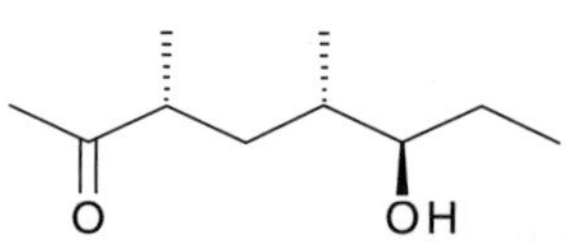

– 2

(1) (4*R*,6*S*,7*R*)-7-hydroxy-4,6-dimethyl-3-nonanone
(2) (3*R*,5*S*,6*R*)-6-hydroxy-3,5-dimethyl-2-octanone

(1*S*, 4*R*, 5*R*)-(+)-α-Acoradiene [1,8-dimethyl-4-(1-methylethenyl)spiro[4.5]dec-7-ene] (I) is shown by its synthesis to be the major component of the male-produced aggregation pheromone of the broad-horned flour beetle (*Gnatocerus cornutus*).[205,206]

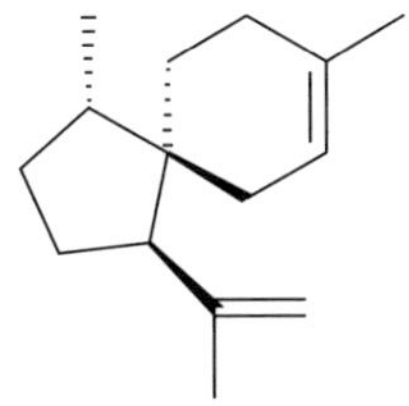

(1*S*,4*R*,5*R*)-(+)-α-acoradiene

Synthesis of the (*S*)-stereoisomers of 3,7-dimethyl-2-heptacosanone and 3,7,15-trimethyl-2-heptacosanone, identified from the locust, *Schistocerca gregaria*, has been reported.[207]

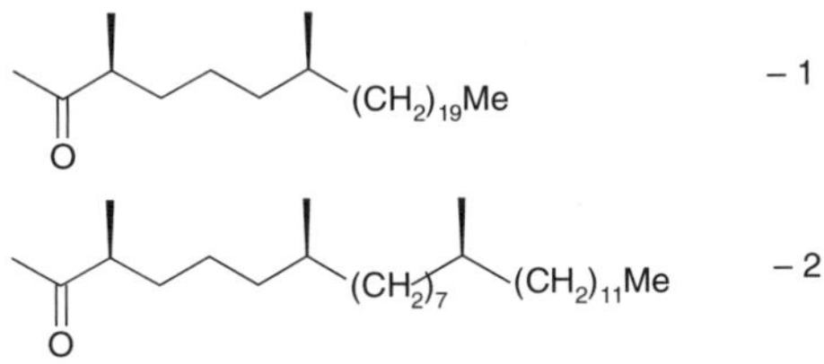

(1) (3*S*,7*S*)-3,7-dimethyl-2-heptacosanone
(2) (3*S*,7*S*,15*S*)-3,7,15-trimethyl-2-heptacosanone

A new synthetic route has been described to prepare matsuone, the pheromone of the pine scale *Matsucoccus matsumurae* and a few other pine scale species. Structure activity is discussed.[208]

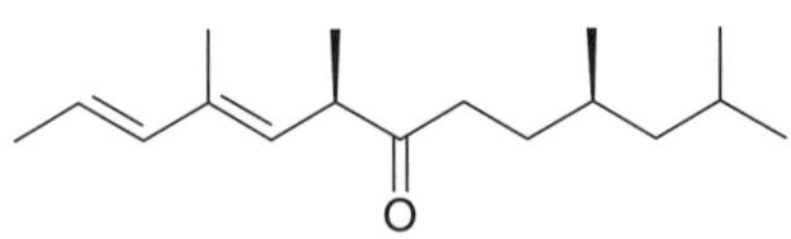

Matsuone

Synthesis of enantiomers of *anti*-2,6-dimethylheptane-1,7-diol monotetrahydropyranyl ether and their conversion to enantiomers of *anti*-10,14-dimethyl-1-octadecene, *anti*-5,9-dimethyloctadecane, and *anti*-5,9-dimethylheptadecane, the sex pheromone components of the apple leafminer, *Lyonetia prunifoliella*, has been reported.[209]

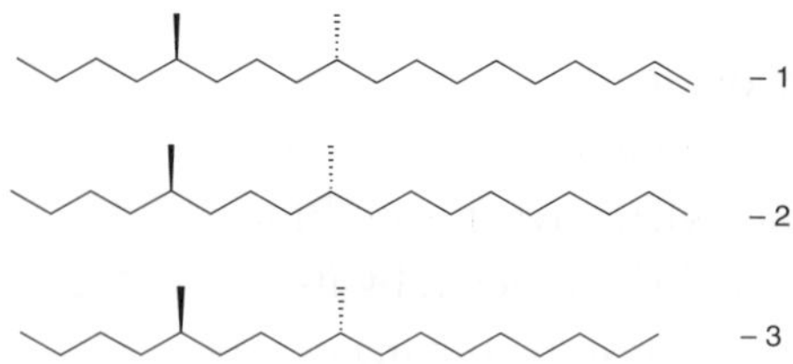

(1) ***anti*-10,14-dimethyl-1-octadecene**
(2) ***anti*-5,9-dimethyloctadecane**
(3) ***anti*-5,9-dimethylheptadecane**

A short route to synthesize vesperal, (*S*)-10-oxoisopiperitenone, the female sex pheromone of the longhorn beetle *Vesperus xatarti* has been reported.[210]

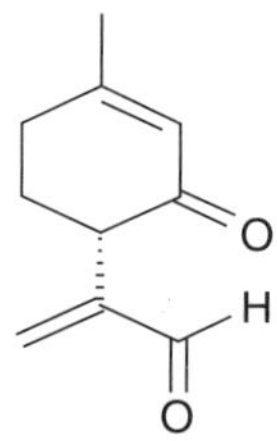

(*S*)-vesperal

Lipase-catalyzed desymmetrization was used to synthesize supellapyrone and its three stereoisomers, the female sex pheromone of the brown-banded cockroach *Supella longipalpa*.[211]

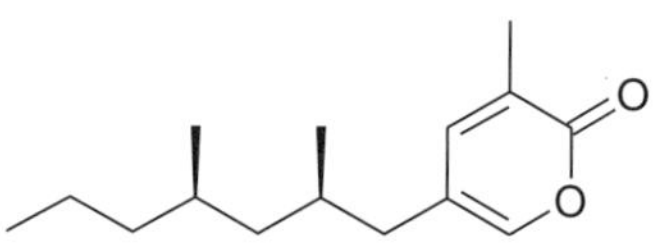

Supellapyrone

Synthesis of posticlure [(6*Z*, 9*Z*, 11*S*, 12*S*)-11,12-epoxyhenicosa-6,9-diene], the female sex pheromone of

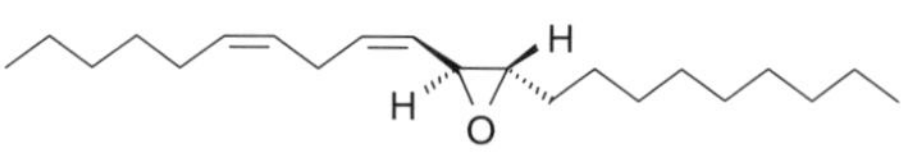

Posticlure

the tussock moth *Orgyia postica* has been reported using two different routes.[212,213]

The male pheromone of a hepialid moth *Endoclita excrescens*, which is (1*R*, 3*S*, 5*S*)-1,3,8-trimethyl-2,9-dioxabicyclo [3.3.1]non-7-ene, and its enantiomer were synthesized.[214,215]

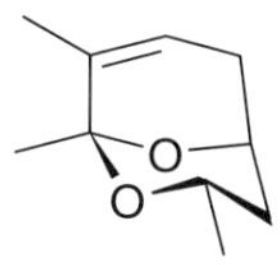

(1*R*,3*S*,5*S*)-1,3,8-trimethyl-2,9-dioxabicyclo[3.3.1]non-7-ene

Lipase PS-C(Amano)-catalyzed asymmetric acetylation was a key step in the synthesis of all four stereoisomers of leucomalure [(3*Z*)-*cis*,6,7-*cis*-9,10-diepoxy-3-henicosene], the female sex pheromone of the satin moth Leucoma salicis.[216]

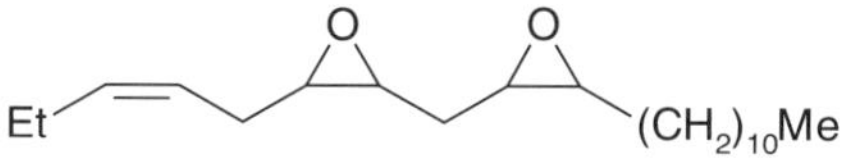

Leucomalure [(3*Z*)-*cis*,6,7-*cis*-9,10-diepoxy-3-henicosene]

The male produced aggregation pheromone of the Colorado potato beetle, *Lepitnotarsa decemlineata*, (*S*)-1,3-dihydroxy-3,7-dimethyl-6-octen-2-one and its (*R*)-isomer were synthesized using lipase-catalyzed asymmetric acetylation of (+)-2, 3-epoxynerol.[217]

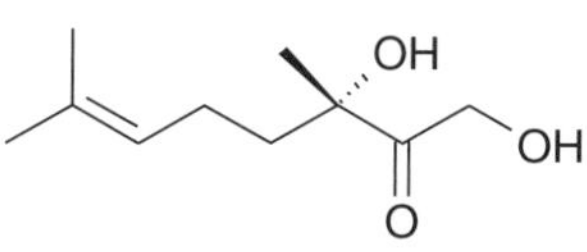

(*S*)-1,3-dihydroxy-3,7-dimethyl-6-octen-2-one

Pheromone components of the female fall webworm *Hyphantria cunea*, which are (3*Z*, 6*Z*, 9*S*, 10*R*)-9,10-epoxy-3,6-henicosadiene and (3*Z*, 6*Z*, 9*S*, 10*R*)-9,10-epoxy-1,3,6-henicosatriene were synthesized.[218]

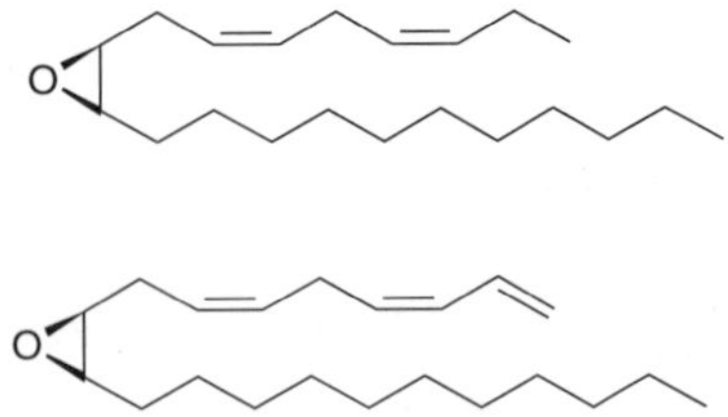

– 1

– 2

(1) (3*Z*,6*Z*,9*S*,10*R*)-9,10-epoxy-3,6-henicosadiene
(2) (3*Z*,6*Z*,9*S*,10*R*)-9,10-epoxy-1,3,6-henicosatriene

Synthetic studies aimed to elucidate the stereostructure of the aggregation pheromone of the male stink bug *Erysarcoris lewisi*, which is 2-methyl-6-(4′-methylenebicyclo [3.1.0] hexyl)hept-2-en-1-ol was reported.[219]

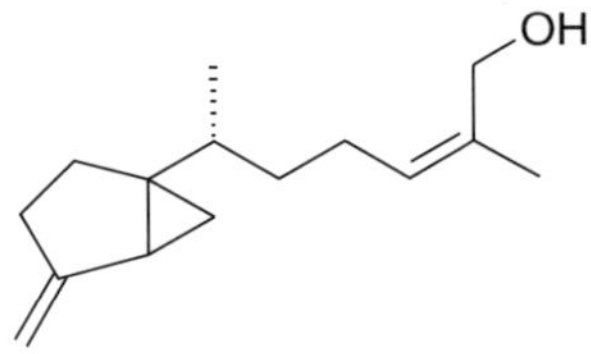

2-methyl-6-(4'-methylenebicyclo[3.1.0]hexyl)hept-2-en-1-ol

Identification of three male-produced sex pheromones of the stink bugs *Chlorochroa ligata* and *C. uhleri*, which are methyl (*R*)-3-(*E*)-6-2, 3-dihydrofarnesoate, methyl (2*E*, 6*E*)-farnesoate, and methyl (*E*)-5-2,6,10-trimethyl-5,9-undecadienoate was confirmed by synthesis.[220]

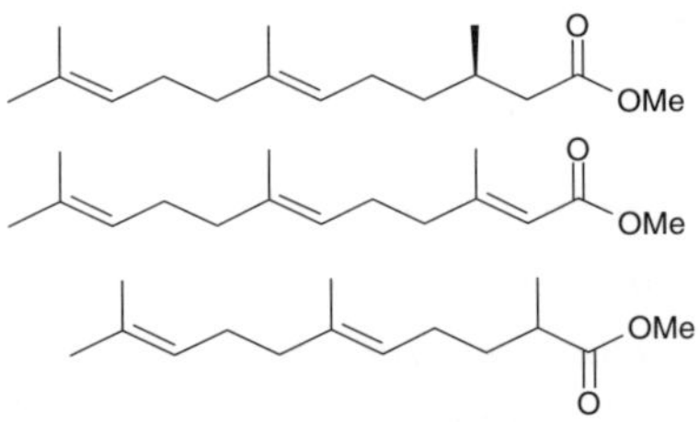

– 1

– 2

– 3

(1) Methyl (*R*)-3-(*E*)-6-2,3-dihydrofarnesoate
(2) Methyl (2*E*,6*E*)-farnesoate
(3) Methyl (*E*)-5-2,6,10-trimethyl-5,9-undecadienoate

Racemic methyl-(*E*)-6-2,3-dihydrofarnesoate was also identified from ***C. sayi*** and confirmed by synthesis.[221]

Two male-produced aggregation pheromone compounds from the eggplant flea beetle *Epitrix fuscula*, were identified by synthesis as (2*E*, 4*E*, 6*Z*)-2,4,6-nontrienal and (2*E*, 4*E*, 6*E*)-2,4,6-nonatrienal.[222]

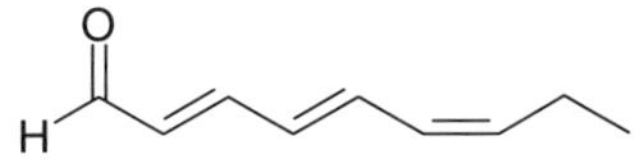

– 1

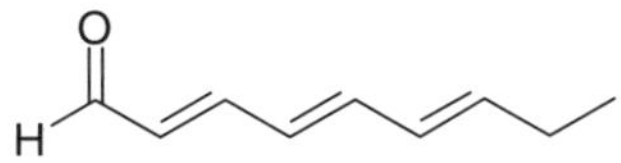

– 2

(1) (2*E*,4*E*,6*Z*)-2,4,6-nontrienal
(2) (2*E*,4*E*,6*E*)-2,4,6-nonatrienal

Two EAD-sensitive compounds,(*S*)-(+)-2,6-dimethyl-5-heptenoic acid and (*E*)-2,6-dimethyl-6-octen-2-ol, of the male dung beetle *Kheper subaeneus*, were synthesized.[223]

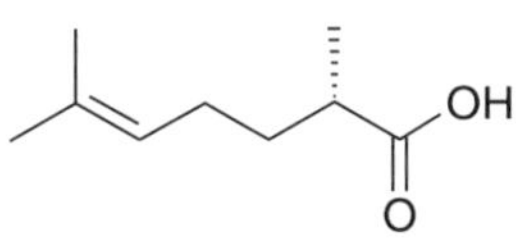

– 1

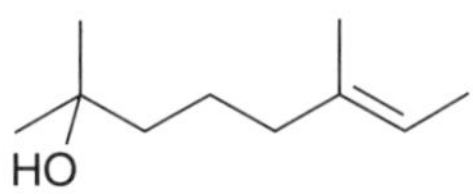

– 2

(1) (*S*)-(+)-2,6-dimethyl-5-heptenoic acid
(2) (*E*)-2,6-dimethyl-6-octen-2-ol

Highly pure (2*R*, 3*R*, 7*R*)- and (2*S*, 3*R*, 7*R*)-stereoisomers of 3,7-di methyl-2-tetradecanol were synthesized. The (2*S*, 3*R*, 7*R*)-isomer was suggested to be the main

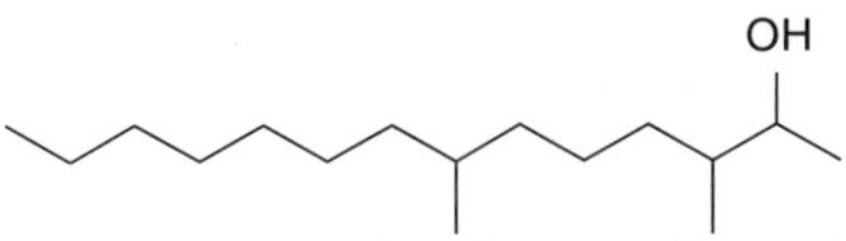

3,7-dimethyl-2-tetradecanol

sex pheromone precursor of the pine sawfly *Gilpinia pallida*.[224]

The natural components of the male aggregation pheromone of the banana weevil *Cosmopolites sordidus*, (1S, 3R, 5R, 7S)-(+)-sordidin and 7-*epi*-(1S, 3R, 5R, 7R)-(−)-sordidin were prepared by asymmetric synthesis.[225]

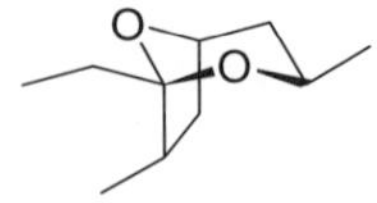
– 1

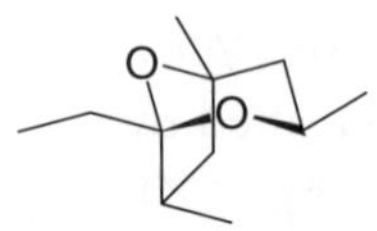
– 2

(1) (1*S*,3*R*,5*R*,7*S*)-(+)-sordidin
(2) 7-epi-(1*S*,3*R*,5*R*,7*R*)-(−)-sordidin

Asymmetric syntheses of (1R, 1′R, 5′R, 7′R) and (1S, 1′R, 5′R, 7′R)-1-hydroxy-exobrevicomin, volatiles of the male mountain pine beetle *Dendroctonus brevicomis* and a synthesis of (+)-exo-brevicomin has been reported, wherein the key steps include the Birch reduction of 2-picoline, selective Wittig olefination and Sharpless asymmetric dihydroxylation.[226] There have been other reports on the synthesis of brevicomin and its derivatives, which are also listed here.[227–231]

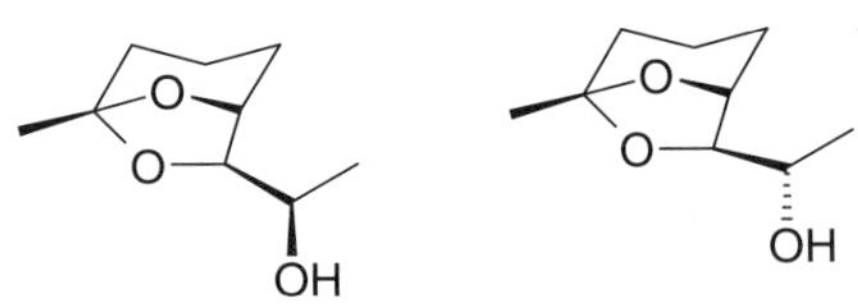

(1) & (2)

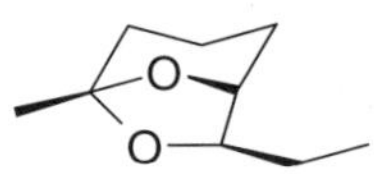

(3)

(1 & 2) 1-hydroxy-exobrevicomins
(3) (+)-exo-brevicomin

Synthesis of the aggregation pheromone of the rice weevil *Sitophilus oryzae* and the maize weevil *S. zeamais*, (+)-sitophilure, was achieved in two steps using NADPH-dependent ketoreductases.[232]

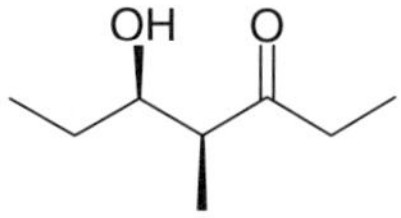

(4*S*,5*R*)-5-hydroxy-4-methyl-3-heptanone [(+)-sitophilure]

The pheromone components of the wasp *Paravespula vulgaris*, (2*R*, 5*S*)- and (2*S*, 5*S*)-2-methyl-1,6-dioxaspiro[4.5]decane, was synthesized diastereoselectively.[233]

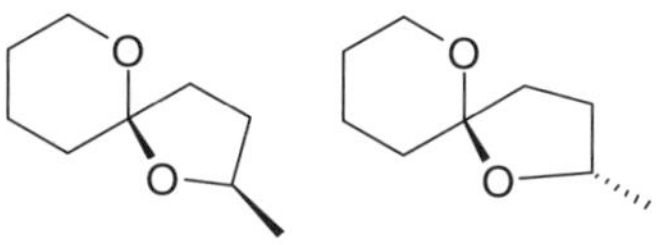

(2*R*,5*S*)- and (2*S*,5*S*)-2-methyl-1,6-dioxaspiro[4.5]decane

A unique semi synthetic pathway in which a genetically modified plant with the ability to produce moth sex pheromones precursors was used to synthesize sex pheromones of the cabbage moth.[234]

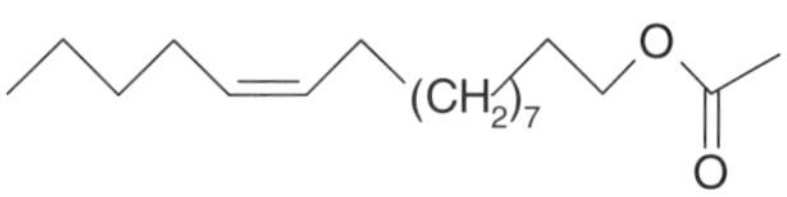

(*Z*11)-hexadec-11-en-1-yl acetate

The use of menthol in the synthesis of important synthons for optically active methyl branched insect pheromones is discussed briefly.[235] Applications of olefin cross metathesis in production of commercial products, including insect pheromones has been discussed.[236]

An improved Horner-Wadsworth-Emmons preparation of α-methyl or α-ethyl-α, β-unsaturated esters from aldehydes, which are useful intermediates in the synthesis of insect pheromones, has been reported.[237]

REFERENCES

1. Butenandt A, Beckman R, Stamm D, Hecher EZ, *Naturforsch*, **14**:283–286, 1959.
2. a) ApSimon John, (ed.) *The Total Synthesis of Natural Products, Volume 9*, John Wiley & Sons. 1992;
   b) Koutek B, Streinz L, Romanuk M, *Collect Czech Chem Commun* **63**:899–954, 1998;
   c) Mori K, *Chirality* **10**:578–586, 1998;
   d) Mori K, *Advances in Asymmetric Synthesis* **1**:211–269, 1995;
   e) Francke W, Schulz S, *Comprehensive Natural Products Chemistry* **8**:197–261, 1999;
   f) Mori Kenji, *Chirality in Natural and Applied Science*, 241–259, 2002;
   g) Jurenka Russell, *Topics in Current Chemistry*, 97–131, 2004;
   h) Mori Kenji, *Topics in Current Chemistry*, 1–50, 2004;
   i) Vogt RG, *Comprehensive Molecular Insect Science* **3**:753–803, 2005;
   j) Baker TC, Heath JJ, *Comprehensive Molecular Insect Science* **6**:407–459, 2005;
   k) Schulz S, *The Chemistry of Pheromone and Other Semiochemicals I. Series: Topics in Current Chemistry, Vol. 239*, Springer, 2004;
   l) Schulz S, *The Chemistry of Pheromone and Other Semiochemicals II. Series: Topics in Current Chemistry, Vol. 240*, Springer, 2004;
   m) Mayer MS, Mclaughlin JR, *Handbook of Insect Pheromones and Sex Attractants*, CRC Press, Boca Raton, 1991;
   n) Morgan DE, Mandava BN, (eds.) *CRC Handbook of Natural Pesticides, Vol. IV, Pheromones, Part A & B*, CRC Press, 1988;
   o) One of the most exhaustive compilations of insect pheromones available online, http://www.pherobase.com/;
   p) Müller-Schwarze D, *Chemical Ecology of Vertebrates*, Cambridge University Press, 2006.
3. Lacey ES, Ginzel MD, Millar JG, Hanks LM, *J Chem Eco* **30**(8):1493–1507, 2004.
4. Yasui H, Wakamura S, Arakaki N, Kishita M, Sadoyama Y, *Chemoecology* **13**:75–80, 2003.
5. Ginzel MD, Millar JG, Hanks LM, *Chemoecology* **13**:135–141, 2003.

6. Toth M, Furlan L, Yatsynin VG, Ujvary I, Szarukan I, Imrei Z, Subchev M, Tollasch T, Francke W, *J Chem Eco* **28**(8):1641–1652, 2002.
7. Hall DR, Cork A, Phythian SJ, Chittamuru S, Jayarama BK, Venkatesha MG, Sreedharan K, Vinod Kumar PK, Seetharama HG, Naidu R, *J Chem Eco* **32**(1):195–219, 2006.
8. Robbins PS, Zhang A, Averill AL, Linn Jr CE, Roelofs WL, Sylvia MM, Villani MG, *J Chem Eco* **32**:1663–1672, 2006.
9. Rochat D, Mohammadpoor K, Malosse C, Avand-Faghih A, Lettere M, Beauhire J, Morin JP, Pezier A, Renou M, Abdollahi GA, *J Chem Eco*, **30**(2):387–407, 2004.
10. Cosse AA, Bartelt RJ, Zilkowski BW, Bean DW, Petroski RJ, *J Chem Eco* **31**(3):657–670, 2005.
11. Bartelt RJ, Cosse AA, Zilkowski BW, Weisleder D, Grode SH, Wiedenmann RN, Post SL, *J Chem Eco* **32**(3):693–712, 2006.
12. Ginzel MD, Moreira JA, Ray AM, Millar JG, Hanks LM, *J Chem Eco* **32**(2):435–451, 2006.
13. Brown AE, Riddick EW, Aldrich JR, Holmes WE, *J Chem Eco*, **32**:2489–2499, 2006.
14. Cosse AA, Bartelt, RJ, *J Chem Eco* **26**(7):1735–1748, 2000.
15. Zhang A, Robbins PS, Averill AL, Weber DC, Linn Jr CE, Roelofs WL, Villani MG, *J Chem Eco* **29**(7):1635–1642, 2003.
16. Toth M, Subchev M, Sredkov I, Szarukan I, Leal W, *J Chem Eco* **29**(7):1643–1649, 2003.
17. Nojima S, Robbins PS, Salsbury GA, Morris BD, Roelofs WL, Villani MG, *J Chem Eco* **29**(11):2439–2446, 2003.
18. Leal WS, Oehlschlager AC, Zarbin PHG, Hidalgo E, Shannon PJ, Murata Y, Gonzalez L, Andrade R, Ono M, *J Chem Eco* **29**(1):15–25, 2003.
19. Ward A, Moore C, Anitha V, Wightman J, John Rogers D, *J Chem Eco* **28**(3):515–522, 2002.
20. Innocenzi PJ, Hall DR, Cross JV, *J Chem Eco* **27**(6):1203–1218, 2001.
21. Zarbin PHG, De Beni Arrigoni E, Reckziegel A, Moreira JA, Baraldi PT, Vieira PC, *J Chem Eco* **29**(2):377–386, 2003.
22. Giblin-Davis RM, Gries R, Crespi B, Robertson LN, Hara AH, Gries G, O'Brien CW, Pierce Jr HD, *J Chem Eco* **26**(12):2763–2780, 2000.
23. Tolasch T, Solter S, Toth M, Ruther J, Francke W, *J Chem Eco* **29**(4):1045–1050, 2003.
24. Arnaud L, Lognay G, Verscheure M, Leenaers L, Gaspar C, Haubruge E, *J Chem Eco* **28**(3):523–532, 2002.
25. Bryning GP, Chambers J, Wakefield ME, *J Chem Eco* **31**(11):2721–2730, 2005.

26. Farine JP, Sirugue D, Abed-Vieillard D, Everaerts C, Le Quere JL, Bonnard O, Brossut R, *J Chem Eco* **33**:405–415, 2007.
27. Hurd LE, Prete FR, Jones TH, Singh TB, Co JE, Portman RT, *J Chem Eco* **30**(1):155–166, 2004.
28. Gries R, Khaskin G, Gries G, Bennett RG, Skip King GG, Morewood P, Slessor KN, Dean Morewood W, *J Chem Eco* **28**(11):2283–2297, 2002.
29. Kah-Wei Hee A, Tan KH, *J Chem Eco* **30**(11):2127–2138, 2004.
30. Choi M, Khaskin G, Gries R, Gries G, Roitberg BD, Raworth DA, Kim D, Bennett RG, *J Chem Eco* **30**(3):659–670, 2004.
31. Hillbur Y, Bengtsson M, Lofqvist J, Biddle A, Pillon O, Plass E, Francke W, Hallberg E, *J Chem Eco* **27**(7):1391–1407, 2001.
32. Gries R, Khaskin G, Bennett RG, Miroshnychenko A, Burden K, Gries G, *J Chem Eco* **31**(12):2933–2946, 2005.
33. Hillbur Y, Celander M, Baur R, Rauscher S, Haftmann J, Franke S, Francke W, *J Chem Eco* **31**(8):1807–1828, 2005.
34. McBrien HL, Millar JG, Gottlieb L, Chen X, Rice RE, *J Chem Eco* **27**(9):1821–1839, 2001.
35. McBrien HL, Millar JG, Rice RE, McElfresh JS, Cullen E, Zalom FG, *J Chem Eco* **28**(9):1797–1818, 2002.
36. Borges M, Birkett M, Aldrich JR, Oliver JE, Chiba M, Murata Y, Laumann RA, Barrigossi JA, Pickett JA, Moraes MCB, *J Chem Eco* **32**:2749–2761, 2006.
37. Zada A, Dunkelblum E, Assael F, Harel M, Cojocaru M, Mendel Z, *J Chem Eco* **29**(4):977–988, 2003.
38. de Assis Marques F, McElfresh JS, Millar JG, *J Chem Eco* **26**(12):2843–2855, 2000.
39. Moraes MCB, Millar JG, Laumann RA, Sujii ER, Pires CSS, Borges M, *J Chem Eco* **31**(6):1415–1427, 2005.
40. Kakizaki M, Sugie H, *J Chem Eco* **27**(12):2447–2458, 2001.
41. Arai T, Sugie H, Hiradate S, Kuwahara S, Itagaki N, Nakahata T, *J Chem Eco* **29**(10):2213–2223, 2003.
42. Millar JG, Midland SL, McElfresh JS, Daane KM, *J Chem Eco* **31**(12):2999–3005, 2005.
43. Boo KS, Choi MY, Chung IB, Eastop VF, Pickett JA, Wadhams LJ, Woodcock CM, *J Chem Eco* **26**(3):601–609, 2000.
44. Goldansaz SH, Dewhirst S, Birkett MA, Hooper AM, Smiley DWM, Pickett JA, Wadhams L, McNeil JN, *J Chem Eco* **30**(4):819–834, 2004.
45. Krokos FD, Konstantopoulou MA, Mazomenos BE, *J Chem Eco* **27**(11):2169–2181, 2001.

46. Kohl E, Holldobler B, Bestmann HJ, *Chemoecology* **13**:113–122, 2003.
47. David Morgan E, Brand JM, Mori K, Keegans SJ, *Chemoecology* **14**:119–120, 2004.
48. Cosse AA, Bartelt RJ, James DG, Petroski RJ, *J Chem Eco* **27**(9):1841–1853, 2001.
49. Clarke SR, Dani FR, Jones GR, Morgan ED, Schmidt JO, *J Chem Eco* **27**(7):1437–1447, 2001.
50. Hunt GJ, Wood KV, Guzman-Novoa E, Lee HD, Rothwell AP, Bonham CC, *J Chem Eco* **29**(2):453–463, 2003.
51. Borg-Karlson AK, Tengo J, Valterova I, Unelius CR, Taghizadeh T, Tolasch T, Francke W, *J Chem Eco* **29**(1):1–14, 2003.
52. Schmitt T, Strohm E, Herzner G, Bicchi C, Krammer G, Heckel F, Schreier P, *J Chem Eco* **29**(11):2469–2479, 2003.
53. Blatrix R, Schulz C, Jaisson P, Francke W, Hefetz A, *J Chem Eco* **28**(12):2557–2567, 2002.
54. Krieger GM, Duchateau, M-J, Van Doorn A, Ibarra F, Francke W, Ayasse M. *J Chem Eco* **32**(2):453–471, 2006.
55. Greenberg L, Aliabadi A, McElfresh JS, Topoff H, Millar JG, *J Chem Eco* **30**(6):1297–1303, 2004.
56. Jarau S, Schulz CM, Hrncir M, Francke W, Zucchi R, Barth FG, Ayasse M, *J Chem Eco* **32**:1555–1564, 2006.
57. Consoli FL, Williams HJ, Vinson SB, Matthews RW, Cooperband MF, *J Chem Eco* **28**(8):1675–1689, 2002.
58. Cosse AA, Bartelt RJ, Weaver DK, Zilkowski BW, *J Chem Eco* **28**(2):407–423, 2002.
59. Sillam-Dusses D, Semon E, Moreay C, Valterova I, Sobotnik J, Robert A, Bordereau C, *Chemoecology* **15**:1–6, 2005.
60. Reinhard J, Lacey MJ, Ibarra F, Schroeder FC, Kaib M, Lenz M, *J Chem Eco*, **28**(1):1–14, 2002.
61. Gibb AR, Suckling DM, Morris BD, Dawson TE, Bunn B, Comeskey D, Dymock JJ, *J Chem Eco* **32**(1):221–237, 2006.
62. Subchev M, Toshova T, Stanimirova L, Stan GH, Embacher G, Francke W, Reckziegel A, Ferreira JT, Priesner E, *J Chem Eco* **26**(2):487–495, 2000.
63. Zhang A, Polavarapu S, *J Chem Eco* **30**(8):1531–1545, 2004.
64. Cork A, Alam SN, Das A, Das CS, Ghosh GC, Farman DI, Hall DR, Maslen NR, Vedham K, Phythian SJ, Rouf FMA, Srinivasan K, *J Chem Eco* **27**(9):1867–1877, 2001.
65. Zhu J, Haynes KF, *J Chem Eco* **30**(10):2047–2056, 2004.
66. McElfresh JS, Hammond AM, Millar JG, *J Chem Eco* **27**(7):1409–1422, 2001.

67. Hung CC, Hwang JS, Hung MD, Yen YP, Hou RF, *J Chem Eco* 27(9):1855–1866, 2001.
68. Chen X, Nakamuta K, Nakanishi T, Nakashima T, Tokoro M, Mochizuki F, Fukumoto T, *J Chem Eco* **32**(3):669–679, 2006.
69. Evenden ML, Gries R, *Chemoecology* **16**:115–122, 2006.
70. Sasaerila Y, Gries R, Gries G, Khaskin G, King S, Takacs S, Hardi, *Chemoecology* **13**:83–89, 2003.
71. Szocs G, Toth M, Karpati Z, Zhu J, Lofstedt C, Plass E, Francke W, *Chemoecology* **14**:53–58, 2004.
72. Gries R, Reckziegel A, Bogenschutz H, Kontzog HG, Schlegel C, Francke W, Millar JG, Gries G, *Chemoecology* **14**:95–100, 2004.
73. Takanashi T, Ohno S, Huang Y, Tatsuki S, Honda H, Ishikawa Y, *Chemoecology* **10**:143–147, 2000.
74. Szocs G, Toth M, Mori K, *Chemoecology* **15**:127–128, 2005.
75. Takacs S, Gries G, Gries R, *Chemoecology* **11**:153–159, 2001.
76. Gibb AR, Jamieson LE, Suckling DM, Ramankutty P, Stevens PS, *J Chem Eco* **31**(7):1633–1644, 2005.
77. Wakamura S, Arakaki N, Yamazawa H, Nakajima N, Yamamoto M, Ando T, *J Chem Eco* **28**(3):449–467, 2002.
78. Jumean Z, Gries R, Unruh T, Rowland E, Gries G, *J Chem Eco* **31**(4):911–924, 2005.
79. Gibb AR, Comeskey D, Berndt L, Brockerhoff EG, El-Sayed AM, Jactel H, Suckling DM, *J Chem Eco* **32**:865–879, 2006.
80. Zhang A, Polavarapu S, *J Chem Eco* **29**(9):2153–2164, 2003.
81. Lofstedt C, Zhu J, Kozlov MV, Buda V, Jirle EV, Hellqvist S, Lofqvist J, Plass E, Franke S, Francke W, *J Chem Eco* **30**(3):643–658, 2004.
82. Zhang A, Leskey TC, Christopher Berg J, Walgenbach JF, *J Chem Eco* **31**(10):2463–2479, 2005.
83. Francke W, Plass E, Zimmermann N, Tietgen H, Tolasch T, Franke S, Subchev M, Toshova T, Pickett JA, Wadhams LJ, Woodcock CM, *J Chem Eco* **26**(5):1135–1149, 2000.
84. Gries R, Khaskin G, Khaskin E, Foltz JL, Schaefer PW, Gries G, *J Chem Eco* **29**(10):2201–2212, 2003.
85. Inomata SI, Watanabe A, Nomura M, Ando T, *J Chem Eco* **31**(6):1429–1442, 2005.
86. Millar JG, Grant GG, McElfresh JS, Strong W, Rudolph C, Stein JD, Moreira JA, *J Chem Eco* **31**(5):1229–1234, 2005.
87. Yasui H, Wakamura S, Arakaki N, Irei H, Kiyuna C, Ono H, Yamazawa H, Ando T, *J Chem Eco* **31**(3):647–656, 2005.

88. Yamamoto M, Kiso M, Yamazawa H, Takeuchi J, Ando T, *J Chem Eco* **26**(11):2579–2590, 2000.
89. Witzgall P, Tasin M, Buser HR, Wegner-Kib G, Marco Mancebon VS, Ioriatti C, Backman AC, Bengtsson M, Lehmann L, Francke W, *J Chem Eco* **31**(12):2923–2932, 2005.
90. Gries R, Khaskin G, Schaefer PW, Hahn R, Gotoh T, Gries G, *J Chem Eco* **31**(1):49–62, 2005.
91. Francke W, Karalius V, Plass E, Lehmann L, Dos Santos A, Buda V, Borg-Karlson AK, Mozuraitis R, *J Chem Eco* **30**(4):805–817, 2004.
92. Kalinova B, Svatos A, Kindl J, Hovorka O, Hrdy I, Kuldova J, Hoskovec M, *J Chem Eco* **29**(2):387–404, 2003.
93. Gries G, Schaefer PW, Gries R, Fan YB, Higashiura Y, Tanaka B, *J Chem Eco* **28**(3):469–478, 2002.
94. Mozuraitis R, Buda V, Liblikas I, Unelius CR, Borg-Karlson AK, *J Chem Eco* **28**(6):1191–1208, 2002.
95. Subchev M, Mircheva A, Pickett J, Wadhams L, Woodcock C, Dos Santos A, Franke S, Francke W, *J Chem Eco* **29**(10):2391–2396, 2003.
96. Downham MCA, Hall DR, Chamberlain DJ, Cork A, Farman DI, Tamo M, Dahounto D, Datinon B, Adetonah S, *J Chem Eco* **29**(4):989–1011, 2003.
97. Gries R, Khaskin G, Gotoh T, Schaefer PW, Gries G, *J Chem Eco* **31**(4):879–891, 2005.
98. Sasaerila Y, Gries R, Gries G, Khaskin G, Skip King, Boo TC, *J Chem Eco* **26**(8):1969–1981, 2000.
99. Gries R, Khaskin G, Clearwater J, Hasman D, Schaefer PW, Khaskin E, Miroshnychenko O, Hosking G, Gries G, *J Chem Eco* **31**(3):603–620, 2005.
100. El Sayed AM, Gibb AR, Suckling DM, Bunn B, Fielder S, Comeskey D, Manning LA, Foster SP, Morris BD, Ando T, Mori K, *J Chem Eco* **31**(3):621–646, 2005.
101. Gries R, Khaskin G, Tan ZX, Zhao BG, Skip King GG, Miroshnychenko A, Lin GQ, Rhainds M, Gries G, *J Chem Eco* **32**:1673–1685, 2006.
102. Naka H, Van Vang L, Inomata SI, Ando T, Kimura T, Honda H, Tsuchida K, Sakurai H, *J Chem Eco* **29**(11):2447–2459, 2003.
103. Gries R, Khaskin G, Daroogheh H, Mart C, *et al. J Chem Eco* **33**:447–448, 2007. (Erratum). Original paper: Gries R, Khaskin G, Daroogheh H, Mart C, *et al. J Chem Eco* **32**:2667–2677, 2006.
104. Inomata SI, Komoda M, Watanabe H, Nomura M, Ando T, *J Chem Eco* **26**(2):443–454, 2000.
105. Foster SP, Thomas WP, *J Chem Eco* **26**(11):2549–2555, 2000.
106. Fang YL, Sun JH, Zhao CH, Zhang ZN, *J Chem Eco* **31**(1):39–48, 2005.

107. Nunez S, De Vlieger JJ, Rodriquez JJ, Persoons CJ, Sctoni I, *J Chem Eco* **28**(2):425–432, 2002.
108. McElfresh JS, Millar JG, Rubinoff D, *J Chem Eco* **27**(4):791–806, 2001.
109. Cabrera A, Eiras AE, Gries G, Gries R, Urdaneta N, Miras B, Badji C, Jaffe K, *J Chem Eco* **27**(10):2097–2107, 2001.
110. Grant GG, Slessor KN, Liu W, Abou-Zaid MM, *J Chem Eco* **29**(3):589–601, 2003.
111. Sasaerila Y, Gries R, Gries G, Khaskin G, Hardi, *J Chem Eco* **26**(8):1983–1990, 2000.
112. Van Vang L, Inomata SI, Kinjo M, Komai F, Ando T, *J Chem Eco* **31**(4):859–878, 2005.
113. Zhang QH, Schneidmiller RG, Hoover DR, Young K, Welshons DO, Margaryan A, Aldrich JR, Chauhan KR, *J Chem Eco* **32**:2163–2176, 2006.
114. Hooper AM, Donato B, Woodcock CM, Park JH, Paul RL, Boo KS, Hardie J, Pickett JA, *J Chem Eco* **28**(4):849–864, 2002.
115. Tschuch G, Lindemann P, Niesen A, Csuk R, Moritz G, *J Chem Eco* **31**(7):1555–1565, 2005.
116. Hamilton JGC, Hall DR, Kirk WDJ, *J Chem Eco* **31**(6):1369–1379, 2005.
117. Schulz S, Fuhlendorff J, Steidle JLM, Collatz J, Franz JT, *Chem Bio Chem* **5**:1500–1507, 2004.
118. Mizoguchi A, Mori N, Nishida R, Kuwahara Y, *J Chem Eco* **29**(7):1681–1690, 2003.
119. Tichy H, Gingl E, Ehn R, Papke M, Schulz S, *J Comp Physiol A* **187**:75–78, 2001.
120. Gemeno C, Yeargan KV, Haynes KF, *J Chem Eco* **26**(5):1235–1243, 2000.
121. Hern A, Dorn S, *Naturwissenschaften* **91**:77–80, 2004.
122. Dickschat JS, Wagner-Dobler I, Schulz S, *J Chem Eco* **31**(4):925–947, 2005.
123. Zada A, Ben-Yehuda S, Dunkelblum E, Harel M, Assael F, Mendel Z, *J Chem Eco* **30**(3):631–641, 2004.
124. Leal WS, Bento JMS, Murata Y, Ono M, Parra JRP, Vilela EF, *J Chem Eco*, **27**(10):2041–2051, 2001.
125. James DG, Petroski RJ, Cosse AA, Zilkowski BW, Bartelt RJ, *J Chem Eco* **29**(10):2189–2199, 2003.
126. Shibata C, Mori K, *Eur J Org Chem* **5**:1083–1088, 2004.
127. Mori K, *Eur J Org Chem* **10**:2040–2044, 2005.
128. Leal WS, Parra-Pedrazzoli AL, Cosse AA, Murata Y, Bento JMS, Vilela EF, *J Chem Eco* **32**(1):155–168, 2006.
129. Moreira JA, McElfresh JS, Millar JG, *J Chem Eco* **32**(1):169–194, 2006.

130. Steinbauer MJ, Ostrand F, Bellas TE, Nilsson A, Andersson F, Hedenstrom E, Lacey MJ, Schiest FP, *Chemoecology* **14**:217–223, 2004.
131. Batista-Pereira LG, Stein K, de Paula AF, Moreira JA, Cruz I, de Lourdes C, Figueiredo M, Perri Jr J, Correa AG, *J Chem Eco* **32**:1085–1099, 2006.
132. Moreira JA, Millar JG, *J Chem Eco* **31**(4):965–968, 2005.
133. Wimalaratne PDC, Slessor K, *J Chem Eco* **30**(6):1225–1244, 2004.
134. Yamazawa H, Nakajima N, Wakamura S, Arakaki N, Yamamoto M, Ando T, *J Chem Eco* **27**(11):2153–2167, 2001.
135. Hedenstrom E, Andersson F, *J Chem Eco* **28**(6):1237–1254, 2002.
136. Zarbin PHG, Reckziegel A, Plass E, Borges M, Francke W, *J Chem Eco* **26**(12):2737–2746, 2000.
137. Nakamura Y, Mori K, *Biosci Biotechnol Biochem* **64**(8):1713–1721, 2000.
138. Ishmuratov G Yu, Yakovieva MP, Tolstikov GA, *Chemistry of Natural Compounds* **36**(2):105–119, 2000.
139. Kashiwagi T, Nakashima T, Tebayashi S, Kim C, *Biosci Biotechnol Biochem* **70**(10):2544–2546, 2006.
140. Mori K, *Tetrahedron: Asymmetry* **17**:2133–2142, 2006.
141. Lamers YMAW, Rusu G, Wijnberg JPBA, de Groot A, *Tetrahedron* **59**:9361–9369, 2003.
142. Das B, Chowdhury N, Banerjee J, Majhi A, *Tetrahedron Letters* **47**:6615–6618, 2006.
143. Das B, Banerjee J, Chowdhury N, Majhi A, Mahender G, *Helvetica Chimica Acta* **89**:876–883, 2006.
144. Das B, Chowdhury N, Banerjee J, Majhi A, Mahender G, *Chemistry Letters* **35**(4):358–359, 2006.
145. Jarvis AP, Liebig J, Holldobler B, Oldham NJ, *Chem Commun*, 1196–1197, 2004.
146. Yadav JS, Reddy EJ, Ramalingam T, *New Journal of Chemistry* **25**:223–225, 2001.
147. Ragoussis V, Panopoulou M, Ragoussis N, *J Agric Food Chem* **52**:5047–5051, 2004.
148. Ebert S, Krause N, *Eur J Org Chem*, 3831–3835, 2001.
149. Prasad KR, Chandrakumar A, Anbarasan P, *Tetrahedron: Asymmetry* **17**:1979–1984, 2006.
150. Chow S, Kitching W, *Tetrahedron: Asymmetry* **13**:779–793, 2002.
151. Chow S, Kitching W, *Chem Commun*, 1040–1041, 2001.
152. Satoh T, Hanaki N, Kuramochi Y, Inoue Y, Hosoya K, Sakai K, *Tetrahedron*, **58**:2533–2549, 2002.

153. Bykov VI, Butenko TA, Egupova EV, Finkelshtein E Sh, *Russian Chemical Bulletin Int Ed* **49**(7):1301–1304, 2002.
154. Zhang A, Nie J, *J Agric Food Chem* **53**:2451–2455, 2005.
155. Montana AM, Garcia F, Batalla C, *Tetrahedron Letters*, **45**:8549–8552, 2004.
156. Hsu JL, Fang JM, *J Org Chem* **66**:8573–8584, 2001.
157. Khrimian A, Klun JA, Hijji Y, Baranchikov YN, Pet'ko VM, Mastro VC, Kramer MH, *J Agric Food Chem* **50**:6366–6370, 2002.
158. Kharisov RYa, Petukhova NI, Davletova AR, Ishmuratova NM, Zorin VV, Ishmuratova G Yu, Tolstikov GA, *Chemistry of Natural Products* **36**(2):210–212, 2000.
159. Ambrosi P, Arnone A, Bravo P, Bruche L, De Cristofaro A, Francardi V, Frigerio M, Gatti E, Germinara GS, Panzeri W, Pennacchio F, Pesenti C, Rotundo G, Roversi PF, Salvadori C, Viani F, Zanda M, *J Org Chem* **66**:8336–8343, 2001.
160. Shakhmaev RN, Vakhidov RR, Zorin VV, Odinokov VN, *Chemistry of Natural Products* **37**(3):282–284, 2001.
161. Stritzke K, Schulz S, Nishida R, *Eur J Org Chem*, 3884–3892, 2002.
162. Muto S, Mori K, *Biosci Biotechnol Biochem* **67**(7):1559–1567, 2003.
163. Dandapani S, Jeske M, Curran DP, *J Org Chem* **70**:9447–9462, 2005.
164. Passaro LC, Webster FX, *J Agric Food Chem* **52**:2896–2899, 2004.
165. Zhang ZB, Wang ZM, Wang YX, Liu HQ, Lei GX, Shi M, *J Chem Soc Perkin Trans 1*, 53–57, 2000.
166. De March P, Figueredo M, Font J, Raya J, *Organic Letters* **2**(2):163–165, 2000.
167. Hamon DPG, Tuck KL, *J Org Chem* **65**(23):7839–7846, 2000.
168. Neokosmidi A, Ragoussis V, Zikos C, Paravatou-Petsotas M, Livaniou E, Ragoussis N, Evangelatos G, *J Agric Food Chem* **52**:4368–4374, 2004.
169. Cho BT, Kim DJ, *Tetrahedron* **59**:2457–2462, 2003.
170. Passaro LC, Webster FX, *Synthesis* **8**:1187–1190, 2003.
171. Ishmuratov G Yu, Yakovleva MP, Galyautdinova AV, Muslukhov RR, Tolstikov GA, *Russian Chemical Bulletin Int Ed* **52**(3):740–744, 2003.
172. Silvestre AJD, Valega M, Cavaleiro JAS, *Industrial Crops and Products*, **12**:53–56, 2000.
173. Diego DG, Cunha RLOR, Comasseto JV, *Tetrahedron Letters* **47**:7147–7148, 2006.
174. Sankaranarayanan S, Sharma A, Chattopadhyay S, *Tetrahedron: Asymmetry* **13**:1373–1378, 2002.
175. Nojima S, Schal C, Webster FX, Santangelo RG, Roelofs WL, *Science* **307**(5712):1104–1106, 2005.
176. Dorit E, Mori K, Takikawa H, Leal WS, Schal C, *J Chem Eco* **30**(9):1839–1848, 2004.

177. Zarbin PHG, Princival JL, de Lima ER, dos Santos AA, Ambrogio BG, de Oliveira ARM, *Tetrahedron Letters* **45**:239–241, 2004.
178. Prokhorevich KN, Kulinkovich OG, *Tetrahedron: Asymmetry* **17**:2976–2980, 2006.
179. Fernandes L, Bortoluzzi AJ, Sa MM, Tetrahedron, **60**:9983–9989, 2004.
180. Maria de Figueiredo R, Berner R, Julis J, Liu T, Turp D, Christmann M, *J Org Chem* **72**:640–642, 2007.
181. Dos Santos AA, Francke W, *Tetrahedron: Asymmetry* **17**:2487–2490, 2006.
182. Ward DE, Jheengut V, Beye GE, *J Org Chem* **71**:8989–8992, 2006.
183. Job A, Nagelsdiek R, Enders D, *Collect Czech Chem Commun* **65**(4):524–538, 2000.
184. Fujita K, Mori K, *Biosci Biotechnol Biochem* **65**(6):1429–1433, 2001.
185. Zlokazov MV, Veselovsky VV, *Russian Chemical Bulletin Int Ed* **51**(8):1600–1603, 2002.
186. Kameda Y, Nagano H, *Tetrahedron* **62**:9751–9757, 2006.
187. Armstrong-Chong RJ, Matthews K, Chong JM, *Tetrahedron* **60**:10239–10244, 2004.
188. Huang PQ, Lan HQ, Zheng X, Ruan YP, *J Org Chem* **69**:3964–3967, 2004.
189. Moreira JA, Correa AGJ, *Braz Chem Soc* **11**(6):614–620, 2000.
190. Nakahata T, Itagaki N, Arai T, Sugie H, Kuwahara S, *Biosci Biotechnol Biochem* **67**(12):2627–2631, 2003.
191. Gowravaram S, Venkata Reddy E, Yadagiri K, Yadav JS, *Synthesis* **19**:3270–3274, 2006.
192. Gowravaram S, Swapna R, Venkata Reddy E, Yadav JS, *Synthesis* **24**:4242–4246, 2006.
193. Mori K, *Tetrahedron: Asymmetry* **16**(3):685–692. Erratum Mori K. *Tetrahedron: Asymmetry* **16**(9):1721, 2005.
194. Tashiro T, Bando M, Mori K, *Synthesis* **13**:1852–1862, 2000.
195. Kurosawa S, Mori K, *Eur J Org Chem* **6**:955–962, 2000.
196. Tashiro T, Mori K, *J Ind Chem Soc* 77(11–12):617–619, 2000.
197. Muto S, Bando M, Mori K, *Eur J Org Chem* **9**:1946–1952, 2004.
198. Shibata C, Furukawa A, Mori K, *Biosci Biotechnol Biochem* **66**(3):582–587, 2002.
199. Kimura T, Carlson DA, Mori K, *Eur J Org Chem* **17**:3385–3390, 2001.
200. Furukawa A, Shibata C, Mori K, *Biosci Biotechnol Biochem* **66**(5):1164–1169, 2002.
201. Mori K, Ohtaki T, Ohrui H, Berkebile DR, Carlson DA, *Biosci Biotechnol Biochem* **68**(8):1768–1778, 2004.
202. Mori K, *Biosci Biotechnol Biochem* **67**(10):2224–2231, 2003.

203. Mori K, Ohtaki T, Ohrui H, Berkebile DR, Carlson DA, *Eur J Org Chem* **5**:1089–1096, 2004.
204. Masuda Y, Fujita K, Mori K, *Biosci Biotechnol Biochem* **67**(8):1744–1750, 2003.
205. Tashiro T, Kurosawa S, Mori K, *Biosci Biotechnol Biochem* **68**(3):663–670, 2004.
206. Kurosawa S, Bando M, Mori K, *Eur J Org Chem* **23**:4395–4399, 2001.
207. Nakamura Y, Mori K, *Eur J Org Chem* 7:1307–1312, 2000.
208. Mori K, *et al. Chem Bio Chem* **1**(1):56–66, 2000.
209. Nakamura Y, Mori K, *Eur J Org Chem* **15**:2745–2753, 2000.
210. Domon K, Mori K, *Eur J Org Chem* **22**:3783–3785, 2000.
211. Fujita K, Mori K, *Eur J Org Chem* **3**:493–502, 2001.
212. Wakamura S, Arakaki N, Yamamoto M, Hiradate S, Yasui H, Yasuda T, Ando T, *Tetrahedron Letters* **42**:687–689, 2001.
213. Muto S, Mori K, *Eur J Org Chem* **24**:4635–4638, 2001.
214. Marukawa K, Mori K, *Chemistry Letters* **1**:40–41, 2002.
215. Marukawa K, Mori K, *Eur J Org Chem* **23**:3974–3978, 2002.
216. Muto S, Mori K, *Eur J Org Chem* 7:1300–1307, 2003.
217. Tashiro T, Mori K, *Tetrahedron: Asymmetry* **16**:1801–1806, 2005.
218. Nakanishi A, Mori K, *Biosci Biotechnol Biochem* **69**(5):1007–1013, 2005.
219. Mori K, *Tetrahedron: Asymmetry* **18**(7):838–846, 2007.
220. Ho HY, Millar JG, *J Chem Eco* **27**(10):2067–2095, 2001.
221. Ho HY, Millar JG, *J Chem Eco* **27**(6):1177–1201, 2001.
222. Zilkowski BW, Bartelt RJ, Cosse AA, Petroski RJ, *J Chem Eco* **32**:2543–2558, 2006.
223. Burger BV, Petersen WGB, Weber WG, Munro ZM, *J Chem Eco* **28**(12):2527–2539, 2002.
224. Hedenstrom E, Edlund H, Wassgren AB, Bergstrom G, Anderbrant O, Ostrand F, Sierpinski A, Auger-Rozenberg MA, Herz A, Heitland W, Varama M, *J Chem Eco* **32**:2525–2541, 2006.
225. Enders D, Breuer I, Nuhring A, *Eur J Org Chem* **13**:2677–2683, 2005.
226. Naveen Kumar D, Rao V, *Tetrahedron Letters* **45**:2227–2229, 2004.
227. Kim SG, Park TH, Kim BJ, *Tetrahedron Letters* **47**:6369–6372, 2006.
228. Gallos JK, Kyradjoglou LC, Koftis TV, *Heterocycles* **55**(4):781–784, 2001.
229. Mayer SF, Mang H, Steinreiber A, Saf R, Faber K, *Canadian Journal of Chemistry* **80**(4):362–369, 2002.
230. Prasad KR, Anbarasan P, *Tetrahedron: Asymmetry* **16**(24):3951–3953, 2005.
231. De Sousa AL, Resck IS, *J Braz Chem Soc* **13**(2):233–237, 2002.
232. Kalaitzakis D, Rozzell DJ, Kambourakis S, Smonou I, *Eur J Org Chem* **10**:2309–2313, 2006.

233. Zarbin PHG, de Oliveira ARM, Delay CE, *Tetrahedron Letters* **44**:6849–6851, 2003.
234. Nesnerova P, Sebek P, Macek T, Svatos A, *Green Chem* **6**:305–307, 2004.
235. Ishmuratov G Yu, Yakovleva MP, Ganieva VA, Amirkhanov DV, Tolstikov GA, *Chemistry of Natural Compounds* **41**(6):719–721, 2005.
236. Pederson RL, Fellows IM, Ung TA, Ishihara H, Hajela SP, *Adv Synth Catal* **344**:728–735, 2002.
237. Petroski RJ, Weisleder D, *Synthetic Communications* **31**(1):89–95, 2001.

Chapter 11

NATURE DERIVED ANTIBIOTICS

Srinivas Kodali and Jun Wang

11.1 INTRODUCTION

Millions of people died of infection before the discovery of antibiotics. During the early twentieth century and before, infectious diseases were cured by using potions and plant extracts. Patients were quarantined to keep the infections from spreading. At first, several scientists started their research by keeping records of the infectious diseases. As a result of this work, pioneers like Joseph Lister, Louis Pasteur, and Robert Koch have proven the existence of microbes and the way they are responsible for infections. Their use of disinfectants, antitoxins, vaccination and anti-infectives laid the foundation for modern day therapy of infectious diseases.

In 1929, Alexander Fleming and his coworkers identified the first natural product antibiotic agent from a fungus, *Penicillium notatum*, that grew as a contaminant on a plate containing bacteria. However, the natural product antibiotic, which inhibited the bacterial growth and was named by Fleming as penicillin, was not isolated till a few more years later. Only in 1932 the first synthetic antibiotic, sulfonamide, has been introduced for human use. In the early 1940s, penicillin G was isolated commercially from fermentation broths of the fungus *Penicillium notatum*. At that time, the pharmaceutical scientific community realized that nature played an irreplaceable role in providing us with antibacterial compounds that led to therapies for the treatment of various infections. It was found that many

Table 1. Nature Derived Antibiotics

Antibiotic	Discovery year	Chemical class	Mechanism of action
Penicillin G	1940	β-Lactam	Bacterial Cell wall
Polymyxin	1942	Glycopeptide	Bacterial Cell Membrane
Streptomycin	1944	Aminoglycoside	Protein Synthesis
Cephalosporin C	1945	β-Lactam	Bacterial Cell Wall
Chloramphenicol	1947	Phenylpropanoid	Protein Synthesis
Chlortetracycline	1948	Tetracycline	Protein Synthesis
Erythromycin	1950	Macrolide	Protein Synthesis
Vancomycin	1955	Glycopeptide	Bacterial Cell Wall
Virginiamycin	1955	Streptogramins	Protein Synthesis
Cycloserine	1955	Cycloserine	Bacterial Cell Wall
Lincomycin	1952	Lincosamide	Protein Synthesis
Novobiocin	1956	Novobiocin	DNA Synthesis
Rifamycin	1959	Ansamycin	RNA Polymerase
Methicillin	1960	β-Lactam	Bacterial Cell Wall
Cephalothin	1962	β-Lactam	Bacterial Cell Wall
Gentamicin	1963	Aminoglycoside	Protein Synthesis
Fosfomycin	1969	Phosphonate	Bacterial Cell Wall
Mupirocin	1971	Pseudomonic Acid	Protein Synthesis
Cefoxitin	1973	Cephamycin	Bacterial Cell Wall
Thienamycin	1976	Carbapenem	Bacterial Cell Wall
Daptomycin	1980	Lipopeptide	Bacterial Cell Membrane

microorganisms, plants and animals produce chemical entities as secondary metabolites. These chemical compounds are not detrimental to the producer but they can prevent growth of other organisms growing in the same niche. Thus, an organized and methodical search to identify antibiotics has begun around the mid-20th century leading to a phenomenal era during which a great number of natural product antibiotics were discovered. Table 1 shows some of the antibiotics with a related clinical drug candidate that has been isolated from natural resources.

These natural product scaffolds identified provided us with most of the antibiotic classes identified to date and they also provided the model for the many semisynthetic antibiotics approved in the recent years. Natural products contributed to more than 78% of the antibiotics approved between 1981 and 2002.[1] Known targets in the clinic are inhibitors of bacterial

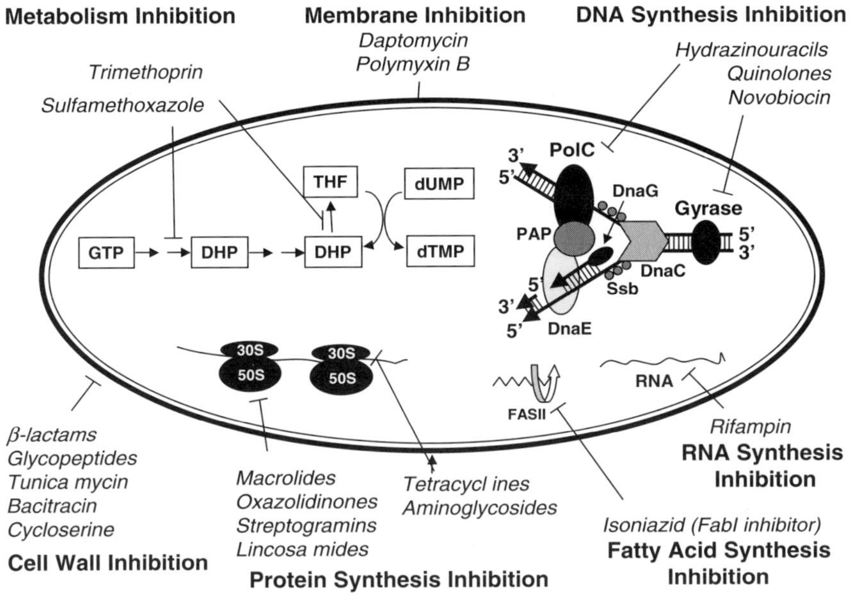

Fig. 1. Mode of action on selected antibiotics. Synthetic antibiotics are underlined.

cell wall, protein biosynthesis, DNA biosynthesis, RNA biosynthesis, cell membrane, fatty acid biosynthesis and metabolism (Fig. 1).

We recommend readers to an informative book on antibiotics.[2] In this chapter, we focus on natural product or natural product derived antibiotics and the discussion of opportunities of antibiotic discovery.

11.2 CELL WALL ANTIBIOTICS

Peptidoglycan allows bacteria, with the exception of mycoplasma, to have rigid shape and structure. Bacteria are generally classified into two groups, Gram-positive and Gram-negative organisms, based on their cell wall composition. Gram-positive bacteria have a thick outer layer of peptidoglycan. Many Gram-positive bacteria also have teichoic and teichuronic acids (sugar alcohol phosphates). In contrast, Gram-negative bacteria have a loosely cross linked thin peptidoglycan layer and an outer asymmetric membrane composed of phospholid and lipopolysaccharide.[2] Peptidoglycan or murein consists of long glycan chain layers cross linked with peptide chains. The glycan layer is made up of repeating units of N-acetylglucosamine linked β-1,4 to N-acetylmuramic acid. A peptide is attached to N-acetylmuramic acid and this peptide composition varies amongst bacteria,[3] but it begins with L-alanine and terminates with D-alanine with either mesodiaminopimelic acid or L-lysine in the middle for cross linking with D-alanine of an adjacent peptide in the majority of bacterial species with some exceptions like Staphylococcus aureus cell wall, where the cross linking is by pentaglycine bridges between the L-lysine and D-alanine of an adjacent peptide.[4] Glycan strands are enzymatically linked by the actions of transglycosylase enzyme and the peptide cross linking is achieved by transpeptidase action resulting in a complex structure that protects the cell from environmental pressures. Biosynthesis of peptidoglycan occurs in three stages[5,6] (Fig. 2). Stage one takes place in the cytoplasm with enzymes MurA through MurF, GlmU, Alr and Ddl to synthesize precursors UDP-*N*-acetylmuramyl-pentapeptide (UDP-MurNAc-pentapeptide) and UDP-*N*-acetylglucosamine (UDP-GlcNAc). Stage two is localized to the cytoplasmic membrane; during this stage precursor lipid intermediates are synthesized. These intermediates carry hydrophilic precursors present in the aqueous environment through the hydrophobic membrane to the growing peptidoglycan outside the cytoplasmic membrane. Lipid I (MurNAc-(pentapeptide)-pyrophosphoryl-undecaprenol) is synthesized by the transfer of phospho-MurNAc-pentapeptide moiety of UDP-MurNAc-pentapeptide to bactoprenol. To synthesize lipid II (GlcNAc-β-(1,4)-MurNAc-(pentapeptide)-pyrophosphoryl-undecaprenol), GlcNAc from UDP-GlcNAc is added to Lipid I. Lipid II is the substrate for the polymerization reactions to synthesize cross-linked peptidoglycan.[7] MraY and MurG are the key enzymes that participate during the second stage of cell wall biosynthesis. Stage three occurs outside the cytoplasmic membrane and is carried out by proteins that have transglycosylase and transpeptidase functions to complete the cell wall biosynthesis. Some transglycosylase and transpeptidase enzymes function as penicillin binding proteins (PBPs). During this stage disaccharide pentapeptides, outside the membrane, are added to the growing peptidoglycan to continue cell wall biosynthesis.[8] The enzymes responsible for the biosynthesis of bacterial cell wall do not have human homologs and because of that it is reasonable to make the cell

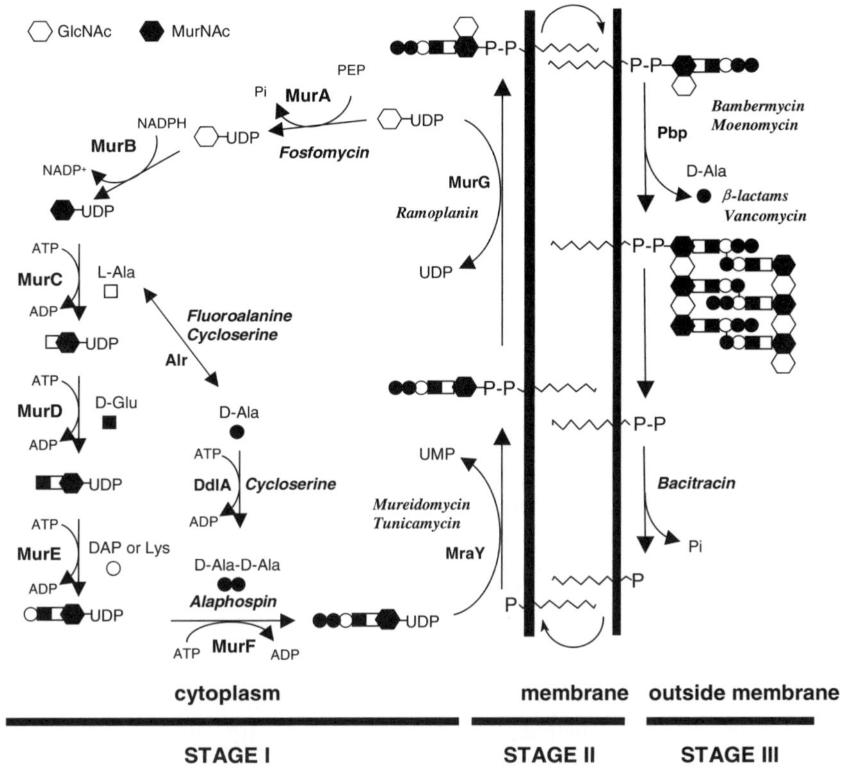

Fig. 2. Cell wall (peptidoglycan or murein) biosynthesis.

wall one of the primary targets for the antibiotic search. Several natural product bacterial cell wall inhibitors have been identified over a period spanning more than six decades. They target enzymes from all three stages of peptidoglycan biosynthesis; beta-lactam antibiotics, bacitracin, cycloserine, glycopeptides, and phosphonates are examples of natural product cell wall biosynthesis inhibitors identified (Fig. 2).

11.2.1 Beta-Lactams

The first natural product antibiotic penicillin, is of a β-lactam type (Fig. 3). It revolutionized the treatment of infections ever since its introduction for clinical use. Since then, other β-lactam containing natural product antibiotics and their derivatives, such as cephalosporins, carbapenems,

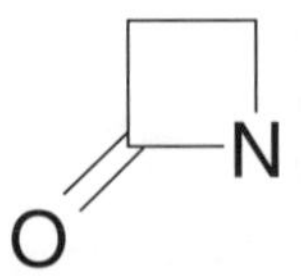

β-lactam ring or basic monobactam structure

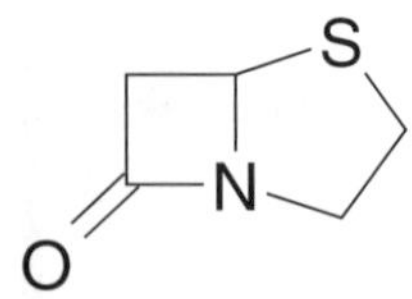

Penicillin basic structure

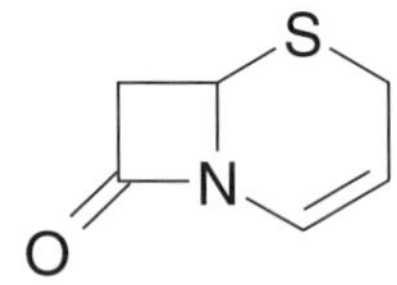

Cephalosporin basic structure

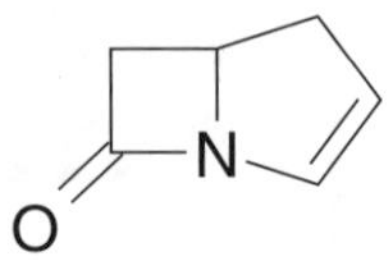

Carbapenem basic structure

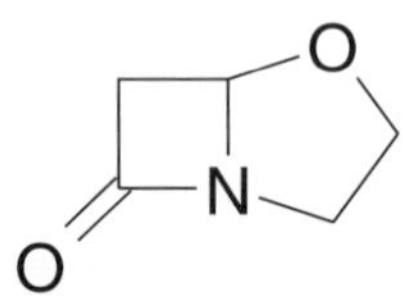

Clavam basic structure

Fig. 3. Naturally occurring β-lactam basic ring structures.

clavulanic acid and monobactams, have been introduced to the market. The mode of action of β-lactam antibiotics has been reviewed in detail.[9,10] It is generally accepted that β-lactam antibiotics bind to PBPs and adversely affect the peptide cross-linking reaction to change the rigid characteristics of the bacterial cell wall. β-lactam containing compounds have a structural similarity to the D-alanyl-D-alanine residues that participates in the transpeptidation reaction. β-lactams target the serine in the active site of the transpeptidase enzyme and irreversibly bind to the enzyme, stopping it from synthesizing cell wall. The exact sequence of events that causes the death of the bacterial cell exposed to β-lactam antibiotics is quite unclear. Functions of the PBPs that seemed to have been previously settled are still in doubt.[11] The mode of action of β-lactam antibiotics and subsequent bactericidal events will be debated for a long time to come.

This antibiotic class has the highest number of antibiotics on the market. All β-lactam antibiotics have a basic β-lactam ring structure and other classes are categorized based on the structure of the ring adjacent to the beta-lactam ring (Fig. 3). The majority of the β-lactams derived from natural products and approved for human use are listed in Table 2.

Table 2. Natural and Semi-synthetic β-lactam Antibiotics

1.	Amdinocillin	41.	Cefteram
2.	Amoxicillin	42.	Ceftibuten
3.	Ampicillin	43.	Ceftin
4.	Apalcillin	44.	Ceftizoxime
5.	Aspoxicillin	45.	Ceftriaxone
6.	Azocillin	46.	Cefuroxime
7.	Aztreonam	47.	Cefuzonam
8.	Benzathine Penicillin	48.	Cefzil
9.	Benzylpenicillin	49.	Cephalexin
10.	Biapenem	50.	Cephalothin
11.	Carbenicillin	51.	Cephapirin
12.	Carumonam	52.	Cephradine
13.	Cefaclor	53.	Clavulanate
14.	Cefazolin	54.	Cloxacillin
15.	Cefbuperazone	55.	Cyclacillin
16.	Cefcapene	56.	Dicloxacillin
17.	Cefdinir	57.	Ertapenem
18.	Cefditoren	58.	Flomoxef
19.	Cefepime	59.	Flucloxacillin
20.	Cefetamet	60.	Fropenam
21.	Cefixime	61.	Hetacillin
22.	Cefmenoxime	62.	Imipenem/Cilastatin
23.	Cefmetazole	63.	Lenampicillin
24.	Cefminox	64.	Loracarbef
25.	Cefodizime	65.	Meropenem
26.	Cefonicid	66.	Methicillin
27.	Cefoperazone	67.	Mezlocillin
28.	Ceforanide	68.	Moxalactam
29.	Cefoselis	69.	Nafcillin
30.	Cefotetan	70.	Oxacillin
31.	Cefotiam	71.	Oxapenem
32.	Cefoxitin	72.	Panipenem
33.	Cefozopran	73.	Phenoxymethylpenicillin
34.	Cefpimizole	74.	Piperacillin
35.	Cefpiramide	75.	Procaine Penicillin
36.	Cefpirome	76.	Sulbactam
37.	Cefpodoxime	77.	Sultamycillin
38.	Cefprozil	78.	Tazobactam
39.	Cefsoludin	79.	Temocillin
40.	Ceftazidime	80.	Ticarcillin

11.2.2 Penicillins

Since the discovery of penicillin, there have been attempts to modify the structure in order to increase their activity against less susceptible bacterial stains or to make it less sensitive to β-lactamases[12,13] that render it inactive. The synthetic modifications to the natural product penicillins can be classified into four different generations.

First generation: Examples are penicillin G (only natural penicillin in clinical use), penicillin V and β-lactamase resistant penicillins like methicillin, nafcillin, oxacillin, cloxacillin, floxacillin etc.

Second generation: Examples are ampicillin, amoxicillin etc.; which have both Gram positive and Gram negative activities, but these antibiotics are susceptible to β-lactamases.

Third generation: Examples are carbenicillin and ticarcillin; which have increased gram negative coverage, but are still sensitive to β-lactamases.

Fourth generation: Examples are piperacillin, azlocillin, and mezlocillin that kill all the bacterial strains covered by the third generation with higher potency. This generation of penicillins is also susceptible to β-lactamases.

11.2.3 Cephalosporins

First generation: Examples are cephalothin, cefazolin, cephalexin, cefadroxil, cephradine, cephaloglycine; which are like penicillins but in general, resistant to majority of the plasmid encoded class A β-lactamases but not to chromosomal encoded class C broader spectrum β-lactamases.[14,15]

Second generation: Examples are cefoxitin, cephamandole, cefmetazole, cefonicid, cefuraxime, cefaclor, cefprozil etc. These drugs in addition to Gram-positive coverage have increased Gram-negative coverage.

Third generation: Examples are cefotaxime, cefoperazone, ceftazidime, ceftizoxime, ceftriaxone, cefdinir etc. This generation has greater Gram-negative activity but loses some Gram-positive potency.

Fourth generation: Cefepime has greatly improved Gram-negative potency and is very resistant to β-lactamases.

11.2.4 Carbapenems

Carbapenems have the broadest antibacterial spectrum, covering both Gram-positive and Gram-negative bacteria and showing high resistance to most β-lactamases but are susceptible to carbapenemases.

11.2.5 Bacitracin

Bacitracin[16] (Fig. 4) is a cyclic peptide antibiotic. The lipid II molecule involved in the bacterial cell wall biosynthesis has a C55 isoprenyl pyrophosphate moiety that must be dephosphorylated so that it can reparticipate in another round of lipid II transfer. Bacitracin binds to the isoprenyl pyrophosphate and prevents the dephosphorylation which, in turn, blocks cell wall growth by interfering with the release of the muropeptide subunits to the outside of the bacterial cell membrane.[17] Bacitracin inhibits similar reactions in eukaryotic cells. So, it is systemically toxic but is an effective and widely used topical antibiotic.

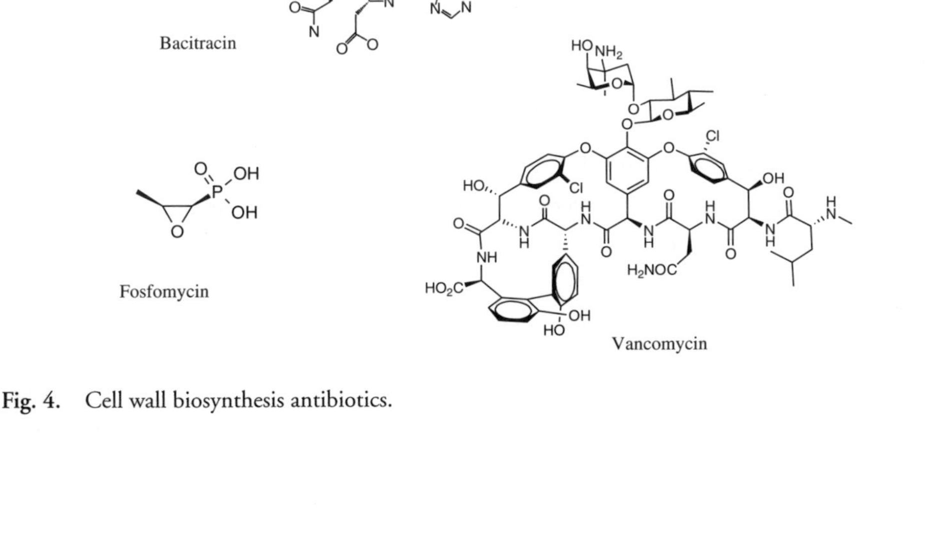

Fig. 4. Cell wall biosynthesis antibiotics.

11.2.6 Cycloserine

Cycloserine[18] (Fig. 4) is produced by several species of *Streptomyces*. One of the basic glycosyl components of the bacterial cell wall, n-acetyl-muramic acid (the product of Mur A and MurB), is modified by the addition of the first three amino acids sequentially by MurC, MurD and MurE enzymes. A dipeptide, D-alanyl-D-alanine is then added to make the pentapeptide. In bacteria, L-alanine is the native form and it is converted to D-alanine form by alanine racemase (Alr). Two D-alanines are joined by D-ala-D-ala ligase (DdlA) to synthesize the dipeptide. Cycloserine resembles the substrate for Alr and Ddl and inhibits their respective reactions in stage I of the peptidoglycan biosynthesis (Fig. 2).

11.2.7 Glycopeptides

Vancomycins[19] (Fig. 4) and Teicoplanin (Coronelli *et al.*, German Patent, 2608216 (1976)) are structurally related examples of glycopeptides. Glycopeptides have rigid cross linked heptapeptide back bone which recognizes the D-ala-D-ala dipeptide[20] present on peptidoglycan intermediates. As the muropeptide subunit of the bacterial cell wall crosses the cell membrane, glycopeptides bind to it cutting off the substrate supply for transglycosylase and transpeptidase enzymes from completing the cell wall assembly.[21]

11.2.8 Phosphonate

Fosfomycin (Fig. 4) is an antibiotic produced by several species of *Streptomyces* grown in the presence of phosphates.[22,23] UDP-n-acetylglucosamine enolpyruvyltransferase (MurA) catalyzes the first committed step in the peptidoglycan biosynthesis by transferring the pyruvyl moiety from phosphoenolpyruvate onto UDP-n-acetyl-glucosamine (Fig. 2). Fosfomycin alkylates a highly conserved cysteine residue essential for binding of phosphoenol pyruvate.[24,25]

11.3 CELL MEMBRANE INHIBITORS

Compounds that disrupt cell membrane integrity either through structural or functional disorganization lead to loss of viability. Polymyxins[26] and daptomycin[27,28] (Fig. 5) are examples of cell membrane inhibitors.

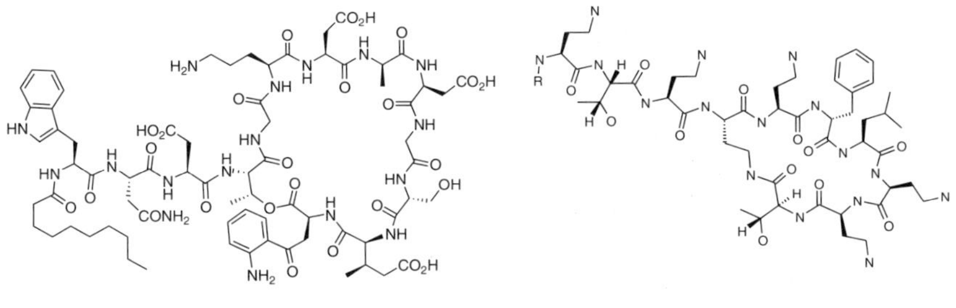

Daptomycin Polymyxin B

Fig. 5. Antibiotics for inhibition of cell mambrane.

11.3.1 Polymyxins

Polymyxins are peptide antibiotics with a hydrophobic side chain[29] that bind to the phospholipid bilayer and interfere with the osmotic equilibrium of the cell. Polymyxins are active against majority of Gram-negative bacteria.

11.3.2 Daptomycin

A novel cyclic lipopeptide antibiotic was isolated from cultures of *Streptomyces roseosporus* grown in the presence of decanioc acid. Daptomycin interacts with the bacterial cell membrane and interferes with membrane potential.[30] Unlike polymyxin B, daptomycin can target majority of clinically relevant Gram-positive pathogens.[31]

11.4 PROTEIN BIOSYNTHESIS INHIBITORS

Bacterial protein biosynthesis is a cascade of events which manufacture chains of amino acids before they are folded into specific structures to carry out various biological functions. Protein biosynthesis is absolutely essential for the survival of prokaryotic and eukaryotic cells. Ribosomes, macromolecular complexes made up of proteins and RNA, participate in decoding the genetic message to synthesize both essential and nonessential proteins to carry out cellular functions.

Bacterial ribosomes consists of a small 30S and a large 50S subunit (Fig. 6).[32]

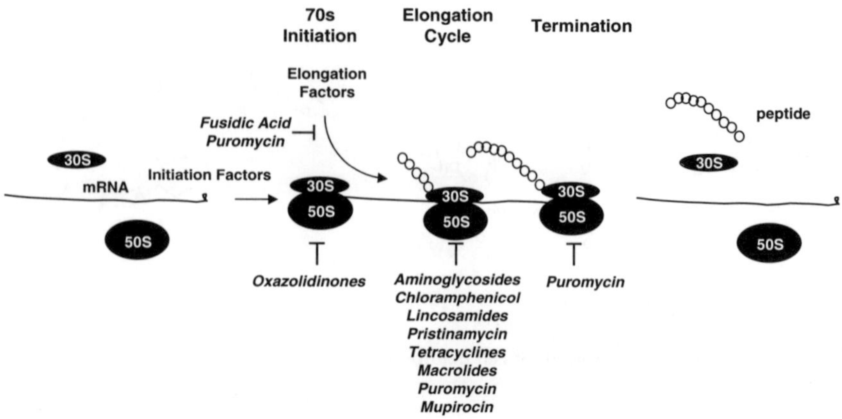

Fig. 6. Protein biosynthesis. All antibiotics in italics are nature derived except oxazolidinone.

The small subunit contains 16S ribosomal RNA (rRNA) and about 20 proteins and the large subunit consists of 23S and 5S rRNAs and about 30 proteins. Protein biosynthesis occurs when messenger RNA (mRNA) is translated into an amino acid sequence by the ribosomal subunits. Before participating in protein biosynthesis, amino acids are attached to specific transfer RNA (tRNA) molecules by aminoacyl-tRNA synthetases. Aminoacyl-tRNA molecules carry a triplet of nucleotide bases called the "anticodon" that specifically pairs with a complimentary triplet "codon" from mRNA. The pairing of codons and anticodons based on the nucleotide sequence of mRNA controls the amino acid sequence of a polypeptide. The nucleotide sequence of mRNA is determined by the DNA strand from which it is transcribed. Thus, mRNA translation is the biosynthesis of a polypeptide using ribosome and all three kinds of RNA. For convenience, protein translation is divided into initiation, elongation, termination and ribosome recycling. The majority of the antibiotics that disrupt protein biosynthesis do so by interfering with the functions of the ribosome (Table 3).

As a group, the protein biosynthesis inhibitors comprise the second largest class of antibiotics available for clinical use. Natural product classes of antibiotics that inhibit the protein biosynthesis are aminoglycosides, tetracyclines, chloramphenicol, macrolides, lincosamides, fusidic acid, streptogramins and mupirocin (Fig. 7).

Table 3. Natural and Semi-synthetic Antibiotics of Protein Biosynthesis Inhibition

1.	Amikacin	20.	Minocycline
2.	Arbekacin	21.	Miokamycin
3.	Astromycin	22.	Mupirocin
4.	Azithromycin	23.	Neomycin
5.	Capreomycin	24.	Netilimicin
6.	Chloramphenicol	25.	Oleandomycin
7.	Chlortetracycline	26.	Oxytetracycline
8.	Clarithromycin	27.	Paromomycin
9.	Clindamycin	28.	Quinupristin
10.	Dalfopristin	29.	Rokitamycin
11.	Dirithromycin	30.	Roxithromycin
12.	Doxycycline	31.	RV-11
13.	Erythromycin	32.	Spectinomycin
14.	Flurithromycin	33.	Streptomycin
15.	Gentamycin	34.	Telithromycin
16.	Isepamicin	35.	Tetracycline
17.	Kanamycin	36.	Tigecycline
18.	Lincomycin	37.	Tobramycin
19.	Micronomicin	38.	Troleandomycin

11.4.1 Aminoglycosides

Some examples of aminoglycoside antibiotics available for clinical use are streptomycin,[33] gentamicin, tobramycin, kanamycin, amikacin, paromomycin, neomycin and netilmicin.[34] The basic structure of aminoglycoside has three or four sugar-like rings (Fig. 7) with amines attached at various locations. Aminoglycosides are bacteriocidal compounds that bind to the 30S subunit of the ribosome to inhibit protein biosynthesis (Fig. 1). They target the A-site, disturbing elongation of the peptide chain. Aminoglycosides induce translational errors resulting in misreading of the mRNA sequence which can lead to premature termination of protein biosynthesis. The entry of the aminoglycosides into the cells is enhanced if the aberrant proteins produced due to misreading are inserted into the cell membrane.[35] Spectinomycin is structurally related to aminoglycosides and similarly, it binds to the A-site, but is not bactericidal[36] and does not cause misreading of mRNA.[37]

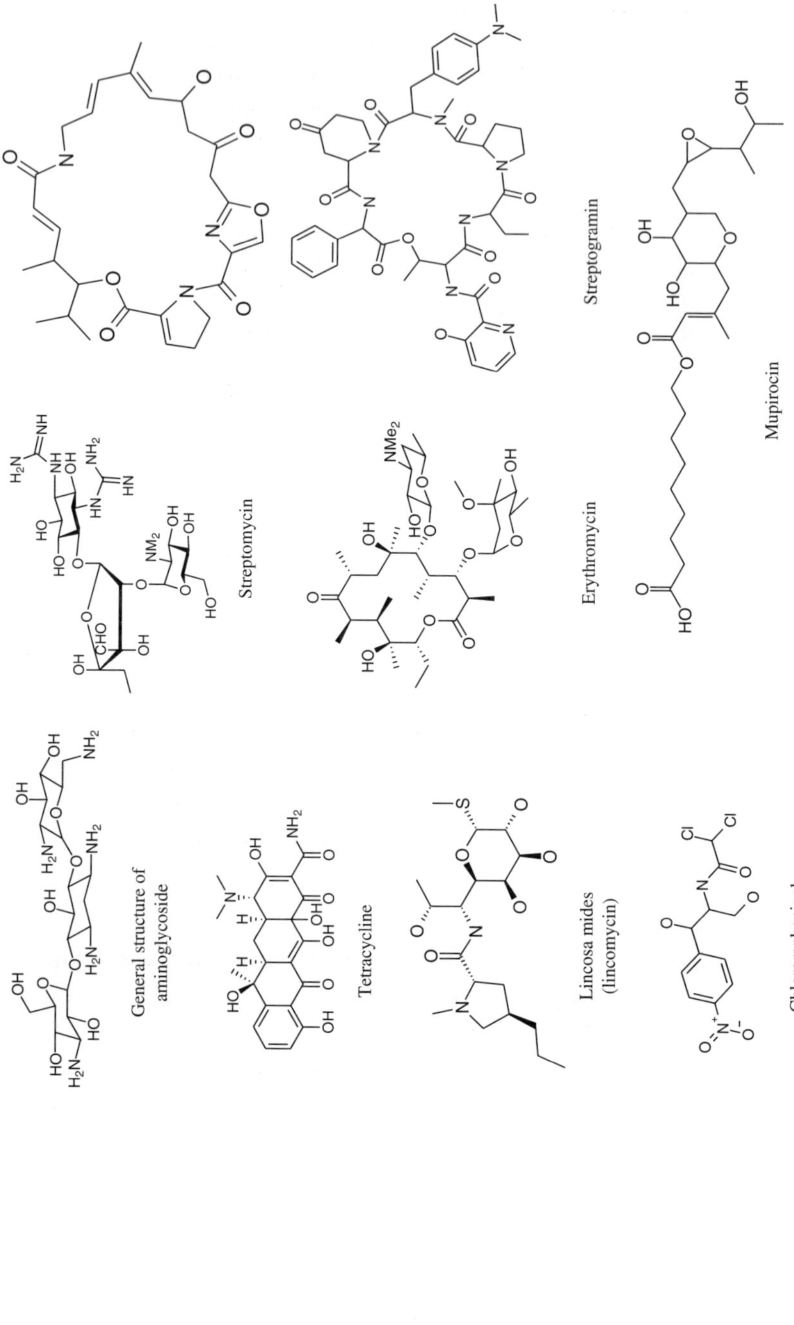

Fig. 7. Antibiotics by inhibition of protein biosynthesis.

11.4.2 Tetracyclines

As the name implies, this class of compounds has four linearly attached six membered rings. Tetracyclines[38] are bacteriostatic and reversibly bind to the 30S ribosomal subunit. They interfere with the binding of aminoacyl tRNA at the A-site of the ribosome.[39]

11.4.3 Macrolides, Lincosamides and Streptogramins

Macrolides, lincosamides and streptogramins are protein biosynthesis inhibitors that bind to 50S subunit of the ribosome and inhibit peptidyl tRNA translocation from the A-site to the P-site.[40–42] Macrolides have a glycosylated 14-, 15- or 16-membered lactone ring structure and are produced by several species of *Streptomyces.*[43] Lincosamide antibiotics were isolated initially from *Streptomyces lincolnensis*[44] but later isolated from different species of *Streptomcyces.* Streptogramins[45] were also isolated from *Streptomyces graminofaciens* and subsequently from several different *Streptomyces* species. There are two structurally different streptogramins, A and B; they are bacteriostatic individually and can be bactericidal when combined.

11.4.4 Mupirocin

Mupirocin, pseudomonic acid, is a product of *Pseudomonas fluorescens* discovered by Fuller *et al.*[46] It is an inhibitor of isoleucyl-transfer ribonucleic acid (isoleucyl-tRNA) synthetase.[47]

11.5 NUCLEIC ACID (DNA AND RNA) INHIBITORS

11.5.1 Novobiocin

Novobiocin (Fig. 8) is a glycosidic antibiotic isolated from *Streptomyces* species.[48] Its mode of action was identified by Smith[49] as DNA biosynthesis inhibitor and it was found to target the B subunit of DNA Gyrase,[50] a DNA replication enzyme (Fig. 1).

11.5.2 Rifampin

Rifampin (Fig. 8) is semisynthetic derivative of Rifamycin B that was isolated from *Sterptomyces mediterranei.*[51,52] Its mode of action was confirmed

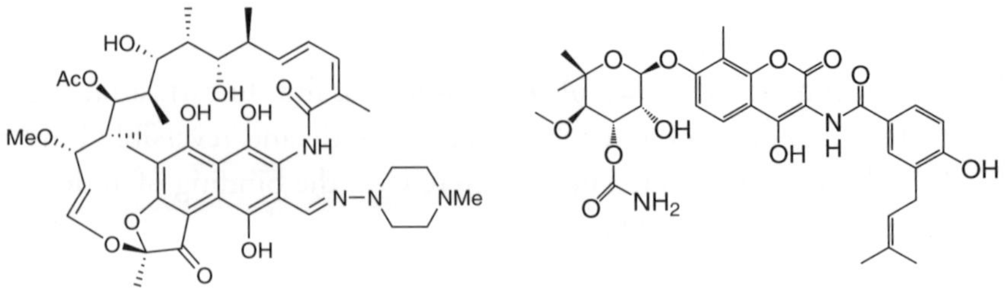

Rifampin   Novobiocin

Fig. 8. Antibiotics by inhibition of RNA or DNA biosynthesis.

as an inhibitor of the initiation of RNA biosynthesis.[53] It targets the β subunit of the RNA polymerase enzyme (RpoB) at an allosteric site.[54]

11.6 FATTY ACID BIOSYNTHESIS INHIBITION

Type II fatty acid biosynthesis (FASII) is essential to bacterial cell viability and highly conserved amongst key pathogens. The significant differences of organization, structure of enzymes and role played by fatty acids between bacteria and humans make this system an attractive target for antibacterial drug discovery[55–61] (Fig. 9). Two marketed synthetic antibacterial agents that target the FabI enzyme are triclosan (antiseptic) and isoniazid (an anti-*Mycobacterium tuberculosis* agent).[62,63]

Although there is no natural product or natural product derived antibiotics that target this pathway in the clinic, this pathway has been extensively investigated in antibiotic discovery. Two natural product inhibitors, cerulenin[64] and thiolactomycin,[65] with poor antibacterial activities were discovered more than two decades ago. Cerulenin targets FabF/B, forming a covalent bond with the catalytic cysteine in the active site with its tail occupying the long hydrophobic cavity that normally contains the growing acyl-chain of the natural substrate.[66,67] Thiolactomycin and its analogs[68,69] target the malonyl binding site of both FabH and FabF/B. Some other inhibitors targeting the bacterial condensing enzymes have been reported but none were suitable drug candidates due to insufficient penetration in whole cells and/or *in vivo* toxicity and/or lack of efficacy and/or

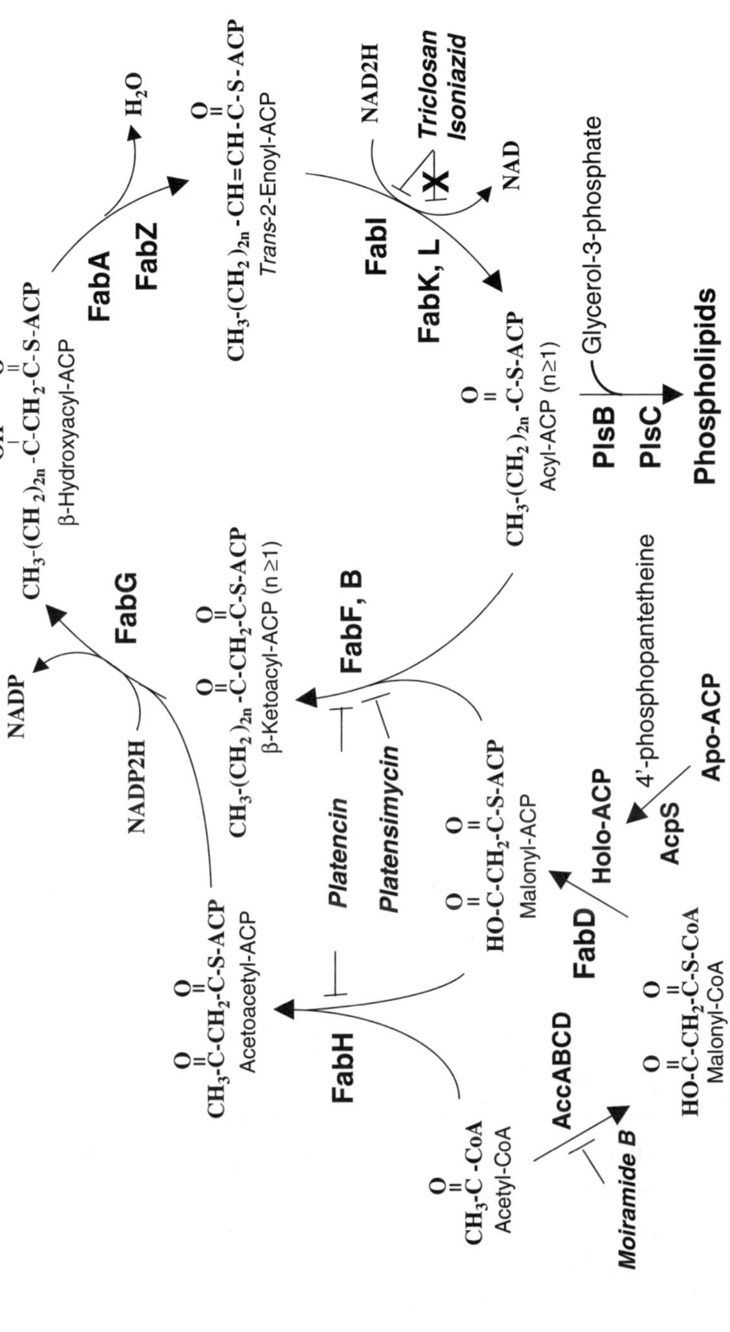

Fig. 9. Bacterial fatty acid biosynthesis (FASII).

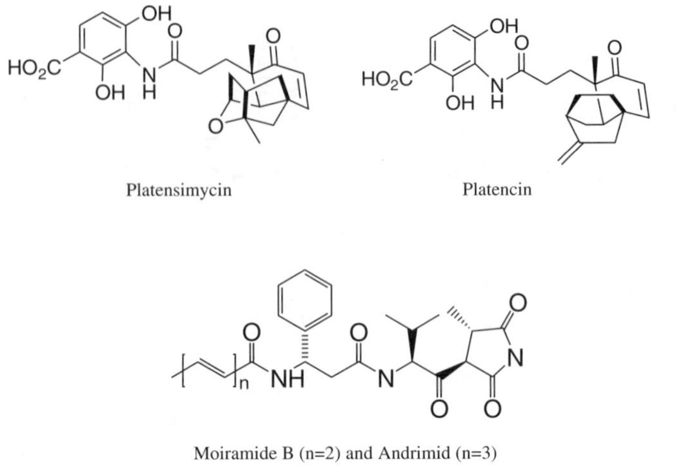

Fig. 10. Naturally occurring fatty acid biosynthesis inhibitors.

poor target selectivity. Several natural products that showed potent *in vitro* and *in vivo* antibacterial properties (Fig. 10) have been discovered recently.

11.6.1 Screening for FabH/F Inhibitors

A team at Merck Research Laboratories developed a two-plate assay,[70] in which one plate was seeded with *Staphylococcus aureus* cells expressing *fabF* antisense RNA (AS plate) and the other with cells lacking antisense RNA expression (control plate). Over 250 000 natural product microbial extracts were screened in the two-plate assay followed by biochemical confirmation.[71] Using this strategy, the Merck team discovered Platensimycin[72] and Platencin[73] with a novel chemical scaffold from strains of *Streptomyces platensis* recovered from soil samples collected in South Africa and Spain, respectively. Platensimycin selectively inhibits FabF while platencin blocks both FabF and FabH condensing enzymes. Direct binding analyses reveal that both platensimycin and platencin interact specially with the acyl-enzyme intermediate of FabF protein. X-ray cocrystallographic studies show that a specific conformational change that takes place on acylation must occur prior to the binding of the inhibitor. In whole-cell labeling experiments, platensimycin and platencin exhibit selective inhibition of bacterial fatty acid biosynthesis without any inhibition of bacterial cell wall, protein, RNA and DNA biosyntheses that most of existing antibiotics target as described above (Fig. 1). Due to the unique mode of action, platensimycin and platencin exhibit no cross-resistance to key antibiotic resistant strains including methicillin-resistant *S. aureus*, vancomycin-intermediate *S. aureus* and vancomycin-resistant *Enterococci*. Both inhibitors present a broad spectrum Gram-positive antibacterial activity and potent *in vivo* efficacy without observed toxicity.[72,73] The natural products, Moiramide B and Andrimid,[74–77] inhibit bacterial acetyl-CoA carboxylase (Acc) which provides unique substrate, malonyl-CoA, for fatty acid initiation and elongation.

11.7 LESSONS LEARNED FROM ANTIBIOTIC DISCOVERY

As described above, the discovery of penicillin followed by its development as the first broad spectrum antibiotic is one of the greatest discoveries of all time. This discovery led to the creation of the field of antibiotic research and, in the next twenty years, culminated in the discoveries of most chemical scaffolds that are in clinical use today.[2,78,79] This glorious time of antibiotic discovery led to the naïve conclusion that these antibiotics or their chemical variants would be sufficient to overcome bacterial infections. However, resistance against all classes of antibiotics has since increased. Certain strains of methicillin resistant *S. aureus* (MRSA) are highly prevalent, accounting for 60%–90% of nosocomial infections. Vancomycin is considered the antibiotic of last resort for the treatment of Gram-positive infections, however 20%–30% of *Enterococci* are found to be resistant to this drug. Chemical modifications of existing chemical scaffolds led to a large number of highly potent broad spectrum antibiotics and continue to deliver newer drugs which provide coverage for resistant pathogens, however, this will surely be a temporary fix. Such efficacious modifications have been extensive and novel classes are increasingly rare. New antibiotic scaffolds are urgently needed.

A number of approaches have been used to discover new antibiotics in past decades. One approach is empirical screening, in which a whole cell assay is used to find antibacterial activities first, followed by the determination of the targets or modes of action to direct further chemical modifications. Another approach is *in vitro* target-based screening, in which a cell free system is used to find inhibitors for an essential enzyme (or enzymes in a pathway) followed by the determination of spectrum of antibacterial activity. However, there have been few successes with the latter approach because of the variation in the ability of the compounds to reach their intracellular targets due to poor penetration[80] and/or active efflux.[81,82] The benefit of the whole cell screening approach is that it ensures antibacterial activity; however, empirical screening may yield toxic compounds or known structures that have been identified during the past half century. Thus, over the years, many target-based whole cell screening methods have been developed to preferentially find certain types of inhibitors. Such screens may compare the responses of paired strains, such as a drug resistant strain and its parent strain, or a wild type strain and a strain over expressing a selected target. Other whole cell screens have monitored changes in cell morphology or specific phenotypes caused by blockage of a particular pathway or measured expression of a reporter gene which is directly or indirectly regulated by essential intracellular proteins. However, most of these approaches have met with limited success.

Most of the antibiotics are natural products or their derivatives. While the widely used antibacterial clasess, fluoroquinolones and oxazolidinones, are synthetic in origin, most screening of corporate chemical collections have turned out to be fruitless exercises. Unfortunately, screening efforts of natural products in the intervening period have allowed the identification of a large number of new natural products with varying degrees of antibiotic activities but none possess adequate activity or desirable properties to be developed as an antibiotic.[83] The biggest challenge in natural product screening by whole-cell empiric assays is to differentiate a known from an unknown, a process called dereplication. Even highly sensitive chemical dereplication methods fail, especially when compounds are highly potent and are present in low concentrations.

Bacterial genome analysis and down-regulation studies have indicated that there are 150–670 essential proteins which are well conserved and could be broad spectrum drug targets.[84–88] However, of those, only a few are proven as therapeutic targets. Recently, a differential sensitivity screening assay using selective expression of antisense RNA against a target was applied to screening natural product extracts to discover target specific antibiotics.[70] The beauty of this discovery method is significantly enhanced when it is combined with target specific and sensitive biochemical assays and coupled with diverse chemistries, such as natural products. This technology is particularly powerful for differentiating the target based antibiotics from nonspecific toxic entities and thus allowing the discovery of novel natural product antibiotics.[72,73,89] The antisense strains are significantly more sensitive than their wild-type counterparts to the targeted inhibitors. This allows the identification of compounds that are present in low concentrations and would have been missed with empiric screen.[73] In addition, antisense technology is a universal approach and could be applied to other targets in bacteria particularly to the ~90% of essential proteins that remain unexplored. The antisense strategy provides a great opportunity for the discovery of novel antibiotic chemical structures from natural products.

11.8 PROSPECTS OF IDENTIFYING NOVEL ANTIBIOTICS

Bacterial infections are the second leading cause of death worldwide accounting for about 17 million deaths despite the success of antibiotics brought in treating infectious diseases for the past 60 to 70 years. During this time, natural products have been responsible for the majority of the backbone and subsequent new designs for most of the clinically used antibiotics. Also, during the same time period, nosocomial pathogens such as *S. aureus* have become resistant to most of the existing antibiotics and multidrug resistant Gram-negative bacteria have emerged. Will it be possible to treat bacterial infections in the future if we keep up with the abysmal record of the last 35 years in identifying new classes of antibiotics? We have to tackle the problem more effectively by identifying new classes of antibiotics from natural resources. During the past decades, the efforts that have gone into identifying novel classes of antibiotics from natural resources diminished significantly as more efforts were dedicated to identifying synthetic compounds.

Natural products can surely provide novel classes of antibiotics. Daptomycin, platensimycin and platencin are some of the examples that may revive interest in natural products as a source of novel structures to provide the medical community with much needed help in the treatment of infectious diseases.

Heading into new era of antibacterial drug discovery, new paradigms are critically needed to identify new chemical leads or antibiotics from natural products. New paradigms that we recommend are to (1) use target-based whole-cell screening approach which up-front ensures antibacterial activity and mode of action; (2) increase sensitivity to enable detection of compounds that have been missed in assays with lower sensitivity[72,73]; (3) increase assay throughput to increase the detection of new rare scaffolds, such as daptomycin — which was found once in 10^7 actinomycete cultures, suggesting many more antibiotics to be discovered at frequency below 10^{-7};[90] (4) increase diversity and size of sample collections, as it is estimated that only 1%–3% of the antibiotics produced by *Streptomyces* species have been discovered;[91] (5) improve technology for compound separation and structure elucidation; (6) combine genomic information and advanced technology including biochemistry, molecular biology, crystallography etc. to set up array format and target multiple essential proteins which are able to decrease frequency of emerging resistance and overcome existing resistant pathogens.[92]

REFERENCES

1. Newman DJ, Cragg GM, Snader KM, *J Nat Prod* **66**(7):1022–1037, 2003.
2. Walsh C, *Antibiotics: Actions, Origins and Resistance*, ASM Press, Washington, DC, 2003.
3. Schleifer KH, Kandler O, *Microbiol Mol Biol Rev* **36**(4):407–477, 1972.
4. Tipper DJ, Strominger JL, Ensign JC, *Biochemistry* **6**(3):906–920, 1967.
5. Rogers HJ, Perkins HR, Ward JB, *Microbial cell walls and membranes*, Chapman and Hall, New York, NY, 1980.
6. Silver LL, *Curr Opin Microbiol* **6**(5):431–438, 2003.
7. van Heijenoort J, *Biosynthesis of the bacterial peptidoglycan unit*, 39–54, Elsevier Science, Amsterdam, The Netherlands, 1994.
8. Scheffers DJ, Pinho MG, *Microbiol Mol Biol Rev* **69**(4):585–607, 2005.
9. Tomasz A, *Annual Review of Microbiology* **33**(1):113–137, 1979.
10. Waxman DJ, Strominger JL, *Annual Review of Biochemistry* **52**(1):825–869, 1983.

11. Pereira SFF, Henriques AO, Pinho MG, *et al.*, *J Bacteriol* JB.00044–00007, 2007.
12. Chapman MJ, Holt RJ, Mattocks AR, *et al.*, *J Gen Microbiol* **36**:215–223, 1964.
13. Holt RJ, Stewart GT, *J Gen Microbiol* **36**:203–213, 1964.
14. Bush K, Jacoby GA, Medeiros AA, *Antimicrob Agents Chemother* **39**(6):1211–1233, 1995.
15. Jacoby GA, Munoz-Price LS, *N Engl J Med* **352**(4):380–391, 2005.
16. Johrfson BA, Anker H, Meleney FL, *Science* **102**:376, 1945.
17. Stone KJ, Strominger JL, *PNAS* **68**(12):3223–3227, 1971.
18. Ravina A, *Presse Med* **63**(26):532, 1955.
19. McCormick MH, McGuire JM, Pittenger GE, *et al.*, *Antibiot Annu* **3**:606–611, 1955.
20. Corti A, Soffientini A, Cassani G, *J Appl Biochem* 7(2):133–137, 1985.
21. Hubbard BK, Walsh CT, *Angewandte Chemie International Edition* **42**(7):730–765, 2003.
22. Stapley EDH, JM, *et al.*, *Antimicrobial Agents Chemother* **9**:284–290, 1969.
23. Hendlin D, EO SM, *et al.*, *Science* **166**(901):122–123, 1969.
24. Kahan F, JS K, PJ, *et al.*, *Ann N Y Acad Sci* **235**(0):364–386, 1974.
25. Thomas AM, Ginj C, Jelesarov I, *et al.*, *European Journal of Biochemistry* **271**(13):2682–2690, 2004.
26. Ainsworth GC, Brown AM, Brownlee G, *Nature* **160**:263, 1947.
27. Debono M, Barnhart M, Carrell CB, *et al.*, Program Abstr 20th, abstr. no. 68. In. *Intersci Conf Antimicrob Agents Chemother*, New Orleans, LA, 1980.
28. Eliopoulos GM, Willey S, Reiszner E, *et al.*, *Antimicrob Agents Chemother* **30**(4):532–535, 1986.
29. Hotchkiuss RD, DuBos RJ, *J Biol Chem* **132**:793–794, 1940.
30. Alborn WE, Jr Allen NE, Preston DA, *Antimicrob Agents Chemother* **35**(11):2282–2287, 1991.
31. Tally FP, DeBruin MF, *J Antimicrob Chemother* **46**(4):523–526, 2000.
32. Garret RA, Douthwaite SR, Liljas A, *et al.*, *The ribosome: Structure, function, antibiotics, and cellular interactions*, ASM Press, Washington, D.C, 2000.
33. Waksman SA, Reilly HC, Schatz A, *Proc Natl Acad Sci USA* **31**(6):157–164, 1945.
34. Kotra LP, Haddad J, Mobashery S, *Antimicrob Agents Chemother* **44**(12):3249–3256, 2000.
35. Davis BD, Chen LL, Tai PC, *Proc Natl Acad Sci USA* **83**(16):6164–6168, 1986.
36. Sparling PF, Davis BD, *Antimicrob Agents Chemother* **1**(3):252–258, 1972.
37. Davis BD, *Microbiol Rev* **51**(3):341–350, 1987.
38. Finlay AC, Hobby GL, *et al.*, *Science* **111**(2874):85, 1950.
39. Chopra I, Roberts M, *Microbiol Mol Biol Rev* **65**(2):232–260, 2001.
40. Vannuffel P, Cocito C, *Drugs* **51 Suppl 1**:20–30, 1996.

41. Cocito C, Chinali G, *J Antimicrob Chemother* **16 Suppl A**:35–52, 1985.
42. Brisson-Noel A, Trieu-Cuot P, Courvalin P, *J Antimicrob Chemother* **22 Suppl B**:13–23, 1988.
43. McGuire JM, Bunch RL, Anderson RC, Boaz HE, Flynn EH, Powell HM, Smith JW, *Schweiz Med Wochenschr* **82**(41):1064–1065, 1952.
44. Mason DJ, Dietz A, Deboer C, *Antimicrobial Agents and Chemotherapy* 554, 1964.
45. Charney JWP, Fischer C, Curran RA, *et al.*, *Antibiot Chemotherapy* **3**:1283–1286, 1953.
46. Fuller ATGMMW, *et al.*, *Nature* **234**(5329):416–417, 1971.
47. Hughes J, Mellows G, *Biochem J* **176**(1):305–318, 1978.
48. Smith C, Dietz A, Sokolski WT, *et al.*, *Antibiotics and Chemotherapy* **6**:135–142, 1956.
49. Smith DH, Davis BD, *J Bacteriol* **93**(1):71–79, 1967.
50. Maxwell A, *Trends Microbiol* **5**(3):102–109, 1997.
51. Sensi P, Timbal MT, Greco AM, *Antibiot Chemother* **12**:488–494, 1962.
52. Colonnello F, Calonghi GF, *Chemotherapia (Basel)* **5**:90–99, 1962.
53. Wehrli W, Knüsel F, Schmid K, *et al.*, *Proc Natl Acad Sci USA* **61**(2):667–673, 1968.
54. Campbell EA, Korzheva N, Mustaev A, *et al.*, *Cell* **104**(6):901–912, 2001.
55. Campbell JW, Cronan JE, *Annual Review of Microbiology* **55**(1):305–332, 2001.
56. Heath R, White S, Rock C, *Prog Lipid Res* **40**(6):467–497, 2001.
57. Zhang YM, Marrakchi H, White SW, *et al.*, *J Lipid Res* **44**(1):1–10, 2003.
58. Heath RJ, Rock CO, *Curr Opin Investig Drugs* **5**(2):146–153, 2004.
59. Smith S, Witkowski A, Joshi AK, *Prog Lipid Res* **42**(4):289–317, 2003.
60. White SW, Zheng J, Zhang YM, *et al.*, *Annu Rev Biochem* **74**:791–831, 2005.
61. Singh SB, Phillips JW, Wang J, *Curr Opin Drug Discov Devel* **10**(2): 160–166, 2007.
62. Heath RJ, Yu YT, Shapiro MA, *et al.*, *J Biol Chem* **273**(46):30316–30320, 1998.
63. Banerjee A, Dubnau E, Quemard A, *et al.*, *Science* **263**(5144):227–230, 1994.
64. Matsumae A, Nomura S, Hata T, *J Antibiot (Tokyo)* **17**:1–7, 1964.
65. Noto T, Miyakawa S, Oishi H, *et al.*, *J Antibiot (Tokyo)* **35**(4):401–410, 1982.
66. Kauppinen S, Siggaard-Andersen M, von Wettstein-Knowles P, *Carlsberg Research Communications* **53**(6):357–370, 1988.
67. Price AC, Choi KH, Heath RJ, *et al.*, *J Biol Chem* **276**(9):6551–6559, 2001.
68. Dolak L, Castle T, Truesdell S, *et al.*, *J Antibiot (Tokyo)* **39**(1):26–31, 1986.
69. Omura SYI, Nakagawa A, Iwata R, *et al.*, *J Antibiot (Tokyo)* **36**(2):109–114, 1983.
70. Young K, Jayasuriya H, Ondeyka JG, *et al.*, *Antimicrob Agents Chemother* **50**(2):519–526, 2006.

71. Kodali S, Galgoci A, Young K, *et al.*, *J Biol Chem* **280**(2):1669–1677, 2005.
72. Wang J, Soisson SM, Young K, *et al.*, *Nature* **441**(7091):358–361, 2006.
73. Wang J, Kodali S, Lee SH, *et al.*, *Proc Natl Acad Sci USA* **104**(18):7612–7616, 2007.
74. Freiberg C, Pohlmann J, Nell PG, *et al.*, *Antimicrob Agents Chemother* **50**(8):2707–2712, 2006.
75. Pohlmann J, Lampe T, Shimada M, *et al.*, *Bioorg Med Chem Lett* **15**(4):1189–1192, 2005.
76. Freiberg C, Brunner NA, Schiffer G, *et al.*, *J Biol Chem* **279**(25):26066–26073, 2004.
77. Fredenhagen A, Tamura SY, Kenny PTM, *et al.*, *J Am Chem Soc* **109**:4409–4411, 1987.
78. Singh SB, Barrett JF, *Biochem Pharmacol* **71**(7):1006–1015, 2006.
79. Butler MS, Buss AD, *Biochem Pharmacol* **71**(7):919–929, 2006.
80. Young K, Silver L, *J Bacteriol* **173**(12):3609–3614, 1991.
81. Poole K, *Current Opinion in Microbiology* **4**(5):500–508, 2001.
82. Markham PN, Neyfakh AA, *Current Opinion in Microbiology* **4**(5):509–514, 2001.
83. Berdy J, *J Antibiot (Tokyo)* **58**(1):1–26, 2005.
84. Akerley BJ, Rubin EJ, Novick VL, *et al.*, *PNAS* **99**(2):966–971, 2002.
85. Hutchison CA, Peterson SN, Gill SR, *et al.*, *Science* **286**(5447):2165–2169, 1999.
86. Ji Y, Zhang B, Van SF, Horn Warren P, *et al.*, *Science* **293**(5538):2266–2269, 2001.
87. Kobayashi K, Ehrlich SD, Albertini A, *et al.*, *PNAS* **100**(8):4678–4683, 2003.
88. Forsyth RA, Haselbeck RJ, Ohlsen KL, *et al.*, *Molecular Microbiology* **43**(6):1387–1400, 2002.
89. Singh SB, Jayasuriya H, Ondeyka JG, *et al.*, *J Am Chem Soc* **128**(36):11916–11920, 2006.
90. Baltz RH, *J Ind Microbiol Biotechnol* **33**(7):507–513, 2006.
91. Watve MG, Tickoo R, Jog MM, *et al.*, *Arch Microbiol* **176**(5):386–390, 2001.
92. Silver LL, *Nat Rev Drug Discov* **6**(1):41–55, 2007.

Chapter 12

NATURAL PRODUCTS AND RELATED COMPOUNDS OF REALIZED AND POTENTIAL USE IN TREATING NEURODEGENERATIVE DISEASE

Peter J. Houghton and Melanie-Jayne Howes

12.1 INTRODUCTION

The relevance of natural products to the treatment of neurodegenerative disease gives good examples of the three ways in which natural products find use in the discovery, development and clinical application of new drugs. Firstly, naturally occurring compounds may be used in their own right, either extracted from plant material or synthesized, according to cost and other relevant factors. Secondly, the naturally occurring compound is derivatized to improve pharmacological function, and also to reduce unwanted effects. The third approach that is commonly employed is the use of the natural product as a template to design and synthesize novel compounds, particularly if the binding properties of the molecule to a receptor or active site of an enzyme have been studied.

The term "neurodegenerative disease" is used for several conditions of chronic, usually progressive, ill health caused by a chronic breakdown and deterioration of the neurons, particularly those of the central nervous system (CNS). Often, a deficiency of one or more neurochemicals, particularly neurotransmitters, is associated with the disease. Neurogenerative disease is also associated with the accumulation of fibrillary materials which causes dysfunction or with the removal of the myelin sheath which encompasses neurons. Parkinson's Disease or Parkinsonism (PD) and Alzheimer's Disease (AD) are the two best known diseases of this type and will be the main diseases considered in this review. However, other diseases which might be classed as "neurodegenerative" include multiple sclerosis and spongiform encephalopathies, but these will not be considered to any great extent.

Most commonly, neurodegenerative disease is seen in the older age groups in society and where life expectancy is long, e.g. in well developed industrialized and postindustrialized societies. This group of conditions imposes severe strains in these communities because of the costs involved in treating and caring for sufferers. Such diseases also, of course, cause much suffering and death, with an estimated nearly 18 million people with dementia in the world in 2004, over 5 million people of these being in Europe.[1] The worldwide figure is expected to increase to 34 million by 2025.[2]

Loss of memory and tremor are common symptoms of neurodegenerative diseases, and have been associated with increasing age for a long time, so that they are recognized and treatments prescribed for them in many traditional medical systems. The modern "named" diseases e.g. Parkinsonism and Alzheimer's disease are a product of scientific medicine and date from the last hundred years. They have attracted increasing attention due to longer life expectancy and the strain on human and financial resources. It is difficult to tell whether the percentage incidence of neurodegenerative disease has increased in various age brackets in recent decades, since in previous years, the shorter life expectancy precluded many surviving to an age where neurodegeneration was likely to affect a significant part of the population. Some would claim that there is an increased incidence due to factors such as unhealthy diets, unhealthy social habits, environmental pollution, radiation and sociological stresses such as the breakdown of extended family support network.

The etiologies of the neurodegenerative diseases are still largely unknown but, especially for AD and PD, postmortem studies have shown clear links between the disease and a deficiency of neurotransmitters in some parts of the brain. Thus, in PD there is a chronic shortage of dopamine

(DA) (**1**) and in AD, a deficit of acetylcholine (ACh) (**2**). Until very recently, the only clinical treatment for these conditions was the reversal of the deficiencies by increasing of transmitter activity by agonists, or by inhibition of enzymes involved in the removal or breakdown of the endogenous transmitters in the immediate locality of the synapse. These approaches are discussed more thoroughly below. This paper gives an overview of natural products, and their derivatives, used clinically because of these properties, as well as others with interesting activity which have not been developed as drugs, or which are still undergoing clinical trials. It should not be forgotten that there are also many traditional medicinal plants with a reputation of alleviating or preventing symptoms of neurodegeneration and that these are used as crude extracts and mixtures.[3,4] Extracts where no active compounds have been isolated are not covered in this chapter.

Increasingly attention is being paid to the excitatory amino acid neurotransmitters and the relevant receptors which are present in the CNS e.g. the N-methyl-D-aspartate (NMDA) receptor. The use of these in designing new drugs or in explaining the use of traditional medicinal plant extracts is in its infancy stage and no significant findings have been made of natural products which either bind to these receptors or affect levels of the transmitters.

12.2 PARKINSON'S DISEASE

12.2.1 Introduction

Parkinson's Disease (PD) is named after the English surgeon who first described a syndrome which he called the "shaking palsy." It affects about 20 000 people in the UK and is most commonly seen in older patients, although an estimated 5% of sufferers are under 40 years old.[5] Its characteristic feature is an increasing tremor in resting limbs and a rigidity, particularly exhibited as a shuffling gait, which is known as dyskinesia, but PD also exhibits degeneration of cognitive function and memory.

In PD, there is a significant loss of neurons which produce DA, thought to be due, at least in part, to oxidative stress in the substantia nigra part of the brain.[6] The result is a characteristic deficiency of DA in the substantia nigra of PD patients when they are examined *post mortem*, the other characteristic observation being the presence of deposits called Lewy bodies in the surviving neurons.

Bearing in mind the associated low levels of DA, the major therapeutic approach to PD has been to elevate the levels by either inhibiting the metabolism of DA to less active compounds by monoamine oxidases (MAO), or by giving L-hydroxyphenylalanine (L-DOPA) **3**, the precursor of DA. An alternative approach is to use compounds which are agonists at the DA receptors.

12.2.2 PD: The Use of Dopaminergic Compounds

12.2.2.1 *General consideration of DA receptor agonists*

The structure-activity relationship for DA receptor agonists are generally considered to involve two OH groups *ortho* to each other on an aromatic ring. These bind with serine residues 505 and 508 on the receptor, the aromatic ring enabling hydrophobic interactions with a phenylalanine at 617. Another necessary feature is a nitrogen with a positive charge, which interacts electrostatically with an aspartate residue. In DA, this is the terminal position of the ethyl amino group attached to the aromatic ring. It is noteworthy that the dopaminergic receptor is very similar in some aspects to the adrenergic receptor.[7] and so it is possible that compounds with a structure favoring binding to one of these receptors may also have some effect on the other.

12.2.2.2 *L-DOPA (refer to Structures 1)*

DA (**1**) itself is quite unstable and cannot cross the blood brain barrier. However, it can be formed within the brain by conversion of its precursor L-DOPA (**3**) and it is this compound which is now used extensively for treating PD patients. (**3**) is present in commercially viable amounts in various species of bean, notably *Mucuna* spp., and these were used as a commercial source when L-DOPA was first investigated, although the drug is now obtained by synthesis. L-DOPA (**3**) is often given together with another analogue of DA, carbidopa (**4**), which inhibits dopadecarboxylase, an enzyme that causes the breakdown of (**3**) and so maintains its levels in the blood, thus prolonging its activity. When (**3**) is given for a long

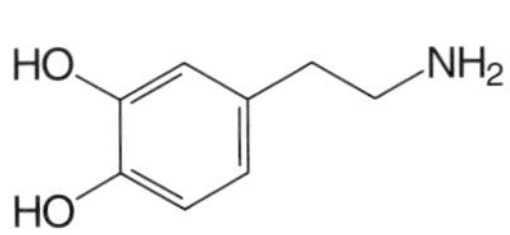

(**1**) Dopamine

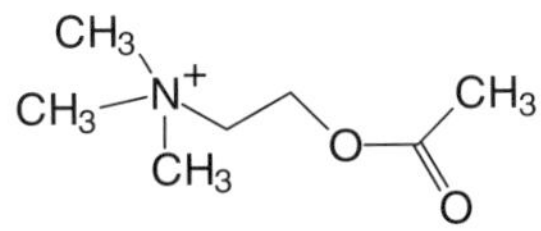

(**2**) Acetylcholine

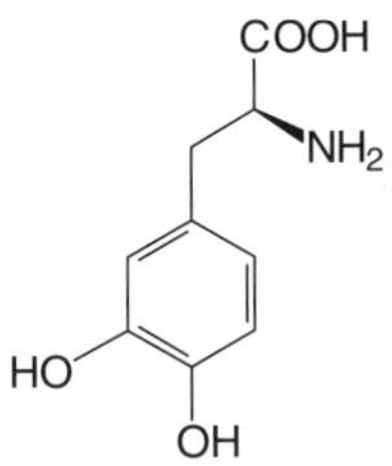

(**3**) LDOPA

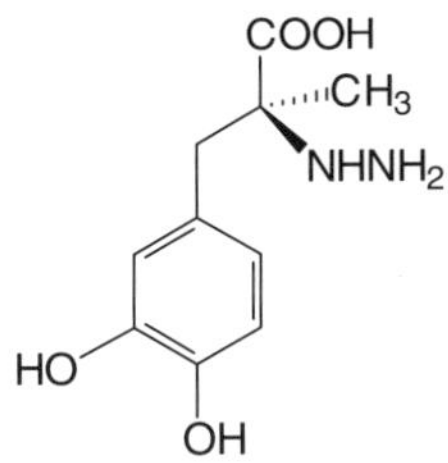

(**4**) Carbidopa

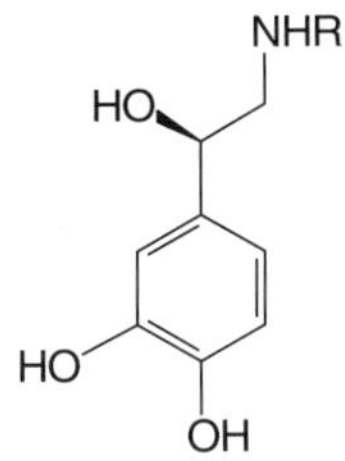

(**5**) R = CH_3 Adrenaline

(**6**) R = H Noradrenaline

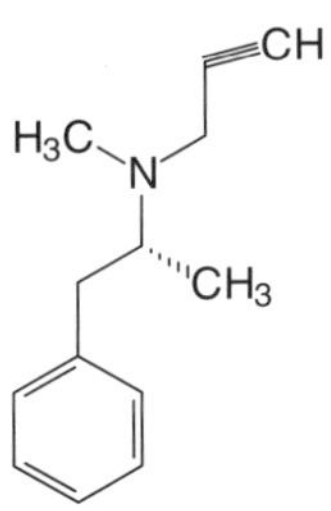

(**7**) Selegiline

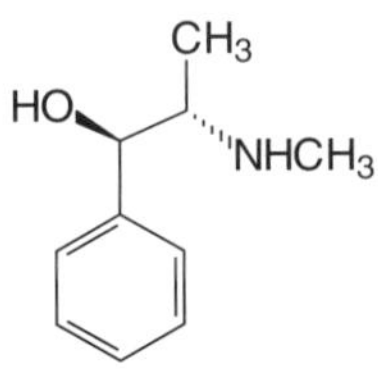

(**8**) Ephedrine

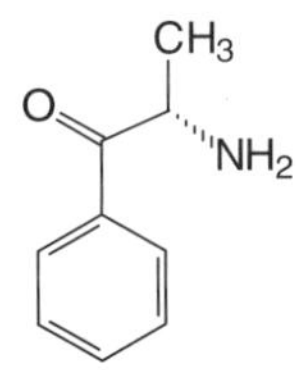

(**9**) Cathinone

Structures 1

time, it is common for a sudden decline in sensitivity to occur occasionally and this is called the "on-off" effect. The causes for this are not completely understood, but are probably due to a number of factors including decrease in the ability of the substantia nigra cells to store DA and the desensitization of the receptors. Other dopaminergic agents, some mentioned below, are often used as adjuvants to reverse the "off" effect.

The powdered seeds of *Mucuna pruriens* L. have been used in Ayurvedic medicine for diseases of the nervous system[8] and have shown to reduce adverse effects in PD patients.[9] HP-200, a commercial extract of *M. pruriens* was shown to be twice as effective as the equivalent dose of L-DOPA in rats,[10] although a later study showed that when the same preparation was given to rats over a 52 weeks period, it did not elevate DA levels in the striatum nigrum, but in the cortex. This calls into question whether the observed improvements in parkinsonian symptoms were due to the hypothesis originally proposed i.e. that the L-DOPA in the *Mucuna* extract was converted to DA and reached the parts of the brain where a deficiency is associated with PD.[8]

12.2.2.3 *Other phenylpropylamines (refer to Structures 1)*

The dopaminergic agonist DA is one of a group of compounds known as phenylpropylamines, which includes some important neurotransmitters such as adrenaline (epinephrine) (**5**) and noradrenaline (norepinephrine) (**6**). Selegiline (**7**) is a synthetic analogue which is widely used as a pharmacological tool and to treat PD, although its mode of action is as an inhibitor of monoamine oxidase B rather than as a dopaminergic compound. The phenylpropylamines also include several naturally-occurring compounds known as protoalkaloids, since their N-atom is not part of a heterocyclic ring. Most protoalkaloids appear to have a greater adrenergic than dopaminergic effect, although they stimulate both types of receptors. The major protoalkaloid in clinical use is ephedrine (**8**), obtained from some species of *Ephedra*, a genus which has been used for many centuries in traditional chinese medicine (TCM). Ephedrine is used principally because of its adrenergic sympathomimetic properties, but it also displays CNS-stimulant side effects, which may be due also to dopaminergic properties. The synthetic phenylpropylamines known as amphetamines are, however, primarily employed as CNS stimulants and some association between their use and alleviation of the dyskinesia often seen in PD sufferers has been noted in recent years. A natural product with structural and and pharmacological similarity to the amphetamines is cathinone (**9**), a major active constituent of *Catha edulis* Forsk. The fresh young leaves of this plant are known as "khat" and are chewed in Ethiopia, Yemen and surrounding countries to achieve a stimulatory effect on the CNS.[11] Cathinone is present in significant amounts only in the young leaves and anecdotal reports of a reduction in PD-like tremors induced by neuroleptic drugs in some regular chewers of khat indicate that cathinone may have a similar effect as amphetamines.[12]

12.2.2.4 *Ergot alkaloids and derivatives (refer to Structures 2)*

Another very important group of compounds used for the treatment of PD are derived from the ergot indole alkaloids. Ergot, *Claviceps purpurea* Tulasne, is a fungus which infects the ears of rye *Secale cereale* L. and is the species which has attracted the most pharmaceutical interest, although several other *Secale* species exist as parasites in cereal crops. Ergot has a long history of poisoning animals and humans through the ingestion of contaminated flour. It has been exploited in traditional medicine in some parts of Europe to aid childbirth, due to its vasoconstriction effects and uterine smooth muscle stimulation. The effects of ergot on the CNS are also well documented and outbreaks of "insanity" involving hallucinations have been shown to correlate with times of heavy contamination of rye flour in the communities affected. The activity of ergot has been shown to be due to the alkaloids present, such as ergotamine (**10**). All of the alkaloids contain the indole ring structure known as lysergic acid and similarities with the three neurotransmitters noradrenaline (norepinephrine), DA and serotonin can be seen (Fig. 1).

This probably explains the wide spectrum of activity of these alkaloids and so it has been found necessary to alter the chemical structure

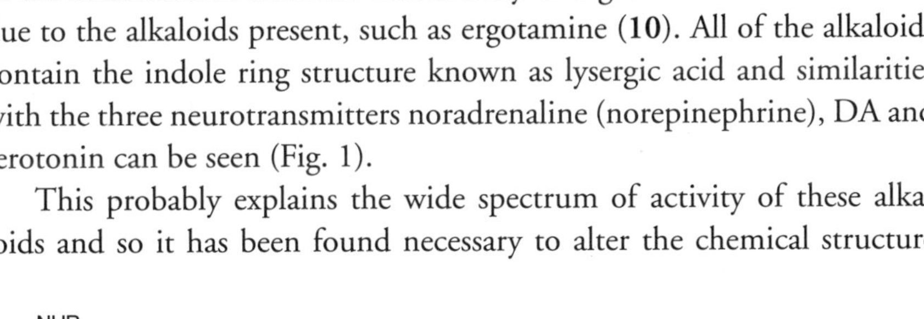

Ergot alkaloid Noradrenaline Dopamine Serotonin (5-HT) Ergot alkaloid

Fig. 1. Structural similarities between ergot alkaloids and CNS neurotransmitters.

of these alkaloids to produce compounds which bind more specifically to only one of the receptors. A large amount of derivatives of ergot alkaloids have been synthesized for differing therapeutic effects, one of these being dopaminergic receptor stimulation for use in treating PD.

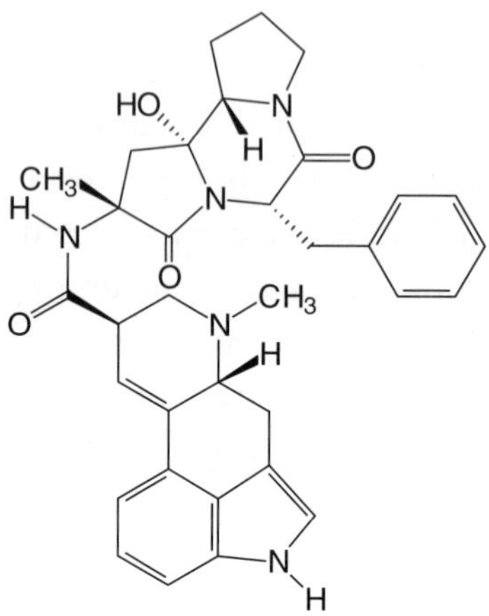

(10) Ergotamine

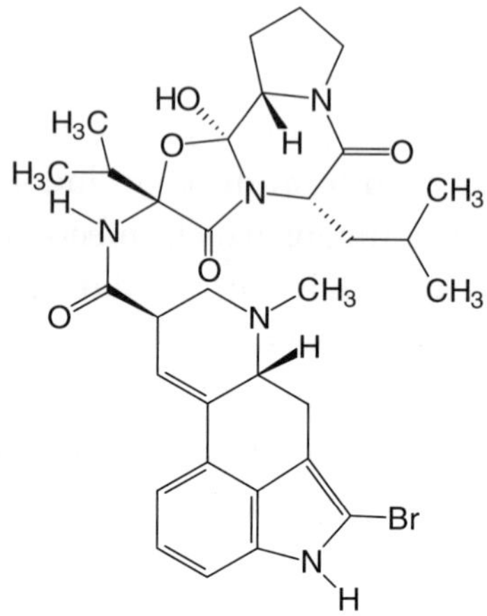

(11) Bromocriptine

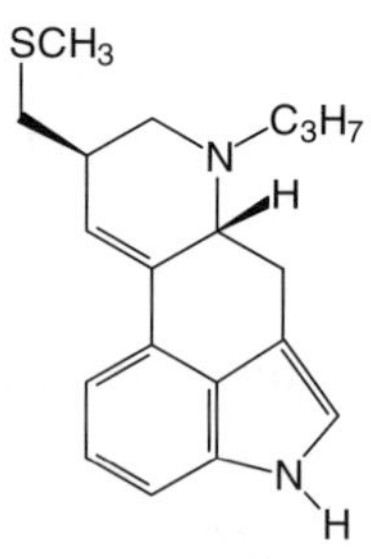

(12) Pergolide

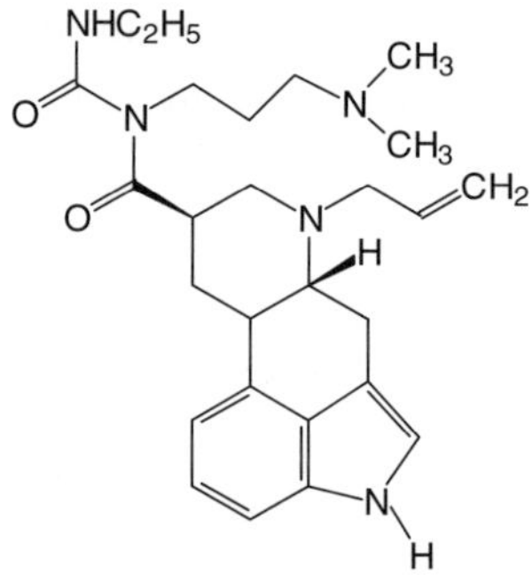

(13) Cabergoline

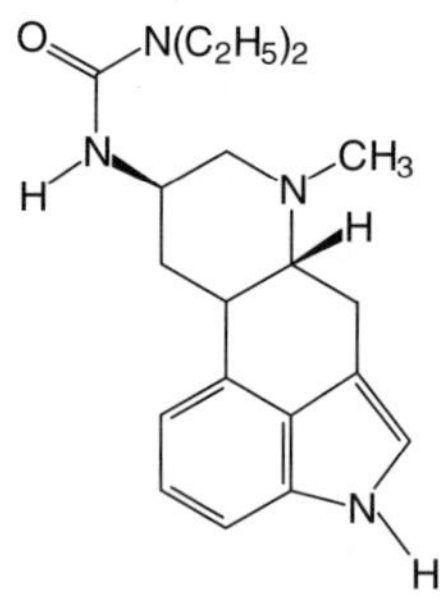

(14) Lisuride

Structures 2

Bromocriptine (**11**), pergolide (**12**), cabergoline (**13**) and lisuride (**14**) are examples of compounds which have been developed in this way and are now used clinically. The pharmacological differences between the compounds is not very great. All are D_2-dopamine receptor agonists, although pergolide also has agonist effects at D_1 and D_3 receptors. All four drugs are well established in the treatment of PD and numerous reviews exist on their clinical efficacy and application.[13,14]

Bromocriptine (**11**) and lisuride (**14**) were the first of these drugs to be introduced and have comparatively short half-lives. Pergolide (**12**) has a half-life of 8 hours[15] and cabergolide (**13**) of 68 hours, making the latter especially useful for administration only once a day.[16] The ergot-derived drugs were originally used in advanced cases of PD but there is increasing interest in their use in the first instance since, unlike L-DOPA, they do not need to be metabolized to the active compound and so are not affected by any degeneration in the dopaminergic terminals.[17]

12.2.2.5 *Other dopaminergic compounds (refer to Structures 3)*

Another alkaloid with agonist activity is apomorphine (**15**), it is derived from the opium alkaloids. It acts on the D_1 and D_2 receptors but, because it is a powerful emetic when given orally, its use is restricted to parenteral administration. It is most commonly given when patients are not exhibiting adverse effects to L-DOPA and is also employed as a diagnostic agent for dopaminergic responsiveness. More recently, other aporphines have been investigated for their dopaminergic effect and pukateine (**16**), originally isolated from the New Zealand species *Laurelia novae-zelandiae* A. Cunn. has received some attention as a possible lead compound for treating PD. This is because pukateine also has antioxidant effects and appears to cause an increase in extracellular DA.[18] The isoquinolinoid salsolinol (**17**) has been investigated extensively over the last ten years. This compound is found in chocolate and is derived from the cocoa bean and the seeds of *Theobroma cacao* L. The addictive properties of chocolate have been ascribed to the dopaminergic activity of this compound,[19] which also occurs as an endogenous catechol in the brain. (**17**) is dopaminergic at the D_2 receptor and has a protective effect against neurodegeneration.[20]

The tropane alkaloids, especially hyoscine (scopolamine) (**18**), have been used to treat PD since they increase DA activity by antagonizing cholinergic activity at the muscarinic receptors in the striatum. The naturally occurring alkaloids, found in various genera of the Solanaceae, are not

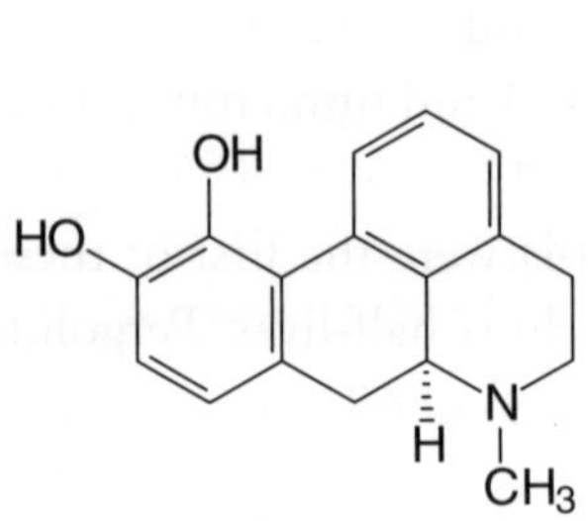

(15) Apomorphine

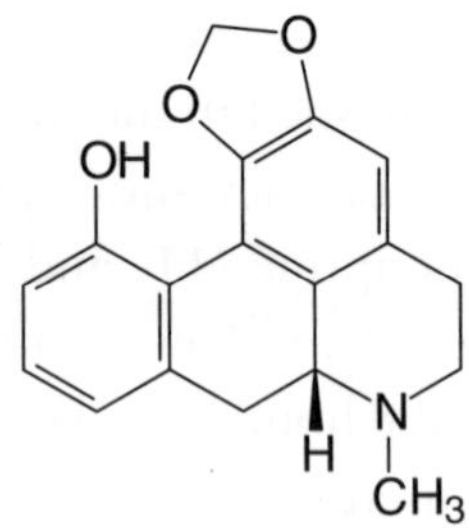

(16) Pukateine

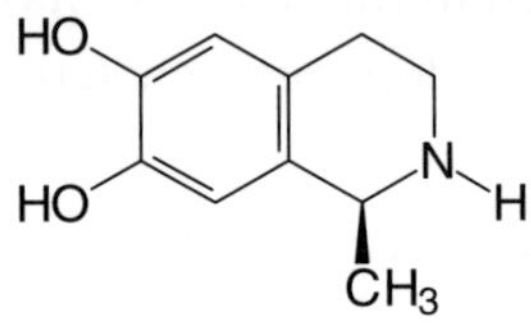

(17) Salsolinol

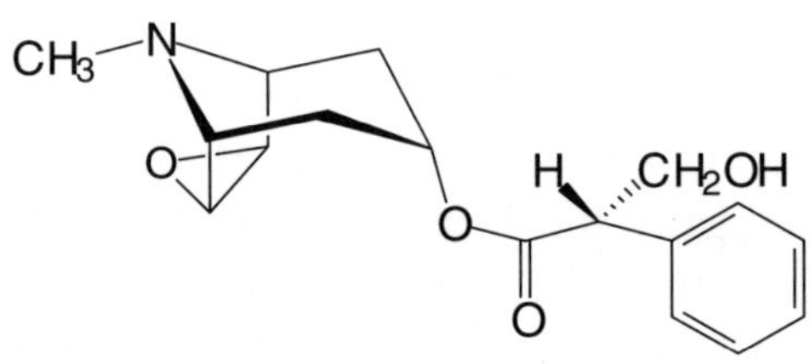

(18) Hyoscine

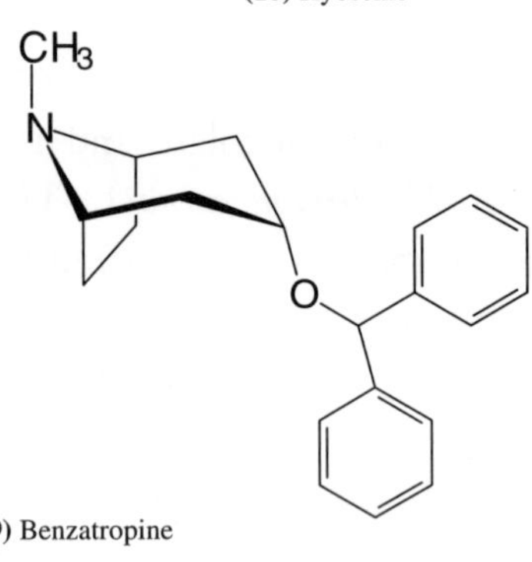

(19) Benzatropine

Structures 3

used much for this purpose but synthetic analogues e.g. benzatropine (**19**) are used since they also inhibit DA reuptake, and the resultant increase in DA compensates for the deficiency of DA associated with PD.

12.2.2.6 *PD: Monoamine oxidase inhibitors (refer to Structures 4 and 5)*

Monoamine oxidase (MAO) is a major enzyme responsible for the fast breakdown of DA and related compounds at the synapse. Inhibitors of this enzyme, known as MAOIs, cause a net increase in DA levels and although their major therapeutic use has been as antidepressants, they have potential use in PD. This has not generally been realized because of the side effects associated with the elevation of peripheral DA levels. The β-carboline alkaloids, harmine (**20**) and harmaline, (**21**) are found in several traditional medicinal plants, including *Banisteriopsis caapi* (Spruce ex Griseb.) Morton, a liana used in Brazil as an ingredient in the hallucinogenic drink "ayahuasca." It has been postulated that the MAOI properties of the *B. caapi* constituents prevent the metabolism of amines in other ingredients of ayahuasca. *B. caapi* root extracts have been reported to be used to treat PD patients in Ecuador, and (**20**) and (**21**) were shown to stimulate the release of DA from striatal cells.[21] These compounds therefore stimulate the release of DA and further elevate its levels by reducing its breakdown, and this may underlie the reputed improvements in PD patients when the extract was taken.

Flavonoids in the diet have been widely promoted as important antioxidant contributors. Their neuroprotective properties, because of this effect, have been demonstrated by several workers. However, they have also been demonstrated to have MAOI activity and this has been proposed as part of the explanation of the use of the common herb, St John's Wort, *Hypericum perforatum* L., as an antidepressant.[22] This dual role has now been proposed for a variety of flavonoids, such as kaempferol (**22**) from the leaves of *Ginkgo biloba* L., a widely used herbal product which has been suggested as a preventative agent against neurodegeneration.[23] Quercetin (**23**), similarly, has also shown to inhibit MAO-B[24] and reverse the effects of induced catalepsy, which mimics the bradykinesia associated with PD.[25] Tangeretin (**24**) also inhibits MAO-B and crosses the blood brain barrier in a rat model,

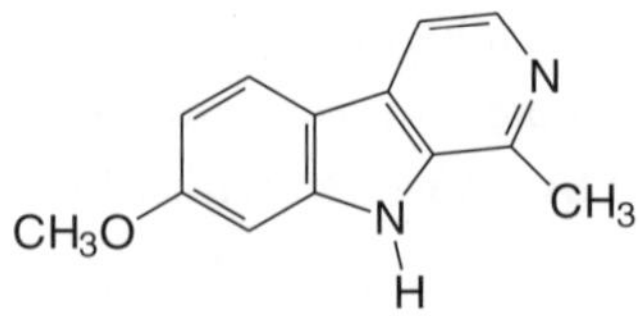

(20) Harmine

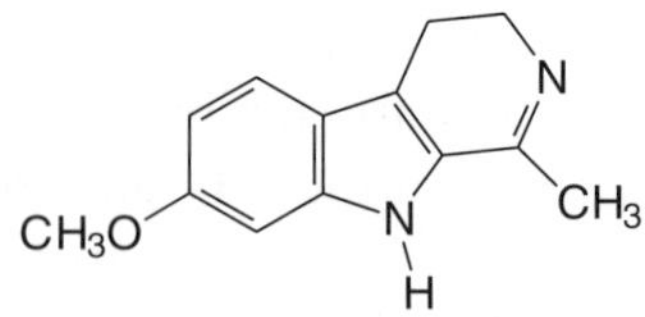

(21) Harmaline

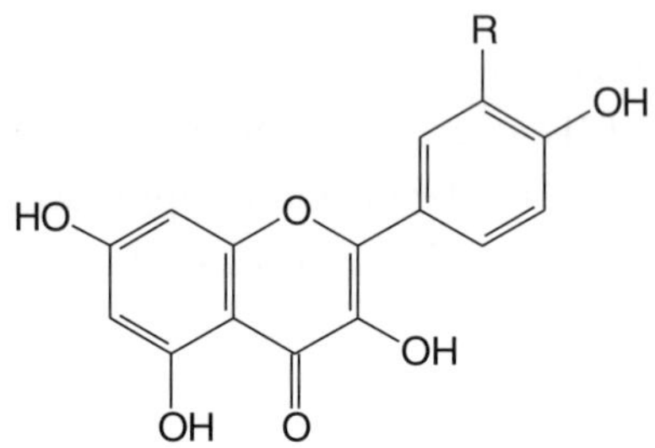

(22) R = H Kaempferol
(23) R = OH Quercetin

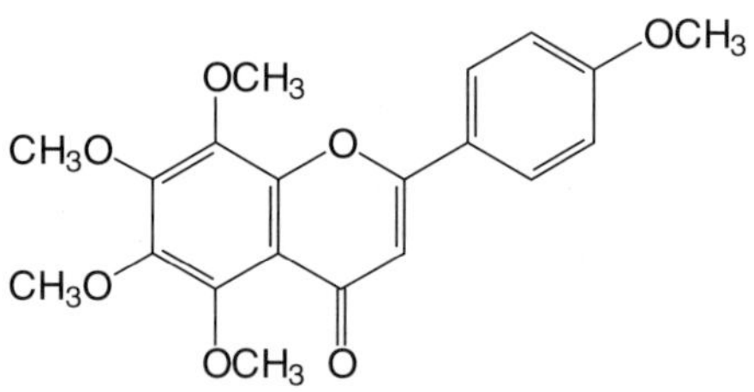

(24) Tangeretin

Structures 4

with consequent reduction in DA depletion, so it may have therapeutic potential.[26] The polyphenol (-)-epigallocatechin-3-gallate (**25**), found in the unfermented ("green") tea leaves from *Camellia sinensis* (L.) Kuntze, has a similar polyvalent activity which may be enough to give a protective effect in PD and other conditions.[27] Since many flavonoids occur in reasonably large amounts in common fruits and vegetables, it is possible that the apparent increase in the incidence of neurodegenerative disease is related

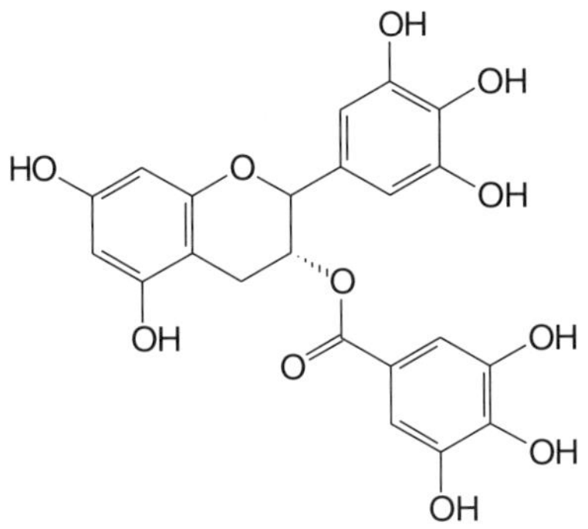

(**25**) Epigallocatechin-3-gallate

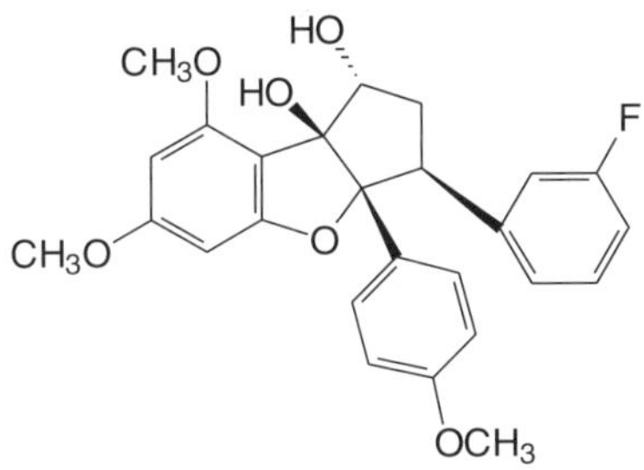

(**26**) Fluororocaglaol

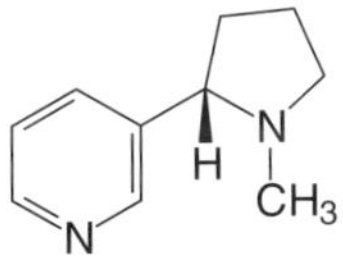

(**27**) Nicotine

Structures 5

to the decline in dietary consumption of these commodities in some sectors of the industrialized world. A different mechanism of protective action apparently occurs with the fluoro derivative of rocaglaoal (**26**), a compound found in *Aglaia odorata* Lour. Fluororocagaol exhibited beneficial effects when given to rats modeling PD and this was ascribed to the reduction of tissue inflammation and of cell death by inhibition of NF-κB.[28]

12.3 ALZHEIMER'S DISEASE

12.3.1 Introduction

Alzheimer's disease (AD) is the major disease of a group which is characterized by the loss of cognitive function leading to dementia. AD is estimated to account for 50% to 60% of dementia cases in persons over 65 years of age.[29] It is a progressive, neurodegenerative disease that primarily affects the elderly population. The main symptoms associated with AD are a decline in cognitive function, which is primarily memory loss,[30] and in the later stages of the disease, language deficits, depression, agitation, mood disturbances and psychosis are often seen.[31]

AD, as a defined medical condition, has only existed for about 100 years, but the loss of memory and cognitive decline as a feature of old age has been documented throughout human history. Ancient writings describe the symptoms but also suggest remedies, based usually on plant extracts, some of which have shown to have relevant activities in recent research and have yielded active compounds of interest.

A feature of AD is the low levels of the neurotransmitter ACh (**2**) revealed by post mortem examination[32] and there also appears to be a depletion of nicotinergic function, since adverse symptoms of AD are reversed with nicotine (**27**), which is reported to increase ACh release by upregulating nicotinic receptors.[33] ACh is certainly associated with cognitive function, since when it is blocked from acting on the nicotinic cholinergic receptors by drugs such as hyoscine (**18**), severe cognitive impairment in the patient is observed. It is still unclear if the low levels of ACh in the CNS are cause or effect as far as AD is concerned, but symptomatic relief by the restoration of ACh levels has been a reasonably successful therapeutic approach in the last 15 years. The synthetic compound, tacrine (**28**), was the first clinically used drug introduced into the clinic. It was the first of several compounds which increase the levels of ACh by inhibition of acetylcholinesterase (AChE), the enzyme responsible for the rapid breakdown of ACh after its release from the nerve ending. This inhibition results in ACh having a longer half-life and therefore its concentration at the synapse increases. (**28**) is no longer used since the more recent introductions, which include at least two based on natural products, are safer and have longer lasting effects. However, it should be stressed that AChE inhibitors only alleviate some of the cognitive symptoms of the disease and do not achieve any permanent improvement. Another approach which has been tried is the employment of cholinergic and, to some extent, nicotinergic, agonists but this has not proved to be as useful therapeutically as the inhibition of cholinesterase.

It should be noted that other types of activity, besides the increase of neurotransmitter levels, are also being explored as leads for chemotherapeutic treatment of AD. These include the prevention of glutamate-mediated neurotoxicity by antagonism of N-methyl-D-aspartate (NMDA) receptors, antioxidants, antiinflammatories and inhibitors of β-amyloid synthesis, but these are not covered to any extent in this chapter.

12.3.2 AD: The Use of Cholinergic Compounds

12.3.2.1 *Introduction (refer to Structures 6)*

The rationale underlying the use of cholinergic compounds is that these are agonists of the nicotinic cholinergic receptor and therefore compensate for the low levels of ACh. Cholinergic compounds have been suggested as

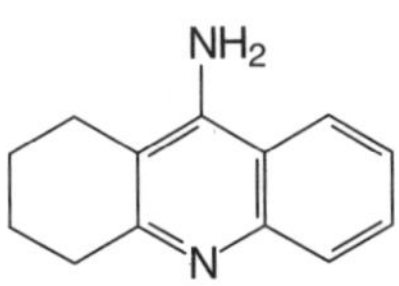

(28) Tacrine

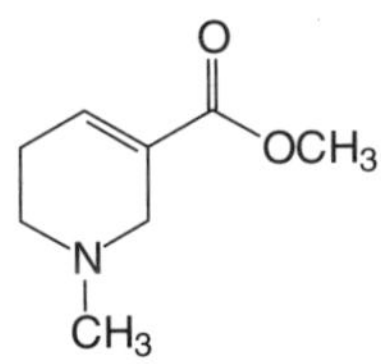

(29) Arecoline

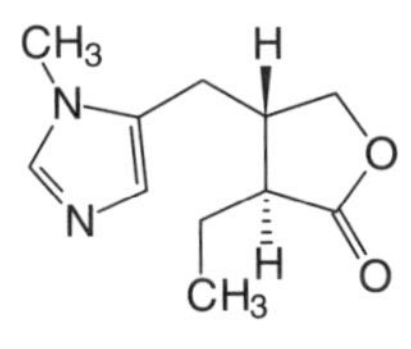

(30) Pilocarpine

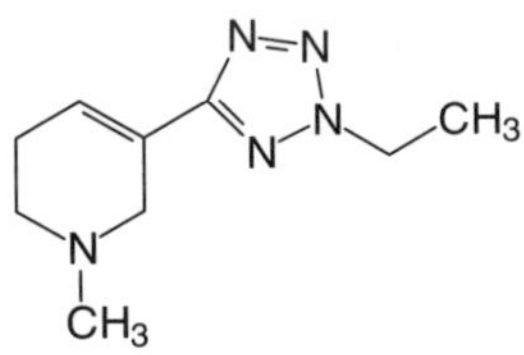

(31) LU 25-109

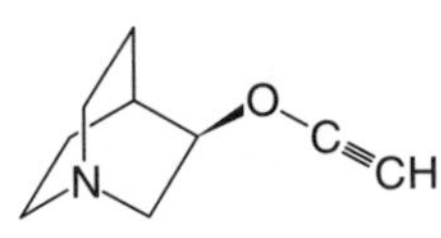

(32) Talsaclidine

Structures 6

valuable agents in treating AD because they also appear to inhibit fibrillary tangle and amyloid production. However, success has been limited as far as clinical studies are concerned, although results in animals were initially promising.[34] Two major alkaloidal natural products that are known to be cholinergic are arecoline (**29**) and pilocarpine (**30**).

12.3.2.2 *Arecoline and pilocarpine (refer to Structures 6)*

Arecoline (**29**) is the major alkaloid present in betel nut, the fruit of the palm tree *Areca catechu* L., which is extensively used as a masticatory throughout the Indian subcontinent and other parts of southeast Asia. It is commonly known in India as "paan" and it is estimated that 500 million people regularly chew it in a form which is usually shredded, mixed with lime and wrapped in a leaf of *Piper betel* L. Excessive salivation occurs, which is a direct result of its cholinergic activity, and a mild CNS stimulation also occurs. There was some interest in (**29**) as a treatment for AD since it showed improvement in memory tests in rats[35] and a small clinical study showed that, when (**29**) was given continually by intravenous infusion to AD patients, it enhanced verbal memory.[36] Derivatives of arecoline have been synthesized in order to improve the selectivity for cortical muscarinic ACh receptors and two examples are Lu 25-109-T (**31**) and talsaclidine (**32**), which are muscarinic M_1 receptor agonists. Although (**31**) showed encouraging results *in vitro*,[37] it failed to improve cognition when tested clinically in patients with mild to moderate AD.[38] Talsaclidine (**32**) has been shown to increase cholinomimetic central activation in animals and humans without some of the side effects seen with AChE inhibitor therapy, but cognitive function did not significantly improve.[39] Tests on rhesus monkeys did however show some improvement in memory-related tasks but at doses which gave unacceptable side effects.[40]

12.3.2.3 *Pilocarpine*

Pilocarpine (**30**) is one of a series of related alkaloids found in the South American plant genus *Pilocarpus*, known commonly as Jaborandi leaf, which was used in traditional medicine to induce sweating and urination. The molecular structure of (**30**) bears similarities to ACh (**2**) since the positively-charged N-atom and the lactone binding to the serine are about the same distance apart. Chewing the leaf results in typical features of cholinergic stimulation such as copious salivation and constriction of the pupils. (**30**) has shown to enhance memory performance in aged rats[41] but no studies on its application in humans for the treatment of AD have been reported. It appears to have been discarded as a potential therapeutic lead because of its undesirable side effects and also because it cannot pass through the blood brain barrier, making it unlikely that it reaches the CNS in significant amounts.

12.3.3 AD: The Use of Cholinesterase Inhibitors

12.3.3.1 *Introduction*

To avoid the undesirable effects of excess cholinergic stimulation, ACh is rapidly hydrolyzed after its release at the synapse by an enzyme named acetylcholinesterase AChE. A similar enzyme, butylcholinesterase BuChE also occurs. If the cholinesterase is inhibited, the ACh is not hydrolyzed so rapidly, and levels of ACh rise.

AChE EC 3.1.1.7 consists of a complex protein of the α/β hydrolase fold type having an overall ellipsoid shape containing a deep groove, usually called the gorge, which is about 20 Å deep. Although the hydrolysis of ACh appears to take place at the bottom of the gorge, the mechanism is fairly complex and it is thought to involve initial binding of ACh by weak nonpolar forces to the rim of the cleft in a region called the "peripheral site." At the bottom of the gorge, where the actual hydrolysis occurs, there are four main subsites, these being the esteratic site, the "oxyanion hole", the "anionic subsite" and the "acyl pocket" and these are shown in Fig. 2.

The substrate enters the enzyme gorge and the product is released in conjunction with defined conformational changes, which have been described in more detail in Ref.[42] The important features required in a molecule which binds to the active site include an amine which becomes positively charged at the pH of the immediate environment and so binds to an aspartate in the receptor. In light of this, it is not surprising that most of the potent AChE inhibitors are alkaloids which exhibit a positively-charged N-atom at the body's pH. Other features associated with binding to the active site are a portion of the molecule being able to form hydrogen bonds with the aspartine domain of the receptor and the small groups being able

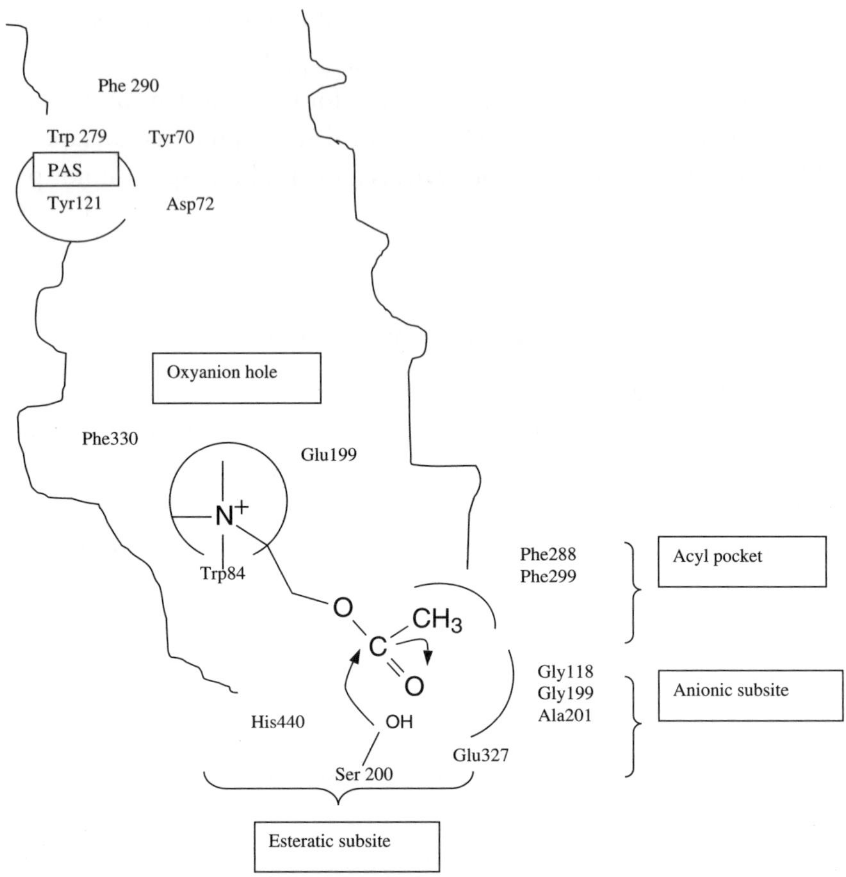

Fig. 2. The acetylcholinesterase "gorge" showing regions where binding takes place and interaction of acetylcholine with active site.

to bind to hydrophobic sites near the aspartate region. In addition, the receptor lies in a pocket which allows only small molecules to enter.

12.3.3.2 *Physostigmine and related compounds (refer to Structures 7)*

The "classic" cholinesterase inhibitor is the alkaloid physostigmine (**33**), also called eserine. Investigations carried out in the nineteenth century on the ordeal poison "esere," which consisted of an extract of the Calabar Bean, the seeds of *Physostigma venenosum* Balf., resulted in the isolation of (**33**).[43]

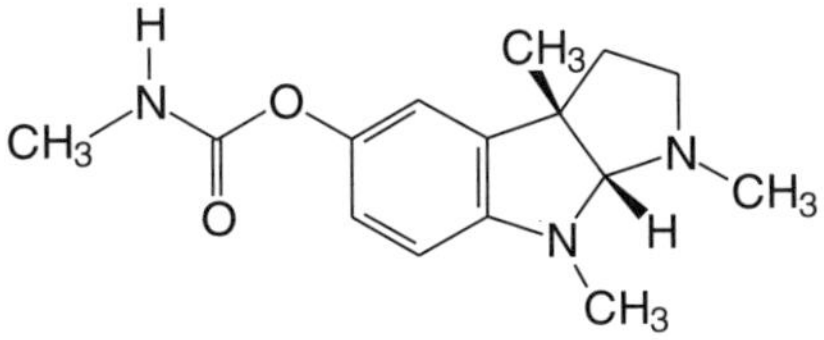

(33) Physostigmine

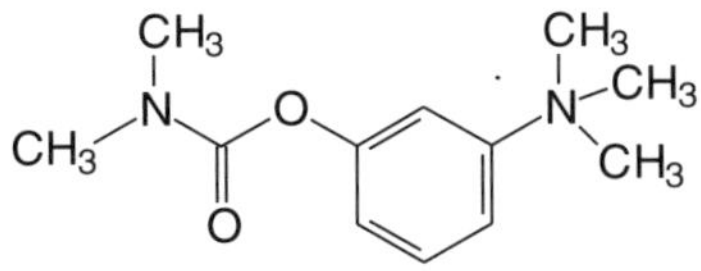

(34) Neostigmine

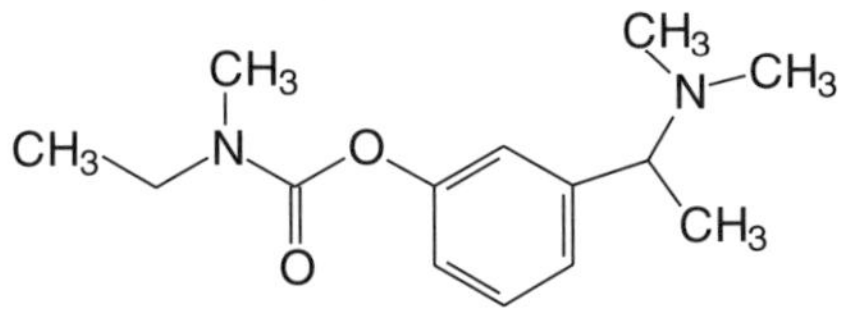

(35) Rivastigmine

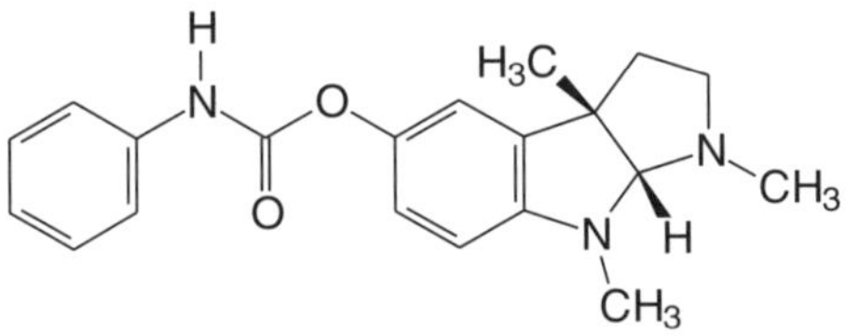

(36) Phenserine

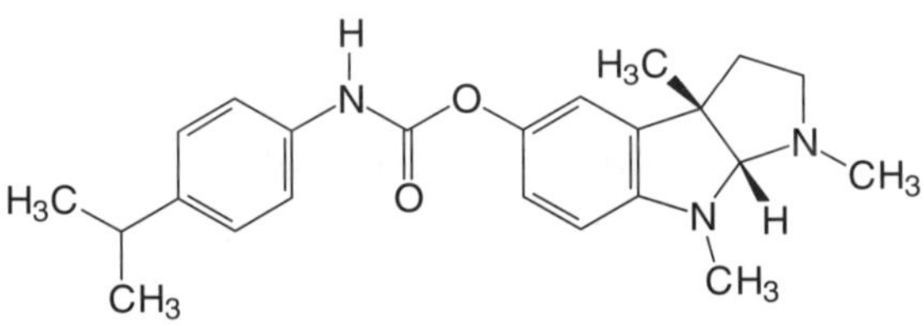

(37) Cymserine

Structures 7

The toxic effects of Calabar Bean extract were found to be due to excessive cholinergic stimulation giving rise to symptoms such as increased salivation, nausea, bradycardia, muscle cramps and respiratory failure, as well as CNS effects such as agitation. The cholinergic excess was found to be caused by the inhibition by (**33**) of the rapid breakdown of acetylcholine, effected naturally by the enzyme known as acetylcholinesterase (AChE).

In therapeutics, cholinesterase inhibition had, until recently, a somewhat limited application in ophthalmology and the treatment of myasthenia gravis. However, the connection between ACh levels and cognitive function led to the findings that early symptoms of AD could be improved upon by the use of cholinesterase inhibitors and this renewed the interest in AChE inhibitors. Physostigmine (**33**) was known to cross the blood brain barrier and several *in vivo* studies showed that it reduced the symptoms of ACh deficiency in the CNS e.g. it was reported to protect mice against cognitive impairment caused by oxygen deficit and it improved learning in rats.[44] Significant cognitive benefits were observed in both normal and AD human patients,[45] but the short half-life of (**33**) prevented its application clinically in AD patients, since this would require multiple daily dosing.

Many analogues of physostigmine have been synthesized to improve its pharmacokinetic profile, and, more recently, to provide compounds with selective BuChE inhibition. Neostigmine (**34**) was developed to treat myasthenia gravis, being the most widely used drug for this disease. Neostigmine is a quaternary amine and this feature severely impairs its ability to cross the blood brain barrier and so it is of little value in treating AD, but is used to treat myasthenia gravis. Rivastigmine (**35**) was produced with the express purpose of engineering a better pharmacokinetic profile for usefulness in AD and it inhibits the G_1 form of AChE in the cortex and hippocampus, which are brain areas involved in cognition,[46] and it has been shown to improve cognition in AD patients.[47,48] Clinical studies have borne out the usefulness of rivastigmine (Exelon®) in mild to moderate AD and it has been licensed as a treatment for symptomatic relief of AD since 2000.

As well as inhibiting AChE, physostigmine also inhibits butylcholinesterase (BuChE) another enzyme found in the CNS. BuChE has recently been implicated in the aetiology and progression of AD.[49] Inhibition of BuChE may therefore prove to be beneficial in treating AD, particularly because BuChE has been shown to produce β-amyloid proteins and to enable them to diffuse into the plaques observed in neurons of AD patients.[50,51] Inhibition of BuChE is therefore likely to stop the production of this toxic protein, so compounds with a strong BuChE inhibitory effect might be useful in treating AD. Over the past 10 years, derivatives of physostigmine have been developed as specific BuChE inhibitors, with phenserine (**36**) and cymserine (**37**) providing starting points for the synthesis of a range of analogues. Phenserine (**36**) was developed primarily as a selective AChE inhibitor and methyl substitution at 2′ was shown to increase AChE selectivity whilst the same substituent at 4′ favored inhibition of BuChE.[52,53] Cymserine (**37**) showed up to 20-fold selective BuChE inhibitory activity compared with AChEI and two of its analogues showed a much greater effect, at least two orders of magnitude in favor of BuChE.[54] The same study showed that the compounds also showed significant improvements in learning experiments in rats and reduced β-amyloid levels in mice, thus supporting the hypothesis that BuChE inhibition might be useful in treating AD.

12.3.3.3 *Galantamine and other amaryllidaceae alkaloids (refer to Structures 8)*

Galantamine (**38**) (sometimes referred to as galanthamine) is found in members of the Amaryllidaceae e.g. the Chinese medicinal herb *Lycoris radiata* Herb. and the European *Galanthus nivalis* L. and *Narcissus* spp.[55] The ethnopharmacological basis for plants containing this compound is not very clear but a full report of the history of the development of galantamine from *Galanthus nivalis* has been recently published.[56] Galantamine has been licensed in Europe for the treatment of AD since 2001. Several good clinical trials have demonstrated good tolerance and significant improvement in cognitive function when galantamine is administered to AD patients.[57–59] These benefits appear to be sustained for at least three years, a much longer time than for other drugs of this type.[60] Galantamine is well absorbed when given orally and is more selective for AChE than BuChE.[61] It also stimulates nicotinic receptors,[62] which may also enhance cholinergic function and memory (see below). Recent studies suggest that galantamine may have therapeutic advantages over other AChE inhibitors and has added value in vascular dementia as well as AD.[63] Several other alkaloids with AChE inhibitory activity have been recently reported from

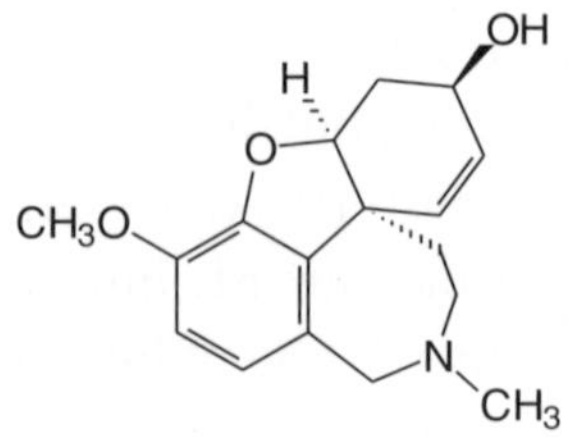

(38) Galantamine

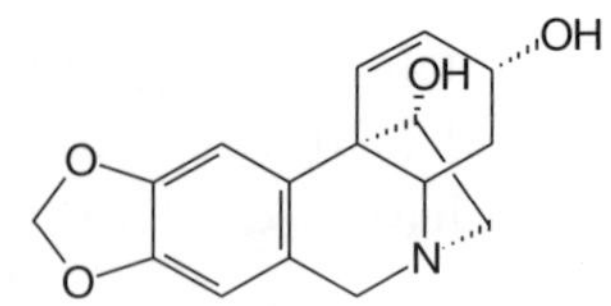

(39) Hamayne

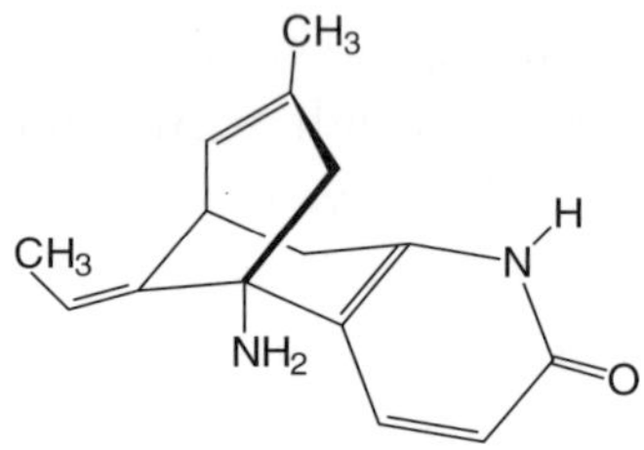

(40) Huperzine A

Structures 8

members of the Amaryllidaceae, including Iberian *Narcissus* species[64] and two Nigerian *Crinum* species used in traditional medicine to help ailing memory.[65] Hamayne (**39**), was the most active alkaloid (IC_{50} value of 250 μM) although it was much weaker than physostigmine.

12.3.3.4 *Huperzine A and analogues (refer to Structures 8 and 9)*

Another natural alkaloidal cholinesterase inhibitor, huperzine A (**40**) is used clinically in China for treating AD. (**40**) is found in the clubmoss *Huperzia serrata* Thunb. and one of its uses in traditional Chinese medicine is to alleviate problems of memory loss.[66] Huperzine A (**40**) reversibly inhibits AChE *in vitro* and *in vivo*[67] and has been shown to improve memory in cognitively impaired rats[68] and to be neuroprotective against β-amyloid peptide fragment 25–53 and free radical-induced cytotoxicity.[69] Clinical evidence for its efficacy also exists e.g. in a multicenter double blind trial, (**40**) significantly improved memory and behavior in AD patients, and was less toxic than the synthetic AChE inhibitors donepezil and tacrine (**28**).[70] There are numerous reports describing the synthesis

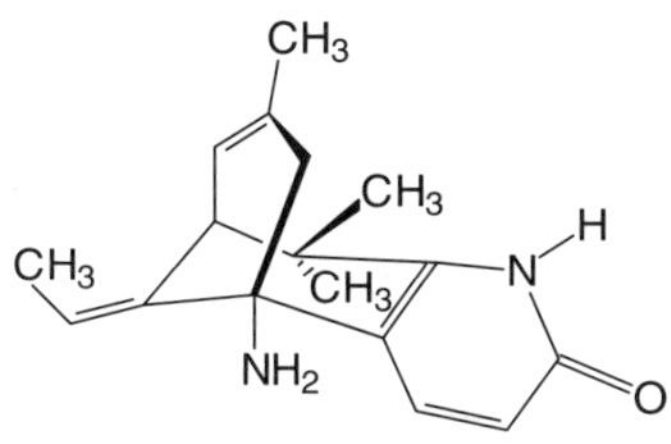

(41) Dimethylhuperzine A

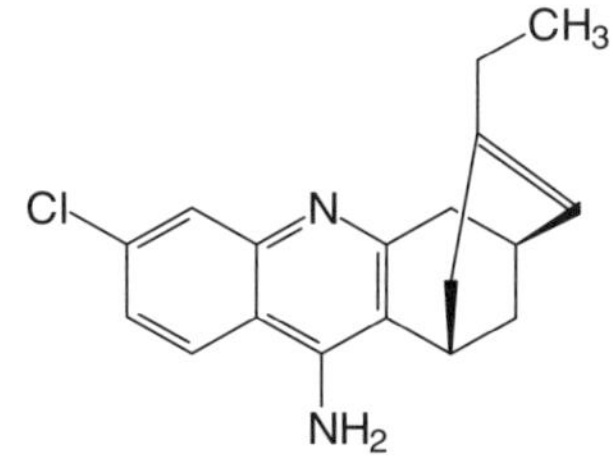

(42) Huprine X

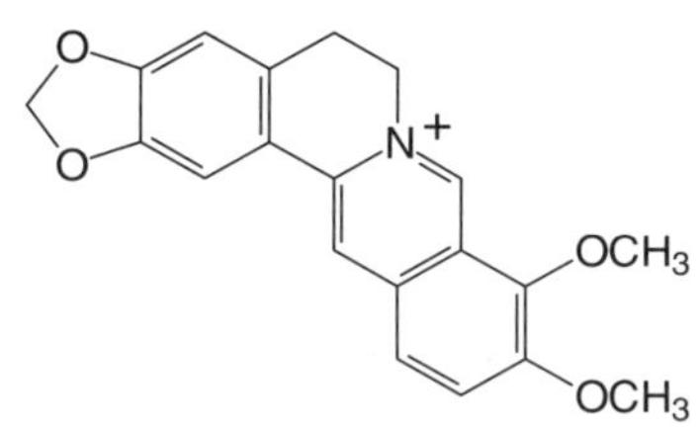

(43) Berberine

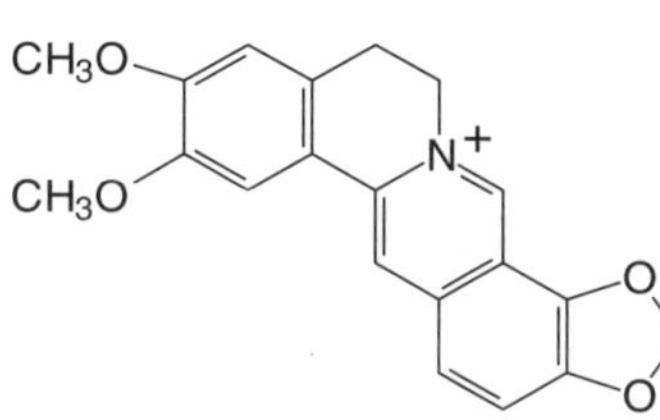

(44) Coptisine

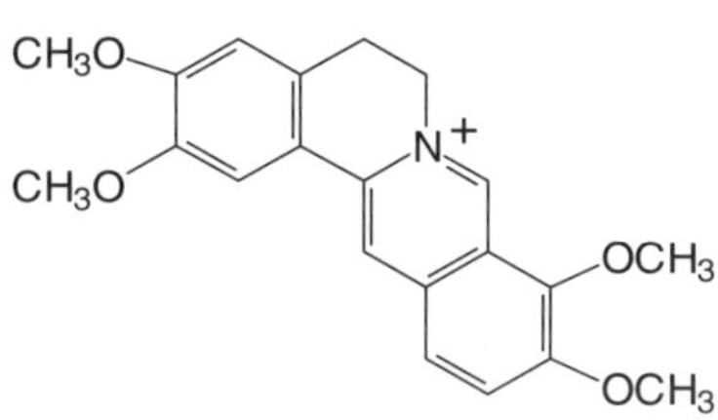

(45) Palmatine

Structures 9

and antiChE activity of a variety of huperzine analogues. With the aim of developing a compound to enhance the H-bond between the C-14 methyl of huperzine A and the backbone carbonyl of His440 on AChE, (±)-14-fluorohuperzine A was synthesized but *in vitro* tests showed the racemic form of 14-fluorohuperzine A to inhibit AChE with a potency of 62 times less than huperzine A.[71] The synthesis of a number of analogues of huperzine A showed the necessity of the C-5 amino group for activity, since it forms a quaternary ammonium under physiological conditions to imitate ACh.[72] (−)-Dimethylhuperzine A (**41**) shows comparable AChE inhibitory activity to (−)-huperzine A, but the (+) enantiomer was inactive, although both enantiomers were equally effective in protecting against glutamate-induced neurotoxicity.[73]

It is interesting that, in addition to its AChE inhibitory effects, huperzine A is reported to inhibit NMDA receptor binding and this is also of use in treating AD.[74] Recently, three compounds, huprine X (**42**) and its F and Br analogues, huprine Y and huprine Z have been synthesized. These compounds combine the carbobicyclic structural feature of huperzine A (**40**) with the 4-aminoquinoline skeleton of tacrine (**28**).[75,76] All three compounds showed a very strong selectivity for AChE over BuChE and also for human as opposed to bovine AChE and this was demonstrated *in vivo* as well as *in vitro*.

12.3.3.5 *Other alkaloids (refer to Structures 9–13)*

Most other alkaloids shown to have AChE inhibitory activity have been isolated from plants used in traditional medicine. *Coptis chinensis* Franch has been used in TCM for several conditions including age-related cognitive and memory decline. Some alkaloids found in this species e.g. berberine (**43**), coptisine (**44**) and palmatine (**45**) are reported to also be antiChE.[77] *Coptis chinensis* extract improved a scopolamine-induced learning and memory deficit in rats and this is likely to be due to the alkaloids present thus raising ACh levels.[78] Berberine (**43**) has been shown to selectively inhibit AChE compared with Bu ChE[79] and it has been shown to improve scopolamine-induced amnesia in rats.[80]

Rutaecarpine (**46**) is the major alkaloid found in *Evodia rutaecarpa* (Juss.) Benth., and activities relevant to AD have been identified with the extract and with rutaecarpine. Dehydroevodiamine (**47**), another alkaloid from the same species, inhibited AChE *in vitro*, and reversed scopolamine-induced memory impairment in rats[81] and increased cerebral blood flow *in vivo* in cats, a property which would supplement its usefulness in AD.[82] The structures of (**46**) and (**47**) and tacrine (**28**) have been used as templates for the development of a series of synthetic compounds which have been evaluated for their antiChE activity. These were found to be inhibitory against both AChE and BuChE with *N*-(2-phenylethyl)-*N*-[(12*Z*)-7,8,9,10-tetrahydroazepino [2,1-*b*]quinazolin-12(6*H*)-ylidene] amine (**48**) showing higher affinity for BuChE.[83]

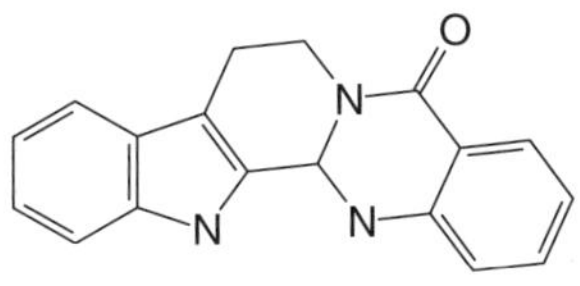

(46) Rutaecarpine

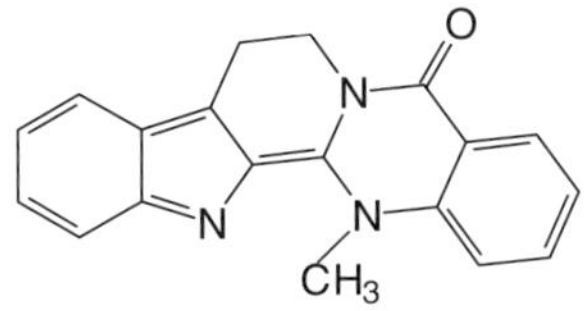

(47) Dehydroevodiamine

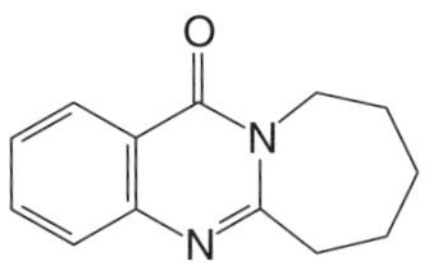

(48) N-(2-phenylethyl)-N-[(12Z)-7,8,9,10-tetrahydroazepino-[2,1-b]quinazolin-12(6H)-ylidene]amine

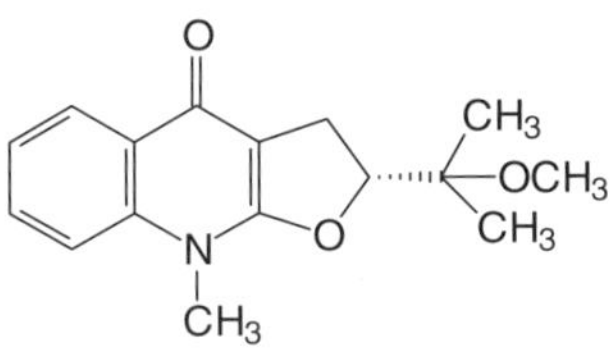

(49) Methyl isoplatydesmine

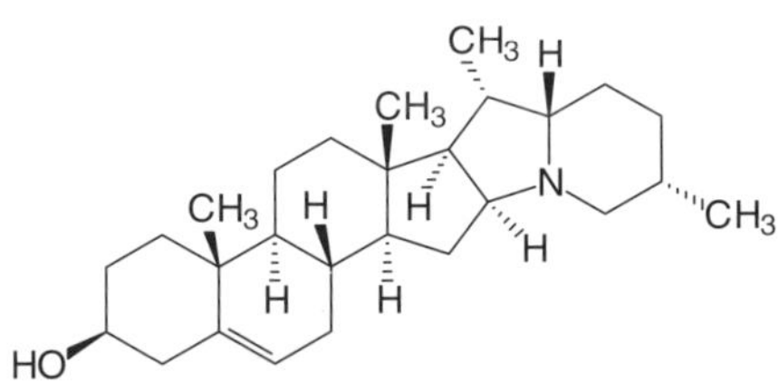

(50) Solanidine

Structures 10

Some quinoline alkaloids from the aerial parts of *Skimmia laureola* (DC.) Dcne have shown antiChE activity *in vitro* and methyl isoplatydesmine (**49**) was shown to be the most active inhibitor of both AChE and BuChE.[84] Solanidine (**50**), related compounds and glycosides are

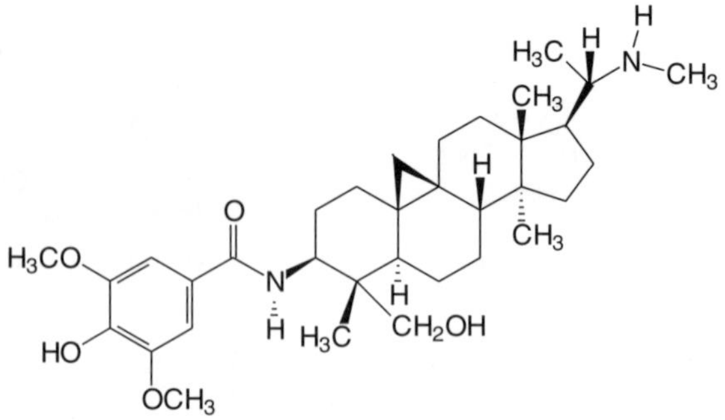

(51) Buxakashmiramine

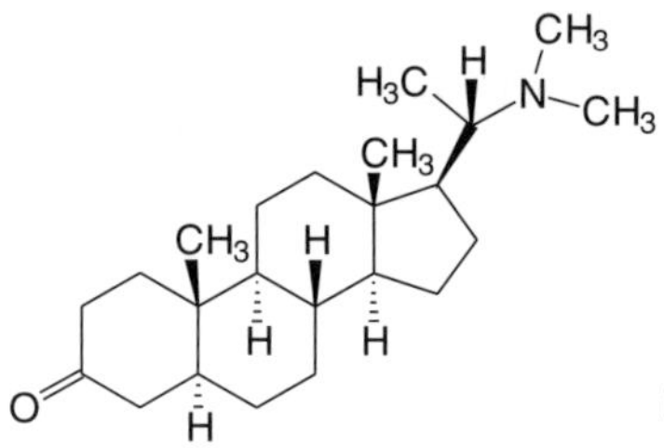

(52) Funtumafrine C

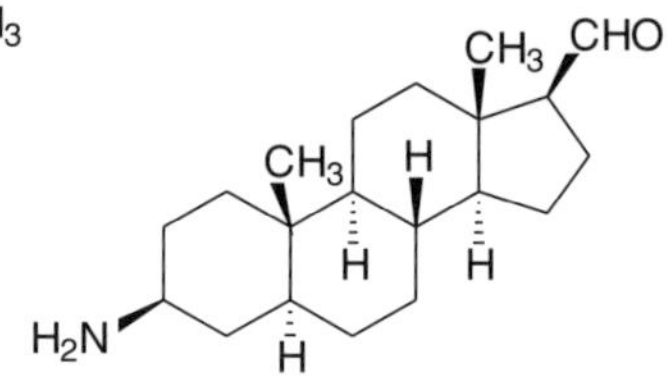

(53) N-methylfuntumine

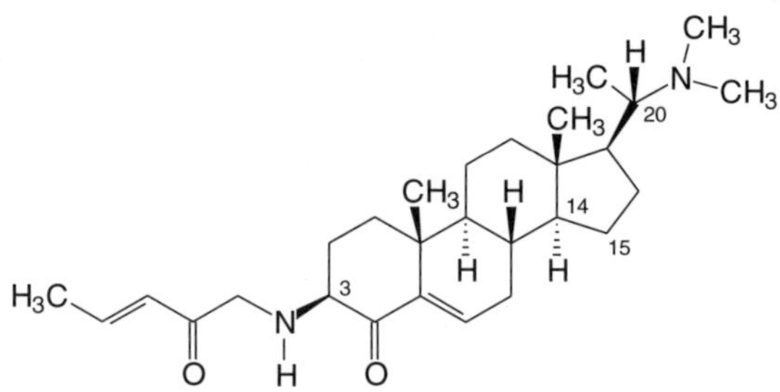

(54) Sarsalignone
(55) $\Delta^{14,15}$ Sarsalignenone

Structures 11

steroidal alkaloids produced in green parts of the genus *Solanum*, which includes the potato. The toxic properties of these alkaloids are due to AChE inhibition, with characteristic signs of cholinergic excess such as sweating, palpitations and CNS disturbances including hallucinations.[85]

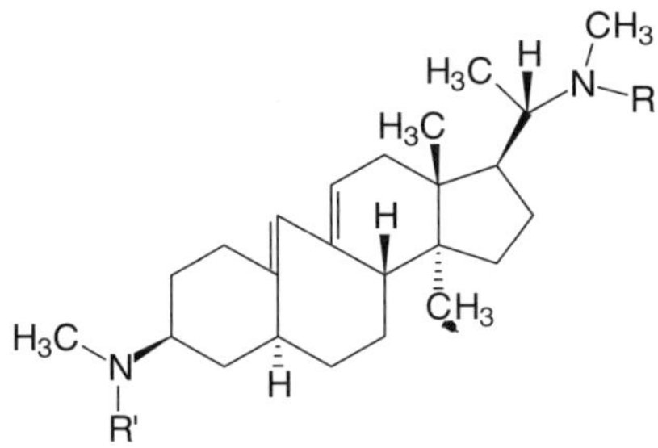

(56) R = CH_3, R′ = H Buxamine B
(57) R = H, R′ = CH_3 Buxamine C

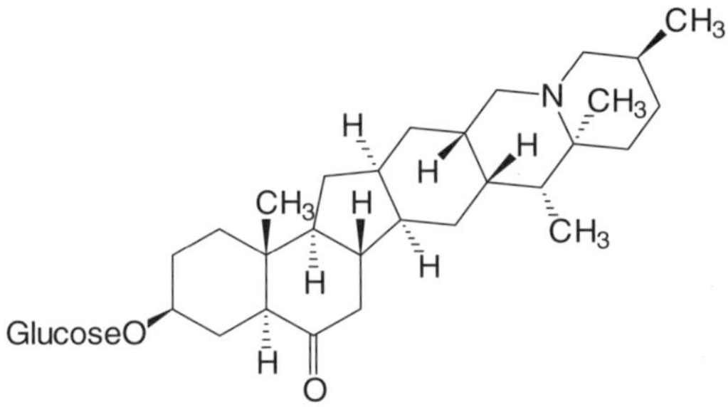

(58) Yibeinoside A

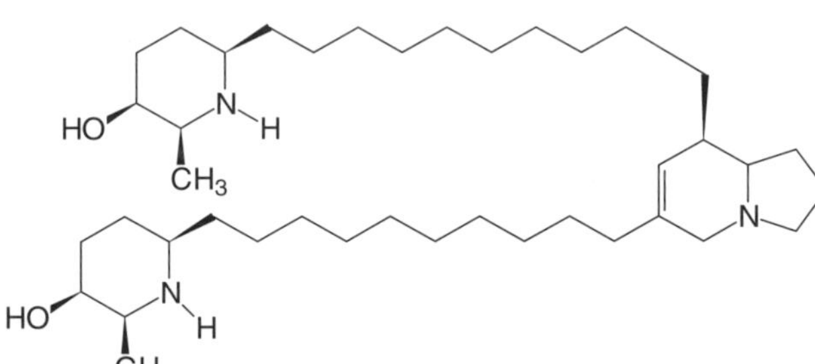

(59) Juliflorine

Structures 12

These toxic symptoms possibly explain why such plants are not reported to be used traditionally for AD or related conditions. The Buxaceae is another family known for its steroidal alkaloids and several of these have been investigated as cholinesterase inhibitors. The first such alkaloid

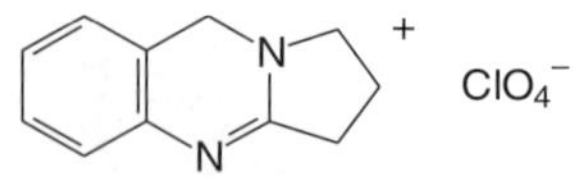

(60) Deoxypeganine perchlorate

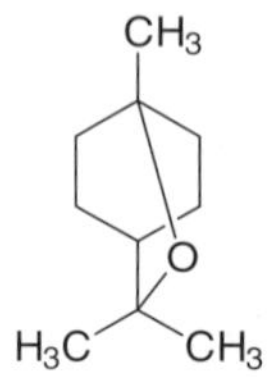

(61) 1,8 Cineole

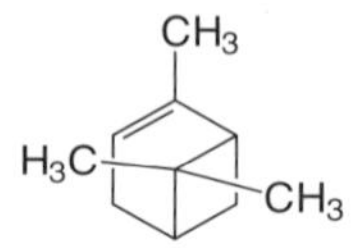

(62) α-Pinene

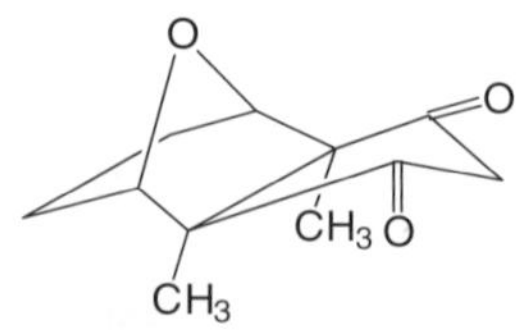

(63) Cantharidin

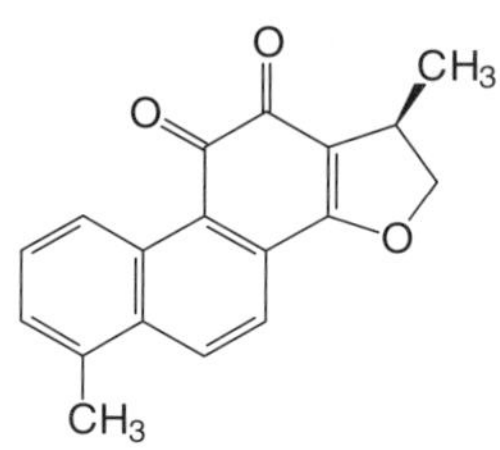

(64) Dihydrotanshinone

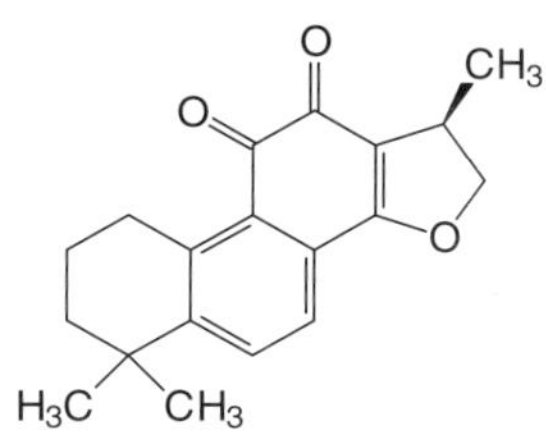

(65) Cryptotanshinone

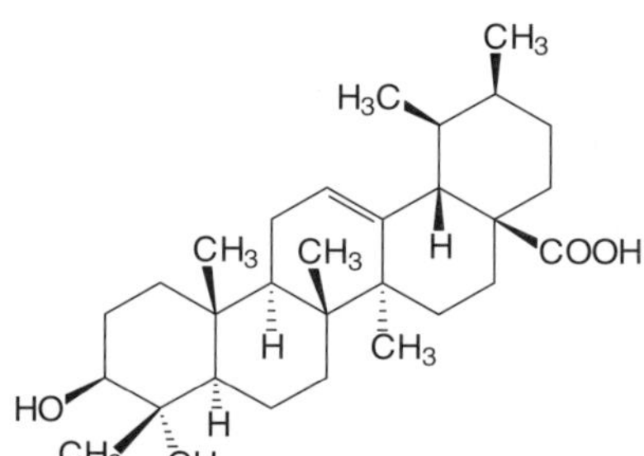

(66) Ursolic acid

Structures 13

reported was buxakashmiramine (**51**) which had moderate activity (IC_{50} 25.4 μM).[86] Steroidal alkaloids from the related genus *Sarcococca coriacea* Sweet have been shown to have similar activity with funtumafrine C (**52**) and N-methylfuntumine (**53**) giving IC_{50} values of 45.8 μM and 97.6 μM respectively.[87] A large number of weakly inhibitory pregnane-type alkaloids from *S. saligna* Müll. Arg. have been isolated.[88] Structure-activity studies suggest that these polycyclic compounds penetrate the aromatic gorge with ring A entering first due to its more hydrophobic character, or due to increased electropositivity of the substituents on ring A. Some of the most active antiChE compounds found are sarsalignone (**54**) and sarsalignenone (**55**). Their potency could be related to the carbonyl group at the C-4 position, which may bind with peripheral site residues such as Tyr-121. Other active steroidal alkaloids have been isolated from *Sarcococca hookeriana* Baill., and also from *Buxus* species. Buxamines B (**56**), C (**57**) from B. *papillosa* C.K. Schneid. and *B. hyrcana* Pojark. inhibited AChE concentration-dependently and noncompetitively (IC_{50}74 μM and 7.5 μM, respectively).[89] (**56**) and (**57**) differ only by the substitution at C-3 and C-20 and it is the amino substituents at C-20 and C-3 in particular that are considered important for the antiChE effects of these compounds. Docking studies suggest that (**57**) penetrates deeper into the AChE gorge and that the positioning of the C-3 tertiary amino group resembles the quaternary ammonium group of ACh more than in (**56**), which may explain the observed differences in its inhibitory activity.

A number of steroidal alkaloids isolated from *Fritillaria* species have been evaluated for their ability to inhibit both AChE and BuChE *in vitro,* with yibeinoside A (**58**) giving the greatest activity (IC_{50} 6.5 μM and 7.3 μM against AChE and BuChE respectively.[90]

Various other types of alkaloids have been shown to inhibit choline esterases. Juliflorine (**59**) from *Prosopis juliflora* DC. appears to block both the peripheral and anionic binding sites in the enzymes, with a slight selectivity for BuChE (IC_{50} values of 0.42 μM and 0.12 μM against AChE and BuChE respectively).[91] An alkaloid from *Peganum harmala* L., a plant used extensively in traditional medicine in lands with an Islamic heritage for CNS and other ailments, is strictly called deoxypeganine (**60**), but it is often called desoxypeganine in the pharmacological literature. This compound is shown to have AChE inhibitory activity twice that of galantamine.[92] In some central Asian countries, it is used for treating patients with lesions in the peripheral nervous system but there are no reports of its clinical use in AD.

12.3.3.6 *Terpenoids (refer to Structures 13–15)*

Although it is comparatively easy to explain the fact that some alkaloids inhibit AChE because of their molecular features, it is not easy to correlate chemical structure and activity for some other phytochemical types for which AChE inhibition has recently been reported.

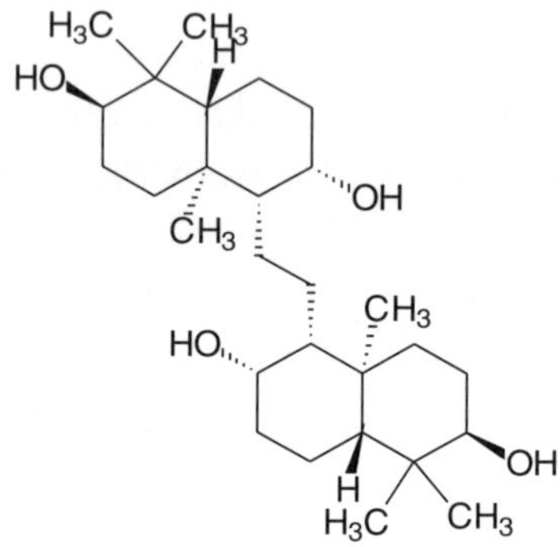

(67) Lyclavatol

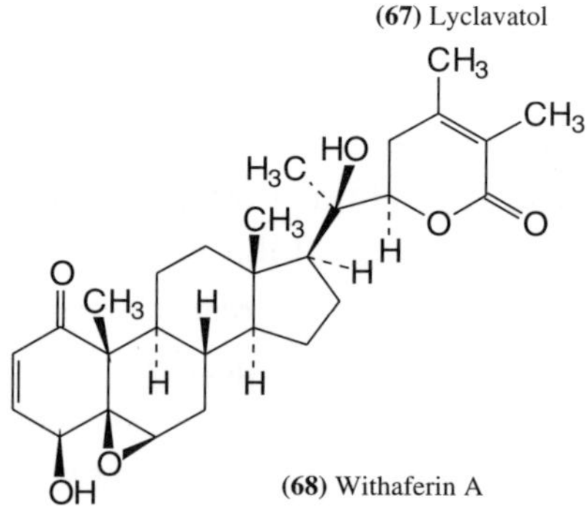

(68) Withaferin A

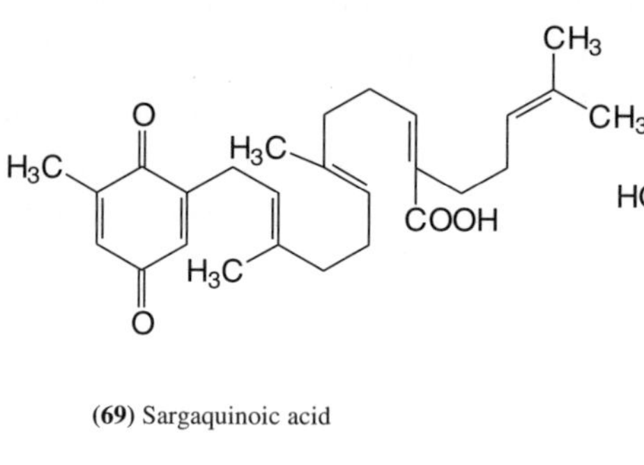

(69) Sargaquinoic acid

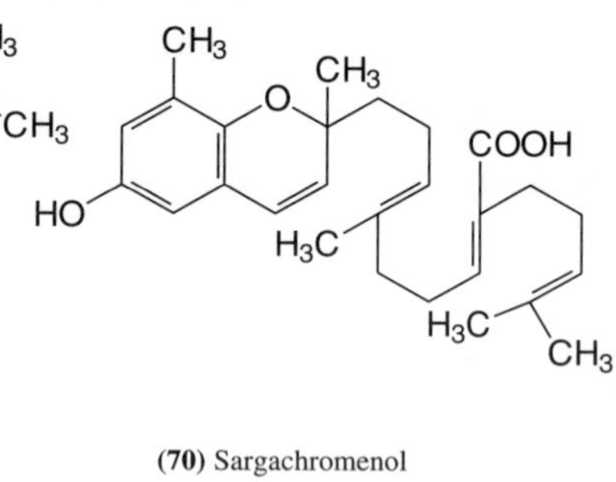

(70) Sargachromenol

Structures 14

Terpenoids comprise a very large group of natural products and tend to be lipophilic, thus they are able to cross the blood brain barrier. In addition, the monoterpenes, and some of the sesquiterpenes, are volatile, and thus effects could occur through inhalation. Investigations into European species of *Salvia*, commonly known as sages. Followed from ethnopharmacological reports that they were "good for the memory," and showed that ethanolic extracts and oils of *S. officinalis* L. and *S. lavandulaefolia* Vahl. inhibits AChE[93] activities. This was shown to be due to the cyclic monoterpenes 1,8-cineole (**61**) and α-pinene (**62**), which were shown to inhibit AChE *in vitro*, with some synergistic contribution from other constituents.[94,95] *In vivo* and clinical studies using the oil showed effects which could be ascribed to AChE inhibition, thus supporting the results of the *in vitro* experiments.[96–98] A related member of the Labiatae, *Melissa officinalis* L. also contains monoterpenes and has a reputation for promoting long life and restoring memory.[99] A recent study showed that an extract of *M. officinalis* administered to patients with mild to moderate AD for four months gave a significantly better outcome on cognitive function than placebo.[100] Another study reports the AChE inhibitory effect of another monoterpenes, the irritant cantharidin (**63**), a constituent of the beetle, *Mylabris phalerata* Pallas, used in traditional Chinese medicine for ailing memory in older people.[101]

The roots of *S. miltiorrhiza* Bunge have been employed for the treatment of cerebral vascular disease, and several reports exist of investigations into possible mechanisms for the protective effect of *S. miltiorrhiza* against cerebral ischemia. An inhibitory effect on AChE has been demonstrated recently. This has been shown to be due to the diterpenes present known as tanshinones.[102] Dihydrotanshinone (**64**) was shown to be the most active inhibitor of AChE ($IC_{50} = 1.0\,\mu M$) with cryptotanshinone (**65**) ($IC_{50} = 7.0\,\mu M$) also showing activity to inhibit AChE.

The triterpene ursolic acid (**66**) was identified as the active component of *Origanum majorana* L.[103] Another triterpene which has been reported to have AChEI effects was detected in the extracts of *Lycopodium clavatum* L. The active compound was originally stated to be α-onocerin but has recently been shown to be lyclavatol (**67**). It had only a weak AChEI effect, being 10^{-4} less that of galantamine.[104]

The withanolides are a unique type of steroid found in some genera of the Solanaceae, notably *Withania somnifera* (L.) Dun. The roots of this plant is known as "ashwagandha" and is most highly regarded in Ayurvedic medicine, being classified among the rejuvenative tonics known as "Rasayanas." Several research groups have described the cognitive enhancing potential of extracts of the roots in experimental animals, but it is only recently that these activities have been linked with the withanolides present, such as withaferin A (**68**).[105] A recent paper has investigated the AChE inhibitory activity of various withanolides thoroughly using enzyme kinetic studies and molecular graphics to visualize docking of the compounds in the enzyme.[106] In addition to their cholinergic activity, the glycowithanolides showed anxiolytic and antidepressant activities in rats,[107] which may be applicable in the symptomatic treatment of AD. Plastoquinones sargaquinoic acid (**69**) and sargachromenol (**70**) from the brown seaweed *Sargassum sagamianum* were shown to possess inhibitory activity against both AChE and BuChE (**69**) showing 1 000 times selectivity for BuChE.[108]

The investigation carried out on some Chinese animals used for CNS symptoms similar to those seen in AD resulted in the isolation of long chain fatty acids from the centipede *Scolopendra subspinipes mutilans* which had weak AChEI activity.[101] A series of fatty acids was tested to discern any structure-activity relationships and it was found that only free acids, not esters, had AChEI activity. The presence of unsaturated bonds and a chain length of not less than 16 C were was necessary for activity. The most active compound tested was linolenic acid (**71**) with IC_{50} of 59.9 μM against AChE.

12.3.3.7 *Phenolic compounds and others derived from the Shikimic pathway (refer to Structures 15–17)*

Of the many types of phenolic compounds, not many have been shown to possess AChE inhibitory activity. The root and stem bark of *Magnolia officinalis* Rehd. Et Wils. contains the biphenolic lignans, honokiol (**72**) and magnolol (**73**). Both lignans increased ChAT activity and inhibited AChE activity *in vitro*, and increased hippocampal ACh release *in vivo*.[109] These two compounds also appeared to have antioxidant antiinflammatory

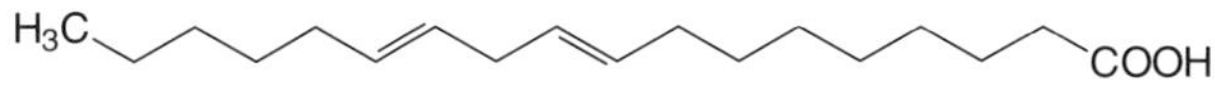

(**71**) ω-6 Linoleic acid

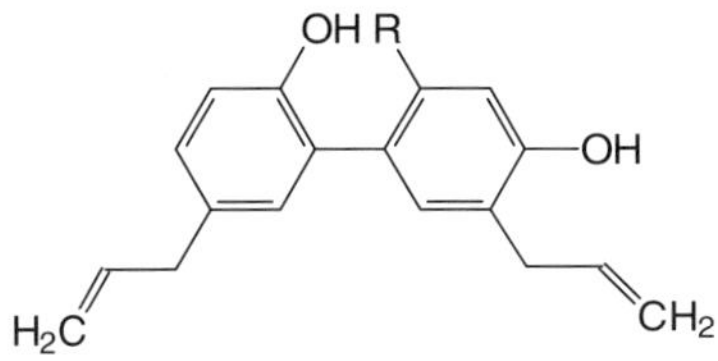

(**72**) R = HHonokiol

(**73**) R = OHMagnolol

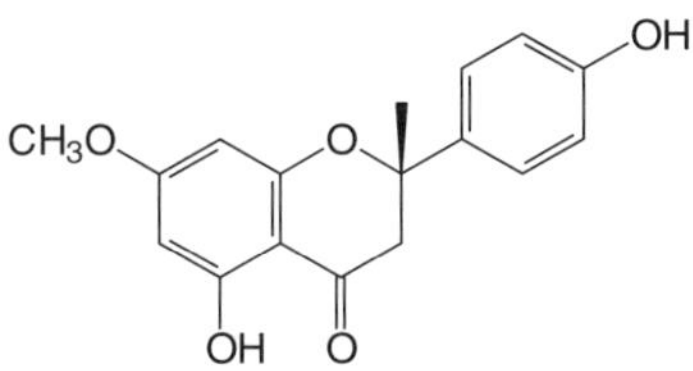

(**74**) Naringenin

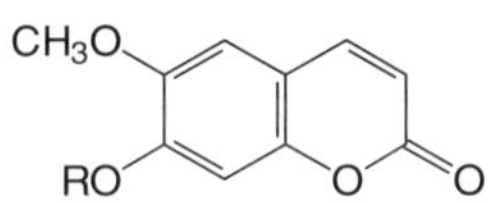

(**75**) R = H Scopoletin

(**76**) R = Glucose Scopolin

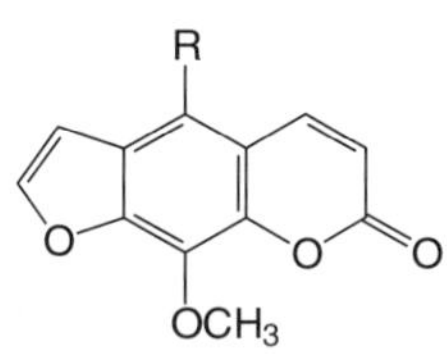

(**77**) R = H Xanthotoxin

(**78**) R = OCH_3 Isopimpinellin

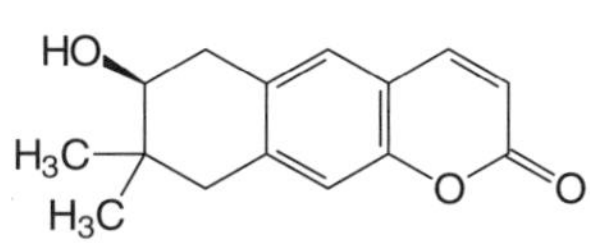

(**79**) Decursinol

Structures 15

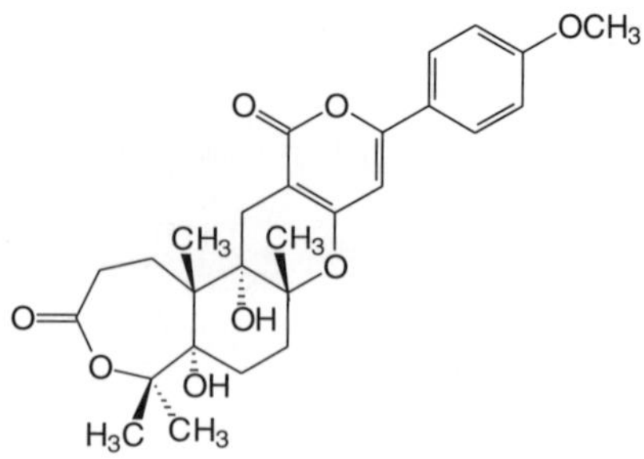

(80) Isoterreulactone A

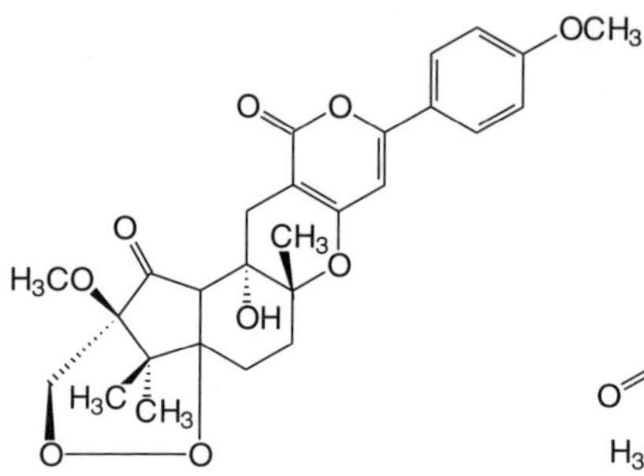

(81) Terreulactone A

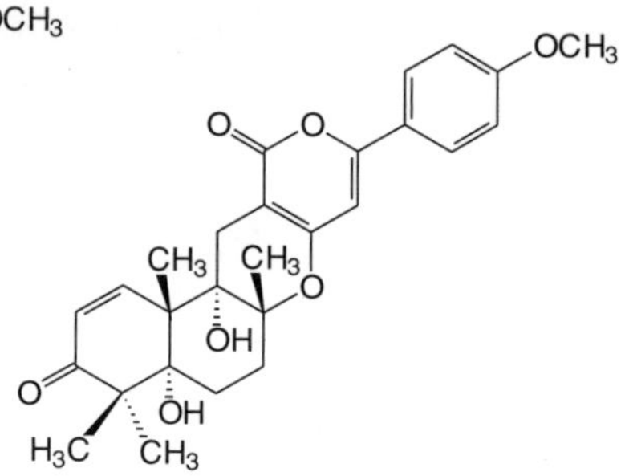

(82) Terreulactone B

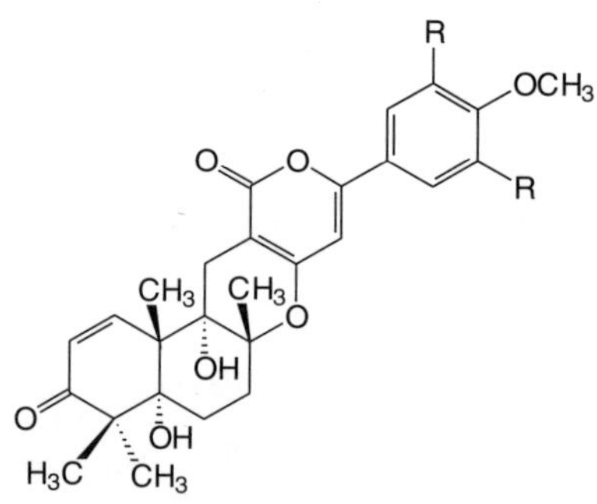

(83) R = H Terreulactone C

(84) R = OCH_3 Terreulactone D

Structures 16

and neuroprotective properties and such a polyvalent activity is of interest in their potential use in treating AD.[110,111]

The biological activity of flavonoids has attracted much interest in the part twenty years and a few compounds of this class have been shown to have AChEI effects. The flavanone naringenin (74) from *Citrus junos* (Rutaceae) ameliorated scopolamine-induced amnesia in mice, which may be related to an antiAChE effect, since naringenin was shown to inhibit AChE *in vitro* dose dependently.[112] A recent theoretical study has shown that flavonoids and xanthones exhibit polyvalent effects such as antioxidant, amyloid reduction and cholinesterase inhibition, which made them interesting candidates for further studies.[113]

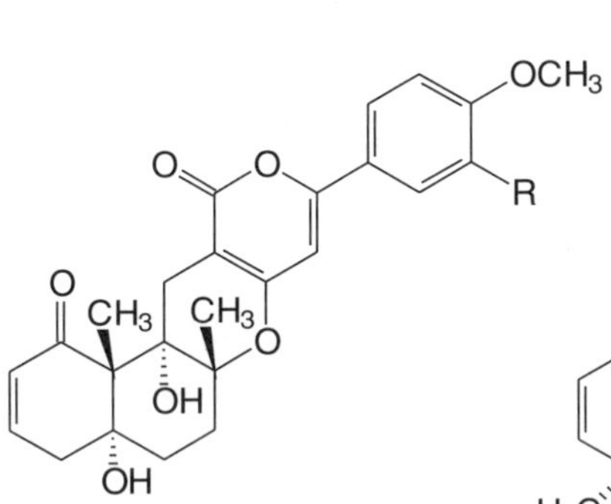

(85) R = H Arisugacin A

(86) R = OCH_3 Arisugacin B

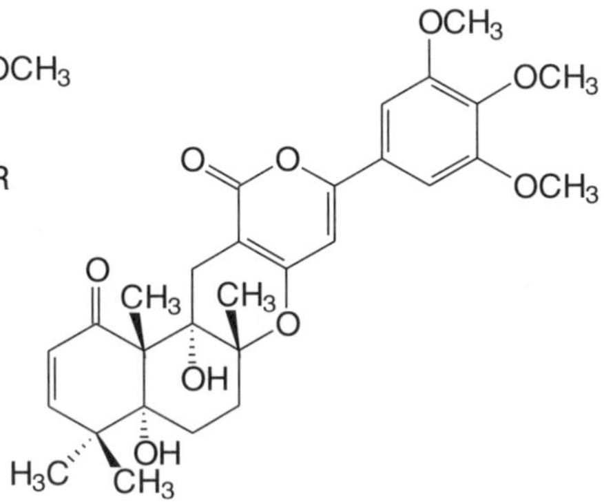

(87) Territrem A

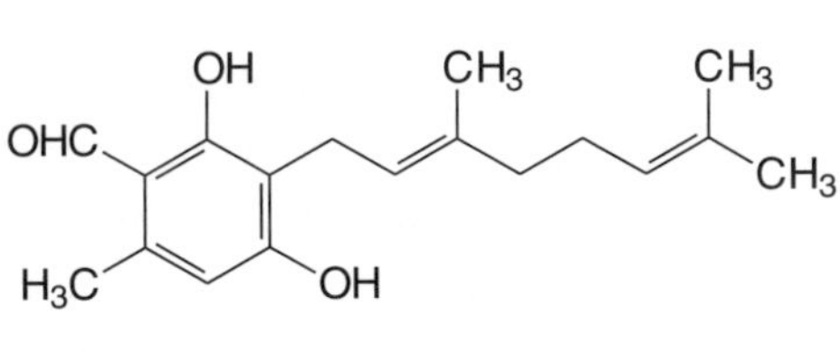

(88) Colletochorin B

Structures 17

The AChE inhibitory properties of the coumarins scopoletin (**75**) and scopolin (**76**) were discovered by an interesting *in silico* approach.[114] A model 3-dimensional pharmacophore was constructed, using a database of known inhibitors and their interaction with the AChE from *Torpedo californica.* The model was then used to predict likely inhibitors from a large database of those whose molecular coordinates were known. (**75**) and (**76**) were predicted and then isolated from *Scopolia carniolica* and tested in the Ellman assay. Results showed that (**75**) was much more active than the glucoside (**76**) but it was 2.5 orders of magnitude weaker than galantamine. Scopoletin also showed activity *in vivo* when given to rats.

The furanocoumarins xanthotoxin (**77**) and isopimpinellim (**78**) from roots of *A. acutiloba* displayed significant AChE inhibitory activity with IC_{50} values of 0.58 μM and 0.32 μM respectively.[115] Prenylated coumarins from *Angelica* spp. had only weak activity, with decursinol (**79**) giving the highest activity (IC_{50} value 28 μM).[116]

Some of the most potent nonalkaloidal AChE inhibitors are isocoumarin meroterpenoids fungal derivatives, especially from *Aspergillus terreus* and *Penicillium* species. Isoterreulactone A (**80**) and terreulactones A–D (**81–84**), meroterpenoids isolated from *Aspergillus terreus*, are potent compounds with selectivity towards AChE.[117,118] Arisugacins A (**85**) and B (**86**) are selective inhibitors of AChE (and were not selective for BChE) and have been shown to be 200-fold more potent than tacrine at inhibiting AChE *in vitro.*[119] Another group of active fungal metabolites obtained from *Aspergillus terreus* are exemplified by territrems A (**87**).[120] Since these antiChE fungal metabolites lack nitrogen, their mechanism for interaction with AChE is likely to differ from that of compounds containing quaternary nitrogen. The suggested mechanism for binding of (**87**) to AChE is an electron-donating/electron-withdrawing interaction; the donation being from the electron density of the dimethoxy group, coupled with electron-withdrawing of the 2-pyrone ring, a moiety which is considered crucial for antiAChE potency. Another group of meroterpenoids have been obtained from liquid cultures of the phytopathogenic fungus, *Nectria galligena* (Hypocreaceae), with colletochlorin B (**88**) being the most active.[121]

12.3.3.8 *AD: Use of nicotinergic compounds (refer to Structures 18)*

A link between smokers and a lower incidence of AD has been noted and this is thought to be associated with increased nicotine intake. Nicotine (27) is reported to have cognition-enhancing effects and these may be due to nicotinic receptor stimulation. It may also protect against AD by

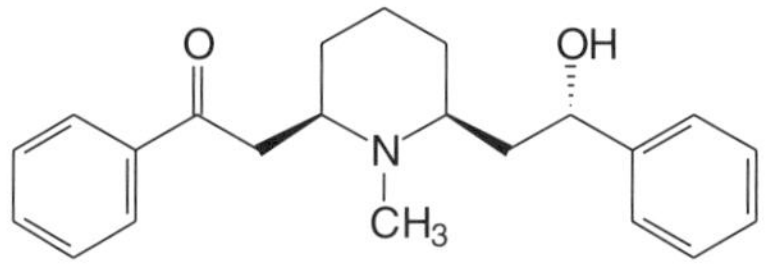

(89) Lobeline

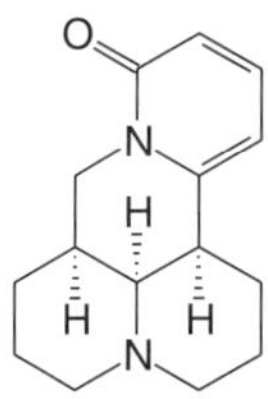

(90) Sophoramine

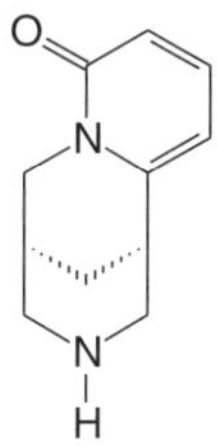

(91) Cytisine

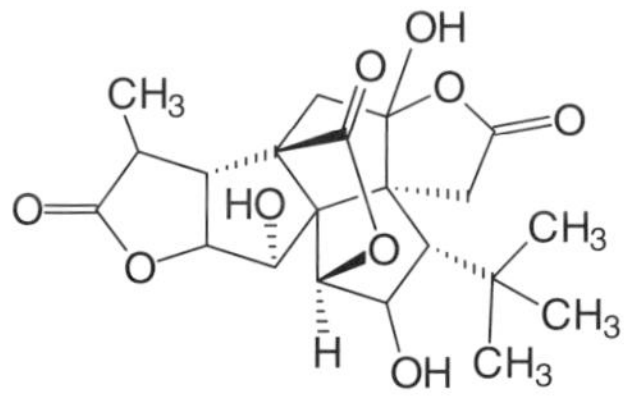

(92) Ginkgolide B

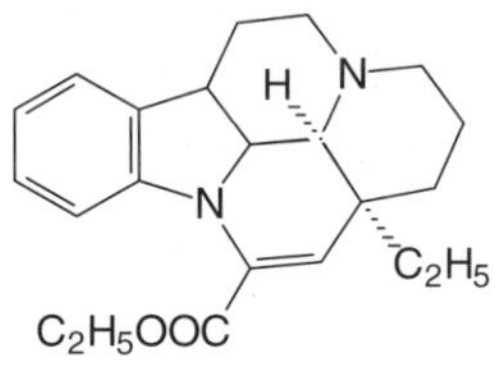

(93) Vincopetine

Structures 18

other mechanisms, such as the inhibition of β-amyloid formation.[122] There are several other alkaloids which are nicotinic agonists at the cholinergic receptor such as lobeline (**89**) from *Lobelia inflata*. *Lobelia inflata* a could be exploited to influence cholinergic function in AD. Sophoramine (**90**) and cytisine (**91**), found in members of the Leguminosae, have nicotinic actions but they do not appear to have been developed for any pharmaceutical purposes, probably because of their toxicity.

12.3.3.9 *AD: Approaches related to improved blood flow through the brain (refer to Structures 18)*

Reactive oxygen species (ROS) damage to CNS neurons is thought to be a factor which results in AD. Therefore the use of antioxidant compounds which might cross the blood brain barrier has been proposed as an approach to the prevention and amelioration of the disease. Ischemic damage caused by ROS is associated with inadequate blood flow due to arterial sclerosis and constriction of capillaries and peripheral blood vessels. The extract of the leaves of *Ginkgo biloba* L. has been promoted and used widely in the recent years for increased cerebral blood flow. Clinical studies have shown that *G. biloba* extract gives beneficial effects in cognitive decline associated with old age.[123] The effect is thought to be polyvalent i.e. various types of compounds having different activities contributing to the overall effect. The diterpenoid ginkgolides e.g. ginkgolide B (**92**), unique to this species, improve blood flow by inhibiting platelet aggregation and thereby formation of thromboses; these compounds also have some antioxidant effect. Other antioxidants are present e.g. bilobalide, and the flavonoids are both antioxidant and improve blood flow due to their relaxant effect on small blood vessels. Several clinical studies on standardized Ginkgo extracts have shown a significant improvement when assessed for cognitive factors in AD patients.[123,124] No individual compounds have been tested since there is some evidence that synergy occurs between the ginkgolides, making it preferable to give them as a mixture.

The indole alkaloid vinpocetine (**93**), has a vasodilatory effect and a clinical study on patients with CNS degenerative function due to vascular deficiency, demonstrated that those who were given vinpocetine, scored significantly better for cognitive function than a placebo group.[125] However, these results are not translated into clinical use yet.

12.4 MULTIPLE SCLEROSIS (MS) (refer to Structures 19)

Multiple sclerosis is a complex disease which consists of a progressive demyelinization and destruction of the neurons. It is characterized by painful spasms, progressive loss of many bodily functions and eventual death. It may progress rapidly, but more commonly, it is marked by periods of stability, or even improvement, before decline once again occurs. Its etiology is not known but appears to involve inflammation and some type of immune response. Treatment is symptomatic rather than curative and is aimed at reducing the progression of the disease and relieving some of its effects such as pain and muscle spasm.

In the past 25 years, anecdotal reports of beneficial effects in MS sufferers of smoking Cannabis (the herb or resin of *Cannabis sativa* L.) have resulted in a considerable amount of research and interest from academia, industry and government. Several products containing cannabis extracts or carefully standardized mixtures of constituents are in development with a number of clinical trials being carried out. In a few countries, products

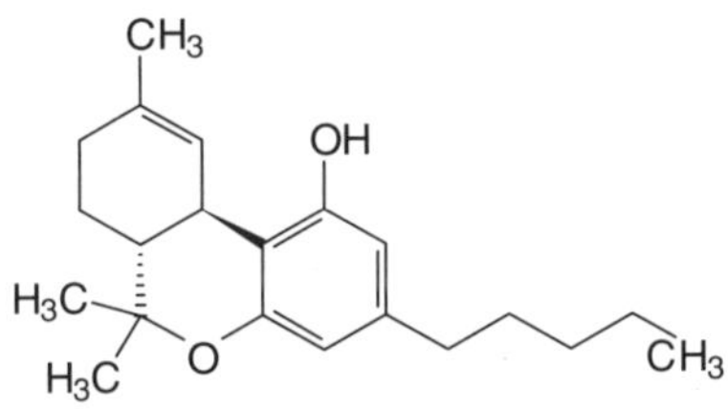

(94) Tetrahydrocannabinol

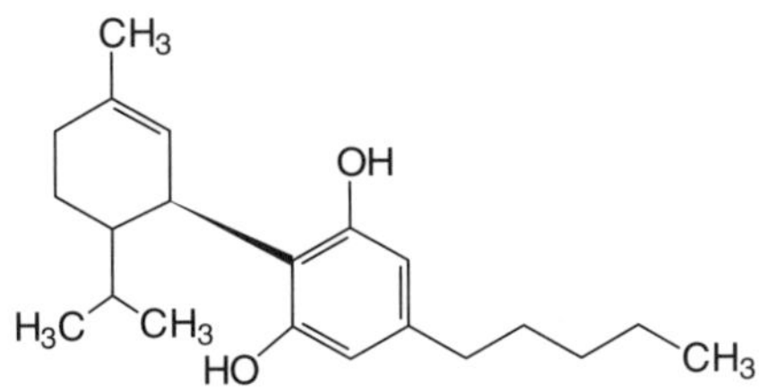

(95) Cannabidiol

Structures 19

derived from Cannabis have been licensed as medicines for treating symptoms of MS.

The two compounds which appear to provide major interest are tetrahydrocannibinol (THC) (**94**) and cannabidiol (CBD) (**95**). THC (**94**) is well known as the major psychotropic compound in Cannabis but has been shown to reduce tremor associated with MS due to its agonist action at the CB1 receptor.[126] (**94**) has also been shown to be analgesic in patients.[127] CBD (**95**) has antiinflammatory effects and also reduces pain to some extent.[128] Studies in MS sufferers using cannabis extracts containing either THC, CBD or a THC:CBD 1:1 mixture showed that the 1:1 mixture gave the best overall results for a number of measured factors, including spasticity, spasm and pain[129] and it is this combination which forms the basis of Sativex®, a product consisting of an oral mucosal spray manufactured by GW Pharmaceuticals, which is licensed in Canada for treatment of the symptoms in patients with multiple sclerosis.

Increased incidence of MS has been noted in those with a diet high in saturated fats and this has led to the proposal that a high intake of polyunsaturated fats, chiefly obtained from vegetable or fish oils, may have beneficial effects in patients and also reduce the risk of contracting MS.[130] The ω-6 fatty acid, linoleic acid (**71**), has been found to be at lower levels than normal in blood and cerebral spinal fluid in MS patients. Consequently, three small clinical trials were carried out where linoleic acid, as an ingredient of sunflower oil, was given to patients. Two of these trials showed trends towards a beneficial effect, but little statistically significant effect, with those given the high doses of **71**.[131] Similar results were obtained with a trial where unsaturated fish oils were given.[131] Although the evidence for any substantial benefit to MS sufferers from a high unsaturated fatty acid diet is not great, it has been suggested that enough is known of the general benefits of such a diet to warrant recommending it.[130]

12.5 CONCLUSIONS

Natural products have provided a useful source of compounds for treating neurodegenerative diseases. Some of these natural products have reached widespread clinical use whilst others provide explanation for the traditional uses of plants. Rational drug design, according to the characteristics of the receptors involved, is now a possibility, and natural molecules have been subjected to "fine-tuning" by chemical derivatization and synthesis of analogues to bind more closely and more exclusively to one type of receptor. However, investigation of traditional plants has revealed that these types of active molecules would have been unlikely to have been predicted from receptor studies.

Chemotherapy of AD with cholinesterase inhibitors is unlikely to develop very much further, since effective agents such as galantamine have been introduced. Future trends could involve the use of a polyvalent "cocktail" of drugs which act in different ways e.g. by mechanisms such as antioxidant and antiinflammatory activity and the inhibition of the formation of fibrillary tangles and β-amyloid plaques. Although the introduction of L-DOPA and other dopaminergic compounds over the past three decades has improved the condition of many sufferers of PD, the side effects and the unpredictability of recurrence highlight the fact that more research is needed. In this context also, the exploitation of compounds derived from plants with other mechanisms, such as monoamine oxidase inhibition, may provide a better treatment experience in due course. It is imperative that clinical trials be carried out to follow up preliminary *in vitro* and *in vivo* results, such as those observed with *Mucuna* bean extract, which indicate an advantage in using an extract over a pure substance.

REFERENCES

1. http://www.alzheimers.org.uk/Facts_about_dementia/index.htm, July 2004.
2. http://www.who.int/mediacentre/factsheets/fs218/en/, August 2004.
3. Howes MJR, Houghton PJ, Plants used in Chinese and Indian traditional medicine for improvement of memory and cognitive function, *Pharm Biochem Behavior* 75:513–527, 2003.
4. Howes MJR, Perry NSL, Houghton PJ, Plants with traditional uses and activities, relevant to the management of Alzheimer's disease and other cognitive disorders, *Phytotherapy Res* 17:1–18, 2003.
5. http://www.parkinsons.org.uk/Templates/
6. Kidd PM, Parkinson's disease as multifactorial oxidative neurodegeneration, implications for integrative management, *Altern Med Rev* 5:502–529, 2000.

7. Houghton PJ, Howes MJR, Natural products and derivatives affecting neurotransmission relevant to Alzheimer's and Parkinson's disease, *Neurosignals* **14**:6–22, 2005.
8. Hussain G, Manyam BV, *Mucuna pruriens* proves more effective than L-DOPA in Parkinson's disease animal model, *Phytotherapy Res* **11**:419–423, 1997.
9. Vaidya RA, Sheth AR, Alookar SD, The inhibitory effect of the Cowhage plant — *Mucuna pruriens* —- and L-DOPA on chlorpromazine-induced hyperprolactinaemia in man *Neurology India* **26**:171–176, 1978.
10. Manyam BV, Dhanasekaran M, Hare TA, Effect of antiparkinson drug HP-200 (*Mucuna pruriens*) on the central monoaminergic neurotransmitters, *Phytotherapy Res* **18**:97–101, 2004.
11. Al-Motarreb A, Baker K, Broadley KJ, Khat, pharmacological and medical aspects and its social use in Yemen *Phytotherapy Res* **16**:403–413, 2002.
12. Ismail M, Personal Communication, 2006.
13. Bonucelli U, Comparing dopamine agonists in Parkinson's disease, *Current Opinion Neurology* **16**:S13–S19, 2003.
14. Sit SY, Dopamine agonists in the treatment of Parkinson's disease — Past, present and future, *Curr Pharm Design* **6**:1211–1248, 2000.
15. Poewe W, Pergolide — A review of its clinical potential, *Aktuelle Neurologie* **22**:71–74, 1995.
16. Pastor P, Tolosa E, Cabergoline in the treatment of Parkinson's disease, *Neurologia* **18**:202–209, 2003.
17. Stocchi F, Dopamine agonists in Parkinson's disease — What is their role in early treatment? *CNS Drugs* **10**:159–170, 1998.
18. Dajas-Bailador FA, Asencio M, Bonilla C, *et al.*, Dopaminergic pharmacology and antioxidant properties of pukateine, a natural product lead for design of agents increasing dopamine neurotransmission, *Gen Pharmacology* **32**:373–379, 1999.
19. Melzig MF, Putscher I, Henklein P, *In vitro* pharmacological activity of the tetrahydroisoquinoline salsinol present in products from *Theobroma cacao* L. like cocoa and chocolate, *J Ethnopharmacol* **73**:153–159, 2000.
20. Antkiewicz-Michaluk L, Wardas J, Michaluk J, Protective effect of 1-methyl-1,2,3,4-tetrahydroisoquinoline against dopaminergic neurodegeneration in the extrapyramidal structures produced by intracerebral injection of rotenone, *Int J Neuropsychopharmacology* 7:153–163, 2004.
21. Schwarz MJ, Houghton PJ, Rose S, Activities of extract and constituents of *Banisteriopsis caapi* relevant to Parkinsonism, *Pharmacol Biochem Behavior* 75:627–633, 2003.

22. Wagner H, Bladt S, MAO inhibition by fractions and constituents of Hypericum extract, *Nervenheilkunde* **12**:349–352, 1993.
23. Sloley BD, Urichuk LJ, Morley P, Identification of kaempferol as a monoamineoxidase inhibitor and potential neuroprotectant in extracts of Ginkgo biloba leaves, *J Pharm Pharmacol* **52**:451–459, 2000.
24. Singh A, Naidu PS, Kulkarni SK, Quercetin potentiates L-DOPA reversal of drug-induced catalepsy in rats, possible COMT/MAO inhibition, *Pharmacology* **68**:81–88, 2003.
25. Naidu PS, Kulkarni SK, Quercetin, a bioflavonoid, reverses haloperidol-induced catalepsy, *Meth Findings Exp Clin Pharmacol* **26**:323–326, 2004.
26. Datla KP, Christidou MA, Widmer WW, Citrus flavonoid tangeretin accumulates in the brain and pre-treatment protects against dopaminergic neuronal loss in a rat model of Parkinson's disease, *Brit J Pharmacol* **135**:350, 2002.
27. Mandel S, Weinreb O, Amit T, Cell signaling pathways in the neuroprotective actions of the green tea polyphenol (−)-epigallocatechin-3-gallate, implications for neurodegenerative diseases, *J Neurochem* **88**:1555–1569, 2004.
28. Fahrig T, Gerlach I, Horvath E, A synthetic derivative of the natural product rocaglaol is a potent inhibitor of cytokine-mediated signaling and shows neuroprotective activity *in vitro* and in animal models of Parkinson's disease and traumatic brain injury, *Mol Pharmacology* **67**:1544–1555, 2005.
29. Francis PT, Palmer AM, Snape M, The cholinergic hypothesis of Alzheimer's disease: A review of progress, *J Neurol Neurosurg Psychiatry* **66**:137–147, 1999.
30. Desgranges B, Baron JC, de la Sayette V, The neural substrates of memory systems impairment in Alzheimer's disease, *Brain* **121**:611–631, 1998.
31. McGuffey EC, Alzheimer's disease, an overview for the pharmacist, *JAMA* **NS37**:347–352, 1997.
32. Perry E, Tomlinson E, Blessed G, Correlation of cholinergic abnormalities with senile plaques and mental test scores in senile dementia, *Brit Med J* **2**:1457–1459, 1978.
33. Balfour DJK, Fagerström KO, Pharmacology of nicotine and its therapeutic use in smoking cessation and neurodegenerative disorders, *Pharmacol Ther* **72**:51–81, 1996.
34. Avery EE, Baker LD, Asthana S, Potential role of muscarinic agonists in Alzheimers disease, *Drugs Aging* **11**:450–459, 1995.
35. Bratt AM, Kelly ME, Domeney AM, Acute and chronic arecoline, effects of a scopolamine-induced deficit in complex maze learning, *Pharm Pharmacol Behavior* **53**:713–721, 1996.

36. Soncrant TT, Raffaele KC, Asthana S, Memory improvement without toxicity during chronic, low-dose intravenous arecoline in Alzheimers disease, *Psychopharmacology* **112**:421–427, 1993.
37. Meier E, Frederiksen K, Nielsen M, Pharmacological *in vitro* characterization of the arecoline bioisostere Lu 25-109-T, a muscarinic compound with M-1-agonistic and M-2/M-3 antagonistic properties, *Drug Dev Res* **40**:1–16, 1997.
38. Tlal LJ, Forreest M, Loft H, Lu-25-109, a muscarinic agent, fails to improve cognition in Alzheimers disease, *Neurology* **54**:421–426, 2000.
39. Wienrich M, Meier D, Ensinger HA, Pharmacodynamic profile of the M-1 agonist talsaclidine in animals and man, *Life Sci* **22**:2593–2600, 2001.
40. Terry AV, Buccafusco JJ, Borsini F, Memory-related performance by aged rhesus monkeys administered the muscarinic M-1-preferring agonist, talsaclidine, *Psychopharmacology* **162**:292–300, 2002.
41. Levin ED, Torry D, Acute and chronic nicotine effects on working memory in aged rats, *Psychopharmacology* **123**:88–97, 1996.
42. Houghton PJ, Ren Y, Howes MR, Acetylcholinesterase inhibitors from plants and fungi, *Nat Prod Rep* **23**:181–199, 2006.
43. Holmstedt B, 'The ordeal bean of Old Calabar: The pageant of *Physostigma venosum* in medicine, in Swain T (ed.), *Plants in the Development of Modern Medicine*, Harvard University Press, Cambridge MS pp. 303–360, 1972.
44. McCaleb R, Nature's medicine for memory loss, *HerbalGram* **23**:15, 1990.
45. Sitaram N, Weingartner H, Gillin JC, Physostigmine, improvement of long-term memory processes in normal humans, *Science* **201**:272–276, 1978.
46. Polinsky RJ, Clinical pharmacology of rivastigmine, a new-generation acetylcholinesterase inhibitor for the treatment of Alzheimer's disease, *Clin Ther* **20**:634–647, 1998.
47. Grossberg G, Desai A, Review of rivastigmine and its clinical applications in Alzheimer's disease and related disorders, *Expert Opin Pharmacother* **2**:653–666, 2001.
48. Spencer CM, Noble S, Rivastigmine — A Review of its use in Alzheimer's disease, *Drugs Aging* **13**:391–411, 1998.
49. Greig NH, Utsuki T, Yu QS, A new therapeutic target in Alzheimer's disease treatment, attention to butyrylcholinesterase, *Curr Med Res Opin* **17**:159–165, 2001.
50. Barber KL, Mesular MM, Kraft GA, Klein WL, Butyrylcholinesterase alters the aggregation rate of β-amyloid, *Proc Soc Neurosci* **72**:1172–1178, 1996.
51. Guillozet A, Smiley JF, Mash DC, Mesulam MM, Butyrylcholinesterase in the life cycle of amyloid plaques, *Ann Neurol* **42**:909–918, 1997.

52. Grieg NH, Pei XH, Soncrant T, Ingram DK, Brossi A, Phenserine and ring C hetero-analogues, Drug candidates for treatment of Alzheimers disease, *Med Res Reviews* **15**:3–31, 1995.
53. Yu QS, Holloway HH, Flippen Anderson JL, *et al.*, Methyl analogues of the experimental Alzheimer drug phenserine, Synthesis and structure/activity relationships for acetyl- and butyrylcholinesterase inhibitory action, *J Med Chem* **44**:4062–4071, 2001.
54. Grieg NH, Utsuki T, Ingram DK, *et al.*, Selective butyrylcholinesterase inhibition elevates brain acetylcholine, augments learning and lowers Alzheimer β-amyloid peptide in rodent, *Prod Natl Acad Sci USA* **102**:17213–17218, 2005.
55. Bores GM, Huger FP, Petko W, Pharmacological evaluation of novel Alzheimer' disease therapeutics, acetylcholinesterase inhibitors related to galanthamine, *Am Soc Pharmacol Exp Therapeut* **277**:728–738, 1996.
56. Heinrich M, Teoh HL, Galanthamine from snowdrop- the development of a modern drug against Alzheimer's disease from local Caucasian knowledge, *J Ethnopharmacol* **92**:147–162, 2004.
57. Wilcock GK, Lilienfeld S, Gaens E, Efficacy and safety of galantamine in patients with mild to moderate Alzheimer's disease, multicentre randomised controlled trial, *Brit Med J* **321**:1445–1449, 2000.
58. Wilkinson D, Murray J, Galantamine, a randomised, double-blind, dose comparison in patients with Alzheimer's disease, *Int J Geriatr Psychiatry* **16**:852–857 2001.
59. Marcusson J, Bullock R, Gauthier S, Galantamine demonstrates efficacy and safety in elderly patients with Alzheimers disease, *Alzheimer Dis Assoc Disord* **17S3**:S86–S91, 2003.
60. Raskind MA, Peskind ER, Truyen L, The cognitive benefits of galantamine are sustained for at least 36 months, a long term extension trial, *Arch Neurol* **61**:252–256, 2004.
61. Fulton B, Benfield P, Galantamine, *Drugs Aging* **1**:60–65, 1996.
62. Woodruff-Pak DS, Vogel RW, Wenk GL, Galantamine, effect on nicotinic receptor binding, acetylcholinesterase inhibition, and learning, *Proc Natl Acad Sci* **98**:2089–2094.
63. Small GW, Rabins RV, Barry PP, Diagnosis and treatment of Alzheimer's disease and related disorders, *JAMA* **278**:1363–1371, 1997.
64. Lopez S, Bastida J, Viladomat F, Acetylcholinesterase inhibitory activity of some Amaryllidaceae alkaloids and Narcissus extracts, *Life Sciences* **71**:2521–2529, 2002.
65. Houghton PJ, Agbedahunsi JM, Adegbulugbe A, Choline esterase inhibitory properties of alkaloids from two Nigerian *Crinum* species, *Phytochemistry* **65**:2893–2896, 2004.

66. Skolnick AA, Old herbal Chinese medicine used for fever yields possible new Alzheimer disease therapy, *JAMA* 277:776, 1997.
67. Bai DL, Tang XC, He XC, Huperzine A, a potential therapeutic agent for treatment of Alzheimers disease, *Curr Med Chem* 7:355–374, 2000.
68. Lu WH, Shou J, Tang XC, Improving effect of huperzine A in aged rats and adult rats with experimental cognitive impairment, *Acta Pharm Sinica* **9**:11–15, 1988.
69. Xiao XQ, Zhang HY, Tang XC, Huperzine A attenuates amyloid β-peptide fragment 25-35-induced apoptosis in rat cortical neurons via inhibiting reactive oxygen species formation and caspase-3 activation, *J Neurosci Res* **67**:30–36, 2002.
70. Zhang Z, Wang X, Chen Q, Clinical efficacy and safety of huperzine alpha in treatment of mild to moderate Alzheimer's disease, a placebo-controlled, double-blind, randomised trial, *Chin Med J* **82**:941–944, 2002.
71. Zeng FX, Jiang HL, Tang XC, Chen KX, Ji RY, Synthesis and acetylcholinesterase inhibitory activity of (+/−)-14-fluorohuperzine A, *Bioorg Med Chem Lett* **8**:1661–1664, 1998.
72. Zhou GC, Zhu DY, Synthesis of 5-substituted analogues of huperzine A, *Bioorg Med Chem Lett* **10**:2055–2057, 2000.
73. Rajendran V, Prakash KRC, Ved HS, *et al.*, Synthesis, chiral chromatographic separation, and biological activities of the enantiomers of 10,10-dimethylhuperzine A, *Bioorg Med Chem Lett* **10**:2467–2469, 2000.
74. Zhang YH, Chen XQ, Yang HH, *et al.*, Similar potency of the enantiomers of huperzine A in inhibition of [H-3]dizocilpine (MK-801) binding in rat cerebral cortex, *Neuroscience Lett* **295**:116–118, 2000.
75. Camps P, Cusack B, Mallender WD, *et al.*, Huprine X is a novel high-affinity inhibitor of acetylcholinesterase that is of interest for treatment of Alzheimer's disease, *Mol Pharmacol* 57:409–417, 2000.
76. Alcala MDM, Vivas NM, Hospital S, *et al.*, Characterisation of the anti-cholinesterase activity of two new tacrine-huperzine A hybrids, *Neuropharmacol* **44**:749–755, 2003.
77. Huang KC, *The Pharmacology of Chinese Herbs*, Boca Raton CRC Press Ltd, 1993.
78. Shigeta K, Ootaki K, Tatemoto H, Potentiation of nerve growth factor-induced neurite outgrowth in PC12 cells by a Coptidis Rhizoma extract and protoberberine alkaloids, *Biosci Biotechnol Biochem* **66**:2491–2494, 2002.
79. Kuznetsova LP, Nikol'skaya EB, Sochilina EE, Inhibition of human blood acetylcholinesterase and butyrylcholinesterase by some alkaloids, *J Evol Biochem Physiol* **38**:35–39, 2002.

80. Peng WH, Hsieh MT, Wu CR, Effect of long-term administration of berberine on scopolamine-induced amnesia in rats, *Jap J Pharmacol* **74**:261–266, 1997.
81. Park CH, Kim S, Choi W, Novel anticholinesterase and antiamnesic activities of dehydroevodiamine, a constituent of *Evodia rutaecarpa, Planta Med* **62**:405–409, 1996.
82. Haji A, Momose Y, Takeda R, Increased feline cerebral blood flow induced by dehydroevodiamine hydrochloride from *Evodia rutaecarpa, J Nat Prod* **57**:387–389, 1994.
83. Decker M, Novel inhibitors of acetyl- and butyrylcholinesterase derived from the alkaloids dehydroevodiamine and rutaecarpine, *Eur J Med Chem* **40**:305–313, 2005.
84. Rahman AU, Khalid A, Sultana N, *et al.*, New natural cholinesterase inhibiting and calcium channel blocking quinoline alkaloids, *J Enz Inhibition Med Chem* **21**:703–710, 2006.
85. Krasowski MD, McGehee DS, Moss J, Natural inhibitors of cholinesterases, implications for adverse drug reactions, *Can J Anaesthesia* **44**:525–534, 1997.
86. Attar-ur-Rahman A, Parveen S, Khalid A, *et al.*, Acetyl- and butyrylcholinesterase inhibiting triterpenoids from *Buxus papillosa, Phytochemistry* **58**:963–968, 2001.
87. Kalauni SK, Choudhary MI, Khalid A, *et al.*, New cholinesterase inhibitory alkaloids from the leaves of *Sarcococca coriacae* of Nepalese origin, *Chem Pharm Bull* **50**:1423–1426, 2002.
88. Khalid A, Zaheer-ul-Haq Anjum S, Khan MR, *et al.*, Kinetics and structure-activity relationship studies on pregnane-type steroidal alkaloids that inhibit cholinesterases, *Bioorg Med Chem* **12**:1995–2003, 2004.
89. Khalid A, Azim MK, Parveen S, *et al.*, Structural basis of acetylcholinesterase inhibition by triterpenoidal alkaloids, *Biochem Biophys Res Commun* **331**:1528–1532, 2005.
90. Lin BQ, Ji H, Li P, Fang W, Jiang Y, Inhibitors of acetylcholine esterase *in vitro* screening of steroidal alkaloids from *Fritillaria* species, *Planta Med* **72**:814–818, 2006.
91. Choudhary MI, Nawaz SA, Zaheer-ul-Haq Azim MK, *et al.*, Juliflorine, A potent natural peripheral anionic-site-binding inhibitor of acetylcholinesterase with calcium-channel blocking potential, a leading candidate for Alzheimer's disease therapy, *Biochem Biophys Res Comm* **332**:1171–1179, 2005.
92. Tuliaganov N, Sadritdinov FS, Suleimanova GA, Pharmacological characteristics of desoxypeganine hydrochloride, *Farmakol Toksikol* **49**:37–40, 1986.
93. Perry N, Court G, Bidet N, European herbs with cholinergic activities, potential in dementia therapy, *Int J Geriatr Psychiatry* **11**:1063–1069, 1996.

94. Perry NSL, Houghton PJ, Theobald AE, *In-vitro* inhibition of human erythrocyte acetylcholinesterase by *Salvia lavandulaefolia* essential oil and constituent terpenes, *J Pharm Pharmacol* **52**:895–902, 2000.
95. Savelev S, Okello E, Perry NSL, Synergistic and antagonistic interactions of anticholinesterase terpenoids in *Salvia lavandulaefolia* essential oil, *Pharmacol Biochem Behavior* 75:661–668, 2003.
96. Perry NSL, Houghton PJ, Jenner P, *Salvia lavandulaefolia* essential oil inhibits cholinesterase *in vivo*, *Phytomedicine* **9**:48–51, 2002.
97. Tildesley NTJ, Kennedy DO, Perry EK *Salvia lavandulaefolia* (Spanish Sage) enhances memory in healthy young volunteers, *Pharmacol Biochem Behavior* 75:669–674, 2003.
98. Akhondzadeh S, Noroozian M, Mohammadi M, *Salvia officinalis* extract in the treatment of patients with mild to moderate Alzheimer's disease, a double-blind, randomized and placebo-controlled trial, *J Clin Pharm Ther* **28**:53–59, 2003.
99. Yarnell E, Lemonbalm *Altern Complement Ther* **4**:417–419, 1998.
100. Akhondzadeh S, Noroozian M, Mohammadi S, *Melissa officinalis* extract in the treatment of patients with mild to moderate Alzheimer's disease, A double blind, randomized, placebo controlled trial, *J Neurol Neurosurg Psychiatry* 74:863–866, 2003.
101. Ren Y, Houghton PJ, Hider RC, Relevant activities of extracts and onstituents of animals used in traditional Chinese medicine for central nervous system effects associated with Alzheimer's disease, *J Pharm Pharmacol* **58**:989–996, 2006.
102. Ren Y, Houghton PJ, Hider RC, Novel diterpenoid acetylcholinesterase inhibitors from *Salvia miltiorhiza*, *Planta Med* **70**:201–204, 2004.
103. Chung YK, Heo HJ, Kim EK, Inhibitory effect of ursolic acid purified from *Origanum majorana* L. on the acetylcholinesterase, *Mol Cells* **11**:137–143, 2001.
104. Rollinger JM, Ewelt J, Seger C, *et al.*, New insights into the acetylcholinesterase activity of *Lycopodium clavatum*, *Planta Med* 71:1040–1043, 2005.
105. Ghosal S, Lal R, Srivastava SK, Immunomodulatory and CNS effects of sitoindosides IX and X, two new glycowithanolides from *Withania somnifera, Phytother Res* **3**:201–206, 1989.
106. Choudhary MI, Nawaz SA, Zaheer-ul-Haq Lodhi MA, *et al.*, Withanolides, a new class of natural cholinesterase inhibitors with calcium antagonistic properties, *Biochem Biophys Res Comm* **334**:276–287, 2005.
107. Bhattacharya A, Ramanathan M, Ghosal S, Effect of *Withania somnifera* glycowithanolides on iron-induced hepatotoxicity in rats, *Phytother Res* **14**:568–570, 2000.
108. Choi BW, Ryu G, Park SH, *et al.*, *Phytother Res* **21**:2007.

109. Hou YC, Chao PD, Chen SY, Honokiol and magnolol increased hippocampal acetylcholine release in freely-moving rats, *Am J Chin Med* **28**:379–384, 2000.
110. Lo YC, Teng CM, Chen CF, Magnolol and honokiol from *Magnolia officinalis* protect rat heart mitochondria against lipid peroxidation, *Biochem Pharmacol* **47**:549–553, 1994.
111. Wang LM, Mineshita S, Preventive effects of Unsei-in and Oren-gedoku-to, Chinese traditional medicines, against rat paw oedema and abdominal constriction in mice, *J Pharm Pharmacol* **48**:327–331, 1996.
112. Heo HJ, Kim MJ, Lee JM, *et al.*, Naringenin from *Citrus junos* has an inhibitory effect on acetylcholinesterase and a mitigating effect on amnesia, *Dement Geriatr Cogn Disord* **17**:151–157, 2004.
113. Ji HF, Zhang HY, Theoretical evaluation of flavonoids as multipotent agents to combat Alzheimer's disease, *J Mol Struct Theochem* **767**:309, 2006.
114. Rollinger JM, Hornick A, Langer T, *et al.*, Acetylcholinesterase inhibitory activity of scopolin and scopoletin discovered by virtual screening of natural products, *J Med Chem* **47**:6248–6254, 2004.
115. Mizayawa M, Tsukamoto T, Anzai J, Ishikawa Y, Insecticidal effect of phthalides and furanocoumarins from *Angelica acutiloba* against *Drosophila melanogaster*, *J Agric Food Chem* **52**:4401–4405, 2004.
116. Kang SY, Lee KY, Sung SH, Park MJ, Kim YC, Coumarins isolated from *Angelica gigas* inhibit acetylcholinesterase: Structure-activity relationships, *J Nat Prod* **64**:683–685, 2001.
117. Kim WG, Cho KM, Lee CK, Yoo ID, Terreulactones A, B, C, and D: Novel acetylcholinesterase inhibitors produced by *Aspergillus terreus* II. Physico-chemical properties and structure determination, *J Antibiot* **56**:351–357, 2003.
118. Yoo ID, Cho KM, Lee CK, Kim WG, Isoterreulactone A, a novel meroterpenoid with anti-acetylcholinesterase activity produced by *Aspergillus terreus*, *Bioorg Med Chem Lett* **15**:353–356, 2005.
119. Otoguro K, Kuno F, Ōmura S, Arisugacins, selective acetylcholinesterase inhibitors of microbial origin, *Pharmacol Ther* **76**:45–54, 1997.
120. Kuno F, Otoguro K, Shiomi K, *et al.*, Arisugacins A and B, novel and selective acetylcholinesterase inhibitors from *Penicillium* sp FO-4259.1. Screening, taxonomy, fermentation, isolation and biological activity, *J Antibiot*, **49**:742–746, 1996.
121. Gutiérrez M, Theodulož C, Rodriguez J, *et al.*, *J Agric Food Chem* **53**:7701–7708, 2005.
122. Salomon AR, Marcinowski KJ, Friedland RP, Nicotine inhibits amyloid formation by the β-peptide, *Biochemistry* **35**:13568–13578, 1996.

123. Barnes J, Anderson LA, Phillipson JD, *Herbal Medicines, a Guide for Healthcare Professionals*, 2nd edn. Pharmaceutical Press London, pp. 250–263, 2002.
124. Le Bars PL, Katz MM, Berman N, *et al.*, for the North American EGb Study Group. A placebo-controlled, double-blind, randomized trial of an extract of *Ginkgo biloba* for dementia, *JAMA* **278**:1327–1332, 1997.
125. Balestreri R, Fontana L, Astengo F, A double-blind placebo controlled evaluation of the safety and efficacy of vinpocetine in the treatment of patients with chronic vascular senile cerebral dysfunction, *J Am Geriatr Soc* **35**:425–430, 1987.
126. Baker D, Pryce G, Croxford JL, *et al.*, Cannabinoids control spasticity and tremor in a multiple sclerosis model, *Nature* **404**:84–87, 2000.
127. Holdcroft A, Smith M, Jacklin A, Pain relief with oral cannabinoids in familial Mediterranean fever, *Anesthesia* **52**:483–486, 1997.
128. Malfait AM, Gallily R, Sumariwalla PF, The non-psychoactive cannabis-constituent cannabidiol is an oral anti-arthritic therapeutic in murine collagen-induced arthritis, *Prod Nat Acad Sci USA* **97**:9561–9566, 2000.
129. Wade DT, Robson PJ, House H, *et al.*, A preliminary controlled study to determine whether whole-plant cannabis extracts can improve intractable neurogenic symptoms, *Clin Rehabil* 17:21–29, 2003.
130. Schwarz S, Leweling H, Multiple sclerosis and nutrition, *Multiple Sclerosis* **11**:24–32, 2005.
131. Bates D, Cartlidge NE, French JM, Jackson MJ, Nightingale S, Shaw DA, A double-blind controlled trial of long chain *n*-3 polyunsaturated fatty acids in the treatment of multiple sclerosis, *J Neurol Neurosurg Psychiatry* **52**:18–22, 1989.

Chapter 13

PHYTOTOXIC COMPOUNDS WITH CALMODULIN INHIBITOR PROPERTIES FROM SELECTED MEXICAN FUNGI AND PLANTS

Rachel Mata, Sergio Martínez-Luis and Araceli Pérez-Vásquez

13.1 INTRODUCTION

Natural products represent an immense reservoir of new molecules for the development of pesticide agents. In particular, natural products based herbicides have become attractive in recent years for a few reasons. One of the reasons is that they are environmentally benign. Furthermore, the novelty of their structures offers the possibility of finding compounds with new mode of actions, thus diminishing the induction of plant resistance, a problem associated with commonly used synthetic herbicides.[1,2]

Traditionally, the majority of the studies in this area have emphasized the analysis of microorganisms, in particular of soil actinomycetes, as the most relevant source of potential herbicide agents.[1,2] The plant kingdom, however, has also been explored for the discovery of new herbicides. So far, bialaphos, originally isolated from different *Streptomyces* strains, is the only commercial herbicide derived from a microbial source. On the other hand, cymethylin, an analog of the monoterpenoid 1,8-cineole, as well as sulcotrione and mesotrione, developed from leptospermone, are examples of herbicides developed from phytochemicals.[1,2]

A country like Mexico, being one of the most megadiverse countries of the world, offers excellent prospects for the discovery of potential herbicides. Its topography, variety of climates and complex geological, biological and cultural history, have contributed to the formation of a mosaic of environmental conditions rich in organisms which are not yet explored. Potentially, these organisms are sources of unique phytotoxic compounds representing novel templates for the development of new herbicides. This chapter will focus on various natural products with phytotoxic and calmodulin (CaM) inhibitor properties recently isolated from selected plants and fungi from Mexico.

Calmodulin is an ubiquitous Ca^{2+} binding protein which regulates many Ca^{2+}-dependent cellular events by interacting with a heterogeneous group of target proteins, both in plant and animal cells. It is a small protein (16 to 18 kD) composed of $\sim$148 amino acids organized in two distinct N- and C-terminal globular domains connected by a flexible central linker. Both domains possess a pair of intimately linked EF hands, each binding a single Ca^{2+} ion.[3,4]

Calmodulin was selected as a molecular target for phytotoxicity for the following three reasons: (1) In higher plants, CaM is a fundamental component of Ca^{2+} signal transduction pathway during plant growth. Indeed, more than 50 enzymes and ion channels are regulated by CaM in plants, some of which are important for plant development. A few of these proteins are found only in plants. Other proteins, although not exclusive from plants, are regulated by CaM only in plants. A third group of proteins are regulated by CaM in all type of organisms.[3,4] In the first two cases, these enzymes could be specific herbicide targets. (2) Some mycotoxins such as ophiobolin A, isolated from several fungus of the genus *Bipolaris*, seem to exert its phytotoxic action by interacting with CaM and inhibiting its ability to activate CaM-sensitive enzymes.[5] (3) Last but not least, unlike animals, higher plants express multiple divergent CaM isoforms, some of which (SCaM-4 and SCaM-5) share 78% of structural identity with vertebrate CaM.[4] These isoforms might differentially regulate CaM-binding enzymes,[3] therefore they could be specific targets for plant-growth inhibitors.

During the 1980s and early 1990s, a few natural products with anti-CaM properties were already discovered using functional enzymatic assays; then, the interest in the topic dwindled. However, more recently, research in several laboratories worldwide revealed a renewed interest for discovering new CaM inhibitors. This concern is probably due to the significant progress in the knowledge of this regulatory protein including its structural features, the discovery of new CaM targets enzymes and isoforms as well as its role in the regulation of several physiological processes such as plant growth and defense,[3,4] muscle contraction and relaxation,[6] learning and memory, immune responses,[7] osteoclastogenesis,[8] as well as mood and anxiety physiology,[9] to mention a few. Furthermore, potential modulation of physiological targets of CaM by natural or synthetic compounds offers great possibilities for the discovery of new drugs, pesticide agents and valuable research tools to understand the complex CaM messenger system in plants and animals.

To discover phytotoxic agents with CaM-inhibitor properties, our studies have been conducted following a bioassay directed isolation and characterization approach. In this endeavor, the phytogrowth-inhibitory activity of organic soluble extracts and pure compounds is initially assessed using a Petri dish assay. In this assay, the effect of the tested materials on seed germination and radicle growth is measured using seedlings of *Amaranthus hypochondriacus* L. and *Echinochloa crus-galli* (L.) Beuv. The effect is compared with that of the known agent 2,2-dichlorophenoxyacetic acid (2,4-D).[10] Bioautography on TLC is used to locate the phytogrowth-inhibitory activity in all fractions obtained from the active extracts. Finally, the action of pure compounds on CaM is analyzed employing a functional enzymatic assay and/or an SDS-PAGE electrophoresis analysis[11,12] (*inter alia*). For the enzymatic assay, bovine brain CaM-sensitive *c*AMP phosphodiesterase (PDE1) is usually used as monitoring enzyme. PDE1 catalyzes the hydrolysis of cyclic nucleotides to nucleotides monophosphates and its activity can be quantified spectrophotocolorimetrically using the Sharma and Wang method.[13] This procedure correlates the activity of PDE1 with the amount of inorganic phosphorous (P*i*) released by the hydrolysis of *c*AMP in the presence of CaM and a snake nucleotidase. Alternatively, we have used CaM-dependent NADK as the target enzyme.[14] This enzyme was selected as the second reporter enzyme because it is a vegetal enzyme and is only modulated by CaM in plants. The use of this enzyme to detect or quantify CaM inhibitors was not previously reported therefore we standardized a procedure for this purpose. NADK catalyzes the conversion of NAD (nicotinamide adenine dinucleotide) into NADP (nicotinamide adenine dinucleotide phosphate), an important coenzyme in many metabolic reactions. The activity of the kinase was correlated with the amount of NADP generated. In turn, the rate of formation of NADP was determined by its rapid conversion to reduced nicotinamide adenine dinucleotide phosphate (NADPH) by glucose-6-phosphate dehydrogenase in the presence of glucose-6-posphate. The rate of NADPH production is monitored by a spectrocolorimetric analysis, following the decrease in absorbance at 600 nm over the time in the presence of the reporter redox dyes. The rate of NADPH formation is proportional to the amount of NADPH present, which in turn correlates with the amount of NADP initially generated.[14] In any case, the effect of phytotoxic agents on CaM-sensitive enzymes is compared with that of chlorpromazine, a well known CaM antagonist.

In the next sections, selected examples of phytotoxic agents with CaM-inhibitor properties stemming from our own work are described. Brief considerations on their natural sources, isolation, structure elucidation and biological properties are discussed.

13.2 PHYTOTOXIC AGENTS FROM SELECTED PLANT SPECIES

13.2.1 *Flourensia cernua*

Flourensia cernua D.C. (Asteraceae) is a bitter tasting shrub with a hop-like odor that grows in the deserts from northern Mexico and the USA southward to the central Mexican states of Zacatecas and Hidalgo. In the USA, it is called "tarbush" while in Mexico, it is referred to as "hojasé," "hojasén" or "hoja ancha." Throughout its Mexican range, an infusion of the leaves is drunk frequently to treat various gastrointestinal ailments; thus, a tea is imbibed to alleviate stomachache, indigestion, diarrhea and dysentery. On the other hand, its uses as a purgative, expectorant and rheumatic remedy are restricted only to a few areas.[15,16]

From the chemical point of view, several authors have investigated this species being flavonoids,[18–20] monoterpenoids,[16,17,21]

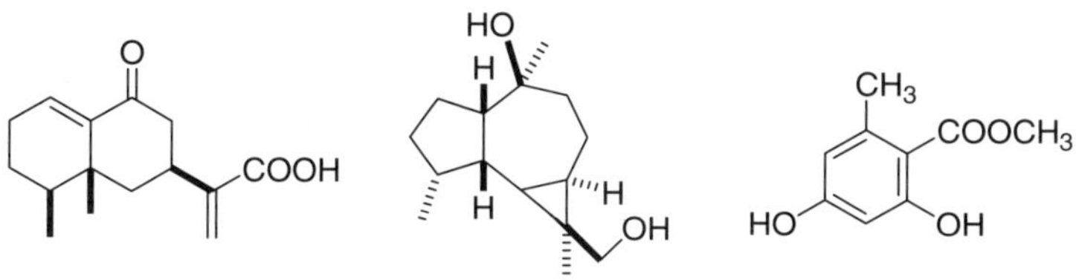

(1) (2) (3)

Fig. 1. Phytotoxic agents from *Flourensia cernua.*

sesquiterpenoids,[16,17,21–24] acetylenes, *p*-acetophenones, benzopyrans and benzofurans,[25] which are the most characteristic metabolites. It has been also demonstrated that crude extracts and fractions from this species possess phytotoxic, antifungal, antialgal and antitermite activities, and reduce the consumption of alfalfa pellets by sheep.[17,21,26]

In our research program, a CH_2Cl_2-MeOH (1:1) extract of *F. cernua* was selected for bioassay-guided fractionation on the basis of its phytogrowth inhibitory activity against *A. hypochondriacus* [IC_{50} (concentration of the testing material inhibiting plant growth by 50%) = 300 μg/mL].[27] Extensive chromatography of the active extract (IC_{50} = 300 μg/mL against *A. hypochondriacus*) led to the isolation of three phytotoxic compounds, namely, dehydroflourensic acid (**1**), fluorensadiol (**2**) and methyl orsellinate (**3**) (Fig. 1).

Dehydroflourensic acid (**1**) turned out to be a new analog of flourensic acid, a sesquiterpenoid previously described by Kingston *et al.* in this species.[22] The structure of compound **1** was established by spectral means. In addition, seven new γ-lactones characterized as tetracosan-4-olide, pentacosan-4-olide, hexacosan-4-olide, heptacosan-4-olide, octacosan-4-olide, nonacosan-4-olide, and triacontan-4-olide were obtained.[27] Compounds **1–3** showed phytotoxic effects when tested against the seedlings of *A. hypochondriacus* and *E. crus-galli*. Compounds **1** [$IC_{50} = 1.96 \times 10^{-4}$ M (*A. hypochondriacus*); $IC_{50} = 6.2 \times 10^{-4}$ M (*E. crus-galli*) and **3** $IC_{50} = 9.2 \times 10^{-4}$ M (*A. hypochondriacus*); $IC_{50} = 3.1 \times 10^{-4}$ M (*E. crus-galli*)] inhibited radicle growth of both target species. However, compound **2** $IC_{50} = 4.12 \times 10^{-4}$ M (*A. hypochondriacus*); [$IC_{50} > 4.2 \times 10^{-3}$ M (*E. crus-galli*)] inhibited only the seedling growth of *Amaranthus* with a potency comparable to 2,4-D [$IC_{50} = 1.8 \times 10^{-4}$ M

(*A. hypochondriacus*); $IC_{50} > 2.3 \times 10^{-4}$ M (*E. crus-galli*)]. According to an SDS-PAGE electrophoresis experiment, compounds **1–3** interacted with bovine brain-CaM since CaM treated with the phytotoxic agents or chlorpromazine had lower electrophoretic mobility than untreated CaM. In addition, the activation of PDE1 was inhibited in the presence of **1–3** and CaM with IC_{50} (concentration of the testing material inhibiting the activity of the enzyme by 50%) values of 23.2 μM, 5.2 μM and 8.1 μM, respectively. Compounds **2** and **3** were more active than chlorpromazine (IC_{50} = 10.2 μM).[27]

13.2.2 *Prionosciadium watsoni*

Prionosciadium watsoni Coulter & Rose ex S.Watson (Umbelliferae) is a medicinal plant found in the pine oak forest from Chihuahua to Hidalgo, Mexico. A tea prepared from the roots of this species is drunk to alleviate grastrointestinal pains, while the fruits are employed to treat diabetes and fever.[15] *P. watsoni* was also selected for activity guided fractionation on the basis of its significant phytotoxicity against *A. hypochondriacus*, *E. crus-galli*, and *Lemna pausicostata* L. (duckweed).[28]

A CH_2Cl_2-MeOH (1:1) extract of the aerial parts of *P. watsoni* provoked notable inhibition of radicle elongation of seedlings of *A. hypochondriacus* (IC_{50} = 0.68 μg/mL) and *E. crus-galli* (IC_{50} = 2.12 μg/mL) when tested by the Petri dish germination and radicle elongation bioassay.[28] On the other hand, the extract caused growth inhibition, chlorophyll reduction (100% at the concentrations of 100 μg/mL and 200 μg/mL) and electrolytic leakage.[28] Bioassay-directed fractionation of the active extract led to the isolation of several coumarins (**4–14**) and pyrones (**15** and **16**) with phytotoxic properties (Fig. 2). Compounds **4**, **5** and **13–16** are novel natural products, and their structures were determined as propionic acid (9*R*, 10*R*)-9-acetoxy-8, 8-dimethyl-9, 10-dihydro-2*H*, 8*H*-benzo[1, 2-*b*:3, 4-*b*′]dipyran-2-one-10-yl ester (**4**), isobutyric acid (9*R*, 10*R*)-9-hydroxy-8,8-dimethyl-9, 10-dihydro-2*H*, 8*H*-benzo[1,2-*b*:3,4-*b*′]dipyran-2-one-10-yl ester (**5**), isobutyric acid (9*R*)-8, 8-dimethyl-9,10-dihydro-2*H*,8*H*-benzo[1,2-*b*:3,4-*b*′]dipyran-2-one-9-yl ester (**13**), 2-methylbut-(2*Z*)-enoic acid (3*R*)-5-methoxy-3,

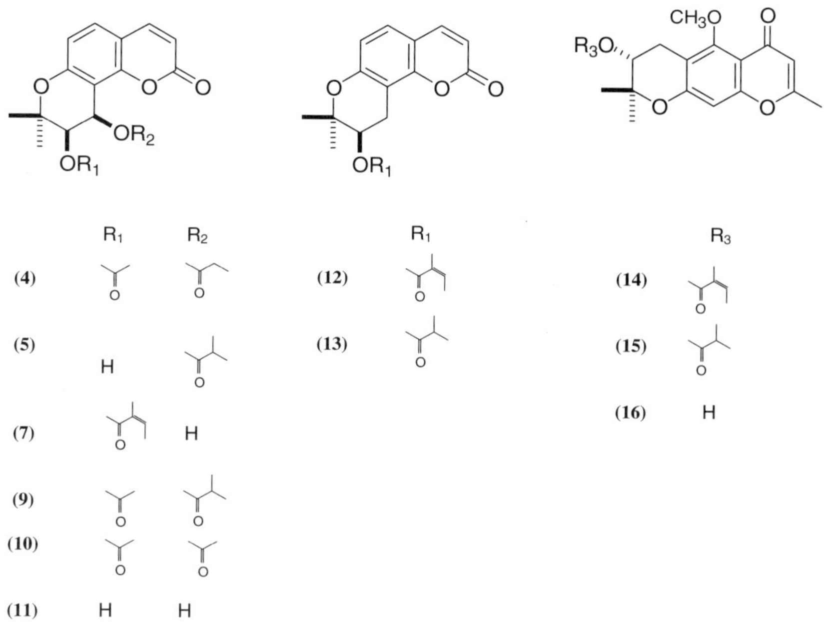

Fig. 2. Phytotoxic compounds from *Prionosciadium watsoni.*

4-dihydro-2, 2, 8-trimethyl-6-oxo-2*H*,6*H*-benzo[1, 2-*b*:5,4-*b*′]dipyran-3-yl ester (**14**), isobutyric acid (3*R*)-5-methoxy-3,4-dihydro-2,2,8-trimethyl-6-oxo-2*H*,6*H*-benzo[1,2-*b*:5,4-*b*′]dipyran-3-yl ester (**15**) and (+)-5-methoxyhamaudol (**16**). In addition, the known compounds: quianhucoumarin A (**6**), 2-methylbut-(2*Z*)-enoic acid (9*R*, 10*R*)-10-hydroxy-8, 8-dimethyl-9,10-dihydro-2*H*, 8*H*-benzo[1,2-*b*:3,4-*b*′]dipyran-2-one-9-yl ester (**7**), 2-methylbut-(2*E*)-enoic acid (9*R*, 10*R*)-10-hydroxy-8,8-dimethyl-9, 10-dihydro-2*H*, 8*H*-benzo[1, 2-*b*:3, 4-*b*′]dipyran-2-one-9-yl ester (**8**), seravshanin (**9**), quianhucoumarin D (**10**), (+)-*cis*khellactone (**11**), and jatamansin (**12**) were obtained.[28]

The compounds were characterized throughout detailed spectroscopic, spectrometric and chemical analyses. The absolute configuration at the stereogenic centers in compounds **4**, **5**, **12** and **16** was established by applying the Mosher ester methodology. Furthermore, the structures of coumarins **4**, **5** and **10** were confirmed by X-ray analysis.[28]

The phytotoxic activity of the isolates was determined on germination and radicle elongation of *A. hypochondriacus* and *E. crus-galli* seedlings. In general, the compounds inhibited radicle growth in a concentration-dependent manner. The IC_{50}'s ranged from 6.19 μM and 863 μM.[28] Compounds **14**, **15**, and **4** were the most active and showed high selectivity against *E. crus-galli* seedlings with IC_{50} of 14.8 μM, 26.4 μM and 6.19 μM, respectively.

Next, the studies were extended to evaluate the effect of the compounds on duckweed which is one of the best characterized models for assessing phytotoxic activity. The duckweed assay system makes it possible to study the toxic effects throughout the plant life cycle, as well as to plant specific toxic effects which target photosynthesis. Of all the isolates, only **12** and **13** showed significant phytotoxicity on duckweed at concentrations of 100 μM and 200 μM inhibiting plant growth by 80% and 100%, respectively and chlorophyll production by 40% and 84%, respectively.[28]

All the phytotoxic agents from *P. watsoni* were found to interact with CaM since at the concentration of 0.033 μg/mL, they modify the electrophoretic mobility of both bovine brain and spinach CaM's.[28]

13.2.3 *Maxillaria densa* and *Epidendrum rigidum*

Maxillaria densa Lindley (Orchidaceae) is an epiphytic orchid widely distributed in Mexico and Guatemala. Our previous chemical investigation of this species allowed the isolation and structure elucidation of several spasmolytic phenanthrene derivatives.[29,30] *Epidendrum rigidum* Jacquin is the most widespread and common *Epidendrum* species in tropical regions of the western hemisphere. During the course of our studies, the extract of both orchids were identified as possessing phytotoxicity against *A. hypochondriacus* and duckweed.[31,32]

The major phytotoxic principles (Fig. 3) from *M. densa*, were shown to be the phenanthrene derivatives erianthridin (**17**; 9,10-dihydro-2,7-dihydroxy-3,4-dimethoxyphenanthrene) and gymnopusin (**18**; 2,7-dihydroxy-3,4,9-trimethoxyphenanthrene).[31]

Erianthridin **17** (200 μM) was highly phytotoxic to duckweed provoking cellular leakage, complete growth inhibition and significant chlorophyll

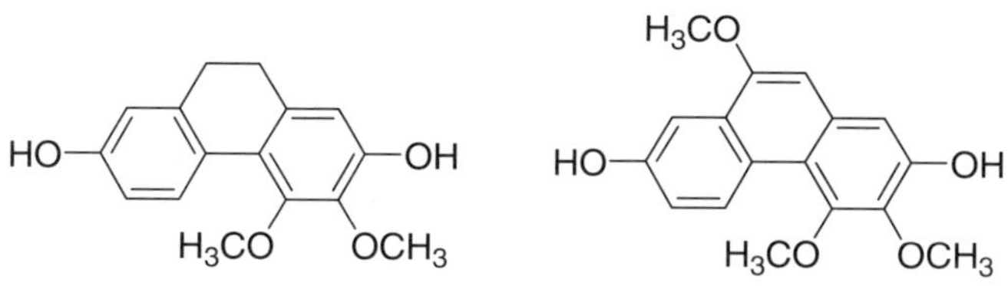

(17) **(18)**

Fig. 3. Major phytotoxic compounds from *Maxillaria densa.*

reduction. Gymnopusin (**18**) was more potent than erianthridin (**17**). The electrolyte leakage induced by 100 μM of **18** began after 12 hours. The level of electrolyte leakage was proportional to the concentration of the phytotoxin with poor leakage at 25 μM, moderate leakage at 50 μM and massive leakage at 100 μM. The effect was more pronounced after 48 hours of treatment.

Gymnopusin (**18**) also caused a significant decrease in chlorophyll content at concentrations ranging from 50 μM to 200 μM (Fig. 4). Finally,

A B
A B
C C
C C

Fig. 4. Effect of gymnopusin (**18**) on duckweed at concentrations ranging from 50 μM to 100 μM. (A) Control; (B) 50 μM; (C) 100 μM. Effects on membrane integrity followed a time course similar to that of electrolyte leakage.

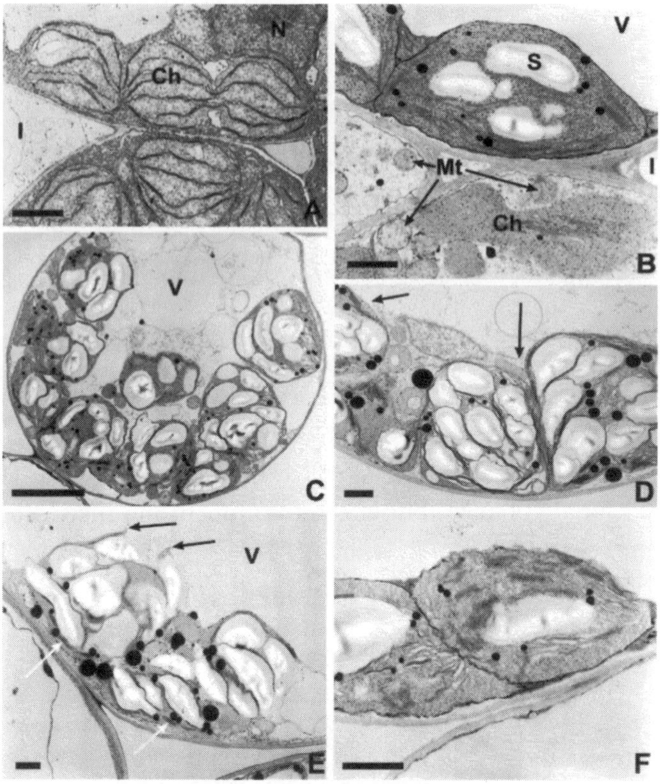

Fig. 5. Transmission electron micrographs of duckweed frond tissue. (A) Control, zero time. The chloroplasts have fully developed thylakoids with small grana stacks. Some small starch grains are present. Ch=chloroplast; I=intercellular space; N=nucleus. Bar=1 μm. (B) Treated with 100 μM gymnopusin for 12 hours. Some damage appears at this time. Pictured here are three adjacent cells, of which the top cell is relatively undamaged. The damaged cells appear to have ruptured tonoplasts, since there are organelles floating freely in the cells. Mt=mitochondrion; S=starch grain; V=vacuole. Bar=1 μm. (C) Treated with 100 μM gymnopusin for 24 hours. Some cells at this time exhibit loss of cytoplasm content. The relatively intact chloroplasts in this example appear to float freely within the cell, again indicating a ruptured tonoplast. Bar=5 μm. (D) Treated with 100 μM gymnopusin for 48 hours. The rate of deterioration of cells exposed to the toxin varies greatly within each sample time. Here, the heavily starch-laden chloroplasts are still contained within a relatively intact, though diffuse cytoplasm. There is evidence, however, that the tonoplast is not intact (arrows). Bar=1 μm. (E) Treated with 100 μM gymnopusin for 60 hour. In this cell, large starch grains and lipid bodies, which may be a sign of membrane deterioration, are present in the chloroplasts and the cytoplasm (black arrows). Grana stacks (white arrows, lower center and center left) are still

18 and **17** inhibited radicle elongation of *A. hypochondriacus* seedlings with IC_{50} values of 330 μM and 58.2 μM, respectively.[31]

Ultrastructural examination of duckweed frond (Fig. 5) and root tissues treated with **18** (100 μM) revealed membrane damage to the tonoplast after 12 hours of exposure. The samples viewed through the transmission electron microscope showed ruptured tonoplasts, free-floating organelles and loss of cytoplasm relative to control tissues. The tonoplast may be the primary target for the phytotoxic effect of **18**, which represents an unusual, if not unique, toxic mechanism among phytotoxic agents.[31]

Bioassay-guided fractionation of the active crude extract of *E. rigidum* furnished four phytotoxic agents, namely, gigantol (**19**), batatasin III (**20**), 2,3-dimethoxy-9,10-dihydrophenanthrene-4,7-diol (**21**) and 3,4, 9-trimethoxyphenanthrene-2,5-diol (**22**) (Fig. 6). Compounds **19**, **20**, **21** and **22** inhibited radicle growth of *A. hypochondriacus* with IC_{50} values of 0.65 μM, 0.1 μM, 0.12 μM and 5.9 μM, respectively. The phytotoxic effect was greater or comparable to that of 2,4-D ($IC_{50} = 0.19$ μM).[32]

Foliar application of gigantol (**19**) at 1 μM to four-week old seedlings of *A. hypochondriacus* reduced shoot elongation (69%) and fresh weight accumulation (54%). At 0.1 μM, the treatments inhibited only fresh weight accumulation (20%); at these two concentrations, the bibenzyl induced necrosis, desiccation and leaf abscission. Finally, at 0.01 μM, no activity was observed. Paraquat (0.1 μM), used as a positive control, had similar effects with the bibenzyl.

Studies on phytotoxicity of bibenzyl derivatives were extended to duckweed axenic cultures. Bibenzyls **19** and **20** as well as the synthetic analogs **23–28** were tested by triplicate in cultures of the small aquatic plant. Analogs **23–28** were synthesized to investigate the effect on phytotoxicity of oxygenated substituents (phenolic *vs.* phenolic methyl ether) and their location on the bibenzyl core structure. All synthesized analogs but

present in the thylakoids. No tonoplast is evident. Bar=1 μm. (F) Treated with 100 μM gymnopusin for 72 hours. The chloroplasts are showing thylakoid disruption. Lamellae are beginning to swell. Bar=1 μm. Reprinted from Phytochemistry, **61** number, Valencia-Islas, N.A., Paul R.N., Shier W.T., Mata R. and Abbas H.K. Phytotoxicity and ultrastructural effects of gymnopusin from the orchid *Maxillaria densa* on duckweed (*Lemna pausicostata*) frond and root tissues, 141–148, 2002, with permission from Elsevier"

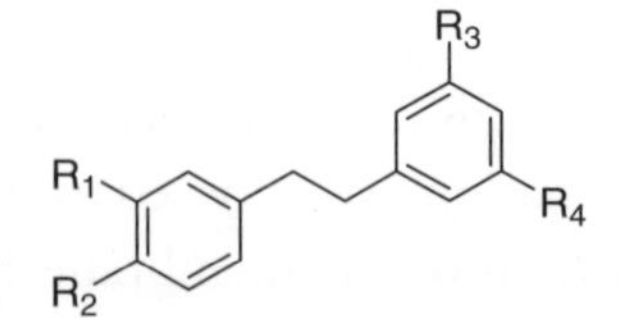

(19) $R_1 = OCH_3$; $R_2 = OH$; $R_3 = OCH_3$; $R_4 = OH$
(20) $R_1 = OH$; $R_2 = H$; $R_3 = OCH_3$; $R_4 = OH$
(21) $R_1 = OCH_3$; $R_2 = OH$; $R_3 = R_4 = OCH_3$
(22) $R_1 = R_2 = R_3 = R_4 = OCH_3$
(23) $R_1 = H$; $R_2 = OH$; $R_3 = R_4 = OCH_3$
(24) $R_1 = OCH$; $R_2 = H$; $R_3 = OCH_3$; $R_4 = OH$
(25) $R_1 = OH$; $R_2 = H$; $R_3 = R_4 = OH$
(26) $R_1 = H$; $R_2 = R_3 = R_4 = OH$

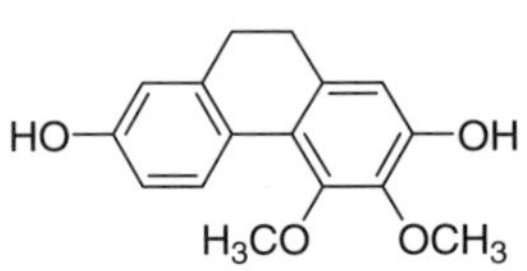

(27)

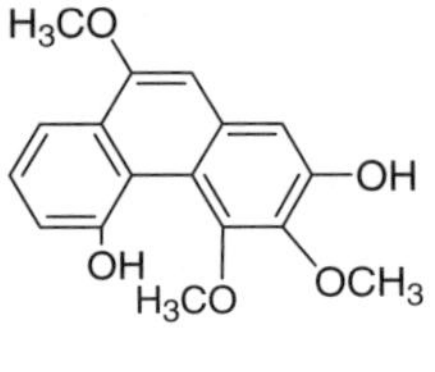

(28)

Fig. 6. Major phytotoxic compounds from *Epidendrum rigidum.*

28 were obtained by the Wittig reaction; compound **28** was synthesized by catalytic reduction of piceatannol.[32]

The data presented in Table 1 show that except for compounds **27** and **28**, all the analogs inhibited growth and caused cellular leakage in duckweed cultures with IC_{50}'s ranging between 89.9 μM–166 μM and 89.9 μ–180 μM, respectively. The most active compounds were **26** and **23**. Compounds **27** and **28**, which possess only free hydroxyl groups in both aromatic rings, were the least active. These results are consistent with bibenzyl derivatives requiring at least one methoxyl group at C-3 or C-5 for strong phytotoxicity.[32] Bibenzyls **19**, **20** and **23–28** inhibited the ability of CaM to activate PDE1 with IC_{50} ranging from 7 μM–13 μM[33] with a similar potency than chlorpromazine ($IC_{50} = 10.3\,\mu M$). Furthermore, they also retarded the electrophoretic mobility of CaM,[32] as an example the results for **19** and **20** are presented in Fig. 7.

Because ideal candidates for commercially-viable herbicides should have both strong phytotoxicity to susceptible weeds and low mammalian

Table 1. Effect of Bibenzyls 19, 20, 23–28 on *Duckweed* Cultures at 72 Hours. (Reprinted from *J Agric Food Chem*, 53, Hernandez-Romero Y, Acevedo L, Sánchez MD, Shier WT, Abbas HK and Mata R, Phytotoxic activity of bibenzyl derivatives from the orchid *Epidendrum rigidum*, 2005, 6276–6280 with permission from ACS.)

	Treatment IC_{50} (μM)	
Compound	Conductivity leakage	Growth inhibition
19	166	180
20	145	159
23	116	89.9
24	148	117
25	144	169
26	89.9	94.7
27	725	985
28	>1000	>1000

Results are the means of three replicates ± standard deviation. All samples exhibited significantly higher phytotoxicity than the controls ($p < 0.005$, student's umpaired t-test).

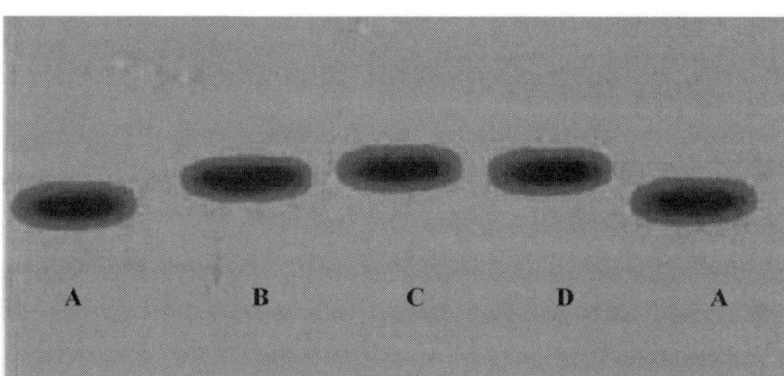

Fig. 7. SDS-PAGE of bovine-brain CaM after treatment with compounds **19** (0.033 μg/mL in DMSO) and **20** (0.033 μg/mL in DMSO). Electrophoresis of 2 μg samples of bovine-brain CaM in the presence of 1 mM $CaCl_2$. Pretreatment of the CaM samples: 1.5 hours at 30°C in the presence of $CaCl_2$; DMSO (A); chlorpromazine in DMSO (B); **19** (C); **20** (D).

toxicity, compounds **17–20** and **23–28** were evaluated for *in vitro* toxicity against four cultured mammalian cell lines representing undifferentiated normal and tumor cells and differentiated kidney and liver lines (Table 2).

Table 2. Cytotoxicity of Compounds 17–20, 23–28 in Cultured Mammalian Cell Lines

	IC_{50} (μM)			
Compound	**NIH3T3**	**MDCK**	**KA31T**	**H4TG**
17	19.3	10.2	13.0	20.0
18	13.0	11.0	21.0	13.0
19	25.8	35.9	57.3	75.0
20	14.7	81.9	49.4	123.7
23	94.2	97.5	141.6	76.5
24	> 200	> 200	108.5	195.6
25	29.3	79.6	22.3	83.8
26	60.6	107.1	45.0	126.2
27	32.3	74.2	57.2	118.1
28	65.1	78.3	43.9	165.8

The concentration of toxin which causes a 50% reduction in cell bound dye after five days in culture. Cell lines used were H4TG, thioguanine-resistant rat hepatoma cells; MDCK, Madin-Darb and canine kidney cells; NIH3T3,NIH Swiss mouse embryo fibroblasts and KA31T, Kirsten strain of Moloney sarcoma virus-transformed 3T3 cells.

The bibenzyls were not toxic but the phenanthrenes showed moderate citotoxicity towards the mammalian cells suggesting that they are not very promising as bioherbicides.[32]

Other orchid metabolites suchs as batatasin I, inhibited the growth of liverworts, algae and oat coleoptiles. Batatasin I also inhibited the CO_2-dependent O_2 evolution and the flow of electrons from water to methylviologen in spinach chloroplasts,[34] and it inhibited the succinate-dependent O_2 uptake in potato tuber mitochondria.[34] Other phenanthrenes such as orchinol, which has a free hydroxyl at the 7-position, inhibit indole-3-acetic acid (IAA) oxidation catalyzed by horseradish peroxidase.[34]

13.2.4 *Hofmeisteria schaffneri*

Hofmeisteria schaffneri (A. Gray) R. M. King & H. Robinson (Asteraceae) is a rare perennial medicinal herb known as "ambar." It grows naturally in the oak and pine oak forested mountains of the central Mexican states of Jalisco and San Luis Potosi. It is cultivated in home gardens in the valley

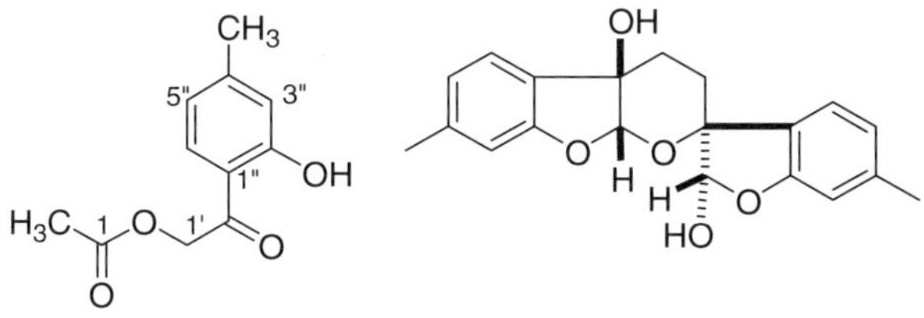

(29) (30)

Fig. 8. Phytotoxic agents from *Hofmeisteria schaffneri*.

of Mexico. The decoction of the fresh, leafy stems and flowers is used for treating skin wounds, fevers and gastrointestinal ailments.[12]

Bioassay-guided fractionation of a phytotoxic extract led to the isolation of two major phytotoxic agents (Fig. 8) characterized as 2′-(2″-hydroxy-4″-methylphenyl)-2′-oxoethyl acetate and designated with the trivial names of hofmeisterin (**29**) and 3′,4′,4*a*′,9*a*′-tetrahydro-6,7′-dimethylspiro [benzofuran-3(2*H*), 2′-pyrano [2, 3-*b*] benzofuran]-2, 4*a*′-diol (**30**), along with several inactive compounds.[12]

Compound **29** was assigned the molecular formula of $C_{11}H_{12}O_4$ (HRMS). The IR spectrum showed bands for phenolic (3500 cm^{-1}), ester (1747 cm^{-1}) and conjugated ketone (1657 cm^{-1}) functionalities. The NMR spectra were consistent with the presence of a trisubstituted benzene ring; a methyl group attached to an aromatic ring; an acetyl moiety, a chelated hydroxyl, a conjugated ketone and a methylene group attached to an oxygenated functionality. The arrangement of the substituents on the aromatic ring was determined by the analysis of the HMBC and NOESY spectra.

In order to verify the structure of **29** and to obtain additional amounts for biological studies, its synthesis was undertaken. The synthetic path to **29** is outlined in Scheme 1.[12] The general strategy was based on a Fries rearrangement of a suitable acetyl derivative of *m*-cresol (**31**). The synthesis started from m-cresol which was treated with Ac_2O to furnish **32** which was subjected to Fries rearrangement at 180°C to generate **33**. The free hydroxyl group of **33** was protected with benzyl bromide to yield the *O*-benzylacetophenone **34** which upon-bromination with Br_2 rendered **35**. The protection of the phenolic group facilitated the α-ketone

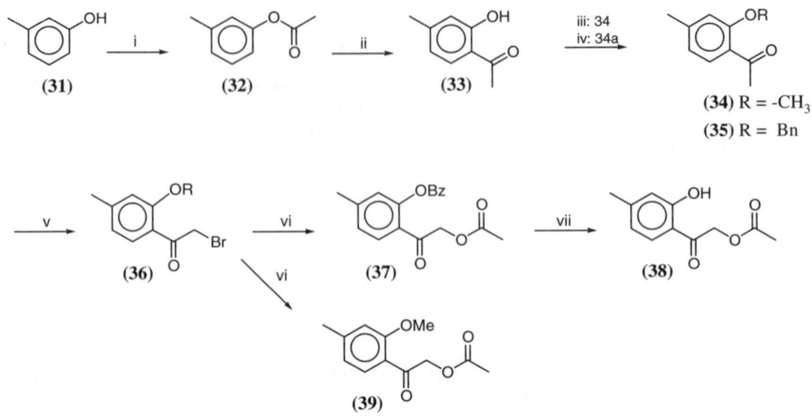

[a] Conditions: (i) Ac_2O; (ii) $AlCl_3$, CS_2, 180°C; (iii) NaOH, $(CH_3)_2SO_4$, bath ice-water; (iv) benzylbromide, K_2CO_3, a: acetone reflux, 5 days; (v) Br_2, benzene, rt, 4h; (vi) AcOH, DBU, THF, rt, 2.5 h; (vii) H_2, 10% Pd-C, 60 psi, 60°, 5h.

Scheme 1. Synthesis of hofmeisterin (Reprinted from *J Nat Prod*, 68, Perez-Vasquez A, Reyes A, Linares E, Bye R and Mata R, Phytotoxic agents from *Hofmeisteria schaffneri*: Isolation and synthesis of 2′-(2″-hydroxy-4″-methylphenyl)-2′-oxoethyl acetate, 959–962, 2005, with permission from ACS).

halogenation reaction. The displacement of bromide from **35** with 1,8-diazabicyclo[5.4.0]undec-7-ene (DBU)-CH_3COOH produced **36**(17% of yield), which was catalytically hydrogenated with palladium-charcoal (10% Pd-C) to afford **29** with an overall yield of 11.8%. The methyl derivative **29a** (Scheme 1) was also synthesized using the same approach; in this case, the intermediate **33** was subjected to methylation with $(Me)_2SO_4$ to yield **34a** which upon treatment with Br_2, rendered **35a**; finally treatment of **35a** with acetic acid/DBU afforded **29a** (15 % of yield).

Natural products **29** and **30** inhibited radical growth of *A. hypochondriacus* (IC_{50} = 380 and 12 μM, respectively). *E. crus-galli* was also sensitive to the treatments with compound **29**. Among the synthetic compounds, only the methyl derivative **34a** interfered (IC_{50} = 320 μM) with the growth of seedlings of *Amaranthus*. None of the tested compounds affected significantly radical growth of *Medicago sativa*.[12]

The action of **29** and **30** on PDE1 was also assessed to check the relationship between phytotoxicity and CaM inhibition. The results indicated that both thymol derivatives inhibited activation of PDE1 in the presence of CaM with IC_{50} values of 4.4 μM and 4.2 μM, respectively, while chlorpromazine had an IC_{50} of 6.8 μM.

Thymol derivatives have been described previously from other Asteraceae, however, northymol derivatives are uncommon. Biogenetically, **29** could be generated from the monoterpenoid thymol which might lose one methyl group from the isopropyl unit through an oxidative decarboxylation; the resulting normonoterpenoid derivative could yield **29** upon the appropriate tailoring reactions.

13.3 PHYTOTOXIC AGENTS FROM SELECTED FUNGI

13.3.1 *Phoma herbarum*

Phoma herbarum Westend (Sphaeropsidaceae) [syn. *Phoma pigmentivora* Massee] has a worldwide distribution and is known from the most varied substrates including herbaceous and woody plants, soil and water. This fungus inhibits the growth of the alga *Chlorella pyrenoidosa in vitro* and is pathogenic to *Avena fatua* and dandelion (*Taraxacum officinale*) seedlings after artificial infection. Furthermore, *P. herbarum* was recovered and characterized from small necrotic lesions on dandelion foliage.[35]

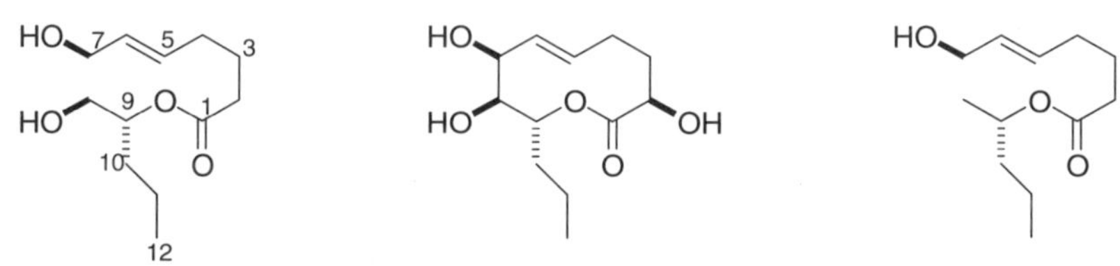

(40) **(41)** **(42)**

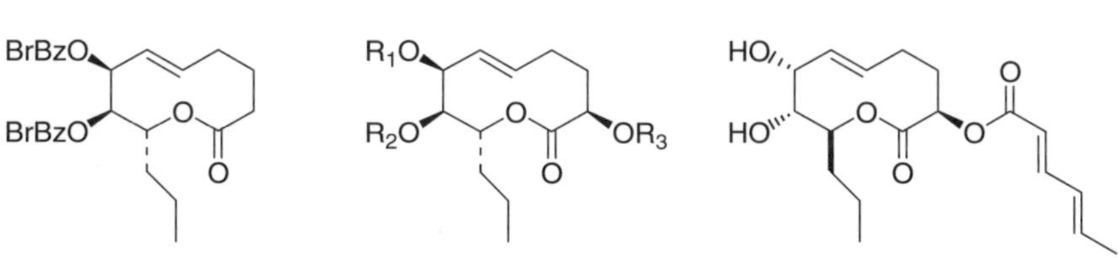

(40) (A)

$R_1 = R_2 = R_3 =$ H (41a)
$R_1 = R_2 = R_3 =$ Ac (41b)
$R_1 = R_2 = R_3 =$ BrBz (41c)

(43)

Fig. 9. Herbarumins I-III (**40–42**) and related compounds.

Mortimer *et al.*,[36] during an experimental study on the mycotoxicosis, provoked by *Phoma herbarum* variety medicaginis in sheep discovered that brefeldin A was the major toxic metabolite isolated from the cultures of the fungus. Cytochalasin B was also isolated from the cultures but at a much lower yield and only caused transient disturbances when given to the animals at doses of 60 mg/kg.[36] Zohri *et al.*[37] confirmed the presence of cytochalasins A and B in several cultures of this fungus. In 1979, Begum and Desphande[38] described a detailed study on the pectolytic enzyme of *P. herbarum*. Later on, MacGahren *et al.*[39] and Chandler and coworkers[40] reported the isolation of the anthraquinones helminthosporin and 1-hydroxyhelminthosporin (cynodontin), and the isomeric chromanones LL-D253α, LL-D253β and LL-D253γ.

From the phytotoxic culture of the fungus *P. herbarum* grown in liquid substrate on modified MD-I medium, we isolated three new nonenolides (Fig. 11), which were given the trivial names of herbarumins I-III (**40–42**), respectively.[41,42]

In general, the structures of these compounds were established on the basis of the spectroscopic methods, in particular, high resolution NMR techniques (COSY, HMQC, NOESY and HMBC). In all cases, the NMR spectra (Table 3) showed the existence of a 10-membered macrolide core, a vicinal diol, a *trans* disubstituted double bond, and a *n*-propyl unit. As an example, the NMR spectra of **40** are shown in Fig. 10. COSY and HMBC experiments indicated the position of the hydroxyl, *n*-propyl and olefin functionalities on the macrolactone core. The stereochemistry and the solution conformation of the ten-membered ring lactones of herbarumins I-III (**40–42**) was determined by molecular mechanics modeling, NOESY data and the comparison of the observed *vs.* the calculated vicinal proton coupling constants. Finally, the absolute configuration at the chiral centers was established using the CD exciton coupling method of Harada and Nakanishi.[43]

The systematic conformational search for herbarumin I (**40**) using the MMX force field as implemented in the PCMODEL program revealed the presence of the minimum energy conformation depicted in Fig. 11 ($E_{MMX} = 9.51$ kcal/mol), which is related to the chair-chair-chair conformation found in cyclodecane.[41] In this conformation, the value of

Table 3. NMR Chemical Shift Assignment of Herbarumins I-III (40–42)

	40		41		42	
C/H	^{13}C δ[a]	^{1}Hδ	^{13}C δ[a]	^{1}Hδ	^{13}C δ[a]	^{1}Hδ[c]
1	176.4	—	177.0	—	176.8	—
2a	34.4	2.32 *ddd* (14.0, 6.0, 2.0)	72.6	3.99 *dd* (8.2, 3.0)	34.6	2.29 *dddd* (13.7, 6.2, 1.6, 0.4)
2b		2.01 *ddd* (14.0, 12.5, 2.0)		2.05 *m*		2.02 *ddd* (13.7, 12.6, 1.5)
3a	24.6	1.89 *m*	33.8	2.10 *m*	25.9	1.99 *ddddd* (13,4, 13.3, 12.6, 3.5, 1.6)
3b		1.73 *m*		2.40 *m*		1.76 *ddddd* (13.4, 6.2, 4.0, 3.3, 1.5)
4a	33.3	2.40 *brdt* (12.0, 3.0)	25.2	2.15 *m*	33.6	2.37 *ddddddd* (14.0, 4.6,3.5, 3.3, 1.7, 1.5, 0.4)
4b		1.97 *m*		5.59 *dddd* (16.3, 7.5, 5.5, 2.5)		1.98 *dddd* (14.0, 13.3, 12.0, 4.0)
5	124.8	5.52 *dddd* (15.5, 10.0, 4.0, 2.5)	122.4	5.63 *dd* (16.3, 2.5)	124.5	5.47 *dddd* (15.8, 12.0, 4.6, 2.3)
6	130.7	5.62 *brd* (15.5, 2.5)	131.9	4.48 *bs*	134.5	5.61 *ddd* (15.8, 4.0, 1.5)
7	73.3	4.43 *quint* (2.0)	68.3	3.56 *dd* (9.75, 2.5)	68.0	4.42 *ddddd* (5.2, 4.0,2.3, 2.2, 1.7)
8a	73.6	3.52 *dd* (9.8, 2.5)	71.8	5.06 *td* (9.75, 2.5)	20.5	1.86 *ddd* (14.6, 5.2, 2.0)
8b						1.79 *ddd* (14.6, 11.2, 2.2)
9	70.2	4.96 *td* (9.8, 2.5)	70.7	1.77 *m*	67.8	5.27 *dddd* (11.2, 8.5,4.2, 2.0)
10a	33.7	1.87 *m*	33.8	1.57 *m*	37.4	1.42 *dddd* (13.5, 9.2, 6.8, 4.2)
10b		1.55 *m*		1.47 *m*		1.55 *dddd* (13.5, 9.2, 8.5, 4.9)
11a	17.9	1.38 *m*	17.8	1.31 m	18.4	1.32 *dddq* (13.5, 9.2, 7.3, 6.8)
11b		1.30 *m*		0.91 *t* (7.5)		1.32 *dddq* (13.5, 9.2, 7.3, 4.9)
12	13.8	0.89 *t* (7.5)	13.8		13.8	0.90 *t* (7.3)

^{1}H NMR (500 MHz) and ^{13}C NMR (125 MHz) spectra were run in $CDCl_3$.
[a]From HMQC and HMBC. [b]Coupling constant in parentheses (Hz). [c]Some coupling constants (Hz) are calculated.

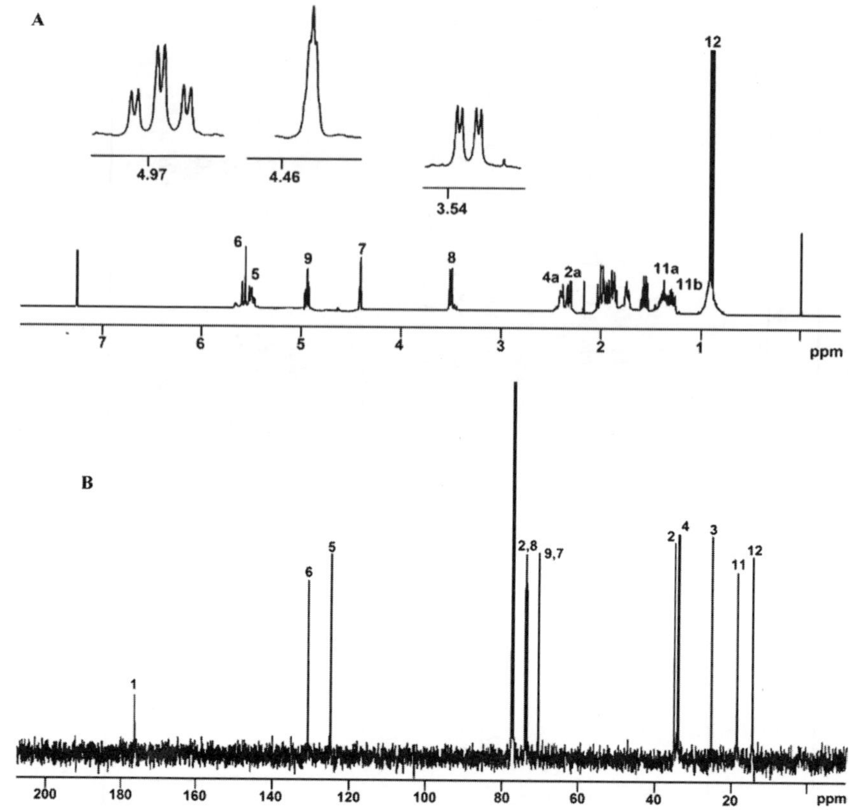

Fig. 10. ^{1}H (A, 500 MHz) and ^{13}C (B, 125 MHz) NMR spectra of herbarumin I (**40**) in $CDCl_3$.

the coupling constant ($J = 2.5$ Hz) between H-7 and H-8 indicated their *cis*-equatorial-axial relationship. On the other hand, the value of the coupling constant ($J = 9.5$ Hz) between H-8 and H-9 revealed their *trans*-diaxial relationship. The absolute configuration at C-7 and C-8 was determined using the CD exciton coupling method of Harada and Nakanishi.[43] For this, **40** was treated with *p*-bromobenzoyl chloride to yield the di-*p*-bromobenzoate derivative **40a**, whose minimum energy conformation ($E_{MMX} = 49.67$ kcal/mol) was similar to that of **40**. The CD spectrum of **40a** indicated a bisignate $[-6.42 \times 10^4(237), 9.26 \times 10^5(255)]$ Cotton effect, corresponding to a positive chirality. Thus, the stereogenic centers

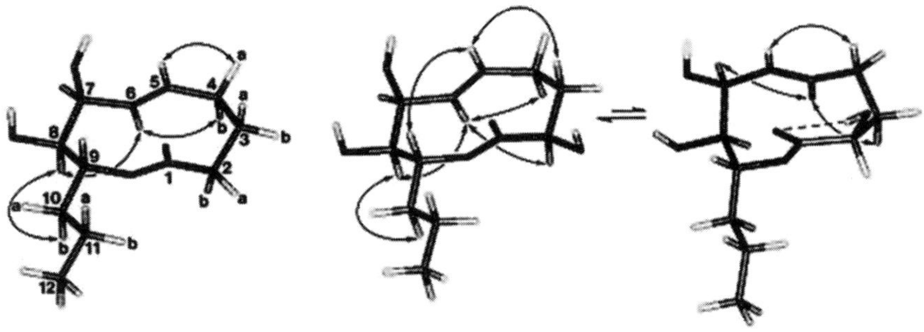

40
$E_{MMX} = 9.51$ kcal/mol

41A
$E_{MMX} = 8.37$ kcal/mol

41B
$E_{MMX} = 8.74$ kcal/mol

Fig. 11. Minimum energy structures of herbarumin I (**40**) and herbarumin II (**41A** and **41B**) (Reprinted from *Tetrahedron*, **56**, Rivero-Cruz JF, García-Aguirre G, Cerda-Garcia-Rojas CM and Mata R, Conformation behavior and absolute stereostructure of two phytotoxic nonenolides from the fungus *Phoma herbarum*, 5337–5344, 2000, with permission from Elsevier).

at C-7 and C-8 were each determined to have the *S*-configuration. The absolute configuration at C-9 was therefore assigned as *R*.

Herbarumin II (**41**) was the C-2 hydroxyl derivative of **40**. Treatment of **41** with pyridine/$(CH_3CO)_2O$ yielded the triacetyl derivative **41a** confirming the presence of three hydroxyl groups in the molecule. In addition, the presence of a vicinal diol in **41** was corroborated by obtaining the acetonide **41b** upon treatment of **41** with anhydrous Me_2CO and $CuSO_4$ (reflux). The NMR spectra of **41** in $CDCl_3$ were almost identical with those of **40** except for the presence of signals for an additional secondary carbinol group (δ_H/δ_C 3.99, dd, $J = 8.2, 3.0$ Hz/72.6) in place of the methylene signals attributed to C-2.

The ten-membered ring in **41** adopts two preferred (Fig. 11) conformations (38A and 38B). In **41A** ($E_{MMX} = 8.37$ kcal/mol), this ring exists in a chair-chair-chair conformation very similar to that found in **40**, while this ring in **41B** ($E_{MMX} = 8.74$ kcal/mol) resembled the twist chair-boat-chair conformation of cyclodecane. The ^{1}H NMR data in $CDCl_3$ is consistent with the formation of the O(2)H–O(1) hydrogen bond which accounts for the stabilization of conformer **41B**. In MeOH-d_4, this intramolecular hydrogen bonding is released by the interaction and exchange of the labile

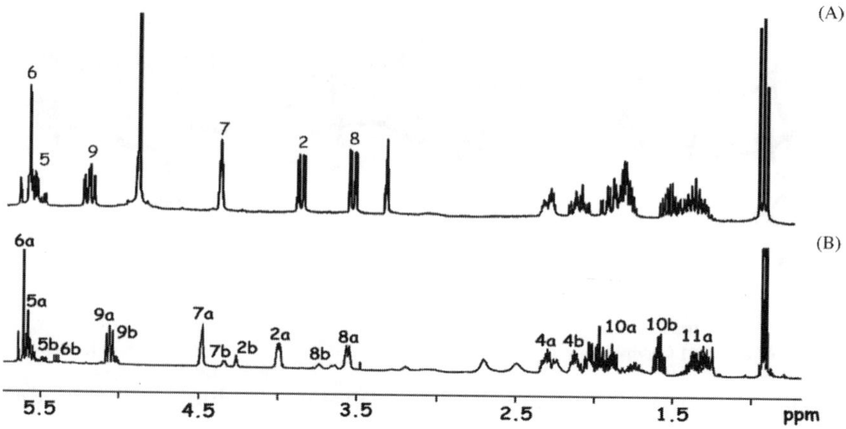

Fig. 12. ^{1}H (500 MHz) NMR spectra of herbarumin II (**41**) in $CDCl_3$ (A) and MeOH-d_4 (B).

hydrogen O(2)H with the solvent atoms, as revealed by the presence of a single conformer (**41A**) in the ^{1}H NMR spectrum (Fig. 12). In MeOH-d_4, the signal for the proton geminal to the C-2 oxygen of **41A** was observed as a sharp *dd* with $J_{2,3a} = 10.2$ Hz and $J_{2,3b} = 3.2$ Hz.

The NOESY spectrum of **41** was also in agreement with the presence of both conformations. In the case of **41A**, the interactions H-5/H-9, and H-2/H-6 were the most relevant; the latter was consistent with a pseudoequatorial (β) disposition of the hydroxyl group at C-2. For **41B**, the most important NOESY interaction was observed between H-7 and H-5 and between H-6 and H-4.

Additional evidence to support the conformational equilibrium in $CDCl_3$ between **41A** and **41B** were obtained through a series of low temperature. ^{1}HNMR (Fig. 13) experiments which allow to propose that the interconversion of **41A** in **41B** might proceed via two non-synchronized conformational movements, involving bond rotations of the C(1)–C(2)–C(3)–C(4)–C(5) and C(4)–C(5)=C(6)–C(7) moieties, which would lead to two intermediate conformations. The presence of these conformations ($E_{MMX} = 8.90$ kcal/mol and $E_{MMX} = 9.90$ kcal/mol, respectively) can be envisaged by the reduction of the *trans* diaxial coupling constants $J_{2b,3a}$ and $J_{4b,5}$ in **41A**, on going from MeOH-d_4, where the interconversion process does not take place to $CDCl_3$.

The absolute configuration of **41** at the stereogenic centers at C-2, C-7, C-8 and C-9 was determined following the same procedure as for compound **40**. The tri-*p*-bromobenzoate **41c**, whose ten-membered ring conformation ($E_{MMX} = 67.34$ kcal/mol) resembled those of **40**, **41A**, **40a** and **41a** showed a typical split CD and this established the chirality of the 7-OBrBz/8-OBrBz as being positive. Therefore, herbarumin II (**41**) was characterized as (2*R*, 7*S*, 8*S*, 9*R*)-2,7,8-trihydroxy-9-propyl-5-nonen-9-olide (**41**).

During the characterization of herbarumin II (**41**), we observed that the H-2 coupling constants of **41A** ($J = 10.0$ and 3.5 Hz) were different to that of the related lactone pinolidoxin (**43**) ($J = 5.6$ and 1.7 Hz) suggesting that the stereochemistry at C-2 in the latter compound was wrongly proposed by Evidente *et al.*[44] In this regard, Furstner *et al.*[45] reported the total syntheses of pinolidoxin (**43**) and herbarumin II (**41**) solving what they called the pinolidoxin puzzle. This work allowed the correction of the stereostructure at C-2 of pinolidoxin which as pointed out in the paper by Rivero-Cruz *et al.*[41] was the opposite to that of herbarumin II (**41**).

The conformation and absolute configuration at the stereogenic centers in herbarumin III (**42**) were established following the same general strategy as for herbarumins I (**40**) and II (**41**).

Since the publication of the structures of herbarumins I-III (**40–43**), many authors have synthesized these medium sized lactones following different approaches. First, Furstner and coworkers from Germany[46] reported the enantioselective total synthesis of herbarumin I (**40**) based on an (E)-selective ring-closing metathesis (RCM) reaction. This group, as already pointed out, also synthesized[45] herbarumin II (**41**) and pinolidoxin (**43**). Sabino *et al.* from Brazil[47] stereoselectively synthesized herbarumin I (**40**) in 17 steps and 6% yield from L-arabinose using the Nozaki-Hiyama-Kishi coupling and modified Yamaguchi macrolactonization as key steps. Later on, herbarumin I (**40**) and (-)-pinolidoxin (**43**) were prepared applying a unique strategy which highlighted the applications of a novel silacyclic alcohol as a common precursor and features the finding of the reversible RCM.[48] Diez and coworkers[49,50] described the total synthesis of herbarumin II (**41**) by using butane diacetal (BDA)-desymmetrized glycolate as building blocks.

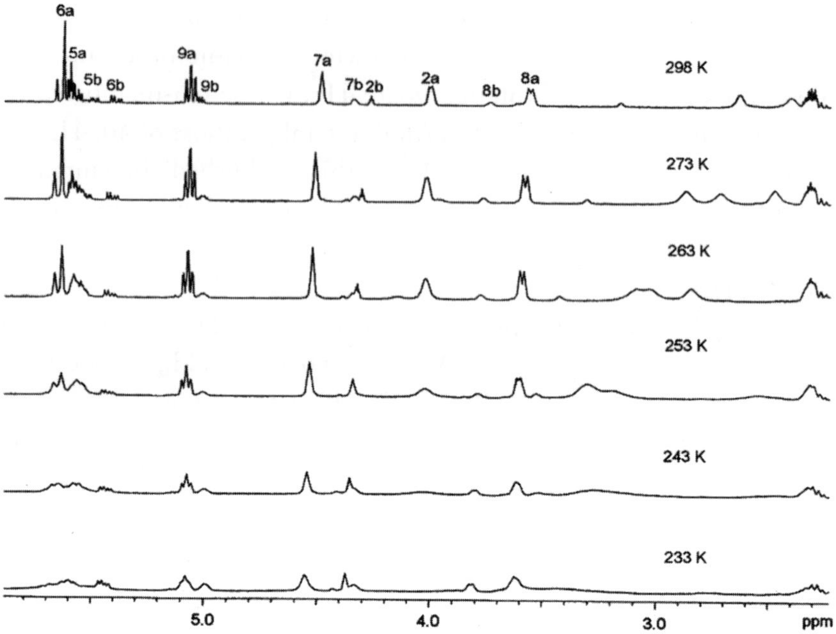

Fig. 13. The δ 4.3-2.5 region of the ^{1}H NMR spectrum of **41** recorded at different temperatures in $CDCl_3$. (Reprinted from *Tetrahedron*, **56**, Rivero-Cruz JF, García-Aguirre G, Cerda-Garcia-Rojas CM and Mata R. Conformation behavior and absolute stereostructure of two phytotoxic nonenolides from the fungus *Phoma herbarum*, 5337–5344, 2000, with permission from Elsevier.)

The first 10-step total asymmetric synthesis of herbarumin III (**42**) in 24% overall was reported in 2004 by Gurjar and coworkers[51,52] who later synthesized the compound using the RCM approach. Thereafter, a chemoenzymatic asymmetric synthesis of **42** which fixed the hydroxyl stereocenters (C7 and C9) by lipase catalyzed irreversible transesterification was described.[53]

Last year, a short enantioselective total synthesis of herbarumin III (**42**) in 11% overall yield was published;[54] the approach applied uses Keck's asymmetric allylation and Sharpless epoxidation to build the key fragment. Esterification with 5-hexenoic acid and a RCM was used to yield **42**. Finally, another asymmetric synthesis of herbarumin III (**42**) was carried out using (R)-cyclohexylidene glyceraldehyde as the chiral template.[55] The key steps of the synthesis were the enantioselective preparation of the

Table 4. Phytotoxic Effect of 40–42, 41a and 41b

Compound	IC_{50} (μM)
Extract*	35.8
40	54.3
41	125.4
42	20
41a	1470
41b	2750
2, 4-D**	192

* Expressed in mg/ml. ** positive control

required homoallylic alcohol an asymmetric dihydroxylation, and a RCM reaction for the macrolactonization.

Natural products **40–43**, as well as derivatives **41a** and **41b**, were evaluated for their ability to inhibit seed germination and seedling growth of *A. hypochondriacus*. Table 4 summarizes the phytotoxic effect of the tested compounds on seedling growth. Compounds **40** and **42** were more potent than 2,4-D, while compound **41** exhibited a similar potency with respect to the standard. Blocking of the diol functionality in compound **41** significantly decreased the pre-emergent phytotoxic activity. Previously, Evidente *et al.*[56] demonstrated that the postemergent phytotoxic effect pinolidoxin (**43**) required not only the integrity of the nonenolide ring and the propyl side chain but also the presence of hydroxyl groups at C-8 and C-9.

Compounds **40–43** interacted with bovine brain-calmodulin as detected in a SDS-PAGE electrophoresis. Calmodulin treated with the lactones had lower electrophoretic mobility than untreated calmodulin. The effect was comparable to that of chlorpromazine, a well known calmodulin inhibitor. In addition, different concentrations of compounds **40** and **41** inhibited calmodulin-dependent PDE1. The inhibitory activity of herbarumins I ($IC_{50} = 14.2\,\mu M$) and II ($IC_{50} = 6.6\,\mu M$) was higher than that of chlorpromazine ($IC_{50} = 9.8\,\mu M$).[42]

More recently, it was demonstrated that in growth room and field conditions, a combination of 2,4-D and *P. herbarum* produced enhanced control of dandelion.[57]

Concerning to other metabolites isolated from this species, an exopolysacharide was obtained from a culture of *P. herbarum* (CCFEE 5080) from the Antarctic.[58] Another polysaccharide (YCP) with a molecular weight of 2.4 × 10^3 kDa was isolated from the mycelium of a marine variety (*P. herbarum* YS4108) using a combination of ion-exchange chromatography on DEAE-32 and gel permeation over Sephacryl S-400. YCP was mitogenic and activated macrophages production in mice and *in vitro* revealing its potential as immunomodulating agent.[59] On the other hand, 3-nitrophthalic acid was isolated from a cell free culture filtrate of *P. herbarum* FGCC no. 75. This compound was phytotoxic against the weed *P. hysterophorus*.[60]

13.3.2 *Guanomyces polytrhix*

Guanomyces polythrix is a species of coprophilous fungus first isolated in 1979, from bat guano obtained from zone III of the cave "Cueva del Diablo" in Tepozotlan, Morelos, Mexico. The fungus forms a membranous ascoma with a long neck that is ornamented with lateral glandular hairs and terminal glandular hairs. When the systematic position of this ascomycete was analyzed, its morphological characters and 18S rDNA sequences revealed that this fungus belongs in the Sordariales and was close to but separate from *Chaetomium*. The hyaline ascospores, without a germ pore, and the possession of an ascoma with a long neck ornamented with lateral glandular hairs is in contrast to other genera of the Chaetomiaceae. For these reasons, the new genus *Guanomyces* was proposed to accommodate the new species *G. polythrix* in the Chaetomiaceae of the Sordariales.[61]

The fungus *G. polythrix* was grown in liquid-substrate fermentation on potato dextrose broth (PDB). The culture broth and the mycelium were extracted with CH_2Cl_2. The combined extract showed phytotoxic activity when evaluated on seedlings of *A.hypochondriacus* and *E. crus-galli* using the Petri dish bioassay. Bioactivity-guided fractionation of this extract led to the isolation of fourteen phytotoxic agents. Compounds **44–50** (Fig. 14) are new natural products and were characterized by spectroscopic and chiroptical methods as (2*S*, 3*S*)-5-hydroxy-6,8-dimethoxy-2,

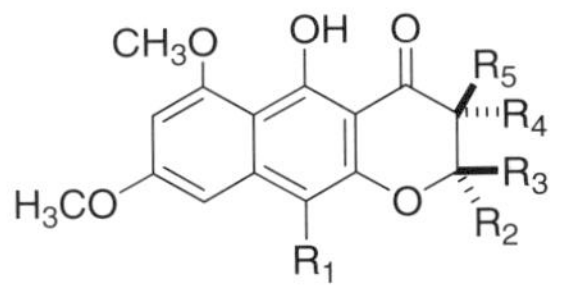

(44) R_1=H, R_2=CH_3, R_3=H, R_4=H, R_5=CH_3
(45) R_1=CH_3O, R_2=CH_3, R_3=H, R_4=H, R_5=CH_3
(46) R_1=H, R_2=CH_3, R_3=H, R_4=H, R_5=H
(47) R_1=CH_3O, R_2=CH_3, R_3=H, R_4=H, R_5=H
(48) R_1=H, R_2=CH_3, R_3=H, R_4=CH_3, R_5=H
(49) R_1=CH_3O, R_2=CH_3, R_3=H, R_4=CH_3, R_5=H

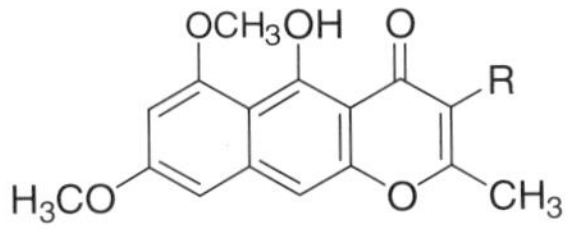

(50) R=CH_3
(51) R=H

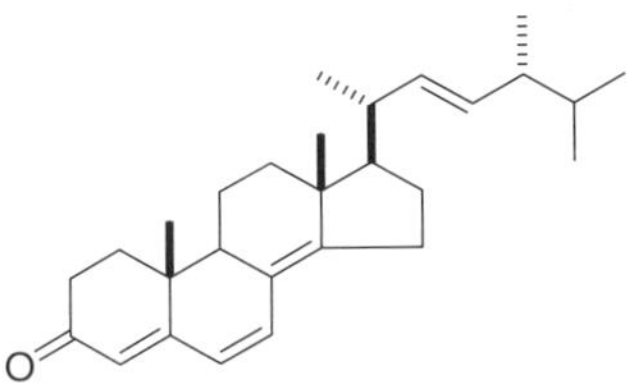

(52)

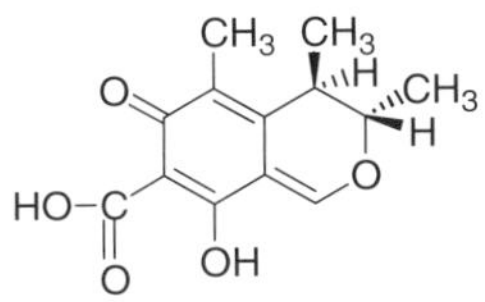

(53)

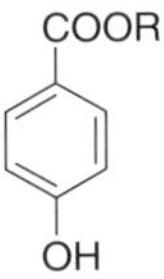

(55) R=H
(57) R=CH_3

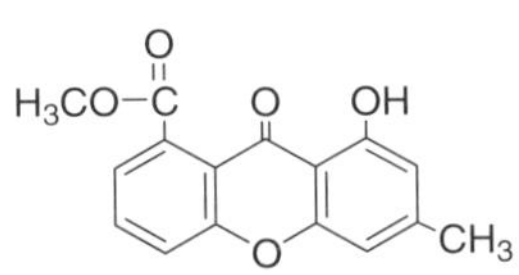

(54)

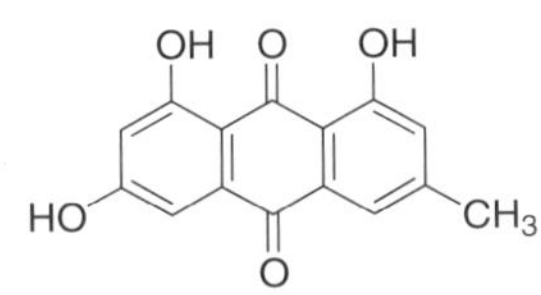

(56)

Fig. 14. Phytotoxic agents from *Guanomyces polythrix.*

3-dimethyl-4*H*-2,3-dihydronaphtho[2,3-*b*]-pyran-4-one (**44**); (2*S*, 3*S*)-5-hydroxy-6, 8, 10-trimethoxy-2,3-dimethyl-4*H*-2,3-dihydronaphtho [2,3-*b*]-pyran-4-one (**45**); (2*S*)-5-hydroxy-6,8-dimethoxy-2-methyl-4*H*-2, 3-dihydronaphtho[2, 3-*b*]-pyran-4-one (**46**); (2*S*)-5-hydroxy-6, 8, 10-trimethoxy-2-methyl-4*H*-2, 3-dihydronaphtho[2, 3-*b*]-pyran-4-one (**47**); 5-hydroxy-6, 8-dimethoxy-2, 3-dimethyl-4*H*-naphtho[2, 3-*b*]-pyran-4-one (**48**); (2*S*, 3*R*)-5-hydroxy-6,8-dimethoxy-2,3-dimethyl-2,3-dihydro-4*H*-naphtho[2,3-b]-pyran-4-one (**49**); and (2*S*, 3*R*)-5-hydroxy-6, 8, 10-trimethoxy-2, 3-dimethyl-2, 3-dihydro-4*H*-naphtho[2, 3-b]-pyran-4-one (**50**). In addition, rubrofusarin B (**51**), ergosta-4, 6, 8(14)22-tetraen-3-one (**52**), citrinin (**53**), 8-hydroxy-6-methyl-9-oxo-9H-xanthene-1-

Table 5. Phytogrowth-Inhibitory Activity (IC_{50}, μM) of Compounds 44–57 from *Guanomyces polythrix*

Treatment	*A. hypochondriacus*	*E. crus-galli*
Extract	54.3*	384.3*
44	320	380
45	65	61
46	23	87
47	13	40
48	80	170
49	110	130
50	75	70
51	13	87
52	80	170
53	55	63
54	15	9.8
55	17	230
56	130	20
57	23	630
2, 4, D[a]	18	88

[a] Positive control (2, 4-dichlorophenoxyacetic acid);
* expressed in μg/mL

carboxylic acid methyl ester (**54**), *p*-hydroxibenzoic acid (**55**), emodin (**56**), and *p*-hydoxybenzoic acid methyl ester (**57**), were obtained. Compound **54** was characterized by X-ray crystallography.[62,63]

Natural products **44–57** were evaluated for their ability to inhibit seed germination and seedling growth of *A. hypochondriacus* and *E. crusgalli*. Table 5 summarizes the phytotoxic effects of the isolates on seedling growth. In general, the tested compounds showed significant phytotoxic effect and were more potent as radicle growth inhibitors than the positive control (2,4-D). In all cases, the target species were affected by the compounds in a concentration-dependent manner.[62,63]

According to a sodium dodecyl sulfate-polyacrylamide gel electrophoresis (SDS-PAGE) and CaM affinity chromatography analyses, compounds **44–53** interacted with both spinach and bovine-brain CaMs. In order to quantify the interaction of the phytotoxic agents **44–53** with

CaM, their effect on the activity of two CaM-dependent enzymes, namely PDE1 and NADK, was investigated. All the tested compounds except citrinin (**53**) showed significant inhibitory effect on PDE activity in a concentration-dependent manner [IC_{50} values for compounds **44–52** were 8.1 μM, 7.2 μM, 6.1 μM, 5.8 μM, 7.6 μM, 6.6 μM, 4.8 μM, 4.7 μM and 5.2 μM, respectively]. The inhibitory activity was higher or comparable to that of chlorpromazine (IC_{50} = 10.6 μM). It is important to point out that none of the tested compounds affected the basal activity of PDE at the concentrations tested.[62,63]

Calmodulin-dependent NADK was selected as the second reporter enzyme because it is a vegetal enzyme and is only modulated by CaM in plants. In addition, NADK seems to be regulated by specific plant-CaM isoforms which could be specific targets for phytotoxic agents. The use of this enzyme to detect CaM inhibitors has not been previously reported. The procedure involved three main steps. The first one was the isolation of a suitable enzymatic preparation free of endogenous CaM from peas according to the method of Harmon. Since the assessment of the NADK activity was performed by a coupled enzymatic reaction involving GPDH,[14] the second step of the assay was the evaluation of the effect of the phytotoxic agents on the basal activity of NADK and GPHD. Finally, in the last step, the ability of the phytotoxic agents to prevent the stimulation of the NADK preparation in the presence of saturating concentrations of spinach CaM, was assessed. After processing the data of the second step, it was found that compounds **46** and **47** inhibited NADK basal activity with IC_{50} values of 49.1 μM and 26.2 μM, respectively. Therefore, these compounds were not further investigated. The remaining compounds did not inhibit NADK or GPDH basal activities at the concentrations tested. As in the case of the experiments with PDE1, all the phytotoxic agents but citrinin (**53**) inhibited the activity of the kinase promoted by spinach CaM. Compounds **50**, **51**, **45** and **49** were the strongest inhibitors of the complex-spinach CaM-NADK with IC_{50} values of 17.1 μM, 13.3 μM, 22.0 μM and 24.3 μM, respectively. The overall results indicate that among the naphtopyrones series, those phytotoxic agents possessing a double bond between C-2 and C-3 are the most active CaM inhibitors (i.e. compounds **50** and **51**).[14]

13.3.3 *Malbranchea aurantiaca*

As the other species of fungi, *Malbranchea aurantiaca* (Fig. 15) was selected as a source of phytotoxic agents considering the noted phytogrowth inhibitory activity against *A. hypochondriacus* seedlings of the extracts prepared from broths and mycelia of the fungus fermented in different conditions.[64]

The first fermentation was accomplished in dark conditions at 25°C. The resulting active organic extract (mycelial and broth) (IC_{50} = 195.0 μg/ml against *A. hypochondriacus*) yielded 1-hydroxy-2-oxoeremophil-1(10), 7(11), 8(9)-trien-12(8)-olide (**58**), a novel natural product and penicillic acid (**59**), a known phytotoxin (Fig. 18).[64]

Compound **58** was isolated as a yellow crystalline, optically active compound. HREIMS data for **58** gave a molecular formula of $C_{15}H_{16}O_4$ with an inherent eight degrees of unsaturation. The IR spectrum was consistent with the presence of conjugated carbonyl functionalities including a γ-lactone (1779 cm^{-1}) and a conjugated ketone (1665 cm^{-1}). The ^{13}C NMR (Fig. 17) spectrum for **58** revealed 15 resonances, interpreted from multiplicity-edited HSQC data as eight quaternary, two methine, two methylene, and three methyl carbons. In the ^{1}H NMR spectrum (Fig. 17)

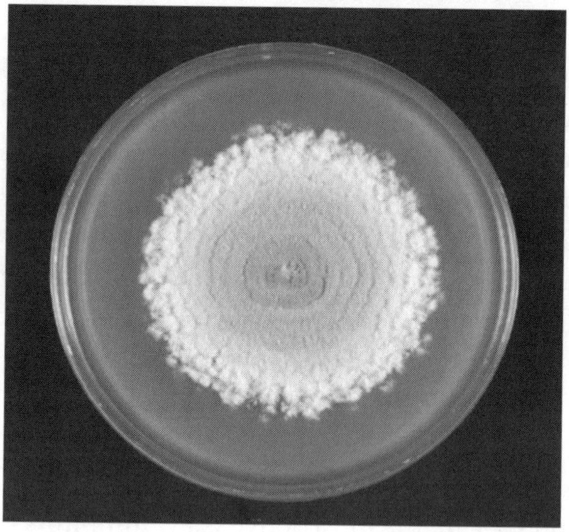

Fig. 15. *Malbranchea aurantiaca.*

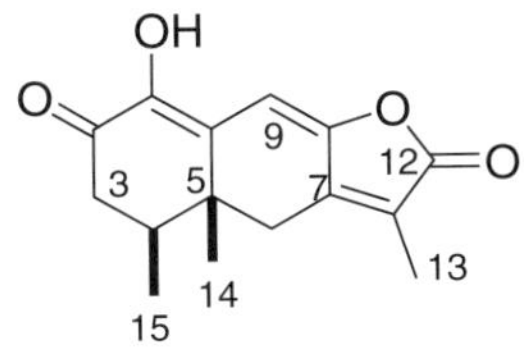

(58)

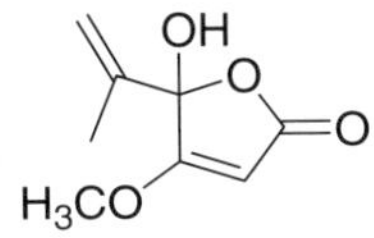

(59)

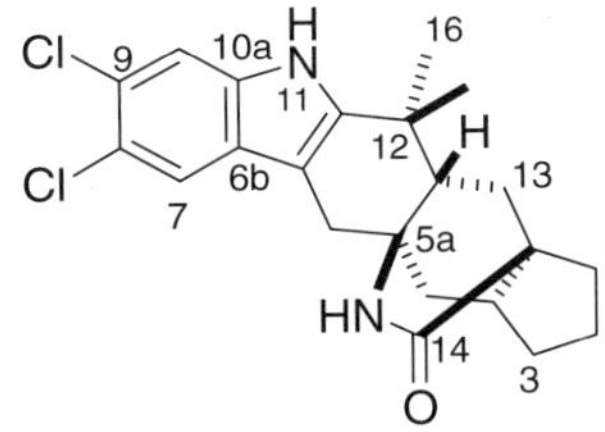

(60)

Fig. 16. Phytotoxic agents from *Malbranchea aurantiaca*.

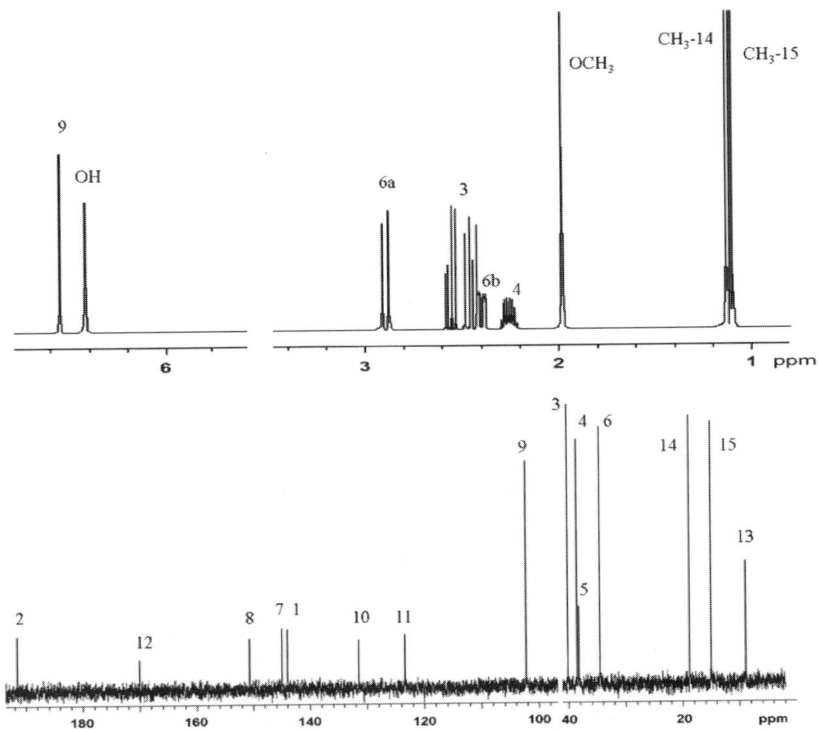

Fig. 17. ^{1}H (A, 500 MHz) and ^{13}C (B, 125 MHz) NMR spectra of **58** in $CDCl_3$.

of **58**, its conjugated unsaturated nature was immediately evident from one olefinic proton resonance (δ_H 6.58, d, J = 0.5 Hz, H-9) accompanied by two deshielded methylene resonances [δ_H 2.56, dd, J = 18, 4.5 Hz ,H-3α; 2.47, dd, J = 18, 7 Hz, H-3β; 2.90, dd, J = 16.5, 0.5 Hz, H-6β; 2.41 ddd,

$J = 16.5, 2, 1$ Hz, H-6α]. Also apparent in the ^{1}H NMR spectrum were signals for three methyl [δ_H 1.98, d,t, $J = 2.5, 1$ Hz, H-13; 1.21,d, $J = 1$ Hz, H-14; and 1.01, d, $J = 6.5$ Hz, H-15] and one hydroxyl (δ_H 6.45, s) groups.

The position of the functional groups along the eremophilane core was confirmed by an HMBC experiment. In particular, the correlations C-10/OH-1, H-6, H-9; C-8/H-9, H-6, H-13; C-2/OH-1, H-4 and C-12/H-13 were consistent with the disposition of the ketone, hydroxyl and lactone functionalities.

The structure of **58** was elucidated as 1-hydroxy-2-oxoeremophil-1(10),7(11),8(9)-trien-12(8)-olide. This proposal was corroborated by an X-ray analysis. Finally, the absolute configuration of compound **58** was proposed as depicted on the basis of the CD spectrum which displayed a strong negative Cotton effect on 378 nm.[23,65] Furthermore, as (-) ligularenolide,[66] whose absolute stereochemistry was determined by chemical correlation, compound **58** possesses a negative optical rotation.

Compounds **58** and **59** showed phytotoxic effects when tested against seedlings of *A. hypochondriacus* using a Petri dish bioassay. Compounds **58** [$IC_{50} = 6.57\,\mu M$] and **59** [$IC_{50} = 3.86\,\mu M$] inhibited radicle growth of this species with a similar potency to 2,2–dichlorophenoxyacetic acid [2,4-D; $IC_{50} = 18\,\mu M$], which was used as a positive control.

Although penicillic acid (**59**) is produced in high quantities in cultures of *Penicillum cyclopium* and *Penicillum canescens*[67] isolated from corn seeds, its toxicity to animals have precluded further development as a herbicidal agent.

The effect of compound **58** on calmodulin (CaM) was also investigated. The results showed that the activation of the calmodulin-sensitive PDE1 was inhibited in the presence of the compound ($IC_{50} = 10.2\,\mu M$) and CaM. Its effect was higher than that of chlorpromazine ($IC_{50} = 18.4\,\mu M$).[64]

The second fermentation of *M. aurantiaca* was carried out in warmer surroundings (~32°C) and under natural day-light cycles.[11] After a usual work out of the phytotoxic extract ($IC_{50} = 195.0\,\mu g/mL$ against *Amaranthus hypochondriacus* seedling) from the mycelium and broth obtained from this second fermentation, a novel indole alkaloid was obtained as a colorless crystalline solid. The compound was characterized as (5a*S*, 12a*S*, 13a*S*)-8,9-dichloro-12,12-dimethyl-2, 3, 11, 12, 12a, 13–hexahydro-1*H*,6*H*-5a,13a (epiminomethano) indolizino [7,6b]carbazol-14–one (Fig. 17), which was given the trivial name of malbrancheamide (**60**).[11] Its molecular formula was determined as $C_{21}H_{23}ON_3Cl_2$ by HRMS. The compound showed a positive color reaction with Dragendroff's and Erlich's reagents. This information as well as the UV absorption maxima at 233 nm and 293 nm suggested that **60** was an indole alkaloid. The presence of two chlorine atoms in the molecule was consistent with the relative abundance of the [M+2] and [M+4] peaks with respect to the molecular ion [M^+] in the mass spectrum (approximately two-third and one-tenth, respectively, of the intensity of the molecular ion peak). The IR spectrum showed typical absorption bands for lactams at 3299 cm^{-1} and 1659 cm^{-1}. In the ^1H NMR spectrum (Fig. 20) of **60**, two singlets at δ_H 7.47 (H-7) and 7.39 (H-10) showed the existence of a tetrasubstituted aromatic ring. Also apparent in the ^1H NMR spectrum were signals for two methyl [δ_H 1.32, s, H-16; and 1.42, s, H-17], one aliphatic methine (δ_H 2.14, m, H-12a) and six methylene groups [δ_H 2.86, m, H-6A; 2.85, m, H-6B; 1.99, m, H-13A; 1.94, m, H-13B; 2.52, m, H-1A; 1.46, m, H-1B; 1.87, m, H-2; 3.05, m, H-3A; 2.15, q, J = 2, 5 Hz, H-3B; 2.25, dd, J = 2, 10 Hz, H-5A; 3.42, d, J = 2 Hz, H-5B]. The ^{13}C NMR spectrum (Fig. 18) revealed 21 resonances, interpreted from multiplicity-edited HSQC data as ten quaternary, three methine, six methylene, and two methyl carbons; this spectrum also supported the presence of a lactam functionality (δ_C 176.67, C-14) and an indole group [δ_C 104.8 (C-6a); 123.3 (C-6b); 119. 6 (C-7); 125.4 (C-8); 128.2 (C-9); 113.1 (C-10); 137.3 (C-10a); 145.2 (C-11a)] in **60**. The chemical shift of the resonances at $_C$ 125.4 and 128.2 were consistent with the placement of the chlorines at C-8 and C-9. The above mentioned structural fragments accounted for seven of the 11 degrees of unsaturation required for the molecular formula, therefore there must be four additional rings in the structure of compound **60**. Detailed 2D-NMR spectra analyses (COSY, HETCOR, HMBC, and NOESY) led to the establishment of the connectivity of functional groups and, in turn, of the molecular structure. The position of the functional groups along the pentacyclic moiety was corroborated by an HMBC experiment. Thus,

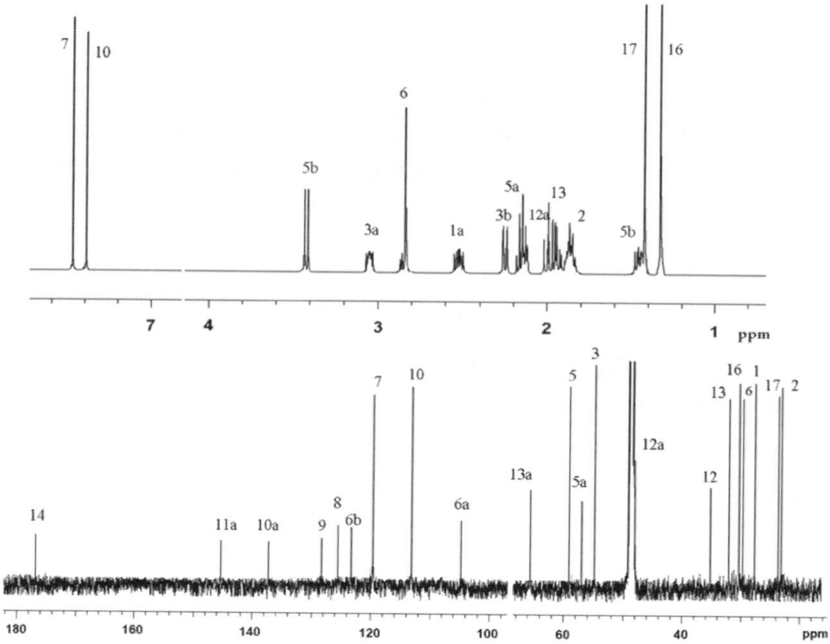

Fig. 18. ^{1}H (A, 500 MHz) and ^{13}C (B, 125 MHz) NMR spectra of malbrancheamide (**60**) in MeOD-d_3.

the correlations C-10a/H-7, C-6b/H-10, C-6a/H-6, H-7, C-5a/H-12a, C-12a/H-6, H-16, H-17 and C-11a/H-6, H-16, H-17 corroborated the position of the chlorine atoms and the fusion of the indole nucleus to the dimethyl cyclohexane ring throughout C-6a and C-11a. On the other hand, the cross peaks C-5a/H-6, H-12a, H-13, H-5; C-12a/ H-5 and C-13a/H-13, H-1, H-3, H-5 indicated that the bicyclo [2.2.2] diazaoctane ring system has the same arrangement than the one observed in alkaloids (-) VM55599 and stephacidin A.[68,69]

The structure of **60** was corroborated by X-ray analysis (Fig. 19). In the bicyclo [2.2.2] diazaoctane system, ring C (carbocyclic) adopts a boat conformation whereas ring F (lactame) an envelope conformation. Ring D was also observed as an envelope while ring E adopts a twisted envelope conformation. The atoms C13a-C1–C2–C3 are located in the ring plane while N4 is out-of the plane. Finally, the absolute configuration was determined applying the Flack's method.[70]

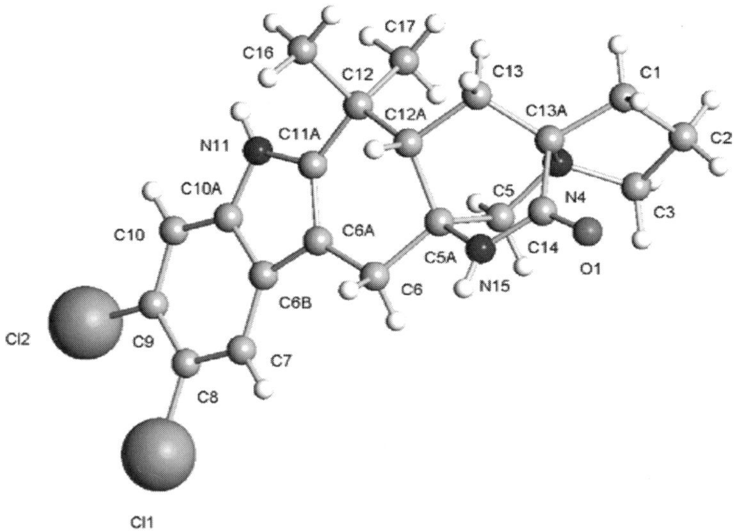

Fig. 19. ORTEP view of malbrancheamide (**60**).

In order to gain a better understanding about the conformational behavior, a high level density functional theory (DFT) investigation was performed. The initial structure was built from standard fragments and minimized using molecular mechanics analysis as implemented in the Spartan'04 program. This initial study revealed a minimum energy conformer (E = 137.77 kcal/mol) which was fully optimized by DFT (B3LYP/G31G*). In the DFT optimized structure (E = − 1973.98 kcal/mol), ring E displays an envelop conformation and the atoms N4-C13a-C1-C2 are located in the plane while the C3 is out-of the ring-plane (Fig. 20).

As previously proposed for the brevianamides, it is highly probable that the biogenesis of malbrancheamide (**60**) proceed from L-tryptophan, L-proline and one isoprene unit, being the bicyclo [2.2.2] ring system arising through an intramolecular Diels-Alder reaction.[68,69] The brevianamides, paraherquamides, aspergamides and macfortines type of compounds possess a spiro- -indoxyl system, while malbracheamide does not. On the other hand, the main difference between malbrancheamide and (-) VM55599 and stephacidin A seems to be the relative configuration at C12a in the

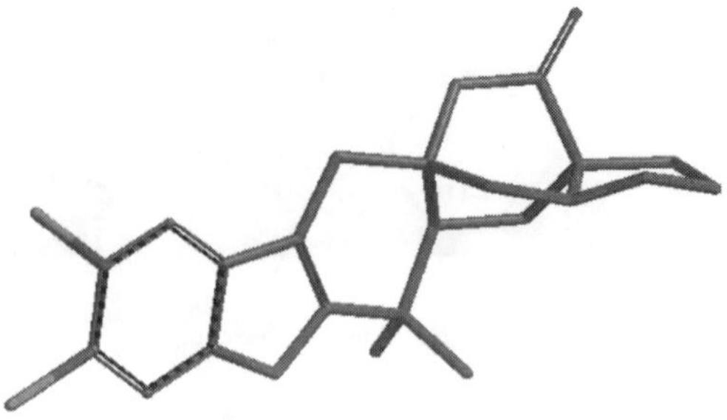

Fig. 20. Optimized structure of compound (**60**) obtained by DFT analysis. (Reprinted from *Tetrahedron,* **66**, Martínez-Luis S, Rodríguez R, Acevedo L, González MC, Lira-Rocha A and Mata R, Malbrancheamide, a new calmodulin inhibitor from the fungus *Malbranchea aurantiaca*, 1012–1016, 2006, with permission from Elsevier).

bicyclo [2.2.2] diazaoctane ring system. Altogether, these features make malbrancheamide[60] unique among the complex indole alkaloid derivatives.

Compound **60** showed phytotoxic effects when tested against seedlings of *A. hypochondriacus* using a Petri dish bioassay.[1] Compound **60** IC_{50} = 0.37 μM inhibited radicle growth of this species with a similar potency to 2,2–dichlorophenoxyacetic acid [2,4–D; IC_{50} = 0.18 μM]. Phytotoxin **60** inhibited the activation of PDE1 in a concentration-dependent manner. The IC_{50} value calculated was 3.65 μM. This effect was comparable to that of chlorpromazine (IC_{50} = 2.75 μM), a well characterized CaM antagonist. During the course of these measurements, **60** was found to have a mild but reproducible effect on the basal PDE1 activity in the presence of bovine-serum albumin (BSA). However, this last effect is unlikely to have physiological significance because its half-amplitude was above 200 μM, and was not pursued.[11]

In order to obtain further evidence of the involvement of CaM in the inhibition of CaM-PDE1, a kinetic analysis of the inhibition of the activity of PDE1 was assessed using different amounts of CaM in the presence of different concentrations of **60** and in the absence of BSA. The BSA was eliminated in order to reduce, though not completely eliminate, the effect of the compound **60** on PDE1 itself. The results were analyzed by means of Dixon plots.[71] In this analysis, the vertical axes are the reciprocal of the PDE1 activity in the presence of each Ca^{2+}-CaM and **60** concentrations, and the horizontal axes are the **60** concentrations.

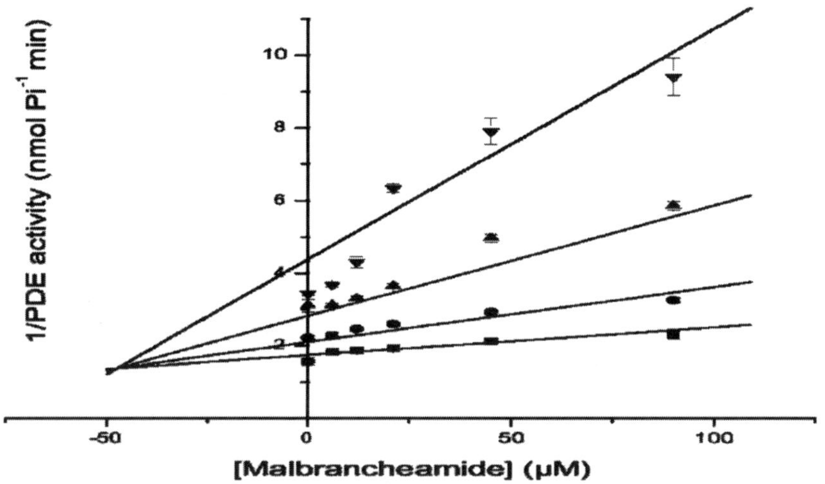

Fig. 21. CaM-dependent PDE activity with constant concentrations of *c*AMP and $CaCl_2$. CaM concentration was held at 12.5 (▼), 25 (▲), 50 (●) and 100 (■) nM, and the concentration of **1** was varied. Error bars represent the standard error. Each point was repeated 6 to 12 times. The lines are the result of fitting all individual readings globally to a competitive inhibition model with CaM as essential activator and (**60**) as the inhibitor. (Reprinted from *Tetrahedron*, (**66**), Martínez-Luis S, Rodríguez R, Acevedo L, González MC, Lira-Rocha A and Mata R, Malbrancheamide, a new calmodulin inhibitor from the fungus *Malbranchea aurantiaca*, 1012–1016, 2006, with permission from Elsevier).

Figure 21 shows the kinetic analysis of **60** induced inhibition of CaM-activated PDE1 by means of Dixon plots. These results suggested that compound **60** acts as competitive antagonist of CaM, thus competing with the formation of the CaM-PDE1 active complex. The estimated *Ki* (inhibition constant) value was 47.4 μM. The difference between the IC_{50} and the Ki values can be explained by the difference in experimental conditions. In fact, an analysis of the impact of **60** on the assay indicated that the Ki value might be as much as three times lower than the IC_{50}, but never higher.[11]

Malbrancheamide (**60**) represents a novel type of CaM inhibitors although other natural indole alkaloids such as the unusual indole-carbazole alkaloids from a culture broth of a *Nocardiopsis* sp, namely K-252a-K252d,[72,73] the brominated β-carbolines eudistomidins A and C, obtained from the Okinawan tunicate *Eudistoma glaucus*,[74,75] pyridindolol from

Streptoverticillium album K-251,[76] the antitumoral bis-indole alkaloids vinblastine and vincristine,[77] verruculogen, a very potent mycotoxin produced by several *Penicillium* species[78] and melatonin, a hormone found in all living creatures from algae to humans,[79] just to mention the most important ones.

M. aurantiaca is very sensitive to light and temperature change. In dark and colder environments, fungi produce sesquiterpenes and poliketides but in warmer surroundings and under natural daylight cycles, malbrancheamide (**60**) is produced. The ecological significance of these variations remains an open question.

13.4 CONCLUSIONS

In conclusion, this short survey shows indeed that Mexican biodiversity offers the prospect for obtaining new and known phytotoxic agents with calmodulin inhibitor properties. Thus, several phytochemicals from *Hofmeisteria schaffneri*, *Prionosciadium watsoni*, and the orchids *Maxillaria densa* and *Epidendrum rigidum* were discovered. In particular, the orchid bibenzyls gigantol and batasin exhibited good phytotoxicity both on *Amaranthus* and duckweed. A series of synthetic analogs showed similar activity to that of the parent compounds, and it was found that for greatest activity, the bibenzyls must have a methoxyl group at C-3 (C-5). These results suggest that these and structurally-related bibenzyls are good lead compounds for the development of novel class of herbicidal agents. Hofmeisterin (**29**), a relatively simple northymol derivative was obtained from an endemic Mexican Asteraceae, showed noted antiCaM activity, thus a reliable synthetic procedure for its synthesis was developed.

The fungi *Malbranchea aurantiaca*, *G. polythrix* and *Phoma herbarum* contain also potent phytotoxic agents with CaM-inhibitor properties. From the structural and biological point of view, the nonenolides **40–42** (herbarumins I-III) as well as malbrancheamide (**60**) were the most interesting. Malbrancheamide is the first chlorinated indole alkaloid possessing a bicyclo [2.2.2] ring. Kinetics studies revealed that this indole is a competitive antagonist of CaM. The studies on the nonenolides **40–42** represent an important contribution to the knowledge of the ten-membered rings conformation and allowed the development of suitable models for the establishment of the absolute configuration at the stereogenic centers of medium-sized lactones. During the investigation of the fungus *G. polythrix*, an assay to detect CaM inhibitors using NADK as monitoring enzyme was standardized. Since the relative order of CaM inhibition by the phytotoxic agents of this fungus was similar for the PDE1 and NADK assays, the latter could be also useful to discover CaM inhibitors. Considering that NADK is only regulated by CaM in plants, this bioassay can be useful to detect CaM isoforms that regulates the activity of NADK.

13.5 ACKNOWLEDGMENTS

This work has been conducted under the auspices of several project grants funded by Consejo Nacional de Ciencia y Tecnología and Dirección General de Asuntos del Personal Académico, UNAM. R. Mata is very grateful to all her colleagues and PhD students whose collaboration has contributed to the results described in this chapter, particularly to Fausto Rivero-Cruz, Martha Macias-Rubalcava and Norma Valencia. Their faith and enthusiasm have been the basis of this work.

REFERENCES

1. Duke SO, Dayan FE, Rimando AM, Schrader KK, Aliotta G, Oliva A, Romagni JG, Chemicals from nature for weed management, *Weed Sci* **50**:138–151, 2002.
2. Singh HP, Batish DR, Kohli RK, Allelopathic interactions and allelochemicals: New possibilities for sustainable weed management, *Crit Rev Plant Sci* **22**:239–311, 2003.
3. Bouché N, Yellin A, Snedden WA, Fromm H, Plant-specific calmodulin-binding proteins, *Ann Rev Plant Biol* **56**:435–466, 2005.
4. Chin D, Means AR, Calmodulin: A prototypical calcium sensor, *Trends Cell Biol* **10**:322–328, 2000.
5. Au TK, Chick WSH, Leung PC, The biology of ophiobolins, *Life Sci* **67**:733–742, 2000.
6. Somlyo AP, Somlyo AV, Ca^{2+} sensitivity of smooth muscle and nonmuscle myosin II: Modulated by G proteins, kinases, and myosin phosphatase, *Physiol Rev* **83**:1325–1358, 2003.

7. Horikawa N, Suzuki T, Uchiumi T, Minamimura T, Tsukada K, Takeguchi N, Sakai H, Cyclic AMP-dependent Cl^- secretion induced by thromboxane A_2 in isolated human colon, *J Physiol* **562**:885–897, 2005.
8. Seales EC, Micoli KJ, McDonald JM, Calmodulin is a critical regulator of osteoclastic differentiation, function, and survival, *J Cell Biochem* **97**:45–55, 2006.
9. Du J, Szabo ST, Gray NA, Manji HK, Focus on CaMKII: A molecular switch in the pathophysiology and treatment of mood and anxiety disorders, *Int J Neuropsychopharmacol* 7:243–248, 2004.
10. Anaya AL, Calera MR, Mata R, Pereda-Miranda R, Allelopathic potential of compounds isolated from *Ipomoea tricolor* Cav. (Convolvulaceae), *J Chem Ecol* 7:2145–2152, 1990.
11. Martínez-Luis S, Rodríguez R, Acevedo L, González MC, Lira-Rocha A, Mata R, Malbrancheamide, a new calmodulin inhibitor from the fungus *Malbranchea aurantiaca Tetrahedron* **66**:1012–1016, 2006.
12. Perez-Vasquez A, Reyes A, Linares E, Bye R, Mata R, Phytotoxic agents from *Hofmeisteria schaffneri*: Isolation and synthesis of 2′-(2″-hydroxy-4″-methylphenyl)-2′-oxoethyl acetate, *J Nat Prod* **68**:959–962, 2005.
13. Greengard P, Robinson GA, (eds.), *Advances in Cyclic Nucleotide Research*, Vol. 10. Raven Press, New York, 1979.
14. Mata R, Gamboa A, Macias M, Santillan S, Ulloa M, Gonzalez MC, Effect of selected phytotoxic agents from *Guanomyces polythrix* on the calmodulin-dependent activity of the enzymes *c*AMP phosphodiesterase and NAD-kinase, *J Agric Food Chem* **51**:4559–4562, 2003.
15. Martínez M, Las Plantas Medicinales de México, Ediciones Botas, México, DF, 1989.
16. Tellez M, Estell R, Fredrickson E, Havstad D, Essential oil of *Fluorensia cernua* DC, *J Ess Oil Res* **9**:619–624, 1977.
17. Tellez M, Estell R, Fredrickson E, Powell J, Wedge D, Schrader K, Kobaisy M, Extracts of *Fluorensia cernua* (L): Volatile constituents and antifungal, antialgal, and antitermite bioactives, *J Chem Ecol* **27**:2263–2285, 2001.
18. Rao MM, Kingston DGI, Spittler TD, Flavonoids from *Fluorensia cernua*, *Phytochemistry* **9**:227–228, 1970.
19. Dillon MO, Mabry TJ, Besson E, Bouillant ML, Chopin J, New flavonoids from *Fluorensia cernua*, *Phytochemistry* **15**:1085–1086, 1976.
20. Wollenweber E, Dietz VH, Occurrence and distribution of free flavonoid aglycones in plants, *Phytochemistry* **20**:869–932, 1981.
21. Estell RE, Anderson DM, Havstad KM, Effects of organic solvents on use of tarbush by sheep, *J Chem Ecol* **20**:1137–1142, 1994.

22. Kingston DGI, Rao MM, Spittler TD, Isolation and structure determination of fluorensic acid, a new sesquiterpene of the eremophilane type, *Tetrahedron Lett* **20**:1613–1616, 1971.
23. Kingston DGI, Rao MM, Spittler TD, Pettersen RC, Cullen DL, Sesquiterpenes from *Fluorensia cernua*, *Phytochemistry* **14**:2033–2037, 1975.
24. Pettersen RC, Cullen DL, Spittler TD, Kingston DGI, The crystal and molecular structure of fluorensadiol, a natural product sesquiterpene isolated from a west Texas shrub, *Acta Crystallogr Sect B: Struct Sci* **31**:1124–1127, 1975.
25. Bohlmann F, Grenz M, Über inhaltsstoffe der gattung *Fluorensia*, *Chem Berich* **110**:295–300, 1977.
26. Dayan FE, Tellez MR, Phytotoxicity of tarbush (*Fluorensia cernua* D. C.), *Allelopathy J* **6**:1–12, 1999.
27. Mata R, Bye R, Linares E, Macias M, Rivero-Cruz I, Perez O, Timmermann B, Phytotoxic compounds from *Flourensia cernua*, *Phytochemistry* **64**:285–291, 2003.
28. Valencia-Islas N, Abbas H, Bye R, Toscano RA, Mata R, Phytotoxic compounds from *Prionosciadium watsoni*, *J Nat Prod* **65**:828–834, 2002.
29. Estrada S, Toscano RA, Mata R, New Phenanthrene Derivatives From *Maxillaria densa*, *J Nat Prod* **62**:1175–1178, 1999.
30. Estrada S, López-Guerrero JJ, Villalobos-Molina R, Mata R, Spasmolytic stilbenoids from *Maxillaria densa*, *Fitoterapia* **75**:690–695, 2004.
31. Valencia-Islas NA, Paul RN, Shier WT, Mata R, Abbas HK, Phytotoxicity and ultrastructural effects of gymnopusin from the orchid *Maxillaria densa* on duckweed (*Lemna pausicostata*) frond and root tissues, *Phytochemistry* **61**:141–148, 2002.
32. Hernandez-Romero Y, Acevedo L, Sánchez MD, Shier WT, Abbas HK, Mata R, Phytotoxic activity of bibenzyl derivatives from the orchid *Epidendrum rigidum*, *J Agric Food Chem* **53**:6276–6280, 2005.
33. Hernández-Romero Y, Rojas J-I, Castillo R, Rojas A, Mata R, Spasmolytic effects, mode of action, and structure-activity relationships of stilbenoids from *Nidema boothii*, *J Nat Prod* **67**:160–167, 2004.
34. Gorham J, *The Biochemistry of the Stilbenoids*, Chapman & Hall, London, 1995.
35. Brebaum SN, Borland GJ, First Report of *Phoma herbarum* and *Phoma exigua* as Pathogens of Dandelion in Southern Ontario, *Plant Dis* **83**:200, 1999.
36. Wyllie TD, Morehouse LG, (eds.), *Mycotoxic Fungi, Mycotoxins, Mycotoxicoses*, Vol. 2, Marcel Dekker, Inc., New York, 1978.
37. Zohri AA, Saber SM, Cytochalasins A and B of *Dematiaceous hyphomycetes*, *Lett Appl Microbiol* **19**:37–39, 1994.

38. Begum S, Deshpande KB, Studies on the pectolytic enzyme of *Phoma herbarum* West. 1. Production of macerating enzyme, *Marathwada Univ J Sci* **18**:29–33, 1979.
39. MacGahren WJ, Ellestad GA, Morton GO, Kunstmann MP, LL-D253α, β and -γ, novel chromanones from the fungus *Phoma pigmentivora*, *J Org Chem* **37**:1636–1639, 1972.
40. Chandler IM, McIntyre CR, Simpson TJ, Structural revision and synthesis of LL-D253 and related chromanone fungal metabolites, *J Chem Soc Perkin Tans I* 2271–2284, 1992.
41. Rivero-Cruz JF, García-Aguirre G, Cerda-Garcia-Rojas CM, Mata R, Conformation behavior and absolute stereostructure of two phytotoxic nonenolides from the fungus *Phoma herbarum*, *Tetrahedron* **56**:5337–5344, 2000.
42. Rivero-Cruz JF, Macias M, Cerda-Garcia-Rojas CM, Mata R, A new phytotoxic nonenolide from *Phoma herbarum*, *J Nat Prod* **66**:511–514, 2003.
43. Harada N, Nakanishi K, Exciton chirality method and its application to configurational and conformational studies of natural products, *Accounts Chem Res* **5**:257–263, 1972.
44. Evidente A, Lanzetta R, Capasso R, Andolfi A, Vurro M, Zonno MC, Putaminoxin, a phytotoxic nonenolide from *Phoma putaminum*, *Phytochemistry* **44**:1637–1641, 1995.
45. Fuerstner A, Radkowski K, Wirtz C, Goddard R, Lehmann CW, Mynott R, Total Syntheses of the Phytotoxic Lactones Herbarumin I and II and a Synthesis-Based Solution of the Pinolidoxin Puzzle, *J Am Chem Soc* **124**:7061–7069, 2002.
46. Furstner A, Radkowski K, Enantioselective total synthesis of the phytotoxic lactone herbarumin I. *Chem Commun* 7:671–672, 2001.
47. Sabino A, Aparecido P, Ronaldo A, Total synthesis of (+)-Herbarumin I via intermolecular Nozaki-Hiyama-Kishi reaction, *Tetrahedron Lett* **43**:2819–2821, 2002.
48. Harmata M, (ed.), *Strategies and Tactics in Organic Synthesis*, Elsevier Ltd., Great Britain, 2004.
49. Diez E, Dixon DJ, Ley SV, Polara A, Rodriguez F, The use of butane diacetals of glycolic acid as precursors for the synthesis of the phytotoxic calmodulin inhibitor herbarumin II, *Helv Chim Acta* **86**:3717–3729, 2003.
50. Diez E, Dixon DJ, Ley SV, Polara A, Rodriguez F, Total synthesis of the phytotoxic agent herbarumin II using butane diacetals of glycolic acid as building blocks, *Synlett* **8**:1186–1188, 2003.
51. Gurjar MK, Karmakar S, Mohapatra DK, First total synthesis of herbarumin III, *Tetrahedron Lett* **45**:4525–4526, 2004.

52. Gurjar MK, Nagaprasad R, Ramana CV, Karmakar S, Mohapatra DK, Ring-closing metathesis mediated total synthesis of microcarpalide and herbarumin III, *ARKIVOC* (Gainesville, FL, U. S.) **3**:237–257, 2005.
53. Nanda S. Chemoenzymatic total synthesis of the phytotoxic lactone herbarumin III, *Tetrahedron Lett* **46**:3661–3663, 2005.
54. Boruwa J, Gogoi N, Barua NC, A short enantioselective total synthesis of the phytotoxic lactone herbarumin III, *Org Biomol Chem* **4**:3521–3525, 2006.
55. Salaskar A, Sharma A, Chattopadhyay S, An asymmetric synthesis of 7-hydroxy-9-propylnonenolide (herbarumin III), *Tetrahedron: Asymmetry* **17**:325–329, 2006.
56. Evidente A, Capasso R, Andolfi A, Vurro M, Zonno MC, Structure-activity relationship studies of putaminoxins and pinolidoxins: Phytotoxic nonenolides produced by phytopathogenic Phoma and Ascochyta species, *Natural Toxins* **6**:183–188, 1998.
57. Schnick PJ, Boland GJ, 2,4-D and *Phoma herbarum* to control dandelion (*Taraxacum officinale*), *Weed Sci* **52**:808–814, 2004.
58. Selbmann L, Onofri S, Fenice M, Federici F, Petruccioli M, Production and structural characterization of the exopolysaccharide of the Antarctic fungus *Phoma herbarum* CCFEE 5080, *Res Microbiol* **153**:585–592, 2002.
59. Yang XB, Gao XD, Han F, Xu BS, Song YC, Tan RX, Purification, characterization and enzymatic degradation of YCP, a polysaccharide from marine filamentous fungus *Phoma herbarum* YS4108, *Biochimie* **87**:747–754, 2005.
60. Vikrant P, Verma KK, Rajak RC, Pandey AK, Characterization of a phytotoxin from *Phoma herbarum* for management of *Parthenium hysterophorus* L, *J Phytopath* **154**:461–468, 2006.
61. González MC, Hanlin RT, Ulloa M, *Guanomyces*, a New Genus of Ascomycetes from Mexico, *Mycologia* **92**:1138–1148, 2000.
62. Macías M, Ulloa M, Gamboa A, Mata R, Phytotoxic compounds from the new coprophilous fungus *Guanomyces polythrix*, *J Nat Prod* **63**:757–761, 2000.
63. Macías M, Ulloa M, Gamboa A, Toscano RA, Mata R, Phytotoxic naphtopyranone derivates from the coprophilous fungus *Guanomyces polythrix*, *Phytochemistry* **58**:751–758, 2001.
64. Martínez-Luis S, González MC, Ulloa M, Mata R, Phytotoxic agents from the fungus *Malbranchea aurantiaca*, *Phytochemistry* **66**:1012–1016, 2005.
65. Kagan HB (ed.), *Stereochemistry, Fundamentals and Methods*, Vol. 2, Georg Thieme Publishers Verlag, Stuttgart, 1977.
66. Jenniskens LHD, Groot A, Enantioselective synthesis of R-(-)-Ligularenolide and the progesterone receptor ligand R-(-)- PF1092C starting from S-(+)-carvone, *Tetrahedron* **54**:5617–5622, 1998.

67. Keromnes J, Thouvenor D, Role of penicillic acid in the phytotoxicity of *Penicillium cyclopium* and *Penicillium canescens* to germination of corn seeds, *Appl Environ Microbiol* **49**:660–663, 1985.
68. Cox RJ, Williams RM, The paraherquamides, brevianamides and asperparaline: Laboratory synthesis and biosynthesis. An interim report, *Acc Chem Res* **36**:127–139, 2003.
69. Stocking C, Williams RW, Chemistry and biology of biosynthetic Diels-Alder reactions, *Angew Chem Int Ed Engl* **42**:3078–3115, 2003.
70. Flack HD, On enantiomorph-polarity estimation, *Acta Crystallogr Sect A: Found Crystallogr* **39**:876–881, 1983.
71. Dixon M, The determination of enzyme inhibitor constant, *Biochem J* **55**:170–171, 1953.
72. Kase H, Iwahashi K, Matsuda Y, K-252a, a potent inhibitor of protein kinase C from microbial origin, *J Antibiot* **39**:1059–1065, 1986.
73. Nakanishi S, Matsuda Y, Iwahashi K, Kase H, K-252b, c and d, potent inhibitors of protein kinase C from microbial origin, *J Antibiot* **39**:1066–1071, 1986.
74. Kobayashi J, Nakamura H, Ohizumi Y, Hirata Y, Eudistomidin-A, a novel calmodulin antagonist from the Okinawan tunicate *Eudistoma glaucus*, *Tetrahedron Lett* **27**:1191–1194, 1986.
75. Kobayashi J, Cheng JF, Ohta T, Nozoe S, Ohizumi Y, Sasaki T, Eudistomidins B, C, and D: Novel antileukemic alkaloids from the Okinawan marine tunicate *Eudistoma glaucus*, *J Org Chem* **55**:3666–3670, 1990.
76. Matsuda Y, Asano K, Kawamoto I, Yasuzawa T, Shirahata K, Sano H, Kase H, K-251 compounds, inhibitors of Ca^{2+} and calmodulin-. dependent cyclic nucleotide phosphodiesterase from *Streptoverticillium album*, *Agric Biol Chem* **52**:3211–3213, 1988.
77. Molnar A, Liliom K, Orosz F, Vertessy BG, Ovadi J, Anti-calmodulin potency of indole alkaloids in *in-vitro* systems, *Eur J Pharmacol* **291**:73–82, 1995.
78. Pala I, Srinivasan A, Vig PVS, Desaiah D, Modulation of calmodulin and protein kinase C activities by *Penicillium* mycotoxins, *Int J Toxicol* **18**:91–96, 1999.
79. Del Rio B, Pedrero JMG, Martinez-Campa C, Zuazua P, Lazo PS, Ramos S, Melatonin, an endogenous-specific inhibitor of estrogen receptor α via calmodulin, *J Biol Chem* **279**:38294–38302, 2004.

Chapter 14

POTENTIAL ANTICANCER NATURAL PRODUCTS FROM PLANT-ASSOCIATED FUNGI

Marilyn T. Marron and A. A. Leslie Gunatilaka

14.1 INTRODUCTION

Cancer is a clinically large group of diseases that varies in the age of onset, invasiveness, the response to treatment, and prognosis. Environmental factors and genetic predisposition play important roles in initiating and promoting carcinogenesis, but cancer knows no geographic boundaries. 6.5 million people worldwide are diagnosed with cancer each year.[1] In the western world, cancer is a leading cause of death with over one million new cases diagnosed each year in the USA alone. For unknown reasons, perhaps due to more advanced detection methods, changes in environment and lifestyle, and increased longevity, there has been a steady increase in certain cancer deaths in the USA over the past 50 years.[2] Fortunately, in recent years, more sophisticated detection processes along with disease management involving both the prevention and treatment of the disease have helped to dramatically increase the number of cancer survivors. Further advances in cancer treatment require the development of novel and improved chemotherapies which may be derived from synthetic or natural sources. The natural products approach to anticancer drug discovery has several advantages, including the potential of finding totally new and unique structural classes and activity types. Among all known producers of natural products, microorganisms represent a rich source of bioactive metabolites and unlike other organisms, microorganisms occupy all living and nonliving niches on earth. Historically, of all microorganisms studied, actinomycetes and fungi have been found to be the most prolific producers of small molecule natural products. It has recently been estimated that less than five percent of fungal species are currently known, suggesting that millions of fungal species remain to be discovered.[3] Microorganisms are known to collaborate with almost all plants to form mutually beneficial associations and produce biologically active metabolites. These include endophytic and rhizosphere microorganisms. Presented in this chapter are some selected small molecule metabolites produced by plant-associated fungi with the potential for anticancer activity. For completeness, endophytic fungal sources of clinically proven plant-based anticancer agents are also included.

14.2 NATURAL PRODUCT-BASED ANTICANCER DRUGS

Over 60% of the anticancer and antiinfective agents that are commercially available or in late stages of clinical development today are of natural origin. It has been estimated that out of the FDA (U.S. Federal Drug Administration) approved anticancer drugs and preNDA (New Drug Application) candidates belonging to natural product and natural product-derived semisynthetic categories, 68 are of microbial origin, 36 are of plant origin and 9 are of marine origin.[4] Some clinically used anticancer agents and/or those currently under clinical evaluation include the microbial-derived natural products adriamycin, daunorubicin, mitomycin C, bleomycin, and the geldanamycin analogue 17-AAG; the plant-derived paclitaxel (taxol®) (**1**) and its analogue docetaxel (taxotere®) (**2**); the podophyllotoxin (**3**) analogues, etoposide (**4**), teniposide (**5**) and etoposide phosphate (**6**), the camptothecin (**7**) analogues, topotecan (**8**) and irinotecan (**9**); and the marine-derived natural products, dolastatin 10, bryostatin, halichondrin B, and discodermolide.

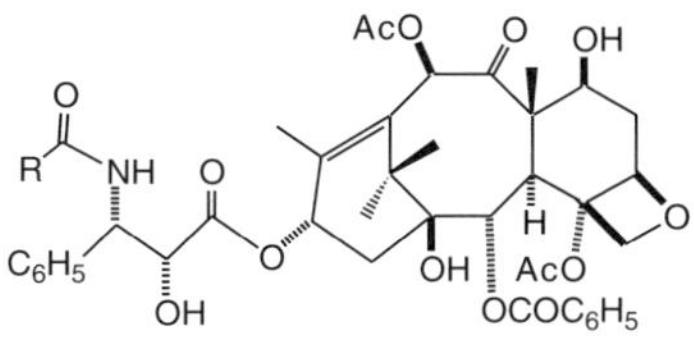

(1) R = Ph
(2) R = t-Butyl

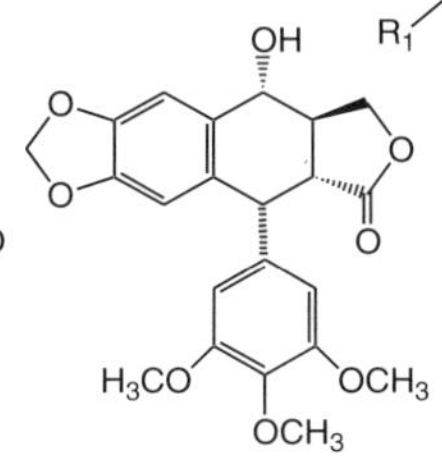

(3)

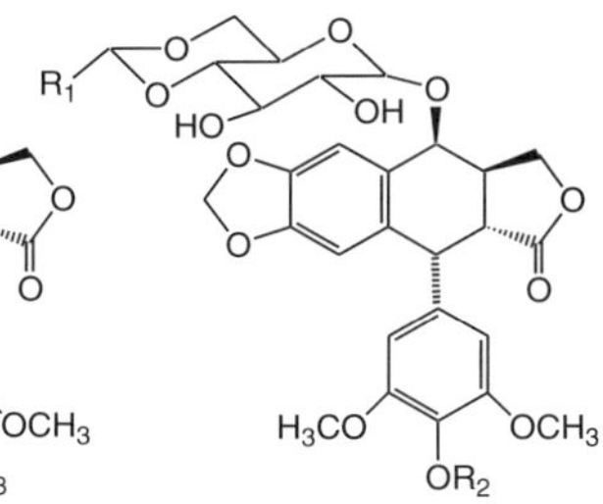

(4) $R_1 = CH_3$; $R_2 = H$

(5) R_1 = 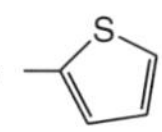 ; $R_2 = H$

(6) $R_1 = CH_3$; $R_2 = OPO_3H_2$

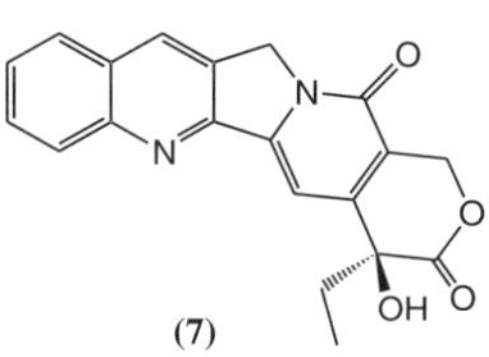

(7)

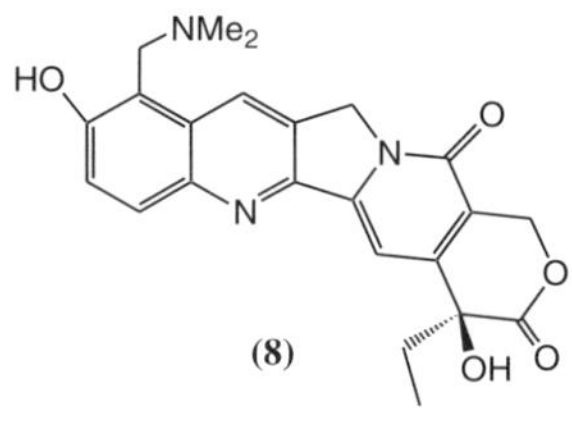

(8)

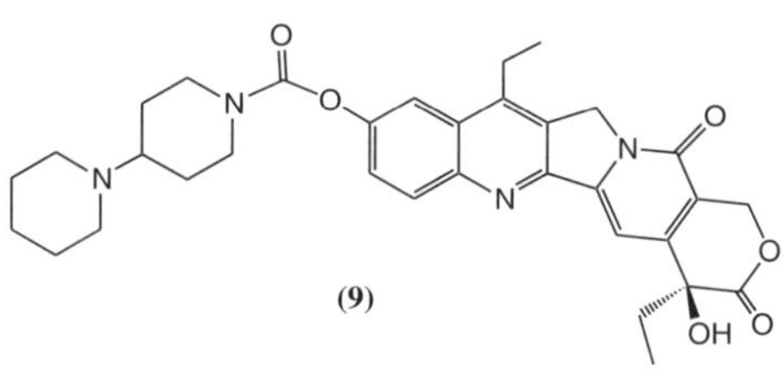

(9)

14.3 PLANT-ASSOCIATED FUNGI AS A SOURCE OF BIOACTIVE COMPOUNDS

Plant-associated microorganisms represent a largely untapped resource of small molecule natural products, some with chemical structures that have been optimized by coevolution for biological and ecological relevance. According to Bode *et al.*,[5] "with more than 20 000 compounds described in the literature, microorganisms must be called metabolic artists superior to any metabolic diversity created by man." Plant-associated microorganisms include endophytes and those which inhabit the rhizosphere of plants. The term endophyte refers to a bacterial or a fungal microorganism that colonizes the interior organs of plants, but does not have pathogenic effects on its host(s). These microorganisms live in a symbiotic relationship with higher plants and produce bioactive secondary metabolites that enhance their host's growth and competitiveness and/or offer chemical protection.[6] Thus, they are known to produce a variety of metabolites with novel structures and interesting biological activities.[7–10] The claimed medicinal properties and biological activities of some plant species have recently been attributed to the microorganisms living in association with these plants.[11,12] Recent studies have illustrated that some important plant-derived anticancer agents such as taxol® (**1**), podophyllotoxin (**3**), and camptothecin (**7**) are also produced by fungal strains endophytic in a number of plant species including those from which these compounds have been reported.[13–15] Plant-associated fungal strains are also known to elaborate many other anticancer agents including cytochalasins[16] and radicicol.[17] It has recently been suggested that even weak anticancer agents derived from these microorganisms may serve as scaffolds for organic synthesis of promising drug-like molecules, as exemplified by the rare cyclopentenone-type metabolites produced by the endophytic fungal strain, *Mitosporic dothideomycete* sp. LRUB20.[18]

Presented in this chapter are some selected metabolites isolated from plant-associated fungal strains that are capable of killing cancer cells due to their cytotoxicity, apoptosis, or disruption of cellular microfilaments, and therefore have potential anticancer activity. It should be noted that most of the reports dealing with anticancer drug discovery from plant-associated fungi have utilized antiproliferation assays in the tumor cell lines recommended by the U.S. National Cancer Institute (NCI). This disease-based approach allows for high-throughput screening for potential anticancer activity since disruption in cancer cell proliferation is considered useful for *in vivo* anticancer activity. More informative mechanism-based approaches such as screening for apoptotic agents offer a route for targeting and identifying specific drug interactions leading to the understanding of their possible mechanism(s) of action. Similarly, microfilament disruption targeted against cancer cells may curb their proliferation and/or reveal novel pathways that can be targeted to limit cancer growth. Several plant-associated fungal strains have been reported to produce secondary metabolites active in these assays and therefore have potential anticancer activity.

14.4 ENDOPHYTES AS SOURCES OF CLINICALLY USED ANTICANCER DRUGS AND THEIR PRECURSORS

Although the natural products approach represents an unparalleled source of molecular diversity that is complementary to other molecular sources such as combinatorial libraries, it faces many challenges. The most significant challenge to the field involves improving its competitiveness with synthetic and combinatorial libraries.[19] Another obstacle is the availability of sufficient quantities of compound(s) required to complete toxicological and clinical studies, as was the case with the well known anticancer drug, taxol® (**1**).[20] Thus, attempts had to be made to develop alternative methods for the production of this compound, such as semisynthetic processes, tissue culture, biotransformation, and the use of endophytic fungal strains.[21] The ability of endophytic fungi to make phytohormones has been suggested to be due to their coevolution with plants and/or as a result of the transfer of biosynthetic genes from their plant hosts.[22] The belief that endophytes, by virtue of their association with plants, may biochemically mimic the host organism, led Strobel and his coworkers to search for paclitaxel (taxol®) producing endophytes in the plant, *Taxus brevifolia* (Western yew; Pacific yew), from which this compound was first isolated. Their pioneering work resulted in the identification of a new taxol-producing endophytic fungus, *Taxomyces andreanae*, inhabiting *Taxus brevifolia*.[20]

14.4.1 Paclitaxel (Taxol®)

Over the past fifteen years, paclitaxel (**1**), a highly functionalized diterpenoid and its structural analogue, docetaxel (taxotere®) (**2**), have emerged as clinically important antitumor agents useful for treating a variety of different cancers including breast, lung, bladder, and cervix cancer. It is likely that the demand for these paclitaxel-based drugs will increase as they are approved for use against additional forms of cancer. Much of the world's supply of paclitaxel has been derived by semisynthesis starting from baccatin III (**10**) obtained by the hydrolysis of a crude mixture of taxanes extracted from the slow-growing yew (*Taxus*) needles.

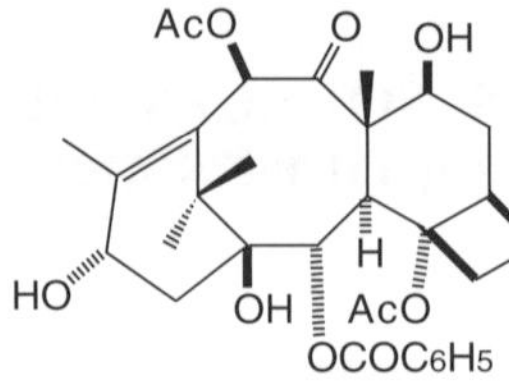

(**10**)

The realization of the potential demand for paclitaxel-based drugs has made microbiologists and natural product chemists to look for alternative sources of taxol, especially microbial sources from which paclitaxel can be obtained by fermentation.[23] Since the first paclitaxel-producing endophyte was discovered by Strobel and coworkers in 1993,[24] several other endophytic fungi inhabiting *Taxus* and other plant species have been described.[23] An interesting example of a paclitaxel producing endophytic fungus is the strain BT2 from *Taxus chinensis* var. *mairei* that is able to produce both paclitaxel and baccatin III (**10**), an important intermediate in the semisynthesis of both paclitaxel (**1**) and docetaxel (**2**). BT2 cultures have been reported to yield 4 μg–7 μg of paclitaxel and 12 μg–18 μg of baccatin III per litre of culture.[25] In general, the yield of paclitaxel from endophytic fungal sources varies between 24 ng/L–50 ng/L of the culture medium.[24] The strain CP-4 of *Pestalotiopsis microspora* produces 50 ng/L–1487 ng/L of paclitaxel, indicating its genetic instability.[26] Despite low and variable yields of paclitaxel from these sources, its unique antitumor activity has provided the impetus for further clinical development. While most anticancer drugs act at the G_1/S phase of the cell cycle, paclitaxel blocks mitosis by inhibiting cell proliferation during the G_2/M phase. Cells treated with paclitaxel are unable to undergo mitosis because their microtubules are stabilized and consequently, do not depolymerize into tubulin.[27]

14.4.2 Podophyllotoxin

The plant, *Podophyllum hexandrum* (Podophyllaceae; Berberidaceae), produces podophyllotoxin (**3**) and its glycoside, both of which are effective anticancer agents. Commercial production of podophyllotoxin from

P. hexandrum was hindered by limited availability of this endangered plant species, low yields of podophyllotoxin and its glycoside,[28] and less successful efforts of metabolic engineering to produce these compounds by plant cell culture.[29] However, very recently *Trametes hirsuta*, an endophyte of *P. hexandrum*, was reported to produce podophyllotoxin.[30] The fungal genus *Trametes* is especially interesting since it is a known producer of several anticancer polysaccharies such as krestin, isolated from the mycelia of *T. versicolor*. Krestin is now approved for use against stomach, esophageal, rectal, lung, and breast cancers.[31] Two strains of the endophytic fungus, *Phialocephala fortinii*, obtained from rhizomes of the plant, *Podophyllum peltatum* (may apple), have also yielded podophyllotoxin. Despite slow growth, these fungal strains were able to produce podophyllotoxin (**3**) at low but measurable yields (0.5 μg/L–189.0 μg/L) in broth cultures.[14]

Podophyllotoxin (**3**) is a precursor to the three clinically used anticancer drugs, etoposide (**4**), teniposide (**5**), and etoposide phosphate (**6**). Its mechanism of action involves inhibition of tubulin polymerization and disruption of mitosis during metaphase of the cell cycle. Both etoposide and teniposide are modifications of podophyllotoxin specifically designed to increase the water solubility, reduce gastric toxicity in addition to inhibiting topoisomerase II, and disrupting the cell cycle.

14.4.3 Camptothecin

The antitumor alkaloid, camptothecin (**7**), was first isolated from the wood of the Chinese plant, *Camptotheca acuminata* Decne (Nyssaceae),[32] and was subsequently found to occur in several other plant species including *Nothapodytes foetida* (Icacinaceae).[33] Unfortunately, camptothecin occurs predominantly in the roots of 50–75 year-old trees of *C. acuminata*. Inconsistent production problems combined with low yields have hindered its further development. Recently, a partially identified endophytic fungus (RJMEF001) belonging to the family Phycomycetes, isolated from the inner bark of *Nothapodytes foetida* growing in India, has been reported to produce camptothecin (**7**).[15] In another study, 52 endophytic fungal strains isolated from young and old twigs of *N. foetida* were grown for 96 hours in shake flasks and one of the isolates, *Entrophospora infrequens*, was found to produce 0.575 mg ± 0.031 mg of camptothecin per 100 g of dry fungal mass. The subsequent growth of this fungal strain in a bioreactor produced an improved yield of camptothecin (7) (4.96 mg ± 0.73 mg per 100 g of dry fungal mass) within 48 hours.[34] As expected, cytotoxicity assays against human cancer cell lines, A-549 (lung carcinoma), HEP-2 (liver carcinoma), OVCAR-5 (ovarian carcinoma), have shown that the fungal-derived camptothecin is significantly toxic to these cell lines.[15] Despite its production difficulties, camptothecin (7) remains an interesting anticancer drug with a unique mechanism of action. It traps the cleavable DNA-topoisomerase I complex and inhibits RNA synthesis by preventing the relaxation of DNA during replication and transcription. Although toxic, camptothecin has demonstrated a strong clinical efficacy against lung, ovarian, and uterine cancers.[32]

14.4.4 Radicicol

Radicicol (**11**), the chlorine containing β-resorcylic acid macrocyclic lactone, was first isolated from the soil borne fungus, *Cylindrocarpon radicicola*,[35] and subsequently from *Penicillium luteo-aurantium*[36] and *Humicola* sp. FO-2942,[37] is known to display *in vitro* anticancer activity due to the inhibition of heat shock protein-90 (Hsp90).[38] To ensure their survival in an acidic, nutrient starved, hypoxic environment, cancer cells are dependent on the protective effect of the heat shock response. Heat shock proteins, especially Hsp90, regulate and stabilize several oncogene products within disordered signal transduction pathways that are responsible for the malignant properties of the cell. The inhibition of Hsp90 activity ultimately leads to an antitumor response, caused by degradation of client proteins involved in promoting tumorigenicity. While radicicol displays *in vitro* anticancer activity, it was found to be devoid of any *in vivo* activity in animal models though some oxime derivatives of it have been shown to possess *in vivo* efficacy.[39] A recent study has demonstrated that the synthetic analog, cyclopropa-radicicol (**13**), in which the oxirane in **11** is replaced with a cyclopropane ring, exhibits strong Hsp90 inhibitory activity. It also demonstrated that the difluorocyclopropyl analog **14** of monocillin I (**12**) was capable of degrading the oncogenic protein HER2 at a concentration of

1.0 μM, whereas the related radicicol analog **15** was found to be less active and degraded HER2 at 10.0 μM.[40] However, in another recent investigation, pochonin D (**16**) was found to be considerably more active than its non-chlorinated analog **17** for affinity towards Hsp90 in a competition assay using geldanamycin (GDA) (**18**).[41] Radicicol (**11**) has recently been encountered in *Chaetomium chiversii* (Chaetomiaceae), an endophytic fungal strain isolated from the stem tissue of Mormon tea (*Ephedra fasciculata* A. Nels) (Ephedraceae). Hsp90 inhibitors such as radicicol (**11**) and GDA (**18**) are attractive agents for anticancer chemotherapy in that they are able to destabilize many substrates involved in carcinogenesis and signal transduction, including the estrogen receptor (ER) and the type 1 insulin-like growth factor receptor (IGF-1R).[17] A partial structure-activity relationship study involving several analogues of radicicol (**11**) and monocillin I (**12**) has suggested some important structural requirements for Hsp90 inhibitory activity of these macrocyclic β-resorcylic acid lactones[17] (Fig. 1).

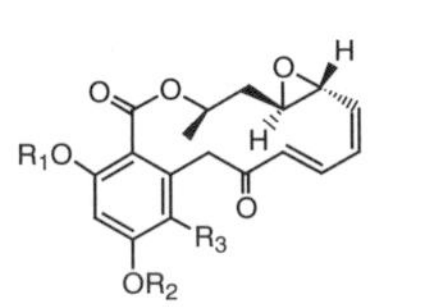

(11) $R_1 = R_2 = H, R_3 = Cl$
(12) $R_1 = R_2 = R_3 = H$

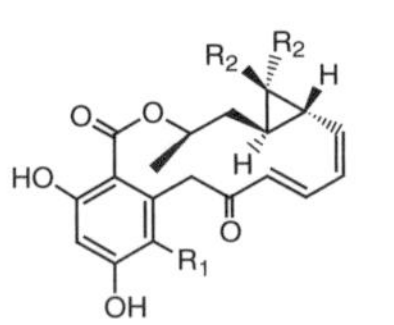

(13) $R_1 = Cl, R_2 = H$
(14) $R_1 = H, R_2 = F$
(15) $R_1 = Cl, R_2 = F$

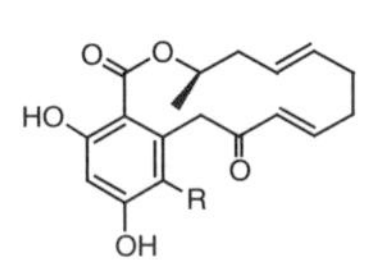

(16) R = Cl
(17) R = H

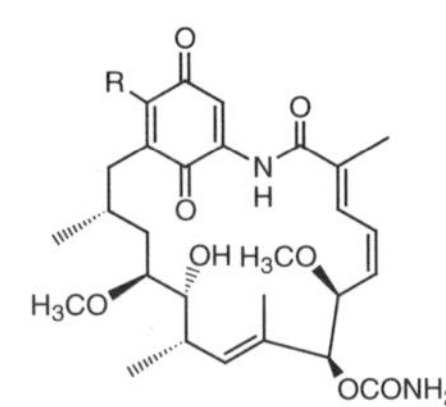

(18) $R = OCH_3$
(19) R = NH

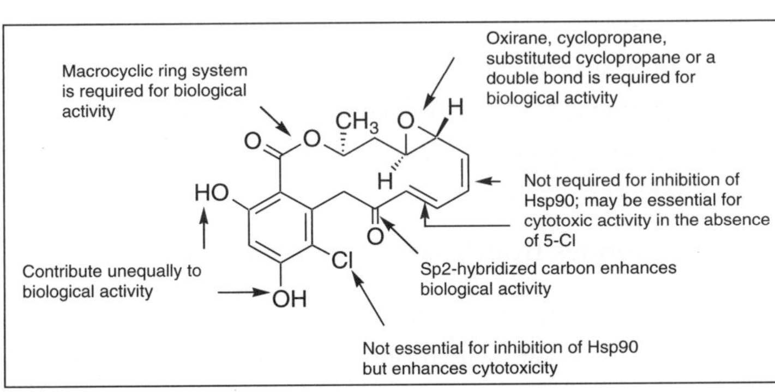

Fig. 1. Partial structure-activity relationships for radicicol (**11**) and monocillin (**12**).

Radicicol (**11**) isolated from *C. chiversii* was able to deplete cellular levels of the known Hsp90 client proteins ER and IGF-1R in MCF-7 human breast cancer cells.[17] An IC_{50} value of 0.03 μM was obtained for the proliferation of MCF-7 cells treated with radicicol for three days. Currently, clinical trials are in progress with a less toxic GDA derivative, allylamino-17-demethoxygeldanamycin (17-AAG) (**19**).[42]

14.5 CYTOTOXIC AGENTS FROM PLANT-ASSOCIATED FUNGI

Most drug discovery efforts utilizing fungal endophytes as sources for potential anticancer agents have used high throughput antiproliferation assays with a panel of NCI cancer cell lines as the initial screen for cytotoxicity. This disease-based approach allows for the rapid screening of libraries of extracts/compounds and may be used to identify agents that demonstrate specific and/or selective cytotoxicity toward cancer cells. Ideally, a toxic small molecule natural product would demonstrate activity at relatively low doses. It would also have a large differential toxicity against cancer cell lines compared to normal cells. This is important because compounds that are nonspecifically toxic would make chemotherapeutic treatment even more destructive to the normal tissue in a patient. Moreover, one must be careful when administering cytotoxic agents since ironically, the very same anticancer agent might also be carcinogenic.[43] Toxicity screening is usually employed as the first step to identify agents that are able to selectively and specifically destroy tumor cells. Once a selective cytotoxic agent is identified, the mechanism leading to this effect can be investigated. Next, modifications to the cytotoxic molecule are made to identify analogues with increased specificity and selectivity against cancer cells.

14.5.1 Globosumones from *Chaetomium globosum*

Chaetomium globosum Ames (Ascomycete) is a fungal endophyte found in the stem tissue of the Mormon tea plant (*Ephedra fasiculata* A.; Ephedraceae). Previous studies on soil-derived *C. globosum* strains have yielded potential anticancer agents such as chetomin,[44] (indole-3-yl-[13])

cytochalasans, chaetoglobosins A–B,[45] C–F,[46] G and J,[47] Q, R, and T,[48] and TAN-1142.[49] The investigation of an endophytic strain of *C. globosum* has recently afforded the orsellinic acid esters globosumones A (**20**), B (**21**), and C (**22**), and the known compounds, orsellinic acid [orsellic acid], orcinol, and trichodion.[50] Antiproliferative activity of these compounds was measured against human cancer cell lines NCI-H460 (non-small cell lung cancer), MCF-7 (breast cancer), SF-268 (CNS glioma), and MIA Pa Ca-2 (pancreatic carcinoma) in addition to normal human fibroblast cells (WI-38) employing the MTT [3-(4,5-dimethylthiazol-2-yl)-2,5-diphenyltetrazolium bromide] assay. Globosumone A (**20**) demonstrated the most cytotoxicity against the NCI-H460 and SF-268 cell cancer lines with IC_{50} values of 6.5 μM and 8.8 μM, respectively. Moderate toxicity of globosumone A against the normal fibroblasts was observed with an IC_{50} of 13.0 μM. Globosumone B (**20**) also demonstrated cytotoxicity, but this activity was limited to higher concentrations and appeared significantly more destructive to normal fibroblasts.[50]

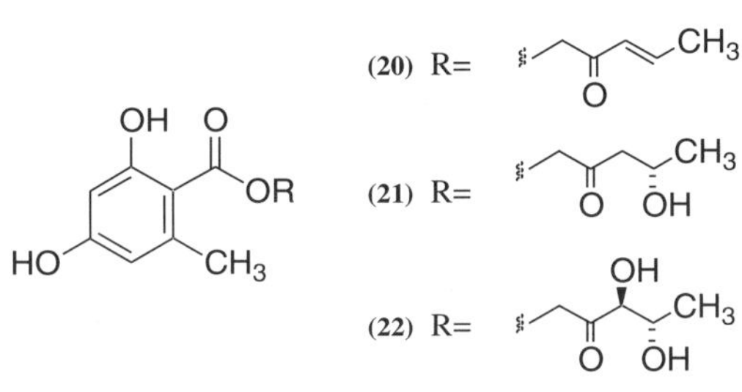

14.5.2 (+)-Epiepoxydon from *Apiospora montagnei*

Apiospora montagnei Saccardo (Apiosporaceae, Ascomycetes) is a fungal endophyte isolated from the inner tissue of the North Sea alga *Polysiphonia violacea* (Roth) Spreng. (Rhodomelaceae, Rhodophyta). New secondary metabolites encountered in this endophyte included the diterpene, myrocin A, the polyketide apiosporic acid, and the monomethyl ester of 9-hydroxyhexylitaconic acid.[51] Additionally, the (−)-enantiomer of the known (+)-hexylitaconic acid and (+)-epiepoxydon (**23**) were isolated. Of the compounds encountered, (+)-epiepoxydon (**23**) demonstrated significant cytoxicity against human cancer cell lines HMO2 (gastric adenocarcinoma), HepG2 (hepatic carcinoma), and MCF-7 with IC_{50} values of 0.70 μg/mL, 0.75 μg/mL, and 0.8 μg/mL, respectively. Significant total growth inhibition was observed with HMO2 at 1.0 μg/mL, HepG2 at 4.6 μg/mL, and MCF-7 at 1.5 μg/mL. At a concentration of 3.6 μg/mL, (+)-epiepoxydon was able to reduce the initial cell number by 50% (LC_{50}) in MCF-7 cells. The activity of (+)-epiepoxydon (**23**) is not surprising, since it has been previously reported to be active against the lymphocytic leukemia cell line P-388 with an IC_{50} of 0.20 μg/mL.[52]

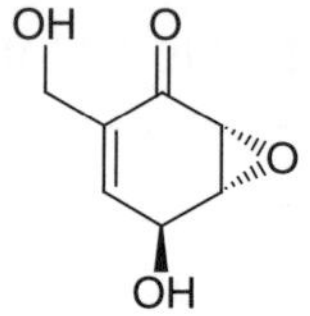

(23)

14.5.3 Myrothecines from *Myrothecium roridum*

Chemical investigation of two endophytic strains of *Myrothecium roridum*, IFB-E009 and IFB-E012, occurring in the Chinese medicinal plants, *Trachelospermum jasminoides* and *Artemisia annua*, respectively, has afforded cytotoxic 10,13-cyclotrichothecane-derived macrolides, myrothecines A–C (**24–26**).[53] Some trichothecenes have previously demonstrated efficacy as anticancer agents.[54] Myrothecines A–C (**24–26**) were tested against the KB human nasopharyngeal epidermoid cell line and demonstrated significant cytotoxicity with IC_{50} values of 15.6 μM, 1.4 μM, and 57.5 μM, respectively.[53]

14.5.4 Curvularins from a *Penicillum* sp. and *Nectria galligens*

Bioactivity-guided fractionation of a cytotoxic extract of an unidentified *Penicillum* sp. (Trichomaceae) inhabiting the rhizosphere of the Apache plume (*Fallugia paradoxa* D. Don; Rosaceae) yielded two anthraquinones and three cytotoxic curvularins identified as α, β-dehydrocurvularin (**27**), 11-hydroxycurvularin (**28**), and 11-methoxycurvularin (**29**).[55] Serial dilutions of each isolated curvularin were applied for 72 hours to human cancer cell lines NCI-H460, MCF-7, and SF-268. The inhibition of proliferation

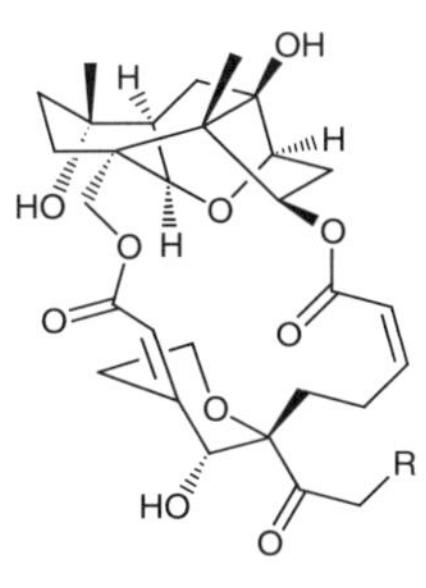

(24) R = H
(25) H = OH

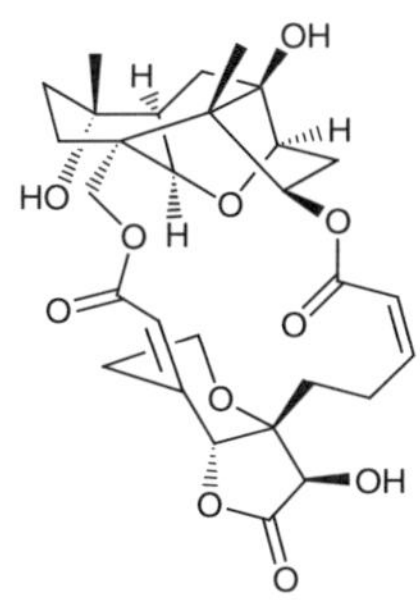

(26)

was then quantified by the MTT assay. The isolated curvularins demonstrated significant cytotoxicity with IC_{50} values ranging from 1.8 μM–13.3 μM. *Nectria galligens* Bres. (Nectriaceae) is a common endophyte found in the apple tree, *Malus x domestica*, and is responsible for the formation of a very common canker disease in hardwoods and shrubs. Infection with this fungus creates swollen dead areas in the bark and causes periodic problems in the agricultural industry due to crop loss. Chemical investigation of a liquid culture of *N. galligens* obtained from the xylem of infected trees in Chile resulted in the isolation of α, β-dehydrocurvularin (27).[56] Other small molecule natural products such as colletochlorin B (**30**), and ilicicolins C (**31**), E (**32**), and F (**33**) were also isolated and were found to be toxic against MRC-5 (human lung fibroblasts) with IC_{50} values in the range of 64 μg/mL–120 μg/mL after a 24 hours exposure. The toxicity for these compounds appeared to be dependent on the presence of the chlorine atom. The cytotoxicity level after exposure to α, β-dehydrocurvularin (27) was found to be much higher with an IC_{50} below 12 μg/mL. Curvularins **27–29** were previously reported to act upon mitotic apparatus components and inhibit cell division in sea urchin embryogenesis,[57] but this is the first time that activity has been shown against human cancer cell lines.

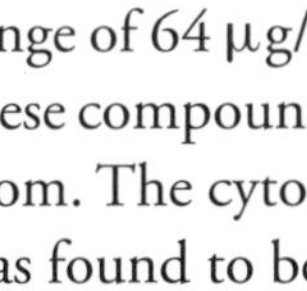

(27)

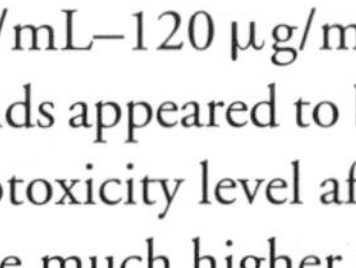

(28) R=OH
(29) R=OCH_3

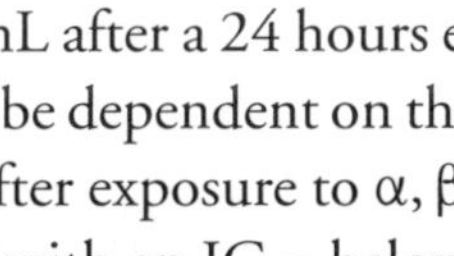

(30)

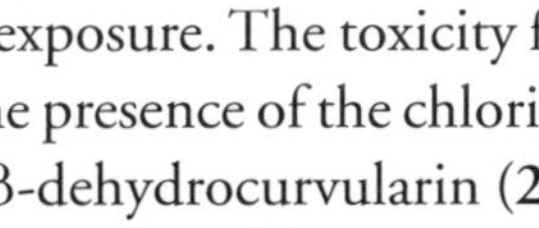

(31) R=H,4′,5′-dihydro
(32) R=H,4″,5″-dehydro
(33) R=OAc

14.5.5 Aspochalasins from *Aspergillus flavipes*

Four known cytochalasins and three new aspochalasins I–K (**34–36**) were obtained from bioassay-guided fractionation of a cytotoxic extract of *Aspergillus flavipes* (Moniliaceae) isolated from the rhizosphere of the desert turpentine bush (*Ericameria laricifolia* Nutt.) (Asteraceae).[58] Aspochalasins are a subgroup of cytochalasins, previously known to have anticancer activity caused in part by inhibiting actin polymerization within the cytoskeleton. After a 48 hours of exposure to the cancer cells, aspochalasins I (**34**), J (**35**), and K (**36**) have demonstrated nonselective weak to moderate antiproliferative activity against the cancer cell lines, NCI-H460, MCF-7, and SF-268, with IC_{50} values ranging from 13.4 μM to 55.2 μM. The cytotoxicity of these molecules appeared to be dependent on an electrophilic α, β-unsaturated carbonyl moiety, as aspochalasin C (**38**), aspochalasin D (**39**), and TMC-169 (**40**) were significantly more active than aspochalasins I (**34**) and J (**35**) (containing an α, β-unsaturated lactone group) or aspochalasins K (**36**) and E (**37**) (with an unconjugated carbonyl group). However, it is unknown whether these cytotoxic aspochalsins are able to inhibit the polymerization of actin, and further work is warranted to elucidate their mechanism of action leading to cell death.

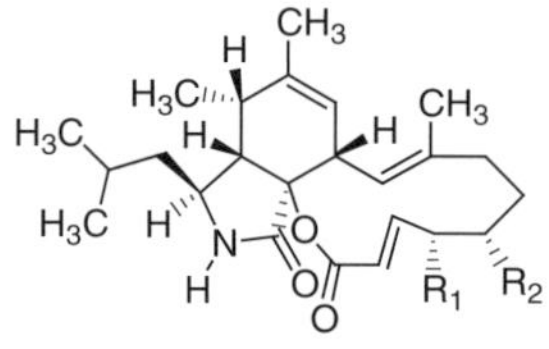

(34) R_1=R_2=OH
(35) R_1=OH, R_2=H

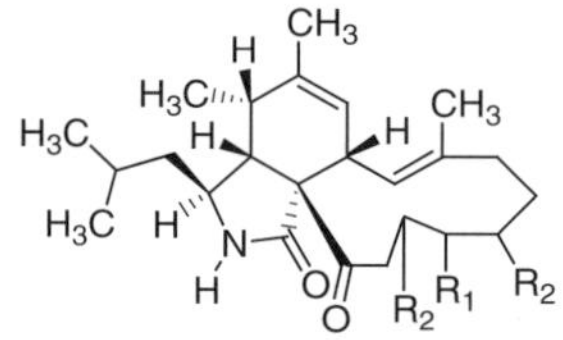

(36) R_1=OCH_3, R_2=OH
(37) R_1=R_2=OH

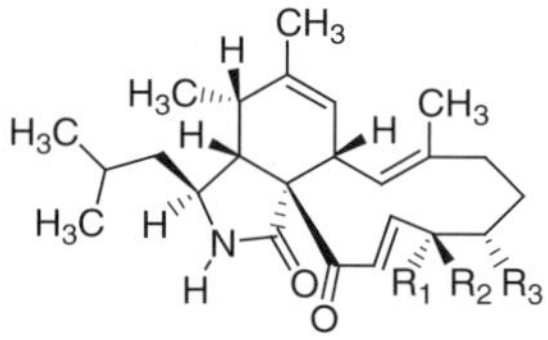

(38) R_1=H, R_2=R_3=OH
(39) R_1=R_3=OH, R_2=H
(40) R_1 or R_2=OH, R_3=H

14.5.6 11-Hydroxycurvularin and Penicillic Acid from *Aspergillus* sp.

Using a bioassay-guided approach, 11-hydroxycurvularin (**28**) and several other known compounds were isolated from an extract of *Aspergillus* *terreus* Thom. occurring in the rhizosphere of an unidentified *Brickellia* sp. (Asteraceae). Extracts of *A. wentii* (Wehmer) found in the rhizosphere of the creosote bush (*Larrea tridentata* DC. Coville) (Zygophyllaceae) and *A. cervinus* from the rhizosphere of the desert honeysuckle (*Anisacanthus thurberi* Torr. Gray) (Acanthaceae), both afforded penicillic acid (**41**).[59] The metabolites isolated from these *Aspergillus* species were subjected to an antiproliferative MTT assay against human cancer cell lines NCI-H460, MCF-7, SF-268, MIA Pa Ca-2 in addition to normal human fibroblast cells (WI-38). Both 11-hydroxycurvularin (**28**) and penicillic acid (**41**) were found to be moderately cytotoxic with selective activity against cancer cell lines compared to the normal fibroblast cells. IC_{50} values ranged from 2.0 μM to 4.1 μM in cancer cells treated with 11-hydroxycurvularin (**28**) and 5.8 μM to 20.0 μM in cancer cells treated with penicillic acid (**41**). IC_{50} values for 11-hydroxycurvularin and penicillic acid on normal human fibroblasts were 11.6 μM and 57.50 μM, respectively, suggesting that these compounds are marginally more toxic to cancer cell lines than the normal fibroblast cells. As noted above, 11-hydroxycurvularin has been reported to act on the mitotic apparatus, inhibiting sea urchin embryogenesis.[57]

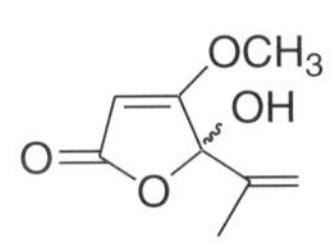

(41)

The activity of penicillic acid is interesting in that it shows selective cytotoxicity against non-small cell lung (NCI-H460) and pancreatic cancer (MIA Pa Ca-2) cell lines with decreased toxicity to normal fibroblasts (WI-38) cells. Previously, penicillic acid has been reported to be cytotoxic to A2780 (human ovarian carcinoma), Chinese hamster ovary, and HeLa cells as well as causing growth inhibition of sarcoma. Its cytotoxicity has been linked to its ability to bind to SH groups in macromolecules, induction of DNA single-strand breaks, and inhibition of DNA synthesis.[60] The derivatives of penicillic acid without the isopropenyl $[-C(CH_3) = CH_2]$ moiety are not cytotoxic, suggesting its requirement for the observed activity.[59]

14.5.7 Terricyclic Acid A (TCA) from *Aspergillus terreus*

Bioactivity-guided fractionation of a cytotoxic extract of a strain of *Aspergillus terreus* isolated from the rhizosphere of *Opuntia versicolor*, a cactus growing in the Sonoran desert, has yielded terrecyclic acid A (TCA) (**42**), a small molecule natural product which demonstrates potent anti-cancer activity in cell-based reporter assays.[61] TCA has shown the ability to induce a heat shock response and affect cellular oxidative and inflammatory stress response pathways. In addition, TCA has demonstrated dose-dependent antiproliferation activity against 3LL (Lewis lung carcinoma) cells, as measured by the MTT assay, and a two-fold induction of the heat shock response over the vehicle (DMSO) control, indicating it to be a possible Hsp90 binding agent. However, a direct interaction of TCA with Hsp90 was ruled out despite its significant antitumor activity.

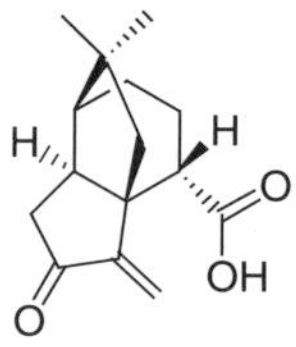

(42)

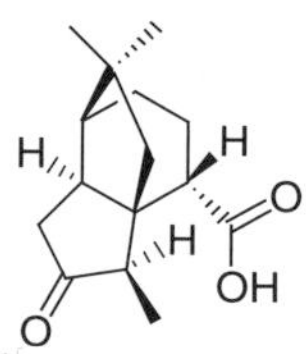

(43)

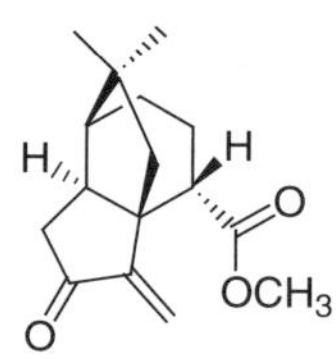

(44)

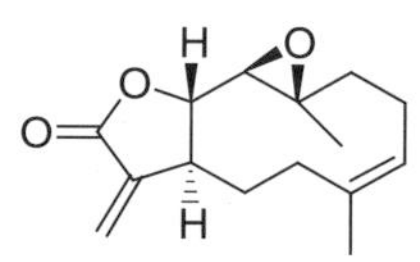

(45)

The presence of an α-methylene ketone group in TCA suggested that it may form adducts with sulfhydryl groups of cellular proteins. Interestingly, the glutathione precursor *N*-acetyl-cysteine was cyto-protective whereas buthionine sulfoximine (a glutathione-depleting agent) enhanced the toxicity of TCA. To examine structure-activity relationships, TCA was converted to dihydro-TCA (**43**) and TCA methyl ester (Me-TCA) (**44**) and compared with the parent compound for their antiproliferative activity. As measured by the MTT assay in 3LL cells, dihydro-TCA (**43**) was nontoxic whereas Me-TCA (**44**) was found to be significantly more cytotoxic. Moreover, Me-TCA was able to induce a 10-fold increase in heat shock response compared to vehicle (DMSO) control whereas dihydro-TCA was shown to be inactive. The generation of reactive oxygen species (ROS) as a response to environmental insults can lead to the activation of HSF-1 (heat shock factor 1), promoting heat shock protein transcription.

Indeed, TCA (**42**) at a concentration of 10 μg/mL, has been shown to elevate levels of ROS, as measured by flow cytometry. Consistent with earlier observations regarding structure-activity relationships, Me-TCA (**44**) showed 3-fold induction of ROS while dihydro-TCA (**43**) had no effect on the cellular levels of ROS.[61] It is noteworthy that parthenolide (**45**), a sesquiterpene natural product structurally related to TCA, has previously been shown to increase the levels of ROS by glutathione depletion in hepatocellular carcinoma cell lines. In a separate study, parthenolide was able to inhibit DNA synthesis, cause cell cycle arrest, and induce apoptosis which are important mechanisms for controlling tumor growth.

Constitutive activation of the nuclear factor κB (NF-κB) pathway is a common feature of cancer cells, promoting their survival, signaling, growth, and invasion.[62] Importantly, TCA was able to inhibit NF-κB while simultaneously increasing ROS. Since tumors with constitutive NF-κB activity are inherently resistant to many chemotherapeutic agents, compounds such as TCA could therefore improve the efficacy of these chemotherapies by blocking the activation of NF-κB. Sesquiterpenes related to TCA have previously shown to increase the level of ROS and also inhibit NF-κB transcriptional activity.[63] TCA not only increased ROS levels in 3LL cells, but also showed concentration dependent inhibition of cytokine-induced NF-κB transcriptional activity in NF-κB reporter constructs.[61] This suggests the modulation of three key stress pathways — those regulating the oxidative, heat shock, and inflammatory responses — by TCA. Since these are the same pathways promoting tumor growth and survival, TCA and other small molecule natural products may be used as probes for discovering the relationships between these pathways, leading to novel anticancer drugs.

14.5.8 Brefeldin A from *Aspergillus clavatus* and *Paecilomyces* sp

The endophytic fungal strains *Aspergillus clavatus* H-037 and *Paecilomyces* sp. were isolated respectively, from the bark of medicinal plants *Taxus maieri* and *Torreya grandis* growing in China. *Paecilomyces* is a common endophyte and *A. clavatus* is usually saprophytic and known to produce cytochalasins. Extracts of both fungi yielded brefeldin A (**46**),[64] a known cytotoxic agent with apoptotic and anticancer activity. The fermentation broths of these fungal strains were tested for cytotoxicity against the human tumor cells lines HL60, KB, HeLa, SPC-A-1 and MCF-7 using the MTT assay after a 3-day exposure to various dilutions of the broths. All fungi demonstrated high cytotoxicity against these cell lines with IC_{50} values ranging between 1:200 to 1:5000. The HL60 cell line was the least sensitive while the most sensitive was SPC-A-1. When tested against the panel of cell lines, brefeldin A (**46**) demonstrated high cytotoxicity with IC_{50} values between 1 ng/mL–10 ng/mL.

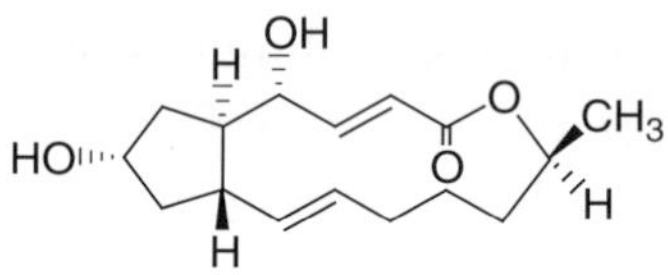

(46)

14.5.9 Rubrofusarin B from *Aspergillus niger*

Fractionation of an extract of the endophytic fungal strain, *Aspergillus niger* IFB-E003, occurring in the leaf tissue of *Cynodon dactylon* (Poaceae) yielded rubrofusarin B (**47**) and aurasperone A (**48**), both of which were found to be cytotoxic against SW1116 (human colon cancer) cells.[65] Previously, naphthopyrones structurally related to **47** and **48** have demonstrated anticancer and anti-signal transduction activity. After a 48 hour-exposure, rubrofusarin B was found to be the most cytotoxic of the isolated

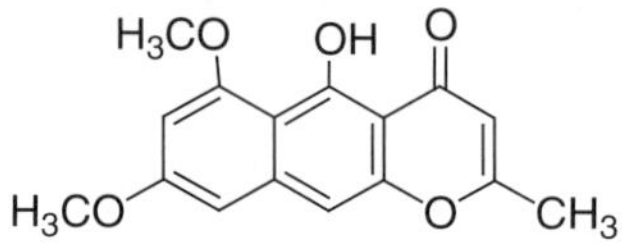

(47)

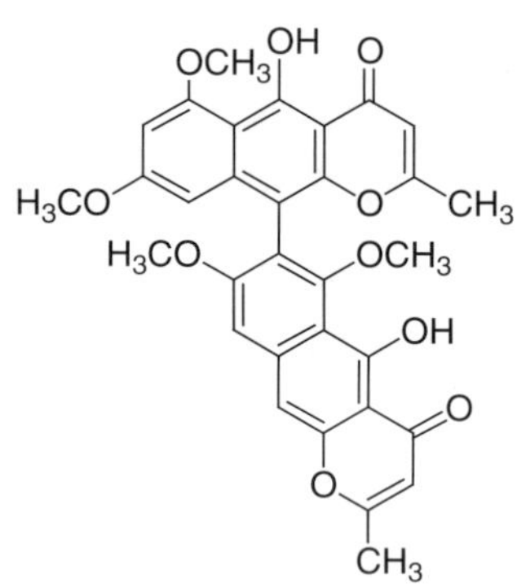

(48)

compounds, with an IC_{50} of 4.5 μg/mL (15.7 μM). The remaining compounds were weakly cytotoxic with only 18.7%–27.9% inhibition at a standard 10 μg/mL concentration.

14.5.10 Sequoaitones from *Aspergillus parasiticus*

Aspergillus parasiticus is an endophyte isolated from the inner bark of the redwood tree, *Sequoia sempervirens* (Taxodiaceae). Previously, several sequoaitones active in the brine shrimp toxicity assay were isolated from this fungal strain and more recently four new sequoaitones C – F (**49–52**) were isolated and found to be toxic to brine shrimp with LD_{50} of 260 μM, 1300 μM, 640 μM, and 260 μM, respectively.[66]

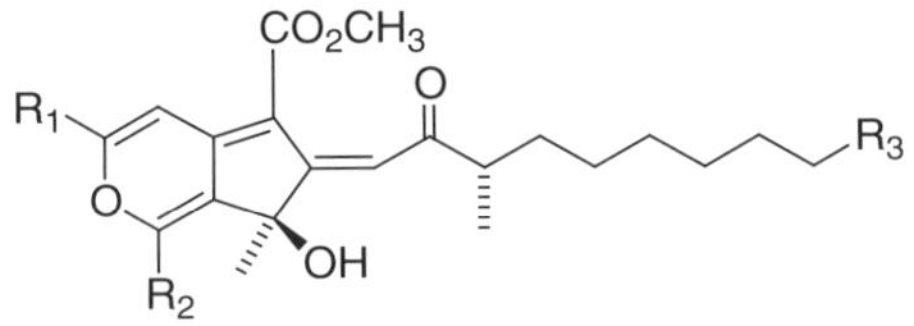

(**49**) R_1=R_3=H, R_2=CH_3
(**50**) R_1=CH_3, R_2=H, R_3=OH
(**51**) R_1=H, R_2=CH_3, R_3=OH
(**52**) R_1=CH_3, R_2=R_3=H, 6, 7-dihydro

14.5.11 Peniprequinolone from *Penicillium janczewskii*

Penicillum janczewskii isolated from the phloem of the Chilean gymnosperm *Prumnopitys andina* (Podocarpaceae) has afforded peniprequinolone (**53**) and gliovictin (**54**).[67] The cytotoxicity of these compounds was measured against normal immortalized cell line MRC-5 (lung fibroblasts) and the cancer cell line AGS (gastric adenocarcinoma) in a neutral red uptake assay. Both peniprequinolone and gliovictin demonstrated weakly differential toxicity with a greater effect against cancer cells as opposed to normal fibroblasts; IC_{50}s for peniprequinolone were 89 μM against the cancer cell line and 116 μM against the fibroblasts. Gliovictin (**54**) was overall less toxic but demonstrated a differential activity with an IC_{50} value of 475 μM against the cancer cell line and 681 μM against the fibroblasts.

In a structure-activity relationship study, it was found that the acetylation of gliovectin (**54**) made it less active with IC_{50} of 1 000 μM.

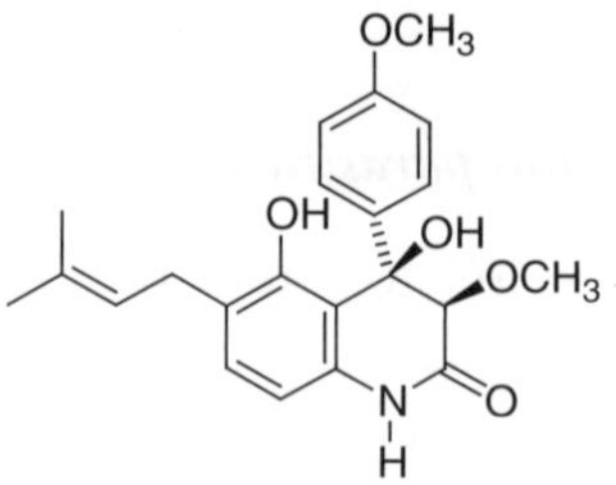

(53)

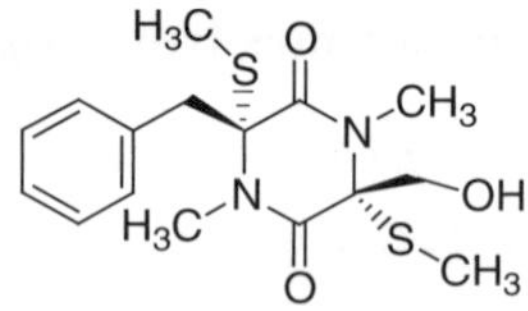

(54)

14.5.12 Phomol from *Phomopsis* sp. EO2O18

The fermentation broths of the endophytic fungal strain, *Phomopsis* sp. E02018, isolated from the dead twigs of *Erythrina crista-galli* (a medicinal plant growing in Argentina) (Fabaceae) has yielded phomol (**55**), a novel antibiotic with weak cytotoxic activity against cancer cells and with anti-inflammatory activity.[68] Cytotoxicity was measured against human cancer cell lines, L-1210, MDA-MB-231, Colo-320, with IC_{50} values of 20 μg/mL, 50 μg/mL, and 50 μg/mL, respectively. Reporter gene assays utilizing STAT1/STAT2 dependent signal transduction, TNF-α, and NF-κB induction were also studied, but phomol (**55**) did not show any activity in these assays. However, **55** did demonstrate anti-inflammatory activity in a mouse ear model assay by inhibiting edema formation by 53.20%.

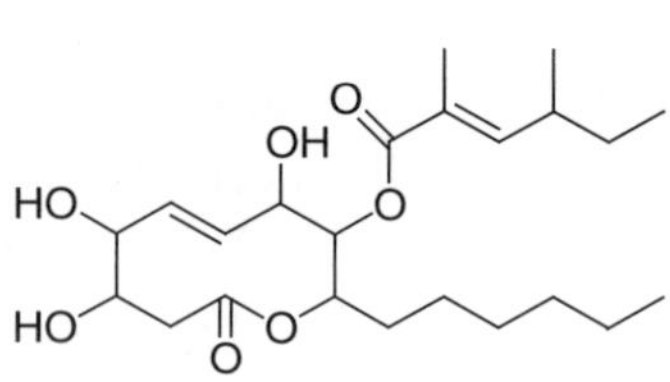

(55)

14.5.13 Cytoskyrin A from *Cytospora* sp. CR200

The endophytic fungal strain, *Cytospora* sp. CR200, was obtained from a branch of the *Conocarpus erecta* (Combretaceae) growing in Costa Rica.

The natural product cytoskyrin A (**56**) was isolated from this fungus, and was found to be highly active in a biochemical induction assay (BIA) for the induction of the SOS response in bacteria.[69] This assay is designed to identify potential anticancer agents that damage DNA or inhibit its synthesis. Cytoskyrin A (**56**) exhibited a strong response for BIA at 12.5 ng whereas cytoskyin B (57) demonstrated no response up to 50 μg. The preliminary data suggested that cytoskyrin A (**56**) is a more powerful inhibitor of *in vitro* DNA synthesis and transcription compared to luteoskyrin or cyotoskyrin B (57).

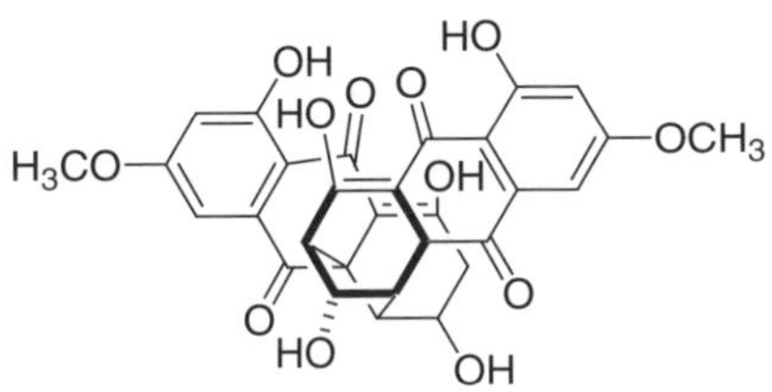

(56)

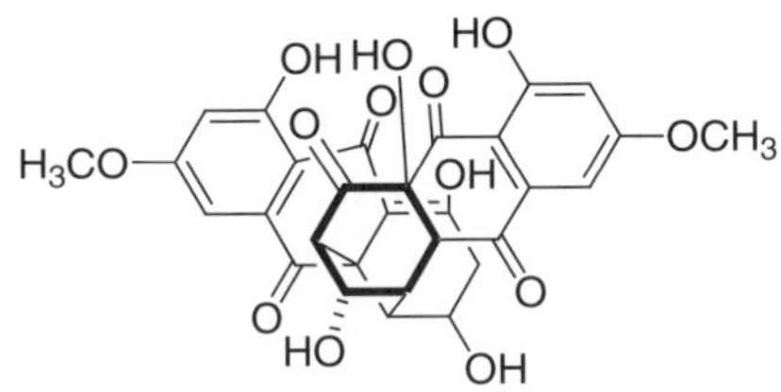

(57)

14.6 APOPTOTIC AGENTS

In its simplest form, apoptosis can be described as programmed cell death. Ultimately, apoptosis is a complex energy-dependent process that requires specific signals to either stimulate its induction or inhibit antiapoptotic factors.[70] Induction of apoptosis involves the activation of cysteinyl aspartate-specific proteases (caspases), the proteolytic enzymes that ultimately signal the cell towards death. Apoptotic signaling occurs mainly via the extrinsic (death receptor) and intrinsic (mitochondrial) pathways, both of which involve activation of caspases and endonucleases.[71] During apoptotic progression, distinct morphological changes within the cell occur, including condensation of chromatin, cellular blebbing, and DNA fragmentation.[70] Unfortunately, some proapoptotic molecules are down-regulated or inactivated in certain types of cancers.[71] Disabled proapoptotic signals combined with a lack of cell cycle regulation are important factors which enable cancer cell survival despite genetic instability. Small molecule natural products that are able to restore the abnormal cell's ability to undergo apoptosis would be important in controlling tumor growth.

14.6.1 Hormonemate from *Hormonema dematioides*

A fermentation broth of the endophytic fungus, *Hormonema dematioides*, (strain E99156) isolated from the needles of a *Pinus* (Pinaceae) species collected in Portugal has yielded hormonemate (**58**) with cytotoxic activity against human colon adenocarcinoma cell lines, COLO-320, DLD-1 and HT-29, in addition to five other human cancer cell lines: HL-60 (promyelocytic leukemia), JURKAT (acute T cell leukemia), HEP-G2, MCF-7, and HeLa S3 (cervical carcinoma).[72] IC_{50} values ranged from 5 μM–10 μM for most of the cell lines tested, with HepG2 and HeLa S3 demonstrating the least sensitivity with IC_{50} of 12 μM–13 μM. Interestingly, hormonemate (**58**) at 7.5 μM was also able to induce 17%–20% of the HL-60 cells to differentiate into granulocytes, which later died by apoptosis. Within two hours of incubation, 13 μM concentration of hormonemate (**58**) induced apoptosis in COLO-320 cells as detected by morphological changes such as cell shrinkage and the formation of apoptotic bodies. The induction of apoptosis was verified by a caspase-3 activity assay which demonstrated the ability of hormonemate (**58**) to increase caspase-3 activity in a dose and time dependent manner that was strongly inhibited by the caspase-3 inhibitor Ac-DEVD-CHO (*N*-acetyl-Asp-Glu-Val-Asp-CHO). When COLO-320 cells were treated with varying concentrations of hormonemate and the incorporation of [^{14}C]-thymidine, [^{14}C]-uridine, and [^{14}C]-leucine was measured to find changes in cellular DNA-, RNA-, and protein synthesis in COLO-320 cells, approximately at 8 μM concentration of **58**, a 50%

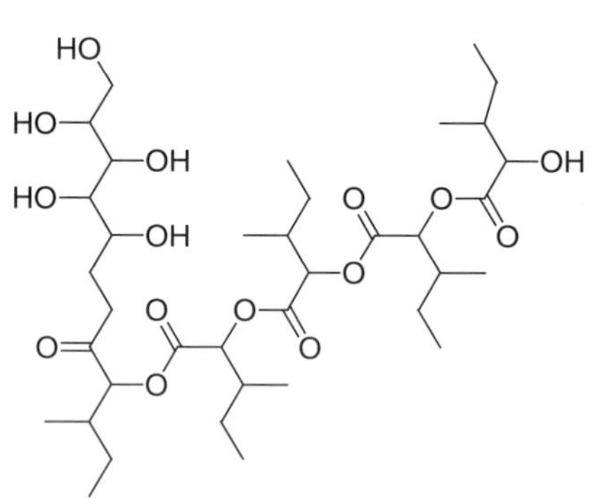

(58)

inhibition of uptake of uridine into RNA was observed. The uptake of thymidine and leucine were also reduced, but only at concentrations above 10.0 μM.

14.7 MICROFILAMENT DISRUPTORS

Cells contain elaborate networks of protein filaments that help establish cell shape, provide strength, allow for locomotion, separate chromosomes during mitosis and meiosis, and are involved in the intracellular transport of organelles. The cytoskeleton itself is made up of actin filaments (microfilaments), intermediate filaments, and microtubules. Microfilaments are involved in anchoring centrosomes at polar ends of the cell during mitosis and linking cell surface receptors to cytosolic proteins. Microtubules are involved in the formation of spindle fibers that drive the migration of chromosomes during mitosis and meiosis. Discovery and development of agents that are able to abrogate mitotic events by disrupting cytoskeletal proteins is a relatively new, but an exciting area of chemotherapeutic research. Since the discovery of paclitaxel (**1**) in 1971,[73] effective antimitotic agents have been in demand. Unlike most other chemotherapies, paclitaxel blocks mitosis at the G_2/M phase of cell proliferation by stabilizing microtubules, rendering them unable to depolymerize into soluble tubulin; ultimately, cell cycle progression is halted and the cell dies.[74] Unfortunately, paclitaxel is considerably toxic in a chemotherapeutic setting. This underlies the need to find new small molecule agents that stop cell cycle progression in a more selective manner in order to slow the growth of cancerous cells.

14.7.1 Microcarpalide I from an Unidentified Endophytic Fungus

The fermentation broth of an unidentified endophytic fungus isolated from plants growing in Hawaii has afforded microcarpalide I (**59**), an alkyl-substituted nonenolide that is weakly cytotoxic as a result of its ability to disrupt microfilaments. Extracts from the strain 112/13 collected from the bark of *Ficus microcarpa* L. (Moraceae) demonstrated a strong abrogation of microfilament activity.[75] In A-10 rat smooth muscle cells, a 5 μg/mL dose was able to induce a 50%–75% loss of actin filaments. Microcarpalide I

(**59**) is the second C_{16} nonenolide natural toxin described to date and is structurally related to achaetolide (**60**) with a similar carbon skeleton. At a concentration of 0.5 μg/mL–1.0 μg/mL, microcarpalide I (**59**) was able to disrupt microfilaments in approximately half of the A-10 cells as detected by immunofluorescence, and its cytotoxicity was observed at a significantly higher dose of 20 μg/mL.

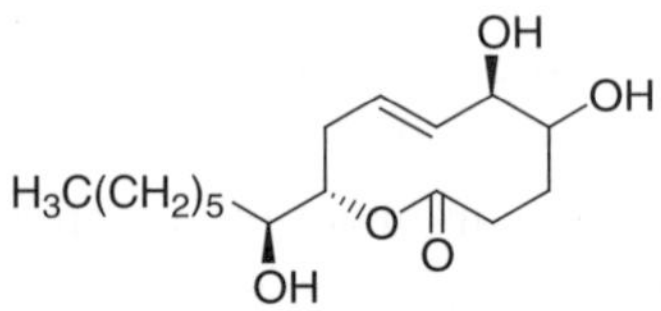

(**59**)

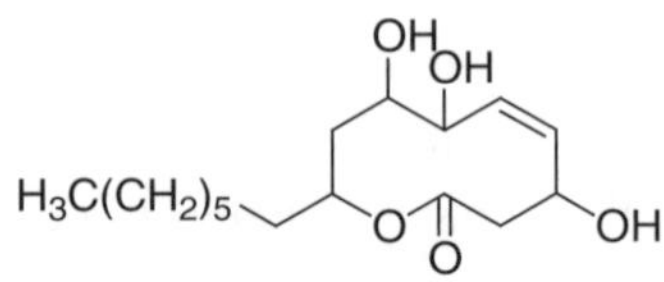

(**60**)

Weak cytotoxicity against human cancer cells, KB and LoVo (colon adenocarcinoma), after treatment with microcarpalide I, was also observed with the IC_{50} values of 50 μg/mL and 90 μg/mL, respectively. The ability of microcarpalide I (**59**) to disrupt microfilaments at subcytotoxic concentrations suggests its usefulness for future cell motility and metastasis studies, since other known microfilament disrupting agents such as cytochalasins are cytotoxic at biologically active concentrations.

14.7.2 Cytochalasins from *Rhinocladiella* sp

The fungi of the genus *Rhinocladiella* are commonly found as saprophytes inside dead tree limbs. An endophytic strain (*Rhinocladiella* sp. 309) of this fungus was recently isolated from the Chinese medicinal plant, *Tripterygium wilfordii* (Celastraceae). The investigation of this endophyte afforded three novel cytotoxic 22-oxa-[12]-cytochalasins (**61–63**), and the known compound cytochalasin E (**64**).[76] Cytochalasins are well known mold metabolites that bind to actin and prevent its polymerization and elongation, thus arresting the cell's ability to divide and ultimately promoting apoptosis. Cytochalasins have been isolated from several fungi including *Helminthosporium* sp., *Phoma* sp., *Xylaria* sp., *Hypoxylon* sp., and *Chalara* sp.[77]

The toxicity of the four purified cytochalasins (**61–64**) was measured against human tumor cell lines 2780S (ovarian carcinoma), SW-620

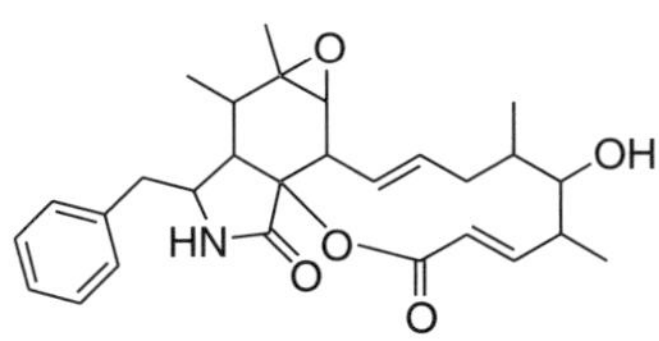

(61)

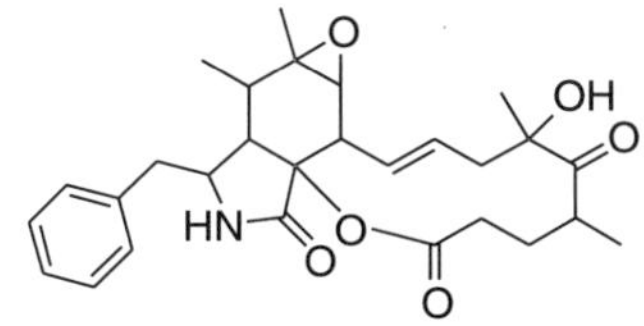

(62)

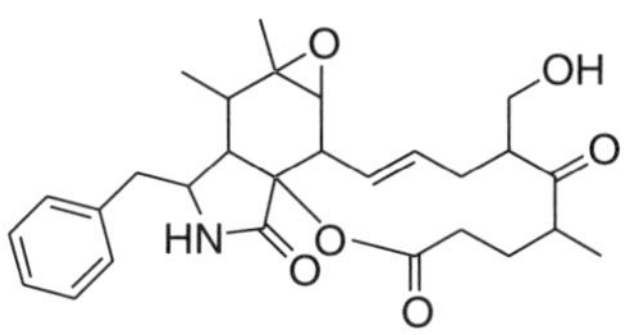

(63)

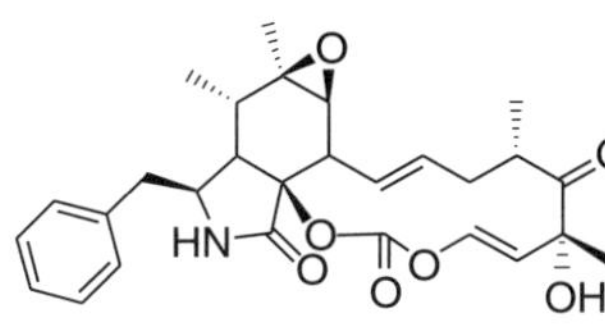

(64)

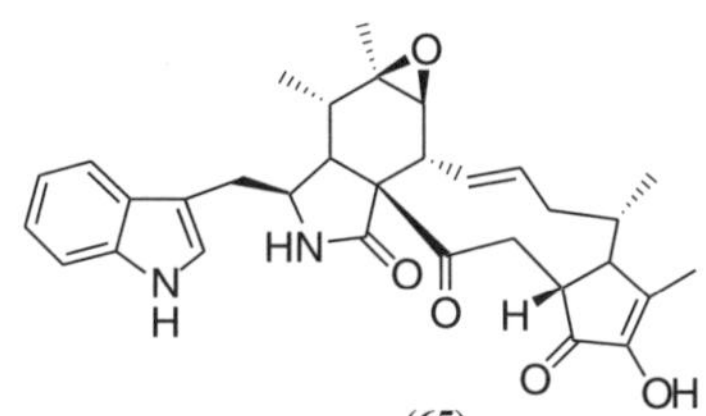

(65)

(colorectal carcinoma), and HCT-116 (colorectal carcinoma). Cytochalasin E (**64**) was found to be the most toxic with IC_{100} values below 1.0 μg/mL whereas the three novel cytochalasins **61–63** were significantly less toxic with IC_{100} values ranging from 3.91 μg/mL to 62.5 μg/mL. Due to their toxicity, the use of cytochalasins for therapeutic applications is limited. Chaetoglobosin U (**65**) was obtained from the extracts of *Chaetomium globosum* IFB-E019 which is an endophytic fungal strain from the stem tissue of *Imperata cylindrica*. It has exhibited cytotoxicity against KB cells with an IC_{50} of 16 μM. Previously known analogues of **65** with moderate cytotoxicity against KB (with IC_{50} values above 30 μM) were also isolated from this fungus.[78]

14.8 ACKNOWLEDGMENTS

We thank U.S. National Cancer Institute (NCI; Grant No. 1RO1CA 90265) and Arizona Biomedical Research Commission for the financial support for some of the work reviewed. AALG is thankful to his collaborators, Drs Luke Whitesell, Stanley H Faeth, Leland S Pierson III, Linda Meade-Tollin, and Hans D VanEtten; and coworkers, Drs E M Kithsiri Wijeratne, Thomas J Turbyville, Bharat P Bashyal, Jixun Zhan, Jian He, Jacqueline Takahashi, Priyani A Paranagama, Guang-Xiong Zhou, G M Kamal B Gunaherath, Mr Christopher J. Seliga, Ms Manping X Liu, Ms Anna M Burns, Ms Anne Fritz, and Ms Libia Luevano, for their valuable contributions to the research programs on Sonoran desert plant-associated microorganisms.

REFERENCES

1. Ruddon R, Cancer Biology 3rd edn., Oxford University Press, Inc., Oxford, 1995.
2. Edwards BK, Brown ML, Wingo PA, Howe HL, Ward E, Ries LAG, Schrag D, Jamison PM, Jemal A, Wu XC, Friedman C, Harlan L, Warren J, Anderson RN, Pickle LW, Annual Report to the Nation on the Status of Cancer, 1975–2002, Featuring Population-Based Trends in Cancer Treatment, *J Nat Cancer Inst* **97**:1407–1427, 2005.
3. Young P, Major microbial diversity initiative recommended, *ASM News* **63**:417–421, 1997.
4. Cragg GM, Newman DJ, Snader KM, Natural products in drug discovery and development, *J Nat Prod* **60**:52–60, 1997.
5. Bode HB, Bethe B, Höfs R, Zeeck A, Big effects from small changes: Possible ways to explore nature's chemical diversity, *Chem Bio Chem* **3**:619–627, 2002.
6. Dreyfuss MM, Chapela IH, *The Discovery of Natural Products with Therapeutic Potential*, Butterworth-Heinemann, Boston, 1994.
7. Tan RX, Zou WX, Endophytes: A rich source of functional metabolites, *Nat Prod Rep* **18**:448–459, 2001.
8. Schulz B, Boyle C, Draeger S, Römmert AK, Krohn K, Endophytic fungi: A source of novel biologically active secondary metabolites, *Mycol Res* **106**:996–1004, 2002.
9. Strobel G, Daisy B, Castillo U, Harper J, Natural products from endophytic microorganisms, *J Nat Prod* **67**:257–268, 2004.
10. Gunatilaka AAL, Natural products from plant-associated microorganisms: Distribution, structural diversity, bioactivity, and implications of their occurrence, *J Nat Prod* **69**:509–526, 2006.
11. Kettering M, Weber D, Sterner O, Ankem T, Secondary metabolites of fungi — Functions and uses, *BIOspectrum* **10**:147, 2004.
12. Chomcheon P, Wiyakrutta S, Sriubolmas N, Ngamrojanavanich N, Isarangkul D, Kittakoop P, 3-Nitropropionic acid (3-NPA), a potent antimycobacterial agent

from endophytic fungi: Is 3-NPA in some plants produced by endophytes? *J Nat Prod* **68**:1103–1105, 2005.

13. Stierle A, Strobel G, Stierle D, Grothaus P, Bignami G, The search for a taxol-producing microorganism among the endophytic fungi of the pacific yew, *Taxus brevifolia*, *J Nat Prod* **58**:1315–1324, 1995.
14. Eyberger AL, Dondapati R, Porter JR, Endophyte fungal isolates from *Podophyllum peltatum* produce podophyllotoxin, *J Nat Prod* **69**:1121–1124, 2006.
15. Puri SC, Verma V, Amna T, Qazi GN, Spiteller M, An endophytic fungus from *Nothapodytes foetida* that produces camptothecin, *J Nat Prod* **68**:1717–1719, 2005.
16. Wagenaar M, Corwin J, Strobel G, Clardy J, Three new cytochalasins produced by an endophytic fungus in the genus *Rhinocladiella*, *J Nat Prod* **63**:1692–1695, 2000.
17. Turbyville TJ, Wijeratne EMK, Liu MX, Burns AM, Seliga CJ, Luevano LA, David CL, Faeth SH, Whitesell L, Gunatilaka AAL, Search for Hsp90 inhibitors with potential anticancer activity: Isolation and SAR studies of radicicol and monocillin I from two plant-associated fungi of the Sonoran desert, *J Nat Prod* **69**:178–184, 2006.
18. Chomcheon P, Sriubolmas N, Wiyakrutta S, Ngamrojanavanich N, Chaichit N, Mahidol C, Ruchirawat S, Kittakoop P, Cyclopentenones, scaffolds for organic syntheses produced by the endophytic fungus *Mitosporic dothideomycete* sp. LRUB20, *J Nat Prod* **69**:1351–1353, 2006.
19. Shu YZ, Recent natural products based drug development: A pharmaceutical industry perspective, *J Nat Prod* **61**:1053–1071, 1998.
20. Stierle A, Stierle D, Strobel G, Bignami G, Grothaus P, Endophytic fungi of pacific yew (*Taxus brevifolia*) as a source of taxol, taxanes and other pharmacophores, in *Bioregulators for Crop Protection and Pest Control*, American Chemical Society, Washington DC, pp. 64–77, 1994.
21. Frense D, Taxanes: Perspectives for biotechnological production, *Appl Microbiol Biotechnol* **73**:1233–1240, 2007.
22. Strobel GA, Hess WM, Ford E, Sidhu RS, Yang X, Taxol from fungal endophytes and the issue of biodiversity, *J Industrial Microbiol* **17**:417–423, 1996.
23. Hoffman A, Khan W, Worapong J, Strobel G, Griffin D, Arbogast B, Berofsky D, Boone RB, Ning L, Zheng P, Daley L, Bioprospecting for taxol in angiosperm plant extracts, *Spectroscopy* **13**:22–32, 1998.
24. Stierle A, Strobel G, Stierle D, Taxol and taxane production by *Taxomyces andreanae*, an endophytic fungus of Pacific yew, *Science* **260**: 214–216, 1993.
25. Guo BH, Wang YC, Zhou XW, Hu K, Tan F, Miao ZQ, Tang KX, An endophytic Taxol-producing fungus BT2 isolated from *Taxus chinensis* var. *mairei*, *African J Biotech* **5**:875–877, 2006.

26. (a) Strobel G, Yang XS, Sears J, Kramer R, Sidhu RS, Hess WM, Taxol from *Pestalotiopsis microspora*, an endophytic fungus of *Taxus wallachiana*, *Microbiol* **142**:435–440, 1996. (b) Li JY, Strobel G, Sidhu R, Hess WM, Ford EJ, Endophytic taxol-producing fungi from bald cypress, *Taxodium distichum Microbiol* **142**:2223–2226, 1996.
27. (a) Schiff PB, Fant J, Horwitz SB, Promotion of microtuble assembly *in vitro* by taxol, *Nature* **277**:665–667, 1979. (b) Schiff PB, Horwitz SB, Taxol stabilizes microtubules in mouse fibroblast cells, *Proc Natl Acad Sci USA* 77:1561–1565, 1980. (c) Manfredi JJ, Parness J, Horwitz SB, Taxol binds to cellular microtubules, *J Cell Biol* **94**:688–696, 1981.
28. Foster S, Medicinal plant conservation and genetic resources: Examples from the temperate Northern hemisphere, *Acta Hort* **330**:67–73, 1993.
29. (a) Berlin J, Bedorf N, Mollenschott C, Wray V, Sasse F, Hofle G, On the podophyllotoxins of root cultures of *Linum flavum*, *Planta Medica* **54**:204–206, 1988. (b) Empt U, Alfermann AW, Pras N, Petersen M, The use of plant cell cultures for the production of podophyllotoxin and related lignans, *J Appl Bot* **74**:145–150, 2001. (c) Giri A, Narasu ML, Production of podophyllotoxin from *Podophyllum hexandrum*: A potential natural product for clinically useful anticancer drugs, *Cytotechnology* **34**:17–26, 2000. (d) Peterson M, Alferman AW, The production of cytotoxic lignans by plant cell cultures, *Appl Microbiol Biotechnol* **55**:135–142, 2001.
30. Puri SC, Nazir A, Chawla R, Arora R, Riyaz-ul-Hasan S, Amna T, Ahmed B, Verma V, Singh S, Sagar R, Sharma A, Kumar R, Sharma RK, Ghulam NQ, The endophytic fungus *Trametes hirsuta* as a novel alternative source of podophyllotoxin and related aryl tetralin lignans, *J Biotechnol* **122**:494–510, 2006.
31. Mizuno TA, Development of antitumor polysaccharides from mushroom fungi, *Food & Food Ingred J* **167**:69–85, 1996.
32. Mallery SR, Shenderova A, Pei P, Begum S, Ciminieri JR, Wilson RF, Casto BC, Schuller DE, Morse MA, Effects of 10-hydroxycamptothecin, delivered from locally injectable poly(lactide-co-glycolide) microspheres, in a murine human oral squamous cell carcinoma regression model, *Anticancer Res* **21**:1713–1722, 2001.
33. Pirillo A, Verotta L, Gariboldi P, Torregiani E, Bombardelli E, Constituents of *Nothapodytes foetida*, *J Chem Soc Perkin Trans 1* 583–587, 1995.
34. Amna T, Puri SC, Verma V, Sharma JP, Khajuria RK, Musarrat J, Spiteller M, Qazi GN, Bioreactor studies on the endophytic fungus *Entrophospora infrequens* for the production of an anticancer alkaloid camptothecin, *Can J Microbiol* **52**:189–196, 2006.
35. Mirrington RN, Ritchie E, Shoppee CW, Sternhell S, Taylor WC, Some metabolites of *Nectria radicicola* Gerlach & Nilsson (Syn. *Cylindrocarpon radicicola* Wr.): the structure of radicicol (monorden), *Aust J Chem* **19**:1265–1284, 1966.

36. Nozawa K, Nakajima S, Isolation of radicicol from *Penicillium luteo-aurantium*, and meleagrin, a new metabolite from *Penicillium meleagrinum*, *J Nat Prod* **42**:374–377, 1979.
37. Arai M, Yamamoto K, Namatame I, Tomoda H, Omura S, New monordens produced by amidepsine-producing fungus *Humicola* sp. FO-2942, *J Antibiotics* **56**:526–532, 2003.
38. (a) Whitesell L, Lindquist SL, Hsp90 and the Chaperoning of Cancer, *Nat Rev Cancer* **5**:761–772, 2005. (b) Neckers L, Hsp90 inhibitors as novel cancer chemotherapeutic agents, *Trends Mol Med* **8**:S55–61, 2002. (c) Maloney A, Workman P, Hsp90 as a new therapeutic target for cancer therapy: the story unfolds, *Expert Opin Biol Ther* **2**:3–24, 2002.
39. (a) Kwon HJ, Yoshida M, Nagaoka R, Obinata T, Beppu T, Horinouchi S, Suppression of morphological transformation by radicicol is accompanied by enhanced gelsolin expression, *Oncogene* **15**:2625–2631, 1997. (b) Soga S, Neckers LM, Schulte TW, Shiotsu Y, Akasaka K, Narumi H, Agatsuma T, Ikuina Y, Murakata C, Tamaoki T, Akinaga S, KF25706, a novel oxime derivative of radicicol, exhibits *in vivo* antitumor activity via selective depletion of Hsp90 binding signaling molecules, *Cancer Res* **59**:2931–2938, 2003. (c) Soga S, Shiotsu Y, Akinaga S, Sharma SV, Development of radicicol analogues, *Curr Cancer Drug Targets* **3**:359–369, 2003.
40. Yang ZQ, Geng X, Solit D, Pratilas CA, Rosen N, Danishefsky SJ, New efficient synthesis of resorcinylic macrolides via ynolides: Establishment of cycloproparadicicol as synthetically feasible preclinical anticancer agent based on Hsp90 as the target, *J Am Chem Soc* **126**:7881–7889, 2004.
41. Moulin E, Zoete V, Barluenga S, Karplus M, Winssinger N, Design, synthesis, and biological evaluation of HSP90 inhibitors based on conformational analysis of radicicol and its analogues, *J Am Chem Soc* **127**:6999–7004, 2005.
42. Zhang H, Chung D, Yang YC, Neely L, Tsurumoto S, Fan J, Zhang L, Biamonte M, Brekken J, Lundgren K, Burrows F, Identification of new biomarkers for clinical trials of Hsp90 inhibitors, *Mol Cancer Therapeutics* **5**:1256–1264, 2006.
43. Blagosklonny MV, Carcinogenesis, cancer therapy and chemoprevention, *Cell Death and Differentiation* **12**:592–602, 2005.
44. Safe S, Taylor A, Sporidesmins. XIII. Ovine III-thrift in Nova Scotia. III. Characterization of chetomin, a toxic metabolite of *Chaetomium cochliodes* and *Chaetomium globosum*, *J Chem Soc Perkin Trans I* **4**:472–479, 1972.
45. Sekita S, Yoshihira K, Natori S, Kuwano H, Structures of chaetoglobosin A and B, cytotoxic metabolites of *Chaetomium globosum*, *Tetrahedron Lett* 2109–2112, 1973.

46. Sekita S, Yoshihira K, Natori S, Kuwano H, Structures of chaetoglobosins C, D, E, and F, cytotoxic indol-3-yl-[13]cytochalasans from *Chaetomium globosum*, *Tetrahedron Lett* 1351–1354, 1976.
47. Sekita S, Yoshihira K, Natori S, Kuwano H, Chaetoglobosins G and J, cytotoxic indol-3-yl-[13]cytochalasans from *Chaetomium globosum*, *Tetrahedron Lett* 2771–2774, 1977.
48. Jiao W, Feng Y, Blunt JW, Cole ALJ, Munro MHG, Chaetoglobosins Q, R, and T, three further new metabolites from *Chaetomium globosum*, *J Nat Prod* **67**:1722–1725, 2004.
49. Tanida S, Tsuboya S, Harada S, Immunomodular and antitumor TAN-1142 and its manufacture with *Chaetomium*, *Jpn. Kokai Tokkyo Koho (Japanese Patent Application JP 91-136729)*, pp. 6, 1992.
50. Bashyal BP, Wijeratne EMK, Faeth SH, Gunatilaka AAL, Globosumones A–C, cytotoxic orsellinic acid esters from the Sonoran desert endophytic fungus *Chaetomium globosum*, *J Nat Prod* **68**:724–728, 2005.
51. Klemke C, Kehraus S, Wright AD, König GM, New secondary metabolites from the marine endophytic fungus *Apiospora montagnei*, *J Nat Prod* **67**:1058–1063, 2004.
52. Iwamoto C, Minoura K, Oka T, Ohta T, Hagishita S, Numata A, Absolute stereostructures of novel cytotoxic metabolites, penostatins A–E, from *Penicillium* species separated from an *Enteromorpha* alga, *Tetrahedron* **55**:14353–14368, 1999.
53. Shen L, Jiao RH, Ye YH, Wang XT, Xu C, Song YC, Zhu HL, Tan RX, Absolute configuration of new cytotoxic and other bioactive trichothecene macrolides, *Chem Eur J* **12**:5596–5602, 2006.
54. (a) Amagata T, Rath C, Rigot JF, Tarlov N, Tenney K, Valeriote FA, Crews P, Structures and cytotoxic properties of trichoverroids and their macrolide analogues produced by saltwater culture of *Myrothecium verrucaria*, *J Med Chem* **46**:4342–4350, 2003. (b) Bondy GS, Pestka JJ, Immunomodulation by fungal toxins. *J Toxicol Environ Health B Crit Rev* **3**:109–143, 2000.
55. Zhan J, Wijeratne EMK, Seliga CJ, Zhang J, Pierson EE, Pierson III LS, VanEtten HD, Gunatilaka AAL, A new anthraquinone and cytotoxic curvularins of a *Penicillium sp.* from the rhizosphere of *Fallugia paradoxa* of the Sonoran desert, *J Antibiot* **57**:341–344, 2004.
56. Gutierrez M, Theoduloz C, Rodriguez J, Lolas M, Schmeda-Hirshmann G, Bioactive metabolites from the fungus *Nectria galligens*, the main apple canker agent in Chile, *J Agric Food Chem* **53**:7701–7708, 2005.
57. Kobayashi A, Hino T, Yata S, Itoh TJ, Sato H, Kawazu K, Unique spindle poisons, curvularins and its derivatives, isolated from *Penicillium* species, *Agric Biol Chem* **52**:3119–3123, 1988.

58. Zhou GX, Wijeratne EMK, Bigelow D, Pierson III LS, VanEtten HD, Gunatilaka AAL, Aspochalasins I, J, and K: Three new cytotoxic cytochalasans of *Aspergillus flavipes* from the rhizosphere of *Ericameria laricifolia* of the Sonoran desert, *J Nat Prod* **67**:328–332, 2004.
59. He J, Wijeratne EMK, Bashyal BP, Zhan J, Seliga CJ, Liu MX, Pierson EE, Pierson III LS, VanEtten HD, Gunatilaka AAL, Cytotoxic and other metabolites of *Aspergillus* inhabiting the rhizosphere of Sonoran desert plants, *J Nat Prod* **67**:1985–1991, 2004.
60. (a) Kawasaki I, Oki T, Umeda M, Saito M, Cytotoxic effect of penicillic acid and patulin on HeLa cells, *Jpn J Exp Med* **42**:327–340, 1972. (b) Dickens F, Jones HEH, Further studies on the carcinogenic and growth-inhibitory activity of lactones and related substances, *Brit J Cancer* **17**:100–108, 1963. (c) Umeda M, Yamamoto T, Saito M, DNA-strand breakage of HeLa cells induced by several mycotoxins, *Jpn J Exp Med* **42**:527–535, 1972.
61. Turbyville TJ, Wijeratne EMK, Whitesell L, Gunatilaka AAL, The anticancer activity of the fungal metabolite terrecyclic acid A is associated with modulation of multiple cellular stress response pathways, *Mol Cancer Therapeutics* **4**:1569–1576, 2005.
62. Aggarwal BB, Nuclear factor-NF-κB: the enemy within, *Cancer Cell* **6**:203–208, 2004.
63. Wen J, You KR, Lee SY, Song CH, Kim DG, Oxidative stress-mediated apoptosis. The anticancer effect of the sesquiterpene lactone parthenolide, *J Biol Chem* **277**:38954–38964, 2002.
64. Wang J, Huang Y, Fang M, Zhang Y, Zheng Z, Zhao Y, Su W, Brefeldin A, a cytotoxin produced by *Paecilomyces* sp. and *Aspergillus clavatus* isolated from *Taxus mairei* and *Torreya grandis*, *FEMS Immun and Med Microbiol* **34**:51–57, 2002.
65. Song YC, Li H, Ye YH, Shan CY, Yang YM, Tan RX, Endophytic naphthopyrone metabolites are co-inhibitors of xanthine oxidase, SW1116 cell and some microbial growths, *FEMS Microbilogy Letters* **241**:67–72, 2004.
66. Stierle A, Stierle D, Bugni T, Sequoiatones C–F, Constituents of the redwood endophyte *Aspergillus parasiticus*, *J Nat Prod* **64**:1350–1353, 2001.
67. Schmeda-Hirschmann G, Hormazabal E, Astudillo L, Rodriguez J, Theoduloz C, Secondary metabolites from endophytic fungi isolated from the Chilean gymnosperm *Prumnopitys andina* (Lleuque), *World J Microbiol Biotechnol* **21**:27–32, 2005.
68. Weber D, Sterner O, Anke T, Gorzalczancy S, Martino V, Acevedo C, Phomol, a new anti-inflammatory metabolite from an endophyte of the medicinal plant *Erythrina crista-galli*, *J Antibiotics* **57**:559–563, 2004.

69. Brady SF, Singh MP, Janso JE, Clardy J, Cytoskyrins A and B, new BIA active bisanthraquinones isolated from an endophytic fungus, *Organic Lett* **2**:4047–4049, 2000.
70. Huerta S, Goulet EJ, Huerta Yepez S, Livingston EH, Screening and detection of apoptosis, *J Surgical Res* **139**:143–156, 2007.
71. Fabregat I, Roncero C, Fernández M, Survival and apoptosis: A dysregulated balance in liver cancer, *Liver International* **2**:155–162, 2007.
72. Filip P, Weber RW, Sterner O, Anke T, Hormonemate, a new cytotoxic and apoptosis-inducing compound from the endophytic fungus *Hormonema dematioides*. I. Identification of the producing strain, and isolation and biological properties of hormonemate, *Z Naturforsch* **58**:547–552, 2003.
73. Wani MC, Taylor HL, Wall ME, Coggon P, McPhail MT, Plant antitumor agents. VI. The isolation and structure of taxol, a novel antileukemic and antitumor agent from Taxus brevifolia. *J Am Chem Soc* **93**:2325–2327, 1971.
74. Schiff PB, Horwitz SB, Promotion of microtuble assembly in vitro by taxol, *Nature* **277**:665–667, 1979.
75. Ratnayake AS, Yoshida WY, Mooberry SL, and Hemscheidt T. The structure of microcarpalide, a microfilament disrupting agent from an endophytic fungus, *Organic Lett.* **3**:3479–3481, 2001.
76. Wagenaar MM, Corwin J, Strobel G, Clardy J, Three new cytochalasins produced by an endophytic fungus in the genus *Rhinocladiella*, *J Nat Prod* **63**:1692–1695, 2000.
77. (a) Buckingham J, *Dictionary of Natural Products*, Chapman and Hall, New York, 1994. (b) Dagne E, Gunatilaka AAL, Asmellash S, Abate D, Kingston DGI, Hofmann GA, Johnson RK, Two New Cytotoxic Cytochalasins from a *Xylaria obovata*, *Tetrahedron* **50**:5615–5620, 1994.
78. Ding G, Song YC, Chen JR, Xu C, Ge HM, Wang XT, Tan RX, Chaetoglobosin U, a cytochalasan alkaloid from endophytic *Chaetomium globosum* IFB-E019, *J Nat Prod* **69**:302–304, 2006.

Chapter 15

PLANT FUNGAL ENDOPHYTES: INTERACTIONS, METABOLITES AND BIOSYNTHESES

John R. Porter

15.1 INTRODUCTION

When people know that I am studying fungi in plants, the general perception is that I must be studying some plant disease that destroys crops or damages ornamental plants. The general view of microorganisms, bacteria, fungi, and protists, is that they cause disease in something about which we care. However, there is a growing appreciation that macroorganisms, from the smallest mite to redwoods, are actually consortia of eukaryotic cells (the host) intermixed or surrounded by a variety of microorganisms (variously referred to as the pathogen, the symbiont, the endophyte, or the epiphyte) living in a symbiotic or parasitic state with the organism we see. Most plants and animals are hosts to hundreds of fungal, bacterial and protistan species, each of which is represented by a few to many billions of individual cells. This leads to the rather disconcerting awareness that there are more microbial cells in me than "my own body" cells in me. Each of these internal species is capable of producing a variety of compounds, some of which serve a nutritional purpose for the host, some of which are harmful to the host, and some of which enhance the ecological fitness of the host in competition with some species or as prey to other species.

The purpose of this chapter is to review the state of knowledge about the endophytic fungal species that inhabit the spaces within and between plant cells and tissues and to examine the array of secondary products produced by these fungal inhabitants. In the few cases in which enough is known, the details of the biosynthesis of these compounds, the medicinal uses to which these compounds have, and the potential for culture of the fungi as industrial producers of the metabolites will be discussed.

There are a number of recent reviews of endophytes, both bacterial[1–3] and fungal, and their secondary metabolism that would provide the interested reader with further details and background. Among these are the excellent reviews concerning the interaction between the endophyte and the host,[4–7] as well as those that specifically address the bioactive natural product production by endophytic fungi.[8–11] The latter collectively list over 330 different compounds produced. To the extent that is reasonable, I have chosen not to repeat the information in these recent reviews, but to summarize their findings (Table 1). This review will address the information from the more recent literature and, insofar as is known, details of biosynthesis.

Following the introductory material, which includes definitions, general ideas, and a discussion of endophyte diversity, the chapter is organized along broad taxonomic lines of the host plants. This should not be taken to imply that a given endophyte inhabits only one species, genus, family or even class of plants. Often, the same endophytic species lives in the tissues of a wide variety of plant species. Some endophytes may be more prevalent in one geographical area than another. The genotype of the host clearly can have a strong influence on the secondary metabolism of the endophyte.[11] The metabolites of a particular fungus will be discussed in the context of the host from which the endophyte was obtained or in which it was studied.

15.2 PRINCIPLES AND DEFINITIONS

15.2.1 Endophytes

The terminology used to describe the relationships among fungi living within host tissues (the symbiont) and the host plant is complex. In fact, the term "symbiont" often connotes a beneficial relationship to at least the host plant, if not to both species, but I am using "symbiont" in the broader sense of "living together." De Bary[12] coined the term "endophyte," literally meaning "internal plant," to describe the presence of fungal tissue within the cells or tissues of a plant, usually when there was no evidence of disease or pathology. This term was proposed at a time when there were only two recognized kingdoms of organisms, the plants and the animals, and the plants were defined as anything not clearly seen as animal. Thus, the bacteria and fungi were seen as plants. A more appropriate set of terms today might be "endomycobiont" or "endobacteriobiont," referring to internal fungi or bacteria, respectively. The usual current usage is "endophyte fungi" or "endophyte bacteria," although it is just as frequently used, and perhaps more grammatically correct to refer to the associations as endophytic fungi or bacteria. Although "endophyte" usually suggests the lack of pathogenicity,[13] with terms such as cryptic, non-obvious, latent, or inconspicuous being commonly used, fungal hyphae of species usually considered pathogenic may be found in tissues showing no signs of disease. Some authors [e.g.,5] have begun referring to such situations as the endophytic phase of a pathogenic relationship.

An endophyte should be distinguished from an epiphyte. An epiphyte[12] ("plant on the surface") is a micro- or macroorganism living on the surface of the plant in question. These could be casually-associated organisms present due to soil contamination, but the term usually refers to an organism actually making its living or occupying a niche on the surface environment of the plant. Leaf, stem and root epiphytes have been studied, including some diazotrophic bacteria that provide nitrogen to the host plant (*e.g., Azospirillium*) or to the local environment of the microorganism (leaf epiphytes). Most bromeliads and some orchids are common examples of macro epiphytes.

It can be seen, then, that the biology of fungal-plant relationships are much more complex than the terminology itself. Pathogenic fungi may be in an endophytic phase in one plant at one point in time, pathogenic at a different time, but can also exist completely endophytically in another species of plant. A widely-distributed endophytic fungal species may become pathogenic only in rare environmental circumstances. An endophytic fungal species may serve to enhance the nutrition of one plant (suggesting a mutualistic mycorrhizal relationship), but then provide no such benefit to a different plant. An endophytic bacterial or fungal species may benefit the plant through suppression of the growth or pathogenicity of other species of fungi or bacteria.[14] Such relationships have led to the terms plant growth-promoting rhizobacteria (PGPR)[1] and plant growth-promoting fungi (PGPF).[15] Finally, there may be no demonstrable benefit or detriment to a plant species due to the presence of an endophytic fungus; this is a neutralistic relationship. Schulz and Boyle[7] have tried to capture the range of endophyte-plant relationships through their term "endophytic continuum," a concept also proposed by Müller and Krauss,[16] although the notion that invasion of the plant tissue by PGPF or the non-pathogenic rhizobia (*Rhizobium, Bradyrhizobium, Sinorhizobium*) is a form of virulence could raise some great confusion in the understanding of these mutualistic relationships. De Bary's original meaning of endophyte, the notation that the fungus was merely found in the plant tissue without obvious disease, may continue to be one of the most useful, particularly for the natural products chemist. The definition of "endophyte" will doubtlessly remain controversial as different researchers disagree on whether "obvious disease" includes only major growth or structural changes (wilts, blights, dieback) or would also includes more subtle changes, such as a small loss of yield or photosynthetic capacity.[17]

15.2.2 Mycorrhizae

Generally distinguished from the endophytes are the mycorrhizae. These are plant-fungus relationships in which the fungus forms a biological link to the soil environment surrounding the plant. The mycorrhizal hyphae generally penetrate much further into the soil than the plant roots or root hairs; form connections to smaller soil particles; have a much larger surface area per weight of tissue; can solubilize poorly-soluble nutrients; and form a large absorptive surface to provide soil nutrients for the benefit of the plant.[18] The nutrient most often studied in mycorrhizal relationships is phosphate, but nitrogen and other nutrients are frequently transferred from the soil as well.[18] The majority of plant species examined to date are facultatively or obligately mycorrhizal. Three major types of mycorrhizae are recognized generally, the ectomycorrhizae, the endomycorrhizae, and the ectendomycorrhizae,[4,19] although individual authors may reject one or more of these classifications.[20] The ectendomycorrhizae are then divided into four subclasses, orchid, ericoid, arbutoid, and monotropoid mycorrhizae.[19]

The ectomycorrhizae form a mat of hyphal tissue (the mantle) surrounding the plant root and have hyphae that radiate away into the soil as well as those that radiate inward to reside between plant cells. The intercellular hyphae usually form a ramifying network in the root cortex (the Hartig's net),[20] and some may have hyphal tips that form a broad pad that makes direct contact with the plant cell (the appressorium)[21,22] to expedite nutrient transfer, but they usually do not invade the plant cell itself. Ectomycorrhizae are often basidiomycete fungi (Basidiomycota).

The endomycorrhizae lack the mantle and the Hartig's net, but reside in greater abundance in intra- and intercellular spaces of the plant. Hyphae radiate away from the plant root to facilitate nutrient collection and transfer. These mycorrhizae enter the plant cells, often forming two, distinguishable structures: a bulbous expansion of the hyphal tip, the vesicle (absent depending on species[20]), and a folded or finger-like in-growth of the hyphae, the arbuscule.[20] These mycorrhizae are often referred to as the vesicular-arbuscular mycorrhizae (VAM)[19] or arbuscular mycorrhizae (AM).[23] VAM are usually aseptate zygomycetes (Zygomycota).

The ectendomycorrhizae, consisting of the four subtypes, have characteristics somewhat intermediate between the other two major classes. There is usually a reduced or absent mantle, there may or may not be a Hartig's net, and the hyphae generally invade the cells to form distinctive features intracellularly. The orchid mycorrhizae form a highly folded and spiraled hyphal tip, the peloton, that distinguishes this group.[19] This association is often obligate for the plant, largely because orchid seeds contain too few cells to become self-sustaining until after considerable growth, if then. The achlorophyllous orchids remain obligately mycorrhizal throughout life.[19] The ericoid mycorrhizae associate with members of the families Ericaceae (northern hemisphere) and Epacridaceae (southern hemisphere).[19] These plants grow in stressed, oligotrophic environments and depend on the fungi for adequate nutrition. The plants have very fine roots, with a much reduced vasculature, and a highly ramifying hyphal network performs much of the work of gathering, transporting and supplying inorganic nutrients to the plant. The hyphal network is both intercellular and intracellular. The arbutoid mycorrhizae form associations with some members of the Ericaceae (Arbutoideae).[19] As in ectomycorrhizae, there may be a well developed mantle around a highly-branched, fine root system or a reduced mantle surrounded by a mucilaginous layer, the morphology dependent on the fungal species.[19] These fungi also invade at least some cells and form a dense hyphal network intracellularly. The monotropoid mycorrhizae form an interlinking connection between parasitic or hemiparasitic plants, such as *Monotropa* spp., and the coniferous host tree that sustains the carbon nutrition of both the fungus and the parasitic plant.[19] The fungus forms a multilaminate hyphal mantle around the plant root, and the well developed intercellular hyphae form small peg-like haustoria which penetrate to the cell interior. The expulsion of hyphal cell contents at the tip of the haustoria and the retention of these contents by plant cells membranes give the haustorium a unique, hydroid appearance at high magnification.[19] Ectendomycorrhizae are often, but not exclusively, basidiomycetes, and the fungi often are ectomycorrhizal with other plant species.[19]

The mycorrhizae are considered below separately from other endophytes. The literature relating to mycorrhizal fungal metabolites is much less rich than that for other endophytes. One reason is a greater difficulty to associate products directly with the fungus, often because the fungi are not culturable away from the plant, especially the AM and VAM. Nor are all non-mycorrhizal endophytes culturable. However, the studies of natural product production in unculturable organisms are much more complicated than the study of culturable organism products, although the nonculturable organisms may be sources of interesting genes and enzymes.[24,25] Novel polyketide synthases and other enzymes could be particularly useful for the "combinatorial biochemistry" approach to the modification of known compounds to new medicinals.[26] Another reason to study mycorrhizae separately from other fungal endophytes is that mycorrhizae generally have been studied as contributors to the plant inorganic nutrition. Again, this is not the sole purview of the mycorrhizae, and the nonmycorrhizal endophytic fungi may also play a role in the transfer of nutrients from the soil[27] leading several to ask the question whether endophytic fungi may also be mycorrhizal fungi,[27–29] at least under some conditions.[4] Because of this trait, some have classified the nonmycorrhizal endophytes as pseudomycorrhizal.[30] The typical mantle and Hartig's net structures of ecto- or ectendomycorrhizae are absent, but there may be some unique structures (microsclerotia) in the endophyte association.[27]

It can also be difficult to position compounds as fungal products because the products do not appear outside the mycorrhizal association. In the future, undoubtedly, many more mycorrhizal compounds will come to light, as this is largely an untapped area of investigation. Likely, such studies will first focus on those compounds that allow communication between the mycorrhizal fungus and the host plant.[23] The suppression of plant pathogens by mycorrhizal fungi is a clear reason to expect bioactive product production by mycorrhizal assocations,[18] but it is not known whether the suppression is due to novel compound syntheses, induced host plant resistance mechanisms, or competition of the mycorrhizal fungus for growth opportunities.

15.2.3 Fungal Morphology and Endophyte Terminology

The details of fungal morphology are complex. Many fungi have very complicated life histories, with different forms being dominant during different stages of reproduction, under different environmental conditions, or associated with particular hosts. Some well known examples are the both hyphal and single-celled forms of many yeasts, including the plant endophytic and free-living *Rhodotorula* spp, the human pathogen/endophyte *Candida albicans*, and even the common *Saccharomyces cerevisiae*.[31,32] Thus, many fungi have multiple form species synonymies, making the clear identification of species difficult. The life form responsible for the production of sexually produced spores is the teleomorph (yeast form in the yeasts). Alternatively, the asexual (conidia, vegetative, *etc.*) reproductive life form is the anamorph (hyphal form of yeasts). The collective fungal hyphae are the mycelium. The hyphal form is the usual endophytic form, but not exclusively. When sexual reproduction ensues, the mycelium may also form a macroscopic fruiting body, the basidiocarp (mushroom) of the Basidiomycota and the ascocarp of some Ascomycota (*e.g.*, morels), becoming obvious to the observer, or they may become weakly to strongly parasitic, replacing plant tissues (seeds) or causing plant disease (*e.g., Epichloë*).

Similarly complex and confusing is the terminology applied to plant endophytic fungi. The most commonly applied term in recent years is "dematiaceous septate endophytes" (DSE), indicating that the hyphae are dark-walled (melanin is deposited[33]), and that the hyphae have cross walls. This is synonymous with "dark septate endophytes" (also DSE). Some current and many older studies use the similar term "*mycelia radicis atrovirens*" (MRA),[4] meaning "dark-green root fungi." If no reproductive spores or structures are seen, the term "*mycelia sterilia*," "sterile fungus," has been applied. DSE and *mycelia sterilia* may apply to fungi found in the roots or in other tissues, and the latter often lack dark pigmentation. It is important to note that hyaline (unpigmented) hyphae have been observed that are continuous with pigmented hyphae,[34] thus, they are a part of the same mycelium. The term "systemic endophytic fungi" (SEF) has been proposed recently[35] to refer to fungi that remain nonpathogenic and colonize a large part of the plant; this term has not been adopted widely. Endophytes have been found to belong to all of the major fungal phyla, Ascomycota, Basidiomycota, Chytridiomycota (=Phycomycota),[36–38] Zygomycota,[19] and the protistan Oomycota.[37,39] The VAM come primarily from the Glomeromycota.[20]

It appears that most endophytes are ascomycete fungi, particularly those that do not sporulate easily for traditional classification techniques. There are some exceptions, however. Studies of *Theobroma cacao*[40] in Mexico, Costa Rica, Brazil, Ecuador, and Cameroon led to a large number of nonsporulating isolates (556 of 854 isolates). These were arranged into 59 morphospecies. When these morphospecies were analyzed using molecular tools, a significant majority (42) were basidiomycetes. This is similar to the results of a survey of the related *T. gileri* in Ecuador,[41] in which 23 of 28 isolates were basiodiomycetes. Both studies contrast with a study of *T. cacao* in Brazil,[42] in which only one of 30 isolates was a basidiomycete. The three studies used similar methods, and this may represent geographic variation.

Several roles of endophytic fungi for the host plant have been postulated. These include acting to increase access to mineral nutrients (a "mycorrhizal" function), to increase access to organic soil N, P and C, to increase drought and stress tolerance, to improve water uptake, protection from herbivory (mammals, insects), and for protection from plant pathogenic fungi, bacteria, nematodes, and other parasites.[34] We should not be surprised that endophytic fungi are such common plant symbioses.

15.2.4 Taxonomy and Molecular Biology

As might be imagined from the foregoing discussions, the identification of a particular endophytic fungus can be a daunting task. Until fairly recently, this has been the realm of the specialist mycologist, and this continues to be necessary in some circumstances. Classical identification studies depend on growing the fungus, determining the conditions that cause or allow sporulation and/or sexual reproduction, and determination of its identity based on morphological, cultural, and nutritional characteristics. Nonsporulating fungi were classified as DSE, MRA, or *mycelia sterilia*, without further affinities.

Studies of endophytes usually begin with a surface sterilization step to remove contamination and epiphytic organisms. [*e.g.*,31,43] The endophyte can then be cultured from sections of the plant tissue laid onto a suitable medium designed for fungal culture (water agar, yeast malt agar, malt agar, corn meal agar),[43–46] although agars containing plant leaf extracts (carnation leaf agar) are sometimes used.[46] Antibiotics to suppress bacterial growth are commonly used. Increasingly, endophytes are being confirmed without culture. This approach originally began with microscopic examination of cleared or uncleared plant tissue and microscopic examination of stained or unstained fungal hyphae. Neutral red, trypan blue, lactophenol blue, aniline blue, acid fuchsin, Rose Bengal, chlorazol black E, nitro blue tetrazolium chloride, diaminobenzidine, wheat germ agglutinin (unconjugated or conjugated to a fluorescent label), or phosphatase substrates are often used.[18,47–50] Some of these preparations target the chitin of the fungal hyphae, and so they do not work well when the endophyte is poorly chitinized (young hyphae) or unchitinized (Oomycota).[35] It has also become popular to stain with a lipid-targeting dye, *e.g.* Sudan IV or Congo Red,[35,49] as the fungi usually accumulate considerable lipid in and on the hyphae. Occasionally, fluorescent dyes are used to observe the hyphae.[18,50,51] Some researchers have confirmed endophytic fungi presence through assays for chitin or ergosterol.[18,49,52] More sophisticated and complicated means of observing the fungi *in situ* include transmission electron microscopy, scanning electron microscopy, and fluorescence microscopy after insertion of GFP into an fungal endophyte genome.[7] Microscopically-observed fungal hyphae cannot usually be identified without culture. Recently, molecular characterization of fungal endophytes has becomes popular (see below). This allows identification of many of the organisms present in a plant tissue, even when the organisms are not culturable.

With the general adoption of molecular methods (PCR, gene sequencing), it has become possible to identify endophytes without a specialist's expertise. The DNA of the large (26 S, 28 S, 36 S) (LSU) or small (5.8 S, 18 S) (SSU) ribosomal RNA subunits (rDNA) have been the most popular targets for PCR, sequencing and identification.[53–57] The sequence data is compared to one or more of a proprietary (Microseq) or public (GenBank, http://www.ncbi.nlm.nih.gov; EMBL Nucleotide Sequence Databank, http://www.ebi.ac.uk/embl/; DNA Data Bank of Japan (DDBJ), http://www.ddbj.nig.ac.jp/) database for identification. Frequently, only a portion of the gene, such as the internal transcribed spacer (ITS) regions, especially the ITS1 region of SSU rDNA[18,30,55,56,58,59] or the variable D2 region of the LSU rDNA,[18,53] are used commonly with primers specific for fungi. Other ITS regions, including ITS2 and ITS4, are sometimes used.[58,60–62] A number of other genes have been used for the identification of species, strains, or related sets of endophytic fungi, singly or in combination. These include the intergenic spacer (IGS) regions,[18] the elongation factor 1-*alpha* (EF1-α) gene,[63,64] β-tubulin,[55,65] 4-(γ, γ-dimethylallyl) tryptophan synthase,[65] RNA polymerase (RPB1, RPB2),[66] and mitochondrial DNA (mtDNA).[66,67] The primary characteristics of genes used for identification are that the gene is widely distributed or universal among the fungi. It is sufficiently variable that there are differences significant at the species or lower levels. It is also sufficiently stable that differences are not excessively subject to over-interpretation (*i.e.*, every isolate considered a new species). They have unique regions that are amenable to primer development, and for which there is enough sequence data that comparison to a reasonably large data set is feasible. A modification of the PCR approach, RAPD-PCR, uses random primers, with no specific gene target, to amplify multiple products, the sizes of which will create a pattern in gel electrophoresis that is unique for each fungal species or strain.[18,58]

Another popular technique for endophyte identification is RFLP (restriction fragment length polymorphism).[18,58,67–70] In this technique, nuclear or mitochondrial DNA is extracted from the endophyte and digested with restriction endonucleases. The digestion products are separated by gel electrophoresis, and a unique pattern emerges for the fungus, similar to the patterns seen in the RAPD-PCR technique.

Even with these powerful techniques, it is often not possible to identify an endophyte unambiguously. It is more powerful to sequence at least two genes, if sufficient comparative data is available, and to combine molecular and classical data (morphology, growth characteristics). Some comprehensive molecular sequence data tools are beginning to become available for some groups of fungi.[58] When such tools are not available, the use of morphology can often "break a tie" when databases return two or more equally likely results. In some cases, the morphology discredits any identification based on the molecular data (none of the database choices are a suitable match for the observed characteristics). The molecular approach will become more robust as more species data are entered. For now, the number of species represented in the databases is a tiny fraction of all the extant species.

One of the major impacts of molecular approaches to fungal identification is the ability to unambiguously place a fungus in relation to related species. Since fungal taxonomy is based on the characteristics of sexual reproduction, many fungi could not be classified, since no teleomorph was known. These fungi were placed into the "catch-all" form-phylum Deuteromycota. Gene sequencing has allowed the placement of many of the deuteromycetes into the natural phyla. Thus, for example, the sterile *Neotyphodium* can now be classified as sterile species or strains of the ascomycete genus *Epichloë*. Such placements allow the examination of genes and the evolution of endophytic systems on a phylogenetic basis, rather than superficial similarities in outcome for the host plant or endophytic fungus.

15.2.5 Transmission of Endophytes

How are endophytes acquired by the host plant? The typical answer for how any fungus, endophytic or parasitic, enters a plant system would be transfer from the environment through contact with hyphae in the soil, wind-borne spores, transfer by an insect or other animal vector, or any of a variety of other means for contact, recognition of the plant as a suitable host, and infection/invasion/inhabitation. These various means of acquisition from the environment are referred to as horizontal transmission.[71] Contrasted with this are the endophytes that are transmitted through the plant seed or other reproductive structures.[72] Thus, the fungus can be "inherited" from the parent plant and develop in the young seedling as a part of that plant individual's inherent characteristics. The transmission from parent to offspring is referred to as vertical transmission.

Although not examined widely for a number of endophytes, the presumption is that most DSE are transmitted horizontally. In part, the geographic, seasonal and species diversity of fungi present in a plant species support this conclusion.[11,71] Many of these endophytes can be cultured away from the plant and used to reinoculate an otherwise naïve or endophyte-free plant (coded as E^- in much of the literature on grass endophytes).[5] Other endophytes are transmitted in plant seeds; they are vertically transmitted. This is true for many of the *Epichloë* endophytes in a number of grass genera. Some endophyte genera, *e.g. Neotyphodium*, may even be obligately transmitted vertically,[71] never spreading by spores or other horizontal means. The genetics of the plant can play a major role in determining the development, secondary metabolite production, and transmission of the endophyte.[73,74] The possibility of vertical transmission is an interesting question to pursue for a wider variety of endophytes.

Equally interesting are the transmission of genes in the endophyte-plant relationship. We generally presume that the genes are transmitted vertically. But, we do know that horizontal transfer is possible through the actions of *Agrobacterium* (=pathogenic *Rhizobium*) bacteria or viruses. Although such studies are in their infancy in regard to acquisition of genes by fungi from plants[75] or by plants from fungi,[76] intriguing possibilities of the outcomes of such transfers arise when two organisms with dramatically different evolutionary histories share the same or very similar metabolic pathways. This is another area of research that will yield interesting solutions in the future.

15.2.6 Diversity of Endophytes

There is considerable uncertainty and controversy over how many endophytic fungi exist in nature, the host range of common or uncommon endophytic species, and the extent to which the host plant controls the range of compounds produced. Environmental conditions play a role in how many endophytes may be obtained through culture or molecular methods, with season,[77] leaf age,[78,79] light exposure, climatic conditions,[80] topography,[77] host plant genetics,[11,77] and plant tannin production[81] all playing a role in the observed diversity. Geographic variation exists for single plant species. *Picea mariana* (black spruce) was studied in four distinctly different habitats in northern Ontario, Canada.[82] Among 64 endophytic taxa isolated, only four (*Meliniomyces variabilis, Mortierella parvispora, Penicillium spinulosum*, and *Phialocephala fortinii*) were found to be endophytic in all four habitats. The isolation technique, whether based on culture of the endophytes or elucidation through molecular techniques, clearly influences the diversity observed.[56,83]

An estimate of the number of endophytic fungal species varies widely. Only a little over 100 species have been studied to any great extent, involving culture and production of these compounds under controlled conditions.[8] Although frequently estimated that most plant species and individuals harbor at least one fungal endophyte,[34] it is common to find multiple endophytes in a single plant. In one recent study, Morakotkarn *et al.*[84] isolated 257 strains representing 71 species from the tissues of two bamboos (*Phyllostachy* spp. and *Sasa* spp.) growing in three sites in Japan. A second study[90] led to the isolation from just the leaves of two tropical trees (*Heisteria concinna* and *Ouratea lucens*) of 242 morphologically distinct strains ("morphospecies") in *H. concinna* and 259 "morphospecies" in *O. lucens*; 418 distinct isolates were isolated *in toto*, meaning that the two species shared 83 isolates. Although hampered somewhat in interpretation by the lack of complete taxonomic data, the authors concluded that tropical endophytes are "hyperdiverse." A study of tanoak (*Lithocarpus densiflorus*) yielded 119 classified taxa and an overall estimated species diversity of 265.[86] Of these, 64% were predicted to be mycorrhizal, 11% endophytic, 3% saprobic, 3% parasitic, and 18% of unknown relationship to the plant (casual or neutral).

Some fungal species may exhibit high host specificity or host preference,[77,87] but others are found on a wide variety of plant species. Grass endophytes seem to exhibit the highest specificity for a single or small range of related hosts,[88] while the endophytes of dicotyledonous plants often have a broad host range. Jumpponen and Trappe[4] reviewed the dark septate endophytes of over 600 plant species. By far, the most frequent isolate was *Phialocephala fortinii*, occurring as an endophyte in 38 of the species. Subsequent studies[7,30,43,60,70,86,89–95] have increased the number of confirmed species inhabited by *P. fortinii* to 75, including six crop species (*Asparagus officinalis, Brassica campestris, Cucumis melo, Fragaria grandiflora, Lycopersicon esculentum*, and *Solanum melongena*) and one liverwort (*Cephalozia varians*). The host range of *Fusarium oxysporum* and its related types is similarly broad, inhabiting at least 95 dicotyledonous and monocotyledous host species as a nonpathogenic or weakly pathogenic endophyte.[96]

The total number of fungal species range from less than 1.5 million to 9.9 million, based in part on assumptions of their presence, with six to 113 fungal species per plant species, including pathogens, endophytes and saprobes.[87] The upper estimate of plant-associated fungi was based on an estimated 100 endophytic fungi, three pathogens and 10 saprobes based on work in palms.[97] Some work suggests "scores" of endophytic fungal species per plant species or even individual plant of a species.[77] Somewhat mid-range in these estimates is the work of Vandenkoornhuyse *et al.*,[56] who used molecular tools to determine affinities of 49 fungal DNA isolates from the grass *Arrhenatherum elatius*.

There are an estimated 150 species of AM fungi and 6000 species of ectomycorrhizal fungi.[34] An estimation of the number of endophytic fungi range from a few hundred species to 1 million unique species.[9,11] If the estimate of the ratio of endophyte-to-other fungal species (88%) holds for plants other than palms, the number of endophytic species may be much higher. Strain diversity gives an even larger estimate of the number of taxa which are potentially biochemically distinct. Twenty-one different *P. fortinii* strains were isolated from one 9 m^2 plot, and a study of twelve 160 m^2 plots gave rise to 48 distinct strains of this endophyte.[98,99] Furthermore, this strain variability appears to remain stable over time.[93] Given that many endophytic species are represented in a single plant from one to 20 or more strains,[43,91,100–102] and each of these strains has a unique molecular and biochemical signature, the potential for interesting and novel chemistries is enormous. Endophytic fungi recently were found in plants in an aquatic habitat,[103] where mycorrhizal and endophytic fungi were thought to be absent. Thus, all plants in every habitat should be expected to harbor one or more endophytic fungal species.

15.2.7 Bioactive Natural Products from Past Literature

As stated previously, the products of endophytic fungi have been the subject of previous reviews. The compounds addressed previously are summarized in Table 1. Clearly, the compounds are chemically diverse and display a wide range of biological activities.

15.3 COMPOUNDS FROM THE RECENT LITERATURE

15.3.1 Gymnosperms and Primitive Plants

Fungal associations with the "lower" land plants, including mosses, liverworts, horsetails, lycopods, ferns, and similar plants have been observed.[70,104,105] These associations are similar to the endomycorrhizal, ectomycorrhizal and endophytic associations found in the more recently evolved gymnosperms and angiosperms. Thus far, there has not been a search for the natural products associated with these fungal symbionts; it is

Table 1. Bioactive fungal products from endophytic fungi in recent literature reviews. The compounds include antifungal, antibacterial, antimycobacterial, antiviral, nematicidal, cytotoxic, antineoplastic, antioxidant, antiinsectant, antifeedant, antidiabetic, herbicidal, algicidal, specific enzyme inhibitory, and immunomodulatory activities.

Plant Species	Fungal Species	Compound(s)	Reference
		Rhodophyta	
Sesquiterpenes (Botryanes)			
Polysiphonia sp.	*Geniculosporium* sp.	7-OH-10-oxodehydro-; 7,10-diOHdehydro-; 7-OH-10-methoxydehydro-; 7-OH-10-ethoxydehydro-; 7-OH-10-dehydroxydehydro-dihydrobotrydial; 7-OH-deacetylbotryenalol; 7-10-diOH-deacetyldihydrobotridial-1(10)-ene; 4,10-dideOH-7-OH-deacetyldihydrobotrydial-1(10),5(9)-diene; 7-OH-10-deOHdeacetyldihydrobotrydial-1(10),5(9)-diene; 15 α-OH-14-aldehyde probotryan-4(5)-ene	246
Cytochalasins			
Polysiphonia sp.	*Geniculosporium* sp.	cytochalasin H and U, 18-deoxycytochalasin H (L-696,474), RKS 1778	246
		Lichens	
Lactones			
Graphis prunicola, G. cognata, G. scripta	Unidentified mycobiont	Alternariol monomethyl ether, graphislactone A	246

		Mosses, Ferns and Allies	
Polyketides			
Selaginella pallescens	*Fusarium* sp.	CR377 (pentaketide)	198, 203
		Gymnosperms	
Aliphatics and Unclassified			
Larix laricina	Unidentified	2-methyloctanoic acid 6-oxo-2-propenyl-3,6-dihydro-2H-pyran-3-yl ester	67, 203, 246
Sequoia sempervirens	*Aspergillus parasiticus*	sequoiatone A-F	203
Amine/Amide Alkaloids			
Prumnopitys andina	*Penicillium janczewskii*	peniprequinolone, gliovictin, gliovictin acetate	67
Taxus mairei, Torreya grandis	*Aspergillus clavatus, Paecilomyces* sp.	brefeldin A	198
Coumarins			
Picea abies, Pinus sylvestris	*Pezicula cinnamomea, P. livida, Pezicula* sp.	mellein	67, 203
Picea glauca	Unidentified	sescandelin, sescandelin B, 5,6,8-triOH-4-(1′-OHethyl)-isocoumarin	67
Picea mariana	*Canoplea elegantula*	(3*R*,4*S*,4α*R*)- & (3*R*,4*S*,4α*R*)-4,8-diOH-3-methyl-3,4,4α,5-tetrahydro-; (3*R*,4α*S*,8*S*,8α*R*)-8-OH-3-methyl-3,4,4α,5,6,7,8,8α-octahydro-; (3*R*,4*S*,6*R*)-3,4,4α,5,6,7-hexahydro-4,8-diOH-3-methyl-; (3*R*,6*R*)-3,4,4α,5,6,7-hexahydro-6,8-diOH-3-methyl-; (3*R*,4*R*,4α*R*,6*R*)-4,8-diOH-6,7-epoxy-3,4,4α,5,6,7-hexahydro-; (3*R*,4*S*)-3,4-dihydro-4,8-diOH-3-methyl-1H-2-benzopyran-1-one; & 4-OH-3-methyl-2-oxabicyclo [3.3.1]non-6-one	67

Table 1. (Continued)

Plant Species	Fungal Species	Compound(s)	Reference
Prumnopitys andina	Unidentified	mellein	67
Taxus brevifolia	*Botrytis* sp.	ramulosin, 6-OH-ramulosin, 8-dihydroramulosin	203
Diterpenoids			
Abies balsaminea	Unidentified	9-α-OH-1,8(14),15-isopimaratrien-3,7,11-trione and -3,11-dione	67
Taxus cuspidata	*Periconia* sp.	periconicin A and B	198
Taxus spp., *Taxodium distichum, Torreya taxifolia, Wollemia nobilis*	*Taxomyces andreanae, Pestalotiopsis microcarpa, P. guepini, Periconia* sp., *Seimatoantlerium nepalense, Sporormia minima, Trichothecium* sp., *Tubercularia* sp.	taxol	67
Lactones			
Ephedra fasciculatus	*Chaetomium chiversii*	monocillin I, radicicol	67, 246
Picea glauca	Unidentified	6,7-diOH-2-propyl-2,4-octadien-4-olide	67
Pilgerodendron uviferum	*Microsphaeropsis olivacea*	graphislactone A, botrallin	67
Sequoia sempervirens	*Aspergillus parasiticus*	sequoiamonascin A-D	203
Monoterpenes			
Taxus brevifolia	*Pestalotiopsis* spp.	(4*S**,5*R**)-(6*Z*,8*E*)-4,5-diOH-6-OHmethyl-6,8-decadiene; (4*S**,5*R**)-(2*Z*,6*Z*,8*E*)-4,5-diOH-6-OHmethyl-2,6,8-decatriene; (4*S**,5*R**)-(2*Z*,6*Z*,8*E*)-4,5-diOH-6-oxomethyl-2,6,8-decatriene	203

Peptides			
Pinus sylvestris	*Cryptosporiopsis* sp.	echinocandin A, B, D, H	203
Taxus baccata	*Acremonium* sp.	leucinostatin A	198
Pyrones			
Torreya taxifolia	*Pestalotiopsis microcarpa*	pestalopyrone, OH-pestalopyrone	203
Quinones			
Abies balsaminea	*Hormonema dematioides*	rugulosin	203
Larix laricina	Unidentified	8,1′, 5′-triOH-3′, 4′-dihydro-1′H-[2,4′]binaphthalenyl-1, 4, 2′-trione	203
Taxus wallachiana	*Phoma* sp.	altersolanol A	198
Torreya taxifolia	*Pestalotiopsis microcarpa*	torreyanic acid	67, 246
T. taxifolia	*Pestalotiopsis* spp., *Monochaetia* sp.	ambuic acid	203
Unidentified?	*Coniothyrium* sp.	preussomerin N_1; palmarumycin $CP_{4\alpha}$ and CP5	203
Phenolics			
Ephedra fasciculata	*Chaetomium globosum*	orsellinic acid, globosumone A-C	203
Taxus wallachiana	*Phoma* sp.	2-OH-6-methylbenzoic acid	198
Torreya taxifolia	Pestalotiopsis microcarpa	pestaloside	246
Sesquiterpenes			
Abies balsamea	*Phyllosticta* sp.	hydro- and heptelidic acid, heptelidic acid chlorhydrin	203
Picea abies, Pinus sylvestris	*Pezicula livida*	(−)-mycorrhizin	198, 203
Taxus brevifolia	*Pestalotiopsis microcarpa*	pestalotipsins A and B; 2α-OH-dimeninol; 5-OH-10-(OH-methyl)-3,7,7-trimethyl-2,9-cycloundecadiene-1,6-dione (humulane derivative)	203

Table 1. (Continued)

Plant Species	Fungal Species	Compound(s)	Reference
Sterols			
Taxus chinensis	*Gliocladium* sp.	(20*S*,22*S*)-4a-homo-22-OH-4-oxoergasta-7,24(28)-dien-3-one	67
Triterpenoids			
Juniperus communis	*Hormonema* sp.	enfumafungin	67, 246
		Dicotyledonous Plants	
Aliphatics and Unclassified			
Adenocarpus foliolosus	*Phomopsis* sp.	phomosine D-G, 6-isopropylcyclohex-1-enecarboxylic acid, (1a*S*,3*R*,4*R*,4a*R*,y*S*,7*R*,8a*S*)-7-chloro-3,6-diOH-3,4a,8,8-tetramethyloctahydro-1a*H*-naphtho[1, *b*]oxirene-4-carboxylic acid	246
Artemisia annua	*Paraphaeosphaeria nolinae*	paranolin	198
Betula pendula, B. pubescens, unidentified tree	*Melanconium betulinum, Phomopsis phaseoli*	3-OH-propionic acid	67
Cinnamomum zeylanicum, Grevillea pteridifolia	*Muscodor albus, M. roseus*	fungitoxic mixture of volatile acids, esters, alcohols, lipids, & ketones	67, 198
Eucryphia cordifolia	*Gliocladium* sp. C-13	fungitoxic mixture of volatile acids, esters, alcohols, and [8]-annulene	203
Fagus sylvatica	*Pezicula livida*	(+)-cryptosporiopsin, 4-*epi*-ethiosolide	246
Melia azedarach	*Aspergillus aculeatus*	fonsecinone A, aurasperone A	203
Paullina paullinioides	*Muscodor vitigenus*	naphthalene	67
Salix spp.	*Phomopsis* spp.	phomodiol, phomopsolide B	203, 246

Amine/Amide Alkaloids			
Quercus variabilis	*Fusarium* sp.	fusaruside, cerebroside D	246
Avicennia marina	Unidentified	[2′, 3′-diOH-tetracosanolyamino]-1,3-diOH-octadecane and [2′,3′-diOH-docosanoylamino]-1,3-diOH-octadecane	198
Bisanthrones			
Conocarpus erecta	*Cytospora* sp.	cytoskyrin A and B	246
Coumarins			
Acer pseudoplatanus, Alnus glutinosa, Betula pendula, Carpinus betulus, Fagus sylvatica, Fraxinus excelsior, Quercus robur, unidentified sp.	*Pezicula* spp.	mellein, (−)-mellein	67
Citrus sinensis	*Aspergillus niger*	orlandin	246
Crassocephalum crepidioides	*Geotrichum* sp.	7-butyl-6,8-diOH-3(*R*)-pent-11-enylisochroman-1-one, 7-but-15-enyl-6,8-diOH-3(*R*)-pent-11-enylisochroman-1-one & 7-butyl-6,8-diOH-3(*R*)-pentylisochroman-1-one	67
Cyclohexenones			
Artemisia annua	*Leptosphaeria* sp.	leptosphaerone	67
Fragraea bodenii, Quercus sp.	*Pestalotiopsis jesteri*	jesterone, OH-jesterone	198, 203
Various tropical species	*Pestalotiopsis microcarpa, Monochaetia* sp.	ambuic acid	67, 203, 246
Cytochalasins			
Salix gracilostyla var.*melanostachys*	*Phomopsis* MF6031	phomopsichalasin	246
Maytenus hookeri	*Chaetonium globosum*	chaetoglobosin A	198, 203
Tripterygium wilfordii	*Rhinocladiella* sp. 309	22-oxa-[12]-cytochalasin 1,2 and 3; cytochalasin E	67, 246

Table 1. (Continued)

Plant Species	Fungal Species	Compound(s)	Reference
Depsides			
Artemisia annua, A. mongolica	*Colletotrichum gloeosporioides*	colletotric acid	198, 203
Quercus sp.	*Cytonaema* F32027	cytonic acid A and B	67
Diterpenoids			
Corylus avellana cv Gasaway, *Maguireothamnus speciosus*	Unidentified, *Se. tepuiense*	taxol, taxanes	67, 203, 246
Daphnopsis americana	Unidentified	guanacastepene A-O	198, 203
Tripterygium wilfordii	*Fusarium subglutinans*	subglutinol A and B	203
Indoles			
Artemisia annua	*Colletotrichum* sp.	6-isoprenylindole-3-carboxylic acid	203
Artemisia annua	*Colletotrichum* spp.	indole-3-acetic acid (IAA), 6-isoprenylindole-3-carboxylic acid	67, 198
Bontia daphnoides	*Nodulisporium* sp.	nodulisporic acid A, A_1 and A_2	203
Cavendishia pubescens	*Phomopsis* sp.	paspalitrem A and C	67, 246
Catharanthus roseus	*Fusarium oxysporum*	vincristine	67, 203
Nicotiana tabacum	*Hypoxylon serpens*	IAA, indole-3-acetonitrile, cytokinins	198, 203
Indolylquinones			
Unidentified	*Pseudomasseria* sp.	L-783,281 (=demethylasterriquinone B1), asterriquinones	203, 246

Lactones			
Buxus sempervirens	*Microsphaeropsis* sp.	lactone S	246
Cistus salviifolius	*Galiella rufa*, unidentified ascomycete spp.	(−)-galiellalactone; (−)-pregaliellalactone; 5(*E*)-buta-1,3-dienyl-3(*E*)-propenyl-5H-furan-2-one; 5(*E*)-but-2-enylidene-3-propyl-5H-furan-2-one; 5(*E*)-but-2-enylidene-3(*E*)-propenyl-5H-furan-2-one; (4*S*,5a*R*,7a*R*,7b*R*)-5,5a,6,7,7a,7b-hexahydro-4-methyl-indeno[1,7-bc]furan-2(4H)-one; (3*R*,5*R*)-5-(3-butenyl)dihydro-3-(1*E*)-1-propenyl-2(3H)-furanone	67, 246
Erythrina crista-galli	*Phomopsis* sp.	phomol	246
Ficus microcarpa	Unidentified	microcarpalide	67
Kandelia candel	Unidentified No. 1893	1893 A and B	67, 246
Malus X domestica	*Nectria galligena*	α, β-dehydrocurvularin	67
Mentha arvensis	*Fusidium* sp.	fusidilactone A-C	67, 246
Trachelospermum jasminoides	*Cephalosporium acremonium*	graphislactone A, G and H; alternariol monomethyl ether	67
Lignans			
Diphylleia sinensis	*Penicillium implicatum*	podophyllotoxin (tentative)	67
Peptides			
Avicennia marina	Unidentified	cyclo-(L-Phe-L-Leu-L-Leu-L-Leu-L-Ile)	203
Fagus sylvatica	*Pezicula* sp.	echinocandin A, B, D, H	198, 203
Tripterygium wilfordii	*Cryptosporiopsis* cf. *quercina* (= *Pezicula cinnamomea*)	cryptocandin	198, 203

Table 1. (Continued)

Plant Species	Fungal Species	Compound(s)	Reference
Phenolics			
Aegiceras corniculatum	*Dothiorella* sp.	cytosporone B, dothiorelone A-D	67
Forsteronia spicata, Conocarpus erecta	*Cytospora* sp., *Diaporthe* sp.	cytosporone A-E	67, 246
Malus X *domestica*	*Nectria galligena*	colletorin B, colletochlorin B, ilicicolin C, E and F	67, 198, 203, 246
Mentha arvensis	*Fusidium* sp.	*cis*-4-OH-6-deoxyscytalone	67, 246
Terminalia morobensis	*Pestalotiopsis microspora*	pestacin, isopestacin	203
Unidentified	*Pezicula* sp.	2-methoxy-4-hydroxy-6-methoxymethylbenzaldehyde	67, 198
Phenylpropanoids			
Spondias mombin	*Guignardia* sp.	(−)-(*S*)-guignardic acid	198, 203
Quinones			
Atropa belladonna	Unidentified *mycelia sterilia*	preussomerin G-L	246
Avicennia sp.	Unidentified No. 1403	anthracenedione, 1,4-diOH-2-methoxy-6-methyl-9,10-anthracenedione, 5,6,7,8-tetrahydro-6,7,10-triOH-2-methoxy-6-methyl-1,4-anthracenedione, 5,6,7,8-tetrahydro-6,7,8,9,10-pentaOH-2-methoxy-6-methyl-1,4-anthracenedione	246
Unidentified	*Penicillium chrysogenum*	xanthoviridicatin E and F	67, 246

Sesquiterpenes			
Artemisia annua	*Leptosphaeria* sp.	leptosphaeric acid	246
Artemisia annua, Trachelospermum jasminoides	*Myrothecium roridum*	myrothecine A-C	246
Baccharis cordifolia	*Certopicnidium baccharidicola*	rodicin A, D, E and H, verrucarin A and J	203, 246
Phellinus robiniae on *Robinia pseudoacacia*	*Chaetomella acutisea* (endophytic on fungal endophyte)	chaetomellic acid A and B	198, 203
Sesquiterpene Benzofurans			
Gaultheria procumbens	Unidentified endophyte	5-OH-2-(1′-OH- & 5-OH-2-(1′-oxo-5′-methyl-4′-hexenyl)benzofuran	67
Sesquiterpene quinones			
Acer pseudoplatanus, Carpinus betulus, Fagus sylvatica	*Pezicula* spp.	mycorrhizin A	67, 246
Spiroquinazolines			
Murraya paniculata	*Eupenicillium* sp.	alanditrypinone A and B, alantryphenone, alantrypinene, alantryleunone	246
Sterols			
Artemisia annua	*Colletotrichum* sp.	3β-OH-ergosta-5-ene; 3-oxo-ergosta-4,6,8(14),22-tetraene; 3β-OH-5α,8α-epidioxy-ergosta-6,22-diene; 3β,5α-diOH-6β-acetoxy-ergosta7,22-diene; 3β,5α-diOH-6β-phenylacetyloxy-ergosta-7,22-diene	67

Table 1. (Continued)

Plant Species	Fungal Species	Compound(s)	Reference
Tetramic acids			
Erigeron annuus	*Phoma* sp.	TAN-1813	246
Euonymus europaeus, Vinca minor	*Alternaria* spp.	altersetin, alternariol, alternariol monomethylether, tenuazonic acid	198, 203, 246
Tripterygium wilfordii	*Cryptosporiopsis* cf. *quercina*	cryptocin	67, 198
Xanthones			
Opuntia leptocaulis	*Chaetomium globosum*	globosuxanthone A-C, OHvertixanthone	67
		Monocotyledonous Plants	
Aliphatics and Unclassified			
Cynodon dactylon	*Aspergillus niger, Aspergillus* sp., *Rhizoctonia* sp.	rubrofusarin B, fonsecinone A, asperpyrone B, aurasperone A, rhizoctonic acid, monomethylsulochrin	203
Phleum pratense	*Epichloë typhina*	ethyl *trans*-9,10-epoxy-11-oxoundecanoate, ethyl 9-oxononanoate, ethyl azelate, hydroxydihydrobovolide, gamahonolide A & B	203
Amine/Amide Alkaloids			
Festuca spp., *Lolium perenne*, other grasses	*Acremonium* spp. (=*Neotyphodium*, anamorph) (*Epichloë*, teleomorph)	peramine; ergobalansine; ergotamine; ergosine; β-ergosine; ergovaline; ergostine; ergoptine; β-ergoptine; ergonine; ergocristine; α-ergocryptine; β-ergocryptine; ergocornine; ergonovine; lysergamide; 8-OH-lysergamide; isolysergamide	67, 246

Cynodon dactylon	*Aspergillus fumigatus, A. niger, Cladosporium herbarum*	asperfumoid, asperfumin, aspernigerin, aspernigrin A	67
Zea maydis	*Acremonium zeae*	pyrrocidine A and B	203
Benzofuranones			
Phleum pratense	*Epichloë typhina*	5-OH-4-phenylfuran-2(5H)-one	67, 246
Coumarins			
Cynodon dactylon	*Cladosporium herbarum*	orlandin, kotanin, 7-OH-4-methoxy-5-methylcoumarin	203
Phleum pratense	*Epichloë typhina*	gamahorin	203
Cytochalasins			
Imperata cylindrica	*Chaetonium globosum*	chaetoglobosin U, chaetoglsin C, E and F, penochalasin	203, 246
Flavonoids			
Poa ampla	*Neotyphodium typhnium*	tricin, 7-O-(β-D-glucopyranosyl)- and 7-O-[α-L-rhamnopyranosyl(1-6)-β-D-glucopyranosyl]-tricin; isoorientin	203
Indoles			
Andropogon virginicus, Sporobolus poiretii, Stipa robusta	*Acremonium* sp., *Balansia* sp.	chanoclavine, agroclavine, elymoclavine	67, 246
Cynodon dactylon	*Aspergillus fumigatus*	fumigaclavine; fumitremorgin lolilline;	203, 246
Lolium perenne, Festuca spp.	*Neotyphodium lolii, Epichloë festucae*	paxilline; lolitrem A, B and E; terpendole C	203, 246
Lolium perenne	*Neotyphodium lolii*	lolitrem C, F and N, lolitriol, lolicine A and B	203

Table 1. (Continued)

Plant Species	Fungal Species	Compound(s)	Reference
Festuca arundinacea	*Acremonium coenophialum, Aureobasidium pullulans*	IAA	246
Lignans			
Phleum pratense	*Epichloë typhina*	chokorin	67, 246
Peptides			
Cynodon dactylon	*Aspergillus fumigatus*	cyclo(Ala-Leu), cyclo(Ala-Ile)	203
Phenolics			
Phleum pratense	*Epichloë typhina*	*p*-hydroxybenzoic acid, *p*-hydroxyphenylacetic acid, tyrosol, *cis*- and *trans-p*-coumaric acid	246
Phenylpropanoids			
Phleum pratense	*Epichloë typhina*	1,3- and 1,2-O-di-trans-p-coumaroylglycerol	203
Pyrrolizidines			
Festuca arundinacea, F. pratensis, F. argentina	*Acremonium* sp., *Neotyphodium coenophialum, N. uncinatum*	lolinine, nor-, N-methyl-, N-formylnor-, N-acetylnor-, N-formyl-, N-acetyl-, & 5,6-dehydro-N-acetyl-loline	67, 246
Sesquiterpenes			
Phleum pratense	*Epichloë typhina*	chokol A-G	67

Quinones			
Imperata cylindrica	*Pleospora* sp.	7-methoxy-2-methyl-3,4,5-triOH-anthraquinone, physcion, macrosporin, deoxybostrycin, altersolanol B, dactylariol, pleospdione	67
Cynodon dactylon	*Aspergillus fumigatus*	physcion	203
Sterols			
Cynodon dactylon	*Aspergillus fumigatus*, *Aspergillus* sp.	3β-OH-ergosta-4,22-diene; ergosterol; 3β-OH-5α,8α-*epi*-dioxyergosta-6,22-diene; helvolic acid; 3β,5α,6β-triOHergosta-7,22-diene	203

not yet known whether the compounds produced are similar and whether they may confer some of the same benefits seen in the "higher" plants.

Molecular techniques were used to isolate endophytic fungi from *Fucus serratus*.[106] Although technically macrophytic, multicellular protists, the brown and red macroalgae have a form, habit and niche similar to marine plants. The same can be said of the macroalgae of the Chlorophyta,

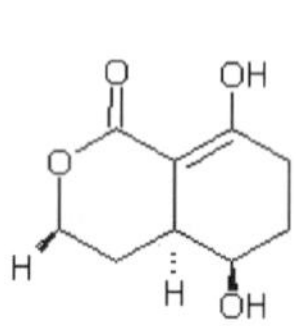

1

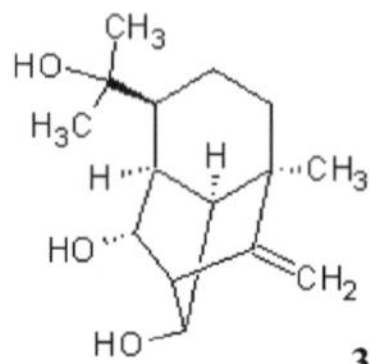

2

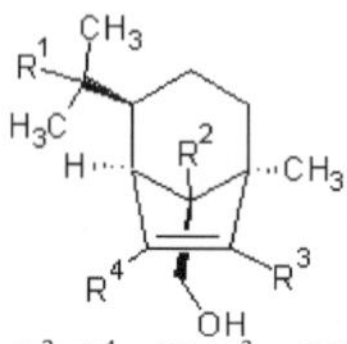

3 R^1, R^2, R^4 = H, R^3 = CH_2OH
4 R^1, R^2 = H, R^3 = CH_3, R^4 = CHO
5 R^1 = OH, R^2 = H, R^3 = CH_3, R^4 = CHO

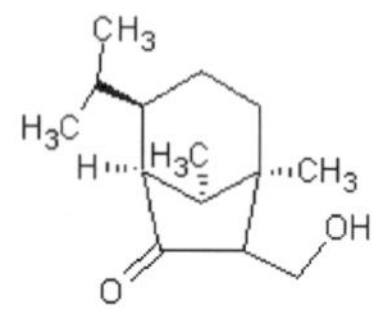

6

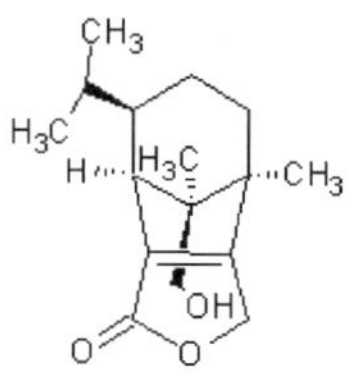

7

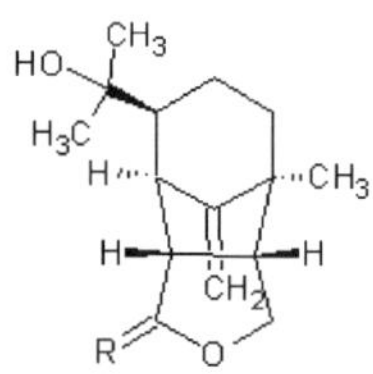

8 R = O
9 R = α H, ß OCH_3

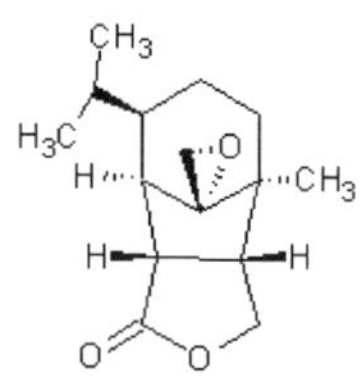

10

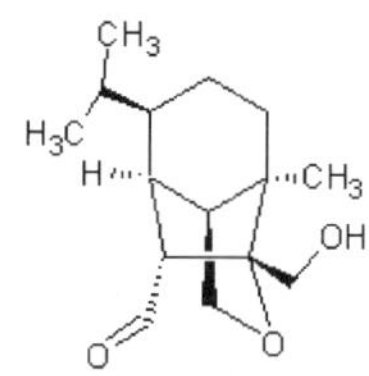

11

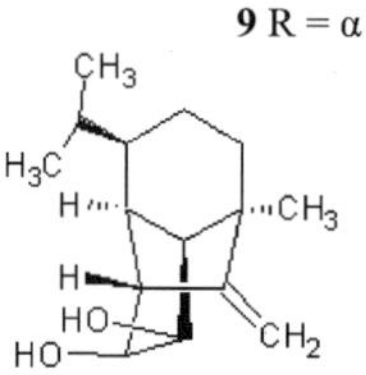

12

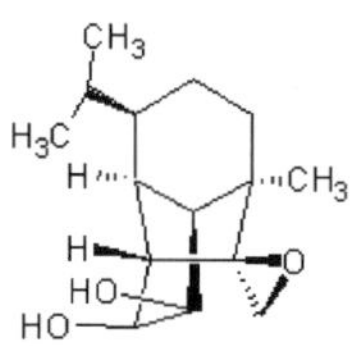

13

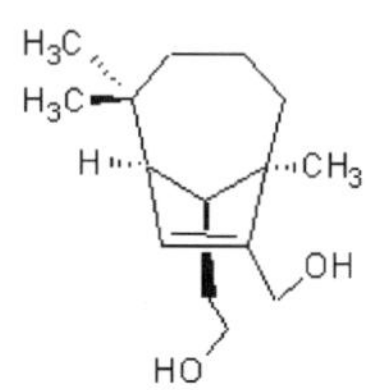

14

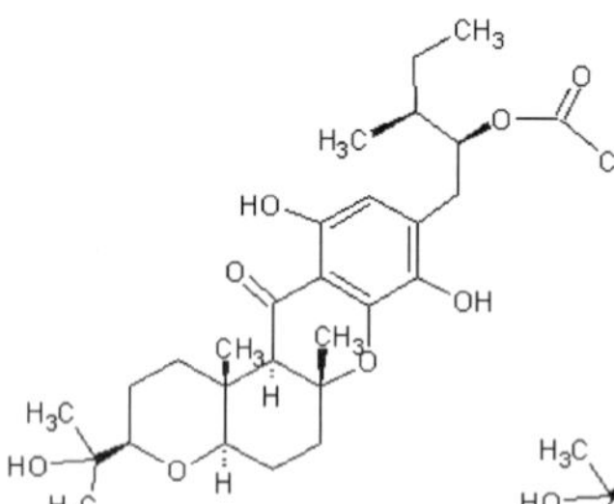

15

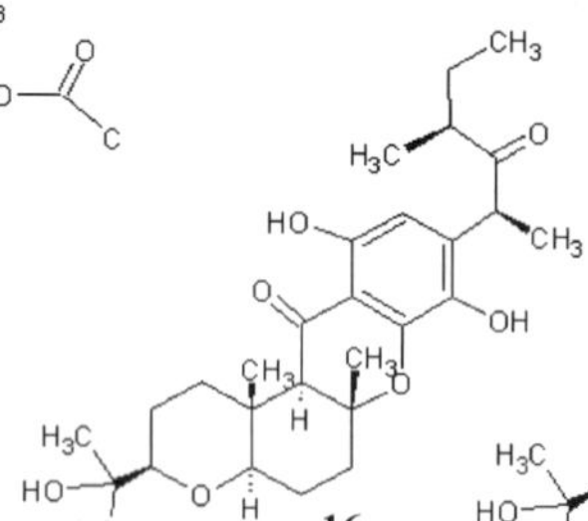

16

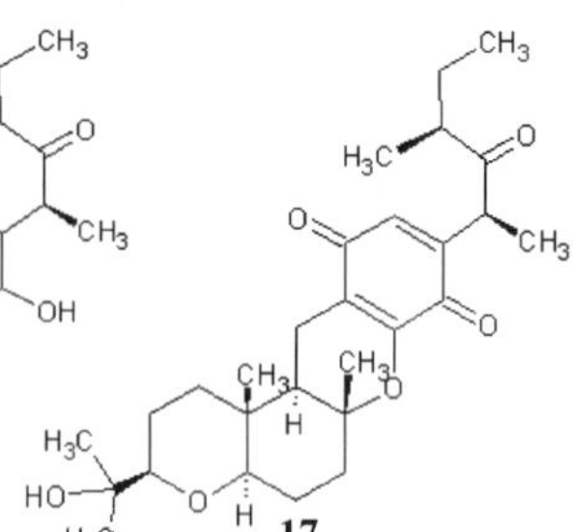

17

although they are true plants. The isolation of the endophytic fungus *Phoma tropica* from the tissues of *F. spiralis* led to a novel isocoumarin product, 5-hydroxyramulosin (**1**).[107] This same group isolated the endophytic fungus *Drechslera dematioidea* from the tissues of *Liagora viscida*, a marine rhodophyte.[108] They isolated 10 new sesquiterpenoids [isosativenetriol (**2**), drechslerine A (**3**), 9-hydroxyhelminthosporol (**5**), drechslerine C (**6**), drechslerine B (**7**), drechslerine D-G (**8-11**), and sativene epoxide (**13**)] and six known compounds [helminthosporol (**4**), *cis*-sativenediol (**12**), seco-longifolene diol (**14**), isochlioquinone (**15**), isocochlioquinone C (**16**), and cochlioquinone B (**17**)]. Compounds **9** and **11** were found to have activity against strains of *Plasmodium falciparum*.

A search was undertaken to discover endophytic fungi in marine algae and plants in the regions of Göttingen and Braunschweig in Germany.[109] In all, 55 endophytic fungi were isolated from rhodophycean, phaeophycean

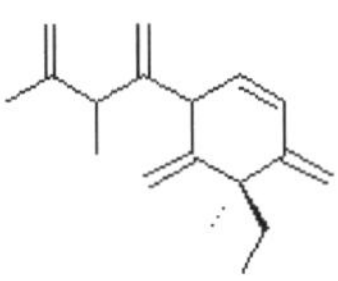

18

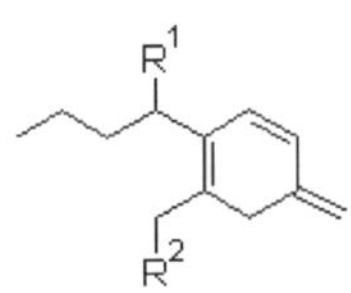

19 R^1 = $=CH_2$, R^2 = CH_3
20 R^1 = CH_3 (α), R^2 = CH_3
21 R^1 = $=CH_2$, R^2 = =

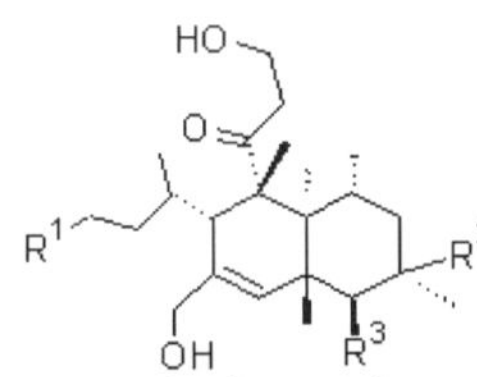

22 R^1, R^2 = H, R^3 = O
23 R^1, R^2, R^3 = H
24 R^1, R^3 = H, R^2 = O
25 R^1 = O R^2 R^3 = H

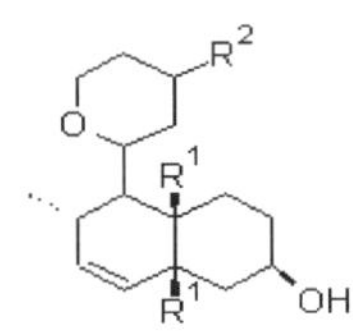

26 R^1 = H, R^2 = =O
27 R1 = CH3, R2 = NCH_2CHOH

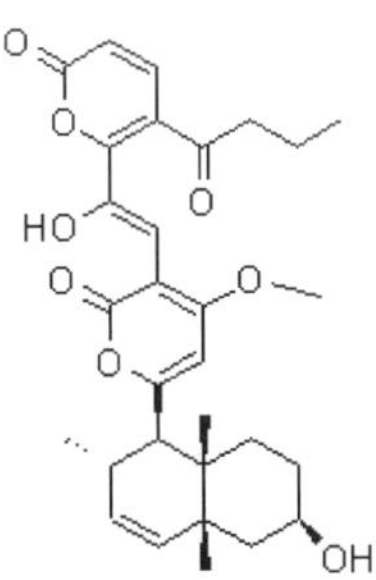

28

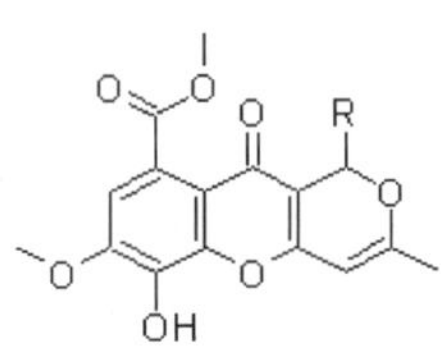

29 R = OCH_3
30 R = H

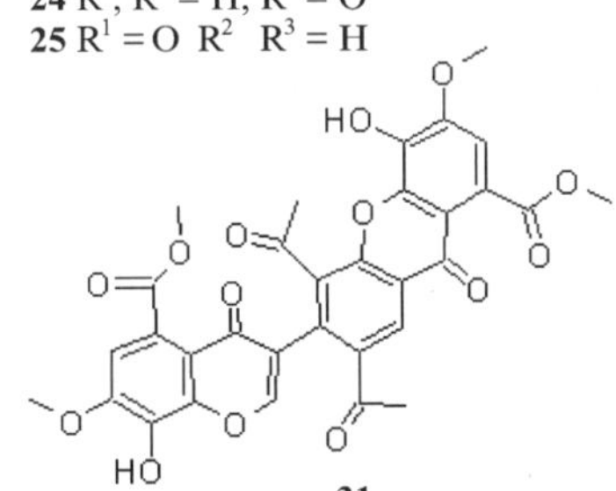

31

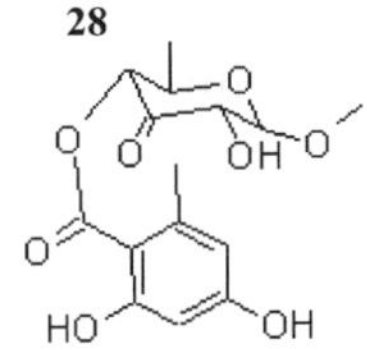

33

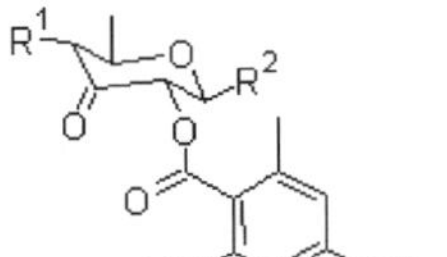

32 R^1 = OH, R^2 = OMe
34 R^1 = H, R^2 = OMe
35 R^1 = OH, R^2 = H
36 R^1, R^2 = H

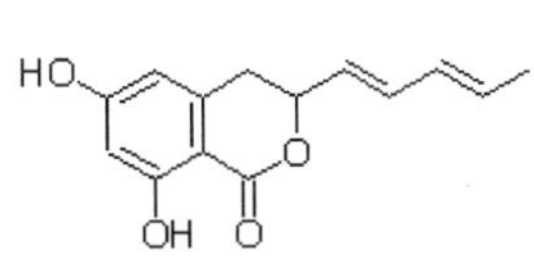

37

and chlorophycean macroscopic algae. The antifungal and antibacterial extracts of these cultures were characterized further. The culture of these fungi in high-salt media led to the isolation of 53 secondary metabolites, 20 of which were new endophyte products. Prospiciferone A (**18**), dehydrotaiwapyrone (**19**), 5,6-dihydrotaiwapyrone (**20**), 7-(1-methyl-ethylidene)-taiwapyrone (**21**), betaenones G-J (**22-25**), and solanapyrones H-J (**26-28**) were isolated from *Microsphaeropsis* sp. in the tissues of the rhodophyte *Ceramium* sp. Chaetocyclinones A-C (**29-31**) and orsellides A-E (**32-36**) were produced by *Chaetomium* sp. isolated from the rhodophyte *Gracilaria tikvahiae*. Arthrolactone (**37**) was isolated from *Aspergillus flavus* from the same rhodophyte species. Several biosynthetic pathways were postulated and some steps were confirmed by feeding with[67]C-acetate and[67]C-methionine. All of the new compounds are synthesized as polyketides. The side groups are derived from methionine, except for the orsellides, in which deoxyglucose is the origin of the modified sugar moieties. A number of the novel and known compounds had significant antifungal (chaetomin, KM-01) and antibacterial [**21**, **26**, **27**, **32-36**, orsellinic acid, 3, 5, 4′-triacetyl-orsellide A, 3, 5, 2′-triacetyl-orsellide B, 3, 5, 4′-triacetyl-orsellide D, (4-hydroxy-pent-2-on-1-yl)-orsellinate, (pent-3-en-2-on-1-yl)-orsellinate, chaetomin, cheatosiminudin, chaetomin B, fuscoatroside, bipolaroxin, dendryphiellin E, and KM-01] activities.[109,110] A number of natural and synthetic orsellinates are known to have cytotoxic[111] and antineoplastic[112] activities, and these new compounds should be investigated further.

Macroscopic chlorophycean algae also harbor endophytic fungi. *Ascochyta salicorniae* was isolated from *Ulva* sp. and cultured to produce two new tetramic acid derivatives and a novel sesquiterpene pyrone.[113] The tetramic acids were ascosalipyrrolidinone A (**38**) and B (**39**), and the other novel compound was ascosalipyrone (**40**). Compound **38** gave moderate activity against *Bacillus megaterium*; low activity against the fungi *Microbotryum violaceum* and *Mycotypha microspora*; inhibition of the tyrosine kinase p56^{lck} (lymphocytic kinase); and moderate-to-good activity against two *P. falciparum* strains. Two known compounds, genistein (**41**) and 2,3-dihydro-2-hydroxy-2,4-dimethyl-5-*trans*-propenyl-furan-3-one (**42**), both with slight antifungal, strong enzyme inhibitory, and slight antiplasmodial activities, were also isolated. Compound **42**, in addition, inhibited the growth of *M. violaceum, Eurotium repens, Trypanosoma brucei* ssp. *rhodisiense*, and *T. cruzi*. Osterhage[114] gives an interesting review of natural products previously isolated from algal endophytes and free-living marine fungi.

Osterhage's work[114] has led to further analysis of bioactive products from the marine fungus *A. salicorniae*. In addition to the novel tetramic acid derivatives (**38, 39**), this fungus also produces two new epimeric lactones, the cycloethers ascolactone A (**43**) and B (**44**), similar to known cephalosporide compounds, and ascochitine (**45**), ascochital (**46**), and halopyrone (**47**), which were previously known.[115] Compound **45** exhibited moderate inhibition of a tyrosine phosphatase B from *Mycobacterium tuberculosis*.

The fungal endophyte *Chaetomium globosum* was isolated from the rhodophyte *Polysiphonia urceolata* in the Yellow Sea coast off east central China.[116] The analysis of the fungus in culture led to moderately active free-radical-scavenging (anti-oxidant) benzaldehyde and anthraquinone derivatives. One benzaldehyde, chaetopyranin (**48**), was new, while the others, 2-(2′,3-epoxy-1′, 3′-heptadienyl)-6-hydroxy-5-(3-methyl-2-butenyl)benzaldehyde (**49**), isotetrahydroauroglaucin (**50**), and erythroglaucin (**51**), were previously known. Six other known compounds were isolated, but they exhibited no activity. Chaetopyranin (**48**) was toxic to human liver carcinoma (SMMC7721), lung carcinoma (A549) and breast cancer (HMEC) cell lines.

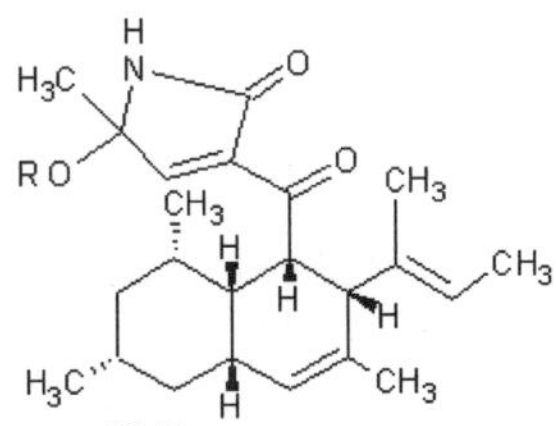

38 R = -$CH_2CH_2CH_2CH_3$
39 R = -CH_3

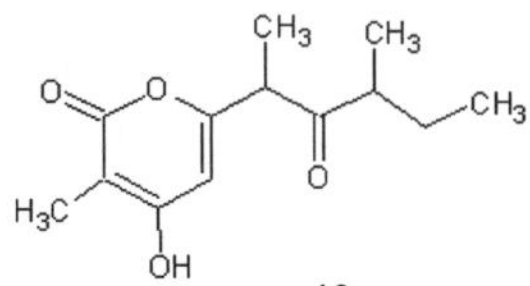

40

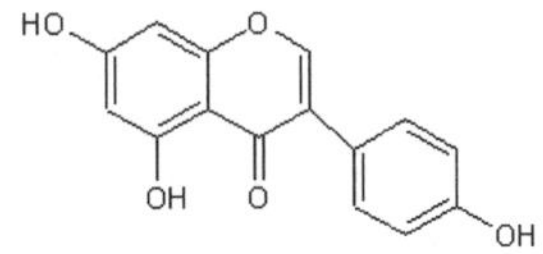

41

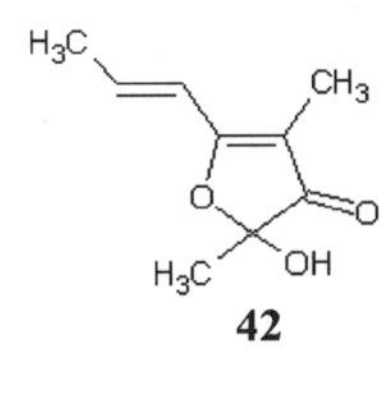

42

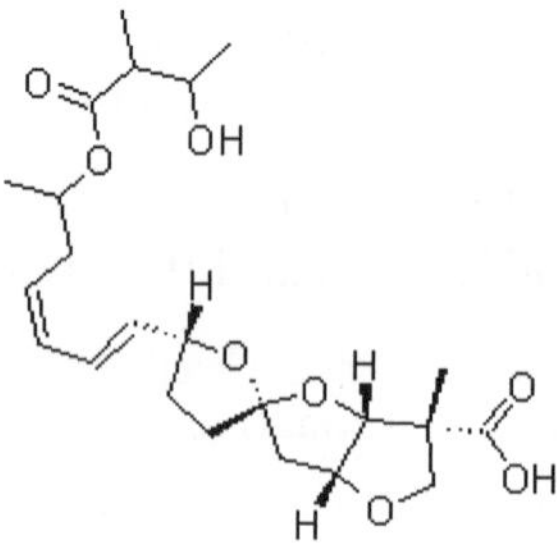

43

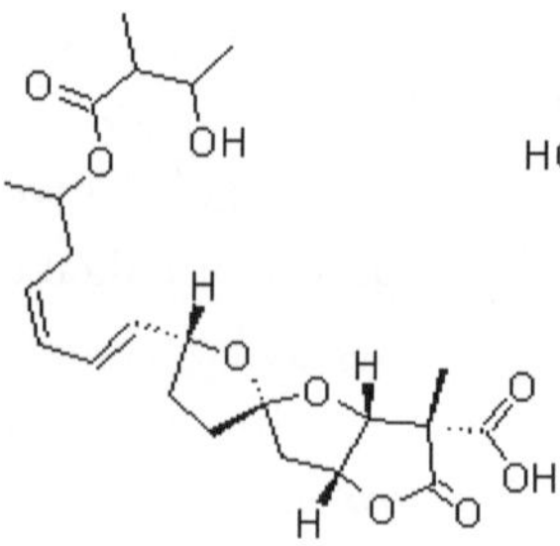

44

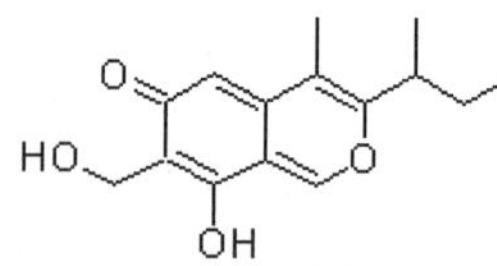

45

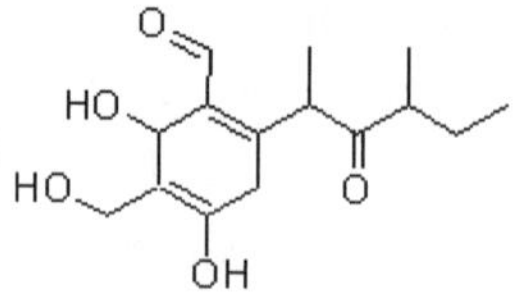

46

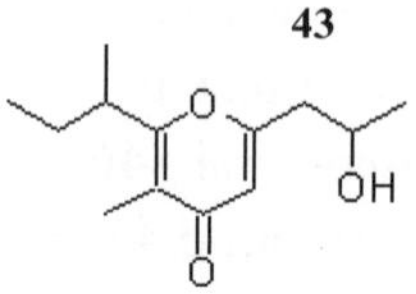

47

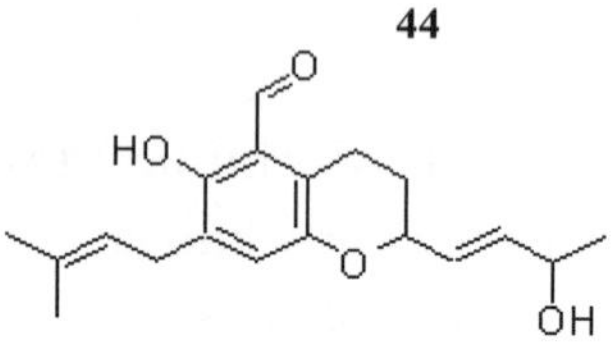

48

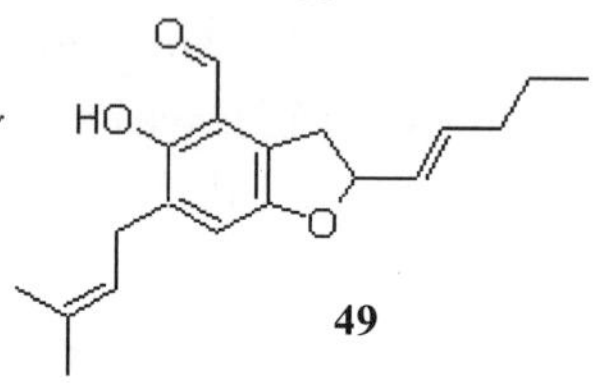

49

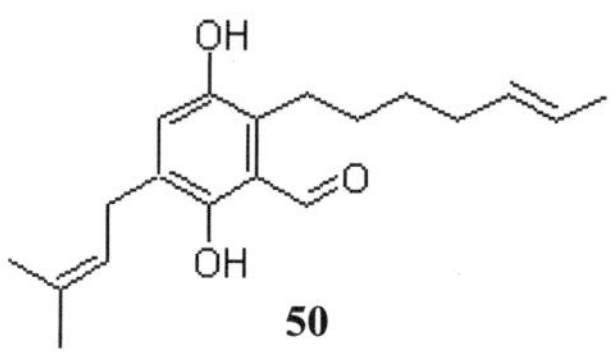

50

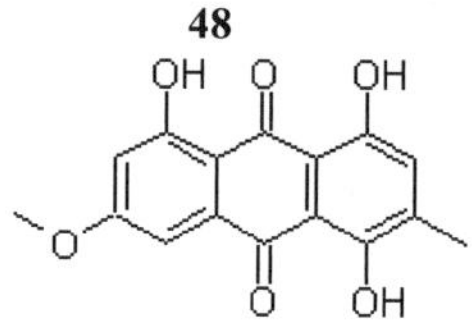

51

Also from the China seas, the phaeophyte *Sargassum* sp. yielded an unidentified fungal endophyte with antibacterial activity. The culture of the fungus (No. ZZF36) led to isolation of two new, 6-oxo-de-*O*-methyllasiodiplodin (**52**) and (*E*)-etheno-lasiodiplodin (**53**), and three known lactones, lasiodiplodin (**54**), de-*O*-methyllasiodiplodin (**55**), and 5-hydroxy-de-*O*-methyllasiodiplodin (**56**).[117] Compounds **54–56** exhibited low to moderate antibacterial and antifungal activities.

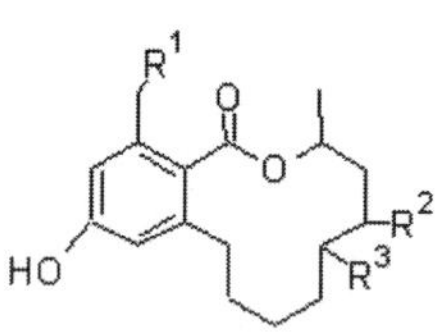

52 R^1, R^2 = H, R^3 = =O
54 R^1 = Me, R^2, R^3 = H
55 R^1, R^2, R^3 = H
56 R^1, R^3 = H, R^2 = OH

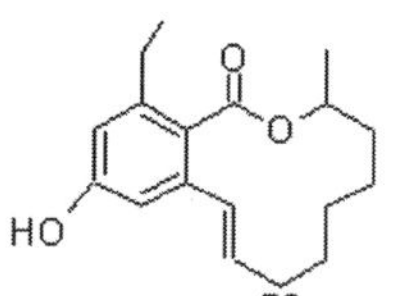

53

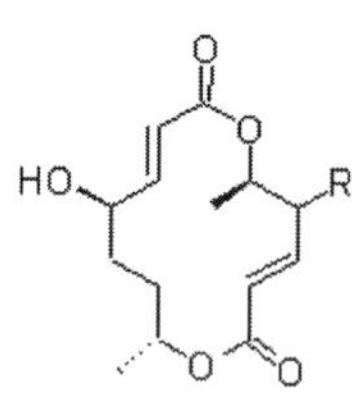

57 R = =O
58 R = OH

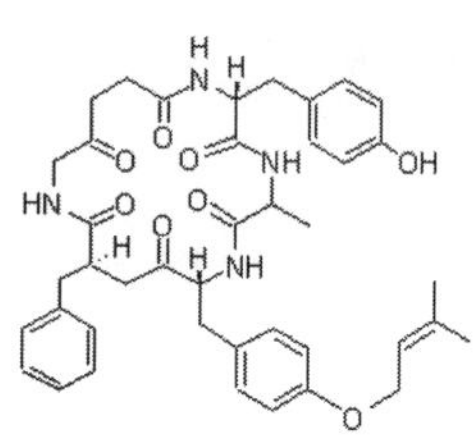

59

Cytotoxic (P388), antibacterial (*Bacillus subtilis*), and antifungal (*Trichophyton mentagrophytes, Cladosporium resinae*) compounds were derived from culture of the fungal endophyte *Gliocladium* sp. from the kelp *Durvillea antarctica*.[118] An HPLC-microtiter plate-bioassay-guided analysis led to two known macrocyclic lactones, 4-keto-clonostachydiol (**57**) and clonostachydiol (**58**), and a new cyclodepsipeptide, gliotide (**59**). Compound **57** was toxic to P388 leukemia cells ($IC_{50} = 0.55\,\mu M$) It was moderately active against the bacterium, and had low activity against the fungi. Two oxidized derivatives of **57** had IC_{50} values of 1.9 and 3.9 μM. Compound **58** had low activity against the leukemia line ($IC_{50} = 25\,mM$) and no antimicrobial activity. Gliotide (**59**) had no activity.

Nine polyketides, including four benzophenones, two of which were new, three known xanthones, and one known indole alkaloid were isolated from the fungus *Emericella nidulans* var.*acristata* isolated from an unidentified chlorophyte alga.[119] The two novel compounds were arugosins G (**60**) and H (**61**). All nine compounds were tested for antimicrobial and antitumor activities, the former against *Mycotypha microspora, Eurotium repens, Microbotryum violaceum, Chlorella fusca, Bacillus megaterium,* and *Escherichia coli*. Compound **61** had low activity against *M. microspora* and the microalga *C. fusca*. A known compound, sterigmatocystin (**62**), had moderate and low activity against *M. microspora* and *C. fusca*, respectively. The known alkaloid, emindole (**63**), gave moderate activity at 10 μg/mL against 92% of a 36-cell line screen including seven tumor types and a mean IC_{50} value of 5.5 μg/mL. Two other known compounds, the arugosins A (no isoprene at position 4) and B (no isoprene at position 2), gave low-to-moderate activity against seven of the cell lines screened. The benzophenone and xanthone compounds isolated were structurally similar, and it was proposed that they are biosynthetically related, arising from a chrysophanol anthrone and proceeding through prenylation, dehydration, reduction, and cyclization to the structures obtained.

Recently,[8,11] little work has been done with the endophytes of gymnosperms. The isolation of *Chaetomium chiversii* from *Ephedra fasciculata* in the Sonoran desert of southwestern USA, led to two new isocoumarins, chaetochiversins A (**64**) and B (**65**); three known isocoumarins eugenetin (**66**), 6-methoxy-methyleugenin (**67**), and 6-hydroxy-methyleugenin (**68**);

and the known macrocyclic isocoumarin derivative radicicol (**69**).[120] Radicicol (**69**) was previously reported as cytotoxic and inhibitory to Hsp90. The activities of the new compounds were not reported.

Sumarah *et al.*[121] developed an immunoassay for a rugulosin-producing endophytic fungus, 5WS22E1, in white spruce (*P. abies*). The fungus colonized 27% of the seedlings exposed. They determined that 90% of the endophytic fungi produced the anti-insectant, quinone dimer rugulosin (**70**). Park *et al.*[122] isolated the well-known antifungal compound griseofulvin (**71**) and the less-active dechlorogriseofulvin (**72**) from *Xylaria* sp. F0010 growing endophytically in *Abies holophylla*.

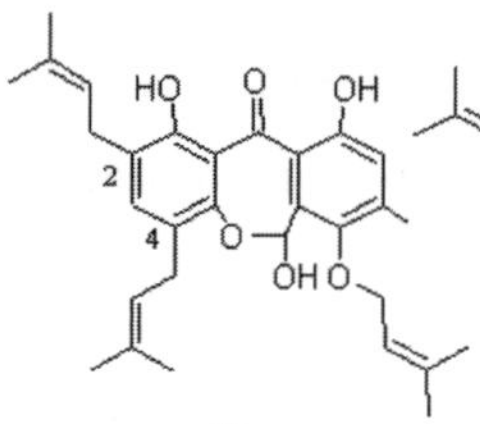

60

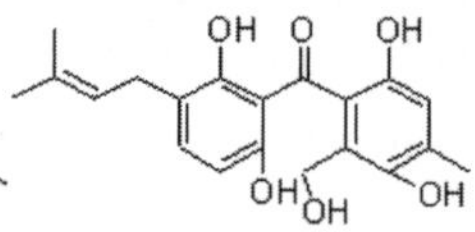

61

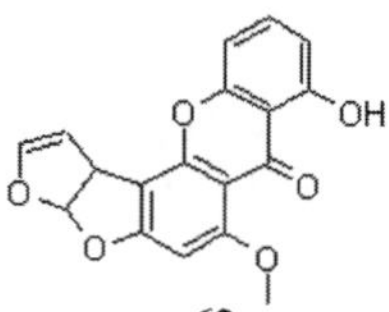

62

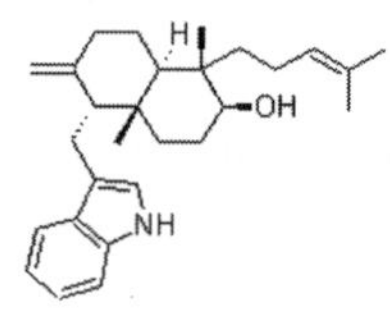

63

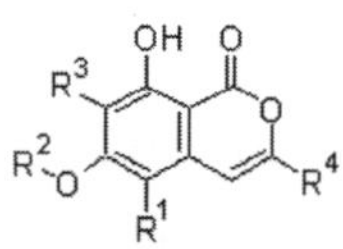

64 R^1 = Cl, R^2, R^3 = H, R^4 = 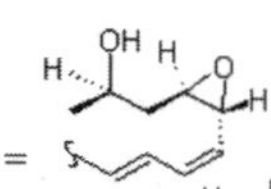

65 R^1 = Cl, R^2, R^3 = H, R^4 = 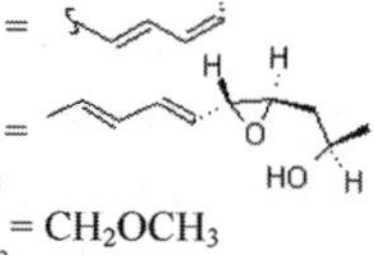

66 R^1 = H R^2, R^3, R^4 = Me
67 R^1 = H, R^2, R^4= Me, R^3 = CH_2OCH_3
68 R^1 = H, R^2, R^4 = Me, R^3 = CH_2OH

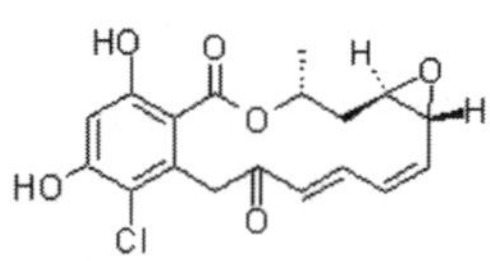

69

General mineral nutrition status improvement for the host plant has been studied frequently to understand the roles of endophytic fungi. Iron nutrition, however, has been studied only occasionally. *P. fortinii* strains obtained from *Pinus sylvestris, Abies alba, Picea abies*, and *Carex curvula* (the last a monocotyledonous plant) were found to produce the cyclic hexapeptide siderophores ferricrocin (**73**), ferrirubin (**74**) and ferrichrome C (**75**).[89] The concentration and pattern of siderophore production was dependent on ferric ion concentration, pH of the medium, and the strain of endophyte.

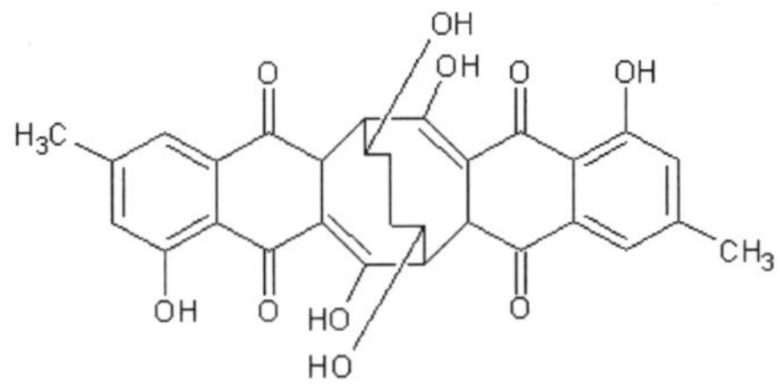

70

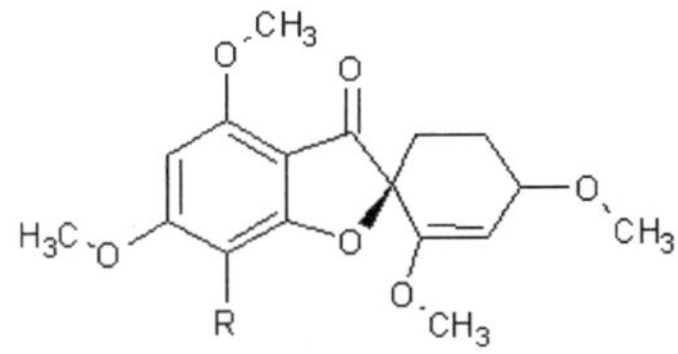

71 R = Cl
72 R = H

15.3.2 Secondary Metabolites from Mycorrhizal Fungi

The secondary metabolites of mycorrhizal fungi deserve special treatment because they are seldom reviewed, especially in the context of endophytic fungi. It is also because a majority of the compounds found have biological activity, in moderating the host defense response toward the mycorrhizal fungus, inhibiting plant pathogens and parasites (bacteria, fungi, nematodes), and in reaching a mutual accommodation between host plant and symbiotic fungus.[123] AM fungi act as enhancers of plant secondary metabolism (terpenoids, phenolics and alkaloids),[124–126] in addition to providing their own products to the plant tissues and rhizosphere. The majority of studies of mycorrhizal natural products have necessarily focused on the ectomycorrhizae, since these are more often culturable. The obligate endomycorrhizae, especially the VAM, AM and orchid mycorrhizae, have seldom been brought into pure culture. The products of these fungi must be studied by difference from the plant host with and without the symbiosis, although this is not always possible, and the producer is often not certain. Molecular techniques are beginning to lead to an elucidation of the biosynthetic capabilities of the host plant and the mycorrhizal

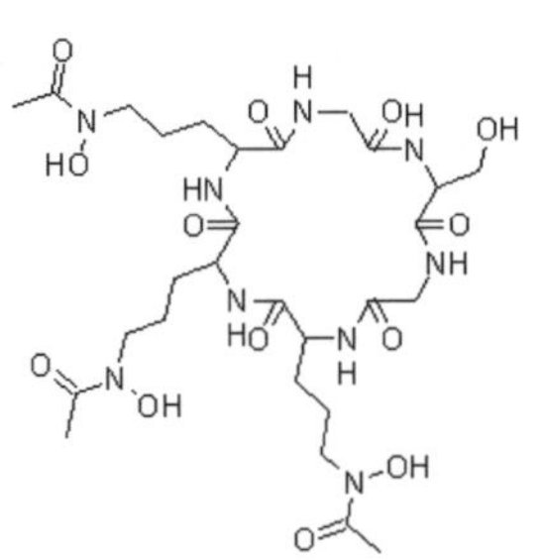

73

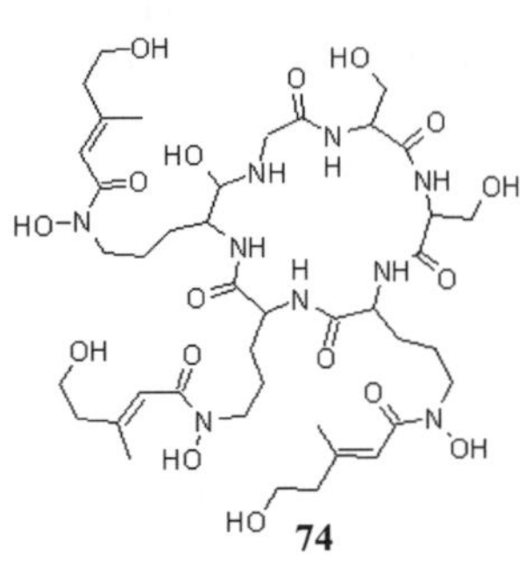

74

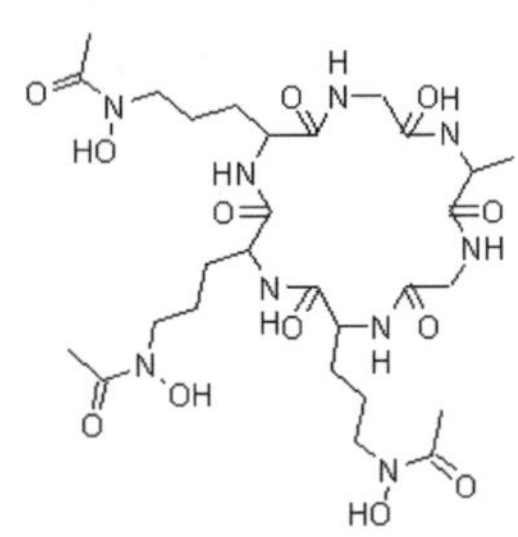

75

fungus;[127,128] the source of metabolites may be assigned unambiguously in the not too distant future. The communications between plant host and the developing symbiotic fungi will similarly be clarified as our understanding of biosynthesis progresses. There has been some study of the increases and decreases of plant secondary metabolites with the onset of mycorrhizal development,[123] but I will focus on the compounds attributable to the fungal component.

Ectomycorrhizal fungi in culture produce compounds with antibacterial or antifungal activities.[129,130] However, many such studies were based on bioassay alone, and little chemical work was done to determine the precise compounds responsible for the activity. Many of the activities reported were against bacteria or fungi that are known root pathogens, again supporting a significant role for these endophytes to reduce plant disease. Although a variety of mechanisms have been proposed for mycorrhizal antibiosis,[130–132] some may be attributable to antibiotic production. A large number of soil pathogens are known to be inhibited by mycorrhizae.[130] Ectomycorrhizae in culture and VAM *in planta* are capable of producing a variety of antifungal compound classes, including terpenes, flavonoids, phenolics, phytoalexins, and other compounds.[129,132]

Specific ectomycorrhizal fungal extracts have shown a wide range of activities, being able to act as broad spectrum antibiotics, act with some selectivity, or cause very specific inhibition of only certain pathogens. Three polyacetylenes, diatretyne nitrile (**76**), diatretyne 3 (**77**), and diatretyne amide (**78**), produced by *Leucopaxillus cerealis* var. *piceina* and *Clitocybe diatreta* in mycorrhizal association with *Pinus echinata*,[129,133] were among the first compounds determined to be of fungal origin in a mycorrhizal relationship and which inhibited pathogenic fungi. Similar to the recent observations of the ability of *M. albus* to produce volatile inhibitory compounds,[134] ectomycorrhizal fungi, *e.g. Boletus variegatus*, can produce

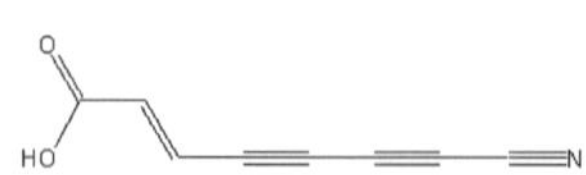

76

78

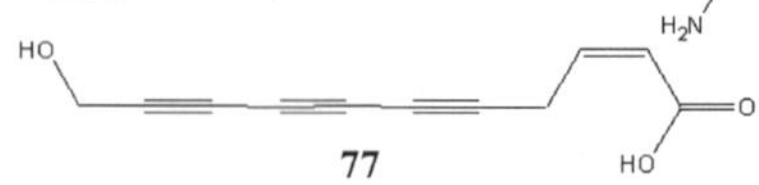

77

volatile products, isobutanol, isobutyric acid, ethanol, acetoin, and isoamyl alcohol, which may play an antifungal role towards pathogens.[129,135]

The antibacterial, antifungal and nematicidal benzofuran, furanone, and benzopyran derivatives mycorrhizin A (**79**), B1 (**80**) and B2 (**81**), chloromycorrhizin A (**82**),[136–138] dechloromycorrhizin A (**83**), lachnumol (**84**), lachnumon A (**85**), B1 (**86**) and B2 (**87**),[136,139] papyracons A-D (**88–91**),[140,141] 6-*O*-methylpapyracon B-C (**92–93**), lachnumfuran A (**94**), lachnumlactone A (**95**), and chloromycorrhizinol (**96**)[140] have been isolated from *mycelia sterilia* mycorrhizae of *Monotropus hypopitys* and *Lachnum papyraceum*, a common mycorrhizal fungus. In addition, compounds **88–90** have shown cytotoxic activity. Some of these compounds have been shown to be produced by free-living, nematode-trapping fungi.[142] The mycorrhizins should not be confused with the ectomycorrhizins[143] which are low molecular mass proteins that are specific to onset of mycorrhizal development. However, this confusion does occur in the literature.[144] Some ectomycorrhizins are similar antigenically to nodulins.[145]

As with the endophytes of conifers and a sedge (see above), some ectomycorrhizal fungi have been shown to produce siderophores in pure culture.[146] The ericoid mycorrhizal fungi, *Hymenoscyphus ericae*, common on ericaceous plants in acidic environments, *Oidiodendron griseum*, both an ericoid mycorrhizal fungus and a saprobe, and an unidentified ericoid mycorrhizal fungus from *Rhodothamnus chamaecistus*, an ericaceous plant found in calcareous environments, were studied in axenic culture. The mycorrhizae from the acidic environments produced ferrichrome (**75**, demethylated central ring), ferricrocin (**73**) and ferrichrome C (**75**), with ferricrocin (**73**) predominating. This is a similar pattern to the results from the endophytic *P. fortinii*.[89] The fungus from the alkaline environment produced these siderophores as well, but, in addition, it produced fusigen (**97**), an ester-type siderophore.

Recently, attention has focused on the interaction between the fungal and plant members of the symbiosis and the establishment of the mycorrhizal association. Plants produce compounds that stimulate fungal spore germination, hyphal growth and branching.[23,123] Mycorrhizal fungi can probably produce plant hormones (*e.g.* cytokinins and gibberellins),[123] which affect the growth of the root tips, growth of individual cells,

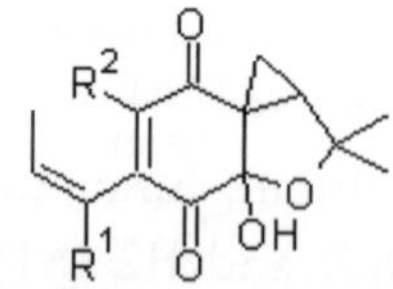

79 R^1 = Cl, R^2 = H
80 R^1 = Br, R^2 = H
81 R^1 = Br, R^2 = Cl
82 R^1 = Cl, R^2 = Cl
83 R^1, R^2 = H

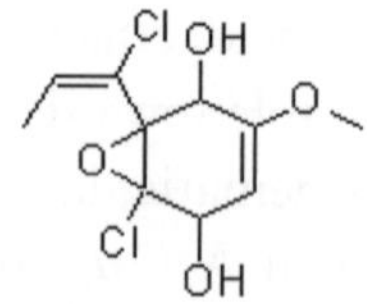

84

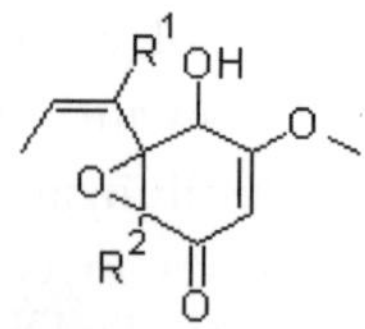

85 R^1 = Cl, R^2 = Cl
86 R^1 = Br, R^2 = H
87 R^1 = Br, R^2 =

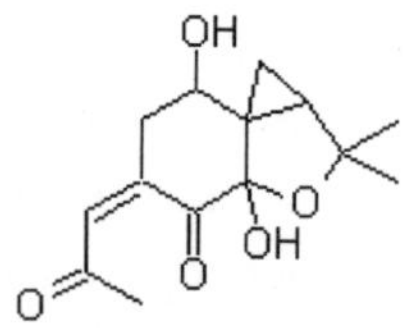

88

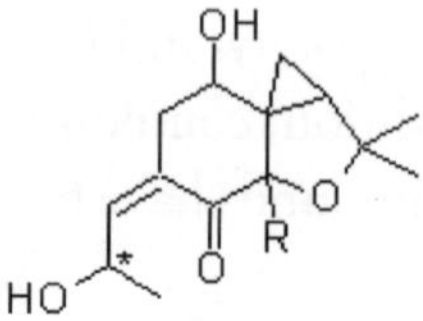

89 R = OH, * = S
90 R = OH, * = R
92 R = OMe, * = S
93 R = OMe, * = R

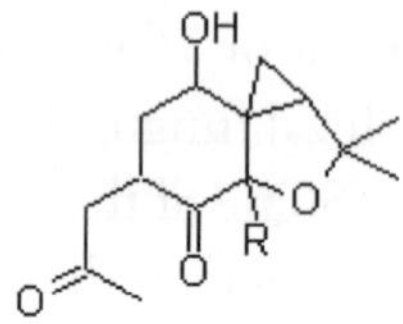

91

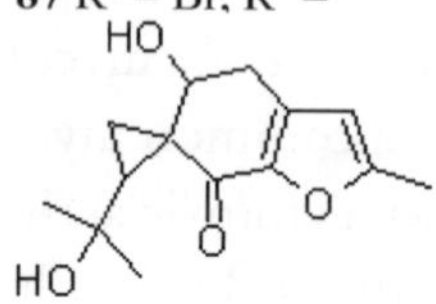

94

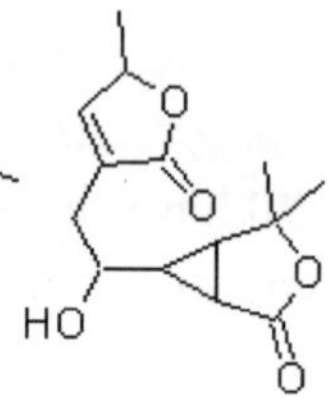

95

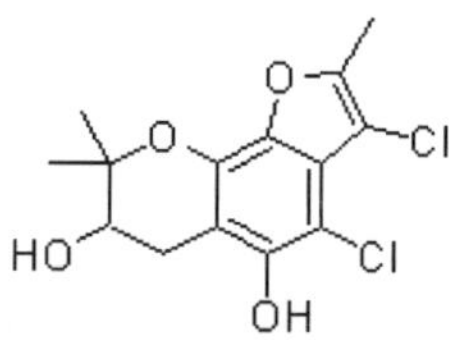

96

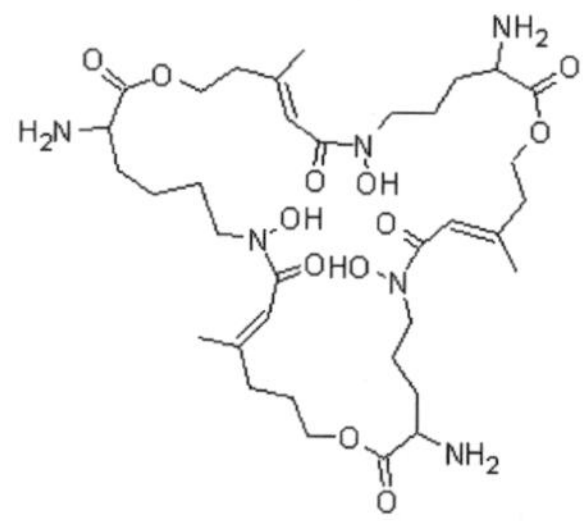

97

branching of roots, and aid the symbiosis development. Generally, ectomycorrhizal fungi produce the auxin IAA (**168**), and several are known to produce cytokinins, gibberellins, and ethylene.[127] Elevated jasmonates,[147] glycosylated cyclohexenones, mycorradicin,[123] flavones,[23] benzoquinones, and volatile organic compounds[148] are common plant products during mycorrhizal establishment. The products specifically produced by the fungal partner are almost completely unknown, except by hypothesis or supposition. This may soon change. Relatively few plant or fungal gene expressions are up- or downregulated during mycorrhizal establishment.[149] The majority of fungal secondary metabolite genes that showed differential expression were downregulated. Only a gene similar to the *Neurospora crassa* pro-tyrosinase was upregulated among the 2,284 genes analyzed. This enzyme may be involved in the production of melanin by the hyphae in the plant root.[150] An effort to identify the key genes that support the establishment and maintenance of VAM and ectomycorrhizal symbioses was undertaken. This led to the analysis of 20 000 expressed-sequence tags, bioinformatic tools for analysis of the data, and web-based databases for the compilation of data from other studies.[128] These and similar studies in future will help elucidate the specific fungal products formed during mycorrhizal establishment and maintenance and the roles that the compounds may play in effecting changes in plant growth and altered pathogen and parasite success. This knowledge will be useful for the establishment of mycorrhizal associations in difficult environments, as well as potential herbicidal, nematicidal, fungicidal, and bactericidal applications, especially in agricultural settings.

15.3.3 Dicotyledonous plant endophyte products

The secondary metabolites of the endophytic fungi associated with dicotyledonous plants (dicots) are chemically diverse (Table 1). There is an equal diversity in the activities of these compounds, including antibacterial, antifungal, nematicidal, phytotoxic, cytotoxic, antineoplastic, anti-insectant, anti-herbivory, and a variety of other activities. The compounds isolated, structurally elucidated, and explored biologically in the short time since previous reviews[8,11] continue to display that same wide array of chemical and biological diversity.

Studies continue to be published in which only the results of bioassay of fungal extracts are presented[151–154] without the identification of active compounds. Some of these studies suggest significant activities for the compounds contained in the extracts. Such studies represent opportunities for the natural products chemist to collaborate with the biology- and molecular biology-oriented groups for more complete studies of the plants and associated endophytic fungi.

Three cyclopentan- and cyclopenten-ones have been reported in the recent literature from endophytic fungal sources.[155,156] Xylobovide (**98**), a previously known compound, was isolated from an unidentified endophyte from the leaves of the Australian tree *Alstonia scholaris*, a known medicinal plant.[156] Xylobovide has been reported previously to inhibit

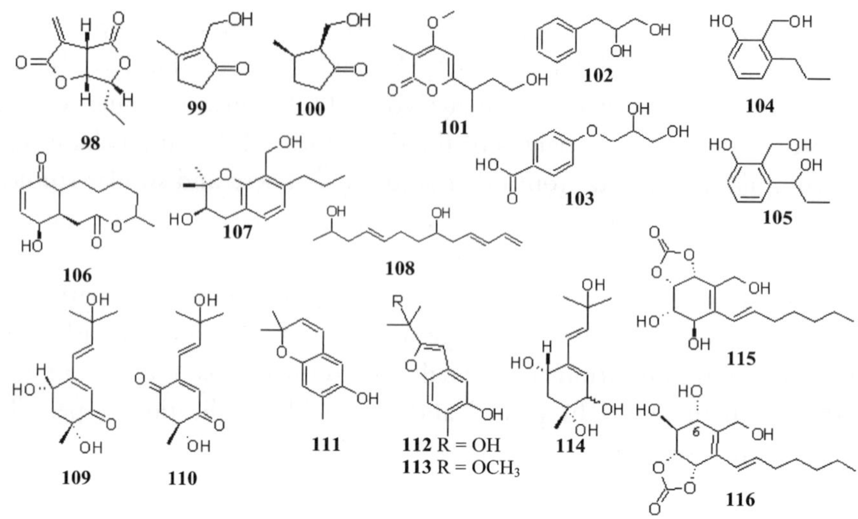

seed germination.[157] A different medicinal plant, *Leea rubra*, yielded a dothidiomycete (Ascomycota) endophyte that gave two new cyclopentane compounds, 2-hydroxymethyl-3-methylcyclopent-2-enone (**99**) and *cis*-2-hydroxymethyl-3-methylcyclopentanone (**100**).[155] Compound **99** showed low activity against *M. tuberculosis* H37Ra and no activity in a cytotocity assay. The hydrazone of compound **100** gave a low toxicity against the Vero cell line. The known compound asterric acid was isolated as well.

Phomopsis sp. strain endophytes of the medicinal plant *Erythrina crista-galli* continue to yield novel compounds (Table 1). New phenylpropane, pyronol, benzoic acid, phenylpyran, macrocyclic, and alkene compounds were discovered, as well as the known compounds clavatol, 4-hydroxymellein, mellein, mevalonolactone, mevinic acid, nectriapyrone, phomol, scytalone, and tyrosol.[158] The new compounds were phomopyronol (**101**), 3-phenylpropane-1,2-diol (**102**), 4-(2,3-dihydroxypropoxy) benzoic acid (**103**), 2-(hydroxymethyl)-3-propylphenol (**104**), 2-(hydroxyl-methyl)-3-(1-hydroxypropyl)phenol (**105**), an octahydrobenzoxecine dione (**106**), 8-(hydroxymethyl)-2,2-dimethyl-7-propylchroman-3-ol (**107**), and (4*E*, 10*E*)-trideca-4,10-12-triene-2,8-diol (**108**). Most of the new compounds (**101–104, 106–108**) had low activity against a battery of bacteria and fungi; some of the compounds gave good-to-moderate activity against the fungi *Paecilomyces islandicum* (**101**), *Absidia glauca* (**102, 103, 106**), and *Sporobolomyces roseus* (**106**). Some of the compounds showed phytotoxicity against *Lepidium sativum* (**101, 105–107**) and *Setaria italica* (**105, 107, 108**). None of the new compounds showed significant activity in a mouse ear edema anti-inflammation assay or a nematicidal activity assay, although one compound (**106**) showed significant cytotoxicity against L1210 ($IC_{50} = 1$ μg/mL), Colo-320 ($IC_{50} = 1$ μg/mL) and MDA-MB-231 ($IC_{50} = 5$ mg/mL) cell lines.

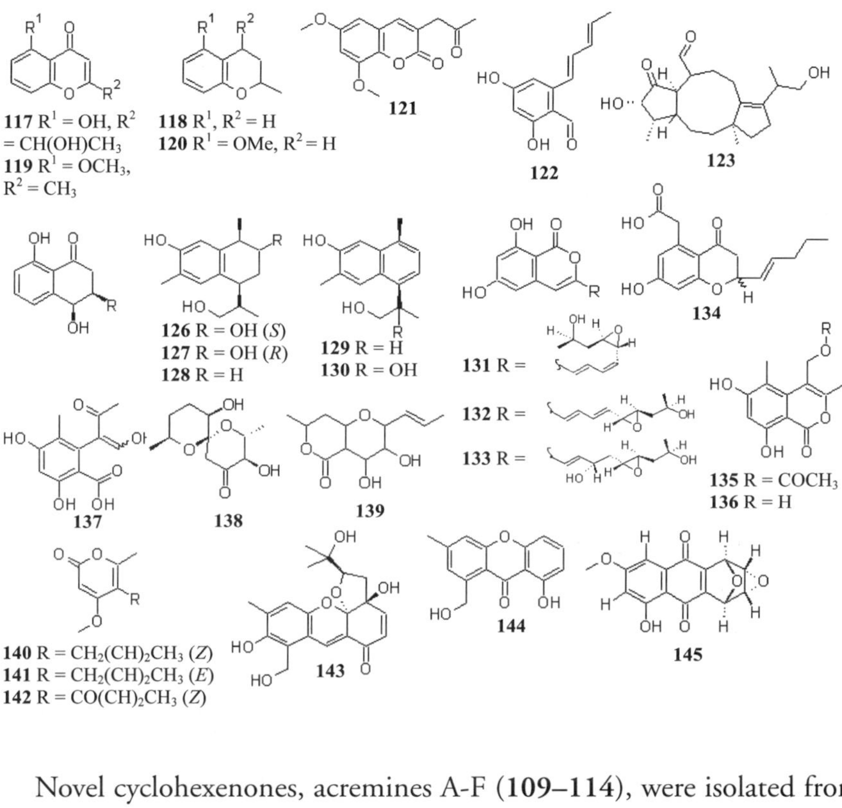

117 R^1 = OH, R^2 = CH(OH)CH_3
119 R^1 = OCH_3, R^2 = CH_3
118 R^1, R^2 = H
120 R^1 = OMe, R^2 = H
121
122
123
126 R = OH (*S*)
127 R = OH (*R*)
128 R = H
129 R = H
130 R = OH
131 R =
132 R =
133 R =
134
135 R = $COCH_3$
136 R = H
137
138
139
140 R = $CH_2(CH)_2CH_3$ (*Z*)
141 R = $CH_2(CH)_2CH_3$ (*E*)
142 R = $CO(CH)_2CH_3$ (*Z*)
143
144
145

Novel cyclohexenones, acremines A-F (**109–114**), were isolated from a mycoparasitic *Acremonium* sp. from the oomycete pathogen *Plasmopara viticola* on *Vitis vinifera* cv. Regina blanca.[159] Compounds **109–112** inhibited germination of the pathogen, and the most activity was shown by acremine C (**111**). This work may lead to a mechanism for the mycocidal activity of mycoparasitic fungi.

Davis *et al.*[160] isolated polyketide cyclohexene compounds, phomoxins B (**115**) and C (**116**) (the latter a stereoisomer of phomoxin A at position 6), that contain a rare cyclic carbonate, and the previously known, structurally-related eupenoxide. The compounds came from the culture of *Eupenicillium* sp., an endophyte obtained from the surface-sterilized outer bark of the tropical tree *Glochidion ferdinandi* growing in Queensland, Australia. Although the structure of these new compounds is interesting, they have not been reported to have any biological activity based on screening against a battery of bacteria and *Candida albicans*.

Several studies have came out of the laboratory of Ângela Regina Araújo, Universidade Estadual Paulista, Araraquara, Brazil. Two new and two known benzopyrans were isolated from culture extracts of the endophyte *Curvularia* sp. from the Brazilian *Ocotea corymbosa*. The new compounds, (2′*S*)-2-(propan-2′-ol)-5-hydroxy-benzopyran-4-one (**117**) and 2,3-dihydro-2-methyl-benzopyran-4,5-diol (**118**), along with the two related and known compounds (**119–120**), were evaluated for antifungal activity and cytotoxicity in HeLa and CHO cell assays.[161] Compound **117** was weakly antifungal against *Cladosporium sphaerospermum* and *C. cladosporioides*, and appeared to act as a mitogen, increasing cell division in the cell assays. From another Brazilian plant, *Xylopia aromatica*, the same group[162] isolated the endophyte *Periconia atropurpurea*. They found two new compounds, a coumarin derivative, 6,8-dimethoxy-3-(2′-oxo-propyl)-coumarin (**121**), and a benzaldehyde, 2,4-dihydroxy-6-[(1′*E*,3′*E*)-penta-1′,3′-dienyl]-benzaldehyde(**122**), as well as the known macrocyclic compound periconicin B (**123**). Compound **122** was significantly antifungal against *C. sphaerospermum* (detection limit of 1 μg on bioautography) and also acted as a mitogen towards HeLa and CHO cells. Periconicin B (**123**) was cytotoxic to these cells ($IC_{50} = 8\,\mu M$). Two previously known compounds, (-)-*cis*-4-hydroxy-6-deoxyscytalone[200] (**124**) and (4*R*)-4,8-dihydroxy-α-tetralone[164] (**125**), were isolated from *Colletotrichum gloeosporoides*, an endophyte of the Brazilian plant *Cryptocarya mandioccana*.[165] In bioautography, the compounds were toxic against the phytopathogenic fungi *C. cladosporioides* and *C. sphaerospermum*. Although the compounds are known from previous work, this appears to be the first report of antifungal activity. Compound **125** had been reported previously as mildly phytotoxic.[164] The Brazilian group also found five cadinane sesquiterpenoids from the cultures of *Phomopsis cassiae*, an endophyte of *Cassia spectabilis*.[166] The compounds were identified as two diastereomers of 3,9,12-trihydroxycalamenene (**126–127**), 3,12-dihydroxycalamenene (**128**), 3,12-dihydroxycadalene (**129**), and 1,11, 12-trihydroxycadalene (**130**). These compounds were tested against the phytopathogens *C. cladosporioides* and *C. sphaerospermum*; compounds **127** and **130** were active, with a detection limit in bioautography about the same as the control, nystatin (1 μg). Compound **129** showed moderate cytotoxicity against HeLa cells ($IC_{50} = 20\,\mu M$), while compounds **128** and **130** were only slightly cytotoxic ($IC_{50} = 100$ and $110\,\mu M$, respectively).

Wijeratne *et al.*[120] discovered three new isocoumarins (**131–133**), a benzopyrone (**134**), and a known compound, monocillin (**69**), from an endophyte, *Paraphaeosphaeria quadriseptata,* from *Opuntia leptocaulis*. No biological activities were reported. Similar compounds were discovered by Sohrab[167] in exploratory work on endophytes in plants in Germany. The endophyte *Scytalidium* sp. was isolated from *Salix* sp., and culture led to three new compounds, (6,8-dihydroxy-3,5-dimethyl-1-oxo-1*H*-isochromen-4-yl)methyl acetate (**135**), 6,8-dihydroxy-4-(hydroxymethyl)-3,5-dimethyl-1*H*-isochromen-1-one (**136**), and 4,6-dihydroxy-2-(1-hydroxy-3-oxobut-1-en-2-yl)-3-methylbenzoic acid (**137**). The isolation and culture of *Dinemasporium strigosum* from *Calystegia sepium* led to the isolation of two new compounds, a spiropyranone, 3,11-dihydroxy-2,8-dimethyl-1,7-dioxaspiro[5,5]undecan-4-one (**138**), and (2*R*, 3*S*, 4*R*, 4a*S*, 7*R*, 8a*S*)-hexahydro-3,4-dihydroxy-7-methyl-2-((*E*)-prop-1-enyl)pyrano[4,3-*b*]pyran-5(7*H*)-one (**139**). Finally, the culture of an unidentified endophytic fungus from *Trifolium dubium* led to the isolation of a further three pyranone compounds, 5-((*Z*)-but-2-enyl)-4-methoxy-6-methyl-2*H*-pyran- 2-one (**140**), 5-((*E*)-but-2-enyl)-4-methoxy-6-methyl-2*H*-pyran-2-one (**141**), and 5-((*Z*)-but-2-enoyl)-4-methoxy-6-methyl-2*H*-pyran-2-one (**142**). A number of known compounds were also isolated. All of the compounds isolated in sufficient quantity were analyzed for antimicrobial activity against *B. megaterium, M. violaceum*, and *C. fusca* to assess bactericidal, fungicidal and algicidal activities, respectively. The bactericidal (**136, 137, 138, 139, 141, 142**), fungicidal (**135, 136, 137,**

138, 139, 141, 142) and algicidal (**136, 137, 138, 139, 141, 142**) compounds had low-to-moderate activities.

Paranolin (**143**) was isolated as a novel xanthene metabolite of *Paraphaeosphaeria nolinae*, an endophyte of *Artemisia annua.*[168] There is continuing interest in *A. annua* and its endophytes since the discovery of artemisinin. Being a xanthene, paranolin (**143**) is novel as well. The compound had low activity against SW116 and HeLa cells ($IC_{50} > 50\,\mu g/mL$). However, as a potentially new class of natural product, biosynthesis could arise from the modification of a fungal metabolite known from *Cyathus intermedius*, 1-hydroxy-6-methyl-8-hydroxymethylxanthone (**144**), through oxidation, isoprenylation, epoxidation, and methylation steps, followed by closure of the five-membered ring prior to reduction and dehydration to **143**. This group also has worked on novel products from monocots (see below).

A tetracyclic epoxyquinone (**145**) has been reported from a leaf endophyte of the locally-medicinal Papua New Guinean plant *Desmodium uncinatum*. So far, this compound has only appeared in short reviews[169] of endophyte products, and no primary literature on the compound has appeared. The compound is reported to be both antifungal and antibacterial and had a rather high IC_{50} against HeLa cells of $0.9\,\mu g/\mu L$.

A compound of a novel class, spirononadienes, was isolated from culture of an unidentified leaf endophyte obtained from the New Zealand plant *Knightia excelsa.*[170] The compound, spiromamakone A (**146**) was toxic to P388 cells ($IC_{50} = 0.33\,\mu M$), and showed significant activities against *Bacillus subtilis, Trichophyton mentagrophytes*, and *Cladosporium resinae* in agar diffusion tests. The biosynthesis of **146** appears to proceed from acetyl-CoA and malonyl-CoA to dihydroxynaphthalene, which dimerizes, oxidizes and rearranges to **146**.

Three new depsidones (**147–149**) were isolated from an endophytic fungus BCC 8616 (dothidiomycete, Ascomycota) obtained from an unidentified leaf in an evergreen forest in Thailand[171] along with the related compound auranticin A. Compound **147** had low ($IC_{50} = 6.5$ and $4.1\,\mu g/mL$, respectively) toxicity against KB and BC (breast cancer) cell lines. Although depsidones are known from plant and free living fungal material, this is the first report of isolation from an endophytic fungus.

Five known xanthones, phomoxanthone A,[172] shamixanthone, 14-methoxytajixanthone-25-acetate, tajixanthone methanoate, and tajixanthone hydrate,[173] were isolated from a *Phomopsis* sp. endophyte from a *Costus* sp. stem in Costa Rica (phomoxanthone A) and the leaf endophyte *Emericella variecolor* from *Croton oblongifolius* in Thailand (the remainder). Phomoxanthone A was shown previously to have activity against malarial parasites, *M. tuberculosis* and cytotoxicity.[174] The authors of the current study of phomoxanthone A[172] determined the stereochemistry of the compound through X-ray spectroscopy and found a low toxicity against *B. megaterium, C. fusca, Ustilago violacea*, and *Phytophthora infestans,* and moderate toxicity against *Botrytis cinerea*, *Pyricularia oryzae,* and*Septoria tritici.* The other xanthones were isolated previously from the free living fungus *Aspergillus variecolor*. Tajixanthone hydrate and 14-methoxytajixanthone-25-acetate were found to have activity similar to doxorubicin in a gastric (KATO3) and breast (BT474) carcinoma cell line assay.[173] Cytotoxic and antimicrobial activities have been reported previously for these compounds.

Endophytic *Penicillium janthinellum* from the fruits of *Melia azadarach* in Brazil yielded known anthraquinones, emodin, citreorosein, and a related compound, citrinin, as well as a novel anthraquinone, janthinone

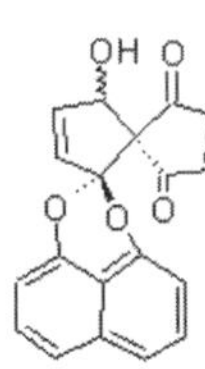

146

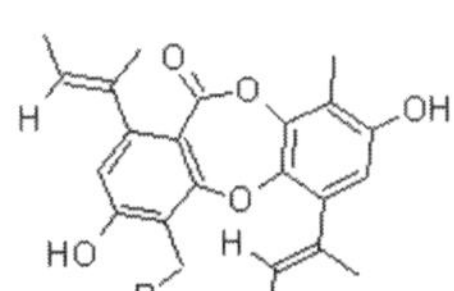

147 R = =O
148 R = H
149 R = OH

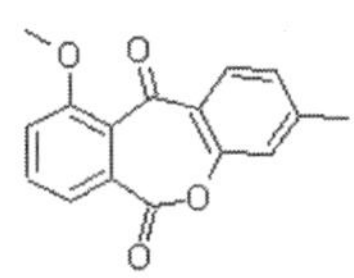

150

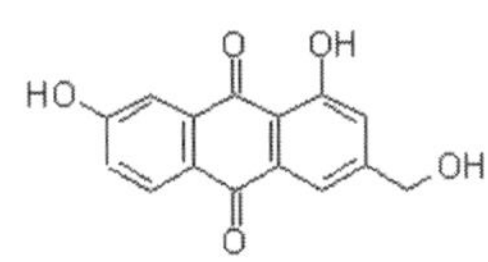

151

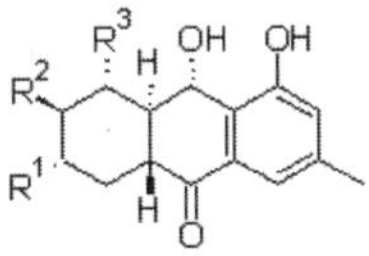

152 R^1 = OH, R^2, R^3 = H
153 R^1 = H, R^2, R^3 = OH

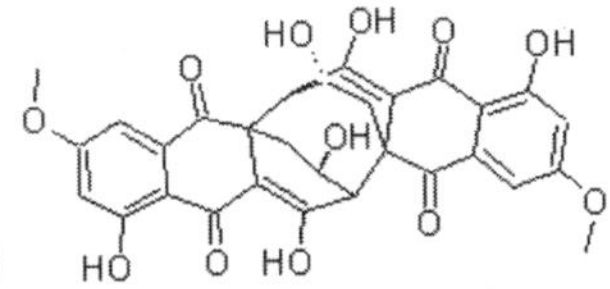

154

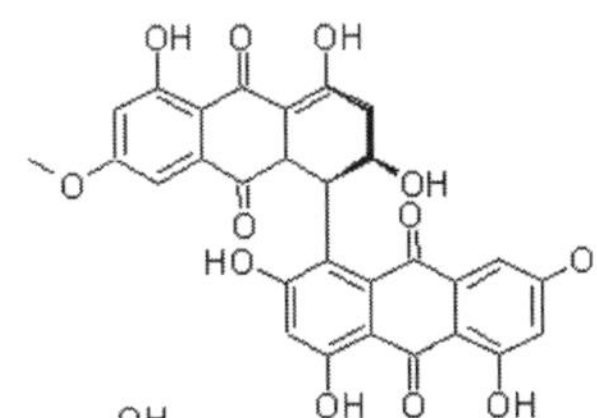

155

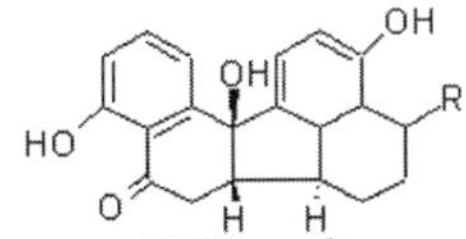

156 R = =O
157 R = OCH_3

(**150**).[175] Janthinone was inactive in antibacterial and antileishmanial assays, but the known compounds, emodin and citrinin, were active against *Pseudomonas aeruginosa* and *B. subtilis* at relatively high concentrations ($\geq$ 8 mg/mL). Citrinin was moderately-to-highly toxic to *Leishmania mexicana* at 10–40 μg/mL. Another Brazilian study led to a new anthraquinone, 1,7-dihydroxy-3-hydroxymethyl-9,10-anthraquinone (**151**), and two, new anthraquinone derivatives, dendryols E (**152**) and F (**153**), isolated from a culture of *Phoma sorghina* obtained from a medicinal plant, *Tithonia diversifolia.*[176] This was a first report of *P. sorghina* as an endophyte. No biological activities were reported, but other, related anthraquinones are both bactericidal and cytotoxic. Two new bisanthraquinones, (+)-epicytoskyrin (**154**) and (+)-1,1′-bislunatin (**155**) were isolated from an endophyte closely related to *Diaporthe phaseolorum* var. *sojae* from *Camelia sinensis.*[177] Compounds **154** and **155** showed cytotoxicity toward KB cells with IC_{50} values of 0.5 μg/mL and 3.5 μg/mL respectively. Another study of *A. annua* endophytes resulted in the fungus *Hypoxylon truncatum.* The culture of the fungus led to two new benzo[*j*]fluoranthenes, daldinones C (**156**) and D (**157**), and two known compounds altechromone A and (4*S*)-5,8-dihydroxy-4-methoxy-α-tetralone,[178] the latter an isomer of **125**. The benzo[*j*]fluoranthenes are uniquely fungal products, and **156** and **157** are the first cytotoxic products of this class (IC_{50} values of 49.5 and 41 μM, respectively, against SW116 cells; the control, 5-fluorouracil, had an IC50 value of 37 μM). The biosynthesis of these compounds could proceed from [1,1'-binaphthalene]-4,4',5,5'-tetrol by conjugation, rearrangement, oxidation, and reduction.

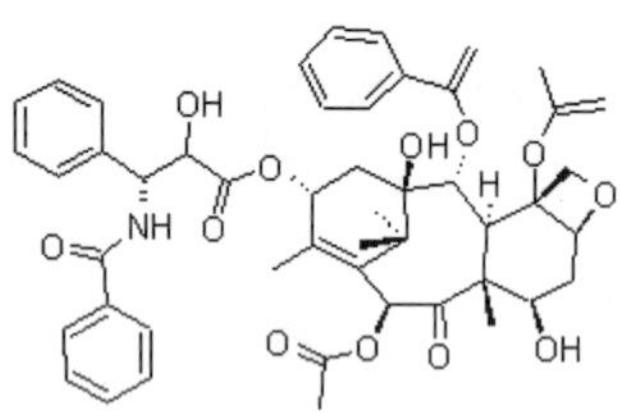

158

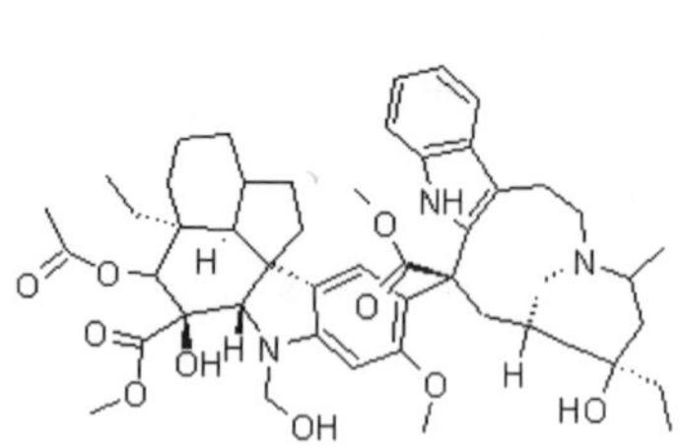

159

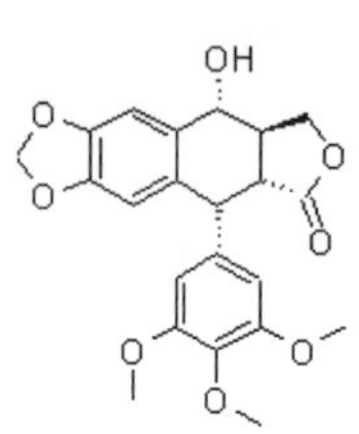

160

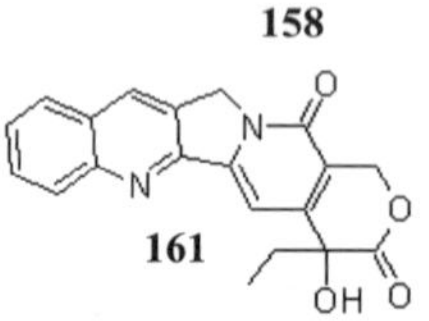

161

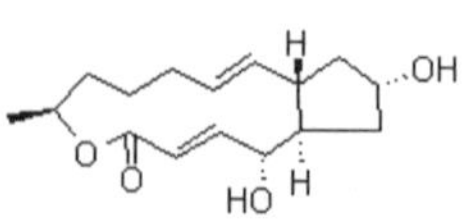

162

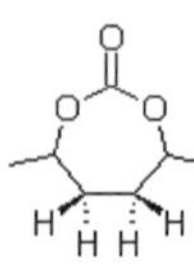

163

Following the groundbreaking work by Stierle *et al.*,[179] in which an endophyte of *Taxus brevifolia, Taxomyces andreanae*, produced paclitaxel (Taxol®) (**158**) and other taxanes, many researchers began to realize the potential of fungal endophytes as sources of rare or source-limited medicinal natural products. Subsequently, Strobel's group found Taxol-producing fungal endophytes in virtually every *Taxus* sp. examined.[9] In the last several months, the remaining major natural product antineoplastic compounds, vincristine (**159**), podophyllotoxin (**160**), and camptothecin (**161**), have been isolated from cultured fungal endophytes of various species. The indolizidine alkaloid camptothecin (**161**) has been obtained from the fungus *Entrophospora infrequens* from the tree *Nothapodytes foetida*,[180–182] which is a normal source for the molecule. Drug production in bioreactors may be an important biotechnological source in the future; the compound can be difficult to obtain in consistent supply from wild collections. The indole alkaloid vincristrine (**159**) has been isolated from the endophytes *Fusarium oxysporum*[183] and *mycelia sterilia* 97CY3,[184] both from *Catharanthus roseus* which is the usual source plant. Again, studies are ongoing to determine whether the fungi may become important sources for the drug compound, although supplies seem to be adequate due to cultivation of the plant. Nearly simultaneously, Puri *et al.*[185] and Eyberger *et al.*[43] reported the production of the lignan podophyllotoxin (**160**), the former from *Trametes hirsuta*, an endophyte of *Podophyllum hexandrum*, and the latter from two strains of *P. fortinii*, endophytes of *P. peltatum*. Both plants are sources for the drug compound which is in decreasing supply due to over-collection. No efficient synthesis has been found, and biotechnological development is in progress. Brefeldin A (**162**) was isolated from an endophyte of *Cephalotaxus fortunei*[186] and *Phoma medicaginis* from *Medicago lupulina* and *M. sativa.*[187] Reports of cytotoxic and antineoplastic compounds from a variety of endophytes have appeared (above and other reviews[8–11]). Endophytic fungi may be important sources of antineoplastic compounds in the future. The normal plant sources of the compounds seem to make good candidates for the recovery of novel endophytic fungal producers.

A new cyclic heptanone, 4,7-dimethyl-1,3-dioxa-cyclohepta-2-one (**163**), was isolated from *Penicillium thiomii* IR7, one of 21 endophytes isolated from leaves, roots and stems of *Terminalia chebula* in Bangladesh.[188]

Methanolic and ethyl acetate extracts of the fungus had low activity against a battery of Gram-positive and Gram-negative bacteria, but the isolated compound was not shown to be the active component.

Three new trichothecene macrolides (**164–166**) and two known analogs, roridin E and mytoxin B, each with significant activity against KB cells, were isolated from *Myrothecium roridum* strains IFB-E009 and IFB-E012 from *Trachelospermum jasminoides* and *A. annua*, respectively. The compounds gave IC_{50} values of 15.6 (**164**), 1.4 (**165**), 57.5 (**166**), 0.034 (roridin E), and 0.0042 μM (mytoxin B). The known compounds are thus substantially more active than the new compounds, although compound **165** shows some promise for development. The trichothecenes are well-known toxic fungal products and their biosynthesis has been studied.[189] The trichothecenes and related products appear to be derived from the C_{15} farnesyl diphosphate followed by conjugation, cyclization and oxidations.

The last of the purely hydrocarbon compounds to be considered in this section are two new deoxypreussomerin derivatives, palmarumycins JC1 and JC2, along with a new naphthoquinone, isodiospyrol A, of uncertain endophytic origins.[190] The researchers isolated the compounds from dry fruits of *Diospyros ehretioides* but not from fresh fruits. They concluded that the compounds must be of fungal, perhaps endophyte, origin. Palmarumycin JC2 had low activity against plasmodia (IC_{50} = 4.5 mg/mL), fungi (IC_{50} = 12.5 mg/mL), *M. tuberculosis* (IC_{50} = 6.25 mg/mL), and toxicity against the NCI-H187 cell line (IC_{50} = 11 mg/mL). It is probably reasonable to presume a fungal origin for these compounds since similar compounds have been isolated from other fungi[191,192] including endophytes. However, since no endophyte was isolated, I will not consider these compounds further here.

A few novel N-containing and alkaloid products have been isolated from dicot plant endophytes recently. Although not nearly as numerous as the hydrocarbon products, the implications of some of these compounds are significant. Simplest among the amino and amide compounds are 3-nitropropionic acid (NPA)(**167**)[172,193] and indole-3-acetic acid (IAA)(**168**).[156,194] NPA is isolated frequently from plant tissues and it has significant biological activity. Chomcheon *et al.*[193] isolated NPA from extracts of endophytic *Phomopsis* sp. and unidentified fungal endophytes isolated from six Thai medicinal plants (*Urobotrya siamensis, Grewia* sp., *Mesua ferrea, Rhododendron lyi, Tadehagi* sp., and *Gmelina elliptica*). The compound had significant activity against *M. tuberculosis* (MIC = 3.3 μM) but no activity against several cancer lines. NPA is a well-known animal toxin. IAA, an auxin hormone normally found in plants, is known to be produced by mycorrhizal and endophytic fungi[195] It is also believed to play a role in the establishment of the symbiosis through its effect on root growth and branching. The production of IAA by endophytic fungi other than mycorrhizae suggests that these fungi may have similar interactions with the host plants in setting up a mutualistic or balanced relationship.

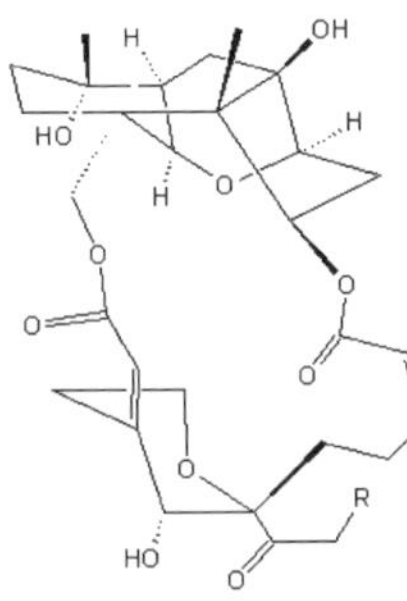

164 R = H
165 R = OH

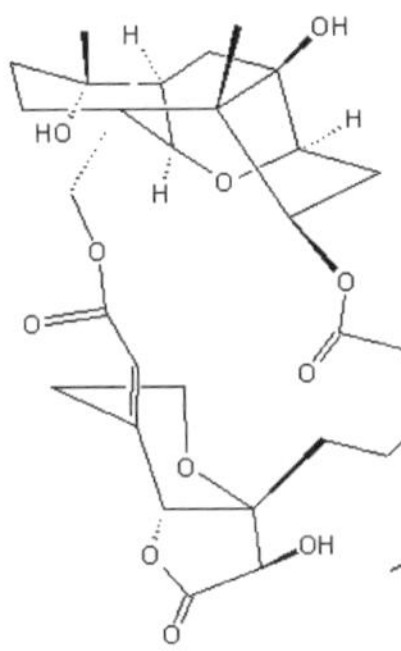

166

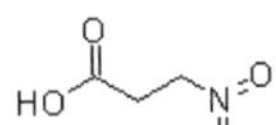

167

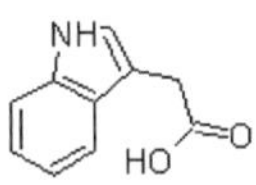

168

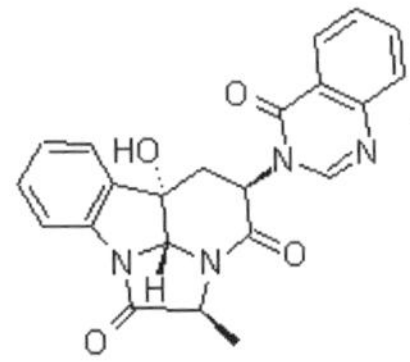

178

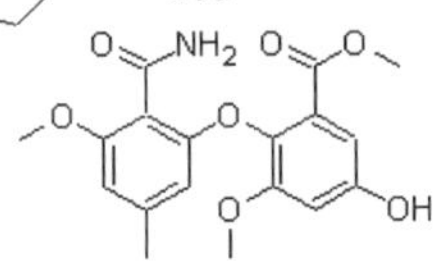

169

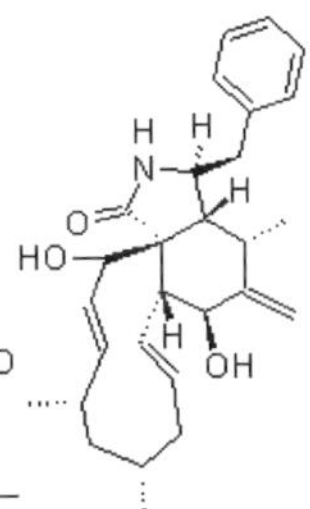

179

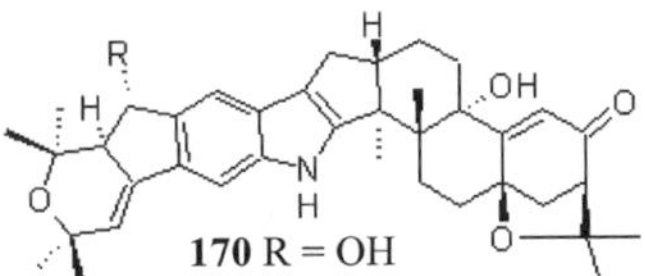

170 R = OH
171 R = OMe

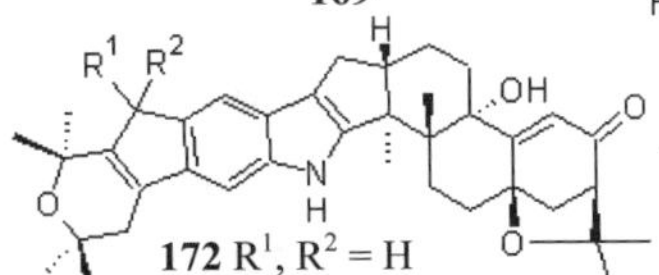

172 R[1], R[2] = H
173 R[1] + R[2] = O

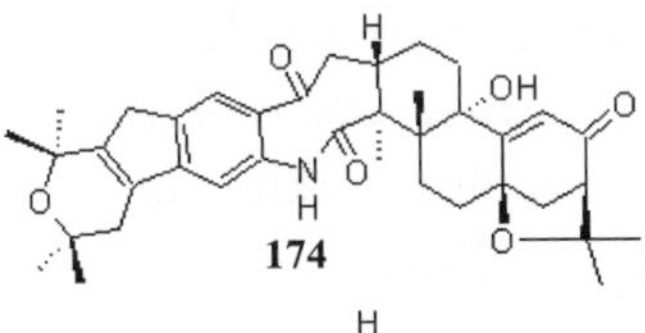

174

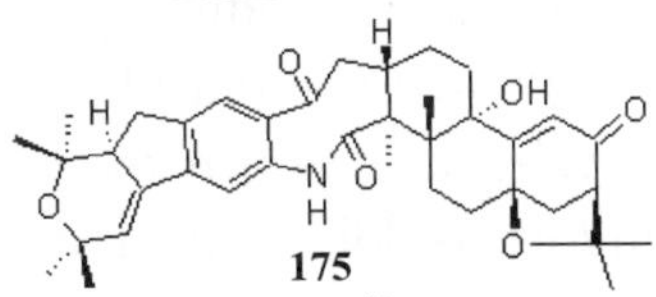

175

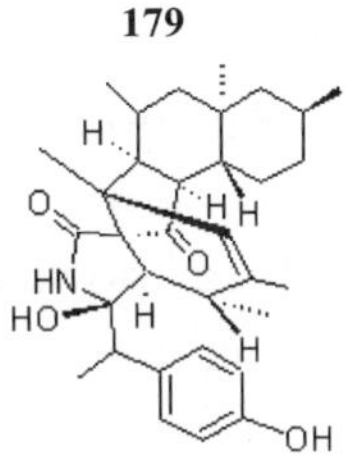

180

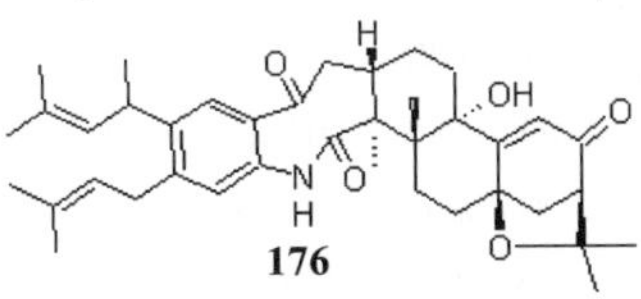

176

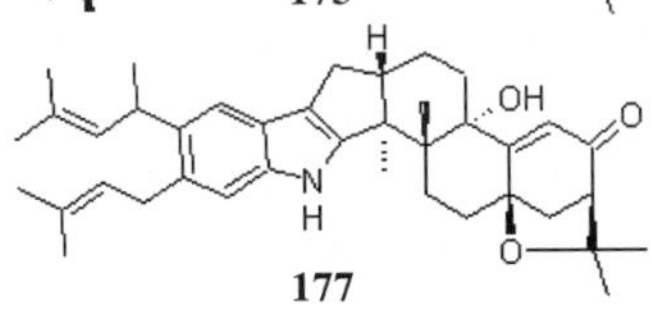

177

An interesting, new modified hydroxybenzoic acid dimer, neoplaether (**169**) was isolated from the endophytic *Neoplaconema napellum* obtained from leaves of *Hopea hainanensis* in China.[196] Diphenyl ethers have been isolated as fungal products in the past with noted biological activities. The current product (**169**) was cytotoxic to KB cells ($IC_{50} = 13.0\,\mu g/mL$), five-fold less than the control 5-fluorouracil, and inhibited growth of *C. albicans* (MIC $= 6.2\,\mu g/mL$) compared to a MIC of $1.5\,\mu g/mL$ for Amphotericin B.

Outside the grasses, the only plants known to significantly accumulate ergoline (ergot) alkaloids are in the Convolvulaceae. *Ipomoea asarifolia* and *Turbina corymbosa* were subjected to endophyte analysis to determine whether the fungi could be responsible for the indole alkaloid accumulation.[65] Twelve endophytic fungi were cultured and one unculturable epiphytic fungus was discovered by microscopic and molecular techniques. Among the culturable endophytic fungi, none was capable of synthesizing in culture the alkaloids found in the plant. Further, none of these had evidence of enzymes at the molecular level that would support that synthesis. One endophyte, *Penicillium roquefortii*, synthesizes alkaloids, but these are different from those isolated in the plants. The epiphytic fungus, however, showed affinities to the Clavicipitaceae, the fungi in grasses known to produce ergot alkaloids. It possesses the genes for ergot alkaloid synthesis (more details of synthesis are discussed below), and was presumed to be the source of the ergot alkaloids in the plants, particularly since fungicidal treatment of the plants "cures" them of alkaloid accumulation. Although an epiphyte, the clavicipitaceous fungus could not be transmitted horizontally among plants, and there is evidence that the fungus is transmitted vertically. There must, then, be an endophytic phase of this epiphyte, and its alkaloid products are accumulated in the plant tissues.

An endophytic *Penicillium* sp. (subgenus *Furcatum*) was isolated from the mangrove plant *Aegiceras corniculatum*.[197] The culture of the fungus led to three known indole alkaloids, shearinine A, paspalitrem, and paspaline, and eight new triterpenoid indole alkaloids, shearinines D-K (**170–177**). All of the alkaloids were studied for the ability to alter voltage-activated potassium (BK_{Ca}) channels from HEK293 cells. The control alkaloid, penitrem A, reduced channel conductance by almost 100% in less than 50 sec at a concentration of 100 nM. Shearinine E (**171**) similarly inhibited the channel conductance, but required about 130 sec. Shearinines D (**170**), G (**173**) and K (**177**), were significantly less potent, requiring about 200 sec, >400 sec and ≫400 sec to disrupt the channel conductance, respectively. The other new compounds had no or slight activities. The BK_{Ca} channels, thought to be mycotoxin targets, are implicated in epilepsy, hypertension, and erectile dysfunction. The new products could be used as channel probes or lead to novel therapeutics. Paspaline is postulated to be a precursor to other alkaloids in the janitrem class, proceeding by isoprenylation, oxidation, hydroxylation, and rearrangement to the other members. Shearinine E (**171**), G (**173**), and H (**174**) are likely metabolic endpoints for this biosynthetic pathway.

A new class of quinazolinone-modified imidazole alkaloids, represented by the product chaetominine (**178**), was isolated from a *Chaetomium* sp. IFB-E015 endophyte from the TCM plant *Adenophora axilliflora.*[198] The novel alkaloid is not seen in the intact plant and is produced by the fungus presumably only in culture or under special conditions. The compound is more cytotoxic to K562 leukemia (IC_{50} = 21 nM) and SW1116 colon cancer (IC_{50} = 28 nM) cells than 5-fluorouracil (IC_{50} = 33 and 76 nM, respectively). Chaetominine (**178**) is similar to the tryptoquivaline-like γ-lactones of *Aspergillus* and *Penicillium* spp., except that it contains a δ-lactone ring presumed to arise from alanine rather than tyrosine or phenylalanine as seen in other products. The compound is postulated to be synthesized by condensation of L-alanine, anthranilic acid and D-tryptophan, accounting for the number of nitrogens in the molecule, and is thus a product of non-ribosomal peptide formation.

Cytochalasins continue to be isolated as products of endophytic fungi. Wright *et al.*[156] isolated two known, bioactive cytochalasins, 19,20-epoxycytochalasin C and D, as well as two known picolinic acid compounds, fusaric acid and 3-(but-3-enyl)picolinic acid, from two endophytic fungi from *Alstonia scholaris* and *Acmena graveolens*. The specific activities of the compounds have not been reported. The cytochalasin L-696,474, an HIV protease inhibitor, previously reported in cultures of an endo-/epiphytic, bark-inhabiting fungus *Hypoxylon fragiforme*, was isolated from an endophytic *Phomopsis* sp. isolated from *Costus* sp. stems, along with its

21-O-deacetyl-derivative (**179**), the latter not known previously as a natural product.[172] L-696,474 is reported also from an endophyte of a rhodophyte *Polysiphonia* sp. (Table 1). The new product showed no activity in antimicrobial screening. Another novel hexacyclic cytochalasin, diaporthichalasin (**180**) (an isomer of phomopsichalasin) and the known compound pycnidione, were isolated from an endophytic *Diaporthe* sp. from *Croton sublyratus* growing in Bangkok, Thailand.[199] The new compound was active ($IC_{50} = 0.626\,\mu M$) in inhibiting the activity of the human cytochrome P450 CYP3A4 compared to pycnidione ($IC_{50} = 465\,\mu M$) and the positive control ketoconazole ($IC_{50} = 0.11\,\mu M$). The biosynthesis of this and other cytochalasins is presumed to follow from condensation of a polyketide and either tyrosine or phenylalanine, with appropriate ring closures.

An interesting new area of work is the biotransformation of plant products to new compounds by their endophytic fungi. A known allelopathic compound, lepidimoide, was synthesized by an endophytic *Colletotrichum* sp. from okra (*Hibiscus esculentus*) polysaccharide [$(1 \rightarrow 4) - O - \alpha$-(D-galactopyranosyluronic acid)-$(1 \rightarrow 2) - O - \alpha$-L-rhamopyranose].[200] The fungus produces a β-galactosidase, a rhamogalacturonan lyase, an acetylesterase, and perhaps two other proteins of unknown type that convert the polysaccharide into the toxic product. Lepidimoide has been reported to be phytotoxic to a variety of plants, but may also promote the growth of the plants in which it is produced.

An endophytic fungus, *Diaporthe* sp., from tea (*C. sinensis*) can convert (+)-catechin into (+)-$(2R, 3S, 4S)$-3, 4, 5, 7, 3′, 4′-hexahydroxyflavan and (−)-epicatechin into (−)-$(2R,3R,4R)$ − 3, 4, 5, 7, 3′, 4′-hexahydroxyflavan,[201,202] both previously known compounds. The isomers (−)-catechin and (+)-epicatechin were not converted. The alkaloids quinine, quinidine, cinchonidine, and cinchonine in *Cinchona pubescens* could be converted to the corresponding 1-*N*-oxides by an endophytic *Xylaria* sp.;[203] again, all of the products are previously known compounds. Finally, the polyamine alkaloid aphelandrine of *Aphelandra tetragona* was metabolized by a majority of 22 shoot and root endophytic fungi.[204] At least three of the fungi isolated could use aphelandrine as a sole nitrogen source. It is interesting to speculate that some of the "break-down" and "precursor" products isolated from plants could, in fact, be a result of the metabolism of the plant's endophytic fungi.

15.3.4 Monocotyledonous Plant Endophyte Products

Compared to the situation with the endophytic fungi of dicots, there are a few recent reports detailing new compounds or novel sources of known compounds in the recent literature. The Australian grass *Neurachne alopecuroidea* yielded an unidentified ectomycorrhizal fungus SDEF 678 that inhibited the growth of the pathogen *Gaeumannomyces graminis* var. *tritici* in coculture experiments.[205] The culture of the fungus led to a novel product, 5-hydroxy-1-(3-methyl-3-buten-1-ynyl)-7-oxabicyclo[4.1.0]-hept-3-en-2-one (**181**), that inhibits growth of the target pathogen and *Phytophthora cinnamoni* (MIC = 0.98 μg/mL), as well as the pathogens *Rhizoctonia solani* (MIC = 7.81 μg/mL), and *Pythium irregulare* and *Alternaria alternata* (MIC = 15.63 μg/mL for both). A regioisomer of **181**, in which the alkyne side group is on C2, was isolated in 1992 as a phytotoxic metabolite of *Pestalotiopsis theae*. In contrast, **181** was growth promoting in barley at concentrations of 0.1 (8.2% increase in stem length) and 1 μg/mL (24.7% increase).

Two new indole alkaloids were isolated from *Phaeosphaeria avenaria* FA39 obtained from an unidentified plant in Florida.[206] The products, phaeosphaerides A and B (**182–183**), were tested in an ELISA assay designed to assess ability to inhibit STAT3 signaling as a new therapeutic target. Compound **182** was active in the ELISA screen (IC_{50} = 0.61 mM), but was even more active in an *in vitro* cell line assay based on STAT3-dependent U266 multiple myeloma cells (IC_{50} = 6.7 μM) and active in a HepG2 cell line assay (IC_{50} = 0.7 μM). Compound **182** showed selectivity for STAT3 signaling and did not significantly inhibit STAT1 or STAT5 pathways. The phaeosphaerides are similar to fungal products obtained previously from *Curvularia pallescens* (curvupallide A) and both *Staphylotrichum coccosporum* and *Drechslera tritici-repentis* (spirostaphylotricins/triticones). The latter products are active as antihyperlidemics.

The production of the siderophores ferricrocin (**73**), ferrirubin (**74**) and ferrichrome C (**75**) by *P. fortinii* endophytes of *Carex curvula* was mentioned above.[89] It appears that the endophytes can play a broad role in plant mineral nutrition, with access to metallic elements as well as the more frequently studied phosphorus, nitrogen and carbon.

An interesting recent study resulted in the discovery of loline alkaloids in an hemiparasitic plant, *Rhinanthus serotinus*, when in association with *Neotyphodium*-infected *Lolium perenne*.[207] *R. serotinus* does not usually have alkaloids, and it was found not to be infected with the *Neotyphodium* endophyte, even though it was in close proximity to E^+ grass hosts. When feeding on E^- hosts, the hemiparasite did not accumulate alkaloids. The hemiparasite acquired the alkaloids through tapping into the phloem-stream of the grass host. The alkaloids were effective in reducing herbivory of the hemiparasite by the aphid *Aulacorhum solani*. This study fits with the data of Spiering *et al.*,[208] in which they showed that the fungal alkaloids are transported in the plant, not necessarily accumulating to the highest levels only in proximity to the endophyte in the plant.

There has been recent interest in the fate of endophytes and their products after the plant is harvested, particularly in silages[209,210] and stored grain.[211] Silages are fermented pasture plant products for the feeding of livestock. Silage containing endophytic fungi when the plant is harvested may accumulate significant quantities of alkaloidal and nonalkaloidal toxic metabolites. The products may be directly toxic, induce immune suppression, or may be metabolized to toxic or restricted products (*e.g.* metabolism of zearalenone to α-zearalanol, a growth stimulant banned in Europe).[209] *Fusarium moniliforme* forms fumonisin toxins in maize, which can be used for feed. A bacterial endophytic *Bacillus subtilis* and a fungal endophyte thought to be *Trichoderma koningii* may provide some control of the toxin-producing endophyte.[211] The mechanism of the control is unknown except that the inhibition "compound" is diffusible in agar. Another control possibility is biofumigation with *Muscodor albus*.[212]

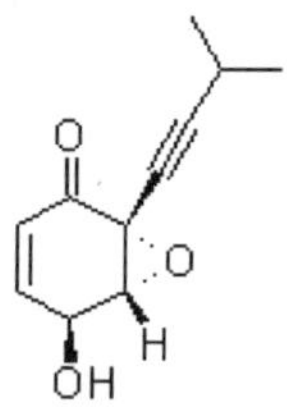

181

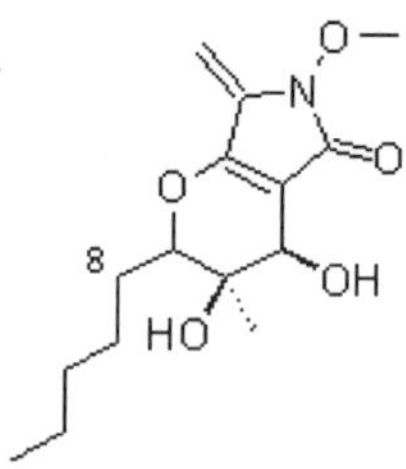

182 C-8 (*R*)
183 C-8 (*S*)

Without a doubt, the interest in endophytic fungal products began through concern about toxicants in pastures and other livestock grazing feeds. Darnel grass (*Lolium temulentum*) toxicosis has been known for at least two (and perhaps four) millennia.[213] Syndromes caused by poisoning of cattle, horses, sheep, and others with grass alkaloids produced by endophytes are known by a variety of names, including fescue toxicosis (tremors, abortion, premature birth, edematous and thickened placenta, agalactia),[214,215] ryegrass staggers,[216] sleepygrass (*Stipa robusta*) syndrome,[217] fescue foot (gangrene and loss of tails, toes or hooves due to loss of blood flow through vasoconstriction by ergot alkaloids),[217] and *tembladera* (an often fatal tremor syndrome).[217] These problems were first recognized to be related to plant endophytic fungi in 1904, studied intensively in the 1930s, and found to be due to endophyte alkaloids in the 1970s.[213] There is thus a rich literature in this area, most of it focused on relatively few alkaloidal compounds. The physiologic bases of the syndromes are well understood,[214,218,219] as are the principal endophytes involved (*Neotyphodium, Acremonium, Epichloë*).[6] The major alkaloid types (with examples of each class) include the clavine alkaloids (chanoclavine), lysergic acid derivatives (ergonovine), ergopeptine alkaloids (ergotamine, ergosine), indole alkaloids (gramine), indolediterpene alkaloids (lolilline, lolitrem B), pyrrolopyrazine alkaloids (peramine), and aminopyrrolizidine alkaloids (loline, norloline).[6,216,220] Some of these alkaloids are primarily anti-insectant (peramine, lolines), while others are tremorgenic (ergot alkaloids, paxilline, lolitrems) in mammalian herbivores. The host genetic and seasonal controls on fungal alkaloid production are beginning to be understood.[73,221–223] These alkaloids and endophyte-host interactions have been reviewed extensively,[6,224,225] and I will not repeat the data here. As would be appropriate for such a mature field, the focus has begun to shift toward a thorough understanding of the biosynthetic steps and regulation of the alkaloid production.

The genes, gene clusters and enzymes responsible for alkaloid synthesis in grass endophytes has become a focus for investigation.[226,227] Recent attention has focused on biosynthesis of the ergot and related alkaloids in *Claviceps purpurea* and *Aspergillus fumigatus*,[227–230] indolediterpene alkaloid biosynthesis in *Penicillium paxillus, Neotyphodium lolii* and *Epichloë*

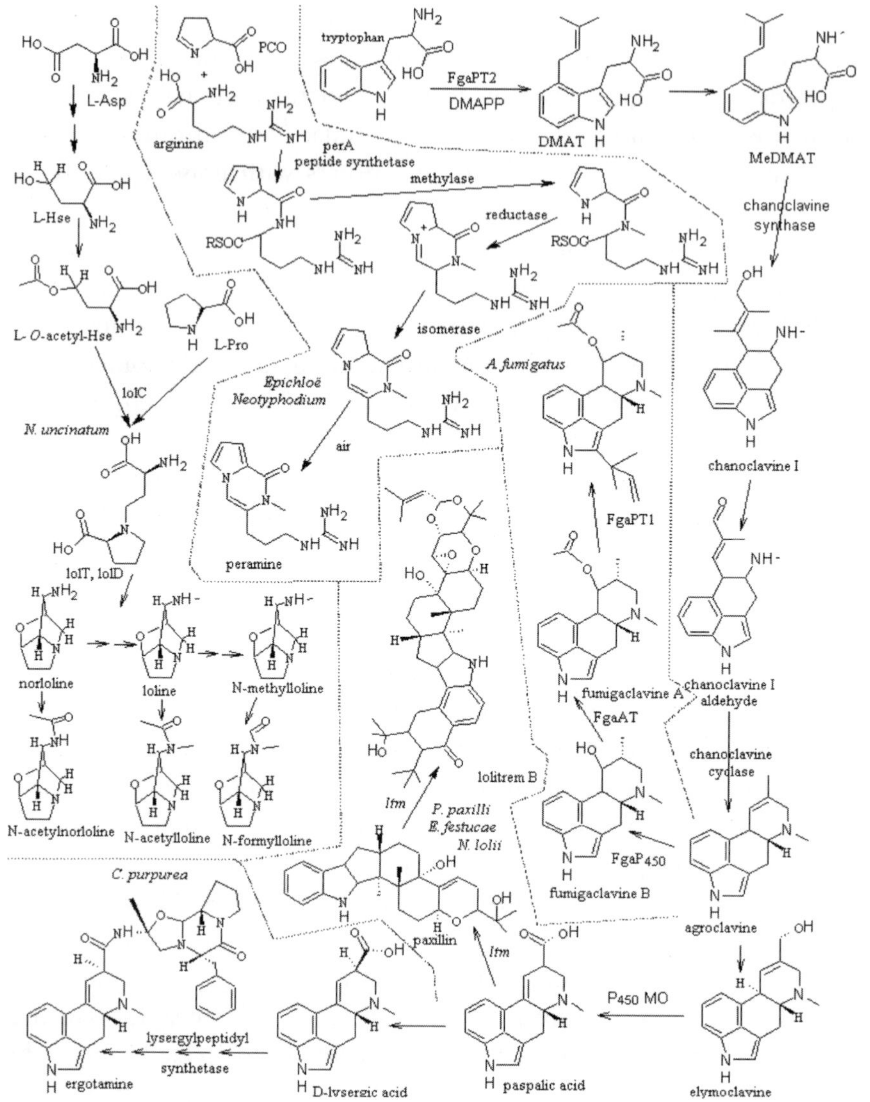

Scheme 1. Pathways to endophyte alkaloids supported by molecular (*A. fumigatus, C. purpurea, Epichloë* spp., *Neotyphodium* spp., *P. paxilli*) and labeling (*N. uncinatum*) studies. Dashed lines surround the reactions for a given organism or group of organisms. The pathway across the top right and down the right side seem to be shared across a wide spectrum of organisms. DMAPP — dimethylallyldiphosphate; DMAT — dimethylallyl tryptophan; Fga — fumigaclavin synthase enzymes; Hse — homoserine; lol — loline synthase enzymes; ltm — lolitrem synthase enzymes; MO — monooxygenase; PCO — 1-pyrroline-5-carboxylate; perA — peramine synthase; PT — prenyl transferase; R — enzyme; TS — tryptophan synthase.

festucae,[231] peramine biosynthesis in *Epichloë* and *Neotyphodium*,[232] and loline biosynthesis in *Neotyphodium uncinatum*.[226,233,234] These studies have led to the identification of gene clusters responsible for the syntheses, characterization of gene and enzyme similarities and functions across species, and the major features of the biosynthetic transformations (Scheme 1). More work needs to be done to complete our picture of alkaloid biosynthesis in grass endophytic fungi, but major progress has been made in the last few years.

One of the ongoing debates concerning the grass endophytes is whether the endophytes are primarily beneficial, detrimental, or have balanced symbioses (net neutral) with respect to their plant hosts. There are strongly held opinions on all sides. Herbivory can result in losses of 10% of the aboveground plant material and 25% below ground.[22] Clearly, the extent to which endophytes and their products reduce or control herbivore destruction confer a significant advantage to the plant. The fact that metabolite concentrations change with herbivore pressure[235] suggests that the protection of the plant is an active system. Endophyte and endophyte product control of herbivory by nematodes,[236] insects,[237,238] and mammals;[239] improved drought tolerance;[16] increased seed germination;[240] increased nutritional status; increased fitness relative to neighboring E^- plants;[241–243] and improved plant growth[225] would all argue for a strong beneficial effect of the endophyte for the plant. At the same time, periodic reproductive parasitism by some endophytes,[71] toxicity to valued livestock, higher seedling mortality,[244] and occasional failures to mitigate herbivory[77] all temper the conclusion, leading to a net neutral result or the parasitism-mutualism continuum frequently discussed in the literature.[7,16,77] An interesting final note in this consideration is that E^+ plants are less digestible by ruminants and lead to higher production of methane,[245] a significant greenhouse gas.[246] Thus, reduced presence or consumption of E^+ grasses may be an important part of the control of global warming.

15.3.5 Gene Transfer

Major questions remain to be resolved in the interaction between plant hosts and their fungal endophytes, particularly when a compound is known both from the plant tissue and from the endophyte in culture. (a) Do biosynthetic genes get transferred from the host to the endophyte? (b) Do biosynthetic genes get transferred from the endophyte to the host? or (c) Did the biosynthetic pathways arise independently? If transfer does not happen at all, then many of the products that we think of as plant metabolites are in fact fungal metabolites that reach appreciable concentrations in the plant tissues. Horizontal gene transfer has been well documented, but it is incompletely understood, especially as it relates to the interaction between endophyte and host.

Jones *et al.*[247] analyzed genes in plant parasitic nematodes and concluded that many genes have been acquired from plants, fungi and bacteria that assist the nematode in adapting to the changing chemical environment of the plant host. Conversely, a genetically engineered insert in poplar (*Populus tremula* X *P. tremuloides* T89) was not transferred to an ectomycorrhizal fungus (*Amanita muscaria*) ($n = 35,000$).[75] A mitochondrial cytochrome *c* (*coxI*) gene intron in *Peperomia polybotrya* was determined to have greater affinities to fungal *coxI* introns than to those of a variety of plants.[76] This intron can function as a ribozyme in the RNA processing, similar to their functions in yeasts and other fungi. This appears to be a case of horizontal gene transfer from a fungus with a recent origin. Some fungi are particularly susceptible to acquiring genetic material.[248] *P. microspora*, an endophyte known to produce Taxol and a number of other natural products in various hosts (Table 1), readily acquired foreign DNA when telomeric repeats were added to the donor sequence. The transfer was stable over time and could reflect a mechanism by which some endophytes acquire host metabolic pathways. The work directed by K. Kitamoto[249–251] demonstrates that *Aspergillus oryzae* and other fungi bear a functional chalcone synthase (CHS) system, previously thought to occur only in plants and some bacteria. There is a high degree of conserved sequence compared to plant CHS systems, and phylogenetic analysis suggests lateral gene transfer as an origin of these genes in fungi. The fungi analyzed were primarily plant pathogens, plant endophytes or close relatives of these. Thus, horizontal transfer of functional material does occur between plants and fungi, and this has occurred at various times. However, it may be rare and occurs primarily when the fungus and plant are in close association over an extended period of time.

What regulates the biosynthesis of metabolites by the endophyte within the plant? Many endophytes appear to produce the metabolites only under very special conditions. For example, many of the endophytes detailed above only produce the bioactive metabolities when grown on specific media, such as plant leaf agars, or directly on maize or cereal products. It is not unusual for secondary metabolite synthesis in cultured fungi to decrease over time, and it would be useful to know the metabolic or environmental triggers that keep metabolite production high. Fungal elicitors prepared from cell walls or extracts are often used to stimulate production of metabolites in tissue, cell and organ cultures,[*e.g.*,252] although elicitation of fungal production by plant tissue preparations is less well studied. The *laeA* system of *Aspergillus* spp.[253] specifically regulates secondary metabolite biosynthesis. Similar systems are likely to occur in endophytic fungi, with activation by molecules or regulators in the plant tissues. Such regulatory systems await discovery.

15.4 CONCLUSIONS

Plant fungal endophytes are a rich source of novel, bioactive metabolites. The diversity of chemistries is enormous, and, if anything, the rate of discovery of new compounds seems to be increasing. There are yet a number of questions waiting to be resolved. (1) Are the biosynthetic pathways, genes and enzymes in endophytic fungi the same as previously described in the plant (suggesting horizontal gene transfer), and what is the ultimate origin of these genes? (2) Are "plant" biosynthetic pathways actually endophytic fungal biosynthetic pathways (at least for some compounds)? (3) Do the biosynthetic genes that reside in endophytes also exist in ancestral nonendophytic fungi, connecting different endophytes with similar biosynthetic pathways evolutionarily? (4) What are the regulatory factors of metabolite production *in planta*, and can these be replicated outside the plant? The answers to these and related questions will enrich our understanding of metabolic potentials and hopefully place the regulation under human control.

REFERENCES

1. Hallmann J *et al.*, Bacterial endophytes in agricultural crops, *Can J Microbiol* **43**:895–914, 1997.
2. Kobayashi DY, Palumbo JD, Bacterial endophytes and their effects on plants and uses in agriculture, in Bacon CW, White JF Jr (eds.), *Microbial Endophytes* Marcel Dekker, Inc., New York, NY, 199–236, 2000.
3. Rosenblueth M, Martinez-Romero E, Bacterial endophytes and their interactions with hosts, *Molec Plant Microbe Interact* **19**:827–837, 2006.
4. Jumpponen A, Trappe JM, Dark septate endophytes:A review of facultative biotrophic root-colonizing fungi, *New Phytol* **140**:295–310, 1998.
5. Saikkonen K *et al.*, Evolution of endophyte-plant symbioses, *Trends Plant Sci* **9**:275–280, 2004.
6. Schardl CL, *Epichloë festucae* and related mutualistic symbionts of grasses, *Fungal Genet Biol* **33**:69–82, 2001.
7. Schulz B, Boyle C, The endophytic continuum, *Mycol Res* **109**:661–686, 2005.
8. Gunatilaka AAL, Natural products from plant-associated microorganisms: Distribution, structural diversity, bioactivity, and implications of their occurrence, *J Nat Prod* **69**:509–526, 2006.
9. Strobel GA *et al.*, Natural products from endophytic microorganisms, *J Nat Prod* **67**:257–268, 2004.
10. Tan RX, Zou WX, Endophytes: a rich source of functional metabolites, *Nat Prod Rep* **18**:448–459, 2001.
11. Zhang HW *et al.*, Biology and chemistry of endophytes, *Nat Prod Rep* **23**:753–771, 2006.
12. De Bary A, Comparative Morphology and Biology of the Fungi, Mycetezoa and Bacteria, Oxford at the Clarendon Press, Oxford, UK, 525, 1887.
13. Stone JK *et al.*, An overview of endophytic microbes: Endophytism defined, in Bacon CW, White JF, Jr (eds.), *Microbial Endophytes* Marcel Dekker, Inc., New York, NY, pp. 3–29, 2000.
14. Kogel KH *et al.*, Endophyte or parasite—what decides?, *Curr Opin Plant Biol* **9**:358–363, 2006.
15. Hyakumachi M, Kubota M, Fungi as plant growth promoter and disease suppressor, *Mycol Ser* **21**:101–110, 2004.
16. Müller CB, Krauss J, Symbiosis between grasses and asexual fungal endophytes, *Curr Opin Plant Biol* **8**:450–456, 2005.
17. Pinto LSRC *et al.*, Symptomless infection of banana and maize by endophytic fungi impairs photosynthetic efficiency, *New Phytol* **147**:609–615, 2000.

18. Godbold DL, Sharrock R, in Schroth G, Sinclair FL (eds.), Mycorrhizas, in Trees, Crops and Soil Fertility. Concepts and Research Methods, CABI Publishing, Cambridge, MA, 271–287, 2003.
19. Smith SE, Read DJ, Mycorrhizal Symbiosis, 2nd ed, Academic Press, San Diego, CA, pp. 605, 1997.
20. Brundrett M, Diversity and classification of mycorrhizal associations, *Biol Rev* **79**:473–495, 2004.
21. Barker SJ *et al.*, Regulation of root and fungal morphogenesis in mycorrhizal symbioses, *Plant Physiol* **116**:1201–1207, 1998.
22. Kyde MM, Gould AB, Mycorrhizal endosymbiosis, in Bacon CW, White JF Jr (eds.), *Microbial Endophytes* Marcel Dekker, Inc., New York, NY, 161–198, 2000.
23. Bécard G *et al.*, Partner communication in the arbuscular mycorrhizal interaction, *Can J Bot* **82**:1186–1197, 2004.
24. Mayer AM, Polyphenol oxidases in plants and fungi:Going places? A review, *Phytochemistry* **67**:2318–2331, 2006.
25. Sauer M *et al.*, Estimating polyketide metabolic potential among non-sporulating fungal endophytes of *Vaccinium macrocarpon, Mycol Res* **106**:460–470, 2002.
26. Oksman-Caldentey KM, Inze D, Plant cell factories in the post-genomic era: New ways to produce designer secondary metabolites, *Trends Plant Sci* **9**:433–440, 2004.
27. Jumpponen A, Dark septate endophytes — are they mycorrhizal?, *Mycorrhiza* **11**:207–211, 2001.
28. Vohnik M *et al.*, The inoculation with *Oidiodendron maius* and *Phialocephala fortinii* alters phosphorus and nitrogen uptake, foliar C:N ratio and root biomass distribution in *Rhododendron* cv. Azurro, *Symbiosis* **40**:87–96, 2005.
29. Zijlstra JD *et al.*, Diversity of symbiotic root endophytes of the Helotiales in ericaceous plants and the grass, *Deschampsia flexuosa, Stud Mycol* **53**:147–162, 2005.
30. Harney SK *et al.*, Molecular characterization of dematiaceous root endophytes, *Mycol Res* **101**:1397–1404, 1997.
31. Larran S *et al.*, Isolation and analysis of endophytic microorganisms in wheat (*Triticum aestivum* L.) leaves, *World J Microbiol Biotechnol* **18**:683–686, 2002.
32. Pirttila AM *et al.*, Seasonal variations in location and population structure of endophytes in buds of Scots pine, *Tree Physiol* **25**:289–297, 2005.
33. Suryanarayanan TS *et al.*, Characterization of the melanin pigment of a cosmopolitan fungal endophyte, *Mycol Res* **108**:974–978, 2004.
34. Mandyam K, Jumpponen A, Seeking the elusive function of the root-colonising dark septate endophytic fungi, *Stud Mycol* **53**:173–189, 2005.

35. Barrow JR, Atypical morphology of dark septate fungal root endophytes of *Bouteloua* in arid southwestern USA rangelands, *Mycorrhiza* **13**:239–247, 2003.
36. Barrow JR *et al.*, Fungal root endophytes in fourwing saltbush, *Atriplex canescens*, on arid rangelands of southwestern USA, *Arid Soil Res Rehab* **11**:177–185, 1997.
37. Schardl CL, Craven KD, Interspecific hybridization in plant-associated fungi and oomycetes: A review, *Mol Ecol* **12**:2861–2873, 2003.
38. Summerbell RC, From Lamarckian fertilizers to fungal castles:recapturing the pre-1985 literature on endophytic and saprotrophic fungi associated with ectomycorrhizal root systems, *Stud Mycol* **53**:191–256, 2005.
39. Jacobson DJ *et al.*, Persistent, systemic, asymptomatic infections of *Albugo candida*, an oomycete parasite, detected in three wild crucifer species, *Can J Bot* **76**:739–750, 1998.
40. Crozier J *et al.*, Molecular characterization of fungal endophytic morphospecies isolated from stems and pods of *Theobroma cacao, Plant Pathol* **55**:783–791, 2006.
41. Evans HC *et al.*, Endophytes and mycoparasites associated with an indigenous forest tree, *Theobroma gileri,* in Ecuador and a preliminary assessment of their potential as biocontrol agents of cocoa diseases, *Mycol Prog* **2**:149–160, 2003.
42. Rubini MR *et al.*, Diversity of endophytic fungal community of cacao (*Theobroma cacao* L.) and biological control of *Crinipellis perniciosa*, causal agent of witches, broom disease, *Intl J Biol Sci* **1**:24–33, 2005.
43. Eyberger AL *et al.*, Endophyte fungal isolates from *Podophyllum peltatum* produce podophyllotoxin, *J Nat Prod* **69**:1121–1124, 2006.
44. Barengo N *et al.*, Diversity of endophytic mycobiota in leaves and twigs of pubescent birch (*Betula pubescens), Sydowia* **52**:305–320, 2000.
45. Lumyong S *et al.*, Endophytes from Thailand:Isolation and screening for tartaric acid and oligosaccharide production, *Biotechnol Sust Util Biol Res Tropics* **12**:394–403, 1998.
46. Metz AM *et al.*, Induction of the sexual stage of *Pestalotiopsis microspora*, a taxol-producing fungus, *Microbiology* **146**:2079–2089, 2000.
47. Ganjali R *et al.*, Classical methods and specific primers in detection of endophytic fungi in some gramineuos plants, *Rostaniha* **5**:15–18, 2004.
48. Hignight KW *et al.*, A clearing technique for detecting the fungal endophyte *Acremonium* sp. in grasses, *Biotech Histochem* **68**:87–90, 1993.
49. Saha DC *et al.*, A rapid staining method for detection of endophytic fungi in turf and forage grasses, *Phytopathology* **78**:237–239, 1988.
50. Vierheilig H *et al.*, An overview of methods for the detection and observation of arbuscular mycorrhizal fungi in roots, *Physiol Plant* **125**:393–404, 2005.

51. Duckett JG, Read DJ, The use of the fluorescent dye, 3, 3′-dihexyloxacarbocyanine iodide, for selective staining of ascomycete fungi associated with liverwort rhizoids and ericoid mycorrhizal roots, *New Phytol* **118**:259–272, 1991.
52. Jurc M *et al.*, Ergosterol content of endophytic fungi from the needles of the Austrian pine (*Pinus nigra* Arn.), *Acta Biol Sloven* **41**:23–33, 1997.
53. Kurtzman CP, Robnett CJ, Identification of clinically important ascomycetous yeasts based on nucleotide divergence in the 5′ end of the large-subunit (26S) ribosomal DNA gene, *J Clin Microbiol* **35**:1216–1223, 1997.
54. Tanaka M *et al.*, Isolation of endophytes from plants in Southeast Asia and Japan, and their identification by 18S rRNA gene, *Biotechnol Sust Util Biol Res Tropics* **13**:227–232, 1999.
55. Untereiner WA *et al.*, Evolutionary relationships of *Hyphodiscus hymeniophilus* (anamorph *Catenulifera rhodogena*) inferred from *beta*-tubulin and nuclear ribosomal DNA sequences, *Can J Bot* **84**:243–253, 2006.
56. Vandenkoornhuyse P *et al.*, Extensive fungal diversity in plant roots, *Science* **295**:2051, 2002.
57. White TJ *et al.*, Amplification and direct sequencing of fungal ribosomal RNA genes for phylogenetics, in *Innis M, Gelfand DH, Sninsky JJ, White TJ (eds.), PCR Protocols: A Guide to Methods and Applications*, Academic Press, San Diego, CA, pp. 315–322, 1989.
58. Druzhinina IS *et al.*, The first 100 *Trichoderma* species characterized by molecular data, *Mycoscience* **47**:55–64, 2006.
59. Luchi N *et al.*, A real-time quantitative PCR assay for the detection of *Sphaeropsis sapinea* from inoculated *Pinus nigra shoots, J Phytopathol* **153**:37–42, 2005.
60. Addy HD *et al.*, Distribution and molecular characterization of the root endophyte *Phialocephala fortinii* along an environmental gradient in the boreal forest of Alberta, *Mycol Res* **104**:1213–1221, 2000.
61. Grünig CR, Sieber TN, Molecular and phenotypic description of the widespread root symbiont *Acephala applanata gen. et sp nov.*, formerly known as dark-septate endophyte Type 1, *Mycologia* **97**:628–640, 2005.
62. Photita W *et al.*, Morphological and molecular characterization of *Colletotrichum* species from herbaceous plants in Thailand, *Fung Divers* **18**:117–133, 2005.
63. Mohali S *et al.*, Two new *Fusicoccum* species from *Acacia* and *Eucalyptus* in Venezuela, based on morphology and DNA sequence data, *Mycol Res* **110**:405–413, 2006.
64. Peterson SW *et al.*, *Penicillium coffeae*, a new endophytic species isolated from a coffee plant and its phylogenetic relationship to *P. fellutanum, P. thiersii* and *P. brocae* based on parsimony analysis of multilocus DNA sequences, *Mycologia* **97**:659–666, 2005.

65. Steiner U *et al.*, Molecular characterization of a seed transmitted clavicipitaceous fungus occurring on dicotyledoneous plants (Convolvulaceae), *Planta* **224**:533–544, 2006.
66. Lutzoni F *et al.*, Assembling the fungal tree of life:Progress, classification, and evolution of subcellular traits, *Am J Bot* **91**:1446–1480, 2004.
67. Attitalla IH *et al.*, A rapid molecular method for differentiating two special forms (*lycopersici* and *radicis-lycopersici*) of *Fusarium oxysporum, Mycol Res* **108**:787–794, 2004.
68. Bougoure DS, Cairney JWG, Assemblages of ericoid mycorrhizal and other root-associated fungi from *Epacris pulchella* (Ericaceae) as determined by culturing and direct DNA extraction from roots, *Environ Microbiol* **7**:819–827, 2005.
69. Grünig CR *et al.*, Evidence for subdivision of the root-endophyte *Phialocephala fortinii* into cryptic species and recombination within species, *Fungal Genet Biol* **41**:676–687, 2004.
70. Jumpponen A *et al.*, Filamentous ascomycetes inhabiting the rhizoid environment of the liverwort *Cephaloziella varians* in Antarctica are assessed by direct PCR and cloning, *Mycologia* **95**:457–466, 2003.
71. Faeth SH, Are endophytic fungi defensive plant mutualists?, *Oikos* **98**:25–36, 2002.
72. Saikkonen K *et al.*, The persistence of vertically transmitted fungi in grass metapopulations, *Proc Roy Soc Lond B Biol Sci* **269**:1397–1403, 2002.
73. Easton HS *et al.*, Ryegrass host genetic control of concentrations of endophyte-derived alkaloids, *Crop Sci* **42**:51–57, 2002.
74. Zhang L *et al.*, Isolation and determination of the anti-cancer substance chaetoglobosin A from endophytic fungus of *Maytenus hookerii*, *Zhongguo Yaoxue Zazhi* **37**:172–175, 2002.
75. Zhang C *et al.*, Investigation of horizontal gene transfer in poplar/*Amanita muscaria* ectomycorrhizas, *Environ Biosaf Res* **4**:235–242, 2005.
76. Vaughan JC *et al.*, Fungal origin by horizontal transfer of a plant mitochondrial group I intron in the chimeric *CoxI* gene of *Peperomia, J Molec Evol* **41**:563–572, 1995.
77. Saikkonen K *et al.*, Fungal endophytes: A continuum of interactions with host plants, *Annu Rev Ecol Syst* **29**:319–343, 1998.
78. Hata K *et al.*, Seasonal and needle age-dependent changes of the endophytic mycobiota in *Pinus thunbergii* and *Pinus densiflora* needles, *Can J Bot* **76**:245–250, 1998.
79. Osono T, Mori A, Seasonal and leaf age-dependent changes in occurrence of phyllosphere fungi of giant dogwood, *Mycoscience* **46**:273–279, 2005.

80. Behnweg G *et al.*, Beech leaf colonization by the endophyte *Apiognomonia errabunda* dramatically depends on light exposure and climatic conditions, *Plant Biol* 7:659–669, 2005.
81. Bailey JK *et al.*, Host plant genetics affect hidden ecological players: Links among *Populus*, condensed tannins, and fungal endophyte infection, *Can J Bot* **83**:356–361, 2005.
82. Summerbell RC, Root endophyte and mycorrhizosphere fungi of black spruce, *Picea mariana*, in a boreal forest habitat: influence of site factors on fungal distributions, *Stud Mycol* **53**:121–145, 2005.
83. Lorenzi E, Picco AM, Isolation of fungal endophytes from *Picea abies* (L.) Karsten with some different methodological approaches, *Micol Ital* **33**:50–58, 2005.
84. Morakotkarn D *et al.*, Molecular diversity of bamboo-associated fungi isolated from Japan, *FEMS Microbiol Lett* **266**:10–19, 2007.
85. Arnold AE *et al.*, Are tropical fungal endophytes hyperdiverse?, *Ecol Lett* **3**:267–274, 2000.
86. Bergemann SE, Garbelotto M, High diversity of fungi recovered from the roots of mature tanoak (*Lithocarpus densiflorus*) in northern California, *Can J Bot* **84**:1380–1394, 2006.
87. Zhou D, Hyde KD, Host-specificity, host-exclusivity, and host-recurrence in saprobic fungi, *Mycol Res* **105**:1449–1457, 2001.
88. Shardl CL, Wilkinson HH, Hybridization and cospeciation hypotheses for the evolution of grass endophytes, in Bacon CW, White JF, Jr (eds.), *Microbial Endophytes* Marcel Dekker, Inc., New York, NY, pp. 63–84, 2000.
89. Bartholdy BA *et al.*, Hydroxamate siderophore synthesis by *Phialocephala fortinii*, a typical dark septate fungal root endophyte, *BioMetals* **14**:33–42, 2001.
90. Jumpponen A, Spatial distribution of discrete RAPD phenotypes of a root endophytic fungus, *Phialocephala fortinii*, at a primary successional site on a glacier forefront, *New Phytol* **141**:333–344, 1999.
91. Menkis A *et al.*, Ecology and molecular characterization of dark septate fungi from roots, living stems, coarse and fine woody debris, *Mycol Res* **108**:965–973, 2004.
92. Narisawa K *et al.*, Suppression of *Verticillium* wilt in eggplant by some fungal root endophytes, *Europ J Plant Pathol* **108**:103–109, 2002.
93. Queloz V *et al.*, Monitoring the spatial and temporal dynamics of a community of the tree–root endophyte *Phialocephala fortinii* sl, *New Phytol* **168**:651–660, 2005.
94. Ruotsalainen AL, Kytoviita M-M, Mycorrhiza does not alter low temperature impact on *Gnaphalium norvegicum, Oecologia* **140**:226–233, 2004.

95. Yu T *et al.*, Characterization of the interaction between the dark septate fungus *Phialocephala fortinii* and *Asparagus officinalis* roots, *Can J Microbiol* **47**:741–753, 2001.
96. Kuldau GA, Yates IE, Evidence for *Fusarium* endophytes in cultivated and wild plants, in Bacon CW, White JF Jr (eds.), *Microbial Endophytes* Marcel Dekker, Inc., New York, NY, 85–120, 2000.
97. Fröhlich J *et al.*, Endophytic fungi associated with palms, *Mycol Res* **104**:1202–1212, 2000.
98. Addy HD *et al.*, Microfungal endophytes in roots, *Can J Bot* **83**:1–13, 2005.
99. Grünig CR *et al.*, Development of single-copy RFLP markers for population genetic studies of *Phialocephala fortinii* and closely related taxa, *Mycol Res* **107**:1332–1341, 2003.
100. Dondapati R, Identification and Molecular Characterization of Endophytic Fungi Isolated from *Podophyllum peltatum*, Department of Biological Sciences, University of the Sciences in Philadelphia, Philadelphia, PA, pp. 63, 2004.
101. Newcome JM, Molecular Identification of Endophyte Fungi of the Medicinal Plant *Podophyllum peltatum*, Department of Biological Sciences, University of the Sciences in Philadelphia, Philadelphia, PA, pp. 52, 2004.
102. Strobel GA, Rainforest endophytes and bioactive products, *Crit Rev Biotechnol* **22**:315–333, 2002.
103. Wang K, Zhao Z, Occurrence of arbuscular mycorrhizas and dark septate endophytes in hydrophytes from lakes and streams in southwest China, *Intl Rev Hydrobiol* **91**:29–37, 2006.
104. Kottke I, Nebel M, The evolution of mycorrhiza-like associations in liverworts:An update, *New Phytol* **167**:330–334, 2005.
105. Read DJ *et al.*, Symbiotic fungal associations in 'lower' land plants, *Philosophical Transactions of the Royal Society of London B Biological Sciences* **355**:815–831, 2000.
106. Zuccaro A *et al.*, Molecular detection of ascomycetes associated with *Fucus serratus, Mycol Res* **107**:1451–1466, 2003.
107. Osterhage C *et al.*, 5-Hydroxyramulosin, a new natural product produced by *Phoma tropica*, a marine-derived fungus isolated from the alga *Fucus spiralis, Plant Med* **68**:1052–1054, 2002.
108. Osterhage C *et al.*, Rare sesquiterpenes from the algicolous fungus *Drechslera dematioidea, J Nat Prod* **65**:306–313, 2002.
109. Schlörke O, Isolierung, Strukturaufklärung und Biosynthese von Sekundärmetaboliten endophytischer Pilze aus Algen und Pflanzen mariner Habitate, Mathematisch-naturwissenschaftliche Fakultäten, Georg-August-Universität, Göttingen, pp. 214, 2005.

110. Schlörke O, Zeeck A, Orsellides A-E: an example for 6-deoxyhexose derivatives produced by fungi, *Europ J Org Chem*, 1043–1049, 2006.
111. Gomes AT *et al.*, Cytotoxic activity of orsellinates *Z Naturforsch C J Biosci* **61**:653–657, 2006.
112. Bashyal BP *et al.*, Globosumones A-C, cytotoxic orsellinic acid esters from the Sonoran Desert endophytic fungus *Chaetomium globosum, J Nat Prod* **68**:724–728, 2005.
113. Osterhage C *et al.*, Ascosalipyrrolidinone A, an antimicrobial alkaloid, from the obligate marine fungus *Ascochyta salicorniae, J Org Chem* **65**:6412–6417, 2000.
114. Osterhage C, Isolation, Structure Determination and Biological Activity Assessment of Secondary Metabolites from Marine-derived Fungi, Technischen Universität Carolo-Wilhelmina zu Braunschweig, Braunschweig, pp. 197, 2001.
115. Seibert SF *et al.*, Ascospiroketals A and B, unprecedented cycloethers from the marine-derived fungus *Ascochyta salicorniae, Org Lett*, in press, 2007.
116. Wang S *et al.*, Chaetopyranin, a benzaldehyde derivative, and other related metabolites from *Chaetomium globosum*, an endophytic fungus derived from the marine red alga *Polysiphonia urceolata*, *J Nat Prod* **69**:1622–1625, 2006.
117. Yang Ry *et al.*, Lactones from a brown alga endophytic fungus (No. ZZF36) from the South China Sea and their antimicrobial activities, *Bioorg Med Chem Lett* **16**:4205–4208, 2006.
118. Lang G *et al.*, Bioactivity profiling using HPLC/microtiter-plate analysis: Application to a New Zealand marine alga-derived fungus, *Gliocladium* sp, *J Nat Prod* **69**:621–624, 2006.
119. Kralj A *et al.*, Arugosins G and H:Prenylated polyketides from the marine-derived fungus *Emericella nidulans var. acristata, J Nat Prod* **69**:995–1000, 2006.
120. Wijeratne EMK *et al.*, Five new isocoumarins from Sonoran desert plant-associated fungal strains *Paraphaeosphaeria quadriseptata* and *Chaetomium chiversii, Tetrahedron* **62**:8439–8446, 2006.
121. Sumarah MW *et al.*, Measurement of a rugulosin-producing endophyte in white spruce seedlings, *Mycologia* **97**:770–776, 2005.
122. Park J-H *et al.*, Griseofulvin from *Xylaria* sp strain F0010, an endophytic fungus of *Abies holophylla* and its antifungal activity against plant pathogenic fungi, *J Microbiol Biotechnol* **15**:112–117, 2005.
123. Strack D *et al.*, Arbuscular mycorrhiza: Biological, chemical, and molecular aspects, *J Chem Ecol* **29**:1955–1979, 2003.
124. Soares ACF *et al.*, Arbuscular mycorrhizal fungi and the occurrence of flavonoids in roots of passion fruit seedlings, *Scient Agric* **62**:331–336, 2005.
125. Zhao X, Yan XF, Effects of arbuscular mycorrhizal fungi on plant secondary metabolism, *Zhiwu Shengtai Xuebao* **30**:514–521, 2006.

126. Zhao X *et al.*, Effects of arbuscular mycorrhiza on camptothecin content in *Camptotheca acuminata* seedlings, *Shengtai Xuebao* **26**:1057–1062, 2006.
127. Debaud JC *et al.*, Genetics and molecular biology of the fungal partner in the ectomycorrhizal symbiosis *Hebeloma cylindrosporum* X *Pinus pinaster*, in Carroll GC and Tudzynski P (eds.), *The Mycota. A Comprehensive Treatise on Fungi as Experimental Systems for Basic and Applied Research. V. Plant Relationships. Part B.*, Springer, Berlin, pp. 95–115, 1997.
128. Küster H *et al.*, Development of bioinformatic tools to support EST-sequencing, *in silico*- and microarray-based transcriptome profiling in mycorrhizal symbioses, *Phytochemistry* **68**:19–32, 2007.
129. Marx DH, Mycorrhizae and feeder root diseases, in Marks GC, Kozlowski TT (eds.), *Ectomycorrhizae: Their Ecology and Physiology*, Academic Press, New York, NY, 351–382, 1973.
130. Singh R *et al.*, Mycorrhiza in control of soil borne pathogens, in Mukerji KG, Chamola BP, and Singh J, (eds.), *Mycorrhizal Biology* Kluwer Academic/Plenum Publ., New York, NY, 173–196, 2000.
131. Gupta V *et al.*, General aspects of mycorrhiza, in Mukerji KG, Chamola BP, Singh J (eds.), *Mycorrhizal Biology*, Kluwer Academic/Plenum Publishers, New York, pp. 27–44, 2000.
132. Whipps JM, Prospects and limitations for mycorrhizas in biocontrol of root pathogens, *Can J Bot* **82**:1198–1227, 2004.
133. Anchel M, Metabolic products of *Clitocybe diatreta*. I. Diatretyne amide and diatretyne nitrile, *Arch Biochem Biophys* **78**:100–110, 1958.
134. Strobel GA, *Muscodor albus* and its biological promise, *J Ind Microbiol Biotechnol* **33**:514–522, 2006.
135. Krupa S, Fries N, Ectomycorrhizae of pine. I. Production of volatile organic compounds, *Can J Bot* **49**:1425–1431, 1971.
136. Stadler M *et al.*, New metabolites with nematicidal and antimicrobial activities from the ascomycete *Lachnum papyraceum* (Karst.) Karst. VII. Structure determination of brominated lachnumon and mycorrhizin A derivatives, *J Antibiot* **48**:158–161, 1995.
137. Stalhandske C *et al.*, Chloromycorrhizin A, *Acta Crystallograph B: Struct Crystallograph Cryst Chem* **B33**:870–873, 1977.
138. Trofast J, Wickberg B, Mycorrhizin A and chloromycorrhizin A, two antibiotics from a mycorrhizal fungus of *Monotropa hypopitys L*, *Tetrahedron* **33**:875–879, 1977.
139. Stadler M *et al.*, Lachnumon and lachnumol A, new metabolites with nematicidal and antimicrobial activities from the ascomycete *Lachnum papyraceum* (Karst.) Karst. I. Producing organism, fermentation, isolation and biological activities, *J Antibiot* **49**:961–967, 1993.

140. Shan R *et al.*, New metabolites with nematicidal and antimicrobial activities from the ascomycete *Lachnum papyraceum* (Karst.) Karst. VIII. Isolation, structure determination and biological activities of minor metabolites structurally related to mycorrhizin A, *J Antibiot* **49**:447–452, 1996.
141. Stadler M *et al.*, Metabolites with nematicidal and antimicrobial activities from the ascomycete *Lachnum papyraceum* (Karst.) Karst. V. Production, isolation and biological activities of bromine-containing mycorrhizin and lachnumon derivatives and four additional new bioactive metabolites, *J Antibiot* **48**:149–153, 1995.
142. Anke H. *et al.*, Secondary metabolites with nematicidal and antimicrobial activity from nematophagous fungi and Ascomycetes, *Can J Bot* **73**:S932–S939, 1995.
143. Hilbert JL *et al.*, Ectomycorrhizin synthesis and polypeptide changes during the early stage of eucalypt mycorrhiza development, *Plant Physiol* **97**:977–984, 1991.
144. Tikhonovich IA *et al.*, Integration of plant and microbial genetic systems in symbiosis, *Uspek Sovrem Biol* **125**:227–238, 2005.
145. Wyss P *et al.*, Vesicular-arbuscular mycorrhizas of wild-type soybean and non-nodulating mutants with *Glomus mosseae* contain symbiosis-specific polypeptides (mycorrhizins), immunologically cross-reactive with nodulins, *Planta* **182**:22–26, 1990.
146. Haselwandter K *et al.*, Isolation and identification of hydroxamate siderophores of ericoid mycorrhizal fungi, *BioMetals* **5**:51–56, 1992.
147. Hause B *et al.*, Jasmonates in arbuscular mycorrhizal interactions, *Phytochemistry* **68**:101–110, 2007.
148. Mandelbaum CI, Piche Y, The role of root exudates in arbuscular mycorrhiza initiation, in Mukerji KG, Chamola BP, Singh J (eds.), *Mycorrhizal Biology* Kluwer Academic/Plenum Publishers, New York, 153–172, 2000.
149. Johansson T *et al.*, Transcriptional responses of *Paxillus involutus* and *Betula pendula* during formation of ectomycorrhizal root tissue, *Molec Plant Microbe Interact* **17**:202–215, 2004.
150. Poma A *et al.*, Effect of tyrosinase inhibitors on *Tuber borchii* mycelium growth *in vitro, FEMS Microbiol Lett* **180**:69–75, 1999.
151. Phongpaichit S *et al.*, Antimicrobial activity in cultures of endophytic fungi isolated from *Garcinia* species, *FEMS Immunol Med Microbiol*, 2006.
152. Sette LD *et al.*, Molecular characterization and antimicrobial activity of endophytic fungi from coffee plants, *World J Microbiol Biotechnol* **22**:1185–1195, 2006.
153. Tejesvi MV *et al.*, Endophytic fungal assemblages from inner bark and twig of *Terminalia arjuna* W. & A. (Combretaceae), *World J Microbiol Biotechnol* **21**:1535–1540, 2005.

154. Tejesvi MV *et al.*, Fungal endophyte assemblages from ethnopharmaceutically important medicinal trees, *Can J Microbiol* **52**:427–435, 2006.
155. Chomcheon P *et al.*, Cyclopentenones, scaffolds for organic syntheses produced by the endophytic fungus mitosporic dothideomycete sp. LRUB20, *J Nat Prod* **69**:1351–1353, 2006.
156. Wright N *et al.*, Structure elucidation of bio-active compounds isolated from endophytes of *Alstonia scholaris* and *Acmena graveolens*, 19 Rocky Mtn Reg Mtg ACS, Tucson, AZ, October 14–18, 2006.
157. Abate D *et al.*, Cytochalasins and phytotoxins from the fungus *Xylaria obovata, Phytochemistry* **44**:1443–1448, 1997.
158. Weber D *et al.*, Metabolites from endophytes of the medicinal plant *Erythrina crista-galli, Zeitschrift für Naturforschung C Journal of Biosciences* **60**:467–477, 2005.
159. Assante G *et al.*, Acremines A-F, novel secondary metabolites produced by a strain of an endophytic *Acremonium*, isolated from sporangiophores of *Plasmopara viticola* in grapevine leaves, *Tetrahedron* **61**:7686–7692, 2005.
160. Davis RA *et al.*, Phomoxins B and C: Polyketides from an endophytic fungus of the genus *Eupenicillium, Phytochemistry* **66**:2771–2775, 2005.
161. Teles HL *et al.*, Benzopyrans from *Curvularia* sp., an endophytic fungus associated with *Ocotea corymbosa* (Lauraceae), *Phytochemistry* **66**:2363–2367, 2005.
162. Teles HL *et al.*, Aromatic compounds produced by *Periconia atropurpurea*, an endophytic fungus associated with *Xylopia aromatica, Phytochemistry* **67**:2686–2690, 2006.
163. Krohn K *et al.*, Fusidilactones, a new group of polycyclic lactones from an endophyte, *Fusidium* sp., *Europ J Org Chem* **2002**:2331–2336, 2002.
164. Venkatasubbaiah P, Chilton WS, Toxins produced by the dogwood anthracnose fungus *Discula* sp., *J Nat Prod* **54**:1293–1297, 1991.
165. Inácio ML *et al.*, Antifungal metabolites from *Colletotrichum gloeosporioides*, an endophytic fungus in *Cryptocarya mandioccana* Nees (Lauraceae), *Biochem Syst Ecol* **34**:822–824, 2006.
166. Silva GH *et al.*, Cadinane sesquiterpenoids of *Phomopsis cassiae*, an endophytic fungus associated with *Cassia spectabilis* (Leguminosae), *Phytochemistry* **67**:1964–1969, 2006.
167. Sohrab H, Isolation and Structure Elucidation of Secondary Metabolites from Endophytic Fungi and the plant *Prismatomeris tetrandra* and Synthesis of (+)-Ochromycinone, Department Chemie, Universität Paderborn, Paderborn, Germany, pp. 141, 2005.
168. Ge HM *et al.*, Paranolin:A new xanthene-based metabolite from *Paraphaeosphaeria nolinae,Helv Chim Acta* **89**:502–506, 2006.

169. Owen NL, Hundley N, Endophytes: The chemical synthesizers inside plants, *Sci Prog* **87**:79–99, 2004.
170. van der Sar SA *et al.*, spiro-Mamakone A: A unique relative of the spirobisnaphthalene class of compounds, *Org Lett* **8**:2059–2061, 2006.
171. Pittayakhajonwut P *et al.*, Depsidones from the endophytic fungus BCC 8616, *J Nat Prod* **69**:1361–1363, 2006.
172. Elsässer B *et al.*, X-ray structure determination, absolute configuration and biological activity of phomoxanthone A, *Europ J Org Chem*, 4563–4570, 2005.
173. Pornpakakul S *et al.*, Cytotoxic activity of four xanthones from *Emericella variecolor*, an endophytic fungus isolated from *Croton oblongifolius, Arch Pharmac Res* **29**:140–144, 2006.
174. Isaka M *et al.*, Phomoxanthones A and B, novel xanthone dimers from the endophytic fungus *Phomopsis* species, *J Nat Prod* **64**:1015–1018, 2001.
175. Marinho AMdR *et al.*, Biologically active polyketides produced by *Penicillium janthinellum* isolated as an endophytic fungus from fruits of *Melia azedarach, J Brazil Chem Soc* **16**:280–283, 2005.
176. Borges WdS, Pupo MT, Novel anthraquinone derivatives produced by *Phoma sorghina*, an endophyte found in association with the medicinal plant *Tithonia diversifolia* (Asteraceae), *J Brazil Chem Soc* **17**:929–934, 2006.
177. Agusta A *et al.*, Bisanthraquinone metabolites produced by the endophytic fungus *Diaporthe* sp, *Chem Pharm Bull* **54**:579–582, 2006.
178. Gu W *et al.*, Cytotoxic benzo[*j*]fluoranthene metabolites from *Hypoxylon truncatum* IFB-18, an endophyte of *Artemisia annua, J Nat Prod*, in press, 2007.
179. Stierle A *et al.*, Taxol and taxane production by *Taxomyces andreanae*, an endophytic fungus of Pacific yew, *Science* **260**:214–216, 1993.
180. Amna T *et al.*, Determination and quantification of camptothecin in an endophytic fungus by liquid chromatography-positive mode electrospray ionization tandem mass spectrometry, *Curr Sci* **91**:208–212, 2006.
181. Amna T *et al.*, Bioreactor studies on the endophytic fungus *Entrophospora infrequens* for the production of an anticancer alkaloid camptothecin, *Can J Microbiol* **52**:189–196, 2006.
182. Puri SC *et al.*, An endophytic fungus from *Nothapodytes foetida* that produces camptothecin, *J Nat Prod* **68**:1717–1719, 2005.
183. Zhang L *et al.*, Isolation of endophytic fungus of *Catharanthus roseus* and its fermentation to produce products of therapeutic value, *Zhongcaoyao* **31**:805–807, 2000.
184. Yang X *et al.*, Preliminary study of a vincristine-producing endophytic fungus isolated from leaves of *Catharanthus roseus, Zhongcaoyao* **35**:79–81, 2004.

185. Puri SC *et al.*, The endophytic fungus *Trametes hirsuta* as a novel alternative source of podophyllotoxin and related aryl tetralin lignans, *J Biotechnol* **122**:494–510, 2006.
186. Fang M *et al.*, Rapid screening and identification of brefeldin A from endophytic fungi, *Fenxi Ceshi Xuebao* **24**:21–24, 2005.
187. Weber RWS *et al.*, Brefeldin A production by *Phoma medicaginis* in dead pre-colonized plant tissue: A strategy for habitat conquest?, *Mycol Res* **108**:662–671, 2004.
188. Rob T *et al.*, A new compound from *Pencillium thiomii*, an endophytic fungus, isolated from the root of *Terminalia chebula* Retz (Haritaki), *J Bangladesh Chem Soc* **18**:64–69, 2005.
189. Tkacz JS, Polyketide and peptide products of endophytic fungi: Variations on two biosynthetic themes of secondary metabolism, in Bacon CW, White JF, Jr (eds.), *Microbial Endophytes* Marcel Dekker, Inc., New York, NY, pp. 263–294, 2000.
190. Prajoubklang A *et al.*, Bioactive deoxypreussomerins and dimeric naphthoquinones from *Diospyros ehretioides* fruits: Deoxypreussomerins may not be plant metabolites but may be from fungal epiphytes or endophytes, *Chem Biodivers* **2**:1358–1367, 2005.
191. Krohn K *et al.*, Biologically active metabolites from fungi. Part 16: New preussomerins J, K and L from an endophytic fungus:Structure elucidation, crystal structure analysis and determination of absolute configuration by CD calculations, *Tetrahedron* **57**:4343–4348, 2001.
192. Polishook JD *et al.*, Preussomerin D from the endophyte *Hormonema dematioides, Mycologia* **85**:62–64, 1993.
193. Chomcheon P *et al.*, 3-Nitropropionic acid (3-NPA), a potent antimycobacterial agent from endophytic fungi:Is 3-NPA in some plants produced by endophytes?, *J Nat Prod* **68**:1103–1105, 2005.
194. Bhuvaneswari V *et al.*, Extraction and estimation of indo-3-acetic acid (IAA) from some endophytic and pathogenic coelomycetes, *Asian J Microbiol Biotechnol Environ Sci* **8**:243–248, 2006.
195. Tudzynski B, Fungal phytohormones in pathogenic and mutualistic associations, in Carroll GC, Tudzynski B (eds.), The Mycota. A Comprehensive Treatise on Fungi as Experimental Systems for Basic and Applied Research. V. Plant Relationships. Part A., Springer-Verlag, Berlin, Germany, pp. 167–184, 1997.
196. Wang FW *et al.*, Neoplaether, a new cytotoxic and antifungal endophyte metabolite from *Neoplaconema napellum* IFB-E016, *FEMS Microbiol Lett* **261**:218–223, 2006.

197. Xu M *et al.*, Shearinines D-K, new indole triterpenoids from an endophytic *Penicillium* sp. (strain HKI0459) with blocking activity on large-conductance calcium-activated potassium channels, *Tetrahedron* **63**:435–444, 2007.
198. Jiao RH *et al.*, Chaetominine, a cytotoxic alkaloid produced by endophytic *Chaetomium* sp. IFB-E015, *Org Lett* **8**:5709–5712, 2006.
199. Pornpakakul S *et al.*, Diaporthichalasin, a novel CYP3A4 inhibitor from an endophytic *Diaporthe* sp, *Tetrahedron Lett* **48**:651–655, 2007.
200. Saranpuetti C *et al.*, Determination of enzymes from *Colletotrichum* sp. AHU9748 essential for lepidimoide production from okra polysaccharide, *J Biosci Bioeng* **102**:452–456, 2006.
201. Agusta A *et al.*, Stereoselective oxidation at C-4 of flavans by the endophytic fungus *Diaporthe* sp isolated from a tea plant, *Chem Pharm Bull* **53**:1565–1569, 2005.
202. Shibuya H *et al.*, Biooxidation of (+)-catechin and (−)-epicatechin into 3,4-dihydroxyflavan derivatives by the endophytic fungus *Diaporthe* sp. isolated from a tea plant, *Chem Pharm Bull* **53**:866–867, 2005.
203. Shibuya H *et al.*, Transformation of *Cinchona* alkaloids into 1-N-oxide derivatives by endophytic *Xylaria* sp. isolated from *Cinchona pubescens, Chem Pharm Bull* **51**:71–74, 2003.
204. Werner C *et al.*, Degradation of the polyamine alkaloid aphelandrine by endophytic fungi isolated from *Aphelandra tetragona, FEMS Microbiol Lett* **155**:147–153, 1997.
205. Kim HJ *et al.*, An antifungal and plant growth promoting metabolite from a sterile dark ectotrophic fungus, *Phytochemistry* **67**:2277–2280, 2006.
206. Maloney KN *et al.*, Phaeosphaeride A, an Inhibitor of STAT3-dependent signaling isolated from an endophytic fungus, *Org Lett* **8**:4067–4070, 2006.
207. Lehtonen P *et al.*, Transfer of endophyte-origin defensive alkaloids from a grass to a hemiparasitic plant, *Ecol Lett* **8**:1256–1263, 2005.
208. Spiering MJ *et al.*, Distribution of the fungal endophyte *Neotyphodium lolii* is not a major determinant of the distribution of fungal alkaloids in *Lolium perenne* plants, *Phytochemistry* **66**:195–202, 2005.
209. Bauer J, Meyer K, Stoffwechselprodukte von Pilzen in Silagen:Einflüse auf die Gesundheit von Nutztieren, *Übersicht Tierernähr* **34**:27–55, 2006.
210. Nielsen KF *et al.*, Production of metabolites from the *Penicillium roqueforti* complex, *J Agric Food Chem* **54**:3756–3763, 2006.
211. Bacon CW *et al.*, Biological control of *Fusarium moniliforme* in maize, *Environ Health Perspec Suppl* **109**:325–332, 2001.
212. Ramin AA *et al.*, *In vitro* effects of *Muscodor albus* and three volatile components on growth of selected postharvest microorganisms, *Hortscience* **40**:2109–2114, 2005.

213. Bacon CW, Toxic endophyte-infected tall fescue and range grasses: Historic perspectives, *J Anim Sci* **73**:861–770, 1995.
214. Blodgett DJ, Fescue toxicosis, *The Veterinary Clinics of North America Equine Practice* **17**:567–577, 2001.
215. Lane GA *et al.*, Coevolution of fungal endophytes with grasses:The significance of secondary metabolites, in Bacon CW and White JF Jr (eds.), *Microbial Endophytes* Marcel Dekker, Inc., New York, NY, 341–388, 2000.
216. Powell RG, Petroski RJ, Alkaloid toxins in endophyte-infected grasses, *Nat Toxins* **1**:163–170, 1992.
217. White JF, Jr *et al.*, Biotrophic endophytes of grasses: A systematic appraisal, in Bacon CW, White JF Jr (eds.), *Microbial Endophytes* Marcel Dekker, Inc., New York, NY, pp. 49–62, 2000.
218. Cross DL *et al.*, Equine fescue toxicosis:Signs and solutions, *J Anim Sci* **73**:899–908, 1995.
219. Paterson J *et al.*, The effects of fescue toxicosis on beef cattle productivity, *J Anim Sci* **73**:889–898, 1995.
220. Cheeke PR, Endogenous toxins and mycotoxins in forage grasses and their effects on livestock, *J Anim Sci* **73**:909–918, 1995.
221. Aiken GE *et al.*, Influence of protein supplementation and implant status on alleviating fescue toxicosis, *J Anim Sci* **79**:827–832, 2001.
222. Faeth SH *et al.*, Temporal and spatial variation in alkaloid levels in *Achnatherum robustum*, a native grass infected with the endophyte *Neotyphodium, J Chem Ecol* **32**:307–324, 2006.
223. Tong DW *et al.*, Seasonal change of loline alkaloids in endophyte-infected meadow fescue, *Agric Sci China* **5**:793–797, 2006.
224. Malinowski DP, Belesky DP, Ecological importance of *Neotyphodium* spp. grass endophytes in agroecosystems, *Grassland Sci* **52**:1–14, 2006.
225. Schardl CL *et al.*, Symbioses of grasses with seedborne fungal endophytes, *Annu Rev Plant Biol* **55**:315–340, 2004.
226. Blankenship JD *et al.*, Biosynthetic precursors of fungal pyrrolizidines, the loline alkaloids, *ChemBioChem* **6**:1016–1022, 2005.
227. Li SM, Unsöld IA, Post-genome research on the biosynthesis of ergot alkaloids, *Plant Med* **72**:1117–1120, 2006.
228. Haarmann T *et al.*, The ergot alkaloid gene cluster in *Claviceps purpurea*: Extension of the cluster sequence and intra species evolution, *Phytochemistry* **66**:1312–1320, 2005.
229. Unsöld IA, Li SM, Overproduction, purification and characterization of FgaPT2, a dimethylallyltryptophan synthase from *Aspergillus fumigatus, Microbiology* **151**:1499–1505, 2005.

230. Unsöld IA, Li SM, Reverse prenyltransferase in the biosynthesis of fumigaclavine C in *Aspergillus fumigatus*: Gene expression, purification, and characterization of fumigaclavine C synthase FGAPT1, *ChemBioChem* 7:158–164, 2006.
231. Young CA *et al.*, Molecular cloning and genetic analysis of a symbiosis-expressed gene cluster for lolitrem biosynthesis from a mutualistic endophyte of perennial ryegrass, *Molec Genet Genomics* **274**:13–29, 2005.
232. Tanaka A *et al.*, A symbiosis expressed non-ribosomal peptide synthetase from a mutualistic fungal endophyte of perennial ryegrass confers protection to the symbiotum from insect herbivory, *Mol Microbiol* **57**:1036–1050, 2005.
233. Spiering MJ *et al.*, Expressed sequence tags and genes associated with loline alkaloid expression by the fungal endophyte *Neotyphodium uncinatum, Fungal Genet Biol* **36**:242–254, 2002.
234. Spiering MJ *et al.*, Gene clusters for insecticidal loline alkaloids in the grass-endophytic fungus *Neotyphodium uncinatum, Genetics* **169**:1403–1414, 2005.
235. Bacon CW, White JF Jr, Physiological adaptations in the evolution of endophytism in the Clavicipitaceae, in Bacon CW, White JF Jr eds. *Microbiol Endophytes*Marcel Dekker, Inc., New York, NY, 237–262, 2000.
236. Dong LQ, Zhang KQ, Microbial control of plant-parasitic nematodes:A five-party interaction, *Plant Soil* **288**:31–45, 2006.
237. Bultman TL *et al.*, Effects of fungal endophyte isolate on performance and preference of bird cherry-oat aphid, *Environ Entomol* **35**:1690–1695, 2006.
238. Tintjer T, Rudgers JA, Grass-herbivore interactions altered by strains of a native endophyte, *New Phytol* **170**:513–521, 2006.
239. Clay K *et al.*, Herbivores cause a rapid increase in hereditary symbiosis and alter plant community composition, *Proc Natl Acad Sci U S A* **102**:12465–12470, 2005.
240. Joost RE, *Acremonium* in fescue and ryegrass:Boon or bane? A review, *J Anim Sci* **73**:881–888, 1995.
241. Clay K, Holah J, Fungal endophyte symbiosis and plant diversity in successional fields, *Science* **285**:1742–1744, 1999.
242. Latch GCM, Grass endophytes as a model, *Sydowia* **50**:213–228, 1998.
243. Spyreas G *et al.*, Effects of endophyte infection in tall fescue (*Festuca arundinacea*: Poaceae) on community diversity, *Intl J Plant Sci* **162**:1237–1245, 2001.
244. Faeth SH, Hamilton CE, Does An asexual endophyte symbiont alter life stage and long-term survival in a perennial host grass?, *Microb Ecol* **52**:748–755, 2006.
245. Vibart RE *et al.*, Varying endophyte status and energy supplementation of fresh tall fescue in continuous culture, *Anim Feed Sci Technol* **132**:123–136, 2007.
246. Wright ADG *et al.*, Reducing methane emissions in sheep by immunization against rumen methanogens, *Vaccine* **22**:3976–3985, 2004.

247. Jones JT *et al.*, Horizontal gene transfer from bacteria and fungi as a driving force in the evolution of plant parasitism in nematodes, *Nematology* 7:641–646, 2005.
248. Long DM *et al.*, *In vivo* addition of telomeric repeats to foreign DNA generates extrachromosomal DNAs in the taxol-producing fungus *Pestalotiopsis microspora, Fungal Genet Biol* **24**:335–344, 1998.
249. Juvvadi PR *et al.*, Genomics reveals traces of fungal phenylpropanoid-flavonoid metabolic pathway in the filamentous fungus *Aspergillus oryzae, J Microbiol* **43**:475–486, 2005.
250. Seshime Y *et al.*, Discovery of a novel superfamily of type III polyketide synthases in *Aspergillus oryzae, Biochem Biophys Res Commun* **331**:253–260, 2005.
251. Seshime Y *et al.*, Genomic evidences for the existence of a phenylpropanoid metabolic pathway in *Aspergillus oryzae, Biochem Biophys Res Commun* **337**:747–751, 2005.
252. Wang JW *et al.*, The preparation of an elicitor from a fungal endophyte to enhance artemisinin production in hairy root cultures of *Artemisia annua* L, *Ch J Biotechnol* **22**:829–834, 2006.
253. Bok JW, Keller NP, LaeA, a regulator of secondary metabolism in *Aspergillus* spp, *Euk Cell* **3**:527–535, 2004.

Index